D1749997

AN ULLMANN'S ENCYCLOPEDIA

INDUSTRIAL ORGANIC CHEMICALS

Starting Materials and Intermediates

VOLUME 1

WILEY-VCH

Weinheim · New York · Chichester · Brisbane · Singapore · Toronto

INDUSTRIAL ORGANIC CHEMICALS

AN ULLMANN'S ENCYCLOPEDIA

VOLUME 1
Acetaldehyde to **Aniline**

VOLUME 2
Anthracene to **Cellulose Ethers**

VOLUME 3
Chlorinated Hydrocarbons to **Dicarboxylic Acids, Aliphatic**

VOLUME 4
Dimethyl Ether to **Fatty Acids**

VOLUME 5
Fatty Alcohols to **Melamine and Guanamines**

VOLUME 6
Mercaptoacetic Acid and Derivatives to **Phosphorus Compounds, Organic**

VOLUME 7
Phthalic Acid and Derivatives to **Sulfones and Sulfoxides**

VOLUME 8
Sulfonic Acids, Aliphatic to **Xylidines**

Index

AN **ULLMANN'S**
ENCYCLOPEDIA

INDUSTRIAL ORGANIC CHEMICALS

Starting Materials and Intermediates

VOLUME 1
Acetaldehyde
to **Aniline**

WILEY-VCH

Weinheim · New York · Chichester · Brisbane · Singapore · Toronto

This book was carefully produced. Nevertheless, authors and publisher do not warrant the information contained therein to be free of errors. Readers are advised to keep in mind that statements, data, illustrations, procedural details or other items may inadvertently be inaccurate.

Library of Congress Card No.: Applied for.
British Library Cataloguing-in-Publication Data: A catalogue record for this book is available from the British Library.

Die Deutsche Bibliothek – CIP-Einheitsaufnahme
Industrial organic chemicals : starting materials and intermediates ;
an Ullmann's encyclopedia. – Weinheim ; New York ;
Chichester ; Brisbane ; Singapore ; Toronto : Wiley-VCH
   ISBN 3-527-29645-X
Vol. 1. Acetaldehyde to Aniline. – 1. Aufl. – 1999.

© WILEY-VCH Verlag GmbH, D-69469 Weinheim (Federal Republic of Germany), 1999
Printed on acid-free and chlorine-free paper.
All rights reserved (including those of translation in other languages). No part of this book may be reproduced in any form – by photoprinting, microfilm, or any other means – nor transmitted or translated into machine language without written permission from the publishers. Registered names, trademarks, etc. used in this book, even when not specifically marked as such, are not to be considered unprotected by law.

Composition and Printing: Rombach GmbH, Druck- und Verlagshaus, D-79115 Freiburg
Bookbinding: Wilhelm Osswald & Co., D-67433 Neustadt (Weinstraße)
Cover design: mmad, Michel Meyer, D-69469 Weinheim
Printed in the Federal Republic of Germany

Preface

This handbook deals with starting materials and intermediates of industrial organic chemistry, alphabetically listed, beginning with acetaldehyde and ending with xylidines.

The information presented is an updated form of that included in the 6th edition of the 36-volume „Ullmann's Encyclopedia of Industrial Chemistry". The wealth of material it contains provides the user with both broad introductory information and in-depth detail of utmost importance in both industrial and academic environments. Due to its sheer size, however, the unabridged Ullmann's is inaccessibile to many potential users, particulary individuals and smaller companies. This is why all the information on organic chemicals and intermediates has been collected together in this convenient 8-volume set.

Users of the handbook of Industrial Organic Chemicals will have the benefit of the most up-to-date professional information on organic starting materials and intermediates, approved by the Ullmann's editorial team, tailored specifically to their needs, and presented in an undiluted and accessible format.

The Publisher

Contents

1 Acetaldehyde

1. Introduction 2
2. Physical Properties 2
3. Chemical Properties and Uses. 4
4. Production 7
5. Quality and Analysis. 17
6. Storage and Transportation 18
7. Economic Aspects. 19
8. Polymers of Acetaldehyde 20
9. Toxicology and Occupational Health 23
10. References. 25

2 Acetic Acid

1. Introduction 30
2. Physical Properties 30
3. Chemical Properties 33
4. Production 34
5. Wastewater and Off-Gas Problems. . 48
6. Quality Specifications 49
7. Chemical Analysis 49
8. Storage, Transportation, and Customs Regulations. 50
9. Uses. 50
10. Derivatives 50
11. Economic Aspects. 57
12. Toxicology and Occupational Health 58
13. References. 59

3 Acetic Anhydride and Mixed Fatty Acid Anhydrides

1. Acetic Anhydride 63
2. Mixed Fatty Acid Anhydrides 81
3. Economic Aspects. 83
4. Toxicology and Occupational Health 83
5. References. 84

4 Acetone

1. Introduction 89
2. Physical Properties 90
3. Chemical Properties 91
4. Production 92
5. Environmental Protection 101
6. Quality Specifications and Analysis . 102
7. Storage and Transportation 102
8. Uses. 103
9. Economic Aspects. 106
10. Toxicology and Occupational Health 108
11. Derivatives 109
12. References. 114

5 Acetylene

1. Introduction 118
2. Physical Properties 119
3. Chemical Properties 123
4. Production 130
5. Safety Precautions, Transportation, and Storage. 171
6. Uses and Economic Aspects 181
7. Propyne 185
8. Toxicology and Occupational Health 189
9. References. 190

6 Acridine

1. Physical Properties 197
2. Chemical Properties 197
3. Production 197
4. Uses. 198
5. Toxicology. 198
6. References. 198

7 Acrolein and Methacrolein

1. Introduction 199
2. Properties 200
3. Production 207
4. Quality and Analysis. 211
5. Handling, Storage, and Transportation 212
6. Uses and Production Data 213
7. Toxicology and Ecotoxicology 215
8. References. 218

8 Acrylic Acid and Derivatives

1. Acrylic Acid and Esters 223
2. Cyanoacrylates 241
3. Acrylamide 243
4. References. 244

9 Acrylonitrile

1. Introduction 249
2. Physical Properties 250
3. Chemical Properties 252
4. Production 252
5. Quality Specifications and Chemical Analysis 255
6. Storage and Transportation 256
7. Uses. 256
8. Economic Aspects. 257
9. Toxicology and Occupational Health 258
10. References. 259

VIII

10 Adipic Acid

1. Introduction 263
2. Physical Properties 263
3. Chemical Properties 264
4. Production 265
5. Byproducts 268
6. Quality Specifications 269
7. Storage and Transportation 269
8. Derivatives 270
9. Uses 272
10. Economic Aspects 273
11. Toxicology and Occupational Health 274
12. References 274

11 Alcohols, Aliphatic

1. Introduction 280
2. Saturated Alcohols 280
3. Unsaturated Alcohols 308
4. Alkoxides 311
5. Toxicology 314
6. References 316

12 Alcohols, Polyhydric

1. General Aspects 321
2. Diols 326
3. Triols 338
4. Tetrols 340
5. Higher Polyols 342
6. Toxicology 343
7. References 344

13 Aldehydes, Aliphatic and Araliphatic

1. Introduction 350
2. Saturated Aldehydes 351
3. Unsaturated Aldehydes 366
4. Hydroxyaldehydes 374
5. Araliphatic Aldehydes 381
6. Dialdehydes 390
7. Acetals 393
8. Quality Control 398
9. Analysis 399
10. Storage, Transportation, and Environmental Regulations 400
11. Economic Aspects 401
12. References 402

14 Allyl Compounds

1. Allyl Chloride 410
2. Allyl Alcohol 419
3. Allyl Esters 425
4. Allyl Ethers 431
5. Allylamines 432
6. Toxicology and Occupational Health 433
7. References 436

15 Amines, Aliphatic

1. Introduction 444
2. General Chemical Properties 444
3. General Production Methods 448
4. Lower Alkylamines 453
5. Cycloalkylamines 459
6. Cyclic Amines 464
7. Fatty Amines 472
8. Diamines and Polyamines 482
9. Toxicology and Occupational Health 490
10. References 494

16 Amines, Aromatic

1. Introduction 503
2. Physical and Chemical Properties . . 506
3. Production 513
4. Economic Aspects 525
5. Toxicology, Occupational Health, and Environmental Protection 525
6. References 530

17 Amino Acids

1. Introduction and History 536
2. Properties 540
3. Production 548
4. Biochemical and Physiological Significance 562
5. Uses 565
6. Chemical Analysis 585
7. Economic Significance 587
8. Toxicology 587
9. References 590

18 Aminophenols

1. Aminophenols 601
2. Aminophenol Derivatives 611
3. References 622

19 Aniline

1. Introduction 627
2. Physical and Chemical Properties . . 628
3. Production 629
4. Quality Specifications 633
5. Handling, Storage, and Transportation 634
6. Aniline Derivatives and Uses 634
7. Economic Aspects 639
8. Environmental Protection, Toxicology, and Occupational Health 640
9. References 641

Acetaldehyde

GERALD FLEISCHMANN, Wacker-Chemie GmbH, Werk Burghausen, Federal Republic of Germany (Chap. 2–8)

REINHARD JIRA, Wacker-Chemie GmbH, Werk Burghausen, Federal Republic of Germany (Chap. 2–8)

HERMANN M. BOLT, Institut für Arbeitsphysiologie an der Universität Dortmund, Dortmund, Federal Republic of Germany (Chap. 9)

KLAUS GOLKA, Institut für Arbeitsphysiologie an der Universität Dortmund, Dortmund, Federal Republic of Germany (Chap. 9)

1.	Introduction	2
2.	Physical Properties	2
3.	Chemical Properties and Uses.	4
3.1.	Addition Reactions	4
3.2.	Derivatives of Aldol Addition.	5
3.3.	Reaction with Nitrogen Compounds	5
3.4.	Oxidation	6
3.5.	Reduction	6
3.6.	Miscellaneous Reactions	6
3.7.	Consumption	7
4.	Production	7
4.1.	Production from Ethanol	8
4.2.	Production from Acetylene	9
4.3.	Production from Ethylene	11
4.3.1.	Direct Oxidation of Ethylene	11
4.3.2.	Acetaldehyde as Byproduct	16
4.3.3.	Isomerization of Ethylene Oxide	16
4.4.	Production from C_1 Sources	16
4.5.	Production from Hydrocarbons	17
5.	Quality and Analysis	17
6.	Storage and Transportation	18
6.1.	Storage	18
6.2.	Transportation	18
6.3.	Other Regulations	19
7.	Economic Aspects	19
8.	Polymers of Acetaldehyde	20
8.1.	Paraldehyde	20
8.2.	Metaldehyde	22
8.3.	Polyacetaldehyde	22
9.	Toxicology and Occupational Health	23
10.	References	25

1. Introduction

Acetaldehyde (ethanal), CH$_3$CHO [75-07-0], was observed in 1774 by Scheele during reaction of black manganese dioxide and sulfuric acid with alcohol. Its constitution was explained in 1835 by Liebig who prepared pure acetaldehyde by oxidation of ethanol with chromic acid and designated this product "aldehyde," a contraction of the term "alcohol dehydrogenatus."

Acetaldehyde is a mobile, low-boiling, highly flammable liquid with a pungent odor. Because of its high chemical reactivity, acetaldehyde is an important intermediate in the production of acetic acid, acetic anhydride, ethyl acetate, peracetic acid, butanol, 2-ethylhexanol, pentaerythritol, chlorinated acetaldehydes (chloral), glyoxal, alkyl amines, pyridines, and other chemicals. The first commercial application was the production of acetone via acetic acid between 1914 and 1918 in Germany (Wacker-Chemie and Hoechst) and in Canada (Shawinigan).

Occurrence. Acetaldehyde is an intermediate in the metabolism of plant and animal organisms, in which it can be detected in small amounts. Larger amounts of acetaldehyde interfere with biological processes. As an intermediate in alcoholic fermentation processes it is present in small amounts in all alcoholic beverages, such as beer, wine, and spirits. Acetaldehyde also has been detected in plant juices and essential oils, roasted coffee, and tobacco smoke.

Commercial production processes include dehydrogenation or oxidation of ethanol, addition of water to acetylene, partial oxidation of hydrocarbons, and direct oxidation of ethylene. In the 1970s, the world capacity of this last process, the Wacker-Hoechst direct oxidation, increased to over $2 10^6$ t/a. However, the importance of acetaldehyde as an organic intermediate is now steadily decreasing, because new processes for some acetaldehyde derivatives have been developed, such as the oxo process for butanol and 2-ethylhexanol and the Monsanto process for acetic acid. In the future, new processes for acetic anhydride (Halcon, Eastman, Hoechst), for vinyl acetate (Halcon), and for alkyl amines (from ethanol) will diminish the use of acetaldehyde as a starting material.

2. Physical Properties

Acetaldehyde, C$_2$H$_4$O, M_r 44.054, is a colorless liquid with a pungent, suffocating odor that is slightly fruity when diluted.

bp at 101.3 kPa	20.16 °C
mp	−123.5 °C
Critical temperature t_{crit}	181.5 °C
other values	187.8 °C [12], 195.7 °C [13]
Critical pressure p_{crit}	6.44 MPa
other values	5.54 MPa [12], 7.19 MPa [13]

Relative density $\quad d^t_4 = 0.8045 - 0.001325 \cdot t$ (t in °C) [3]
Refractive index $\quad n^t_D = 1.34240 - 0.0005635 \cdot t$ (t in °C) [14]
Molar volume of the gas
  at 101.3 kPa and 20.16 °C $\quad$ 23.40 L/mol
  at 25.0 °C $\quad$ 23.84 L/mol
    For dependence on T (293.32 – 800 K) and p (0.1 – 30 MPa), see [15].
Specific volume of the vapor
  at 20.16 °C $\quad$ 0.531 m^3/kg
  at 25.0 °C $\quad$ 0.541 m^3/kg
Vapor density (air = 1) $\quad$ 1.52

Vapor pressure

t, °C	-20	-0.27	5.17	14.76	50	100
p, kPa	16.4	43.3	67.6	82.0	279.4	1014.0

For further values between –60 and +180 °C, see [14]

Viscosity of liquid η
  at 9.5 °C $\quad$ 0.253 mPa · s
  at 20 °C $\quad$ 0.21 mPa · s
Viscosity of vapor η
  at 25 °C $\quad$ 86$10^{-4}$ mPa · s
    For further values between 35.0 and 77.8 °C and between 0.13 and 0.40 kPa, see [16].
Surface tension γ at 20 °C $\quad$ 21.2$10^{-2}$ mN cm^{-1}
Dipole moment (gas phase) $\quad$ 2.69±2% D [12]
Dielectric constant
  of liquid at 10 °C $\quad$ 21.8
  of vapor at 20.16 °C, 101.3 kPa $\quad$ 1.0216
Heat capacity of liquid
  c_p (l) at 0 °C $\quad$ 2.18 J g^{-1} K^{-1}
  at 20 °C $\quad$ 1.38 J g^{-1} K^{-1}
    For further values between –80 °C (c_p = 1.24 J g^{-1} K^{-1}) and +120 °C (c_p = 1.50 J g^{-1} K^{-1}), see [17].
Heat capacity of vapor
  c_p (g) at 25 °C, 101.3 kPa $\quad$ 1.24 J g^{-1} K^{-1}
    For dependence on temperature (nonlinear) between 0 °C (c_p = 1.17 J g^{-1} K^{-1}) and 1000 °C (c_p = 2.64 J g^{-1} K^{-1}), see [17].
c_p/c_v (= $\varkappa$) at 30 °C, 101.3 kPa $\quad$ 1.145 [18]
Thermal conductivity
  of liquid at 20 °C $\quad$ 0.174 J m^{-1} s^{-1} K^{-1};
  for more values, see [19]
  of vapor at 25 °C $\quad$ 1.09$10^{-2}$ J m^{-1} s^{-1} K^{-1}
  for further values, see [20].
Cubic expansion coefficient per K (0 – 20 °C) $\quad$ 0.00169
Heat of combustion of liquid at constant p $\quad$ 1168.79 (1166.4 [12]) kJ/mol
Heat of solution in water (infinite dilution) $\quad$ 17 906 J/mol
Latent heat of fusion $\quad$ 3246.3 J/mol
Latent heat of vaporization at 20.2 °C $\quad$ 25.73 kJ/mol
  other values $\quad$ 27.2 [21], 30.41, 27.71 [12], 26.11 [22] kJ/mol
    For dependence on temperature (nonlinear) between -80 °C (32.46 kJ/mol) and 182 °C (0 kJ/mol), see [17].
Heat of formation ΔH from the elements at 25 °C for gaseous acetaldehyde
$\quad$ -166.47 (–166.4 [21]) kJ/mol
    For dependence of heat of formation for gaseous and liquid acetaldehyde, and enthalpy of vaporization on temperature up to 800 K and 30 MPa, see [15].
Gibbs free energy of formation ΔG from elements
  at 25 °C for gaseous acetaldehyde $\quad$ -133.81 kJ/mol
  other values $\quad$ 133.72 [12], 132.9 [21] kJ/mol

Entropy for gaseous acetaldehyde at 25 °C	265.9 J mol^{-1} K^{-1}
Entropy for liquid acetaldehyde at 20.16 °C	172.9 J mol^{-1} K^{-1}
Entropy of vaporization at 20.16 °C	91.57 J mol^{-1} K^{-1}
First ionization potential	10.5 eV
Dissociation constant at 0 °C	0.710^{-14} mol/L (H$_3$CCHO·H$_2$CCHO + H$^+$)

For the second virial coefficient of the equation of state for gaseous acetaldehyde at 31 °C, 66 °C, and 85 °C, see [23].

Acetaldehyde is completely miscible with water and most organic solvents. It forms no azeotrope with water, methanol, ethanol, acetone, acetic acid, or benzene. Binary azeotropes are formed with butane (*bp* −7 °C, 84 wt% of butane) and diethyl ether (*bp* 18.9 °C, 23.5 wt% of ether).

Other physical data: compressibility and viscosity at higher pressure are given in [24], vapor pressure of aqueous acetaldehyde solutions in [25]. For solubility of carbon dioxide, acetylene, and nitrogen in acetaldehyde, see [11]; for freezing points of aqueous acetaldehyde solutions, see [11]; for vapor–liquid equilibria of binary systems of acetaldehyde with water, ethanol, acetic acid, and ethylene oxide, see [26, pp. 392, 561, 565, and 570], with vinyl acetate, see [27].

Safety Data. Flash point (Abel–Pensky; DIN 51755; ASTM 56–70) −20 °C (−40 °C according to the safety regulations of the Berufsgenossenschaft der Chemischen Industrie, Federal Republic of Germany). Ignition temperature (DIN 51794; ASTM D 2155–66) 140 °C; for ignition retardation when injected into a hot air stream, see [28]. Explosive limits in air: 4–57 vol%; for influence of pressure on explosive limits, see [29].

3. Chemical Properties and Uses

Acetaldehyde is a highly reactive compound showing all of the typical aldehyde reactions as well as those of an alkyl group in which hydrogen atoms are activated by the carbonyl group in the α position. When heated above 420 °C acetaldehyde decomposes into methane and carbon monoxide.

3.1. Addition Reactions

With water, acetaldehyde forms an unstable hydrate; isolable solid hydrates are known only with chlorinated acetaldehydes. *Alcohols* add to acetaldehyde giving hemiacetals, which form acetals (→ Aldehydes, Aliphatic and Araliphatic) with additional alcohol in the presence of acids by removal of water. Diols give cyclic acetals; for example, 2-methyl-1,3-dioxolane is obtained from ethylene glycol and acetaldehyde, and 2-methyl-1,3-dioxane from 1,3-propanediol.

Aqueous *sodium bisulfite* solution and acetaldehyde give a crystalline adduct from which acetaldehyde can be liberated. Dry *ammonia* forms crystalline acetaldehyde ammonia. Acetaldehyde and *hydrocyanic acid* react to give lactonitrile (α-hydroxypropionitrile), a possible intermediate in acrylonitrile production [30].

Acetaldehyde reacts with *acetic anhydride* to give ethylidene diacetate, an intermediate in the vinyl acetate process of Celanese Corp. [31] (→ Vinyl Esters).

3.2. Derivatives of Aldol Addition

Two molecules of acetaldehyde combine in the presence of alkaline catalysts or dilute acids at room temperature or with moderate heating to form acetaldol [*107-89-1*], $CH_3CH(OH)CH_2CHO$. At increased temperatures, water is cleaved easily from this acetaldol, forming crotonaldehyde (→Aldehydes, Aliphatic and Araliphatic). Further condensation under more stringent conditions to form aldehyde resins (e.g., synthetic shellac) now has no industrial importance.

Urea and acetaldehyde condense in the presence of H_2SO_4 to form crotonylidenediurea (6-methyl-4-ureidohexahydropyrimidin-2-one [*1129-42-6*]), which is used as a long-term nitrogen fertilizer.

Acetaldehyde is also an intermediate in the butadiene synthesis starting from acetylene and proceeding via acetaldol and its hydrogenation product, 1,3-butanediol [32]. This process was introduced around 1918 and is still carried out on a commercial scale in some Eastern European countries.

Acrolein is obtained by aldol condensation of acetaldehyde and formaldehyde and subsequent water elimination, analogous to the formation of crotonaldehyde. This method is also without commercial importance today, whereas the production of pentaerythritol from acetaldehyde and a fourfold amount of formaldehyde in the presence of $Ca(OH)_2$ or NaOH is very important industrially (→ Alcohols, Polyhydric).

3.3. Reaction with Nitrogen Compounds

With primary amines, Schiff bases, $CH_3CH=NR$, are formed. Nitrogen compounds such as hydroxylamine, hydrazine, phenylhydrazine, and semicarbazide react with acetaldehyde to give easily crystallizable compounds that are used for the analytical determination and characterization of aldehydes (semicarbazone, *mp* 162–163 °C; *p*-nitrophenylhydrazone, *mp* 128.5 °C; 2,4-dinitrophenylhydrazone, *mp* 168 °C; oxime, *mp* 47 °C). Many other aldehydes and ketones can be characterized in the same way because their analogous derivatives generally have sharp melting points.

The synthesis of pyridine and pyridine derivatives is of increasing importance. 5-Ethyl-2-methylpyridine is obtained in the presence of fluoride ions by the reaction of aqueous ammonia with acetaldehyde (or with paraldehyde, which slowly releases the monomer). In the added presence of formaldehyde or acrolein, mixtures of pyridine and alkylpyridines form (→ Pyridine and Pyridine Derivatives).

3.4. Oxidation

The major part of the acetaldehyde produced commercially is used for manufacturing acetic acid by oxidation with oxygen or air (→ Acetic Acid). Acetaldehyde monoperacetate is formed as an intermediate and decomposes into peracetic acid and acetaldehyde at elevated temperatures and in the presence of catalytic amounts of iron or cobalt salts. In the presence of Mn^{2+} salts, acetic acid is obtained from acetaldehyde monoperacetate, and in the presence of Co^{2+} and Cu^{2+} salts, acetic anhydride can be formed.

Oxidation with nitric acid gives glyoxal (→ Glyoxal and → Glyoxylic Acid). Halogenated acetaldehydes are prepared by halogenation.

$$2\,CH_3CHO \xrightarrow{O_2} CH_3\!-\!\underset{O\!-\!O}{\overset{O\cdots HO}{C}}\!\!\diagup\!\!\underset{}{CH\!-\!CH_3} \begin{array}{l} \xrightarrow{Co^{2+},\,Cu^{2+}} (CH_3CO)_2O + H_2O \\ \xrightarrow{Mn^{2+}} 2\,CH_3COOH \\ \xrightarrow{Mn^{2+}} CH_3CHO + CH_3CO_3H \end{array}$$

Mono-, di-, and trichloroacetaldehydes (→Chloroacetaldehydes) and tribromoacetaldehyde (bromal) are useful for producing insecticides (e.g., DDT, DDD), pharmaceuticals, and dyes.

3.5. Reduction

Acetaldehyde is hydrogenated readily to ethanol. Prior to 1939, that is, before petrochemically produced ethylene became available in Europe, this reaction was used industrially to produce ethanol from acetaldehyde and, therefore, from acetylene.

Mono-, di-, and triethylamine [75-04-7], [109-89-7], [121-44-8] can be produced from acetaldehyde, ammonia, and hydrogen in the presence of a hydrogenation catalyst [33] (→ Amines, Aliphatic).

3.6. Miscellaneous Reactions

The Tishchenko reaction of acetaldehyde gives the commercially important solvent ethyl acetate (→ Acetic Acid); it is catalyzed by aluminum alcoholate.

Table 1. Consumption of acetaldehyde (10^3 t) 1993

Product	U.S.A	Mexico	Western Europe	Japan	Total
Acetic acid/acetic anhydride		139	307	97	543
Acetate esters	32	5	133	158	328
Pentaerithritol	22		37	12	71
Pyridine and pyridine bases	45		44	17	106
Peracetic acid	19			*	19
1,3-Butylene glycol	13			*	13
Others	16	105	85	65	271
Total	147	249	606	349	1351

* Included in others (glyoxal/glyoxalic acid, crotonaldehyde, lactic acid, n-butanol, 2-ethylhexanol).

As a "radical trapping agent," acetaldehyde is used to control chain length in the polymerization of vinyl compounds.

Oligomers of acetaldehyde are treated in Chapter 8.

3.7. Consumption

The consumption of acetaldehyde has changed during the last few years. Since 1993 in the USA, acetaldehyde is no longer used for the production of acetic acid, butanol, or 2-ethylhexanol, which are now produced by other routes (Table 1). The consumption of acetaldehyde for some other chemicals like peracetic acid or pyridine bases is increasing.

4. Production

Raw materials that have been used for the production of acetaldehyde are:

1) Ethanol from fermentation of carbohydrates or from hydration of ethylene
2) Acetylene
3) Ethylene
4) Lower hydrocarbons
5) Carbon monoxide and hydrogen
6) Methanol

The economy of the commercial processes depends essentially upon prices and the availability of raw materials. In highly industrialized countries maintaining high prices for ethanol by fiscal measures or where petrochemical ethanol was not available, as in Germany or Japan before 1939, acetylene was the favored starting material for acetaldehyde. The acetylene process is still operated in some Eastern European countries and also by companies where cheap acetylene is available. Petrochemically produced ethanol, however, was the favored raw material in the other countries, whereas ethanol made by fermentation was and still is used on a small scale in countries with less chemical industry.

In Western countries, including Japan, all these processes have now been almost completely replaced by the direct oxidation process developed in the late 1950s by Wacker-Chemie and Hoechst. This is because ethylene is available at a lower price than acetylene.

Even the two-stage processes using ethanol from ethylene as starting material are no longer competitive because of the decreasing importance of acetaldehyde as an organic intermediate (see Chap. 7).

Generally, all processes based on acetylene, ethylene, and ethanol are more selective than the oxidation of saturated hydrocarbons. This is because, in the latter case, other oxidation products are formed in addition to acetaldehyde. Because of the great expense of separating the product mixture, such processes are economical only in large units and when all main and secondary products obtained in the process are utilized.

4.1. Production from Ethanol

For the production of acetaldehyde, ethanol can either be dehydrogenated or oxidized in the presence of oxygen. Between 1918 and 1939, dehydrogenation took precedence over oxidation because of the simultaneous production of hydrogen. Later, however, the catalytic vapor-phase oxidation of ethanol became the preferred process, probably because of the long catalyst life and the possibility of recovering energy.

Dehydrogenation of Ethanol. In the first work on ethanol dehydrogenation, published in 1886, ethanol was passed through glass tubes at 260 °C.

$$CH_3CH_2OH\,(l) \rightarrow CH_3CHO\,(l) + H_2\,(g) \quad \Delta H = +82.5 \text{ kJ/mol}$$

Improved yields are obtained in the presence of catalysts such as platinum, copper, or oxides of zinc, nickel, or cobalt. In later patents, zinc and chromium catalysts [34], oxides of rare earth metals [35], and mixtures of copper and chromium oxides [36] have been reported. The lowest amounts of decomposition products are obtained using copper catalysts. Frequent regeneration of the catalysts is required, however.

Process Description. Ethanol vapor is passed at 260–290 °C over a catalyst consisting of copper sponge or copper activated with chromium oxide in a tubular reactor [37]. A conversion of 25–50% per run is obtained. By washing with alcohol and water, acetaldehyde and ethanol are separated from the exhaust gas, which is mainly hydrogen. Pure acetaldehyde is obtained by distillation; the ethanol is separated from water and higher-boiling products by distillation and flows back to the reactor. The final acetaldehyde yield is ca. 90%. Byproducts include butyric acid, crotonaldehyde, and ethyl acetate.

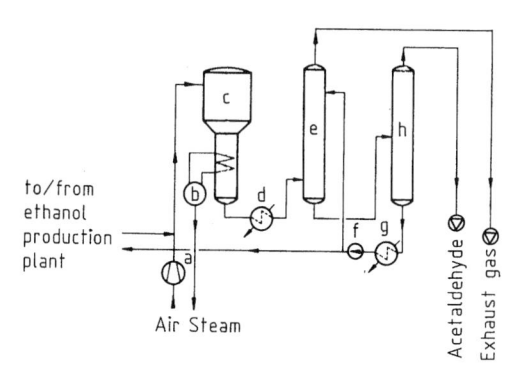

Figure 1. Acetaldehyde production by the Veba-Chemie process
a) Air compressor; b) Heat recovery system; c) Reactor; d) Cooler; e) Waste-gas scrubber; f) Washing-alcohol and return pump; g) Cooler; h) Acetaldehyde rectification

Oxidation of Ethanol. Oxidation of ethanol is the oldest and the best laboratory method for preparing acetaldehyde. In the commercial process, ethanol is oxidized catalytically with oxygen (or air) in the vapor phase.

$$CH_3CH_2OH\,(g) + 1/2\,O_2\,(g) \rightarrow CH_3CHO(l) + H_2O\,(l) \quad \Delta H = -242.0 \text{ kJ/mol}$$

Copper, silver, and their oxides or alloys are the most frequently used catalysts [38]. For an example of a simultaneous oxidation–dehydrogenation process, see [39].

Veba-Chemie Process (Fig. 1). Ethanol is mixed with air and passed over a silver catalyst at 500–650 °C (c). The temperature depends on the ratio of alcohol to air and the flow rate of the gas through the catalyst. Alcohol conversion varies between 50 and 70 % and the yield is between 97 and 99 % depending on the reaction conditions. Acetaldehyde and unreacted alcohol are removed from the waste gas by washing with cold alcohol (e) and separated by fractional distillation (h); after concentration the alcohol returns to the reactor. Heat formed in the reaction is utilized for steam production using a waste-heat recovery system immediately after the reaction zone.

The waste gas consists mainly of nitrogen, hydrogen, methane, carbon monoxide and carbon dioxide; it is burned as lean gas with low calorific value in steam generators. Small amounts of acetic acid are obtained as a byproduct.

4.2. Production from Acetylene

The most important catalysts for the industrial water addition (hydration) are mercury compounds:

$$C_2H_2\,(g) + H_2O\,(l) \xrightarrow[H_2SO_4]{Hg^{2+}} CH_3CHO\,(l) \quad \Delta H = -138.2 \text{ kJ/mol}$$

This method only suceeds industrially when the polymerization and condensation products of acetaldehyde formed in the acid medium are eliminated. To achieve this, the Consortium für elektrochemische Industrie in 1912 proposed a process using excess acetylene at an elevated temperature and removing the acetaldehyde product immediately from the reaction liquid. At the same time, the heat of reaction is removed by distilling an appropriate amount of water. Secondary reactions, such as the oxidation of acetaldehyde to form acetic acid and carbon dioxide, result in reduction of Hg^{2+} to metallic mercury. In Western countries, acetaldehyde production from acetylene has now been discontinued.

Wet Oxidation Process (Hoechst). The wet oxidation process avoided direct handling of the toxic mercury compounds. It was operated, among others, by Wacker-Chemie until the changeover to ethylene as the starting material in 1962.

In this method, iron(III) sulfate is added to reoxidize the mercury metal to the mercury(II) salt, thus ensuring sufficient concentrations of active catalyst. The acetylene reacts at 90–95 °C with the aqueous catalyst solution; between 30 and 50% of the injected acetylene reacts in one run. The gas emerging from the reactor is cooled; mainly water and traces of mercury are separated and returned to the reactor. Acetaldehyde and water are condensed in additional coolers and the acetaldehyde finally is washed out with water from the cycle gas which has been cooled to 25–30 °C. An 8–10% aqueous acetaldehyde solution is obtained. Nitrogen is introduced with the feed gas while carbon dioxide is formed as a byproduct; to avoid excessive accumulation, these gases are removed by withdrawing a small stream of the cycle gas. Iron(II) sulfate is formed in the reaction and is oxidized in a separate reactor with 30% nitric acid at 95 °C. Pure acetaldehyde is obtained by fractional distillation of the aqueous solution at about 200 kPa. For further details of this process, see [40].

Chisso Process [41]. The Chisso process also uses sulfuric acid/mercury sulfate solution as a catalyst. The acetylene reacts completely with the catalyst solution at 68–78 °C and a gauge pressure of 140 kPa. A combination of pressure and vacuum process stages at low temperature and without excess acetylene are used; pure acetaldehyde can be isolated and distilled by utilizing the heat of the reaction. As in the Hoechst process, the catalyst can be regenerated with nitric acid. Production of acetaldehyde by this method was discontinued at Chisso Corp. more than a decade ago.

Production via Vinyl Ether. REPPE at BASF developed the process using vinyl ether [40]; it was operated in a pilot plant between 1939 and 1945. The use of toxic mercury compounds is avoided altogether. Methanol is added to acetylene at 150–160 °C and 1600 kPa in the presence of potassium hydroxide to form methyl vinyl ether [107-25-5]. The methyl vinyl ether is then hydrolyzed with dilute acid:

$$C_2H_2 + CH_3OH \xrightarrow{KOH} CH_3OCH=CH_2 \xrightarrow{H_2O, H^+} CH_3OH + CH_3CHO$$

Production via Ethylidene Diacetate. Addition of acetic acid to acetylene in the presence of mercury(II) salts yields ethylidene diacetate [542-10-9], $CH_3CH(OCOCH_3)_2$, which decomposes into acetaldehyde and acetic anhydride at 130–145 °C in the presence of acid catalysts (e.g., $ZnCl_2$). This process was developed by the Societe

Chimique des Usines du Rhône on an industrial scale in 1914 but is now without importance.

4.3. Production from Ethylene

Ethylene is now the most important starting material for the production of acetaldehyde. Most of the present capacity works by the direct oxidation of ethylene (Wacker process).

4.3.1. Direct Oxidation of Ethylene

This process was developed between 1957 and 1959 by Wacker-Chemie and Hoechst [42]. Formally, the reaction proceeds as follows:

$C_2H_4 + 1/2\,O_2 \rightarrow CH_3CHO \quad \Delta H = -244\ kJ/mol$

An aqueous solution of $PdCl_2$ and $CuCl_2$ is used as catalyst. Acetaldehyde formation had already been observed in the reaction between ethylene and aqueous palladium chloride. This reaction is almost quantitative:

$C_2H_4 + PdCl_2 + H_2O \rightarrow CH_3CHO + Pd + 2\,HCl$

In the Wacker-Hoechst process, metallic palladium is reoxidized by $CuCl_2$, which is then regenerated with oxygen:

$Pd + 2\,CuCl_2 \rightarrow PdCl_2 + 2\,CuCl$
$2\,CuCl + 1/2\,O_2 + 2\,HCl \rightarrow 2\,CuCl_2 + H_2O$

Therefore only a very small amount of $PdCl_2$ is required for the conversion of ethylene. The reaction of ethylene with palladium chloride is the rate-determining step.

One- and two-stage versions of the process are on stream. In the *one-stage method*, an ethylene – oxygen mixture reacts with the catalyst solution. During the reaction a stationary state is established in which "reaction" (formation of acetaldehyde and reduction of $CuCl_2$) and "oxidation" (reoxidation of CuCl) proceed at the same rate. This stationary state is determined by the degree of oxidation of the catalyst, as expressed by the ratio $c_{Cu^{2+}}/(c_{Cu^{2+}} + c_{Cu^+})$. In the *two-stage process* the reaction is carried out with ethylene and then with oxygen in two separate reactors. The catalyst solution is alternately reduced and oxidized. At the same time the degree of oxidation of the catalyst changes alternately. Air is used instead of pure oxygen for the catalyst oxidation.

Reaction Mechanism. The first step of the reaction is the complexation of ethylene to give a palladium ethylene complex

$[PdCl_4]^{2-} + H_2C{=}CH_2 \rightleftharpoons [(C_2H_4)PdCl_3]^- + Cl^-$

Kinetic studies of this reaction show that chloride ions have a inhibiting effect which is explained by the following substitution of a further chloride ligand by a water molecule:

$[(C_2H_4)PdCl_3]^- + H_2O \rightleftharpoons [(C_2H_4)PdCl_2(H_2O)] + Cl^-$

Dissociation of hydrogen ions explains the inhibiting effect of acids:

$[(C_2H_4)PdCl_2(H_2O)] \rightleftharpoons [(C_2H_4)PdCl_2(OH)]^- + H^+$

It is assumed that the hydroxyl complex has the *trans* geometry. Evidence for the *trans–cis* isomerisation of this complex was revealed by a detailed kinetic study of the reaction. π-Bonded ethylene ligands in the *trans* position weaken the metal–chlorine bonds, so that the chloro ligand can be easily substituted by a water molecule from which a hydrogen ion dissociates.

trans-$[(C_2H_4)PdCl_2(OH)]^- + H_2O \rightleftharpoons [(C_2H_4)PdCl(OH)_2]^- + H^+ + Cl^-$

The OH⁻ ligand in the *trans* position is replaced by a Cl⁻ ion, leading to a *cis* complex:

$[(C_2H_4)PdCl(OH)_2]^- + H^+ + Cl^- \rightleftharpoons$ *cis*-$[(C_2H_4)PdCl_2(OH)]^- + H_2O$

The next reaction step is the formation of a σ-bonded hydroxyethylpalladium species. This reaction has been regarded as a *cis* ligand insertion reaction in which the OH⁻ ligand attacks the π-bonded olefin:

cis-$[(C_2H_4)PdCl_2(OH)]^- \rightleftharpoons [HOCH_2CH_2PdCl_2]^-$

Hydride transfer to give an α-hydroxyethylpalladium complex is followed by reductive elimination, which is the rate-determining step:

$[HOCH_2CH_2PdCl_2]^- \rightleftharpoons [CH_3CH(OH)PdCl_2]^-$

$[CH_3CH(OH)PdCl_2]^- \rightarrow CH_3CHO + Pd + H^+ + 2\,Cl^-$

For a detailed description, see [43]. The rate of reaction can be given by the following equation [44]:

$$\frac{-d\,c_{C_2H_4}}{dt} = k\,\frac{c_{PdCl_4^{2-}} \cdot c_{C_2H_4}}{c_{H^+} \cdot c_{Cl^-}^2}$$

The rate of reaction is diminished by the acid formed in the reduction of palladium chloride. This can be prevented by buffering the acid with basic copper salts (copper oxychloride, copper acetate). Reformation of the basic copper salts takes place during catalyst oxidation.

One-Stage Process (Fig. 2). Ethylene and oxygen are charged into the lower part of the reaction tower (a); the catalyst is circulated via the separating vessel (b) by the airlift principle and thoroughly mixed with the gas. Reaction conditions are about 130 °C and 400 kPa. An acetaldehyde–water vapor mixture, together with unreacted gas, is with-

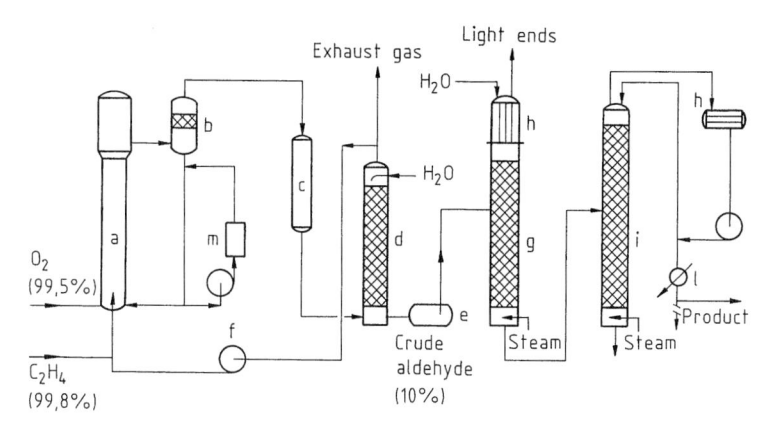

Figure 2. One-stage process
a) Reactor; b) Separating vessel; c) Cooler; d) Scrubber; e) Crude aldehyde tank; f) Cycle-gas compressor; g) Light-ends distillation; h) Condensers; i) Purification column; l) Product cooler; m) Regeneration

drawn from the separating vessel; from this mixture the reaction products are separated by cooling (c) and washing with water (d); unreacted gas is returned to the reactor. A small portion is discharged from the cycle gas as exhaust gas to prevent accumulation of inert gases in the cycle gas; these inert gases are either introduced as contamination of the feed gas (nitrogen, inert hydrocarbons) or formed as byproducts (carbon dioxide). A partial stream of catalyst is heated to 160 °C (m) to decompose byproducts that have accumulated in the catalyst.

Crude acetaldehyde obtained during washing of the reaction products is distilled in two stages. The first stage (g) is an extractive distillation with water in which lights ends having lower boiling points than acetaldehyde (chloromethane, chloroethane, and carbon dioxide) are separated at the top, while water and higher-boiling byproducts, such as acetic acid, crotonaldehyde, or chlorinated acetaldehydes, are withdrawn together with acetaldehyde at the bottom. In the second column (i) acetaldehyde is purified by fractional distillation.

Two-Stage Process (Fig. 3). Tubular reactors (a), (d) are used for both "reaction" and "oxidation". The gases react almost completely in the presence of the catalyst. Reaction of ethylene takes place at 105–110 °C and 900–1000 kPa. Catalyst solution containing acetaldehyde is then expanded in a flash tower (b) by reducing the pressure to atmospheric level. An acetaldehyde–water vapor mixture distills overhead while catalyst is sent via the pump (c) to the oxidation reactor (d), in which it reacts with oxygen at about 1000 kPa. As oxidation and reaction are carried out separately, no high-purity starting gas is required. Generally, air is used instead of oxygen. Oxygen conversion is almost complete; the exhaust air from (e) can be used as inert gas for plant use. The oxidized catalyst solution separated from exhaust air in the separator (e) is reused for the reaction with ethylene in (a).

Acetaldehyde–water vapor mixture from the flash tower (b) is preconcentrated in column (f) to 60–90% acetaldehyde by utilizing the heat of reaction. Process water discharged at the bottom of (f) is returned to the flash tower to maintain a constant

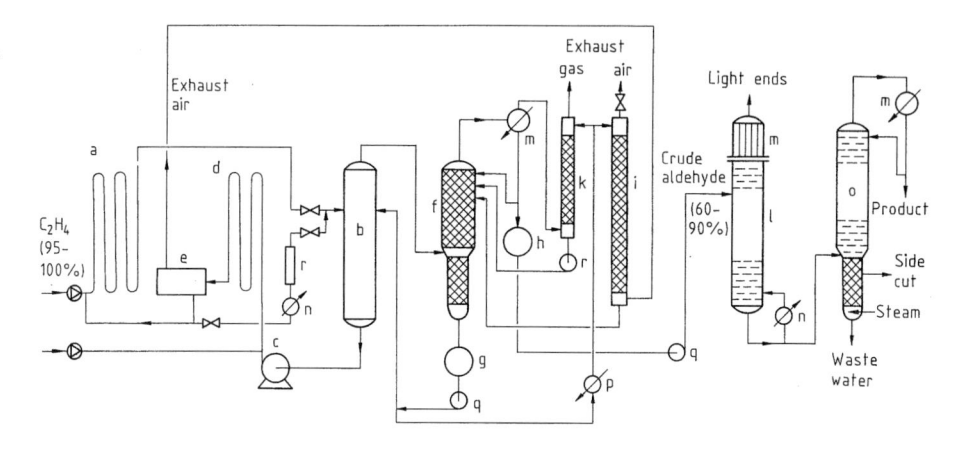

Figure 3. Two-stage process
a) Reactor; b) Flash tower; c) Catalyst pump; d) Oxidation reactor; e) Exhaust-air separator; f) Crude-aldehyde column; g) Process-water tank; h) Crude-aldehyde container; i) Exhaust-air scrubber; k) Exhaust-gas scrubber; l) Light-ends distillation; m) Condensors; n) Heater; o) Purification column; p) Cooler; q) Pumps; r) Regeneration

catalyst concentration. A portion of the process water is used for scrubbing exhaust air (nitrogen from the "oxidation") in (i) and exhaust gas (inert gas from the "reaction") in (k) free of acetaldehyde. Scrubber water then flows to the crude aldehyde column (f).

A two-stage distillation of the crude acetaldehyde follows. In the first stage (l), low-boiling substances, such as chloromethane, chloroethane and carbon dioxide, are separated. In the second stage (o), water and higher-boiling byproducts, such as chlorinated acetaldehydes and acetic acid, are removed from acetaldehyde, and the latter is obtained in pure form overhead. Chlorinated acetaldehydes become concentrated within the column as medium-boiling substances and are discharged laterally. From this mixture, monochloroacetaldehyde can be obtained as the hemihydrate. Residual byproducts can be returned to the catalyst for oxidative decomposition. This oxidative self-purification is supported by thermal treatment of a partial stream of catalyst at about 160–165 °C (regeneration, r).

When gas mixtures obtained in naphtha cracking processes are used as raw material, conventional towers are used as reactors instead of coiled pipes; such mixtures contain 30–40% ethylene in addition to inert hydrocarbons and hydrogen [45].

Comparison of the Two Methods. In both one- and two-stage processes the acetaldehyde yield is about 95% and the production costs are virtually the same. The advantage of using dilute gases in the two-stage method is balanced by higher investment costs. Both methods yield chlorinated hydrocarbons, chlorinated acetaldehydes, and acetic acid as byproducts. Generally, the choice of method is governed by the raw material and energy situations as well as by the availability of oxygen at a reasonable price.

Balance of Reaction and Side Products. The yield in both type of process is nearly the same. The balance of the two stage process is as follows :
100 parts of ethylene gives:

95 parts acetaldehyde
1.9 parts chlorinated aldehydes
1.1 parts unreacted ethylene
0.8 parts carbon dioxide
0.7 parts acetic acid
0.1 parts chloromethane
0.1 parts ethyl chloride
0.3 parts ethane, methane, crotonealdehyde and other minor side products

The chlorinated aldehydes consist of chloroacetaldehyde, dichloroacetaldehyde, trichloroacetaldehyde, and 2-chloro-2-butenal.

Process Variant. An interesting variant of the process, although so far of no technical importance, uses glycol as the reaction medium. The cyclic acetal of acetaldehyde, namely the easily hydrolyzable 2-methyl-1,3-dioloxane, is obtained. An advantage of this method is the high rate of reaction [46].

Construction Materials. During process development, serious problems have been caused by the extremely corrosive aqueous $CuCl_2 - PdCl_2$ solution. These problems have been solved in the two-stage process either by constructing parts in contact with the catalyst solution entirely from titanium or by lining those parts with the metal. In the one-stage process, the reactor is lined with acid-proof ceramic material, the tubing is made of titanium, and certain other parts are of tantalum.

Waste Air. The waste air from the oxidation process contains small amounts of unreacted ethylene, some acetaldehyde, and side products from the reaction such as ethane, chloromethane, chloroethane, and methane. In Germany the waste air must be purified of these side products to meet the criteria of TA-Luft [47]. The byproducts are oxidized over a chromium oxide catalyst. The hydrogen chloride generated is removed by washing, so that the waste air contains mainly carbon dioxide.

Wastewater. Side products of the oxidation process that enter the wastewater are acetic acid, crotonaldehyde and chlorinated aldehydes. Some of the chlorinated aldehydes are highly toxic and show high antimicrobial activity. Therefore they must be treated before entering the wastewater plant to render them biologically degradable. Cleavage of organic chlorine by alkaline hydroyis is a possible method. If the chlorinated compounds can not be destroyed the wastewater has to be incinerated.

4.3.2. Acetaldehyde as Byproduct

Acetaldehyde is also formed in the production of vinyl acetate from ethylene or acetylene (→ Vinyl Esters). It is separated by distillation and is normally converted to acetic acid for reuse. In one version of the method starting from ethylene, vinyl acetate and acetaldehyde are obtained in a molar ratio of 1:1. This makes the process nearly self-sufficient in acetic acid.

4.3.3. Isomerization of Ethylene Oxide

Research was carried out on this process [48] before adopting the direct oxidation of ethylene. Catalysts were Al_2O_3, SiO_2, and acid salts of mineral acids, such as sulfuric acid, phosphoric acid, or molybdic acid. Yields of 90–95 % have been reported, but the process has not gained industrial importance.

4.4. Production from C_1 Sources

Since the increases in oil price in 1973/74 and 1977, C_1 material has gained interest as a feedstock for organic chemicals and as a substitute for petrochemicals. However, for acetaldehyde production, C_1 material seems to be of minor importance because most of the classical acetaldehyde derivatives can be made from C_1 sources.

Production Directly from Synthesis Gas. Acetaldehyde is formed with low selectivity and a yield of ca. 30 % from synthesis gas, together with acetic acid, ethanol, and saturated hydrocarbons, mainly methane. Catalysts are cobalt and rhodium compounds activated by iodine compounds or magnesium chloride and supported on a silicate carrier [49]. There is so far no industrial use of this process.

Production via Methanol, Methyl Acetate, or Acetic Anhydride. *Hydroformylation of methanol* with CO/H_2 has been well known since the discovery of the oxo process [50]. It takes place in the presence of hydroformylation catalysts, such as cobalt, nickel, and iron salts (e.g., $CoBr_2$, CoI_2) or the corresponding metal carbonyls, at increased temperature (180–200 °C) and high pressure (30–40 MPa). Because of the increase in price for petrochemicals, research work in this field recently has been reinstituted by several companies. An acetaldehyde selectivity of 80 % or more has been claimed using an iron–cobalt carbonyl or alternatively a cobalt–nickel catalyst in the presence of tertiary amines, phosphines, or nitriles as the catalyst [51], [52].

Similarly, high selectivity has been claimed for the *hydrocarbonylation of methyl acetate* with palladium or rhodium catalysts in the presence of tertiary phosphines and iodine compounds as well as cobalt–ruthenium catalysts in the presence of methyl and sodium iodides [53]. So far, these processes have not been developed industrially.

Recent patents describe the formation of acetaldehyde by *reduction of acetic anhydride* with hydrogen over palladium or platinum on a carrier at low pressures and moderate temperatures [54]. If the importance of acetaldehyde as an organic intermediate were to decrease in the future as outlined in Chapter 8, this method might be of some interest for the economical production of small quantities of acetaldehyde.

4.5. Production from Hydrocarbons

Acetaldehyde is a byproduct of the production of acrolein, acrylic acid, and propene oxide from propene. It is also formed in the oxidation of saturated hydrocarbons (e.g., propane or butane) in the gas phase as operated by Celanese in the United States [55] (→Acetic Acid).

5. Quality and Analysis

High demands on acetaldehyde purity generally are made; a typical specification is:

Color	practically colorless
Acetaldehyde	more than 99.5 wt%
Acid (as acetic acid)	less than 0.1 wt%
Water	less than 0.02 wt%
Chlorine	less than 30 mg/kg
Dry residue	less than 10 mg/kg

The *acid content* is determined directly by titration, *water content* using the Karl Fischer reagent or empirically from the cloud point of a carbon disulfide-acetaldehyde mixture, and *chlorine content* (mostly in the form of organic chlorine compounds) by combustion in a hydrogen stream and determination of the hydrochloric acid in the condensate.

6. Storage and Transportation

6.1. Storage

For storage of acetaldehyde, the national regulations must be observed. In the *Federal Republic of Germany* the regulation for flammable liquids (Verordnung über brennbare Flüssigkeiten) applies, under which acetaldehyde belongs to: Gefahrenklasse B, Explosionsklasse 1, Temperaturklasse T 4. In the *United States*, loading and storage of acetaldehyde currently are governed by the U.S. Environmental Protection Agency (EPA).

In *Japan* the Fire Defense Law and the supplement "Cabinet Order for Control of Dangerous Articles" applies. According to these regulations, outer storage tanks must not be made of copper, magnesium, silver, mercury, or alloys of these metals. They must be equipped with cooling facilities in order to keep the temperature below 15 °C, and also with an inert-gas sealing system.

For safety data, see Chapter 2.

6.2. Transportation

International Regulations. Acetaldehyde is classified as a flammable liquid. Transportation is governed by the IMDG-Code, no. D 3102, class 3.1, UN-no. 1089, RID, ADR, and ADNR, class 3, no. 5, Rn 301, 2301, and 6301, respectively. Air transportation: IATA-DGR, class 3, packing group PAC forbidden, CAC 304, or ICAO; RAR: art no.6, flammable liquid.

National Regulations. *Federal Republic of Germany:* GGVS (road); GGVE (rail); Gefahrgut V See (sea). *United States:* Regulations of the U.S. Department of Transportation: CFR 49, no. 172 101. *United Kingdom:* Blue Book, flammable liquid, IMDG-Code E 3019. *Japan:* Fire Defense Law: Cargo Transfer Rule; Rule for Transportations by Ship and Storing of Dangerous Materials.

Containers. Types and sizes of containers are recommended by the regulations mentioned above, for example:

1) Gas pressure bottles, maximum size 1 L, with suitable closure to withstand pressure buildup, placed in authorized outer containers.
2) Metal drums, stainless steel or equivalent phenolic resin-lined drums.
3) Insulated tank trucks of stainless steel or phenolic resin-lined.
4) Insulated tank cars, phenolic resin-lined.

In the *Federal Republic of Germany*, under the regulations for dangerous materials, acetaldehyde is classified according to Anhang Nr. 1.1. All containers must be marked with the appropriate symbols and phrases. In the *United States*, besides name and producer label, the Manufacturing Chemists Association recommends labels with the appropriate warning and safety recommendations.

6.3. Other Regulations

In *Germany*, any accidental release of acetaldehyde into the air, water, or soil must be reported to the appropriate authorities according to the regulations for flammable liquids (Verordnung über brennbare Flüssigkeiten, VbF) or the regulations on units for storing, filling, and moving materials dangerous to water supplies (Störfallverordnung, 1980).

In the *United States*, a similar regulation applies according to the Comprehensive Environmental Response, Compensation, and Liability Act of 1980. Incidents must be reported to the National Response Center for Water Pollution.

In *Japan*, acetaldehyde is specified as an Offensive Odor Material by the Offensive Odor Control Law. Under this law a district of dense population is specified, for which the concentration of the material in the air is limited. The maximum level of acetaldehyde allowed at the boundary of the factory or place where acetaldehyde is handled is 0.5 ppm.

Environmental problems are dealt with by the respective national laws (see above). If acetaldehyde is highly diluted with water it can easily be degraded biologically. In higher concentrations it kills bacterial flora.

7. Economic Aspects

Today the most important production process worldwide is the direct oxidation of ethylene. In Western Europe there is also some capacity for the production of acetaldehyde by oxidation of ethanol ($<$ 15%) and hydration of acetylene (2 %). In Eastern Europe the hydration of acetylene is even more important. It is estimated that in Eastern Europe about 235 000 t (44 % of the anual capacity) of acetaldehyde can be produced by this route. The worldwide production of acetaldehyde has been nearly constant since the early 1980s, although in the USA the production of acetic acid from acetaldehyde ceased in 1991. Table 2 gives production data for acetaldehyde.

Table 2. Production of acetaldehyde (10^3 t)

Year	USA	W. Europe	Mexico	Japan	Total
1983	259	554	212	256	1281
1987	270	564	193	285	1312
1990	274	596	190	384	1444
1993	147	606	249	349	1351

Important producers and their production capacities (10^3 t) are listed in the following [56]:

Azot PO, Ukraine	100
BP Chemical Ltd., Italy	140
Doljchim, Romania	45
Eastman Chemical Company, U.S.	250
Erkimia, SA , Spain	75
Hoechst AG, Germany	290
Hüls AG, Germany	80
Japan Aldehyde Company Ltd., Japan	69
Kyowa Yuka Copany Ltd., Japan	62
Mitsui Petrochemical Industries Ltd., Japan	53
Morelos, Mexico	150
Neftechim, Bulgaria	90
Omskiy Zavod Synth. Caoutch., Russia	100
Pajaritos, Mexico	44
Petroleos Mexicanos (PEMEX), Mexico	100
Salavatnephteorgsyntez, Russia	100
Showa Denko K.K., Japan	140
Societe Francaise Hoechst SA, France	96
Tokuyama Petrochemical Company, Ltd., Japan	100
Wacker Chemie GmbH, Germany	65
Zacklady Chemiczne, Poland	90

8. Polymers of Acetaldehyde

8.1. Paraldehyde

Paraldehyde, 2,4,6-trimethyl-1,3,5-trioxane [*123-63-7*], $C_6H_{12}O_3$, M_r 132.161, is a cyclic trimer of acetaldehyde:

Properties. Paraldehyde is colorless and has an ethereal, penetrating odor.

bp	124.35 °C
mp	12.54 °C
Critical temperature t_{crit}	290 °C
Solubility in 100 g water at 13 °C	12 g
at 75 °C	5.8 g

In the solid state, four different crystal forms exist; transition points: 230.3 K, 147.5 K, 142.7 K. Paraldehyde is miscible with most organic solvents.

Density d_4^{20}	0.9923
Refractive index n_D^{20}	1.4049
Viscosity at 20 °C	1.31 mPa · s
Heat of combustion at constant pressure	3405 kJ/mol
Heat capacity c_p at 25 °C	1.947 J g^{-1} K^{-1}
Entropy (l) at 25 °C	2.190 J g^{-1} K^{-1}
Free energy (l) at 25 °C	276.4 J/g
Heat of vaporization	41.4 kJ/mol
Latent heat of melting	104.75 J/g
Heat of formation from acetaldehyde (calculated from combustion enthalpies)	−113.0 kJ/mol

The equilibrium 3 acetaldehyde ⇌ paraldehyde is 94.3 % on the paraldehyde side at 150 °C.

Production. Paraldehyde is produced from acetaldehyde in the presence of acid catalysts, such as sulfuric acid, phosphoric acid, hydrochloric acid, or acid cation exchangers. In the homogeneous reaction, acetaldehyde is added, with stirring and cooling, to paraldehyde containing a small amount of sulfuric acid. After the addition is completed, stirring is continued for some time to establish the equilibrium; the sulfuric acid is exactly neutralized with a sodium salt, such as sodium acetate, sodium carbonate, or sodium bicarbonate; the reaction mixture is separated into acetaldehyde, water, and paraldehyde by fractional distillation [57].

For continous production, liquid acetaldehyde at 15 – 20 °C or acetaldehyde vapor at 40 – 50 °C is passed over an acid cation exchanger [58]. Conversion is greater than 90 %. Acetaldehyde and paraldehyde are separated by distillation. For depolymerization, acetaldehyde is slowly distilled off in the presence of acid catalysts. Paraldehyde also can be decomposed in the gas phase. Catalysts are HCl, HBr, H_3PO_4, or cation exchangers. The reaction is first order. Other catalysts described in the literature are Al_2O_3, SiO_2, $ZnSO_4$, and $MgSO_4$ [59].

Uses. Paraldehyde is used in chemical synthesis as a source of acetaldehyde whereby resin formation and other secondary reactions are largely eliminated. Such synthetic reactions are, for instance, used for the production of pyridines and chlorination of chloral. Between 1939 and 1945 paraldehyde was used as a motor fuel.

8.2. Metaldehyde

Metaldehyde [*9002-91-9*], $C_8H_{16}O_4$, M_r 176.214, is the cyclic tetramer of acetaldehyde:

$$\begin{array}{c} CH_3 \\ | \\ O-CH-O \\ CH_3-CH \qquad CH-CH_3 \\ O-CH-O \\ | \\ CH_3 \end{array}$$

Properties. Metaldehyde forms tetragonal prisms, *mp* (closed capillary) 246.2 °C, sublimation temperature (decomp.) 115 °C, heat of combustion at constant volume 3370 kJ/mol.

Metaldehyde is insoluble in water, acetone, acetic acid, and carbon disulfide.

Depolymerization of metaldehyde to acetaldehyde begins at 80 °C, and is complete above 200 °C. Depolymerization takes place faster and at lower temperatures in the presence of acid catalysts, such as dilute H_2SO_4 or H_3PO_4. Metaldehyde does not show the typical acetaldehyde reactions. It is stabilized by ammonium carbonate or other weakly basic compounds which neutralize acidic potential catalysts.

Production. Metaldehyde is obtained in addition to large amounts of paraldehyde during polymerization of acetaldehyde in the presence of HBr and alkaline earth metal bromides, such as $CaBr_2$, at temperatures below 0 °C. However, yields are scarcely higher than 8 %. Yields of 14 – 20 % have been reported when working in the presence of 7 – 15 % of an aliphatic or cyclic ether at 0 – 20 °C [60]. Insoluble metaldehyde is filtered out. Acetaldehyde is then distilled from the filtrate following depolymerization of the paraldehyde and is returned to the polymerization. Recycling of the large amounts of acetaldehyde results in losses that increase the process costs.

Uses. Metaldehyde in pellet form is marketed as a dry fuel (Meta). Mixed with a bait, metaldehyde is used today as a molluscicide.

8.3. Polyacetaldehyde

Polyacetaldehyde [*9002-91-9*] is a high-molecular-mass polymer with an acetal structure (polyoxymethylene structure):

$$\begin{array}{c} -CH-O-CH-O-CH-O- \\ | \qquad | \qquad | \\ CH_3 \quad CH_3 \quad CH_3 \end{array}$$

By using cationic initiators, mainly an amorphous polymer is obtained. Temperatures below –40 °C are preferred in this case. Above –30 °C, mainly paraldehyde and metaldehyde are produced. The initiator activity also depends on the solvent used. Suitable initiators include H_3PO_4 in ether and pentane, as well as HCl, HNO_3, CF_3COOH, $AlCl_3$ in ether, and particularly BF_3 in liquid ethylene [61]. Al_2O_3 and SiO_2 also seem to be good initiators [62].

The polymer has a rubber-like consistency and is soluble in common organic solvents. It depolymerizes at room temperature, liberating acetaldehyde. It evaporates completely within a few days or weeks. Acidic compounds accelerate depolymerization, and amines (e.g., pyridine) stabilize polyacetaldehyde to a certain extent. A complete stabilization (as, for instance, in the case of polyformaldehyde) has not yet been achieved, so the polymer is still of no practical importance.

Copolymers with propionaldehyde, butyraldehyde, and allylacetaldehyde also have been produced [63]. Crystalline, isotactic polymers have been obtained at low temperatures (for example, –75 °C) by using anionic initiators [64]. Suitable initiators are alkali metal alkoxides, alkali metals, or metal alkyls in hydrocarbon solvents. The products are insoluble in common organic solvents but have an acetal structure like the amorphous polymers' [65]. Polymerization of acetaldehyde to poly(vinyl alcohol), which in contrast to polyacetaldehyde has a pure carbon backbone, has not yet been achieved [66].

9. Toxicology and Occupational Health

Acetaldehyde. At higher concentrations (up to 1000 ppm), acetaldehyde irritates the mucous membranes. The perception limit of acetaldehyde in air is in the range between 0.07 and 0.25 ppm [67], [68]. At such concentrations the fruity odor of acetaldehyde is apparent. Conjunctival irritations have been observed after a 15-min exposure to concentrations of 25 and 50 ppm [69], but transient conjunctivitis and irritation of the respiratory tract have been reported after exposure to 200 ppm acetaldehyde for 15 min [69], [70]. The penetrating odor, the low perception limit, and the irritation that acetaldehyde causes, give an effective warning so that no serious cases of acute intoxication with pure acetaldehyde have been reported. Acute acetaldehyde intoxication can also be observed following combined ingestion of disulfiram (Antabuse) and ethanol [73]. In animal experiments at high concentrations (3000 – 20 000 ppm), pulmonary edema and a narcotic effect become evident. The clinical course is similar to alcohol intoxication. Death occurs by breath paralysis or — with retardation — by pulmonary edema.

For rats, the LC_{50} (30 min inhalation) is 20 500 ppm [71]. No studies on subchronic or chronic toxicity of acetaldehyde in humans are available. Investigations of sister chromatid exchange in cell cultures [72] and in human lymphocytes [74] and studies of single- and double-strand breaks in human lymphocytes incubated with acetaldehyde [75] have revealed mutagenic effects of acetaldehyde. Long-term exposures of Syrian golden hamsters to concentrations in the range 1650–2500 ppm have resulted in inflammatory hyperplastic and metaplastic alterations of the upper respiratory tract with an increase in carcinomas of the nasal mucosa and the larynx [76]. Male and female Wistar rats were exposed to aldehyde concentrations of 0 ppm, 750 ppm, 1500 ppm for 6 h daily, 5 d per week for 52 weeks. In the highest dose group, the initial aldehyde concentration of 3000 ppm was reduced to 1000 ppm in the course of the study due to toxic effects. The dose-dependent effects found were increased mortality in all dose groups and delayed growth in the middle- and highest-dose group. Adenocarcinomas were observed at all three investigated concentrations. An increased rate of squamous epithelial carcinomas was only seen at 1500 ppm or more. Histological signs of irritation were observed in the larynx region in most animals from the medium- and high-dose groups [77]. Concomitant exposure to acetaldehyde also considerably increases the number of tracheal carcinomas induced by instillation of benzo[a]pyrene [72]. This suggests that chronic tissue injury is a prerequisite for tumor formation by acetaldehyde. Tumors probably do not develop if doses are not sufficient to cause tissue necrosis. In male Wistar rats exposed to 150 ppm or 500 ppm for 6 h per day, 5 d per week for 4 weeks, morphological changes in the olfactory epithelium were observed in the high-dose group [78]. However, long-term toxicity data at lower exposure concentrations are not yet available.

Currently, the TLV is 25 ppm (STEL/ceiling value) [79], and the MAK is 50 ppm [80]; the latter value is preliminary. At 50 ppm acetaldehyde, no irritation or local tissue damage in the nasal mucosa is observed. Because the mechanism of action is assumed to be analogous to that of formaldehyde, acetaldehyde is regarded as a suspected carcinogen [80].

When taken up by the organism, acetaldehyde is metabolized rapidly in the liver to acetic acid. Only a small proportion is exhaled unchanged. After intravenous injection, the halflife in the blood is approximately 90 s [81].

Paraldehyde acts as a sedative with few side effects. Ingested paraldehyde partly is metabolized to carbon dioxide and water and partly is exhaled unchanged. It generates an unpleasant odor in the expired air and therefore is not much used.

Metaldehyde decomposes slowly to acetaldehyde in the presence of acids, so ingestion may cause irritation of the gastric mucosa with vomiting. As characteristic signs of a metaldehyde intoxication, especially in children, heavy convulsions (sometimes lasting several days) have been reported, with lethal outcomes, after ingestion of several grams of metaldehyde [82]. For these reasons, those molluscicides and solid fuels which contain metaldehyde must be kept away from children.

10. References

General References

Information on United States and Japanese regulations was submitted to the author by Dr. Joe L. Sadler, Celanese Chemical Co. and by Mr. Shinichi Oka, Celanese Chemical Co. and by Mr. Shinichi Oka, Mitsubishi Chemical Ind., Tokyo, respectively. Their contributions are gratefully acknowledged.

[1] *Beilstein* **1**, 594; **1**, 1st suppl., 1, 321; **1**, 2nd suppl., 1, 654; **1**, 2nd suppl., 1, 654; **1**, 3rd suppl., 1, 2617; **1**,4th suppl., 3094.
[2] S. A. Miller (ed.): *Acetylene*, E. Benn, London 1965.
[3] S. A. Miller (ed.): *Ethylene*, E. Benn, London 1969.
[4] R. Jira in [3] : pp. 639 and 650.
[5] R. Sieber in [3] : pp. 659 and 668.
[6] R. Page in [3] : p. 767.
[7] W. Reppe: *Chemie und Technik der Acetylen-Druck-Reaktionen*, Verlag Chemie, Weinheim 1951.
[8] *Review on Manufacturing Processes of Wacker, Lonza, IG-Farben factories Hüls, Gendorf, Schkopau and Ludwigshafen, Stickstoff AG, Knapsack,* BIOS Final Report 1049, item no. 22 (1946).
[9] *Report on the IG Works Hüls and Gendorf,* FIAT Final Report 855 (1946).
[10] *Acetaldehyde from Acetylene,* BIOS Final Report 370.
[11] *Kirk-Othmer,* 3rd ed., **1**, 97–112.

Specific References

[12] R. C. Weast, H. J. Astle (eds.): *CRC Handbook of Chemistry and Physics,* 60th ed., CRC Press, Boca Raton, Florida 1979–1980.
[13] D'Ans-Lax, *Taschenbuch für Chemiker und Physiker,* 3rd ed., Springer Verlag, Berlin 1964.
[14] Th. F. Smith, R. F. Bonner, *Ind. Eng. Chem.* **43** (1951) 1169.
[15] T. R. Das, N. R. Kuloor, *J. Indian Inst. Sci.* **50** (1968) 45.
[16] P. M. Craven, J. D. Lambert, *Proc. R. Soc. (London)* **205 A** (1951) 439 and 444.
[17] R. W. Gallant, *Hydrocarbon Process.* **47** (1968) no. 5, 151.
[18] A. K. Shaka, *Indian J. Phys.* **6** (1931) 449.
[19] L. P. Filippov, *Vestn. Mosk. Univ.* **9**, no. 12, *Ser. Fiz. Mat. Estestv. Nauk.* (1954) no. 8, 45–48.
[20] J. D. Lambert, E. N. Staines, S. D. Woods, *Proc. R. Soc. (London)* **200 A** (1950) 262.
[21] *Landolt-Börnstein,* 6th ed., vol. **2**, part 4, Springer Verlag, Berlin 1961.
[22] M. M. Brazhnikov, A. D. Peshchenko, O. V. Ral'ko, *Zh. Prikl. Khim (Leningrad)* **49** (1976) no. 5, 1041; *Chem. Abstr.* **85** (1976) 45 851.
[23] J. D. Lambert, G. A. H. Roberts, J. S. Rowlington, V. J. Wilkinson, *Proc. R. Soc. (London)* **200 A** (1960) 262.
[24] P. M. Chaudhuri, R. A. Stanger, G. P. Mathur, *J. Chem. Eng. Data* **13** (1968) 9–11.
[25] A. A. Dobrinskaya, V. G. Markovich, M. B. Neiman, *Isv. Akad. Nauk. S.S.S.R., Otd. Khim. Nauk.* 1953, 434–441;*Bull. Acad. Sci. U.S.S.R. Div. Chem. Sci. (Engl. Transl.)* 1953, 391–398; *Chem. Abstr.* **49** (1955) 4378.
[26] *Landholt-Börnstein,* 6th ed., vol. **2**, part 2, Springer Verlag, Berlin 1961.
[27] G. O. Morrison, T. P. G. Shaw, *Trans. Electrochem. Soc.* **63** (1933) 425.
[28] B. P. Mullins, *Fuel* **32** (1953) 481.
[29] F. C. Mitchell, H. C. Vernon, *Chem. Met. Eng.* **44** (1937) 733.

[30] K. Sennewald, *Erdöl Kohle* **12** (1959) 364.
[31] *Petroleum Refiner* **40** (1961) no. 11, 308.
[32] *Ullmann*, 4th ed., **9**, 6.
[33] *Ullmann*, 4th ed., **7**, 374.
[34] SU 287 919, 1970.
[35] Heavy Minerals Co., US 2 884 460, 1955 (V. I.Komarevsky).
[36] Knapsack-Griesheim, DE 1 097 969, 1954 (W. Opitz, W. Urbanski); DE 1 108 200, 1955 (W. Opitz, W. Urbanski).
[37] W. L. Faith, D. B. Keyes, R. L. Clarks: *Industrial Chemicals*, 3rd ed., J. Wiley & Sons, New York 1965, p. 2.
[38] Shell Development Co., US 2 883 426, 1957 (W. Brackman). Eastman Kodak Co., US 3 106 581, 1963 (S. D.Neely). Veba-Chemie, DE 1 913 311, 1969 (W. Ester, W. Hoitmann).
[39] *Petroleum Refiner* **36** (1957) 249.
[40] *Ullmann*, 3rd ed., **3**, 4.
[41] K. Kon, T. Igarashi, *Ind. Eng. Chem.* **48** (1956) 1258.
[42] Consortium für elektrochemische Industrie, DE 1 049 845, 1957 (W. Hafner, J. Smidt, R. Jira, R. Rüttinger, J. Sedlmeier). DE 1 061 767, 1957 (W. Hafner, J. Smidt, R. Jira, J. Sedlmeier); DE 1 080 994, 1957 (W. Hefner, J. Smidt, R. Jira). J. Smidt, W. Hafner, R. Jira, J. Sedlmeier, R. Sieber, R. Rüttinger, H. Kojer, *Angew. Chem.* **71** (1959) 176; *ibid.* **74** (1962); *Angew. Chem. Int. Ed. Engl.* **1** (1962) 80; *Chem. Ind.* 1962, 54.
[43] R. Jira: "Oxidations," in W. A.Herrmann (ed.): *Applied Homogeneous Catalysis with Organometallic Compounds*, vol. **1** VCH, Weinheim 1996.
[44] I. I. Moiseev, M. N. Vargaftik, Y. K. Sirkin, *Dokl. Akad. Nauk SSSR* **153** (1963) 140. P. M. Henry, *J. Am. Chem. Soc.* **86** (1964) 3246.
[45] Consortium für elektrochemische Industrie, DE 1 215 677, 1964 (W. Hefner, R. Jira, J. Smidt).
[46] Lummus Co., BE 668 601, 1965. W. G. Lloyd, *J. Org. Chem.* **34** (1969) 3949.
[47] Erste Allgemeine Verwaltungsvorschrift zum Bundes-Immissionsschutzgesetz (Technische Anleitung zur Reinhaltung der Luft TA-Luft) vom 27.02.1986, GMBL 1986, S. 95.
[48] Sicedison, FR 1 316 720, 1961 (J. Herzenberg, P. Gialtoni). British Petroleum Co., FR 1 367 963, 1963. Hoechst, US 3 067 256, 1959 (K. Fischer, K. Vester).
[49] Rhône-Poulenc, EP 0 011 043, 1979 (J. Gauthier-Lafage, R. Perron).Union Carbide, US 4 235 798, 1980 (W. J. Bartley, T. P. Wilson, P. C. Ellgen). Hoechst, DE-OS 2 814 365, 2 825 495, 2 825 598, 1978 (H. J. Arpe, E.-J. Leupold, F. A. Wunder, H.-J. Schmidt).
[50] Du Pont, US 2 457 204, 1946 (R. E. Brooks). US 3 356 734, 1963 (M. Kuraishi, S. Asano, A. Takahashi).
[51] Exxon Research and Eng. Co., EP 0 027 000, 1980 (G. Doyle). Celanese Corp., US 4 201 868, 1980 (W. E. Slinkard). Union Rheinische Braunkohlen Kraftstoff, DE-OS 2 913 677, 1979 (J. Korff, M. Fremery, J. Zimmermann).
[52] Gulf Research & Development Co., US 4 239 704/5, 1980 (W. R. Pretzer, T. P. Kobylinsky, J. E. Bozik). British Petroleum Co., EP 0 029 723, 1980 (M. T.Barlow).
[53] Rhône-Poulenc, EP 0 046 128/9, 1981 (J. Gauthier-Lafage, R. Perron). Halcon Research & Development Corp., US 4 302 611, 1981; BE 890 376, 1982 (R. V. Porcelli).
[54] Halcon SD Group, FR 8 118 437/8, 1981 (D. Moy). Kuraray Co., EP 0 040 414, 1981 (S. Nakamura, M. Tamura).
[55] *Ullmann*, 3rd ed., suppl. vol., p. 172.
[56] W. K. Johnson, A. Leder, Y. Sakuma: Acetaldehyde, Chemical Economics Handbook, SRI International 1995.

[57] Eastman Kodak Co., US 2 318 341, 1937 (B. Thompson).
[58] Publicker Ind., US 2 479 559, 1947 (A. A. Dolnick et al.). Melle-Bezons, DE-OS 1 927 827, 1969 (M. G.Gobron, M. M. Repper).
[59] T. Kawaguchi, S. Hasegawa, *Tokyo Gakugei Daigaku* **21** (1969) 63.
[60] Publicker Ind., US 2 426 961, 1944 (R. S. Wilder).
[61] O. Vogl, *J. Polym. Sci., Part A* **2** (1964) 4591; H. Staudinger, *Trans. Faraday Soc.* **32** (1936) 249;H. A. Rigby, C. J. Danby, C. N. Hinshelwood, *J. Chem. Soc.* 1948, 234.
[62] J. T. Furukawa, T. Saegusa, T. Tsuruta, H. Fujii, T. Tatana, *J. Polym. Sci.* **36** (1959) 546. J. T. Furukawa, T. Saegusa, T. Tsuruta, H. Fujii, A. Kawazaki, *Makromol. Chem.* **33** (1960) 32. Consortium für elektrochemische Industrie, DE 1 106 075, 1958 (J. Smidt, J. Sedlmeier).
[63] Consortium für elektrochemische Industrie, DE 1 292 394, 1959 (J. Smidt, J. Sedlmeier).
[64] Bridgestone Tire Co., FR 1 268 322 and 1 268 191, 1960 (J. Furukawa, T. Tsuruta, T. Saegusa, H. Fujii). J. T. Furukawa, *Makromol. Chem.* **37** (1969) 149.G. Natta, *Makromol. Chem.* **37** (1960) 156; *J. Polym. Sci.* **51** (1960)505.
[65] O. Vogl, *J. Polym. Sci., Part A* **2** (1964) 4607.
[66] T. Imoto, T. Matsubara, *J. Polym. Sci.* **56** (1962) 5.
[67] C. P. McCord: *Odors, Physiology and Control*, McGraw-Hill, New York 1949.
[68] W. Summer: *Odour Pollution of Air Causes and Control*, Chemical and Process Engineering Series, Leonard Hill, London 1971.
[69] L. Silverman, H. F. Schulte, M. W. First, *J. Ind. Hyg. Toxicol.* **28** (1946) 265.
[70] V. M. Sim, R. E. Pattle, *J. Am. Med. Assoc.* **165** (1957) 1908.
[71] E. Skog, *Acta Pharmacol.* **6** (1950) 299.
[72] G. Obe, H. J. Ristow, *Mutat. Res.* **58** (1978) 115.
[73] J. Becker, H. Desel, H. Schuster, G.F. Kahl, *Ther. Umsch.* **52** (1995) 183.
[74] G. Obe, R. Jonas, S. Schmidt, *Mutat. Res.* **174** (1986) 47.
[75] N. P. Singh, A. Kahn, *Mutat. Res.* **337** (1995) 9.
[76] V. J. Feron, A. Kruysse, R. A. Woutersen, *Eur. J. Cancer Clin. Oncol.* **18** (1982) 13.
[77] R. A. Woutersen, L. M. Appelman, A. van Garderen-Hoetmer, V. J. Feron, *Toxicology* **41** (1986) 213.
[78] L. M. Appelman et al., Report No. V84.382/140327, CIVO Institutes TNO, NL-3700AJ Zeist, The Netherlands, 1985.
[79] *1997 TLVs and BEIs*. American Conference of Governmental Industrial Hygienists Inc., Cincinnati, Ohio 1997.
[80] DFG Deutsche Forschungsgemeinschaft: *Occupational Toxicants: Critical Data for MAK Values and Classification of Carcinogens*, vol. **3**, Commission for the Investigation of Health Hazards of Chemical Compounds in the Work Area, VCH Verlagsgesellschaft, Weinheim 1992.
[81] K. J. Freundt, *Naunyn-Schmiedeberg's Arch. Exp. Pathol. Pharmakol.* **260** (1968) 111; *Beitr. Gerichtl. Med.* **27** (1970) 368.
[82] O. R. Klimmer: *Pflanzenschutz- und Schädlingsbekämpfungsmittel, Abriß einer Toxikologie und Therapie von Vergiftungen*, Hundt-Verlag, Hattingen 1972.

Acetic Acid

ADOLFO AGUILÓ, Celanese Chemical Company Technical Center, Corpus Christi, Texas 78469, United States

CHARLES C. HOBBS, Celanese Chemical Company Technical Center, Corpus Christi, Texas 78469, United States

EDWARD G. ZEY, Celanese Chemical Company Technical Center, Corpus Christi, Texas 78469, United States

1.	Introduction	30
2.	Physical Properties	30
3.	Chemical Properties	33
4.	Production	34
4.1.	Carbonylation of Methanol	34
4.2.	Direct Oxidation of Saturated Hydrocarbons	39
4.3.	Acetaldehyde Process	43
4.4.	Other Processes	46
4.5.	Concentration and Purification	47
4.6.	Materials of Construction	48
5.	Wastewater and Off-Gas Problems	48
6.	Quality Specifications	49
7.	Chemical Analysis	49
8.	Storage, Transportation, and Customs Regulations	50
9.	Uses	50
10.	Derivatives	50
10.1.	Salts	51
10.1.1.	Aluminum Acetate	51
10.1.2.	Ammonium Acetate	52
10.1.3.	Alkali-Metal Salts	52
10.2.	Esters	52
10.2.1.	Methyl Acetate	53
10.2.2.	Ethyl Acetate	53
10.2.3.	Butyl Acetate	54
10.2.4.	2-Ethylhexyl Acetate	54
10.2.5.	Other Esters	55
10.3.	Acetyl Chloride	55
10.4.	Amides	56
10.4.1.	Acetamide	56
10.4.2.	N,N-Dimethylacetamide	56
10.5.	Phenylacetic Acid	57
11.	Economic Aspects	57
12.	Toxicology and Occupational Health	58
13.	References	59

1. Introduction

Acetic acid [64-19-7], CH$_3$COOH, M_r 60.05, is a colorless, corrosive liquid. It has a pungent odor and is a dangerous vesicant. It is found in dilute solutions in many plant and animal systems. Vinegar (4–12% acetic acid solutions produced by the fermentation of wines) has been known for more than 5000 years.

The major producers of synthetic acid are currently the United States, Western Europe, Japan, Canada, and Mexico. The total capacity in these countries is close to 4×10^6 t/a and production is 3×10^6 t/a. The largest end uses are in the manufacture of vinyl acetate and acetic anhydride. Vinyl acetate is used in the production of latex emulsion resins for application in paints, adhesives, paper coatings, and textile treatments. Acetic anhydride is used in the manufacture of cellulose acetate textile fibers, cigarette filter tow, and cellulosic plastics.

2. Physical Properties

Acetic acid, *mp* 16.66 °C [1], *bp* 117.9 °C at 101.3 kPa [2] (this value differs somewhat from that in Table 4), is a clear, colorless liquid. Normal impurities in acetic acid are trace amounts of acetaldehyde, other oxidizable substances, and water. Glacial acid (acetic acid containing <1% water) is very hygroscopic. The presence of 0.1 wt% water lowers the melting point by about 0.2 °C [3].

Acetic acid has a pungent odor and taste and solidifies into colorless, lamellar, icelike crystals on freezing. The freezing point can be used to determine the purity. An example of this for acetic acid–water mixtures is given in Table 1.

The density of aqueous acetic acid (Table 2) goes through a maximum between 77 and 80 wt% at 15 °C. This maximum corresponds to the monohydrate (77% acetic acid). The density of acetic acid as a function of temperature also has been determined [4] (Table 3).

The vapor pressure of pure acetic acid is given in Table 4 [5]. The density of the vapor corresponds to approximately twice the molecular mass because of vapor-phase hydrogen bonding [4]. Both dimeric and tetrameric hydrogen bonded species have been proposed:

Table 1. Freezing points for various acetic acid–water mixtures

wt % CH$_3$COOH	fp, °C	wt % CH$_3$COOH	fp, °C
100	16.75	96.8	11.48
99.6	15.84	96.4	10.83
99.2	15.12	96.0	10.17
98.8	14.49	93.46	7.1
98.4	13.86	80.6	-7.4
98.0	13.25	50.6	-19.8
97.6	12.66	18.11	-6.3
97.2	12.09		

Table 2. Densities of aqueous acetic acid solutions at 15 °C

wt % CH$_3$COOH	ϱ, g/cm^3	wt % CH$_3$COOH	ϱ, g/cm^3
1	1.007	60	1.0685
5	1.0067	70	1.0733
10	1.0142	80	1.0748
15	1.0214	90	1.0713
20	1.0284	95	1.0660
30	1.0412	97	1.0625
40	1.0523	99	1.0580
50	1.0615	100	1.0550

Table 3. Dependence of the density of pure acetic acid on temperature

t, °C	ϱ, g/cm^3	t, °C	ϱ, g/cm^3
26.21	1.0420	97.42	0.9611
34.10	1.0324	106.70	0.9506
42.46	1.0246	117.52	0.9391
51.68	1.0134	129.86	0.9235
63.56	1.0007	139.52	0.9119
74.92	0.9875	145.60	0.9030
85.09	0.9761	156.40	0.8889

Table 4. Vapor pressure of pure acetic acid

t, °C	p, mbar	t, °C	p, mbar
0	4.7	150.0	2 461.1
10	8.5	160	3 160
20	15.7	170	4 041
30	26.5	180	5 091
40	45.3	190	6 333
50	74.9	200	7 813
60	117.7	210	9 612
70	182.8	220	11 733
80	269.4	230	14 249
90	390.4	240	17 057
100	555.3	250	20 210
110	776.7	260	23 854
118.2	1013	270	28 077
130.0	1386.5	280	32 801
140.0	1841.1		

Vapor – liquid equilibria for the acetic acid – chloroacetic acid system confirm the association of the acid molecules [6]. Vapor – liquid equilibria of other acetic acid binary and multicomponent systems have been studied [7].

Heat is evolved when mixing water with acetic acid (at 15 – 18 °C), up to 32 % acid; at higher acid concentrations heat is absorbed [8]. The measured values of the heat of mixing are consistent with the calculated values based on the dimers and tetramers described above [9].

Other physical properties of acetic acid are listed below.

Specific heat capacity
Gaseous acid, c_p 1.110 J g^{-1} K^{-1} at 25 °C [10]
Liquid acid, c_p 2.043 J g^{-1} K^{-1} at 19.4 °C
Crystalline acid, c_p 1.470 J g^{-1} K^{-1} at 1.5 °C
  0.783 J g^{-1} K^{-1} at –175.8 °C
Heat of melting 195.5 J/g
Heat of vaporization 394.5 J/g at *bp*

Viscosity				11.83 mPa · s at 20 °C [11]		
				10.97 mPa · s at 25 °C [12]		
				8.18 mPa · s at 40 °C [11]		
Dielectric constant				6.170 at 20 °C (liquid) [14]		
				2.665 at −10 °C (solid) [15]		
Refractive index n_D^{20}				1.3719 [16]		
Enthalpy of formation						
$\Delta H°$ (l, 25 °C)				−484.50 kJ/mol [17]		
$\Delta H°$ (g, 25 °C)				−432.25 kJ/mol [17]		
Heat of combustion ΔH_c (l)				−874.8 kJ/mol [18]		
Normal entropy						
$S°$ (l, 25 °C)				159.8 J mol^{-1} K^{-1} [17]		
$S°$ (g, 25 °C)				282.5 J mol^{-1} K^{-1} [17]		
Flash point				43 °C (closed cup) [19]		
Autoignition point				465 °C [19]		
Flammability				4.0 to 16.0 vol% in air [19]		
Critical data						
p_c				5.786 MPa [20]		
T_c				592.71 K		
Surface tension [13]:						
t, °C	20.1	23.1	26.9	42.3	61.8	87.5
σ, mN/m	27.57	27.25	26.96	25.36	23.46	20.86

Acid dissociation constants (in water) [16]:

t, °C	0	25	50
pK_a	4.78	4.76	4.79

3. Chemical Properties

Although acetic acid is not unusually reactive, many useful and commercially valuable materials can be prepared from it. Several of these compounds are discussed in greater detail (see Chap. 9). Acetic acid reacts with alcohols or olefins to form various esters [21], [22]. Acetamide is prepared by the thermal decomposition of ammonium acetate [23]. Acetic acid also can be converted to acetyl chloride using chlorinating agents, such as phosphorus trichloride or thionyl chloride [23].

Acetic acid is a raw material for a number of commercial processes. It can be converted to vinyl acetate on treatment with ethylene in the presence of noble metal catalysts [24]. Acetic acid also is used in the manufacture of acetic anhydride (→Acetic Anhydride) (via ketene generation) [25], and chloroacetic acid (→Chloroacetic Acids) [26].

4. Production

Table vinegar is still occasionally made by fermentation. However, the most important synthetic routes to acetic acid are methanol carbonylation and the liquid-phase oxidation of *n*-butane, naphtha, or acetaldehyde. In addition some acetic acid is recovered as a byproduct, mainly from poly(vinyl alcohol) production. *Methanol carbonylation* has been the technology of choice for new capacity in the last 10 years [27], and it will be the preferred route in the near future because of its favorable raw material and energy costs. The synthesis gas raw material required for this process can be obtained from a variety of sources, ranging from natural gas to coal.

Processes that have been studied but have not been commercialized are the vapor-phase oxidation of *n*-butenes, ethylene, and saturated hydrocarbons [28], and the liquid-phase oxidation of *sec*-butyl acetate (from *n*-butenes).

4.1. Carbonylation of Methanol

The manufacture of acetic acid from methanol [67-56-1] and carbon monoxide [630-08-0] at high temperature and high pressure was described by BASF as early as 1913 [28].

$$CH_3OH + CO \longrightarrow CH_3COOH \quad \Delta H = -138.6 \text{ kJ}$$

The extreme conditions of temperature and pressure and the highly corrosive substances (iodides) used in the process impeded commercialization. In 1941 Reppe's work at BASF demonstrated the efficiency of group VIII metal carbonyls as catalysts for carbonylation reactions, including hydroformylation [29], [30]. This work led to the development of a high-pressure, high-temperature process, 70 MPa (700 bar) and 250 °C, with a cobalt iodide catalyst. The process was commercialized in 1960 by BASF in Ludwigshafen, Federal Republic of Germany, [29], [31]–[33]. The initial capacity of 3600 t/a was expanded to 10 000 t/a in 1964 and in 1970 to 35 000 t/a. The 1981 capacity was 45 000 t/a [27]. In 1966 Borden Chemical Co. started up an acetic acid unit in Geismar, Louisiana, United States, based on BASF technology [29], [31]. The original capacity of 45 000 t/a grew to 64 000 t/a in 1981 [27].

In 1968 Monsanto reported the discovery of a new iodide-promoted rhodium catalyst with remarkable activity and selectivity for methanol carbonylation to form acetic acid. Methanol can be carbonylated even at atmospheric pressure with yields of 99 % based on methanol and 90 % based on carbon monoxide [34]. This catalytic process was commercialized by Monsanto in 1970 at Texas City, Texas. The initial capacity was 135 000 t/a, expanded to 180 000 t/a since 1975. Operating conditions in the reactor are much milder (3 MPa and 180 °C) than in the BASF process [35].

Chemistry and Reaction Conditions. The chemistry of both the BASF and the Monsanto processes is similar, but the kinetics are different, indicating different rate-determining steps. In both systems there are two important catalytic cycles, one that involves the metal carbonyl catalyst and one that involves the iodide promoter [36].

The BASF process uses a cobalt carbonyl catalyst with an iodide promoter. Cobalt(II) ionide [15238-00-3] is used for in situ generation of $Co_2(CO)_8$ [10210-68-1] and hydrogen iodide [10034-85-2]. As mentioned above, severe conditions are required to give commercially acceptable reaction rates. The rate of reaction depends strongly on both the partial pressure of carbon monoxide and the methanol concentration. Acetic acid yields are 90% based on methanol and 70% based on carbon monoxide. The mechanism of the reaction can be represented by the following reaction sequence [29]:

$$Co_2(CO)_8 + H_2O + CO \longrightarrow 2\ Co(CO)_4H + CO_2 \qquad (1)$$

$$CH_3OH + HI \rightleftharpoons CH_3I + H_2O \qquad (2)$$

$$HCo(CO)_4 \rightleftharpoons H^+ + [Co(CO)_4]^- \qquad (3)$$

$$[Co(CO)_4]^- + CH_3I \longrightarrow CH_3Co(CO)_4 + I^- \qquad (4)$$

$$CH_3Co(CO)_4 \longrightarrow CH_3\overset{O}{\overset{\|}{C}}-Co(CO)_3 \qquad (5)$$

$$CH_3\overset{O}{\overset{\|}{C}}-Co(CO)_3 + CO \rightleftharpoons CH_3\overset{O}{\overset{\|}{C}}-Co(CO)_4 \qquad (6)$$

$$CH_3\overset{O}{\overset{\|}{C}}-Co(CO)_4 + HI \longrightarrow CH_3COI + H^+ + [Co(CO_4)]^- \qquad (7)$$

$$CH_3COI + H_2O \longrightarrow CH_3COOH + HI \qquad (8)$$

Equation (1) can be considered a water-gas shift reaction (see Eq. 9) in which $Co_2(CO)_8$ is a catalyst and the hydrogen formed is dissociated via the hydridocarbonyl complex. Subsequently the methyl iodide undergoes nucleophilic attack by the $[Co(CO)_4]^-$ anion, Equation (4). Iodide facilitates this reaction because it is a better leaving group than OH^-. The CH_3I reacts with a coordinatively saturated d^{10} complex, and therefore oxidative addition of CH_3I is more difficult than addition to $[Rh(CO)_2I_2]^-$ (see below), a coordinatively unsaturated d^8 complex. Also the initial adduct with Co, $CH_3-Co(CO)_4$, is a five-coordinate d^8 species, which is the preferred configuration of cobalt(I). Therefore, the methyl migration step represented by Equation (5) is less favored than the same process for the rhodium(III) species.

Once formed, the acyl intermediate on cobalt, Equation (6), cannot undergo simple reductive elimination to acetyl iodide because iodide is not coordinated to cobalt. All of the individual steps involved in the otherwise similar mechanisms can be assumed to be of lower rate for cobalt than for rhodium. This explains the higher temperature needed for the BASF process. In addition, higher carbon monoxide partial pressures are required to stabilize the [Co(CO)$_4$]$^-$ complex at the higher reactor temperatures.

Byproducts in the BASF process are CH$_4$, CH$_3$CHO, C$_2$H$_5$OH, CO$_2$, C$_2$H$_5$COOH, alkyl acetates, and 2-ethyl-1-butanol [37]–[39]. About 3.5% of the methanol reactant leaves the system as methane, 4.5% as liquid byproducts, and 2% is lost as off-gas. Some 10% of the CO feed is converted to CO$_2$ through the water-gas shift reaction:

$$CO + H_2O \longrightarrow CO_2 + H_2 \qquad (9)$$

The Monsanto process with rhodium carbonyl catalyst [38255-39-9] and iodide promoters operates under much milder conditions than the BASF process. As a consequence, methanol efficiency is 99% and carbon monoxide efficiency is probably close to 90% [34]. The system is not as sensitive to H$_2$ as the BASF process and therefore reduction products, if present, are very insignificant [37].

Kinetic studies of the rhodium-catalyzed methanol carbonylation reaction [35] show the reaction to be zero order in carbon monoxide and methanol, and first order in rhodium and iodide promoter. Many different types of rhodium compounds act as effective catalysts at common reaction temperatures of 150–200 °C. The iodide promoter is normally methyl iodide [74-88-4], but iodide also can be used in several different forms without marked differences in reaction rates.

Spectroscopic investigations [35] have shown that rhodium(III) halides can be reduced in aqueous or alcoholic media to [Rh(CO)$_2$X$_2$]$^-$. Moreover, when different rhodium(I) complexes are charged to the reaction medium, [Rh(CO)$_2$I$_2$]$^-$ becomes the predominant rhodium species, which strongly suggests this anion is the active catalytic species.

The catalytic cycle shown in Figure 1 is based on kinetic and spectroscopic studies [35]. The [Rh(CO)$_2$I$_2$]$^-$ reacts in the rate-determining step with methyl iodide by oxidative addition to make the transient, methylrhodium(III) intermediate [35]. Insertion of carbon monoxide or, more correctly, methyl migration gives the pentacoordinate acyl intermediate. The acyl intermediate eliminates acetyl iodide and regenerates the [Rh(CO)$_2$I$_2$]$^-$. The acetyl iodide reacts with water to regenerate HI and produce acetic acid. Hydrogen iodide reacts with methanol to form methyl iodide. In this way both the original rhodium complex and the methyl iodide promoter are regenerated.

The main byproducts of this process are carbon dioxide and hydrogen via the water-gas shift (Eq. 9), which is catalyzed by rhodium compounds [35], [40], [41].

$$\begin{aligned}
[Rh(CO)_2I_2]^- + 2\,HI &\longrightarrow [Rh(CO)I_4]^- + H_2 + CO \\
[Rh(CO)I_4]^- + 2\,CO + H_2O &\longrightarrow [Rh(CO)_2I_2]^- + CO_2 \\
& + 2\,HI \\
\hline
CO + H_2O &\longrightarrow CO_2 + H_2
\end{aligned}$$

Other transition-metal complexes have been investigated as catalysts for the carbonylation reaction, but none has been commercialized. Of these, the iridium complexes

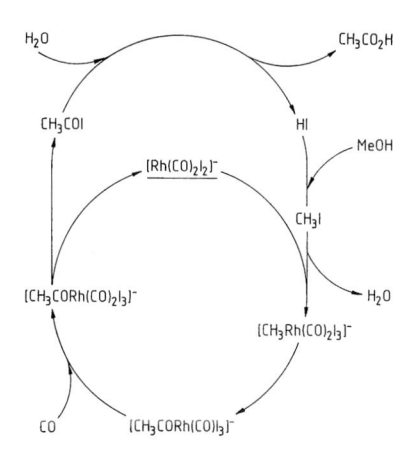

Figure 1. Reaction cycle proposed for the rhodium catalyzed methanol carbonylation reaction (Monsanto process)

studied by Monsanto [35], and the nickel complexes by Halcon and Rhône-Poulenc [37] are the most promising.

BASF Process [5] (Fig. 2).

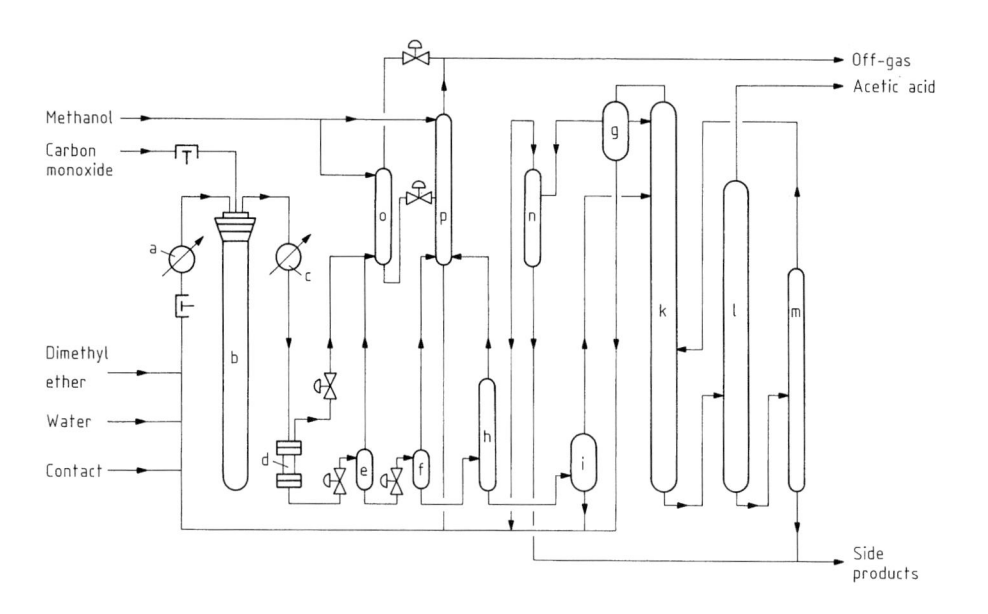

Figure 2. Production of acetic acid (BASF process)
a) Preheater; b) Reactor; c) Cooler; d) High-pressure separator; e) Intermediate pressure separator; f) Expansion chamber; g) Separation chamber; h) Degasser column; i) Catalyst separation column; k) Drying column; l) Pure acid column; m) Residue column; n) Auxiliary column; o) Wash column; p) Scrubbing column

Carbon monoxide, methanol (containing up to 60% dimethyl ether), catalyst recycle, catalyst makeup, and methyl iodide recycle (from the wash column) are sent to the high-pressure reactor (b) (stainless steel lined with Hastelloy). Part of the relatively low heat of reaction is used to preheat the feed and the rest ultimately is dissipated through the reaction vent. The reaction product is cooled and sent to the high-pressure separator (d). The off-gas goes to the wash column (o) and the liquid is expanded to a pressure of 0.5 – 1.0 MPa (5 – 10 bar) in the intermediate pressure separator (e). The gas released also is sent to the wash column; the liquid from the intermediate pressure separator is sent to the expansion chamber (f). The gas from the chamber goes to the scrubber (p). The gas from the scrubber and the wash column is discarded as off-gas. Both scrubber and wash column use the methanol feed to recover methyl iodide and other iodine-containing volatile compounds; this methanol solution is returned to the reactor. The off-gas composition in vol% is 65 – 75 CO, 15 – 20 CO_2, 3 – 5 CH_4, and the balance CH_3OH_g. The raw acid from the expansion chamber contains 45 wt% acetic acid, 35 wt% water, and 20 wt% esters, mainly methyl acetate. The acid is purified in five distillation towers. The first column (h) degasses the crude product; the off-gas is sent to the scrubber column. The catalyst is then separated as a concentrated acetic acid solution by stripping the volatile components in the catalyst separation column (i). The acid is then dried by azeotropic distillation in the drying column (k). The overhead of the drying column contains acetic and formic acids, water, and byproducts that form an azeotrope with water. This overhead is a two-phase system that is separated in the chamber (g). Part of the organic phase, composed mainly of esters, is returned to (k), where it functions as an azeotroping agent; the remainder is sent to the auxiliary column (n) where heavy ends are separated in the bottom of the column and light esters from the overhead are recycled to the reactor. The aqueous phase and the catalyst solution are returned to the reactor. The base of the drying column is sent to a finishing tower (1), in which pure acetic acid is taken overhead. The bottom stream of the finishing tower is sent to the residue column (m). The overhead of this residue column is sent back to the dehydration column. The bottom of the residue column contains about 50 wt% propionic acid, which can be recovered.

Monsanto Process [5], [28] (Fig. 3).

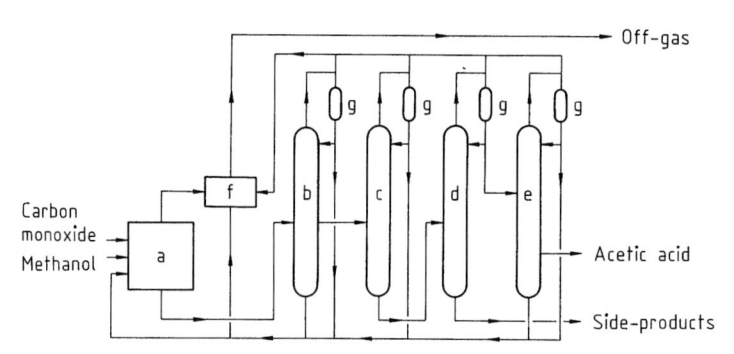

Figure 3. Production of acetic acid (Monsanto process)
a) Reaction system; b) Light ends column; c) Drying column; d) Heavy-ends column; e) Finishing column; f) Scrubber system; g) Distillation receiver

Carbon monoxide and methanol react in (a) to produce acetic acid. The reaction and purification system vent gases are combined and scrubbed (f) to recover light ends, including organic iodides, for recycle to the reactor. Crude acetic acid is sent to a light-ends column (b). The overhead light ends and

Table 5. Butane liquid-phase oxidation processes

Company	Location	Reported acetic acid capacity, t/a	Startup date
Celanese	Pampa, Texas	263 000 [43]	1952
	Edmonton, Alberta, Canada	64 000 [99]	1966
Union Carbide	Brownsville, Texas	295 000 [43]	1961
AKZO Zout Chemie	Europoort, The Netherlands	110 000 [100]	1965
Chemische Werke Hüls	Marl, Federal Republic of Germany	24 000 [45]	1961
Russian Refinery	Moscow, USSR	n.a.*	1962

* Not available

the residue are sent back to the reaction system (a), while acetic acid is removed as a side stream and sent to the drying column (c), where water is removed by conventional distillation. The overhead of the drying column, which is an acetic acid-water mixture, is sent back to the reaction system. The dry acetic acid from the base of the drying column is sent to column (d), where propionic acid is separated as a heavy end. Overhead acetic acid is sent to a finishing column (e) to produce high-purity acetic acid as a vapor side stream. Overhead and residue from the finishing towers are recycled.

4.2. Direct Oxidation of Saturated Hydrocarbons

The liquid-phase oxidation (LPO) of many aliphatic hydrocarbons, particularly those with straight-chain structures, can be used to produce carboxylic acids. n-Butane [106-97-8] is especially suitable for the production of acetic acid. This process is or has been operated on a large scale by the companies shown in Table 5 [42], [43] – [46]. The initial operation dates and reported acetic acid capacities also are given.

Reaction Mechanism. The chemistry of butane LPO is quite complex and has been the subject of various interpretations [43], [47], [48].

The predominant initial step of chain propagation is probably the abstraction of a secondary hydrogen atom (Eq. 10). Sufficient oxygen is dissolved in the liquid to scavenge radicals bearing the odd electron on carbon [49]; the sec-butyl radicals are converted rapidly to sec-butylperoxy radicals (Eq. 11). Alkylperoxy radicals are, in general, not highly reactive hydrogen abstractors [50]. Their concentrations therefore increase so that they undergo not only relatively selective hydrogen abstractions (Eq. 12) but a variety of bimolecular radical reactions as well (Eqs. 13, 14).

The hydroperoxide also can be generated through reaction of the peroxy radical with a lower valent catalyst ion (Eq. 15) [51]. The hydroperoxide is a source of new radicals to replenish the chain (Eqs. 16, 17).

$$CH_3-CH_2-CH_2-CH_3 + R\cdot \longrightarrow CH_3-CH_2-\overset{\cdot}{C}H-CH_3 \quad (10)$$
$$+ RH$$

$$CH_3-CH_2-\overset{\cdot}{C}H-CH_3 + O_2 \longrightarrow CH_3-CH_2-\overset{\overset{OO\cdot}{|}}{C}H-CH_3 \quad (11)$$

$$CH_3-CH_2-\overset{\overset{OO\cdot}{|}}{C}H-CH_3 + RH \longrightarrow CH_3-CH_2-\overset{\overset{OOH}{|}}{C}H-CH_3 + R\cdot \quad (12)$$

$$CH_3-CH_2-\overset{\overset{OO\cdot}{|}}{C}H-CH_3 + ROO\cdot \longrightarrow \quad (13)$$
$$CH_3-CH_2-\overset{\overset{O\cdot}{|}}{C}H-CH_3 + RO\cdot + O_2$$

$$CH_3-CH_2-\overset{\overset{OO\cdot}{|}}{C}H-CH_3 + R-\overset{\overset{OO\cdot}{|}}{\underset{R'}{C}}-R'' \longrightarrow \quad (14)$$

[cyclic transition state structure]

$$\downarrow$$

$$CH_3-CH_2-\overset{\overset{O}{\|}}{C}-CH_3 + O_2 + R-\overset{\overset{OH}{|}}{\underset{R'}{C}}-R''$$

$$CH_3-CH_2-\overset{\overset{OO\cdot}{|}}{C}H-CH_3 + M^{n+} \longrightarrow \quad (15)$$
$$CH_3-CH_2-\overset{\overset{OO^-}{|}}{C}H-CH_3 + M^{(n+1)+} \xrightarrow{H^+} CH_3-CH_2-\overset{\overset{OOH}{|}}{C}H-CH_3$$

$$CH_3-CH_2-\overset{\overset{OOH}{|}}{C}H-CH_3 \longrightarrow CH_3-CH_2-\overset{\overset{O\cdot}{|}}{C}H-CH_3 + OH\cdot \quad (16)$$

$$CH_3-CH_2-\overset{\overset{OOH}{|}}{C}H-CH_3 + M^{n+} \longrightarrow \quad (17)$$
$$CH_3-CH_2-\overset{\overset{O\cdot}{|}}{C}H-CH_3 + OH^- + M^{(n+1)+}$$

$$RO\cdot + RH \longrightarrow ROH + R\cdot \quad (18)$$

$$R-\overset{\overset{O\cdot}{|}}{\underset{R'}{C}}-R'' \longrightarrow R-\overset{\overset{O}{\|}}{C}-R' + R''\cdot \quad (19)$$

Reaction (13) converts the weakly reactive peroxy radicals to highly reactive alkoxy radicals that can abstract even relatively unreactive hydrogen atoms (Eq. 18). However, the alkoxy radicals also can undergo a β-scission reaction to generate a carbonyl compound and an alkyl radical (Eq. 19). The bulkier alkyl group tends to become the departing radical. If the alkoxy radical is tertiary, the carbonyl product is a ketone; if it is secondary, the carbonyl product is an aldehyde; and if the radical is primary, formaldehyde is generated. In butane LPO the major fate of secondary and tertiary (from isobutane) alkoxy radicals is the β-scission reaction, whereas primary alkoxy radicals are predominantly converted to alcohols via hydrogen abstraction [49].

Equation (14), sometimes called the "Russell chain termination mechanism," explains the conversion of two peroxy radicals, at least one of which must be secondary or tertiary, directly to an alcohol, a carbonyl compound, and molecular oxygen [52]. Because chain termination in butane LPO seems to occur rapidly under commercial conditions, the pathway in Equation (14) contributes significantly to the product distribution and may provide the major source of 2-butanone [78-93-3] and 2-butanol 78-92-2].

Contrary to literature reports, however, 2-butanone is not the most significant source of acetic acid and is less important in this respect than acetaldehyde. This conclusion is based on the fact that about 25% of all the carbon in the consumed butane appears as ethyl alcohol (with some catalysts) as the first isolable, nonperoxidic intermediate [43], [49]. The ethyl alcohol most probably arises from the decomposition of sec-butoxy radicals to form ethyl radicals and acetaldehyde (Eq. 19). Ethyl radicals are converted efficiently to ethyl alcohol under the conditions of this reaction. Over 50% of the carbon in the consumed butane must go through sec-butoxy radicals that have undergone β-scission. Ethyl alcohol is not a significant product from 2-butanone oxidation and only the ethyl groups in sec-butoxy radicals are converted readily to ethyl alcohol.

Acetaldehyde also is derived from the further oxidation of ethanol as well as from β-scission of sec-butoxy radicals. Acetaldehyde is, therefore, a major intermediate in butane LPO. It reacts rapidly to produce acetic acid, the main product.

The major byproduct is 2-butanone. Although it does so more slowly than acetaldehyde, some of it is further oxidized to acetic acid. Minor byproducts are propionic acid (from oxidation of propionaldehyde produced by β-scission of sec-butoxy radicals) and butyric acid (from n-butyl radicals).

Isobutane, a contaminant in all commercial butane (1–5%), oxidizes by a mechanism similar to that for n-butane with initial attack predominantly on the tertiary hydrogen atom. The tert-butoxy radical undergoes β-scission to give methyl radical and acetone:

$$\underset{\underset{CH_3}{|}}{\overset{\overset{O\bullet}{|}}{CH_3-C-CH_3}} \longrightarrow CH_3-\overset{\overset{O}{\|}}{C}-CH_3 + CH_3\bullet$$

Higher hydrocarbons oxidize in a manner similar to butane with the added complication that intromolecular hydrogen abstraction by alkylperoxy radicals can lead to significant amounts of shorter chain methyl ketones and difunctional intermediates [53]. Therefore, in areas of the world where naphtha is cheaper and more readily available than butane (mainly regions outside of oil producing areas, e.g., Europe,

Japan), naphtha-based acetic acid processes, developed by British Petroleum [54] and others [55], have proved quite useful. In these cases, the number of byproducts is much larger and the problems in product recovery and purification are proportionately greater.

Industrial Operation (Fig. 4). The reaction section (a) of a liquid-phase oxidation unit consists of a sparged vessel for the reactor and one or more phase separation vessels for separating the gas and two liquid phases formed by the reactor exit streams. Either air or oxygen enriched air can be used as the oxidant. The reactors have been described as "columns" or, in the case of naphtha oxidation, as tubular reactors [46], [54]. Several reactors may be staged to increase production of intermediates, especially 2-butanone [48]. The reported reaction temperatures are generally 150–200 °C (the critical temperature of n-butane is 152 °C [56]). Reported reactor pressures cover a relatively broad range but the range generally includes 5.6 MPa (56 bar). The pressure for naphtha oxidation is somewhat lower. Low-pressure steam can be generated from the heat of reaction. The reaction solvent consists of acetic acid, varying amounts of intermediates, water, and dissolved hydrocarbons. Control of water concentration below some maximum level appears to be critical [43], [48]. Varivalent metal ions, such as Mn, Co, Ni, and Cr, are used as catalysts [42], but noncatalytic operation is reported also [48].

The hydrocarbon vapor in the vent gas from (a) must be recovered for recycling. One means to accomplish this is to expand the vent gas through a turbine (b) to recover work (which may be applied to compress air for use in the reactor). The consequent reduction in temperature condenses the hydrocarbon [54]. The upper, organic layer from the phase separation vessel (d) is rich in hydrocarbons and is recycled to the reactor. The lower, aqueous phase is distilled to recover hydrocarbons for recycle.

The residual, hydrocarbon-free product consists of: volatile, neutral oxygenated derivatives (mostly aldehydes, ketones, esters, and alcohols), water, volatile monocarboxylic acids (formic, acetic, propionic, and butyric from butane), and essentially nonvolatile material (difunctional acids, γ-butyrolactone, condensation products, catalyst residues, etc.).

The volatile, neutral materials can be recovered as mixtures or individually. They are used for derivatives, sold, or recycled to the reactor. Most of these components generate acetic acid on further oxidation. In the case of butane oxidation, 2-butanone usually is isolated as a pure component for sale.

The separation of water is perhaps the most difficult and costly step of the purification process. This is achieved by azeotropic distillation using entrainment agents, such as ethers [45], or by extraction methods. Extractive distillations employing basic extractants also have been reported [5], [57].

The separation of formic acid from the anhydrous residue can be accomplished by careful fractionation; but, generally, an azeotroping agent, such as chlorinated hydrocarbons or aromatic hydrocarbons, is used. The remaining acids are separated by

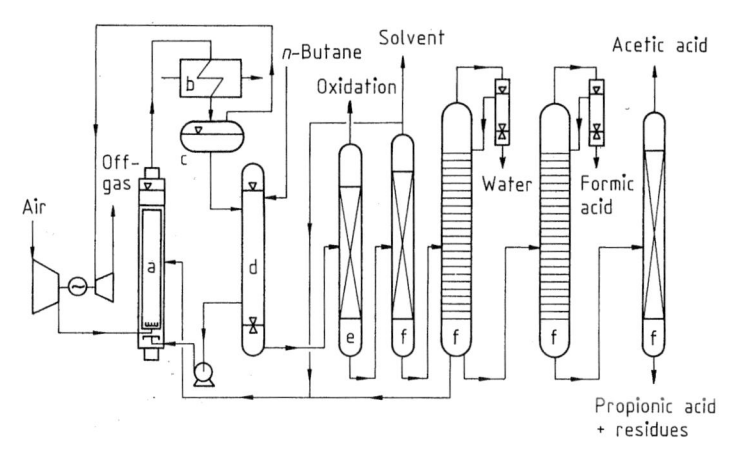

Figure 4. Oxidation of *n*-butane in the liquid phase (Chemische Werke Hüls process)
a) Reactor; b) Air cooler; c) Collector; d) Separation vessel; e) Pressure column; f) Distillation column

fractionation. After further chemical treatment [46], excellent quality acetic acid (>99% pure) is isolated in yields that can exceed 60% [44].

Particularly with naphtha oxidation, some of the diacids (especially succinic acid) are recovered for sale [54], [55]. The nonvolatile residue can be burned to recover energy.

4.3. Acetaldehyde Process

The oxidation of acetaldehyde [75-07-0] to acetic acid proceeds through a free-radical chain, which produces peracetic acid [79-21-0] as an intermediate (Eq. 22):

$$CH_3-\overset{O}{\underset{\|}{C}}-H + R\cdot \longrightarrow CH_3-\overset{O}{\underset{\|}{C}}\cdot + RH \qquad (20)$$

$$CH_3-\overset{O}{\underset{\|}{C}}\cdot + O_2 \longrightarrow CH_3-\overset{O}{\underset{\|}{C}}-OO\cdot \qquad (21)$$

$$CH_3-\overset{O}{\underset{\|}{C}}-OO\cdot + CH_3-\overset{O}{\underset{\|}{C}}-H \longrightarrow \qquad (22)$$

$$CH_3-\overset{O}{\underset{\|}{C}}-OOH + CH_3-\overset{O}{\underset{\|}{C}}\cdot$$

Peracetic acid then can react with acetaldehyde to generate acetaldehyde monoperacetate (AMP) [7416-48-0]. The AMP decomposes efficiently to acetic acid by a hydride shift in a Baeyer-Villiger reaction. A methyl migration (dashed line, below) to produce methyl formate also appears to occur [58]:

$$CH_3-\underset{\underset{O}{\|}}{C}-OOH + CH_3-\underset{\underset{O}{\|}}{C}-H \longrightarrow CH_3-\underset{\underset{O}{\|}}{C}\underset{O-O}{\overset{CH_3\ \ OH}{\underset{H}{\diagdown C \diagup}}} \quad (23)$$

$$CH_3-\underset{\underset{O^-}{\|}}{C} + \overset{+}{\underset{CH_3-O\ \ H}{C\diagup^{OH}}} \qquad CH_3-\underset{\underset{O^-}{\|}}{C} + \underset{HO}{\overset{CH_3}{\diagdown}}\overset{OH}{\underset{}{C+}}$$

$$CH_3-\underset{\underset{O}{\|}}{C}-OH + CH_3-O-\underset{\underset{O}{\|}}{C}-H \qquad 2\ CH_3-\underset{\underset{O}{\|}}{C}-OH$$

The alkyl migration becomes more pronounced with higher aldehydes, particularly those having a branch at the α-position.

Chain termination occurs primarily via bimolecular reactions of acetylperoxy radicals going through an intermediate tetroxide [59].

$$2\ CH_3-\underset{\underset{O}{\|}}{C}-OO\bullet \longrightarrow \left[CH_3-\underset{\underset{O}{\|}}{C}\underset{O-O-O-O}{\diagdown\diagup}\underset{\underset{O}{\|}}{C}-CH_3 \right] \longrightarrow \quad (24)$$

$$2\ CH_3-\underset{\underset{O}{\|}}{C}-O\bullet + O_2$$

$$CH_3-\underset{\underset{O}{\|}}{C}-O\bullet \longrightarrow CH_3\bullet + CO_2$$

$$2\ CH_3\bullet + 2\ O_2 \longrightarrow 2\ CH_3-OO\bullet \underset{\diagdown}{\overset{\diagup 2\ CH_3O\bullet + O_2}{}}$$
$$CH_3OH + HCHO + O_2$$

$$CH_3O\bullet + RH \longrightarrow CH_3OH + R\bullet$$

Reaction (24) is the source of most of the carbon dioxide, methanol, formaldehyde, and formic acid byproducts of the acetaldehyde oxidation. The uncatalyzed oxidation can be fairly efficient provided the conversion of acetaldehyde is low enough to maintain a significant concentration of acetaldehyde in the reaction solvent. This keeps the steady-state concentration of acetylperoxy radicals low by facilitating reaction (22), thus supressing the bimolecular reaction (24). In uncatalyzed reactions, however, special precautions must be taken to prevent the concentration of AMP from reaching explosive levels [42], [46].

Another free-radical decomposition reaction can become important at high temperatures and/or at low oxygen concentrations:

$$CH_3-\underset{\underset{O}{\|}}{C}\bullet \longrightarrow CH_3\bullet + CO$$

The rate of this decarbonylation increases with increasing temperature; however, it becomes important only when insufficient oxygen is present to scavenge the acetyl radicals (Eq. 21).

Catalysts can play several important roles in aldehyde oxidations. They generally decompose the peroxides and so minimize the explosion hazards noted previously; however, the situation is more complex than this. An important function of manganese (the preferred catalyst) is the reduction of acetylperoxy radicals [51]:

$$CH_3-\overset{O}{\underset{\|}{C}}-OO\cdot + Mn^{2+} \longrightarrow CH_3-\overset{O}{\underset{\|}{C}}-OO^- + Mn^{3+} \quad (25)$$

$$CH_3-\overset{O}{\underset{\|}{C}}-H + Mn^{3+} \longrightarrow CH_3-\overset{O}{\underset{\|}{C}}\cdot + Mn^{2+} + H^+ \quad (26)$$

Reaction (25) assists reaction (22) in supressing the concentration of acetylperoxy radicals. The manganese(III) ion generated in reaction (25) can perform the function of acetylperoxy radicals in reaction (22) (Eq. 26) but does not contribute to the inefficiency generating reaction (24).

Manganese also greatly increases the rate of reaction of peracetic acid and acetaldehyde to produce acetic acid [60]. The reaction in the presence of manganese is first order with respect to peracid, aldehyde, and manganese. The rate is much higher than the uncatalyzed rate of formation of AMP; therefore, manganese may catalyze the formation of AMP, and possibly the decomposition of AMP as well. Alternatively, manganese may catalyze a different mechanism not involving AMP (AMP is not detected in manganese-catalyzed oxidations) [60].

Copper can interact synergistically with a manganese catalyst [61]. Manganese has some negative aspects as a catalyst as well as the benefits noted above. The major problem is that it greatly increases the reaction rate. Part of the reason for this may be that manganese can create new reaction centers or chain initiators [manganese(III) ion] by reaction with peracetic acid:

$$CH_3-\overset{O}{\underset{\|}{C}}-OOH + 2\ Mn^{2+} \longrightarrow CH_3-\overset{O}{\underset{\|}{C}}-O^- + OH^- + 2\ Mn^{3+}$$

The resulting very high rate exacerbates any oxygen starvation problem and, by raising steady-state radical concentrations, at least partly restores the participation of reaction (24) [43]. This is particularly the case with higher aldehydes but is also true to some extent with acetaldehyde.

Copper(II) ion, however, can oxidize acetyl radials very rapidly [62]:

$$CH_3-\overset{O}{\underset{\|}{C}}\cdot + Cu^{2+} \longrightarrow CH_3-\overset{O}{\underset{\|}{C}}^+ + Cu^+$$

The acetylium ion can react as shown below:

$$CH_3-\overset{O}{\underset{\|}{C}}{}^+ \begin{array}{l} \xrightarrow{H_2O} CH_3-\overset{O}{\underset{\|}{C}}-OH + H^+ \\ \xrightarrow{CH_3COOH} CH_3-\overset{O}{\underset{\|}{C}}-O-\overset{O}{\underset{\|}{C}}-CH_3 + H^+ \\ \xrightarrow{ROH} CH_3-\overset{O}{\underset{\|}{C}}-OR + H^+ \end{array}$$

Production The copper(I) ion can be reoxidized by peroxide or manganese(III) ion. In this fashion, copper diverts a fraction of the reaction through a nonradical path and, in essence, provides a termination path that does not result in inefficiency. The product is acetic acid (or an acetic acid precursor).

Many other catalysts are mentioned in the literature [63]–[65]. Cobalt is cited frequently but seems to be somewhat inferior to manganese [60]. Some effects of cobalt do not appear to be well understood. A cobalt and copper mixture is preferred when maximum anhydride production is desired [66].

Additional minor products encountered are ethylidene diacetate (from acetaldehyde and acetic acid), crotonic acid, succinic acid, etc.

Industrial Operation. A typical acetaldehyde oxidation unit is depicted in Figure 5. The reactor (a) is sparged with air or oxygen-enriched air. Temperatures are typically 60–80 °C with pressures of 0.3–1.0 MPa (3–10 bar). The reaction mixture is circulated rapidly through an external heat exchanger to remove the heat of reaction. The vent gas is cooled and then scrubbed with recirculated crude product (which goes to the reactor) and finally with water (which goes to the aldehyde recovery column). The reactor product is fed to the aldehyde recovery column (b) (the aldehyde is recycled) and then to a low-boilers column (c) where methyl acetate is removed. The next column is the acetic acid finishing column (d), where water is removed overhead by azeotropic distillation and finished product comes off as a vapor side stream. Yields are generally in excess of 90%; purity is excellent (>99%).

4.4. Other Processes

Research and development efforts on new acetic acid processes have been significant. Routes that have been investigated are the oxidation of *n*-butenes in the vapor-phase [67], [68], the liquid-phase oxidation of *sec*-butyl acetate (from *n*-butenes) [69], the oxidation of ethylene in the vapor-phase [70], and the vapor-phase oxidation of saturated hydrocarbons, especially *n*-butane [71], [72]. None of these routes has been commercialized and only the oxidation of butenes has been developed through pilot plant scale.

The *Chemische Werke Hüls process* is based on the vapor-phase oxidation of *n*-butenes [25167-67-3] at 180–245 °C and superatmospheric pressure (200–3000 kPa) over a titanium–vanadium oxide catalyst. The acetic acid yield for this process is only 46%. The oxidation is carried out in the presence of air and steam in a fixed-bed tubular reactor. The heat of reaction is recovered through process steam generation. Purification is carried out in four towers: lightends column, drying column, formic acid separation column, and finishing column.

The *Bayer process* is based on the liquid-phase oxidation of *sec*-butyl acetate [105-46-4]. The *n*-butenes first react with acetic acid at 100–120 °C and 1.9 MPa (19 bar), to

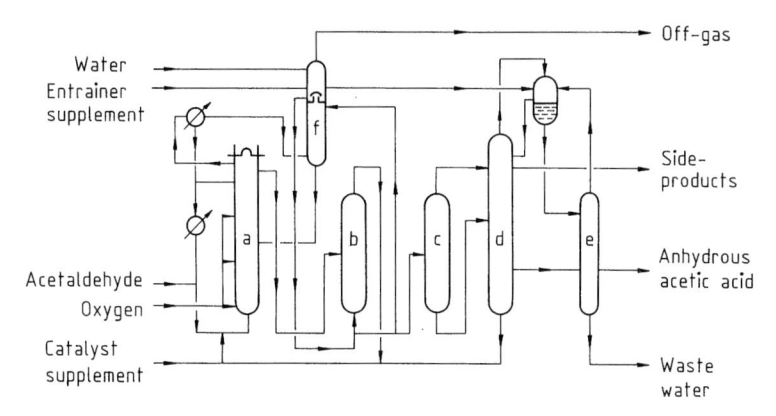

Figure 5. Oxidation of acetaldehyde to acetic acid
a) Reactor; b) Acetaldehyde column; c) Methyl acetate column; d) Finishing column; e) Column for recovering entrainer; f) Off-gas scrubber column

produce *sec*-butyl acetate. The *sec*-butyl acetate is then catalytically oxidized in the liquid phase at 200 °C and 6.3 MPa (63 bar). The overall process yield of acetic acid is 58%. Purification is by conventional methods using four towers [5].

4.5. Concentration and Purification

Aqueous solutions of acetic acid are produced as byproducts of important industrial processes, e.g., cellulose acetate production. Recovery of the acid from these aqueous streams can be accomplished by common distillation, azeotropic distillation, solvent extraction, and extractive distillation. Direct distillation requires a significant number of plates and a high reflux ratio in the columns. In extractive distillation, the vapor streams of acetic acid and water are scrubbed with a high-boiling solvent that preferentially dissolves one of the components.

Azeotropic Distillation. In azeotropic distillation, a compound that decreases the boiling point of water is used. If the "azeotropic agent" is immiscible with water, it can be separated easily from the water and recycled. Esters, ethers, benzene, and chlorinated hydrocarbons have been proposed as azeotropic agents [5].

Solvent Extraction. In this method, contact is made between the acetic acid solution in water and a countercurrent organic solvent of low miscibility with water. Because water is a very good solvent for acetic acid, the organic solvent to water feed ratio is generally high, ranging from 2:1 to 5:1. The compounds mentioned as azeotropic agents usually can be used in liquid-liquid extraction. Dow Chemical has developed an interesting extractant that is a solution of the weak base, tri-*n*-octylphosphine oxide [*78-50-2*] (TOPO), in high-boiling hydrocarbons [73]. Badger has disclosed the use of 1-methyl-2-pyrrolidinone [*872-50-4*] in extractive distillation [57].

The selection of the method to remove water depends on the economics. In general the more dilute the acid solution the more attractive solvent extraction becomes.

Removal of other impurities has been described in discussing the different processes.

4.6. Materials of Construction [39]

The selection of materials of construction, heat treatments, welding techniques, and other technical aspects of each acetic acid process are proprietary knowledge. Corrosive attack on metals by liquid streams increases with acetic acid, halide, and/or formic acid concentrations and with increasing temperature. In the absence of reducing agents and halides, AISI 316 stainless steel [12597-68-1] is widely used (Cr 16–18 wt%, Ni 10–14 wt%, Mo 2–3 wt%). If a reducing condition can exist, then AISI 321 stainless steel is required (Cr 17–19 wt%, Ni 9–12 wt%, Ti >0.4 wt%). If halides are present, Hastelloy B [12605-84-4] or C [12605-85-5], and/or other exotic materials of construction are necessary. At ambient temperatures and high acid concentration, aluminum [7429-90-5] and AISI 304 stainless steel (Cr 18–20 wt%, Ni 8–10.5 wt%) are used in storage tanks, pumps, and piping.

5. Wastewater and Off-Gas Problems

The acetic acid contained in waste water is diluted, neutralized, and then degraded biologically. In the Federal Republic of Germany, acetic acid belongs to class 2 materials (TA-Luft) as an off-gas problem [74]. Under this class, a limit of 150 mg/m^3 acid is permitted for emissions ≥ 3 kg/h of off-gas. Acetic acid can be removed from the off-gas by cooling or by washing with water.

Methyl acetate also belongs to class 2, TA-Luft. Ethyl acetate, butyl acetate, and phenyl acetate belong to class 3. Off-gas emissions (≥ 6 kg/h) containing esters in class 3 may contain a maximum of 300 mg/m^3 of these substances. Removal of esters from gas emissions can be accomplished by one or more of the following methods: thermal condensation, washing with nonvolatile organic solvents, physical adsorption, or burning.

In the United States, emissions are governed by the United States Clean Air Act [75], with each state, having enforcement responsibilities. Typically, acetic acid and other volatile organic compounds (like the acetic acid esters) are limited to emissions of less than 22.675×10^3 kg/a (<2.6 kg/h) in the various states.

Table 6. Sales specifications of acetic acid, glacial

Sales specifications		Limits
Acetic acid, wt%	min.	99.85
Specific gravity, 20/20 °C	–	1.0505 – 1.0520
Distillation range, °C	max.	1.0
Initial *bp*	min.	117.3
Dry point	max.	118.3
Freezing point, °C	min.	16.35
Color, Pt – Co units	max.	10
Water content, wt%	max.	0.15
Reducing substances, as formic acid, wt%	max.	0.05
Aldehydes, as acetaldehyde, wt%	max.	0.05
Iron, ppm	max.	1
Heavy metals, as Pb, ppm	max.	0.05
Chlorides, ppm	max.	1
Sulfates, ppm	max.	1
Sulfurous acid, ppm	max.	1
Permanganate time	min.	2 h
Appearance	–	clear and free of suspended matter

6.   Quality Specifications

There is essentially only one commercial grade of acetic acid sold in the United States. Elsewhere, three or more grades are available [5], differing as to specifications for heavy metals, chloride, sulfate, arsenic, and iron. One of the most critical requirements is "permanganate time." For this test, a solution of 5 mL of product, 10 mL of water, and 0.25 mol of 0.1 N potassium permanganate must retain its color for up to 2 h. One manufacturer's sales specifications are given in Table 6 [76].

7.   Chemical Analysis

Gas chromatography is the usual technique for determining the acetic acid content of a volatile mixture [77]. Packed columns can be used, but capillary columns are especially effective. Carbowax 20 M, terephthalic acid terminated [*41479-14-5*], is the preferred substrate.

8. Storage, Transportation, and Customs Regulations [5], [78]

Acetic acid can be stored and transported in containers lined with stainless steel, glass, or polyethylene. Aluminum is also resistant to glacial acetic acid (99.7%). Aluminum in contact with acetic acid is slowly attacked to form a layer of aluminum oxide. This prevents further corrosion; but some of the oxide may be suspended in the acid, giving it a cloudy appearance. Concentrated acetic acid crystallizes at $\approx 16\,°C$. Storage containers, tank cars, tank trucks, and pipes for concentrated acid should be equipped with a heating coil that can be connected to a steam line and steam trap. All storage tank vents must be steam traced to prevent plugging by acetic acid crystals.

German Legal Requirement for Foodstuffs. Vinegar containing more than 11 wt% acetic acid may be marketed commercially in closed containers made from materials resistant to acetic acid. The containers must be labeled clearly with a warning "Handle with Care, Do Not Swallow Undiluted." Acetic acid concentrations higher than 25 wt% can be handled only by dealers who are not the consumers.

German Customs Requirements (Vinegar Law) [5]. According to Paragraph 1 of the Vinegar Law, the transfer of synthetic vinegar on the free market is subject to the Vinegar Tax. Imports are also subject to the tax. According to Paragraph 2, the vinegar tax is recalculated each time the price of vinegar changes.

Acetic acid that is suitable only for commercial purposes, or that is suitable for human consumption but is used only for commercial purposes, is exempt from the tax.

9. Uses

Acetic acid has a broad spectrum of applications. An outline of these is shown in Table 7 [27], [28]. Over 60% of the acetic acid produced goes into polymers derived from either vinyl acetate, or cellulose. Most of the poly(vinyl acetate) is used in paints and coatings, or used for making poly(vinyl alcohol) and plastics. Cellulose acetate is used to produce acetate fibers. Acetic acid and acetate esters are used extensively as solvents.

10. Derivatives

This section contains information about acetic acid derivatives that are not discussed in detail elsewhere in this encyclopedia (→Acetic Anhydride, →Chloroacetic Acids).

Table 7. Acetic acid end-use applications

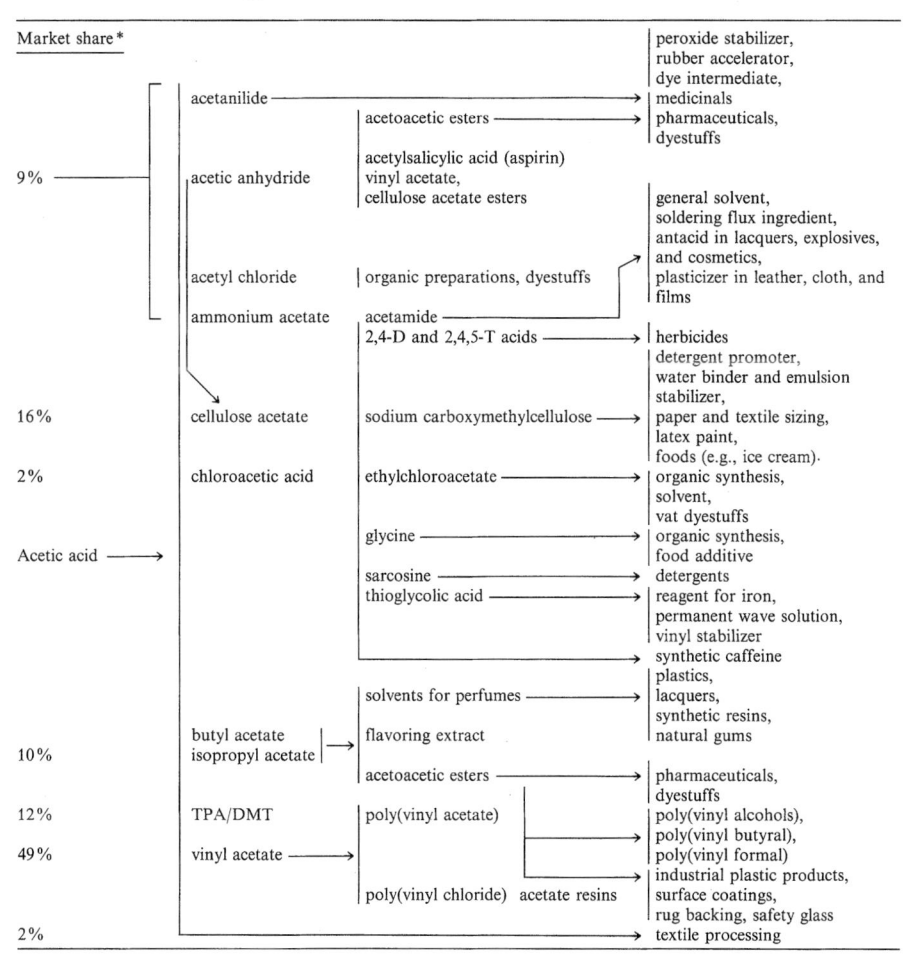

Market share*			
	acetanilide		peroxide stabilizer, rubber accelerator, dye intermediate, medicinals
		acetoacetic esters	pharmaceuticals, dyestuffs
9%	acetic anhydride	acetylsalicylic acid (aspirin) vinyl acetate, cellulose acetate esters	
	acetyl chloride	organic preparations, dyestuffs	general solvent, soldering flux ingredient, antacid in lacquers, explosives, and cosmetics, plasticizer in leather, cloth, and films
	ammonium acetate	acetamide	
		2,4-D and 2,4,5-T acids	herbicides
16%	cellulose acetate	sodium carboxymethylcellulose	detergent promoter, water binder and emulsion stabilizer, paper and textile sizing, latex paint, foods (e.g., ice cream).
2%	chloroacetic acid	ethylchloroacetate	organic synthesis, solvent, vat dyestuffs
Acetic acid →		glycine	organic synthesis, food additive
		sarcosine	detergents
		thioglycolic acid	reagent for iron, permanent wave solution, vinyl stabilizer
			synthetic caffeine
		solvents for perfumes	plastics, lacquers, synthetic resins, natural gums
10%	butyl acetate isopropyl acetate	flavoring extract	
		acetoacetic esters	pharmaceuticals, dyestuffs
12%	TPA/DMT	poly(vinyl acetate)	poly(vinyl alcohols), poly(vinyl butyral), poly(vinyl formal)
49%	vinyl acetate		industrial plastic products, surface coatings, rug backing, safety glass
2%		poly(vinyl chloride) acetate resins	textile processing

* Reproduced with permission from [27].

10.1. Salts

10.1.1. Aluminum Acetate

Three aluminum acetates are known, two basic and one neutral: Al(OH)$_2$CO$_2$CH$_3$ [*24261-30-1*], Al(OH)(CO$_2$CH$_3$)$_2$ [*142-03-0*], and Al(CO$_2$CH$_3$)$_3$ [*139-12-8*].

Properties. Aluminum acetate, Al(CO$_2$CH$_3$)$_3$, M_r 204.1, is a white, watersoluble powder that decomposes at about 130 °C to give acetic anhydride along with basic aluminum acetates [79].

Production. Neutral aluminum acetate is made from aluminum metal and glacial acetic acid that contains 0.1 – 1 % acetic anhydride [80] which scavenges any water present. The anhydrous triacetate cannot be made from aqueous solutions. The basic acetates can be prepared from $Al(OH)_3$ and aqueous acetic acid solutions [81].

Uses. Aluminum acetate is used as a dye mordant in the dyeing of fabrics.

10.1.2. Ammonium Acetate

Properties. Ammonium acetate [*631-61-8*], $CH_3CO_2NH_4$, M_r 77.08, *mp* 114 °C, forms colorless, hygroscopic needles. The solubilities of this salt in 100 g of water or methyl alcohol are, respectively, 148 g (4 °C) and 7.9 g (15 °C).

Production. Ammonium acetate is manufactured by neutralizing acetic acid with ammonium carbonate or by passing ammonia gas into glacial acetic acid [5]. Acidic ammonium acetate, $CH_3CO_2NH_4 \cdot CH_3CO_2H$ [*25007-86-7*], is manufactured by dissolving the neutral salt in acetic acid.

Uses. Ammonium acetate is used in the manufacture of acetamide, an excellent solvent, and as a diuretic and diaphoretic in medical applications. The wool industry also uses this salt as a dye mordant.

10.1.3. Alkali-Metal Salts

Properties. All of the alkali metals (Li, Na, K, Rb, Cs) are known to form acetates. The aqueous solubilities of these salts increase with increasing atomic mass of the alkali metal [82]. The potassium salt [*127-08-2*], M_r 98.14, *mp* 292 °C, and sodium salt [*127-09-3*], M_r 82.04, *mp* 324 °C, are the most common. The former crystallizes from water as white columns. The latter can be purified by crystallization from acetic acid. Several acetic acid solvates of the alkali metal salts are known to exist [83].

Production. Potassium and sodium acetates normally are manufactured from glacial acetic acid and the corresponding hydroxides. The salts also can be prepared from acetic acid on treatment with the metal, the metal carbonate, or the metal hydride [81].

Uses. Potassium acetate is used to purify penicillin. It is also used as a diuretic and as a catalyst to make polyurethane. The sodium salt is used frequently in water as a mild alkali.

10.2. Esters (→Esters, Organic)

Production. Most of the esters of acetic acid are manufactured by the liquid-phase esterification of the acid with alcohols. Such catalysts as sulfuric acid, toluenesulfonic acid, or acid ion-exchanger resins normally are used. Instead of secondary and tertiary alcohols, the corresponding olefins also can be used to esterify acetic acid. Ethyl acetate can be produced directly from acetaldehyde.

The formation of esters from alcohols and acids is an equilibrium process. Therefore it is necessary to remove either the water byproduct or the esters to drive the reaction to completion. The esters normally are taken off by azeotropic distillation.

Uses. Acetic acid esters are used in large quantities as solvents for plastics, lacquers, resins, and gums.

10.2.1. Methyl Acetate

Properties. Methyl acetate [79-20-9], M_r 74.08. n_D^{20} 1.3594, d_{20}^{20} 0.9330, bp 57 °C, fp −98.1 °C, is colorless and has a pleasant odor. Methyl acetate forms azeotropes with water (3.5 wt% H_2O, bp 56.5 °C) and methanol (19 wt% MeOH, bp 54.0 °C).

Production. Most commercial methyl acetate is obtained as a byproduct from acetic acid manufacture. When this ester is produced via esterification of acetic acid with methanol [811-98-3] using a sulfuric acid catalyst, the product ester can be removed by means of the methanol-methyl acetate azeotrope. Methyl acetate occurs as an intermediate in the acetic acid synthesis from methanol and carbon monoxide.

Uses. Very little methyl acetate is sold commercially. Methyl acetate is used as a solvent for cellulose nitrates, esters, and ethers.

10.2.2. Ethyl Acetate

Properties. Ethyl acetate [141-78-6], M_r 88.10, n_D^{20} 1.3723, d_{20}^{20} 0.9003, bp 77.1 °C, fp −83.6 °C, forms azeotropes with water (8.2 wt% H_2O, bp 70.4 °C), ethanol (30.8 wt% EtOH, bp 71.8 °C), and methanol (44.0 wt% MeOH, bp 60.25 °C). A ternary ethyl acetate − water − ethanol azeotrope also is formed (7.8 wt% H_2O, 9.0 wt% EtOH, bp 70.3 °C).

Production. Ethyl acetate can be made from ethanol [64-17-5] and acetic acid in either a batch or a continuous process. In the continuous process, the key to production is removal of the ester by the alcohol − ester − water azeotrope. After the water is decanted, the organic layer contains 93 wt% ester. Further purification gives ethyl acetate in yields greater than 95% (based on acetic acid).

Ethyl acetate also can be made by the Tischtschenko reaction [83]:

$$2\ CH_3CHO \xrightarrow[0-5\,°C]{Al(OR)_3} CH_3CO_2CH_2CH_3$$

Uses. Ethyl acetate is used extensively as a solvent for nitrocellulose, coatings, inks, and polymers. It is also used as an extraction solvent.

Table 8. Binary azeotropes with the butyl acetates

Component A	Component B	bp, °C	wt % B
n-Butyl acetate	Water	90.2	26.7
sec-Butyl acetate	Water	86.6	19.4
Isobutyl acetate	Water	87.4	–
n-Butyl acetate	n-Butanol	117.2	47
sec-Butyl acetate	sec-Butanol	99.6	86.3

Table 9. Ternary azeotropes with the butyl acetates

Component A	Component B	Component C	bp, °C	wt % B	wt % C
n-Butyl acetate	Water	n-Butanol	89.4	37.3	37.3
sec-Butyl acetate	Water	2-Butanol	86.0	23.0	45.0
Isobutyl acetate	Water	Isobutyl alcohol	86.8	30.4	23.1

10.2.3. Butyl Acetate

Properties. n-Butyl acetate [123-86-4] has M_r 116.16, n_D^{20} 1.3951, d_{20}^{20} 0.882, bp 126 °C, and fp -73.5 °C. All four of the esters, n-butyl [123-86-4], sec-butyl [105-46-4], isobutyl [110-19-0], and tert-butyl acetate [540-88-5], are colorless compounds with pleasant odors. Azeotropic data for the butyl acetates are given in Tables 8 and 9 [5].

Production. The butyl acetates are prepared from acetic acid and the corresponding alcohol in the presence of sulfuric acid catalyst. The reaction is driven to completion by azeotropic removal of water. The acid catalyst is neutralized and the ester purified by distillation [5].

Uses. The butyl acetates are used primarily as solvents in the lacquer and enamel industry. These acetates are used also in the photographic industry, in adhesives, and as extraction solvents.

10.2.4. 2-Ethylhexyl Acetate

2-Ethylhexyl acetate [103-09-3], M_r 172.26, n_D^{20} 1.4204, d_{20}^{20} 0.8734, bp 199.3 °C, fp -93 °C, is made by esterifying acetic acid with 2-ethyl-1-hexanol in the presence of an acid catalyst and a low-boiling azeotroping solvent. This acetate is used in lacquers, in silk-screen inks, and as a coalescing aid in certain paints.

Table 10. Physical properties of other acetate esters

Ester		M_r	n_D^{20}	d_{20}^{20}	bp, °C	fp, °C
n-Propyl	[109-60-4]	102.13	1.3844	0.887	101.6	−92.5
Isopropyl	[108-21-4]	102.13	1.3773	0.8718	90	-73.4
n-Pentyl	[628-63-7]	130.18	1.4023	0.8756	149.2	-70.8

10.2.5. Other Esters

There are many other esters of commercial or academic interest. Several of these are listed in Table 10 [84]. These esters have a mild fruity smell and are colorless. These materials also are used as solvents in the production of cellulose and elastomers. The propyl compounds are used as ink solvents.

10.3. Acetyl Chloride

Properties. Acetyl chloride [75-36-5], M_r 78.5, n_D^{20} 1.3871, d_4^{20} 1.1051, bp 51.8 °C, mp -112.86 °C, is colorless and a strong irritant. It is very reactive and hydrolyzes in the presence of moist air to hydrogen chloride and acetic acid. Great care should be used when working with it.

Production. The normal industrial method of acetyl chloride preparation involves treatment of acetic anhydride [108-24-7] with hydrogen chloride [7647-01-0]. On a smaller scale the chloride can be prepared from the acid using reagents, such as thionyl chloride [7719-09-7] or phosphorus trichloride [7719-12-2] [23]. Moisture must be excluded carefully from these reactions.

Large-scale users normally manufacture the compound themselves, because transportation and storage are difficult. Suitable handling materials are glass, enamel, porcelain, clay, or polytetrafluoroethylene.

Uses. Acetyl chloride is an efficient acetylating agent for alcohols and amines to produce the corresponding esters and amides. Industrially acetyl chloride is used in the synthesis of dyes and pharmaceutical chemicals; also in the acylation of benzene to make acetophenone. Many other reactions also can be carried out with this reactive material [85].

10.4. Amides

10.4.1. Acetamide

Properties. Acetamide [60-35-5], CH_3CONH_2, M_r 59.07, *mp* 81.5 °C, *bp* 222 °C, d_4^{20} 1.161, forms deliquescent hexagonal crystals that are odorless when pure (musty when impure). This material dissolves in a wide variety of solvents, such as water and lower molecular mass alcohols.

Production. The heating of ammonium acetate to produce acetamide is well known [23]:

$$CH_3CO_2NH_4 \xrightarrow{120\,°C} CH_3CONH_2 + H_2O$$

Acetamide also can be prepared by acylation of ammonia with acetyl chloride, acetic anhydride, or alkyl acetates. The acetates are preferred because they are easy to handle:

$$CH_3CO_2R + NH_3 \longrightarrow CH_3CONH_2 + HOR$$

Uses. Acetamide is used extensively as a solvent. It is also used in the manufacture of methylamine and as a plasticizer.

10.4.2. *N,N*-Dimethylacetamide

Properties. *N,N*-Dimethylacetamide, $CH_3CON(CH_3)_2$, M_r 87.1, *mp* −20 °C, d_4^{20} 0.943, is a colorless, high-boiling solvent. This hygroscopic material boils at 165.5 °C and dissolves in a wide variety of solvents. This amide is a powerful solvent and frequently exhibits catalytic properties.

Production. *N,N*-Dimethylacetamide can be made by most of the classical methods for producing amides. The three routes commonly used are the following:

$$CH_3COOH + NH(CH_3)_2 \longrightarrow \qquad (27)$$
$$CH_3CON(CH_3)_2 + H_2O$$

$$(CH_3CO)_2O + NH(CH_3)_2 \longrightarrow \qquad (28)$$
$$CH_3CON(CH_3)_2 + CH_3COOH$$

$$CH_3COOCH_3 + NH(CH_3)_2 \longrightarrow \qquad (29)$$
$$CH_3CON(CH_3)_2 + CH_3OH$$

Normally Reaction (27), is carried out at elevated temperature and pressure. The amide is initially isolated as an azeotropic mixture with acetic acid. In the case of Reaction (28), acetic anhydride is treated with dimethylamine to give the amide and acetic acid byproduct. *N,N*-Dimethylacetamide also can be obtained upon reaction of methyl acetate with dimethylamine (Eq. 29). Normally sodium methoxide catalyst is employed in this latter route.

Uses. As mentioned above this material is a very effective reaction solvent. In many cases this solvent also catalyzes the reaction (e.g., halogenation, cyclization, alkylation reactions).

N,N-Dimethylacetamide is a very good polymer solvent, dissolving polyacrylates and polyesters. It can be used also as a spinning solvent for polyimides. Polyester raw materials such as terephthalic acid can be recrystallized from this amide.

10.5. Phenylacetic Acid

Properties. Phenylacetic acid [*103-82-2*], $C_6H_5CH_2CO_2H$, M_r 136.14, *bp* 265.5 °C, *mp* 78 °C, $d_4^{79.8}$ 1.0809, forms leaflets on vacuum distillation. This material dissolves readily in hot water but is only slightly soluble in cold water. It is soluble in chloroform (4.4 mol/L, 25 °C).

Production. Most phenylacetic acid is produced by acid hydrolysis of the corresponding nitrile, which is prepared from benzyl chloride, a cyanide salt, and a base [23].

Uses. Phenylacetic acid is a starting material for synthetic perfumes. It is also used in the synthesis of penicillin G.

11. Economic Aspects [5], [27]

Synthetic acetic acid was produced at levels of 3×10^6 t in 1981. This included 1.4×10^6 t/a in the United States, 9×10^5 t/a in Western Europe, 4×10^5 t/a in Japan, and 1×10^5 t/a in both Canada and Mexico. The total world capacity was about 4×10^6 t/a, with 2×10^6 t/a in the United States, 1.2×10^6 t/a in Western Europe, 5.5×10^5 t/a in Japan, and 1.3×10^5 t/a in both Canada and Mexico. The annual capacity of the Federal Republic of Germany in July 1981 was estimated at 4×10^5 t/a. The major producers in that country in order of importance are: Hoechst AG (including Knapsack) with a capacity of 2.25×10^5 t/a, Wacker-Chemie GmbH with 8×10^4 t/a, BASF with 4.5×10^4 t/a, Chemische Werke Hüls with 4×10^4 t/a, and Lonza with 1.2×10^4 t/a.

In the United States about one-half of the acetic acid is consumed in the manufacture of vinyl acetate, whereas 20% is used in the manufacture of cellulose acetate. The remaining 30% is used in the manufacture of terephthalic acid and dimethyl terephthalate, esters of acetic acid, acetic anhydride (other than that used for cellulose acetate manufacture), and a variety of other minor uses.

Acetic acid growth during the 1970s was 8% per year; however, in 1980 a decline of 9% in consumption occurred because of the general business recession. Stanford Research Institute [27] predicted a growth in acetic acid consumption of only 2.0–2.5% per year between 1980 and 1985. This is because growth of vinyl acetate consumption is not expected to be higher than 3% per year and also because the new Eastman Chemical coal-based acetic anhydride process will reduce the amount of acetic acid required to make anhydride for cellulose acetate [27].

There is very little doubt that the technology of the future is methanol carbonylation. In 1972 low-pressure methanol carbonylation in the United States was only 10% of the total acetic acid capacity, whereas in 1982 the share had increased to 40% [86]. In the United States the main producers using this technology are Celanese, Clear Lake, Texas, with 3.6×10^5 t/a; National Distillers (USI), Deer Park, Texas, with 2.7×10^5 t/a; and Monsanto, Texas City, Texas, with 1.8×10^5 t/a. Other companies that have licensed the methanol carbonylation technology are BP Chemicals International and also Lummus Company, which has constructed an acetic acid plant in the Soviet Union [87].

In Europe, Rhône-Poulenc has built a unit of 2.25×10^5 t/a at Pardies in Southern France [27], [87].

In Japan methanol carbonylation has led to a reorganization of acetic acid producers [27]. In 1979 there were seven companies producing acetic acid, all by oxidation of acetaldehyde. By the middle of 1981 there were only three acetic acid manufacturers operating. The largest, Kyodo Sakusan K. K., with a capacity of 2.2×10^5 t/a, uses the methanol carbonylation process.

Acetaldehyde oxidation units are the most vulnerable because of the high price of the ethylene used to produce acetaldehyde. As a result acetaldehyde capacity is being cut back [88]. Other technologies are also being idled, such as the high-pressure methanol carbonylation plant of Borden in Geismar, Louisiana, USA, and the liquid-phase oxidation of butane by Union Carbide at Brownsville, Texas, USA [89].

12. Toxicology and Occupational Health

The acetate ion is quickly consumed in the body and is considered to be a normal metabolite. One radioactive labeling study [90] using [1–^{14}C] acetate gave results that indicate acetic acid incorporation into liver and brain cholesterol of rats and guinea pigs. However, there are no reports of cumulative toxicity, presumably because of its incorporation into intermediary metabolism.

Vinegar normally contains 3–6% acetic acid. Solutions of acetic acid in the 5% range cause human mucous membrane irritation that can lead to weight loss. For persons more than 2 years old, estimated daily acetic acid intakes up to 2.1 g are possible [91]. Diluted acetic acid can act more strongly on the skin than some diluted mineral acids because the diluted acetic acid is readily miscible with the lipids.

Acetic acid can irritate the eyes, nose, and throat at 10 ppm (8 h). At 100 ppm, possible damage to these organs can occur. However, repeated inhalation leads to

Table 11. Exposure limits to acetic acid and its derivatives

Substance	TLV				MAK	
	TWA		STEL			
	ppm	mg/m^3	ppm	mg/m^3	mL/m^3	mg/m^3
Acetic acid	10	25	15	37	10	25
n-Butyl acetate	150	710	200	950	200	950
sec-Butyl acetate	200	950	250	1190	200	950
Dimethylacetamide (skin)	10	35	15	50	10	35
Ethyl acetate	400	1400	–	–	400	1400
Methyl acetate	200	610	250	760	200	610

habituation and daily concentrations up to 60 ppm can be tolerated [92]. Table 11 gives established TLV and MAK for acetic acid and its derivatives [93], [94].

Oral acetic acid poisoning leads to severe mouth and digestive tract pain. In extreme cases vomiting, respiratory and circulatory distress, and sometimes death may follow [95].

Most of the lower molecular mass *acetate esters* have a distinct aromatic odor and slightly irritate the mucous membranes. They act as mild narcotics when ingested or adsorbed through the skin. Ethyl acetate is less potent than the butyl and amyl acetates [96]. Methyl acetate poses a special hazard because of the generation of methanol after ingestion or absorption.

Ethyl acetate has a pleasant fruity flavor. The LD$_{50}$ (oral) for ethyl acetate is similar for rats (5.6 g/kg) and rabbits (4.9 g/kg) [97]. Irritation of the human nose and throat occurs at about 400 ppm.

In humans 1,3-dimethylbutyl acetate has the lowest reported irritating level at 100 ppm. For most of the other acetates, irritating levels are between 100 and 200 ppm. An odor detection level of 0.002 ppm. in water has been recorded for hexyl acetates.

The *amides* pose more of a safety problem than the esters. Acetamide is known to cause cancer in rats (2.5% acetamide fed for 12 months). Methylacetamides show teratogenic and neoplastic potentials. N,N-Dimethylacetamide causes hallucinogenic effects similar to LSD at doses of 400 mg/kg [98]. Employees exposed to 20–25 ppm of N,N-dimethylacetamide showed liver damage [5]. A TLV of 10 ppm has been recommended with no skin exposure.

13. References

[1] A. P. Kudchadker, G. H. Alani, B. J. Zwolinski: *Chem. Rev.* **68** (1968) no. 6, 696.
[2] J. A. Riddick, W. B. Bunger: *Organic Solvents: Physical Properties and Methods of Purification,* 3rd ed., Wiley-Interscience, New York 1970.

[3]   D. D. Perrin, W. L. F. Armarego, D. R. Perrin: *Purification of Laboratory Chemicals*, Pergamon Press, New York 1966, p. 56.
[4]   H. L. Ritter, J. H. Simon, *J. Am. Chem. Soc.* **67** (1945) 757–762.
[5]   *Ullmann*, 4th ed., vol. **11,** p. 57–74.
[6]   T. Grewer, A. Schmidt, *Chem. Ing. Tech.* **45** (1973) 1063–1067.
[7]   J. Gmehling, U. Onken, P. Grenzheuser: *Vapor-Liquid Equilibrium Data Collection*, Schon E. Wetzel, Frankfurt 1982, p. 54–191.
[8]   C. Sandonnini, *Atti Accad. Naz. Lincei Cl. Sci. Fis. Mat. Nat. Rend.* **4** (1926) 63–68.
[9]   E. Sebastiani, L. Lacquaniti, *Chem. Eng. Sci.* **22** (1967) 1155–1162.
[10] J. A. Dean: *Lange's Handbook of Chemistry*, 12th ed., McGraw-Hill, New York 1978, Sec. 9, p. 65.
[11] S. P. Miskidzh'yan, H. A. Trifonof: *Zh. Obshch. Khim.* **17** (1947) 1033. *Chem Abstr.* **42** (1948) 3651 c.
[12] M. Usanovich, L. N. Vasil'eva, *Zh. Obshch. Khim.* **16** (1946) 1202. *Chem. Abstr.* **41** (1947) 2976 c.
[13] A. I. Vogel, *J. Chem. Soc.* 1948 1814–1819.
[14] R. J. W. LeFévre, *Trans. Faraday Soc.* **34** (1938) 1127.
[15] R. Philippe, A. M. Piette: *Bull. Soc. Chim. Belg.* **64** (1955) 600.
[16] J. A. Dean: *Lange's Handbook of Chemistry*, 12th ed., McGraw-Hill, New York 1978, Sec. 10, p. 103; Sec. 5, p. 42.
[17] D. D. Wagman, W. H. Evans, V. B. Parker, R. H.Schumm et al., *J. Phys. Chem. Ref. Data* **11** (1982) Suppl. 2.
[18] N. D. Lebedeva: *Russ. J. Phys. Chem. (Engl. Transl.)* **38** (1964) 1435–1437.
[19] N. I. Sax: *Dangerous Properties of Industrial Materials*, 5th ed., Van Nostrand Reinhold Co., New York 1979, p. 333.
[20] D. Ambrose, J. H. Ellender, C. H. S. Sprake, R. Townsend, *J. Chem. Thermodyn.* **9** (1977) 735–741.
[21] E. K. Euranto in S. Patai (ed.): *The Chemistry of Carboxylic Acids and Esters*, Interscience Publishers, London 1969, p. 505–588.
[22] I. T. Harrison, S. Harrison: *Compendium of Organic Synthetic Methods*, Wiley-Interscience, New York 1971, Chap. 8, p. 271.
[23] A. I. Vogel: *Practical Organic Chemistry*, 3rd ed.,Longmans, Green & Co., London 1957.
[24] W. Schwerdtel: *Hydrocarbon Process.* **47** (1968) no. 11, 187.
[25] A. V. G. Hahn: *The Petrochemical Industry*, McGraw-Hill, New York 1970, Chap. 5, p. 251.
[26] W. L. Faith, D. B. Keyes, R. L. Clark: *Industrial Chemicals*, 3rd ed.,J. Wiley & Sons, New York 1965, p. 257.
[27] S. F. Dickson,J. Bakker, A. Kitai: "Acetic Acid," in *Chemical Economics Handbook*, SRI International, Menlo Park, Calif. March 1982, revised July 1982, 602.5020, 602.5021.
[28] R. P. Lowry, A. Aguiló, *Hydrocarbon Process.* **53** (1974) no. 11, 103–113.
[29] H. Hohenschutz et al., *Hydrocarbon Process.* **45** (1966) no. 11, 141–143.
[30] J. W. Reppe: *Acetylene Chemistry*, Charles A. Meyer & Co., Boston, Mass. 1949.
[31] BASF, *Hydrocarbon Process.* **46** (1967) no. 11, 136.
[32] BASF, *Hydrocarbon Process.* **52** (1973) no. 11, 92.
[33] N. V. Kutepow, W. Himmele, H. Hohenschutz, *Chem. Ing. Tech.* **37** (1965) 383–388.
[34] F. E. Paulik, J. F. Roth, *Chem. Commun.* (1968) 1578.
[35] D. Foster in F. C. A. Stone, R. West (eds.): *Advances in Organometallic Chemistry* vol. **17,** Academic Press, New York 1979, p. 255–267.
[36] G. W. Parshal: *Homogeneous Catalysis*, Wiley & Sons, New York 1980, Chap. 5, p. 80–81.

[37] D. L. King, K. K. Ushiba, T. E. Whyte, Jr., *Hydrocarbon Process.* **61** (1982) no. 11, 131–136.
[38] P. Ellwood, *Chem. Eng. (N.Y.)* **76** (1969) no. 11, 148–150.
[39] J. J. McKetta, W. A. Cunningham (eds.): *Encyclopedia of Chemical Processing Design,* Marcel Dekker, New York 1976, p. 216–246.
[40] C. H. Cheng, D. E. Hendrickson, R. Eisenberg, *J. Am. Chem. Soc.* **99** (1977) 2791.
[41] B. R. James, G. L. Rempel, *J. Chem. Soc., A*1969 78–84.
[42] *Kirk-Othmer,* 3rd ed., vol. **1,** p. 124–147.
[43] *Kirk-Othmer,* 3rd ed., vol. **12,** p. 826.
[44] F. Broich, *Chem. Ing. Tech.* **36** (1964), 417–422.
[45] H. Höfermann, *Chem. Ing. Tech.* **36** (1964) no. 5, 422–429.
[46] *Acetic Acid, Report 37, 37 A,* SRI Internationals Process Economics Program, private report, Feb. 1968, March 1973.
[47] F. R. Mayo, *Prepr. Div. Pet. Chem. Am. Chem. Soc.* **19** (1974) 627.
[48] J. B. Saunby, B. W. Kiff, *Hydrocarbon Process.* **55** (1976) no. 11, 247.
[49] C. C. Hobbs, T. Horlenko, H. R. Gerberich, F. G.Mesich et al., *Ind. Eng. Chem. Process Des. Dev.* **11** (1972) 59.
[50] K. U. Ingold: *Acc. Chem. Res.* **2** (1969) 1.
[51] W. J. de Klein, E. C. Kooyman: *J. Catal.* **4** (1965) 626.
[52] G. A. Russell, *J. Am. Chem. Soc.* **79** (1957) 3871.
[53] R. K. Jensen, S. Korcek, L. R. Mahoney, M. Zinbo, *J. Am. Chem. Soc.* **101** (1979) 7574.
[54] F. J. Weymouth, A. F. Millidge, *Chem. Ind. (London)* 1966, 887–893.
[55] T. Yamaguchi, *Jpn. Chem. Q.* **IV – 1** (1968) 27–33.
[56] G. M. Forker, N. A. Lange (ed.): *Handbook of Chemistry,* 10th ed., McGraw-Hill, New York 1961, p. 1494.
[57] Badger BV, EP 71 298, 1983.
[58] J. Royer, M. Beugelmans-Verrier, *C. R. Hebd. Seances Acad. Sci. Ser. C* **272** (1971) 1818.
[59] N. A. Clinton, R. A. Kenley, T. G. Traylor, *J. Am. Chem. Soc.* **97** (1975) 3746, 3752, 3757.
[60] G. C. Allen, A. Aguiló, *Adv. Chem. Ser.* **76** (1968) 363.
[61] Celanese Corp., NL 8004 525, 1980.
[62] R. A. Sheldon, J. K. Kochi: *Metal Catalyzed Oxidations of Organic Compounds,* Academic Press, New York 1981, p. 361.
[63] S. Kakutani, *Rep. Imp. Ind. Res. Inst. Osaka Jp.* **19** (1939) no. 11, 68. *Chem. Abstr.* **33** (1939) 8566.
[64] Degussa, FR 844 531, 1939.
[65] Showa Acetyl Chem K, JP-Kokai 55 167 242, 1979.
[66] Hoechst, EP 2 696, 1981.
[67] J. Morrison, *Oil Gas Int.* **11** (1971) 26–29.
[68] S. Writers, *Chem. Econ. Eng. Rev.* **18** (1972) 22–26.
[69] E. G. Bayer, *Hydrocarbon Process.* **52** (1973) no. 11, 93.
[70] Halcon International Inc., US 3240805, 1966 (A. N. Nagliery).
[71] M. D. Mikkal, *Tr. Tallin. Politekh. Inst.* Ser. **A 228** (1965) 49–60.
[72] H. T. Raudsepp, M. D. Mikkal, *Tr. Tallin. Politekh. Inst. Ser.* A 228 (1965) 61–70.
[73] Dow Chemical Co., US 3980701, 1975 (R. R.Ginstead).
[74] T. A. Luft: vol. 28, Gemeinsames Ministerialblatt Ausgabe A, 25. Jahrgang Nr. 24, Bonn Aug. 1974.
[75] United States Clean Air Act, 42. U.S.C. 7401, as amended by U.S. Public Law 98-45 July 12, 1983, Bureau of National Affairs, Washington, D.C. 20037.

[76] J. L. Sadler: *Private Communication,* Celanese Chemical Co., Inc., Dallas, Tex., July 11, 1983.
[77] B. Byars, G. Jordan, *J. Gas Chromatogr.* **2** (1964) 304.
[78] *Chemical Safety Data Sheet SD-41,* Manufacturing Chemists Association, Inc., Washington, D.C. 1951.
[79] A. I. Grigor'ev, E. G. Pogudilova, A. V. Novoselova, *Zh. Neorg. Khim.* **10** (1965) 772.
[80] A. G. für Stickstoffdünger, US 2141477, 1937 (J. Losch).
[81] J. C. Bailar, H. J. Emeleus, R. Nyholm, A. F. Trotman-Dickenson: *Comprehensive Inorganic Chemistry,* vol. **1,** Pergamon Press, Oxford 1973 p. 462–464.
[82] H. Stephen, T. Stephen: *Solubilities of Inorganic and Organic Compounds,* vol. **1,** Pt 1, Macmillan Publ. Co., New York 1963, 107–212.
[83] *Houben-Weyl,* 4th ed., **8,** p. 558.
[84] *Kirk-Othmer,* 3rd ed., vol. **9,** p. 328.
[85] M. F. Ansell, R. H. Gigg in S. Coffey (ed.): *Rodd's Chemistry of Carbon Compounds,* vol. **1C,** Elsevier, Amsterdam 1965,p. 92-249.
[86] A. Aguiló, J. S. Alder, D. N. Freeman, R. J. H. Voorhoeve, *Hydrocarbon Process.* **62** (1983), no. 3, 57–65.
[87] *Chem. Mark. Rep.* **213** (1978) March 6, 5.
[88] *Chem. Mark. Rep.* **222** (1982) Dec. 20, 5, 12.
[89] *Chem. Week* **132** (1983) April 13, 16.
[90] H. J. Nicholas, B. E. Thomas, *Brain* **84** (1961) 320–328.
[91] *NTIS,* no. PB-274670, Life Sci. Res. Office, Fed. Am. Soc. Exp. Biol., Bethesda, Md, 1977.
[92] E. L. Vigliani, *Arch. Gewerbepathol. Gewerbehyg.* **13** (1955) 528.
[93] American Conference of Governmental Industrial Hygienists (ACGIH) (ed.): *Threshold Limit Values (TLV) 1982,* Cincinnati, Ohio, 1982, p. 10.
[94] Deutsche Forschungsgemeinschaft (ed.): *Maximale Arbeitsplatzkonzentrationen (MAK) 1983,* Verlag Chemie, Weinheim 1983, p. 15.
[95] D. Guest, G. V. Katz, B. D. Astill in G. D. Clayton, F. E. Clayton (eds.): *Patty's Industrial Hygiene and Toxicology,* 3rd ed., vol. **2C,** Wiley-Interscience, New York 1982, 4909–4911.
[96] J. C. Munch, *Ind. Med. Surg.* **41** (1972) 31–33.
[97] G. D. Clayton, F. E. Clayton (eds.): *Patty's Industrial Hygiene and Toxicology,* 3rd ed, vol. 2A, Wiley-Interscience, New York 1982, p. 2270–2282.
[98] A. J. Weiss, *Science* **136** (1962) 151.
[99] *Chem. Age* (London) **107** (1973) Aug. 24, 15.
[100] *Eur. Chem. News* **40** (1983) no. 1068, no. 14.

Acetic Anhydride and Mixed Fatty Acid Anhydrides

HEIMO HELD, Wacker-Chemie GmbH, Werk-Burghausen, Burghausen, Federal Republic of Germany (Sections 1.1–1.3.2, 1.4–1.7, and Chaps. 2 and 3)

ALFRED RENGSTL, Wacker-Chemie GmbH, Werk-Burghausen, Burghausen, Federal Republic of Germany (Section 1.3.3)

DIETER MAYER, Pharma Forschung Toxikologie, Hoechst Aktiengesellschaft, Frankfurt, Federal Republic of Germany (Chap. 4)

1.	Acetic Anhydride	63
1.1.	Physical Properties	64
1.2.	Chemical Properties	65
1.2.1.	Acetylation	66
1.2.2.	Dehydration	70
1.2.3.	Reactions of the α-Protons	70
1.2.4.	Reactions of a Single Carbonyl Group	71
1.2.5.	Production of Silver Ketenide	71
1.3.	Production	71
1.3.1.	Ketene Process	72
1.3.1.1.	Production of Ketene	72
1.3.1.2.	Reaction of Ketene with Acetic Acid	73
1.3.1.3.	Pure Anhydride Distillation	74
1.3.1.4.	Environmental Problems	75
1.3.2.	Oxidation of Acetaldehyde	75
1.3.3.	Carbonylation of Methyl Acetate	76
1.4.	Analysis	79
1.5.	Quality Specifications	79
1.6.	Storage and Transportation	80
1.7.	Uses	80
2.	Mixed Fatty Acid Anhydrides	81
2.1.	Physical Properties	81
2.2.	Chemical Properties	81
2.3.	Production	82
2.4.	Uses	83
3.	Economic Aspects	83
4.	Toxicology and Occupational Health	83
5.	References	84

1. Acetic Anhydride

Acetic anhydride [108-24-7], $(CH_3CO)_2O$, was first prepared by C. GERHARDT in 1852 by the reaction of benzoyl chloride with molten potassium acetate. Today it is one of the most important organic intermediates and is used widely in both research and industry.

1.1. Physical Properties

Acetic anhydride, $C_4H_6O_3$, M_r 102.09, mp −73.1 °C, bp 139.5 °C (at 101.3 kPa), is a colorless liquid with a pungent odor and is strongly lachrymatory. The most important physical data are given below.

Critical pressure						4680 kPa
Critical temperature						296 °C
Vapor pressure						
t, °C	20	40	60	80	100	120
p, kPa	0.4	1.7	5.2	13.3	28.7	53.3

Density	d_{20}^{20} 1.0838; d_{20}^{195} 1.3290; d_{20}^{79} 1.277; d_4^{15} (1.0870) up to d_4^{50} (1.0443), see [1]
Refractive index	
n_D^{20}	1.39038
n_{322nm}^{16}	1.4174
$n_{667.8nm}^{16}$	1.3897, see [2]
^{1}H NMR absorption	see [3]
UV absorption maximum	217 nm
Specific heat capacity (23–122 °C)	1.817 J/g
Heat of vaporization	
at 18.5 °C	496.5 J/g
at 137 °C	276.7 J/g
Heat of combustion at constant volume at 25 °C	1804.5 kJ/mol
Thermal conductivity at 25 °C	2.215 mJ cm^{-1}s^{-1}K^{-1} [4]
Electric conductivity at 20 °C	2.3×10^{-8} Ω^{-1}cm^{-1} [5]
Dielectric constant at 20 °C	20.5
Viscosity	
at 15 °C	0.971 Pa s
at 30 °C	0.783 Pa s
(for temperature dependence, see [2], [6])	
Cubic expansion coefficient at 18 °C	1.13×10^{-9} K^{-1}
Adiabatic compressibility constant	4.86×10^8 cm^2/N
Molecular refraction	22.38 cm^3
Surface tension	

t, °C	15	20	40	60	100	139.5
σ, mN/m	33.4	32.7	30.0	28.1	23.3	18.6

Acetic anhydride is miscible with polar sol-vents and dissolves in cold alcohol with slow decomposition. The solubility of acetic anhydride in water at 20 °C is 2.6 wt %, with slow decomposition; the solubility of water in acetic anhydride at 15 °C is 10.7 wt %, with gradual decomposition. An overview of the solubilities of several anhydrous organic and inorganic compounds is given in [7].

Data on the kinetics of hydrolysis in water or aqueous mixtures of acids, bases, or organic solvents, with or without additives and at different temperatures, are given in [8]. Rate constants for the solvolysis in alcohols or alcoholic mixtures of organic solvents, with or without additives, are also given.

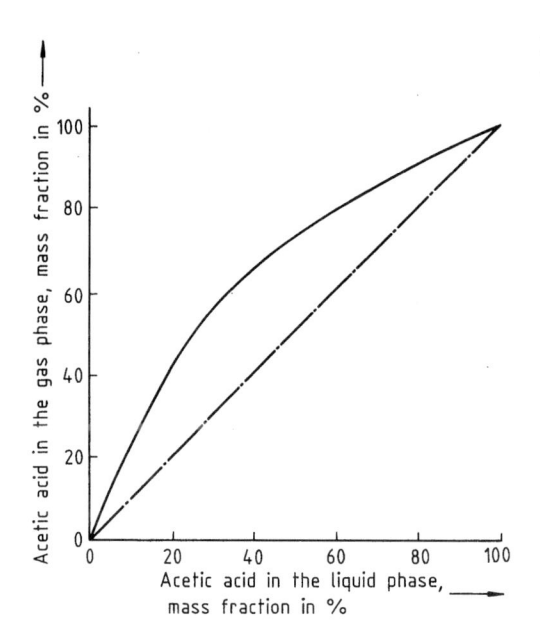

Figure 1. Equilibrium curve: acetic acid – acetic anhydride

Melting points of acetic acid – acetic anhydride mixtures are as follows:

% acetic acid	100	80	60	40	20	10
mp,°C	16.7	5.4	−5.8	−19.8	−44.8	−68.1

The vapor-liquid equilibrium curve for mixtures of acetic acid and acetic anhydride is shown in Figure 1; this is of great importance in industry. Further vapor – liquid equilibria, for example, with water, benzene, diketene, propionic acid, pyrimidine, and water + acetic acid are given in [8].

Safety data:

Flash point 52.5 – 53 °C (closed cup, ASTM 56–70, DIN 51755)
Ignition temperature 315 °C
Explosive limits in air (20 °C, 101.3 kPa) 2 – 10.2 vol%

1.2. Chemical Properties

Acetic anhydride undergoes a large variety of chemical reactions and is by far the most researched aliphatic carboxylic acid anhydride. In about half of the several thousand relevant publications and patents over the last 25 years, acetic anhydride was used for the acetylation of OH or NH groups (at least for the primary step). Many special reactions of acetic anhydride are summarized in [9].

1.2.1. Acetylation

O-Acetylation. Acetic anhydride is particularly suitable for the esterification of alcohols, a reaction that is difficult or impossible with acetic acid. Acetic acid is set free in the course of the reaction. Bases and strong acids as well as salts, such as sodium acetate, are suitable as catalysts. Examples of the reaction with hydroxyl groups are the formation of acetyl cellulose, acetylsalicylic acid (Aspirin), and glycerol triacetate (→ Salicylic Acid, → Glycerol).

Carboxylic Acid reactions are described in Chap. 2 (→ Carboxylic Acids, Aliphatic; → Carboxylic Acids, Aromatic).

Hydrogen peroxide reacts with acetic anhydride to give peracetic acid [*79-21-0*] or diacetyl peroxide [*110-22-5*], depending on the molar ratio of the reactants:

$(CH_3CO)_2O + 2 H_2O_2 \rightleftharpoons 2 CH_3CO_3H$

$(CH_3CO)_2O + H_2O_2 \rightleftharpoons CH_3COOH + CH_3CO_3H$

$(CH_3CO)_2O + CH_3CO_3H \rightleftharpoons CH_3CO\text{-}OO\text{-}COCH_3 + CH_3COOH$

N-Acetylation. Acetylation of compounds containing NH groups yields acetamides following the general equation:

$RR'NH + (CH_3CO)_2O \rightarrow RR'N\text{-}COCH_3 + CH_3COOH$

where R, R' = H or alkyl.

Amines. Aliphatic amines usually react without heating. Aniline gives acetanilide (→ Aniline), the acetyl group of which prevents oxidation during subsequent nitration. N-Acetylation generally occurs faster than the acetylation of OH groups. Therefore partial acetylation of compounds with several functional groups is possible. Examples of such reactions are the production of N-acetylamino acids, such as *N*-acetylmethionine-*S*-oxide (**1**) [10] and of N-acetylanthranilic acids (**2**) [11]

$CH_3SOCH_2CH_2\underset{NHCOCH_3}{CHCOOH}$

1

(structure **2**: aromatic ring with R, COOH, R', NHCOCH$_3$ substituents)

2

Amides and Carbamides. Aliphatic and aromatic carboxylic acid amides as well as carbamides with a free NH group can be acetylated with acetic anhydride. Strong acids, such as sulfuric acid, are most often used as catalysts. The breadth of application of this reaction can be seen from the following products: *N,N,N',N'*-tetraacetylethylenediamine [*10543-57-4*] [12],

$(CH_3CO)_2NCH_2CH_2N(COCH_3)_2$

and 2,4,6,8-tetraacetylazabicyclo [3.3.1] nonane-3,7-dione [13]:

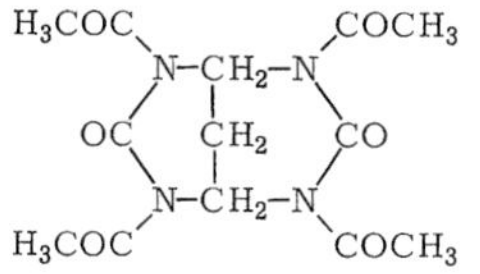

$$\text{H}_3\text{COC}\diagdown\text{N}-\text{CH}_2-\text{N}\diagup\text{COCH}_3$$
$$\text{OC}\quad\text{CH}_2\quad\text{CO}$$
$$\text{H}_3\text{COC}\diagup\text{N}-\text{CH}_2-\text{N}\diagdown\text{COCH}_3$$

C-Acetylation. Compounds with reactive CH bonds can be acetylated with acetic anhydride, with the aid of a catalyst if necessary. Examples are the production of ethyl α-cyanoacetoacetate [634-55-9], $CH_3COCH(CN)COOC_2H_5$, (catalyst: K_2CO_3) [14] and the Friedel-Crafts reaction of acetic anhydride with aromatic hydrocarbons, for example, with benzene to form acetophenone (→ Ketones), as well as with unsaturated hydrocarbons. An interesting example of the latter group is the acetylation of 2,3-dimethylbutadiene tricarbonyl iron in the presence of aluminum chloride [15] to give the following complex:

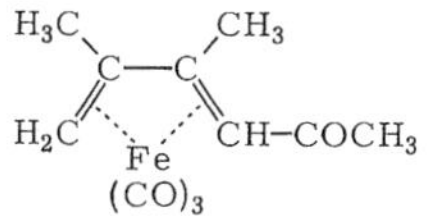

The reaction of ketones with acetic anhydride using boron trifluoride as a catalyst gives β-diketones [16]. Triacetylmethane [815-68-9] can be synthesized directly from isopropenyl acetate, acetic anhydride, and aluminum chloride catalyst [17]. In the same way, unsaturated methyl ketones can be produced by acetylation of olefins with acetic anhydride in the presence of zinc chloride as catalyst [18]:

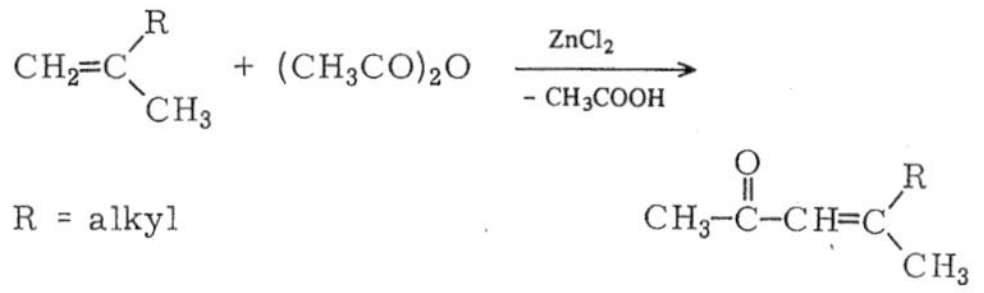

Acetylation of Mineral Acids. The reaction of nitric acid with acetic anhydride gives acetyl nitrate [591-09-3], which is used often as a nitrating agent in organic chemistry. For example, it is used in the production of 2-nitrocyclohexanone [4883-67-4], a precursor of ε-caprolactam [19]. Acetyl nitrate can be synthesized also from dinitrogen pentoxide and acetic anhydride.

Other strong acids that form mixed anhydrides include sulfuric acid [20], sulfonic acids [21], and hydrochloric acid. The reaction of acetic anhydride with phosphorous acid leads to 1-hydroxyethane-1,1-diphosphonic acid [22]:

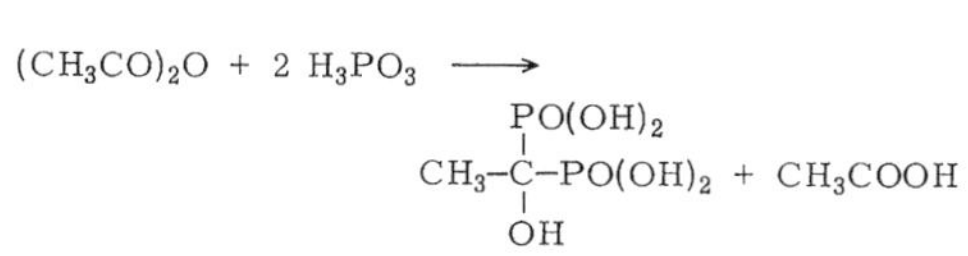

Acetylation of Oxides. Acetic anhydride reacts with antimony trioxide to give antimony triacetate, Sb(OCOCH$_3$)$_3$ [*5692-86-4*] [23]. The corresponding reaction with chromium trioxide leads to chromyl acetate solutions [24], which often are used to oxidize olefins and hydrocarbons and which may occasionally explode [25].

Acetylation of Salts. Various salts react with carboxylic acid anhydrides in the same way as the corresponding free acids [26]:

$$\text{Li-C} \equiv \text{CR} + (\text{CH}_3\text{CO})_2\text{O} \rightarrow \text{CH}_3\text{CO-C} \equiv \text{CR} + \text{CH}_3\text{COOLi}$$

With barium peroxide, diacetyl peroxide [*110-22-5*] is obtained [27]:

$$\text{BaO}_2 + (\text{CH}_3\text{CO})_2\text{O} \longrightarrow (\text{CH}_3\text{CO})_2\text{O}_2 + \text{BaO}$$

A general method of producing vinyl ketones is the reaction of vinyl magnesium bromides and acetic anhydride [28]:

$$2\ \text{CHR} = \text{CHMgBr} + 2\ (\text{CH}_3\text{CO})_2\text{O} \xrightarrow{\text{THF}}$$
$$2\ \text{CHR} = \text{CHCOCH}_3 + \text{MgBr}_2 + (\text{CH}_3\text{COO})_2\text{Mg}$$

Production of Acetoxy Silanes. Acetic anhydride reacts with silanes according to the equation:

$$-\overset{|}{\underset{|}{\text{Si}}}-\text{X} + (\text{CH}_3\text{CO})_2\text{O} \longrightarrow -\overset{|}{\underset{|}{\text{Si}}}-\text{OCOCH}_3 + \text{CH}_3\text{COX}$$

where X = H, Cl [29]; OR [30]; NR$_2$ [31].

Addition to Heterocyclic Compounds with Ring Cleavage. These reactions differ only formally from those described above. Examples are the production of ethylene glycol diacetate [*111-55-7*] from ethylene oxide in the presence of strongly acidic [32] or basic [33] catalysts and the production of oxymethylene diacetates from trioxane [34].

Oxidative Addition to Carbon-Carbon Double Bonds. Oxidative addition leads to the corresponding diacetates, as for example in the addition of ethylene, which leads to ethylene glycol diacetate [35]:

$$\text{CH}_2 = \text{CH}_2 + (\text{CH}_3\text{CO})_2\text{O} \xrightarrow[1/2\text{O}_2]{\text{Re}_2\text{O}_7}$$
$$\text{CH}_3\text{CO} - \text{OCH}_2\text{CH}_2\text{O} - \text{COCH}_3$$

1,4-Diacetoxy-2-butene [*18621-75-5*] can be obtained from butadiene in a similar manner [36].

Production of Mixed Diacyl Peroxides. Diacyl peroxides can be made by the reaction of oxygen with mixtures of aliphatic aldehydes and acetic anhydride in the presence of sodium acetate [37]:

$$2\,RCHO + (CH_3CO)_2O \xrightarrow{2O_2} 2\,RCO-OO-COCH_3 + H_2O$$

Reaction with N-Oxides. Reactions with N-oxides lead to a variety of products, depending on the type of N-oxide. From pyridine-*N*-oxide, 2-acetyloxypyridine is obtained [38]. However, 4-picoline-*N*-oxide yields a mixture of 4-acetyloxymethylene-pyridine and 3-acetyloxy-4-methyl- pyridine [39]. In the Polonovski reaction of N-oxides, formaldehyde and unsaturated aldehydes are formed as well as the acid amides [40], as in the following example:

$$(CH_3)_2\overset{O\uparrow}{N}(CH_2)_n-CH=CH_2 \xrightarrow{(CH_3CO)_2O}$$

$$\underset{|}{\overset{COCH_3}{CH_3N}}(CH_2)_n-CH=CH_2 + CH_2O$$

$$+ CH_3(CH_2)_n-CH=CH-CHO$$

Reaction with S-Oxides. The reduction of sulfoxides to sulfides with acetic anhydride is known as the Pummerer reaction [41]:

$$CH_3SOR + (CH_3CO)_2O \rightarrow CH_3COOCH_2SR + CH_3COOH$$

An interesting use of this reaction is the oxidation of primary and secondary alcohols, even when sterically hindered, by mixtures of dimethylsulfoxide (or tetramethylene sulfoxide) and acetic anhydride, to give the corresponding carbonyl compounds [42]:

$$CH_3SOCH_3 + (CH_3CO)_2O \rightarrow CH_3COO-\overset{+}{S}\underset{CH_3}{\overset{CH_3}{\diagup}}$$

$$\xrightarrow{>CH-OH} >CH-O-\overset{+}{S}\underset{CH_3}{\overset{CH_3}{\diagup}} \rightarrow >C=O$$

Production of Acylals and Vinyl Acetates. Aldehydes react with acetic anhydride in the presence of acid catalysts to form acylals:

$$RCHO + (CH_3CO)_2O \longrightarrow RCH(OCOCH_3)_2$$

If R has an α-hydrogen atom available, the corresponding vinyl acetate forms by elimination of acetic acid. The temperature required for the reaction can be lowered by adding catalysts [43]–[45].

$$>CH-CH\underset{OCOCH_3}{\overset{OCOCH_3}{\diagup}} \longrightarrow >C=CH-OCOCH_3 + CH_3COOH$$

A similar process is particularly suitable for the synthesis of 2,2-dichlorovinyl acetate [*36597-97-4*]. Here, in the absence of an α- proton, cleavage of the acetic acid moiety is facilitated by the presence of zinc [46]:

$$Cl_3CCHO + (CH_3CO)_2O \rightarrow Cl_3CCH(OCOCH_3)_2 \xrightarrow{Zn}$$
$$Cl_2C=CH-OCOCH_3 + ClZnOCOCH_3$$

Boron Trifluoride. Acetic anhydride and substituted acetic anhydrides can be converted into acetyl ketones via a boron trifluoride complex that decomposes in warm water [16], [47].

1.2.2. Dehydration

Acetic anhydride is used as a dehydrating agent, for example, in the explosives industry. The most important example of dehydration is in the production of hexogen (1,3,5-trinitrohexahydro-1,3,5-triazine) [*121-82-4*] [16]:

$$3\ CH_2O + 3\ NH_4NO_3 + 6\ (CH_3CO)_2O \longrightarrow$$

[structure of 1,3,5-trinitrohexahydro-1,3,5-triazine] $+\ 12\ CH_3COOH$

In the production of the nitroester of 1,2,4,5-tetrahydroxy-3,6-dinitrocyclohexane [48], the water-binding property of the anhydride is used again:

[structure of 1,2,4,5-tetrakis(nitrooxy)-3,6-dinitrocyclohexane]

Alkyl cyanides can be made by dehydrating aldoximes with acetic anhydride [49].

$$RCH=NOH + (CH_3CO)_2O \longrightarrow RCN + 2\ CH_3COOH$$

Acetic anhydride serves as a dehydrating agent in a large number of cyclization reactions.

1.2.3. Reactions of the α-Protons

The Perkin reaction is used for the production of α, β-unsaturated acids from aromatic aldehydes, such as benzaldehyde, in the presence of potassium acetate or sodium acetate. This reaction is also suitable for synthesizing cinnamalacetic acid [50]:

$$C_6H_5CH=CHCHO + (CH_3CO)_2O \xrightarrow[140\,°C]{CH_3COOK}$$
$$CH_3COOH + C_6H_5CH=CHCH=CHCOOH$$

Other examples of this type of reaction are the oxidative carboxymethylation with acetic anhydride and oxidizing agents [51] and the conversion of long-chain alkenes to the corresponding carboxylic acid derivatives [52].

1.2.4. Reactions of a Single Carbonyl Group

In some cases, it is possible to make just one of the carbonyl groups of acetic anhydride participate in a reaction. Examples are the reaction with hydrogen cyanide, in the presence of a base, to form 3 [53] and in the presence of Grignard reagent (RMgBr) to form 4 [54]:

$$CH_3CO-O-\underset{OH}{\underset{|}{\overset{CN}{\overset{|}{C}}}}-CH_3 \qquad CH_3CO-O-\underset{OMgBr}{\underset{|}{\overset{R}{\overset{|}{C}}}}-CH_3$$

$$\quad\quad 3 \qquad\qquad\qquad\qquad 4$$

1.2.5. Production of Silver Ketenide

Silver acetate and acetic anhydride react at room temperature in pyridine to form a pyridine complex of silver ketenide. When an excess of anhydride is used, silver ketenide is obtained after fractional distillation of pyridine and acetic acid [55], [56]:

$$CH_3COOAg + (CH_3CO)_2O + \text{pyridine} \xrightarrow{20\,°C} \text{[pyridine complex of silver ketenide]}$$

1.3. Production

History. The oldest process for making acetic anhydride is based on the conversion of sodium acetate with an excess of an inorganic chloride, such as thionyl chloride, sulfuryl chloride, or phosphoryl chloride. In this process, half of the sodium acetate is converted to acetyl chloride, which then reacts with the remaining sodium acetate to form acetic anhydride:

$CH_3COONa + X\text{-}Cl \longrightarrow CH_3COCl + XONa$

$CH_3COONa + CH_3COCl \longrightarrow (CH_3CO)_2O + NaCl$

where $X = SOCl, SO_2Cl, POCl_2$.

A further development, the conversion of acetic acid with phosgene in the presence of aluminum chloride, has the advantage that it allows continuous operation:

$2\,CH_3COOH + COCl_2 \longrightarrow (CH_3CO)_2O + 2\,HCl + CO_2$ 7

Two other methods also were used in the past: the cleavage of ethylidene diacetate to form acetaldehyde and acetic anhydride in the presence of acid catalysts, such as zinc chloride, and the reaction of vinyl acetate with acetic acid on palladium(II) contacts to form acetaldehyde and acetic anhydride [16]. Not one of these processes is now of any industrial importance.

Today, acetic anhydride is made mostly by either the ketene process or the oxidation of acetaldehyde. Production by another process, the carbonylation of methyl acetate (Halcon process), was begun in 1983. In Western Europe, 77% of acetic anhydride is made by the ketene process and 23% by the oxidation of acetaldehyde. Since production by the Halcon process began at the Tennessee-Eastman plant, 25% of acetic anhydride in the United States has been made by this process and 75% by the ketene process.

1.3.1. Ketene Process

The ketene process for the production of acetic anhydride proceeds in two steps: the thermal cleavage of acetic acid to form ketene and the reaction of ketene with acetic acid:

$$CH_3COOH \rightarrow CH_2=C=O + H_2O \qquad \Delta H = 147 \text{ kJ/mol}$$
$$CH_2=C=O + CH_3COOH \rightarrow (CH_3CO)_2O \qquad \Delta H = -63 \text{ kJ/mol}$$

1.3.1.1. Production of Ketene

Thermal Cleavage of Acetic Acid. Hot acetic acid vapor is broken into ketene and water at 700–750 °C in the presence of traces of phos-phoric acid catalyst. The pressure in the reactor is generally reduced so that the ketene can be isolated before it reacts with acetic acid or with water. The cleavage takes place in a multicoil reactor with coils made of highly heat-resistant steel alloys. The alloy Sicromal, containing 25% Cr, 20%Ni, and 2% Si, is particularly suitable. Triethyl phosphate is employed as the catalyst [57].

Figure 2 shows how the vacuum process operates. Acetic acid is fed continuously into the evaporator (a). The acetic acid vapor leaving the evaporator passes over the catalyst evaporator (b), which is supplied continuously with catalyst, and thence into the ketene oven (c). The acetic acid is heated (d) and then cleaved (e). As soon as the cracked gases have left the oven, ammonia is added to prevent the ketene from reacting with the water or with the remaining acetic acid. For the same reason, and also to remove condensable gases, the hot gas is cooled in combined water and brine condensers (f), (g). A 40–48% aqueous solution of acetic acid condenses at about 0 °C and is separated (h) from the cleavage gas (0–10 °C). About 4–8% of this acetic acid results from the reaction of the condensed acetic anhydride with water. This process has been developed by the Consortium für Elektrochemische Industrie, the research institute of Wacker-Chemie.

The construction of the ketene oven depends strongly on the power required. In small ovens (ca. 50 t/month), separate preheating and cleaving ovens are preferred. The acetic acid is cleaved in a spiral tube which is usually electrically heated. Ovens of medium size (up to ca. 700 t/month), however, are often constructed as single-chamber ovens, generally heated with gas or oil. For capacities of 700 t/month, product loss is to be expected. This is because of the unsatisfactory pressure conditions in the cleavage zone. Ovens with much greater capacities work satisfactorily with three- and four-chamber systems and partial gas flows [58]. It is also possible to operate several ovens with acetic acid supplied from a central evaporator.

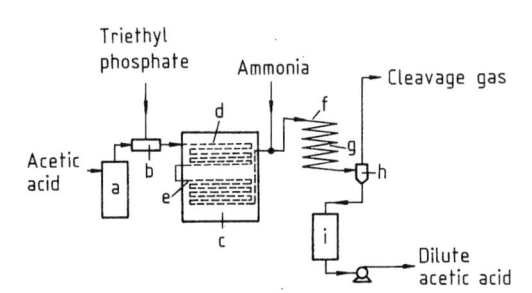

Figure 2. Cleavage of acetic acid by the vacuum process a) Acetic acid evaporator; b) Catalyst evaporator; c) Ketene oven; d) Convection zone; e) Cleavage zone; f) Water cooler; g) Brine cooler; h) Separator; i) Receiver

Thermal Cleavage of Acetone. This variation of the ketene process (cleavage of acetone into ketene and methane [59]) has no economic significance for the manufacture of acetic anhydride today.

1.3.1.2. Reaction of Ketene with Acetic Acid

Two processes are available for the reaction of ketene with acetic acid. The classical *scrubber process* is being replaced gradually by the *Wacker process* with liquid-ring pumping [60]. A further process [61], in which the reaction occurs at higher temperatures, has not yet gained industrial importance.

Scrubber Process (see Fig. 3). The ketene emerging from the separator (Fig. 2) is absorbed by glacial acetic acid circulating in scrubbers (a) and (b), which are filled with Raschig rings [62]. The circulation is maintained by centrifugal pumps and the heat of reaction removed by coolers (f). Most of the ketene is absorbed in scrubber (a), the rest in scrubber (b). The liquid mixture leaving scrubber (a) contains 85–90% raw anhydride and is collected in (g), cooled and used in scrubber (c) to wash the off-gas from (b). The off-gas from (c) is finally washed with brine-cooled, dilute acetic acid in scrubber (d) in order to remove acid.

Wacker Process (Liquid-ring pump process) (see Fig. 4). The ketene emerging from the separator (Fig. 2) is pumped through a Nash Hytor liquid-ring pump (a) at about 13–20 kPa. It reacts in the pump, at about 45–55°C, with acetic acid. At the same time, the liquid-ring pump provides the vacuum required for the acetic acid cleavage. Reaction and compression heat are removed by passing the raw anhydride through a cooler (c). This serves simultaneously as separator for the off-gas from the cleavage reactor. The off-gas is burnt in the ketene oven after passing the measuring point (d) or is led to a flare system. The raw anhydride (90%) is collected continuously. Part of the anhydride is returned to the reaction pump, where it is used as the reaction medium. Advantages of this process stem from the impressive simplicity of the apparatus and operation, which lead to high reliability and the possibility of producing raw anhydride of higher purity [60].

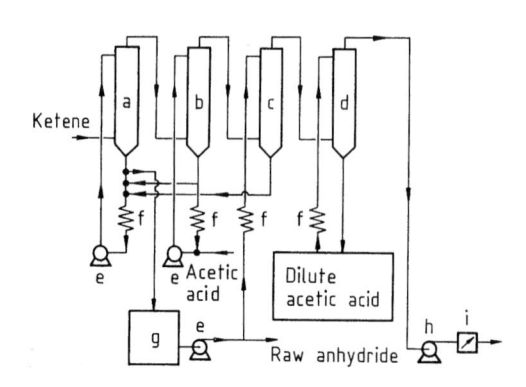

Figure 3. Scrubber process
a), b), c), d) Scrubbers; e) Centrifugal pumps; f) Coolers; g) Collecting vessel; h) Vacuum pump; i) Ring–balance manometer

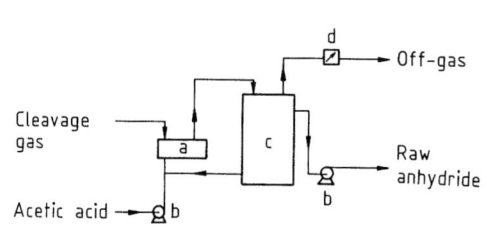

Figure 4. Wacker process (Liquid–ring pump process)
a) Liquid–ring pump; b) Pump; c) Cooler; d) Measuring point

1.3.1.3. Pure Anhydride Distillation

The raw anhydride can be distilled either continuously or discontinuously. In discontinuous distillation, three fractions are obtained:

Forerun (1%) is about 40% acetic acid; the rest consists of volatile liquids, particularly methyl acetate and acetone.

Middle run (about 10%). Depending on the quality of the column, it contains 30–60% acetic anhydride; the rest is acetic acid.

Pure Anhydride. Using discontinuous distillation at normal pressures, the purity of the anhydride cannot exceed 99% (the remainder is acetic acid) because, at the still temperature required, a certain amount of decomposition occurs. These difficulties can be overcome by operating under reduced pressure.

For continuous distillation, two or three columns are used, depending on the required number of fractions (forerun and middle run are separated). The middle run portion is returned to the reactor as reagent acid. The residue of the distillation contains small amounts of tarry products and is worked up continuously in a Sambay evaporator.

A newer continuous and energy-saving vacuum process works with two evaporators and only one column [63].

The *Wacker process*, described above (see Fig. 4), is also particularly suitable as a workup process. Waste acids, such as those from cellulose acetylation, can be processed after concentration without difficulty. No auxiliary materials are necessary, only energy in the form of gas, oil, or electric current. The process has been adopted by many firms. For the production of 100 kg acetic anhydride, about 122 kg acetic acid is required, taking account of the reconcentrated dilute acetic acid. The yield is over 96% at about 75% cleavage.

1.3.1.4. Environmental Problems

No significant environmental problems arise with the processes described above. The off-gas occuring in the production of ketene consists of ca. 45% carbon monoxide, 15–20% carbon dioxide, ca. 15% ethylene, ca. 10% methane, ca. 7% propylene, and less than 1% acetic acid and ethane. The remainder is air. The off-gas can be burnt in the ketene oven saving energy (ca. 10%), or led to a flare system. There are no wastewater problems.

1.3.2. Oxidation of Acetaldehyde

Acetic anhydride can be obtained directly by liquid-phase oxidaton of acetaldehyde. The peracetic acid formed from oxygen and acetaldehyde reacts under suitable conditions with a second molecule of acetaldehyde to form acetic anhydride and water [64]:

$$CH_3CHO + O_2 \rightarrow CH_3CO_3H \xrightarrow{CH_3CHO} (CH_3CO)_2O + H_2O$$

Rapid removal of the reaction water and the use of suitable catalysts are essential in this process. Mixtures of acetic acid and acetic anhydride are always obtained; their ratio can be varied within wide limits by changing the reaction conditions. Generally, the highest possible anhydride yield is sought.

Because of the rapid hydrolysis of acetic anhydride above 60 °C, the process is operated preferably between 40 °C and 60 °C [65]. Suitable catalysts are combinations of metal salts [66]. Particularly important are mixtures of manganese acetate and copper acetate [67], of cobalt acetate and nickel acetate, and of cobalt and copper salts of higher fatty acids [68]. Manganese acetate should hinder the formation of explosive amounts of peracetic acid during the oxidation of acetaldehyde. For increasing the rate of oxidation, the use of pure oxygen at a pressure of several hundred kilopascals instead of air has been proposed [69].

The strongly exothermic reaction requires efficient cooling. For this purpose, the addition of low-boiling solvents has been found to be of assistance. Methyl and ethyl acetates are favored because they form azeotropic mixtures with water (but not with acetic acid or acetic anhydride) and hence allow a rapid, continuous separation of the water formed in the reaction. The ratio of acetic anhydride to acetic acid in the product depends on the ratio of ethyl acetate to acetaldehyde in the initial mixture (Table 1).

Table 1. Formation of acetic anhydride by oxidation of acetaldehyde

Ethyl acetate: acetaldehyde in starting mixture	Acetaldehyde conversion, %	Acetic anhydride yield based on acetaldehyde, %
20:80	80	13.5
30:70	80	57
60:40	80	64
70:30	80	68.5

In practice, a 1:2 mixture of acetaldehyde and ethyl acetate is oxidized with the addition of 0.05 to 0.1% cobalt acetate and copper acetate at 40 °C; the ratio of Co:Cu is 1:2. The ratio of acetic anhydride to acetic acid obtained is 56:44, whereas on oxidizing in the absence of ethyl acetate this ratio is only 20:80 [16]. The optimization of other reaction conditions can also lead to an increase in the acetic anhydride-acetic acid ratio. For example, at 55 °C and atmosphericpressure, a ratio of 80:20 was achieved [70]. At a higher temperature (62–90 °C, 200–300 kPa, acetaldehyde concentration in the final mixture of up to 40%) a ratio of 75:25 was obtained at high aldehyde conversion [71].

Other suitable low-boiling solvents are methylene chloride, diisopropyl ether, cyclohexanone, or ethylidene diacetate. Nonvolatile esters also can be used as diluents, provided they do not have to be removed from the reaction zone. These include alkyl benzoates and alkyl phthalates [72].

The acetaldehyde oxidation is illustrated in Figure 5 by the process of Usines de Melle [73]. The gas mixture containing oxygen and acetaldehyde is pumped into the reactor (a). The oxidation takes place in the liquid phase and in the presence of catalysts. The reactor effluent is sent through a water-cooled condenser (b) constructed as a separator; non-condensable gases are sent to the packed column (c). Fresh acetaldehyde is introduced at the top of this column. The condensates from both the cooler (b) and the column (c) are distilled to obtain the product. Acetaldehyde is recovered from the branch stream (d) of the non-condensable gas. The other part of the gas flow is supplemented with air and returned to the reactor.

Both towers and vessels are suitable as reactors if the heat of reaction can be dissipated. The process of Distillers Co. [69] is shown in Figure 6 as an example. The off-gas contains combustible low-boiling products, such as acetaldehyde, and solvents, such as methyl acetate and ethyl acetate. These can be flared off.

1.3.3. Carbonylation of Methyl Acetate

The thermal decomposition of acetic acid to form ketene requires a large amount of energy, a disadvantage of the conventional process for the production of acetic anhydride. Moreover, processes based on synthesis gas have been developed in recent years that allow the manufacture of products from coal that were hitherto produced from oil. An important example is the acetic acid production process developed by Monsanto (→ Acetic Acid).

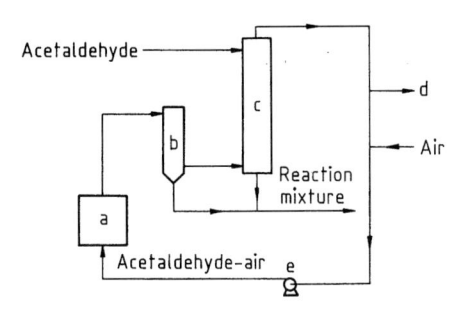

Figure 5. Acetaldehyde oxidation
a) Reactor; b) Condenser; c) Column; d) Branch stream (acetaldehyde recovery from non-condensable gas); e) Pump

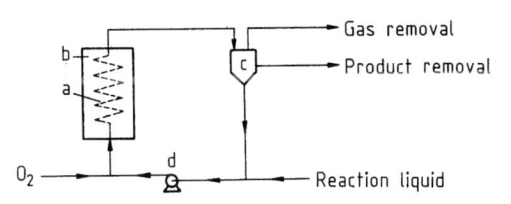

Figure 6. Reactor for the acetaldehyde oxidation (Distillers Co.)
a) Reactor tubing; b) Cooling bath; c) Separator; d) Circulation pump

In 1973 Halcon patented the carbonylation of methyl acetate in the presence of a rhodium catalyst to form acetic anhydride [74]. However, the first plant (Eastman-Kodak Co., in Kingsport, Tenn., USA) using this process was not put into operation before 1983. Because no exact description of the process has so far been available, the following details are based upon information from the patent literature.

Methyl acetate is carbonylated to acetic anhydride in the liquid phase at a temperature of 160–190 °C and at a carbon monoxide partial pressure of 2–5 MPa:

$$CH_3COOCH_3 + CO \underset{}{\overset{catal.}{\rightleftharpoons}} (CH_3CO)_2O \quad \Delta H = -94.8 \text{ kJ/mol}$$

The starting material, methyl acetate, can be produced by esterification of acetic acid with methanol. However, the methyl acetate obtained as a byproduct of the acetic acid synthesis from methanol and carbon monoxide is used preferably.

Catalysts. Rhodium [74] and nickel compounds [75], activated by CH_3I, HI, LiI, I_2, or other iodides, are particularly appropriate as catalysts. Rhodium catalysts have about a tenfold higher activity than nickel catalysts. The selectivity is higher than 95% for both catalysts. Their activities and lifetimes are increased by mixing the carbon monoxide used for the synthesis with 2–7% hydrogen. Chromium compounds have been used to shorten the induction phase of the reaction [74]. Kinetic investigations on the rhodium system have shown that the reaction is zero order with respect to the methyl acetate and carbon monoxide concentrations [76]. As possible byproducts, only acetic acid and ethylidene diacetate are mentioned.

The process also can start from dimethyl ether [74]. In this case, dimethyl ether is first carbonylated to methyl acetate, which is then converted to acetic anhydride by using more carbon monoxide in the same reactor:

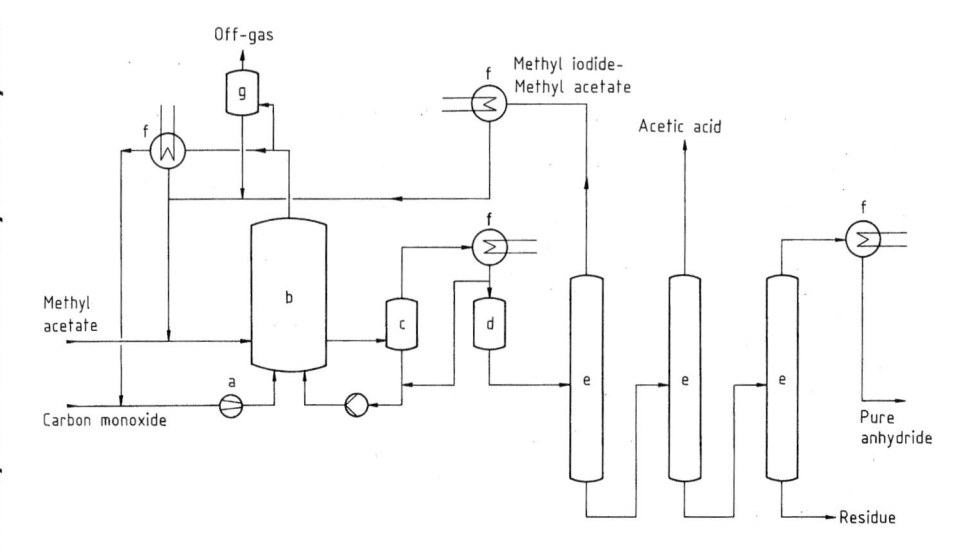

Figure 7. Halcon process for the production of acetic anhydride
a) Compressor; b) Carbonylation reactor; c) Evaporator; d) Adsorber; e) Distillation column; f) Condenser; g) Scrubber

$$CH_3OCH_3 \overset{CO}{\rightleftharpoons} CH_3COOCH_3 \overset{CO}{\rightleftharpoons} (CH_3CO)_2O$$

Process Description. The carbonylation method is illustrated by the Halcon process, shown schematically in Figure 7.

The methyl acetate is dried with acetic anhydride [77] and is sent to a reactor (b) lined with Hastelloy. Carbon monoxide is compressed (a) to the reaction pressure and then added. The reaction proceeds continuously at 175 °C in the presence of a catalyst consisting of $RhCl_3 \cdot 3\,H_2O$, CH_3I, and LiI. The considerable amount of heat generated by the reaction is removed by heat exchange and is used both for preheating methyl acetate and for the production of low-pressure steam. The unreacted carbon monoxide leaves the top of the reactor, is freed of condensable gases (methyl iodide, methylacetate, acetic anhydride, acetic acid, and ethylidene diacetate) by cooling (f), and is then recirculated. However, part of the circulating gas is separated from the main stream to avoid buildup of inert gases, which may be present in the carbon monoxide. The side stream is washed (g) with a countercurrent of pure acetic anhydride; in this way, the loss of methyl iodide in the off-gas can be kept below 0.1 % [78]. This acetic anhydride is combined with the top condensate from the reactor (b), supplemented with fresh catalyst as required, and recirculated. The liquid reaction product leaving the reactor is expanded and subjected to flash distillation (c) [78]. To prevent catalyst decomposition, this distillation is performed in a carbon monoxide-hydrogen atmosphere at about 500 kPa [79]. At the bottom of the evaporator (c) a stream of liquid containing the catalyst is separated and recirculated into the carbonylation reactor. The vapor leaving the top of the evaporator is condensed and passed over adsorbers (d) to remove traces of rhodium and iodine compounds.

The raw anhydride is purified by distillation in three consecutive and continuously operating columns (e). In the first column, methyl iodide and methyl acetate are distilled overhead and recirculated to the carbonylation reactor. In the second column, acetic acid is distilled overhead. The bottom product is distilled in a third column to acetic anhydride of 99% purity. The bottom product of the third column contains ethylidene diacetate and unidentified high-boiling components. To further reduce the iodide content of the pure anhydride, a solution of potassium acetate in acetic anhydride is added to the top of the column [80].

Other processes. Other processes and catalysts for the carbonylation of methyl acetate have been patented by various companies: Hoechst [81] (rhodium catalyst), Air Products and Chemicals Co. [82] (rhodium catalyst), Mitsubishi Gas and Chemical Co. [83] (nickel catalyst), and Rhône-Poulenc [84] (nickel catalyst). Several patents also describe the carbonylation of esters of higher alcohols and carboxylic acids. This reaction results in the formation of the corresponding mixed anhydrides [74], [83].

Environmental problems do not arise. The off-gas from the acetic anhydride production contains large quantities of carbon monoxide, some inert gases (nitrogen, rare gases, and carbon dioxide), and traces of hydrogen, methane, methyl iodide, and methyl acetate. It can be burned. There are no problems with waste water.

1.4. Analysis

A very reliable method for determining the content of acetic anhydride consists of mixing stoichiometric quantities of the sample and water, then heating carefully to reflux temperature. After completion of the reaction, two drops of concentrated sulfuric acid are added, and the mixture is boiled for 20 min to insure that the last traces of the anhydride have reacted. The anhydride content is calculated from the unused water, which is determined by the Karl Fischer method.

In the *aniline method* [85], the total acid content is first determined. After addition of aniline to a second sample, the aniline number is established with alkali and the acetic anhydride content determined from the difference.

A rapid determination of the anhydride content can be obtained from the refractive index. If the temperature remains constant to within ±0.05 °C, the precision obtainable is ±0.2%. Gas chromatography also is recommended for purity determination. Analytical methods are discussed in detail in [86].

1.5. Quality Specifications

Acetic anhydride is marketed with more than 95% purity; the normal product is over 98% pure, but it is marketed also as over 99% pure. The color number (APHA) should be below 10 (DIN 53409). The nonvolatile part should not exceed 0.003%. The product also should contain as few substances as possible that reduce permanganate. According to American Chemical Society specifications, for example, a 2 g sample should not decolorize 0.4 ml of a 0.1 N potassium permanganate solution within 5 min. In particular applications, the impurities that can be oxidized by potassium chromate are also of interest. They may not consume more than 200 ppm of oxygen. The contents of phosphate, sulfate, chloride, aluminum, and iron may not exceed 1 ppm each. Heavy metals should be absent.

1.6. Storage and Transportation

For storage and transportation of pure acetic anhydride, tanks made of aluminum, stainless steel (18% Cr, 8% Ni, and 2% Mo), or polyethylene normally are used, although glass or enamel containers also may be employed. Iron is highly resistant to acetic anhydride, provided moisture is excluded. Therefore it is possible to use iron in the production and workup in certain instances, for example, in pumps and tanks.

Because there are no international arrangements for the storage of dangerous goods, the specifications of individual countries must be observed. Also, acetic anhydride (EG-no. 607–008–00–9) is subject to various industrial working regulations, for example, Appendix I, no. 1.1 of the EEC Guidelines.

Transportation. IMDG-Code, class 8, UN-no. 1715; United Kingdom: Blue Book: Corrosive, IMDG-Code E 8018; United States: DOT Safety Act, Title 46 and Title 49; Cor. M; Europe: RID, ADR, and ADNR: class 8, no. 21 e (from 1985: class 8, no. 32 e), RN 801, 2801, and 6801, respectively. International air transportation: IATA-DGR, class 8, UN-no. 1715, RAR art. no. 9, Cor. M.

1.7. Uses

Acetic anhydride is used chiefly as an acetylating and dehydrating agent; it is used on a large scale for the acetylation of cellulose. Other areas of application for acetic anhydride are:

1) The production of poly(methylacrylimide) hard foam, where acetic anhydride is used for binding the ammonia that is liberated on conversion of two amide groups to an imide group.
2) Acetylated plastic auxiliaries, such as glycerol triacetate, acetyl tributyl citrate, and acetyl ricinolate.
3) Explosives, particularly hexogen production (see Section 1.2.2.).
4) The production of certain types of brake fluids.
5) The production of auxiliaries for drilling fluids.
6) The detergent industry, for the production of cold–bleaching activators such as tetraacetylethylenediamine [12].
7) The dyeing industry, where acetic anhydride is used chiefly in mixtures with nitric acid as a nitrating agent. Here, the solvent and dehydrating properties of acetic anhydride are used.
8) In the preparation of organic intermediates, such as chloroacetylchloride, diacetyl peroxide, higher carboxylic anhydrides, acetates, and the boron trifluoride complex.

9) In the production of pharmaceuticals, such as acetylsalicylic acid, *p*-acetylaminophenol, acetanilide, acetophenacetin, theophyllin, acetylcholine chloride, sulfonamides, a number of hormones and vitamins, and the x-ray contrast agent 2,4,6-triiodo-3,5-diacetyl-amidobenzoic acid.
10) In the food industry, mainly in the acetylation of animal and plant fats, in order to obtain the desired solubilities; in the production of acetostearins, the edible packing materials; and to clarify plant oils.
11) Flavors and fragrances (production of esters and cumarin).
12) Herbicides such as metolachlor (Dual) and alachlor (Lasso).

2. Mixed Fatty Acid Anhydrides

2.1. Physical Properties

Important mixed anhydrides, that is, anhydrides with two different fatty acid radicals, are compiled in Table 2. The lower anhydrides can be distilled partly undecomposed at reduced pressure. At high temperature they disproportionate fairly rapidly into the symmetrical anhydrides. Only acetoformic anhydride distills (at 127 – 130 °C) at normal pressure, although with partial decomposition to acetic acid and carbon monoxide. The first four mixed anhydrides have odors very similar to pure acetic anhydride; acetic isovaleric anhydride has a fruitlike odor.

2.2. Chemical Properties

Acetoformic anhydride [2258-42-6] acts as a formylation agent in acylation reactions. In the other mixed acetic – fatty acid anhydrides, the higher acyl group is more reactive; for example, in the reaction with benzene in the presence of aluminum chloride, compound **5** is formed preferably.

$$RCOOCOCH_3 \xrightarrow[(AlCl_3)]{C_6H_6} \underset{5}{C_6H_5COR} + \underset{6}{C_6H_5COCH_3}$$

where R = alkyl.

Branching of the higher acyl group at its α position leads to a decrease in the yield of the acylbenzene **5** in favor of acetophenone (**6**) [87]. From acetic chloroacetic anhydride, mainly ω-chloroacetophenone is formed [87]. Changes in reaction conditions have little influence on the ratio of reaction products.

Table 2. Mixed fatty acid anhydrides

Anhydride	Formula	bp or mp,°C	Preferred production method (Section 2.3)
Acetoformic anhydride	HCOOCOCH$_3$	bp 32 (at 2.7 kPa)	3
Acetic propionic anhydride	CH$_3$COOCOC$_2$H$_5$	bp 25–27 (at 0.15 kPa) (decomposes at 30 °C)	3
Acetic butyric anhydride	CH$_3$COOCO(CH$_2$)$_2$CH$_3$	–	3
Acetic isobutyric anhydride	CH$_3$COOCOCH(CH$_3$)$_2$	–	3
Acetic valeric anhydride	CH$_3$COOCO(CH$_2$)$_3$CH$_3$	(decomposes above 130 °C)	3
Acetic isovaleric anhydride	CH$_3$COOCOCH$_2$CH(CH$_3$)$_2$	–	3
Acetic 2,2-dimethylpropanoic anhydride	CH$_3$COOCOC(CH$_3$)$_3$	–	3
Acetic hexanoic anhydride	CH$_3$COOCO(CH$_2$)$_4$CH$_3$	–	3
Acetic octanoic anhydride	CH$_3$COOCO(CH$_2$)$_6$CH$_3$	–	3
Acetic hexadecanoic anhydride	CH$_3$COOCO(CH$_2$)$_{14}$CH$_3$	mp 62.5 (decomposes on crystallization from polar solvents)	2,3
Acetic octadecanoic anhydride	CH$_3$COOCO(CH$_2$)$_{16}$CH$_3$	–	3
Butyric tetradecanoic anhydride	CH$_3$(CH$_2$)$_2$COOCO(CH$_2$)$_{12}$CH$_3$	mp 52.7	2
Hexanoic dodecanoic anhydride	CH$_3$(CH$_2$)$_4$COOCO(CH$_2$)$_{10}$CH$_3$	mp 42.4	2
Octanoic decanoic anhydride	CH$_3$(CH$_2$)$_6$COOCO(CH$_2$)$_8$CH$_3$	mp 16	2

2.3. Production

Three processes are generally employed for producing mixed anhydrides:

1) The reaction of acetic anhydride with a higher organic acid [88].
2) Heating an acyl chloride with a fatty acid salt, usually a sodium or potassium salt [89]–[91].
3) Reacting ketene with a carboxylic acid [92]–[94].

The latter two processes are more useful than the first. According to the patent literature, mixed anhydrides also can be produced by carbonylating the corresponding esters (Section 1.3.3).

2.4. Uses

Acetoformic anhydride often is used as a formylating agent. Acetic propionic [13080-96-1] and acetic butyric [7165-13-1] anhydrides are used in the production of acylated cellulose.

3. Economic Aspects

The production data of recent years, as far as they are available, are given in Table 3. Recent annual capacity data for acetic anhydride are as follows: Western Europe (1978) 357000 t, (1980) 362000 t; United States (1980) 905000 t, (1982) 929000 t. The start of operation at the new Tennessee-Eastman plant (Halcon Process) with a capacity of 227000 t should not change the total capacity in the United States significantly, because it is expected that a corresponding capacity from the ketene process will be shut down in the next few years. No data for the mixed anhydrides are available.

4. Toxicology and Occupational Health

In acute oral studies in rats, an LD_{50} value of 1780 mg/kg was determined for acetic anhydride. The dermal LD_{50} in rabbits is 4000 mg/kg [97]. Inhalation of 2000 ppm for 4 h causes death in rats [98]. The lowest published lethal concentration (LCLo) for a 4-h exposure of rats is 1000 ppm [99].

In rabbits, skin contact for 24 h causes only mild irritation [97]. These findings are in accordance with observations in humans. Tingling sensations followed by slight erythema 30 min after skin contact have been reported. Pain or vesication has not been observed [100]. In contrast to these observations, acetic anhydride is reported to cause wrinkling, whitening, and peeling, if it is not removed from the skin at once [101]. Acetic anhydride is a severe eye irritant [102]. In rabbits, 250 µg causes severe irritation. Human occupational exposure can cause lacrimation, conjunctivitis, photophobia [103], corneal burns with loss of vision [104], and iritis [105]. It is assumed that acetic anhydride penetrates the corneal epithelium rapidly without hydrolyzing and reaches the iris in concentrations high enough to cause iritis.

Inhalation can lead to asthmoid bronchitis with lung edema [106]. Human exposure to concentrations of 800 ppm cannot be tolerated for longer than 3 min because of the burning sensation in the nose and throat. Humans are believed to be generally more sensitive to the irritant effects of inhalation of vapors than animals [107]. *Ingestion* is followed by a burning pain in the stomach, nausea, and vomiting [108]. The acute

Table 3. Production of acetic anhydride (in kt)

	1961	1971	1974	1979	1980	1981	1982
United States	571	686	741			567	481
Federal Republic of Germany	32	47	74	91	85	77	76
Japan	33	96	115	114	150		

irritant effects are caused partly by the hydrolysis of acetic anhydride to acetic acid, which itself is an irritant. The rate of hydrolysis depends on the water content of the tissues.

Systemic effects after repeated exposure are unlikely to occur because of the potent warning properties of acetic anhydride. The MAK and TLV are both 5 ppm (20 mg/m^3) [109], [110].

5. References

[1] D. T. Lewis, *J. Chem. Soc.* 1940, 33.
[2] W. J. Lewis, E. J. Evans, *Phil. Mag.* **13** (1932) 268.
[3] L. H. Meyer, A. Saika, H. S. Gutowsky, *J. Am. Chem. Soc.* **75** (1953) 4569; L. H. Allred, E.G. Rochow, *J. Am. Chem. Soc.* **79** (1957) 5361.
[4] L. P. Filippov, *Vestnik Moskov. Univ.* **9** (1954) 45; *Chem. Abstr.* **15430** (1955).
[5] G. Jander, H. Surawski, *Z. Elektrochem.* **65** (1961) 469.
[6] T. V. Malkova, *Zh. Obshch. Khim.* **24** (1954) 1157; Engl. ed. 24 (1954) 1151; K. N. Kovalenko et al., *Zh. Obshch. Khim.* 26 (1956) 403; Engl. ed. 26 (1959) 427.
[7] G. Jaudes, E. Rüsberg, H. Schmidt, *Z. Anorg. Chem.* **255** (1948) 238.
[8] *Beilstein*, **4**, 2nd suppl., 387.
[9] D. H. Kim, *J. Heterocycl. Chem.* **13** (1976) 179.
[10] K. Murihara, *Bull. Chem. Soc. Japan* **37** (1964) 1787.
[11] S. S. Parmar, R. C. Arora, *J. Med. Chem.* **13** (1970) 135.
[12] BASF DE-OS 2 118 281, 1971; NL 7 204 899, 1972 (C. Palm, G. Matthias).
[13] Henkel u. Cie., DE-OS 2 112 557, 1971.
[14] J. Hori, H. Midorikawa, *Sci. Pap. Inst. Phys. Chem. Res. (Tokyo)* **56** (1962) 216.
[15] A. N. Nesmejanov, K. N. Anisimov, G. K. Magornedov, *Izv. Akad. Nauk SSSR, Ser. Khim.* **4** (1970) 959.
[16] *Ullmann*, 3rd ed., **6**, 804.
[17] F. Merenyi, N. Nilsson, *Acta Chem. Scand.* **18** (1964) 1368.
[18] Houben-Weyl: *Methoden der organischen Chemie*, 4th ed., vol. **VII/2a**, Thieme Verlag, Stuttgart 1983, p. 457.
[19] Techn.-Chem., DE-OS 1 940 809, 1969 (D. Sheehan, W. P. Hegarty, A. F. Vellturo, W. A. Gay, D. D.Threlkeld).
[20] A. Cassadevall, A. Commeyras, *Bull. Soc. Chim. Fr.* **5** (1970) 1850.
[21] A. Cassadevall, A. Commeyras, *Bull. Soc. Chim. Fr.* **5** (1970) 1856.
[22] B. Blaser, K.-H. Worms, H. G. Germscheid, K. Wollmann, *Z. Anorg. Allg. Chem.* **381** (1971) 247.
[23] G. Gattow, H. Schwank, *Z. Anorg. Allg. Chem.* **382** (1971) 49.

[24] J. G. Dawber, *Chem. Ind.* **23** (1964) 973.
[25] J. Leleu, *Cah. Notes Doc.* **83** (1976) 281.
[26] R. Finding, U. Schmidt, *Angew. Chem.* **82** (1970) 482; *Angew. Chem. Int. Ed. Engl.* **9** (1970) 456.
[27] J. D'Ans, J. Mattner, *Chem. Ztg.* **74** (1950) 435.
[28] M. S. Newman, A. S. Smith, *J. Am. Chem. Soc.* **67** (1945) 154; *J. Org. Chem.* **13** (1949) 592.
[29] H. Kelling, *J. Chem. Soc.* **8** (1968) 391.
[30] R. C. Mehrotra, *Pure Appl. Chem.* **13** (1966) 111.
[31] J. L. Speier, C. A. Roth, J. W. Ryan, *J. Org. Chem.* **36** (1971) 3120.
[32] V. F. Shvets, I. Al-Vakhib, *Kinet. Katal.* **13** (1972) 98.
[33] V. F. Shvets, I. Al-Vakhib, *Kinet. Katal.* **16** (1975) 785.
[34] J. Tomiska, *Collect. Czech. Chem. Commun.* **28** (1963) 1612.
[35] Union Oil, US 3 393 225, 1965 (D. M. Fenton).
[36] Toyo Soda Mfg Co., JA 7 319 293, 1969 (T. Ono, T. Yanagihara, H. Okada, T. Koga).
[37] J. A. Ol'Dekop, A. N. Sevcenko, J. P. Zjat'Kov, A. P. El'Nickij, *Zh. Obshch. Khim.* **33** (1963) 2771.
[38] C. Rüchardt, S. Eichler, O. Krätz, *Tetrahedron Lett.* **4** (1965) 233.
[39] H. Iwamura, M. Iwamura, T. Nishida, S. Sato, *J. Am. Chem. Soc.* **92** (1970) 7474.
[40] M. Ferles, M. Jankovsky, *Collect. Czech. Chem. Commun.* **36** (1971) 4103.
[41] C. R. Johnson, J. C. Sharp, W. G. Phillips, *Tetrahedron Lett.* **52** (1967) 5299.
[42] J. D. Albright, L. Goldman, *J. Am. Chem. Soc.* **87** (1965) 4214.
[43] R. P. Arganbright, R. J. Evans, *Hydrocarbon Process.* **43** (1964) 159.
[44] American Home Products Corp., US 3 663 605, 1970 (R. J. McCaully, G. L. Conklin).
[45] Y. Masada, *Kogyo Kagaku Zasshi* **74** (1971) 1149.
[46] A. N. Mirskova, E. F. Zorina, A. S. Atavin, *Izv. Sib. Otd. Akad. Nauk SSSR, Ser. Khim. Nauk* **6** (1971) 72.
[47] H. Musso, K. Figge, *Justus Liebigs Ann. Chem.* **668** (1963) 1.
[48] ICI, GB 1 107 907, 1965 (A. H. Dinwoodie, G. Fort).
[49] J. S. Buck, W. S. Ide in: *Organic Synthesis,* collective vol. **II,** Wiley-Interscience, New York 1943, p. 622.
[50] M. Tsuda, H. Tanaka, K. Ikeda, *J. Chem. Soc. Japan, In. Chem. Sect.* **73** (1970) 1888.
[51] P. L. Southwick, *Synthesis* **12** (1970) 628.
[52] G. T. Nikisin, J. N. Ogibin, J. A. Palanuer, *Izv. Akad. Nauk SSSR* **11** (1967) 2478.
[53] B. F. Goodrich Co., DE-AS 1 086 683, 1953 (L. F. Arnold).
[54] W. R. Edwards, K. P. Kamman, *J. Org. Chem.* **29** (1964) 913.
[55] H. Eck, H. Spes in: J. Falbe (ed.), *Methodicum Chimicum,* vol. **5,** p. 498.
[56] E. T. Blues, D. Bryce-Smith, H. Hirsch, M. J.Simons, *Chem. Commun.* **11** (1970) 699.
[57] Consortium für Elektrochemische Industrie, DE 408 715, 1922; DE 417 731, 1924; DE 475 885, 1926; DE 488 573, 1926 (R. Meingast, M. Mugdan); DE 634 438, 1933; DE 687 065, 1933; DE 734 349, 1934; US 2 249 543, 1937 (M. Mugdan, J. Sixt).
[58] Eastman Kodak, US 3 403 181, 1966 (E. S. Painter, R. C. Petrey, J. H. Jensen).
[59] *Ullmann,* 4th ed., **11,** 80.
[60] Wacker-Chemie, DE 1 076 090, 1959 (Th. Altenschöpfer, H. Spes, L. Vornehm).
[61] BASF, DE-OS 2 005 970, 1970 (G. Matthias, G. Schulz, C. Palm).
[62] Consortium für Elektrochemische Industrie, DE 403 863, 1922 (R. Meingast, M. Mugdan).
[63] Wacker-Chemie, DE 2 505 471, 1975 (H. Eck, H. Schwarzbauer, E. Bethe, K. Kaiser, H. Spes).
[64] Wacker-Chemie, DE 867 689, 1940 (A. Krug, J. Sixt).
[65] Carbide & Carbon Chem. Co., US 2 225 486, 1940 (H. L. Reichart).
[66] A. G. f. Stickstoffdünger, DE 699 709, 1934 (J. Lösch, F. Walter, H. Behringer, O. Schlöttig).

[67] A. G. f. Stickstoffdünger, DE 708 822, 1934 (J. Lösch, F. Walter, H. Behringer, O. Schlöttig).
[68] H. Dreyfus, GB 510 959, 1938.
[69] Distillers Co., US 2 514 041, 1946 (A. Elce, H. M.Stanley, K. H. W. Tuerck).
[70] D. V. Musenko, G. N. Gvozdovskij, *Zh. Vses. Khim. Ova.* **14** (1969) 263.
[71] Hoechst, DE-OS 2 757 222, 1977; EP 0 002 696, 1978; US 4 252 983, 1981 (H. Erpenbach, K. Gehrmann, A. Hauser, K. Karrenbauer, W. Lork).
[72] Les Usines de Melle, US 2 658 914, 1950 (L. Rigon). Hoechst, DE 2 757 173, 1977 (H. Erpenbach, K. Gehrmann, A. Hauser, K. Karrenbauer, W. Lork).
[73] Les Usines de Melle, DE-AS 1 142 857, 1963 (L. Alheritierc).
[74] Halcon International, DE-OS 2 441 502, 1974 (C. Hewlett).
[75] Halcon International, US 4 002 678, 1975 (A. N.Naglieri, N. Rizkalla).
[76] M. Schrod, G. Luft, *Ind. Eng. Prod. Res. Dev.* **20** (1981) 649.
[77] Halcon Research and Development Corp., GB 2 033 385 A, 1980 (C. G. Wan).
[78] Halcon Research and Development Corp., US 4 241 219, 1979 (C. G. Wan).
[79] Halcon Research and Development Corp., DE-OS 2 940 752, 1979 (R. V. Procelli, V. S. Bhise, A. J. Shapiro).
[80] Halcon Research and Development Corp., GB 2 033 901 A, 1980 (P. L. Szecsi).
[81] Hoechst, DE-OS 2 836 084, 1978 (H. Erpenbach, K. Gehrmann, H. K. Kübbeler).
[82] Air Products and Chemicals, US 4 333 885, 1981 (D. Feitler).
[83] Mitsubishi Gas Chemical Co., DE-OS 2 844 371, 1978 (T. Isshiki, Y. Kijima, Y. Miyauchi).
[84] Rhône-Poulenc Industries, EP 0 050 084, 1980 (J. Gauthier-Lafaye, R. Perron).
[85] A. Menschutkin, B. Wasiljeff, *Z. Anal. Chem.* **60** (1921) 425; T. Ellerington, J. J. Nichols, *Analyst London* **82** (1957) 233.
[86] E. F. Joy, A. J. Barnard in: F. D. Snell, C. L. Hilton (eds), *Encyclopedia of Industrial Chemical Analysis,* vol. **4,** J. Wiley & Sons, New York 1967, p. 102.
[87] W. R. Edwards, E. C. Sibille, *J. Org. Chem.* **28** (1963) 674.
[88] W. Autenried, G. Thomae, *Ber. Dtsch. Chem. Ges.* **34** (1901) 168; **57** (1924) 423.
[89] A. W. Ralston, R. A. Reck, *J. Org. Chem.* **11** (1946) 625.
[90] J. B. Polya, T. M. Spotswood, *J. Amer. Chem. Soc.* **71** (1949) 2938.
[91] BASF, EP 0 029 176, 1980 (K. Blatt, H. Naarmann).
[92] De Bataafsche, GB 389 049, 1932.
[93] W. Stevens, A. van Es, *C. R. Acad. Sci. Ser. B* **83** (1964) 863.
[94] J. Dickert, A. Krynitsky, *J. Am. Chem. Soc.* **63** (1941) 2511.
[95] *Chem. Eng. News* **61** (1983) no. 24, 29.
[96] *Chemfacts Japan,* Chemical Data Services, IPC Industrial Press, Sutton, UK 1981.
[97] *Registry of Toxic Effects of Chemical Substances,* U.S. Department of Health, Education, and Welfare, Public Health Service, Center for Disease Control, NIOSH, Cincinnati, Ohio 1980.
[98] H. F. Smyth Jr., C. P. Carpenter, U. C. Pozzani, *A. M. A. Arch. Ind. Hyg. Occup. Med.* **10** (1954) 61.
[99] *Toxic and Hazardous Industrial Chemicals Safety Manual,* The International Technical Information Institute, Tokyo 1975, p. 3.
[100] H.-J. Oettel, *Naunyn-Schmiedebergs Arch. Exp. Pathol. Pharmakol.* **183** (1936) 641.
[101] E. R. Plunkett, *Handbook of Industrial Toxicology,* Chemical Publ. Co., New York 1968, p. 4.
[102] F. A. Patty, *Industrial Hygiene and Toxicology,* 2nd ed., vol. **II,** Interscience, New York 1962, p. 1817.
[103] A. Hamilton, H. L. Hardy, *Industrial Toxicology,* 2nd ed., P. B. Hoeber, New York 1949, p. 338.
[104] R. S. McLaughlin, *Am. J. Ophthalmol.* **29** (1946) 1355.

[105]　J. Doull, C. D. Klaassen, M. O. Amdur, *Toxicology. The Basic Science of Poisons,* 2nd ed., Macmillan Publ. Co., New York 1980, p. 284.
[106]　S. Moeschlin, *Klinik und Therapie der Vergiftungen,* Thieme Verlag, Stuttgart 1980.
[107]　K. B. Lehmann, J. Wilke, J. Yamada, J. Wiener, *Arch. Hyg.* **67** (1908) 57.
[108]　N. I. Sax, *Dangerous Properties of Industrial Materials,* 3rd ed., Reinhold Publ. Co., New York 1968, p. 367.
[109]　Deutsche Forschungsgemeinschaft (ed.), *Maximale Arbeitsplatzkonzentrationen (MAK) 1982,* Verlag Chemie, Weinheim 1982.
[110]　American Conference of Governmental Industrial Hygienists (ed.), *Threshold Limit Values (TLV) 1982,* ACGIH, Cincinnati, Ohio 1982.

Acetone

STYLIANOS SIFNIADES, Allied Signal Inc., Morristown, New Jersey 07962, United States
ALAN B. LEVY, Allied Signal Inc., Morristown, New Jersey 07962, United States

1.	Introduction	89
2.	Physical Properties	90
3.	Chemical Properties	91
4.	Production	92
4.1.	Cumene Oxidation (Hock Process)	93
4.2.	Dehydrogenation of 2-Propanol	97
4.3.	Propene Oxidation	98
4.4.	Oxidation of 2-Propanol	99
4.5.	Oxidation of p-Diisopropyl Benzene	100
4.6.	Fermentation of Biomass	100
5.	Environmental Protection	101
6.	Quality Specifications and Analysis	102
7.	Storage and Transportation	102
8.	Uses	103
8.1.	Methyl Methacrylate	103
8.2.	Bisphenol A	105
8.3.	Aldol Chemicals	105
8.4.	Solvent Uses	106
9.	Economic Aspects	106
10.	Toxicology and Occupational Health	108
11.	Derivatives	109
11.1.	Acetone Cyanohydrin	109
11.2.	Diacetone Alcohol	111
11.3.	Miscellaneous Derivatives	113
12.	References	114

1. Introduction

Acetone, 2-propanone, dimethyl ketone, CH_3COCH_3, [67-64-1], is the first and most important member of the homologous series of aliphatic ketones. It is a colorless, mobile liquid widely used as a solvent for various polymers. Its largest application, however, is as an intermediate in the synthesis of methyl methacrylate, bisphenol A, diacetone alcohol, and other products.

Acetone was first manufactured by the dry distillation of calcium acetate, [62-54-4]. Calcium acetate was originally a product of wood distillation, and later was obtained by fermentation of ethanol. Carbohydrate fermentation directly to acetone and butyl and ethyl alcohols displaced these processes in the 1920s. The carbohydrate route, in turn, was replaced in the 1950s and 1960s by the 2-propanol dehydrogenation process and by the oxidation of cumene to phenol [108-95-2] plus acetone. Together with direct propene oxidation, these methods account for over 95% of the acetone produced worldwide.

2. Physical Properties

Acetone has the following physical properties: M_r 58.081; bp at 101.3 kPa, 56.2 °C; mp − 94.7 °C; relative density, d_4^0 0.81378, d_4^{15} 0.79705, d_4^{20} 0.7908; relative vapor density (air = 1) 2.0025; refractive index n_D^{20} 1.35868; critical temperature 235.0 °C, critical pressure 4.6 MPa (46 bar), critical density 0.278 g/cm^3; cubic expansion coefficient (18 °C) 1.43×10^{-3} K^{-1}; compressibility coefficient (18 °C) 1.286×10^{-6} kPa^{-1} (1.286×10^{-4} bar^{-1}).

Viscosity in mPa·s: 1.53 (− 80 °C), 0.71 (− 40 °C), 0.40 (0 °C), 0.32 (20 °C), 0.27 (40 °C). Surface tension in mN·m^{-1}: 38.1 (− 91.09 °C), 23.9 (15 °C), 23.3 (20 °C), 23.0 (24.8 °C), 22.0 (30 °C), 21.6 (42 °C).

Thermal properties: Specific heat capacity, c_p (20 °C) 2.135 kJ kg^{-1} K^{-1}; heat of fusion (− 95 °C) 98.47 kJ kg^{-1}; heat of vaporization (30 °C) 545.2 kJ kg^{-1} , (0 °C) 588.2 kJ kg^{-1}; molar entropy 0.2001 kJ mol^{-1} K^{-1}; heat of combustion 1804 kJ mol^{-1}; heat of formation (20 °C) 235.3 kJ/mol; thermal conductivity of the liquid 1.976 W m^{-1} K^{-1}.

Vapor Pressure in kPa: 24 (20 °C), 37.3 (30 °C), 56.0 (40 °C), 82.8 (50 °C), 114.8 (60 °C), 214.8 (80 °C), 372.8 (100 °C), 929.6 (140 °C).

Electrical properties: Electric conductivity (20 °C) 5.5×10^{-8} Ω^{-1} cm^{-1}; dipole moment (20 °C) 2.69 Debye; dielectric constant of the liquid 21.58 (0 °C), 22.64 (10 °C), 20.70 (25 °C), 19.38 (40 °C); dielectric constant of the vapor 1.0235 (24.8 °C), 1.0277 (29.8 °C).

At ambient temperature acetone is a clear, colorless liquid with a characteristic odor. It is miscible in all proportions with water and polar organic solvents, such as the lower molecular mass alcohols, carboxylic acids, and ethers. It is miscible in limited proportions with nonpolar solvents, such as hydrocarbons. Some azeotropic mixtures are shown in Tables 1 and 2 [3], [4].

Acetone dissolves many synthetic resins, e.g., nitrocellulose, acetylcellulose, poly(acrylate esters), and alkyd resins. It also dissolves most natural resins, fats, and oils.

Table 1. Acetone binary azeotropes *Azeotropes, with acetone

Second component	Acetone, wt%	bp (101.3 kPa), °C
Carbon tetrachloride	88.5	56.08
2-Butylchloride	80	55.75
Hexane	53.5	49.7
Methyl acetate	49	55.65
Diethylamine	38	51.55
Carbon disulfide	33	39.25
tert-Butylchloride	25	49.2
Isoprene	20	30.5
n-Propylchloride	15	45.8
Methanol	14	55.59

* Source and further examples [3].

Table 2. Acetone ternary azeotropes *

Components (A is acetone)	Composition, wt%	bp (101.3 kPa), °C
B water	0.81	38.04
C carbon disulfide	75.21	
B water	0.4	32.5
C isoprene	92.0	
B chloroform	46.7	57.5
C methanol	23.4	
B chloroform	70.2	55.0
C ethanol	6.8	
B methanol	16	51.1
C cyclohexane	40.5	
B methyl acetate	5.6	49.7
C hexane	43.3	

* Source and further examples [4].

3. Chemical Properties

Pure acetone is essentially inert to air oxidation and to diffuse sunlight under ambient conditions. Its chemical stability diminishes significantly in the presence of water. Acetone may react violently and sometimes explosively, especially in a confined vessel [5]. For example it is particularly sensitive to oxidizing agents, such as nitrosyl chloride [6] [2696-92-6], chromium trioxide [7] [1333-82-0], and hydrogen peroxide [8] [7722-84-1], or organic peroxides [9]. Mixtures of acetone with chloroform [67-66-3] may react violently in the presence of alkali [10]. Reaction even may be initiated by surface alkali on new glassware [11]. Acetone has a flash point of –17 °C (closed cup). Flammability limits in air are: lower 2.13 vol%, upper 13 vol%; autoignition temperature 465 °C. The flammability of acetone can be reduced by mixing it with less flammable and/or less volatile solvents [12]. Fires have been started during recovery of acetone from air by adsorption on activated carbon when air flow was too low to effectively remove the heat generated by surface oxidation [13].

Acetone undergoes typical carbonyl reactions with particular ease. Acid- or base-catalyzed self-condensation produces the dimers diacetone alcohol and mesityl oxide and the cyclic trimer isophorone.

Under strongly basic conditions hydrogen cyanide adds to acetone to form 2-cyano-2-propanol (acetone cyanohydrin), an important intermediate in the manufacture of methyl methacrylate and other methacrylate esters (Section 8.1).

In liquid ammonia solution acetone condenses with acetylene [74-86-2] in the presence of catalytic amounts of alkali metals to form 2-methyl-3-butyn-2-ol [115-19-5], an intermediate in the synthesis of isoprene [14] [563-46-2]. Catalytic hydrogenation of acetone yields 2-propanol [67-63-0]. Pyrolysis produces methane [74-82-8] and ketene [463-51-4], a powerful acetylating agent. A more economical source of ketene, however, is the pyrolysis of acetic acid, which produces ketene and water.

Reductive ammonolysis of acetone yields isopropylamine [75-31-0]. Condensation with 2 mol phenol in the presence of an acidic catalyst yields bisphenol A (Section 8.2), an important monomer used in the manufacture of polycarbonate resins.

Perchlorination yields hexachloroacetone [116-16-5], which is cleaved into chloroform [67-66-3] and sodium trichloroacetate [650-51-1] upon treatment with sodium hydroxide.

4. Production

Approximately 83% of the acetone produced worldwide today is manufactured from cumene as a coproduct with phenol. In the United States and Western Europe dehydrogenation of 2-propanol is also important, whereas in Japan catalytic oxidation of propene is used as a second process. Cumene, 2-propanol, and propene together as starting materials account for over 95% of the acetone produced worldwide. Because propene is used in the manufacture of both cumene and 2-propanol, propene is the ultimate raw material for the production of acetone.

Small amounts of acetone are made by oxidation of p-diisopropyl benzene and of p-cymene. Coproducts from these reactions are hydroquinone and p-cresol, respectively. Acetone is also produced by propene oxidation and as a byproduct of acetic acid manufacture.

Fermentation of cornstarch and molasses to acetone and 1-butanol was important in the past. It is believed to be practiced today to a limited extent in several countries.

4.1. Cumene Oxidation (Hock Process)
(Fig. 1)

Propene [*115-07-1*] is added to benzene [*71-43-2*] to form cumene [*98-82-8*], which is then oxidized by air to cumene hydroperoxide (**1**), and cleaved in the presence of an acid catalyst. Phenol [*108-95-2*] and acetone produced in the process are recovered by distillation.

The alkylation of benzene by propene proceeds under typical Friedel–Crafts conditions. In 1996, a number of processes using zeolite catalysts came on-stream. The cumene produced is purified by chemical means and refined by distillation to 99.9 % minimum purity. Oxidation-grade cumene must meet strict quality standards. The newer zeolite-based processes have led to slightly tighter specifications (Table 3) [15].

Cumene Oxidation. The oxidation of cumene is a free-radical chain reaction [16]. The chain initiator is cumene hydroperoxide, the main product of the reaction. The rate of oxygen consumption can be approximated by the following expression:

$$\frac{-dc_{O_2}}{dt} = k_p \cdot c_{RH} \sqrt{\frac{2k_i c_{ROOH}}{k_t}}$$

c_{RH} and c_{ROOH} are the concentrations of cumene and cumene hydroperoxide, respectively; k_i, k_p, and k_t are the rate constants for chain initiation, propagation, and termination.

The expression shows that the rate of oxidation is zero in the absence of cumene hydroperoxide. This is not exactly true, because the expression is only an approximation; but the oxidation of cumene does require long induction periods when starting with pure cumene. Consequently, the industrial oxidation always is carried out in a series of continuous reactors; the concentration of cumene hydroperoxide is at least 8 wt % in the first reactor. Because the sum of c_{RH} and c_{ROOH} remains roughly constant during the reaction, the rate of reaction cannot increase indefinitely as c_{ROOH} increases. The maximum rate is achieved at approximately 35 wt % cumene hydroperoxide.

Besides cumene hydroperoxide, both dimethylphenylmethanol and acetophenone are also formed as byproducts during this oxidation. These arise from a secondary chain reaction that proceeds in parallel with the main chain. Byproduct formation is accelerated as the concentration of cumene hydroperoxide increases. For these reasons, most plants operate between 25 and 40 wt % in the last oxidation reactor.

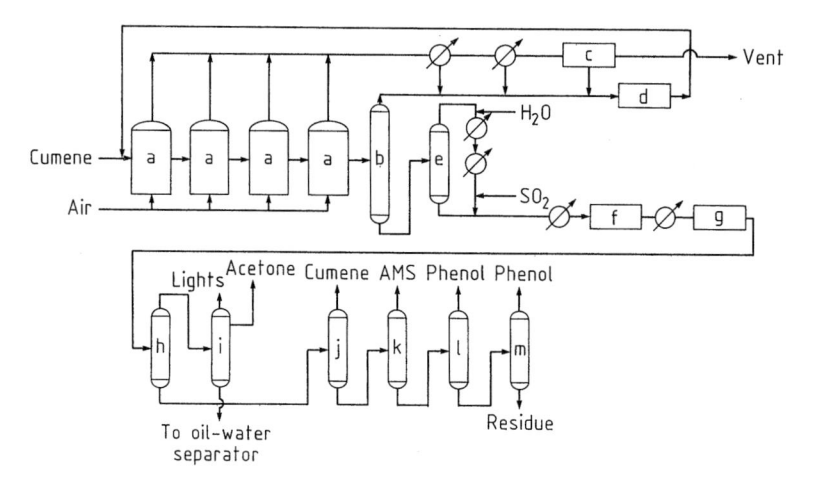

Figure 1. Cumene phenol–acetone process (Allied)
a) Oxidizers; b) Flash column; c) Carbon adsorber; d) Alkaline extraction and wash; e) Cumene hydroperoxide decomposer; f) Dicumyl peroxide decomposer; g) Ion exchange; h) Crude acetone column; i) Acetone-refining column; j) Cumene column; k) α-Methylstyrene column; l) Phenol column; m) Phenol residue topping column
AMS = α-methylstyrene

Table 3. Specifications for oxidation-grade cumene (zeolite process) [15]

Property	ASTM test	Specification
Appearance		Clear, colorless liquid
Color, Pt–Co scale	D1209-79	15 max.
$d_{15.5}^{15.5}$	D891-59	0.864–0.867
Acid wash color, W scale	D848-62	2 max.
Sulfur compounds	D853-47	Free from H_2S and SO_2
Copper corrosion	D849-47	No iridescence, gray or black
Distillation range	D950-56	1.0 °C max.
Cumene content		99.93 % min.
Phenolics content		5 ppm max.
Cumene hydroperoxide content		200 ppm max.
Sulfur content		0.1 ppm max.

A minor but significant byproduct of the oxidation is dicumyl peroxide [80-43-3]. This arises during the termination of the chain reaction. Dicumyl peroxide also contributes to chain initiation [17], but to a much lesser degree than cumene hydroperoxide. Other minor byproducts are formaldehyde and formic acid, which are produced along with acetophenone by methyl group degradation.

The oxygen needed for cumene oxidation is supplied by air. Use of pure oxygen has been suggested [18] but is disfavored by both economic and safety considerations [19]. At low initiation rates, the rate of the reaction is essentially independent of the oxygen concentration at a partial pressure of oxygen over 33 kPa (0.33 bar) [20]. A detailed study of the rate of oxygen uptake in a bubble column as a function of temperature and partial oxygen pressure has been made [21]. The study served as a basis for a mathematical model of the oxidation [22], [23].

There are currently two cumene oxidation processes in use in the United States, which with minor variations are practiced also in the rest of the world [19], [22]. One process was developed by Hercules and is currently licensed by Kellogg (previously BP/Hercules) and GE/Lummus [24]. The other process was developed by Allied and is currently licensed by Allied/UOP [25]–[28].

In both processes several reactors are employed in series. Fresh and recycled cumene are fed to the first reactor, which may operate at 8–12 wt% cumene hydroperoxide. The concentration increases by 4–8 wt% in each successive reactor; the last reactor may operate at 25–40 wt% cumene hydroperoxide. Fresh air is pumped in parallel to each reactor and vented at the top after removal of organic vapors.

In the *Hercules process*, the oxidation of cumene is carried out at approximately 620 kPa (6 bar)/90–120 °C, in the presence of a sodium carbonate buffer [29]. Under these conditions the residence time in the oxidizer train is 4–8 h and the hydroperoxide molar selectivity 90–94%. The spent air is first passed through water cooled and refrigerated condensers in series to remove organic vapors, and is finally vented. The condensate is returned to the oxidizers after treatment [29].

In the *Allied/UOP process* (Fig. 1) the oxidation is carried out at atmospheric pressure. No buffer or promoter is added, but great care is taken to wash all streams recycled to the oxidizer with alkali and water [27]. Temperature is maintained at 80–100 °C. Residence time in the oxidizer train is 10–20 h and hydroperoxide molar selectivity is 92–96%. Spent air is vented after organic vapors are removed by condensation followed by activated carbon adsorption. The recovered materials are washed with aqueous sodium hydroxide and water, then returned to the oxidizers.

The oxidation of cumene generates approximately 116 kJ of heat per mole of cumene oxidized [30]. Part of this heat is carried to the condensers by organic vapors (this part is larger in the Allied/UOP process because of the lower operating pressure). The rest is removed by heat exchangers.

In both processes cumene hydroperoxide is concentrated to over 80 wt% by evaporation of excess cumene. In the Hercules process the oxidate is washed with water prior to distillation in order to remove the buffer added during oxidation.

Cumene Hydroperoxide Cleavage. Cumene hydroperoxide [*80-15-9*] is cleaved to phenol and acetone in the presence of catalytic amounts of a strong acid. The acid most commonly used is sulfuric acid. Sulfur dioxide is used as catalyst in the Allied/UOP process. Several patents claim the use of solid acids as catalysts for the decomposition [31]. Strongly acidic resins have been used to that effect in the Soviet Union [32]. However, today all commercial units use strong mineral acids or SO_2, which generates sulfuric acid in situ, as catalysts.

The cleavage proceeds through an ionic mechanism and releases approximately 252 kJ/ mol of cumene hydroperoxide decomposed [30]. The reaction rate accelerates rapidly with increasing temperature. Consequently, decomposition of cumene hydroperoxide commonly is carried out in a continuously stirred reactor in which the steady-state concentration of cumene hydroperoxide is maintained at a low level. The heat released by the reaction can be used to estimate the concentration of hydroperoxide present in the reactor at any time [33].

The molar selectivity of the cleavage to phenol and acetone is higher than 99.5% at temperatures below 70 °C, but it decreases at higher temperatures as increasing amounts of dimethylphenylmethanol and acetophenone (in addition to those present in the cumene oxidate) are formed (Table 4) [28].

Acetone produced during the cleavage of cumene hydroperoxide can react further. Oxidation by cumene hydroperoxide forms hydroxyacetone [34] to the extent of 0.2–0.5% of acetone present. Self-condensation catalyzed by acid results in diacetone alcohol and mesityl oxide. Conversion of acetone to these condensates is normally below 0.1% but may increase upon protracted exposure to strong acid. For example, when the cumene hydroperoxide cleavage was carried out with refluxing acetone using a sulfonic acid resin as catalyst, approximately 1.7% of acetone was transformed to diacetone alcohol and mesityl oxide [32].

Under the conditions of the cumene hydroperoxide cleavage, dimethylphenylmethanol is dehydrated to α-methylstyrene (**2**) and also forms undesirable condensates.

$$C_6H_5\underset{CH_3}{\overset{CH_3}{C}}-OH \xrightarrow{H^+} C_6H_5\underset{CH_3}{\overset{CH_2}{C}} + H_2O$$

2

Compound **2** may be either hydrogenated [35] to cumene and recycled, or recovered and sold.

In the *Hercules process*, the cumene hydroperoxide decomposition is carried out in a constant-flow, stirred tank reactor in the presence of sulfuric acid or another strong mineral acid [19], [28]. The acid is added to the reactor as an acetone solution. The reactor temperature is maintained below 95 °C by refluxing approximately 2.8 kg acetone per kilogram cumene hydroperoxide. The ratio of the quantity of reflux to the quantitiy of hydroperoxide fed is used as a monitor of the cleavage reaction [19].

In the *Allied/UOP process* the cumene hydroperoxide cleavage is carried out at 60–80°C in a pressurized, constant-flow, back-mixed reactor. Temperature is controlled by means of heat exchangers in the loop. The catalyst is either sulfuric acid or sulfur dioxide. Up to 5 wt% cumene hydroperoxide remains unreacted. Under these conditions, dimethylphenylmethanol combines with cumene hydroperoxide to form dicumyl peroxide (**3**) [*80-43-3*], which upon subsequent heating to 110–140 °C in a short-residence-time plug-flow reactor is cleaved into phenol, acetone, and α-methylstyrene.

Table 4. Formation of byproducts during cumene hydroperoxide decomposition *

Temp. °C	Molar ratio (byproducts/phenol)×100	
	DMPM equivalents**	Acetophenone
70	0.36	0.06
90	0.61	0.06
110	1.24	0.15
122	2.19	0.25
146	5.04	0.69

* Pure cumene hydroperoxide added to phenol–acetone–cumene solution containing initially 0.5 wt% water and 100 ppm sulfuric acid; data from [28].
** Sum of dimethylphenylmethanol, α-methylstyrene, and their condensation products.

$$C_6H_5\underset{CH_3}{\overset{CH_3}{C}}-OOH + C_6H_5\underset{CH_3}{\overset{CH_3}{C}}-OH \xrightarrow[-H_2O]{H^+} C_6H_5\underset{CH_3}{\overset{CH_3}{C}}-O-O-\underset{CH_3}{\overset{CH_3}{C}}-C_6H_5$$
$$\mathbf{3}$$

$$\xrightarrow{H^+} C_6H_5OH + \mathbf{2} + CH_3COCH_3$$

This sequence suppresses the formation of condensates by approximately 50%. Variants of this two-stage process have recently been patented [36].

Product Separation. In the *Hercules process* the cleavage mixture is neutralized with base and then fed to a separation column. The overheads from this column contain acetone, α-methylstyrene, and cumene; acetone is recovered by distillation, and α-methylstyrene is hydrogenated without prior separation from cumene. This cumene stream is then recycled to the oxidizers. Phenol from the bottoms of the separation column is recovered by distillation.

In the *Allied/UOP process* the cleavage mixture is treated with an ion-exchange resin to remove the acid catalyst and then is distilled. Acetone is removed first in a crude acetone column and purified by distillation with steam in an acetone-refining column. Cumene, α-methylstyrene, and phenol are recovered by sequential distillation of the bottoms from the crude acetone column. Cumene is recycled to the oxidizers after it has been washed with alkali, and α-methylstyrene is marketed.

4.2. Dehydrogenation of 2-Propanol

The hydration of propene [*115-07-1*] gives 2-propanol [*67-63-0*], which is then dehydrogenated to acetone. In the United States a C_3 stream containing 40–60% propene is used for the manufacture of 2-propanol (**4**).

$$CH_3CH=CH_2 \xrightarrow{H_2O} CH_3CH(OH)CH_3 \longrightarrow$$
$$\mathbf{4} \quad CH_3COCH_3 + H_2$$

The dehydrogenation of **4** is endothermic by 66.6 kJ/mol at 327 °C. The equilibrium constant, K_p (bar), obeys the following equation [37]:

$$\log K_p = -2764/T + 1.516 \log T + 1.765$$

The main side reaction is the dehydration of 2-propanol to propene. Other competing reactions are the self-condensation of acetone to diacetone alcohol, which leads to further condensation products.

A large number of catalysts for 2-propanol dehydrogenation have been studied, including copper, zinc, and lead metals, as well as metal oxides, e.g., zinc oxide, copper oxide, chromium-activated copper oxide, manganese oxide, and magnesium oxide. Inert supports, such as pumice, may be used.

Highly active catalysts are the precious metals platinum and ruthenium [39] or 0.25 % platinum on sodium-activated alumina [40]. These catalysts are particularly effective for the dehydrogenation of aqueous 2-propanol, which is obtained by hydration of propene.

All catalysts gradually lose activity because of a buildup of carbon deposits, so the operating temperature is increased as the catalyst ages. The catalyst is regenerated periodically by burning out the deposits. A good catalyst lasts for several months.

In a typical process, the azeotropic mixture of water and 2-propanol (87.8 wt % 2-propanol) is evaporated (sometimes using steam as carrier) and fed to a catalyst bed in a reactor specially designed for effective heat transfer. Hydrogen, produced downstream, may be mixed with the feed to prevent catalyst fouling. The reactor consists of a multitude of 2.5-mm steel tubes heated by oil, high-pressure steam, hot gases, or molten salts. The reaction produces hydrogen (> 99 % purity) as a valuable byproduct. This is separated by condensing all other components. Acetone is separated by distillation. The process is illustrated in Figure 2. Typical operating conditions are shown in Table 5.

4.3. Propene Oxidation [43], [44]

Direct oxidation of propene (Wacker – Hoechst process) currently is practiced only in Japan. A mixture of acetone (92 % selectivity) and propionaldehyde (2 – 4 % selectivity) is produced.

$$CH_3CH=CH_2 + 1/2\ O_2 \xrightarrow{catalyst}$$
$$CH_3COCH_3 + CH_3CH_2CHO$$

The process is analogous to the oxidation of ethylene to acetaldehyde by the Wacker process. The catalyst solution typically contains 0.045 M palladium(II) chloride, 1.8 M

Table 5. Gas-phase dehydrogenation of 2-propanol

Company	Catalyst	Temperature, °C	Pressure, kPa	Conversion, %	Selectivity, %	Yield, %	Reference
Standard Oil	ZnO/ZnO$_2$	400	201–304	98.2	90.2	88.6	[38]
Knapsack-Griesheim	CuO/Cr$_2$O$_3$/Na$_2$O pumice	300		89.5	99.0	88.6	[39]
Toyo-Rayon	CuO/NaF/SiO$_2$	300		93.4	100	93.4	[40]
Engelhard Industries	5% Pt/C	310				92.4	[41]
Usines de Melle	CuO/Cr$_2$O$_3$/SiO$_2$	220	151	75	98.2	73.7	[42]

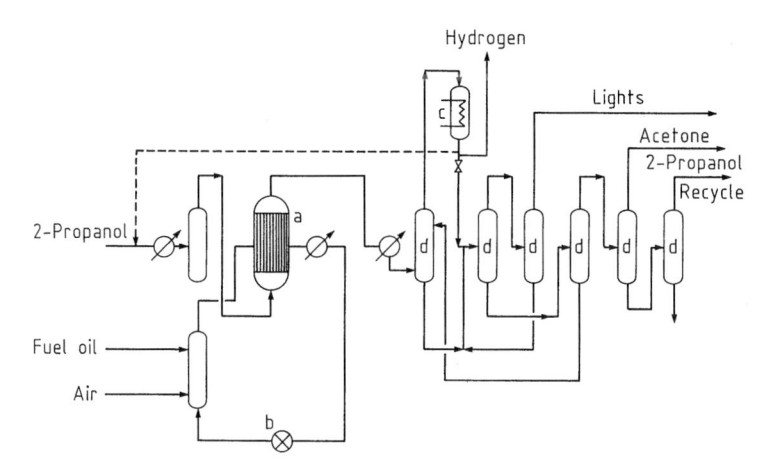

Figure 2. Acetone production via 2-propanol dehydrogenation
a) Reactor; b) Heating loop; c) Refrigeration; d) Distillation columns

copper(II) chloride, and acetic acid [45]. The reaction usually is carried out in two alternating stages. In the first stage, air is used to oxidize the metal ions to the +2 oxidation state. In the second, air is removed and propene added. Palladium(II) oxidizes propene, and the resulting palladium(I) is reoxidized by the pool of copper(II). Reaction conditions are 1–1.4 MPa (10–14 bar) and 110–120 °C. Propene conversion is higher than 99%.

Besides propionaldehyde, chlorinated carbonyl compounds and carbon dioxide also are formed. Acetone and the byproducts are removed from the catalyst solution by flash evaporation with steam and separated by fractional distillation. Propionaldehyde (*bp* 49 °C) distills in one column and acetone (*bp* 56 °C) distills in the other.

4.4. Oxidation of 2-Propanol [46]

In the absence of catalysts 2-propanol reacts with oxygen via a free-radical reaction to form acetone and hydrogen peroxide.

$$\text{CH}_3\text{CH(OH)CH}_3 \xrightarrow{\text{O}_2} \text{CH}_3\text{COCH}_3 + \text{H}_2\text{O}_2$$

Until the mid-1980s the Shell process used hydrogen peroxide for the manufacture of glycerol from propene. The theoretical yield of acetone based on glycerol produced is 1.26 kg/kg. Acetone yields of about 90 % of theoretical were obtained.

4.5. Oxidation of *p*-Diisopropyl Benzene

Acetone is coproduced with hydroquinone [123-31-9] from *p*-diisopropylbenzene [100-18-5] in a process analogous to the phenol–acetone production from cumene.

$$\underset{\text{H}_3\text{C}}{\overset{\text{H}_3\text{C}}{\text{HC}}}\!\!-\!\!\bigcirc\!\!-\!\!\underset{\text{CH}_3}{\overset{\text{CH}_3}{\text{CH}}} \xrightarrow{\text{O}_2} \text{HOO}-\underset{\text{H}_3\text{C}}{\overset{\text{H}_3\text{C}}{\text{C}}}\!\!-\!\!\bigcirc\!\!-\!\!\underset{\text{CH}_3}{\overset{\text{CH}_3}{\text{C}}}-\text{OOH}$$
$$\mathbf{5}$$
$$\xrightarrow{\text{H}^+} \text{HO}-\bigcirc-\text{OH} \;+\; 2\,\text{CH}_3\text{COCH}_3$$

In the Goodyear process [47] *p*-diisopropylbenzene is oxidized by oxygen in the presence of caustic. The *p*-diisopropylbenzene dihydroperoxide (**5**) [3159-98-6] formed is crystallized and washed with benzene. It is then dissolved in acetone and cleaved to hydroquinone and acetone in the presence of sulfuric acid. Next the acid is neutralized with ammonia and the ammonium sulfate formed is filtered. Acetone is recovered by distillation from the reaction mixture. Some of this acetone is recycled to the cleavage section while the rest passes through a finishing column for purification to at least 99.5 %. Eastman Chemical and Goodyear Tire & Rubber Company use this process in the United States. Annual US capacity is estimated to be 18 – 20 t/a. Sumitomo Chemical Company and Mitsui Petrochemical Industries of Japan use a similar process to produce *p*-cresol from cymene. Their annual capacity of acetone byproduct is 48 000 t.

4.6. Fermentation of Biomass

The fermentation of cornmeal or molasses by various members of the *Clostridium* genus yields a mixture of 1-butanol, acetone, and ethanol in 2 % overall concentration. The products are recovered by steam distillation and then fractionated.

The process was started during World War II to provide acetone needed for the manufacture of cordite. The last operating plant in the United States (Publicker Industries) closed in 1977.

The mixture of butanol, acetone and ethanol produced has been considered for use as a gasoline substitute in France [48]. Research aimed at increasing the concentration of useful products obtained in the process was carried out in the United States in the early 1980s [49]. The future of the fermentation process is tied to the availability of petrochemical feedstocks. High oil prices during the oil crises of the mid to late 1970s led to renewed interest in the process. Given the low oil prices of the 1990s and the ready availability of feedstocks at reasonable prices, it does not appear that these processes can compete under current conditions.

5. Environmental Protection

Because approximately 70% of acetone is produced from cumene, a close examination of this process is warranted. Potential pollution sources in a phenol–acetone plant are emissions to the atmosphere and liquid discharge. Atmospheric emissions from the phenol–acetone process in the late 1970s have been estimated [29]. However over the past 20 years, and particularly in the 1990s with the renewal of the Clean Air Act, these emissions have been reduced significantly.

Aqueous streams containing significant amounts of organic substances arise from the various wash operations and sumps at the plant. Insoluble material is recovered by decantation. Phenol and acetone (0.5–3 wt%) each) are the most abundant organic compounds remaining in the water after decantation. There are also minor quantities (0.001–0.1 wt%) of cumene, α-methylstyrene, dimethylphenylmethanol, acetophenone, formaldehyde, formic acid, and various condensates. Of these compounds, phenol, formaldehyde, and formic acid are listed as hazardous substances in the U.S. Federal Water Pollution Control Act [50], but only phenol is present in sufficient quantities to require removal. Phenol is removed from the aqueous solution by solvent extraction, steam stripping, or adsorption on carbon or resins [51] and subsequently is recovered. The recovered phenol is valuable enough to pay for the capital and operating expenses of phenol abatement. Residual phenol in the water (10–500 ppm) is destroyed by biological degradation.

The federal regulatory status of acetone has recently changed. Acetone was granted VOC-exempt status by EPA on June 16, 1995 [53]. As of August, 1997, forty-four states had promulgated similar state rules. In states that have not yet promulgated state exemptions, acetone may technically still be regulated as a VOC. Acetone is not listed as a hazardous air pollutant (HAP) under section 112(b) of the Clean Air Act (CAA), or as an extremely hazardous substance under EPCRA Section 302. Acetone is also not listed as a priority pollutant under the Clean Water Act. It has been approved under the CAA as a substitute for ozone-depleting substances. Acetone was removed from the Federal Emergency Planning and Community Right-to-Know Act (EPCRA Section 313) list in June of 1995. Acetone is not regulated as a known or suspected carcinogen, and the National Toxicology Program (NTP) has recommended against testing for carcinogenicity because of its low toxicity and absence of any evidence supporting the carcinogenic potential of acetone.

Acetone is listed as a "U" waste under the Resource Conservation and Recovery Act (RCRA) based on its ignitability. "U" wastes are commercial chemicals that must be treated as hazardous wastes when discarded. Because of its RCRA listing it is included in the list of hazardous substances in the Superfund statute (Comprehensive Environmental Response, Compensation, and Liability Act).

6. Quality Specifications and Analysis

Acetone is produced industrially in relatively high purity, the main impurity being water. Table 6 summarizes the quality requirements for commercial 99.5% acetone. Methods for preparing very high purity acetone from the commercial material are given in reference [54].

Gas chromatography is the most widely used method for the quantitative analysis of acetone. For example, good separation of acetone from other low-boiling organic compounds can be obtained on a 30 m × 0.32 mm Carbowax capillary column. Extensive data on packed column separations are compiled in [55]. Infrared (carbonyl absorption, 1711 cm^{-1}) and ^{1}H NMR (singlet at ca. 1.05 ppm) spectroscopy may be used for both qualitative and quantitative analysis.

7. Storage and Transportation

Acetone has a low flash point; therefore, all shipping and storage containers must carry a red, diamond shaped "flammable liquid" label. Strict precautions should be taken to guard against fire hazards whenever acetone is handled. All wiring should be installed as described in Article 500 of the U.S. National Electrical Code or corresponding regulations in other countries. Explosion-proof motors, switches, etc., should be used. Accumulation of static electricity should be prevented by grounding and humidity control. Use of spark-resistant tools is recommended. Small fires may be controlled by use of carbon dioxide or dry chemical extinguishers. "Alcohol"-type foam should be used on larger fires; water spray will reduce the intensity of the flame.

Contact of acetone with oxidants should be avoided because it may lead to explosion [5]. Contamination with chlorinating agents may lead to the formation of toxic chloroketones. Prolonged exposure to direct sunlight may result in the formation of carbon monoxide. Packaging requirements for acetone are described in paragraph 49 CFR 173.242 (bulk), Bulk Packaging with Packaging for Certain Medium Hazard Liquids and Solids, Including Solids with Dual Hazards [56]. Transportation of acetone is covered in paragraph 49 CFR 172.101, Table of Hazardous Materials, of the Department of Transport Regulations [57]. The international transportation codes are IMDG

Table 6. Standard specifications for acetone, ASTM D329-90

Property	ASTM test	Specification
Relative density	D268	
20/20 °C		0.7910 – 0.7930
25/25 °C		0.7865 – 0.7885
Color	D1209	≤ 5 on platinum-cobalt scale
Distillation range	D1078	1.0 °C, including 56.1 °C
Nonvolatile matter	D1353	≤ 5 mg/100 mL
Odor	D1296	characteristic, nonresidual
Water	D1364	0.5 wt%*
Acidity (as acetic acid)	D1613	0.002 wt%
Water miscibility	D1722	passes test
Alkalinity (as ammonia)	D1614	0.001 wt%
Permanganate time	D1363	30 min at 25 °C

* This water limit ensures that the material is miscible without turbidity with 19 volumes of 99% heptane at 20 °C (ASTM D1476).

Code D 3102; UN no. 1090; CFR 49, 172.101; RID (ADR, ADNR): Class 3, IATA: flammable liquid. The quantity of acetone in one package may not exceed 5 L in plastic, metal or aluminum, 1 L in glass, or 0.5 L in a glass ampoule in a passenger aircraft. The quantity of acetone in one package may not exceed 60 L in a cargo plane.

8. Uses

The main uses of acetone are as a chemical intermediate and as a solvent. The estimated 1995 acetone consumption by area of application in the USA is shown in Table 7 [59].

8.1. Methyl Methacrylate

Acetone is condensed with hydrogen cyanide to form acetone cyanohydrin (**6**) (see Section 11.1), which is next hydrolyzed with sulfuric acid to methacrylamide sulfate (**7**).

Further reaction with methanol yields methyl methacrylate (**8**) [*80-62-6*]. Approximately 0.70 kg of acetone is required per kilogram of methyl methacrylate produced.

Table 7. Estimated 1995 US acetone consumption by area of application

Use	Acetone used, 10^3 t
Acetone cyanohydrin/methacrylates	500
Bisphenol A	203
Aldol chemicals	140 (total)
Methyl isobutyl carbinol	35
Methyl isobutyl ketone	76
Others	22
Solvent use	191
Other uses	90
Total	1124

$$CH_3COCH_3 + HCN \longrightarrow CH_3\underset{CN}{\overset{OH}{C}}CH_3 \xrightarrow{H_2SO_4}$$

$$\underset{6}{}$$

$$\underset{CH_3}{CH_2{=}CCONH_2} \cdot H_2SO_4 \xrightarrow{CH_3OH} \underset{CH_3}{CH_2{=}CCOOCH_3} + NH_4HSO_4$$

$$\underset{7}{} \qquad \underset{8}{}$$

Higher methacrylate esters may be produced either by transesterification of methyl methacrylate or by esterification of methacrylic acid (**9**) [79-39-0]; the latter is made by hydrolysis of methacrylamide sulfate:

$$\underset{CH_3}{CH_2{=}CCONH_2} \cdot H_2SO_4 \xrightarrow{H_2O} \underset{CH_3}{CH_2{=}CCOOH} + NH_4HSO_4$$

$$\underset{9}{}$$

At the end of 1995 there were 22 plants manufacturing MMA in the United States, Western Europe, and Japan. Five basic process routes have been commercialized: The acetone cyanohydrin route; two-stage oxidation of isobutylene to methacrylic acid followed by esterification; two-stage oxidation of *tert*-butyl alcohol to methacrylic acid followed by esterification; hydroformylation of ethylene to propionaldehyde, condensation with formaldehyde to methacrolein, oxidation, and esterification (BASF); and ammoxidation of *tert*-butyl alcohol to methacrylonitrile, which is hydrolyzed to methacrylamide sulfate and then esterified to MMA (Asahi). One new route has been announced by Mitsubishi Gas Chemicals, which is a recycle version of the acetone cyanohydrin route. A 41×10^3 t/a plant to make MMA and MAA started up in 1997. Worldwide production of MMA in 1996 was about 1682×10^3 t [60].

8.2. Bisphenol A (→ Phenol Derivatives)

Bisphenol A (**10**), 4,4′-isopropylidenediphenol [*80-05-7*] is manufactured by condensation of 2 mol phenol with 1 mol acetone in the presence of an acid catalyst:

$$2 \text{ } C_6H_5OH + (CH_3)_2C=O \xrightarrow{H^+} HO\text{-}C_6H_4\text{-}C(CH_3)_2\text{-}C_6H_4\text{-}OH + H_2O$$
$$\textbf{10}$$

Approximately 0.28 kg of acetone is required per kilogram of bisphenol A. In the 1990s, bisphenol A has had the fastest growing demand of the phenol derivatives. Four US. companies produce bisphenol A: Shell (Deer Park, Texas), General Electric (Mount Vernon, Indiana), Dow (Freeport, Texas), and Aristech (Haverill, Ohio). Estimated worldwide usage in 1995 was 1600×10^3 t [61].

8.3. Aldol Chemicals (see Section 11.2, also → Ketones)

These chemicals are produced by condensation of acetone. Two moles of acetone form 1 mol of diacetone alcohol, 4-hydroxy-4-methyl-2-pentanone (**11**) [*123-42-2*]. Subsequent dehydration yields mesityl oxide, 4-methyl-3-penten-2-one (**12**) [*141-79-7*]. Hydrogenation of **11** yields 2-methyl-2,4-pentanediol (**13**) [*107-41-5*]. By hydrogenation of **12** methyl isobutyl ketone (**14**) [*108-10-1*] is available; further hydrogenation produces 4-methyl-2-pentanol (**15**) [*108-11-2*]. Three moles of acetone are condensed to 1 mol of isophorone, 3,5,5-trimethyl-2-cyclohexen-1-one (**16**) [*78-59-1*].

$$2 \text{ } CH_3COCH_3 \xrightarrow{OH^-} CH_3COCH_2\underset{OH}{C}(CH_3)_2 \xrightarrow{-H_2O}$$
$$\textbf{11}$$

$$CH_3COCH=C(CH_3)_2$$
$$\textbf{12}$$

$$\textbf{11} + H_2 \longrightarrow CH_3\underset{OH}{C}HCH_2\underset{OH}{C}(CH_3)_2$$
$$\textbf{13}$$

$$\textbf{12} + H_2 \longrightarrow CH_3COCH_2CH(CH_3)_2$$
$$\textbf{14}$$

$$\xrightarrow{H_2} CH_3\underset{OH}{C}HCH_2CH(CH_3)_2$$
$$\textbf{15}$$

$$3 \text{ } CH_3COCH_3 \xrightarrow{-2 H_2O} \text{(isophorone)}$$
$$\textbf{16}$$

Approximately 1.25 kg of acetone is used per kilogram of methylisobutyl ketone produced. The US 1996 production of ca. 100×10^3 t of MIBK consumed ca 75×10^3 t of acetone. The US manufacturers are: Eastman (Kingsport, Tennessee), Shell (Deer Park, Texas), and Union Carbide (Institute, West Virginia) [62]. Methyl isobutyl ketone is used as a solvent for nitrocellulose lacquers, vinyl polymers, and acrylic resins [62].

Diacetone alcohol, mesityl oxide, and isophorone are used mainly as solvents. Their use currently is diminishing in the USA because of their status as photochemically reactive solvents under Rule 66 of Los Angeles County [52]. The primary use of 4-methyl-2-pentanol is for ore flotation, and 2-methyl-2,4-pentanediol is used in hydraulic fluids and printing inks.

8.4. Solvent Uses

Acetone is used as a solvent for paints, varnishes, and lacquers. It is also used as a wash solvent for these materials and as a spinning solvent in the manufacture of cellulose acetate. A small amount of acetone is used as a solvent for acetylene. Approximately 191×10^3 t of acetone was consumed in direct solvent applications. The major solvent market for acetone is in paints and coatings. Consumption of acetone in these applications increased by 9×10^3 t in 1995 because of its delisting as a VOC.

The pharmaceutical industry is also a large consumer of acetone for the manufacture of pharmaceuticals, vitamins, and cosmetics. In 1995 acetone consumption in pharmaceutical and cosmetic applications was $(36-43) \times 10^3$ t.

The removal of acetone from the VOC list has made it more attractive as a solvent, particularly for replacing other chemicals on the VOC list. Following the EPA's August 1995 action, eight states automatically delisted acetone as a volatile organic compound. As of August, 1997, forty-four states had promulgated similar state rules.

9. Economic Aspects

The United States acetone capacity by manufacturer and production process can be found in [63]. World capacity data are given in Table 8 [59]. The United States acetone production for the last two decades is summarized in Table 9 [59]. Worldwide production in 1994 was: USA 1281×10^3 t/a; Western Europe 1200×10^3 t/a; Asia 746×10^3 t/a.

In 1995 the phenol process accounted for 83% of all acetone made and 9% was derived from 2-propanol; only 8% was produced by all other processes (66×10^3 t/a from propene oxidation in Japan).

The economics of acetone are unusual. The bulk of acetone is made as a coproduct with phenol. Consequently, phenol demand determines to a large extent the availability of acetone. Fortunately acetone serves to some extent the same markets as phenol does. These are mainly the automotive and housing markets. As a result, when economic

Table 8. World acetone capacity, 1995

Location	Capacity, 10^3 t
United States	1281
Mexico	22.3
Western Europe	1200
Japan	475
Other Asia	271.4
Others	592
World total	3842

Table 9. United States acetone production, 10^3 t

Year	From 2-propanol	From cumene	Other	Total
1970	379	329	26	734
1975	312	433		745
1980	250	693		943
1985	41	768	3	812
1990	70	972	17	1059
1994	73	1110	20	1203

conditions place a demand on phenol, acetone demand also increases. However, an unusually steep demand for phenol may render acetone an overabundant byproduct [63].

Because the single largest use of acetone is as an intermediate in the manufacture of methacrylates, alternate routes to methacrylates, such as the oxidation of C_4 hydrocarbons, are a potential threat to acetone. In the USA, C_4 hydrocarbon stocks are currently in demand for the manufacture of gasoline additives and, therefore, they are not likely to be used for acrylate manufacture. However, production of methacrylate by oxidation of C_4 hydrocarbons started in Japan in 1982 [65].

Dehydrogenation of 2-propanol accounted for approximately 80% of US acetone produced in 1960. As the cumene-based process expanded, the 2-propanol contribution shrank to 52% in 1970, 27% in 1980, and 6% in 1994. If acetone supply does outstrip demand, it is likely that production based on 2-propanol will be further curtailed.

Phenol manufacturing processes that do not coproduce acetone have been developed partly because of fears that there will be a supply/demand imbalance between phenol and acetone. DSM and its licensees produce phenol by the air oxidation of toluene via benzaldehyde and benzoic acid. Subsequently, benzoic acid is decomposed to phenol with a copper catalyst. This route provides phenol, benzaldehyde, and benzoic acid. Mitsui Petrochemical has developed a recycle scheme for converting the acetone byproduct from the cumene hydroperoxide rearrangement back to propylene for feed to the front end of the cumene process. Solutia (formerly Monsanto) and the Boreskov Institute of Catalysis have developed a catalyst capable of oxidizing benzene in high yield with nitrous oxide to give directly phenol. The key to this process is the inexpensive nitrous oxide available as a byproduct from the manufacture of adipic acid. Asahi Chemical has patented a process in which benzene is partly hydrogenated to cyclohexene. The cyclohexene is hydrolyzed to cyclohexanol or oxidized to cyclohexanone; dehydrogenation then gives phenol [66].

10. Toxicology and Occupational Health

Acetone is one of the least toxic industrial solvents [67]. However, exposure to vapor at high concentration should be avoided because it can produce temporary narcosis and cause slight eye irritation. Repeated skin contact with the liquid defats the skin and may cause dermatitis. The liquid is also irritating to the eyes and may cause moderate corneal injury.

The ACGIH [68] has adopted a time-weighted average threshold limit value (TLV-TWA) of 750 ppm, 1.78 g/m^3, and a short-term exposure limit (TLV-STEL) of 1000 ppm, 2.375 g/m^3, for acetone. OSHA regulations [69] set a limit of 2.4 g/m^3. However, the ACGIH has issued a notice of intended change to lower the TLV to 500 ppm (1.19 g/m^3), and the STEL to 750 ppm (1.78 g/m^3) [70]. Exposure limits (TLV-TWA) adopted by the main industrial countries are shown in Table 10 [71]. The odor threshold of acetone is 48 mg/m^3 provided desensitization has not occurred. Animal studies have shown acetone to be relatively nontoxic [67], [71].

LD_{50} (oral, mouse) 4–8 g/kg [67]
LD_{50} (oral, rabbit) 5.3 g/kg [67]
LD_{50} (intraperitoneal, mouse) 1.3 g/kg [67]
LD_{50} (dermal, rabbit) 20 g/kg [71]
Nonteratogenic at 39 or 78 mg per chicken egg [67]
Nonmutagenic in the *Salmonella*/microsome (Ames) test [67]
Nononcogenic on skin of mice, three times a week for 1 year [67]
Moderate corneal injury on rabbit eye [67]
Environmental toxicity: LC_{50} (rainbow trout, 96 h) 5540 mg/L; LC_{50} (bluegill sunfish, 96 h) 8300 mg/L.

Minimum lethal concentration in air (LC_{50}) was 50.1 g/m^3 (21 100 ppm) for rats exposed to acetone vapor for 8 h and 44 g/m^3 (18 500 ppm) for mice exposed for 4 h. Human exposure to acetone has been studied [72]–[77]. Eye and nasal irritation were observed at 1.2 g/m^3. Other effects are similar to those of ethanol, but the anesthetic potency is greater. Prolonged or repeated skin contact may defat the skin and can produce dermatitis. Direct contact of acetone with the eyes can produce corneal injury. Acetone is a solvent of comparatively low acute and chronic toxicity. However it does not have sufficient warning properties to prevent repeated exposures to vapors, which may have adverse effects. There have been no reports that prolonged inhalation of low vapor concentrations result in any serious chronic effects in humans.

Table 10. Time-weighted average threshold limits for acetone

Country	Limit, mg/m^3
Australia	1780
Belgium	1780
Germany	2400
Italy	1000
Japan	470
Netherlands	1780
Former Soviet Union	475
United States	1780

Cases of acetone poisoning are rare [67]. In one case, a solvent mixture containing 90% acetone and 9% pentane was used to set a cast for a broken leg on a 10-year-old boy [78]. The boy became ill and collapsed 12 h later. After the cast was removed the boy became comatose but recovered completely in 4 days. In another case, a 42-year-old man ingested 200 mL of acetone and became comatose for 12 h [79]. Subsequently, hyperglycemia was diagnosed and attributed to acetone ingestion.

Acetone does not cause neurotoxicity, an occupational disorder caused by exposure to some higher aliphatic ketones and related compounds [80]. Acetone vapor is absorbed with 75% efficiency by the lungs [76]. The half-life for the elimination of acetone by expired air is approximately 5 h. The metabolism of acetone may proceed through 1,2-propanediol [67], [81], [82].

11. Derivatives

11.1. Acetone Cyanohydrin

Acetone cyanohydrin, 2-hydroxy-2-methylpropanenitrile, CH$_3$C(OH)(CN)CH$_3$ [75-86-5] is an important chemical intermediate for the manufacture of methacrylates (→ Methacrylic Acid and Derivatives). Small amounts of acetone cyanohydrin are used in insecticide manufacture.

Physical Properties. Acetone cyanohydrin is a colorless liquid. The pure compound is practically odorless but usually has an odor of bitter almonds because of traces of hydrogen cyanide. It is very soluble in water and polar solvents and sparingly soluble in hydrocarbons.

M_r 85.11, mp −19 °C, relative density d_4^{25} 0.9267, relative vapor density (air = 1) 2.96, refractive index n_D^{25} 1.3980, flash point 73 °C.

Vapor pressure

p, kPa	5.3	3.1	1.3	1.2
t, °C	95	82	74	72

Chemical Properties. Acetone cyanohydrin exhibits the combined characteristics of a nitrile and an alcohol. Under neutral and particularly under alkaline conditions it decomposes to acetone and hydrogen cyanide. The decomposition is inhibited by the addition of small amounts of sulfuric or phosphoric acid; consequently, technical-grade material is stabilized by addition of 0.01 wt % of either acid. Reaction with concentrated sulfuric acid converts acetone cyanohydrin to methacrylamide sulfate; subsequent neutralization with ammonia yields methacrylamide [79-39-0]; alcoholysis yields methacrylate esters; alternatively, hydrolysis gives methacrylic acid [79-41-4]. The yield of methacrylic acid is improved if 3 – 10 % oleum is used instead of 100 % sulfuric acid in the reaction with acetone cyanohydrin [83].

Production. Acetone cyanohydrin is manufactured by the base-catalyzed condensation of acetone with hydrogen cyanide according to the following mechanism [84]:

$$\begin{array}{c}CH_3\\ \diagdown\\CH_3\end{array}C=O + CN^- \rightleftharpoons \begin{array}{c}CH_3\\ \diagdown\diagup O^-\\ C\\ \diagup\diagdown\\CH_3CN\end{array} \xrightleftharpoons{HCN} \begin{array}{c}CH_3\\ \diagdown\diagup OH\\ C\\ \diagup\diagdown\\CH_3CN\end{array} + CN^-$$

The reaction is reversible but formation of the cyanohydrin is quite favorable; the equilibrium constant is 28 L/mol at 20 – 25 °C [85]. The reaction usually is carried out in the liquid phase. Representative catalysts used industrially are sodium hydroxide [86], potassium hydroxide [87], potassium carbonate [88], and anion-exchange resins. A schematic flowsheet of the Rohm & Haas [89] process is shown in Figure 3. Acetone and liquid hydrogen cyanide are fed continuously to a cooled reactor along with an alkaline catalyst. The catalyst is next neutralized with sulfuric acid and the resulting salt is removed by filtration. The crude product is then distilled in a two-stage process. The overheads from the first column consist mainly of acetone and hydrogen cyanide, which are recycled to the reactor. The second column removes water overhead and leaves 98 % pure acetone cyanohydrin at the bottom. Nitto Chemical claims a two-column distillation system that delivers acetone cyanohydrin of 99.1 % purity [90]. The manufacture of acetone cyanohydrin produces no byproducts other than small amounts of sulfate salts formed during catalyst neutralization. However, the conversion of acetone cyanohydrin to methacrylate in the classical process produces a large amount of ammonium sulfate byproduct, which is usually pyrolyzed to sulfuric acid. A recent alternative process developed by Mitsubishi Gas Chemical recycles the HCN via formamide. In this process HCN is not directly consumed and no ammonium sulfate is formed.

Uses. By far the largest use of acetone cyanohydrin is as an intermediate in the synthesis of methyl methacrylate [80-62-6], methacrylic acid, and higher methacrylate esters. A small amount is converted to methacrylamide. The estimated amount of acetone used in the manufacture of methacrylate esters in 1995 in the United States was

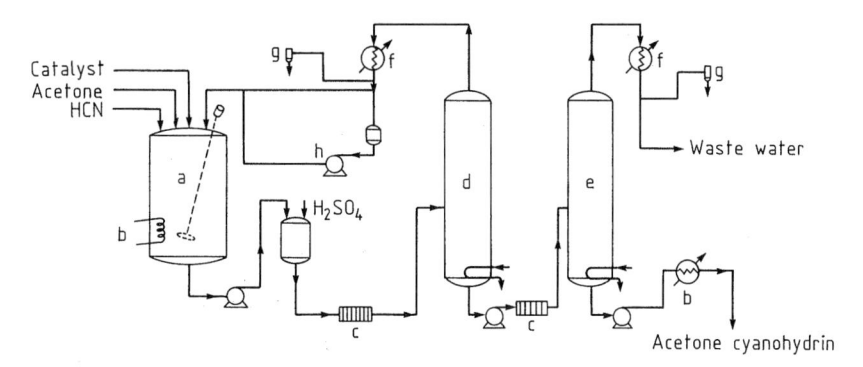

Figure 3. Rohm and Haas acetone cyanohydrin process [89]
a) Reactor; b) Cooling; c) Filter press; d) Concentrator; e) Concentrator; f) Condenser; g) Vacuum jet; h) Pump for acetone and HCN recycle

500×10^3 t (Table 7). Based on this estimate, the quantity of acetone cyanohydrin produced as an intermediate was approximately 714×10^3 t.

Certain esters of acetone cyanohydrin, such as 2-chloroethyl-α-cyanoisopropylsulfite [91] and α-cyanoisopropyl-2,6-dichlorobenzoate [92] have strong fungicidal, herbicidal, and insecticidal properties. Nitrilurethanes made from acetone cyanohydrin and substituted phenylisocyanates are intermediates for 4-iminooxazolidin-2-ones, which are plant growth inhibitors [93].

Transportation and Toxicology. Most acetone cyanohydrin is consumed on site for the manufacture of methacrylates. Because of its high toxicity, acetone cyanohydrin is classified as a poison B and all shipping containers must carry a "poison inhalation hazard" label [57]. Transport aboard a passenger-carrying or cargo aircraft is forbidden [58].

Acetone cyanohydrin has the following toxicologic properties [94]: LD_{50} (oral, rat) 18.6 mg/kg; LD_{50} (oral, rabbit) 14 mg/kg; LD_{50} (dermal, rabbit) 17 mg/kg; LD_{50} (dermal, guinea pig) 150 mg/kg; aquatic toxicity rating (TLm96) 10 to 1 ppm. Threshold limit values have been established as cyanide: TLV-STEL, 4.7 ppm ceiling limit (5 mg/m^3) [68].

11.2. Diacetone Alcohol

Diacetone alcohol, 4-hydroxy-4-methyl-2-pentanone [*123-42-2*]

$$CH_3-\underset{\underset{OH}{|}}{\overset{\overset{CH_3}{|}}{C}}-CH_2-\underset{\underset{O}{\|}}{C}-CH_3$$

is a dimer of acetone that is used as a solvent and as an intermediate for the manufacture of mesityl oxide, methyl isobutyl ketone, and hexylene glycol.

Physical Properties. Diacetone alcohol is a colorless liquid of mild odor. It is miscible with water and polar solvents and is an excellent solvent for cellullose acetate and various oils and resins. M_r 116.16, mp −47 °C, relative density, d_4^{20} 0.9387, refractive index n_D^{20} 1.4235, heat of vaporization at the boiling point at 101.3 kPa 357.1 kJ/kg, specific heat capacity (20 °C) 18.84 kJ kg^{-1} K^{-1}, thermal expansion coefficient (20 °C) 0.00099 K^{-1}, viscosity at 20 °C 2.9 mPa · s, surface tension (20 °C) 31.0 mN/m, dielectric constant (25 °C) 18.2, heat of combustion 3544.5 kJ/mol, flash point 58 °C, lower explosion limit in air 2.6 vol%, auto ignition temperature 624 °C.

Vapor pressure

p, kPa	101.3	1.7	0.108
t, °C	168.1	61.7	20.0

Azeotropic mixture with water: bp 98.8 °C/ 101.3 kPa, 12.7 wt% diacetone alcohol.

Chemical Properties. Diacetone alcohol dehydrates readily in the presence of acids to form mesityl oxide. Catalytic hydrogenation [95] yields hexylene glycol. In the presence of bases, diacetone alcohol reverts to acetone. The reaction is first order in diacetone alcohol and first order in base. The dissociation is accompanied by volume increase; consequently, the reaction is inhibited by pressure [96]. The second-order rate constant is 7.33×10^{-4} L mol^{-1} s^{-1} at 101.3 kPa and decreases to 2.38×10^{-4} L mol^{-1} s^{-1} at 4.05×10^5 kPa. In neutral aqueous solution, dissociation to acetone is very slow at room temperature; it reaches approximately 0.1% in 1 year.

Diacetone alcohol may be acylated by acetic anhydride under mild conditions to form diacetone alcohol acetate [1637-25-8], 4-methyl-4-acetyloxy-2-pentanone, which is claimed as an octane-improving gasoline additive [97]. Condensation with urea in the presence of sulfuric acid yields diacetone-monourea, 3,4-dihydro-4,4,6-trimethyl-2(1H)-pyrimidone [4628-47-1] [98], [99].

The compound is claimed to improve the egg-laying capacity of hens [98].

Production. Diacetone alcohol is manufactured by self-condensation of acetone in the presence of a basic catalyst. The reaction is exothermic by 14.65 kJ/mol and is easily reversible. The equilibrium concentration of diacetone alcohol is 23.1 wt% at 0 °C and decreases with increasing temperature [100]. Kinetic considerations dictate, however, a higher temperature for the manufacture of diacetone alcohol. An optimum temperature range is 10–20 °C.

The self-condensation of acetone is carried out in continuous-flow reactors containing a solid alkaline catalyst, such as barium hydroxide or calcium hydroxide [101], [102]. Anion-exchange resins have been investigated [103], [104], but are not believed to be used commercially. Catalyst performance deteriorates with time, but may last up to 1 year. A patent [105] describes how addition of small amounts of methanol, ethanol, or 2-propanol to the reaction mixture retards catalyst deterioration. The selectivity for diacetone alcohol is 90–95%. Mesityl oxide and higher condensates, such as triacetone alcohol, are the main products. The acetone solution of the crude product is neutralized, e.g., with phosphoric acid [106], prior to concentration under reduced pressure. The recovered acetone is recycled to the condensation reactor and the acidity adjusted by subsequent addition of a base, such as triethylamine. After such treatment, diacetone alcohol of 99.68% purity was obtained by vacuum distillation [106].

Uses. Diacetone alcohol is an excellent solvent for many natural and synthetic resins. It is used in the coatings industry, especially for hot lacquers, and is also used as a solvent for nitrocellulose, cellulose acetate, and epoxy resins. However, its use has been diminishing in recent years because it is not exempt from restrictions under Rule 66 and related federal regulations [52]. Therefore, US sales have declined more than 50% since 1978, reaching 10 000 t/a for the past several years and are expected to remain flat. A large portion of diacetone alcohol is used as an intermediate for the manufacture of mesityl oxide, methyl isobutyl ketone, and hexylene glycol.

Transportation and Toxicology. Diacetone alcohol has a flash point of 58 °C, and all transport containers must carry a "flammable liquid" label. Threshold limit values (TLV) for diacetone alcohol vapor at the workplace are: 50 ppm, 240 mg/m^3 (TLV-TWA) [69]; MAK is 50 mL/m^3, 240 mg/m^3. These limits also apply to the other major industrial countries [71]. The toxicologic properties of diacetone alcohol are as follows [107]: LD_{50} (oral, rat) 4 g/kg; LD_{50} (intraperitoneal, mouse) 933 mg/kg; LD_{50} (dermal, rabbit) 13.5 g/kg; aquatic toxicity rating (TLm96) 1000–100 ppm.

11.3. Miscellaneous Derivatives

Acetone is used in the production of methyl amyl ketone (MAK, 2-heptanone). MAK is produced by the condensation of *n*-butyraldehyde and acetone. In 1995, US consumption of acetone for MAK was approximately 12×10^3 t, producing ca. 17.5×10^3 t of MAK. Eastman Chemical is currently the sole US producer. Acetone is also used to produce methyl isoamyl ketone (MIAK) by condensation of acetone with isobutyraldehyde. Eastman Chemical is the sole US producer of MIAK, which is primarily used in lacquers and surface coatings. Other minor uses of acetone include the manufacture of DuPont triazine herbicide Bladex [$(3.5-3.6) \times 10^3$ t/a]; Ethoxyquin, a Monsanto antioxidant (2.7×10^3 t/a), as a raw material in the production of hexafluoroacetone, methylbutynol, and pseudoionone; and as an auxiliary blowing agent for the production of flexible polyurethane foam.

12. References

General References

[1] *Beilstein* **1** 635, **1 (1)** 335, **1 (2)** 692. **1 (3)** 2696, **1 (4)** 3180. (Acetone); **3** 316, **3 (2)** 224, **3 (3)** 597 (Acetone cyanohydrin).

[2] *Ullmann*, 4. Aufl., **7:25.**

Specific References

[3] *Beilstein* **1** (3) 2707.
[4] L. H. Horsley: *Azeotropic Data-III.* Advances in Chemistry Series No. 116, American Chemical Society, Washington, D.C. 1973.
[5] L. Bretherick: *Handbook of Reactive Chemical Hazards,* CRC Press, Cleveland 1981, p. 362.
[6] G. B. Kaufmann, *Chem. Eng. News* **35** (1957) no. 43, 60.
[7] R. Delhez, *Chem. Ind.* (London) 1956, 931.
[8] H. Seidl, *Angew. Chem. Intern. Ed. Engl.* **3** (1964) 640; *Angew. Chem.* **76** (1964) 716.
[9] A. Naponen, *Chem. Eng. News* **55** (1977) no. 8, 5.
[10] H. K. King, *Chem. Ind. (London)* 1970, 185.
[11] D. H. Grant, *Chem. Ind. (London)* 1970, 919.
[12] United States Dept. of the Navy, U.S. Appl. 109 692, 1980 (B. E. Douda, C. F. Parrish, J. E. Short Jr.); *Chem. Abstr.* **93** (1980) 240 992.
[13] D. A. Boiston, *Br. Chem. Eng.* **13** (1968) 85.
[14] *Hydrocarbon Process.* **44** (1965) no. 11, 231; *Chem. Eng. (N.Y.)* **71** (1964) no. 20, 78.
[15] Allied Signal Corp., Specifications for Oxidation Grade Cumene (Partial Listing), 1983.
[16] J. A. Howard in: G. H. Williams (ed.): *Advances in Free Radical Chemistry,* vol. **4,** Academic Press, New York 1972, p. 49.
[17] H. C. Bailey, G. W. Godin, *Trans. Faraday Soc.* **52** (1956) 68.
[18] A. K. Roby, J. P. Kingsley, *Chem. Tech.* (1996) 41.
[19] J. B. Fleming, J. R. Lambrix, J. R. Nixon, *Hydrocarbon Process* **55** (1976) no. 1, 185.
[20] H. W. Melville, S. Richards, *J. Chem. Soc.* 1954, 944.
[21] K. Hattori, Y. Tanaka, H. Suzuki, T. Ikawa, H. Kubota, *J. Chem. Eng. Jpn.* **3** (1970) 72.
[22] C. G. Hagberg, F. X. Krupa, *Chem. React. Eng. Proc. Int. Symp.* **4** (1976) 408.
[23] P. Andrigo, A. Caimi, P. Cavalieri d'Oro, A. Fait, L. Roberti, M. Tampieri, V. Tartari, *Chem. Eng. Sci.* **47** (1992) 2511.
[24] Hercules Powder Co., US 2 484 841, 1949 (E. J.Lorand).
[25] Allied Corp., US 2 613 227, 1950 (G. G. Joris).
[26] Allied Corp., US 2 757 209, 1956 (G. G. Joris).
[27] Allied Corp., US 3 404 901, 1975 (R. L. Feder, R. Fuhrmann, J. Pisanchyn, S. Elishewitz, T. H. Insinger, C. T. Mathew).
[28] Allied Corp., US 4 358 618, 1982 (S. Sifniades, A. A.Tunick, F. W. Koff).
[29] J. L. Delaney, T. W. Hughes: *Source Assessment: Manufacture of Acetone and Phenol from Cumene,* Environmental Protection Agency Report No. EPA-600/2-79-019d, 1979. Available NTIS PB80-150592.
[30] P. R. Pujado, J. R. Salazar, C. V. Berger, *Hydrocarbon Process,* **55** (1976) no. 3, 91.
[31] J. F. Knifton, J. R. Sanderson, *Appl. Catal. A* **161** (1997) 199.
[32] V. A. Galegov, J. E. Pokrovskaya, V. R. Rakhimov, *Int. Chem. Eng.* **16** (1976) no. 3, 454.

[33] Mitsui Petrochemical Industries, Ld. JP 74 46 278, 1974 (T. Akira); *Chem. Abstr.* **84** (1976) 38 444.
[34] G. Messina, L. Lorenzoni, O. Cappellazzo, A. Gamba, *Chim. Ind.* **65** (1983) no. 1, 10.
[35] J. C. Bonacci, R. M. Heck, R. K. Mahendroo, G. R. Patel, E. D. Allan, *Hydrocarbon Process.* **59** (1980) no. 11, 179.
[36] General Electric, US 5 254 75, 1993 (V. M. Zakoshansky).
[37] H. J. Kolb, R. L. Burwell, Jr., *J. Am. Chem. Soc.* **67** (1945) 1084.
[38] Standard Oil Development Co., US 2 549 844, 1951 (H. O. Mottern).
[39] Knapsack-Griesheim AG, GB 804 132, 1958.
[40] Toyo Rayon Co., Ltd., JP 68 03 163, 1968 (T. Miyata, M. Sato); *Chem. Abstr.* **70** (1969) 19 573.
[41] Engelhard Industries, GB 823 514, 1959; *Chem. Abstr.* **54** (1960) 7562d.
[42] Usines de Melle, GB 1 097 819, 1968; *Chem. Abstr.* **68** (1968) 63105.
[43] J. Smidt, H. Krekeler, *Hydrocarbon Process Pet Refiner.* **42** (1963) no. 7, 149.
[44] *Chem. Eng. (N.Y.)* **70** (Sept. 30, 1963), 48.
[45] Hoechst, US 3 149 167, 1964 (L. Hornig, E. Paszthory, R. Wimmer).
[46] T. Kunugi, T. Matsuura, S. Oguni, *Hydrocarbon Process.* **44** (1965) no. 7, 116.
[47] A. H. Olzinger, *Chem. Eng. (N.Y.)* **82** (June 9, 1975) 50.
[48] R. Marchal, *Rev Inst Fr Pet* **37** (1982) no. 3, 389.
[49] D. I .C. Wang, C. L. Coney, A. L. Demain, R. F. Gomez, A. J. Sinskey, *Degradation of Cellulosic Biomass and Its Subsequent Utilization for the Production of Chemical Feedstocks.* Report 1979, Department of Energy ET/ 20030-1; Chem. Abstr. 95 (1981) 153 746.
[50] U.S. Code of Federal Regulations 40 117.3, 1981.
[51] C. R. Fox, *Hydrocarbon Process.* **57** (1978) no. 11, 269.
[52] Rules and Regulations, County of Los Angeles Air Pollution Control District, Los Angeles, CA, Rule 66, amended Aug. 31, 1921; Rule 442, amended March 5, 1982.
[53] 40 CFR Part 51 (1998).
[54] J. A. Riddick, W. B. Bunger: *Organic Solvents,* Wiley-Interscience, New York 1970, p. 722.
[55] G. Zweig, J. Sherma (eds.): *Handbook of Chromatography,* vol. **1,** CRC Press, Cleveland, Ohio 1972, p. 56.
[56] 49 CFR 173.242, 1997.
[57] 49 CFR 172.100, 172.101, 1997.
[58] IATA Dangerous Goods Regulations 38th ed., Jan. 1997.
[59] "Acetone", SRI Consulting, *Chemical Economics Handbook*, April 1996.
[60] *Chem. Mark. Rep.*1997 (March 17), SR4.
[61] *Chem. Week,* 1997 (Aug. 27), 68.
[62] *Chem. Mark. Rep.*1996 (Aug. 5), 53.
[63] *Chem. Mark. Rep.*1996 (Jan. 22), 4.
[64] Chemical Products Synopsis 1995 (March), A Reporting Service of: Mannsville Chemical Prod. Corp, Adams, NY.
[65] *Chem. Mark. Rep.*1983 (Jan. 17), 3, 17.
[66] Asahi Chemical JP Kokai 02 188 542, 1990, (M. Furuya, H. Nakajima, Y. Fukuoka)
[67] W. J. Krasavage, J. L. O'Donoghue, G. D. DiVincenzo: *Patty's Industrial Hygiene and Toxicology,* 4th ed., vol. **2A,** Wiley-Interscience, New York 1993, p. 149.
[68] American Conference of Governmental Industrial Hygienists (ed.): *Threshold Limit Values (TLV),* Cincinnati, Ohio 1995.
[69] 29 CFR 1910, 1000, 1997.

[70] Acetone (HAZARDTEXT Hazard Management). In: Hall AH @ Rumack BH (Eds.): TOMES System Micromedex, Inc., Englewood, Colorado Edition expires 1/31/98.

[71] *Registry of Toxic Effects of Chemical Substances,* vol. **1,** U.S. Department of Health and Human Services, Washington, D.C. 1985–1986, p. 120.

[72] K. W. Nelson, J. F. Ege Jr., N. Ross, L. E. Woodman, L. Silverman, *J. Ind. Hyg. Toxicol.* **25** (1943) 282.

[73] L. Parmeggiani, C. Sassi, *Med. Lav.* **45** (1954) 431.

[74] T. Matsushita, T. Yoshea, A. Yoshimune, T. Inoue, F. Yamata, H. Suzuki, *Jpn. J. Ind. Health* **11** (1969) 477.

[75] R. L. Raleigh, W. A. McGee, *J. Occup. Med.* **14** (1972) 607.

[76] G. D. DiVincenzo, F. J. Yanno, B. D. Astill, *Am. Ind. Hyg. Assoc. J.* **34** (1973) 329.

[77] A. P. Lupulesku, D. J. Birmingham, H. Pinkus, *J. Invest. Dermatol.* **60** (1973) 33.

[78] L. C. Harris, R. H. Jackson, *Br. Med. J.* **2** (1952) 1024.

[79] S. Gitelson, A. Werczberger, J. B. Herman, *Diabetes* **15** (1966) 810.

[80] P. S. Spencer, M. C. Bischoff, H. H. Schaumburg, *Toxicol. Appl. Pharmacol.* **44** (1978) 17.

[81] G. A. Mourkides, D. C. Hobbs, R. E. Koeppe, *J. Biol. Chem.* **234** (1959) 27.

[82] T. D. Price, D. Rittenberg, *J. Biol. Chem.* **185** (1950) 449.

[83] A. A. Michurin, E. A. Sivenkov, E. N. Zilberman, T. I. Tretyakova, *J. Appl. Chem. USSR* **47** (1974) 1383.

[84] P. A. S. Smith: *Open Chain Nitrogen Compounds,* vol. **1,** W. A. Benjamin, New York 1965, p. 217.

[85] J. Hine: *Structural Effects on Equilibria in Organic Chemistry,* Wiley & Sons, New York 1975, p. 259.

[86] US 2731490, 1953 (G. Barsky).

[87] Du Pont, US 2101823, 1954 (H. R. Dittmar).

[88] American Cyanamid Co., US 2537814, 1951 (H. S. Davis).

[89] M. Salkind, E. H. Riddle, R. W. Keefer, *Ind. Eng. Chem.* **51** (1959) 1232.

[90] Nitto Chemical Co. JP-Kokai 7511020, 1975 (K. Nakai, H. Owa, S. Kezuka); *Chem. Abstr.* **84** (1976) 43369.

[91] Rohm & Haas Co., US 3052702, 1962 (H. F. Wilson).

[92] Tenneco Chemicals, US 3371107, 1968 (J. F. DeGaetano).

[93] Etat Français, BE 644178, 1964 (J. Boileau, M. Faidutti, J. P. Konrat, R. Billaz); *Chem. Abstr* **63** (1965) 11567b.

[94] [71], vol. **3a,** p. 3025.

[95] Società Italiana Serie Acetica Sintetica, BE 869056, 1978; *Chem. Abstr.* **90** (1979) 151571.

[96] A. Gronlund, B. Andersen, *Acta Chem. Scand.,* Ser. A 33 (1979) 329.

[97] Texaco, US 3181 938, 1959 (G. W. Eckert, H. Chafetz).

[98] Harvey Research Corp., US 2782197, 1957 (M. T. Harvey).

[99] T. Inoi, T. Okamoto, Y. Koizumi, *J. Org. Chem.* **31** (1966) 2700.

[100] F. C. Craven, *J. Appl. Chem.* **13** (1963) 71.

[101] *Hydrocarbon Process.* **48** (1969) no. 11, 205.

[102] R. A. Garcia, J. V. Sinistera, J. M. Marinas, *React. Kinet. Catal. Lett.* **18** (1981) 33.

[103] H. Matyschok, S. Ropuszynski, *Chem. Stosow. Ser. A* **12** (1968) 283; *Chem. Abstr.* **69** (1968) 95877.

[104] Z. N. Verkhovskaya, M. Ya. Klimenko, E. M. Zalevskaya, I. N. Bychkova, *Khim. Prom.* **43** (1967), 500; *Chem. Abstr.* **68** (1968) 29204 g.

[105] Mitsui Petrochemical Industries, JP-Kokai 80108831, 1980; *Chem. Abstr.* **95** (1981) 97067.

[106] J. Przondo, E. Bielous, I. Franek, *Przem. Chem.* **59** (1980), 436; *Chem. Abstr.* **94** (1981) 30132.

[107] [71], vol. **1,** p. 244.

Acetylene

PETER PÄSSLER, BASF Aktiengesellschaft, Ludwigshafen, Federal Republic of Germany (Chaps. 1, 3, 4.1, 4.2, 4.4.1, 5.1 (in part), 6.2, 6.3)

WERNER HEFNER, BASF Aktiengesellschaft, Ludwigshafen, Federal Republic of Germany (Chaps. 1, 3, 4.1, 4.2, 4.4.1, 5.1 (in part), 6.2, 6.3)

KLAUS BUCKL, Linde AG, Höllriegelskreuth, Federal Republic of Germany (Chaps. 2, 4.4.2, 5.1 (in part), 7)

HELMUT MEINASS, Linde AG, Höllriegelskreuth, Federal Republic of Germany (Chap. 5.1 (in part), 5.2

HANS-JÜRGEN WERNICKE, Linde AG, Höllriegelskreuth, Federal Republic of Germany (Chap. 6.1)

GÜNTER EBERSBERG, Chemische Werke Hüls AG, Marl, Federal Republic of Germany (Chaps. 4.3.1 – 4.3.3)

RICHARD MÜLLER, Chemische Werke Hüls AG, Marl, Federal Republic of Germany (Chaps. 4.3.1 – 4.3.3)

JÜRGEN BÄSSLER, Uhde GmbH, Dortmund, Federal Republic of Germany (Chap. 4.3.4)

HARTMUT BEHRINGER, Hoechst Aktiengesellschaft, Werk Knapsack, Federal Republic of Germany (Chap. 4.3.4)

DIETER MAYER, Hoechst Aktiengesellschaft, Pharma-Forschung, Toxikologie, Frankfurt, Federal Republic of Germany (Chap. 8)

1.	Introduction	118
2.	Physical Properties	119
3.	Chemical Properties........	123
3.1.	Industrially Important Reactions	124
3.2.	Other Reactions; Derivatives .	129
4.	Production	130
4.1.	Thermodynamic and Kinetic Aspects	130
4.2.	Partial Combustion Processes .	132
4.2.1.	BASF Process (Sachsse-Bartholome)	133
4.2.2.	Other Partial Combustion Processes	141
4.2.3.	Submerged Flame Process	143
4.2.4.	Partial Combustion Carbide Process...................	144
4.3.	Electrothermic Processes	146
4.3.1.	Production from Gaseous and/or Gasified Hydrocarbons (Hüls Arc Process)	147
4.3.2.	Production from Liquid Hydrocarbons (Plasma Arc Process)	153
4.3.3.	Production from Coal (Arc Coal Process)	154
4.3.4.	Production from Calcium Carbide	158
4.3.4.1.	Wet Generators	159
4.3.4.2.	Dry Generators	160
4.3.4.3.	Acetylene Purification	162
4.4.	Other Cracking Processes....	163
4.4.1.	Thermal Cracking By Heat Carriers	163
4.4.2.	Acetylene as a Byproduct of Steam Cracking............	167
5.	Safety Precautions, Transportation, and Storage..	171
5.1.	General Safety Factors and Safety Measures...........	171
5.2.	Acetylene Storage in Cylinders	177

6.	Uses and Economic Aspects	181	7.	Propyne	185
6.1.	Use in Metal Processing	181			
6.2.	Use as Raw Material in Chemical Industry	183	8.	Toxicology and Occupational Health	189
6.3.	Competitive Position of Acetylene as Chemical Feedstock	184	9.	References	190

1. Introduction

Acetylene [74-86-2] is the simplest hydrocarbon with a triple bond. In the days before oil gained widespread acceptance as the main feedstock of chemical industry, acetylene was the predominant building block of industrial organic chemistry. The calcium carbide process was the sole route for acetylene production until 1940, when thermal cracking processes using methane and other hydrocarbons were introduced. At first, these processes used an electric arc; then, in the 1950s, partial oxidation and regenerative processes were developed.

However, along with the expansion of the petroleum industry there was a changeover from coal chemistry to petrochemistry, in the 1940s in the United States and in the 1950s in Europe. As a consequence, acetylene lost its competitive position to the much cheaper and more readily available naphtha-derived ethylene and other olefins. This competition between acetylene and ethylene as feedstocks for chemical industry has been much discussed over the last 20 years [1], [2]. The few hopes, such as BASF's contribution to the submerged flame process, Hoechst's crude oil cracking (HTP), or Hüls' plasma process, have not halted the clear trend toward ethylene as a basic chemical. With the first oil price explosion in 1973, the development of crude cracking processes suffered a setback, and the new processes, such as the Kureha/Union Carbide process, DOW's PCC process (PCC = partial combustion cracking), or the Kureha/Chiyoda/Union Carbide ACR process (ACR = advanced cracking reactor), raise little hope for a comeback of acetylene chemistry. Acetylene production peaked in the United States at 480 000 t in the 1960s, and in Germany at 350 000 t in the early 1970s [3]. Since then, acetylene production has decreased steadily. In both countries the losses were principally in carbide-derived acetylene; in fact, Germany has produced acetylene for chemical purposes almost exclusively from natural gas and petrochemical sources since 1975.

Now, however, the decline in importance of acetylene as a fundamental chemical seems to be slowing. There are two main reasons for this. First, for some chemicals, such as 1,4-butanediol and special vinyl esters, the acetylene route has always been the major commercial process. Second, the oil price rises of 1973 and 1980 made naphtha, the main feedstock for olefins in Europe, much more expensive than the acetylene feedstocks natural gas and coal. As a result, syntheses with acetylene can compete with ethylene in the cases of vinyl chloride and vinyl acetate and even with propene in the case of acrylic acid [4], [5] (see Sections 6.2 and 6.3). Today the United States and Western Europe are the two largest producers of hydrocarbon-derived acetylene for chemical purposes. Other countries, especially the German Democratic Republic, India, Japan, and South Africa, still produce acetylene from calcium carbide for chemical syntheses [3].

All acetylene processes, including carbide processes, are high-temperature processes, requiring a large amount of energy. They differ essentially only in the manner in which the necessary energy is generated and transferred. They can be classified into three groups: partial combustion processes, electrothermic processes, and processes using heat carriers. Finally, the use of byproduct acetylene from olefin plants is economically viable in many cases. For each group of acetylene processes several variants have been developed using various feedstocks and techniques. Today, only three processes remain for the commercial production of acetylene: the *calcium carbide route*, in which the carbide is produced electrically, the *arc process*, and the *partial oxidation of natural gas*. Other once popular processes have become uneconomical as the price of naphtha has increased.

Some processes were shelved in the experimental or pilot-plant stage as the importance of acetylene declined. However, other new processes involving the use of coal, sulfur-containing crude oil, or residues as feedstocks for acetylene production are in the pilot-plant stage.

However, the position of acetylene in chemical industry may improve because of the variety of valuable products to which acetylene can be converted with known technology and high yields.

2. Physical Properties

Due to the carbon–carbon triple bond and the high positive energy of formation, acetylene is an unstable, highly reactive unsaturated hydrocarbon. The C–C triple bond and C–H σ bond lengths are 0.1205 and 0.1059 nm, respectively. For the electronic structure of acetylene and a molecular orbital description, see [6]. The acidity of acetylene (pK_a = 25) permits the formation of acetylides (see Section 3.2).

Under normal conditions acetylene is a colorless, nontoxic but narcotic gas; it is slightly lighter than air. The main physical properties are listed in Table 1. The critical temperature and pressure are 308.32 K and 6.139 MPa. The triple point at 128.3 kPa is 192.4 K. The vapor pressure curve for acetylene is shown in Figure 1. The formation of acetylene is strongly endothermic (ΔH_f = + 227.5 kJ/mol at 298.15 K).

Self-decomposition can be initiated when certain pressure limits above atmospheric pressure are exceeded (for details see Section 5.1).

Table 1. Physical properties of acetylene

Molecular mass	26.0379
Critical temperature	308.32 K (35.17 °C)
Critical pressure	6.139 MPa
Critical volume	0.113 m^3/kmol
Triple point	192.4 K (− 80.75 °C)
Triple point pressure	128.3 kPa
Normal sublimation point and normal boiling point	189.15 K (− 84.0 °C)
Crystal transition point	133.0 K (− 140.15 °C)
Enthalpy of transition	2.54 kJ/mol
Density	760.2 kg/m (131 K)
	764.3 kg/m^3 (141 K)
Density (liquid C_2H_2)	465.2 kg/m^3 (273.15 K)
Enthalpy of vaporization (calculated)	10.65 kJ/mol (273.15 K)
Enthalpy of sublimation	21.168 kJ/mol (5.55 K)
Enthalpy of formation	227.5±1.0 kJ/mol (298.15 K)
Gibbs free energy of formation	209.2±1.0 kJ/mol (298.15 K)
Entropy of formation	200.8 J mol^{-1} K^{-1} (298.15 K)
Enthalpy of combustion	−1255.6 kJ/mol (298.15 K)
Vapor pressure	2.6633 MPa (273.15 K)
Heat capacity (ideal gas state)	43.990 J mol^{-1} K^{-1} (298.15 K)

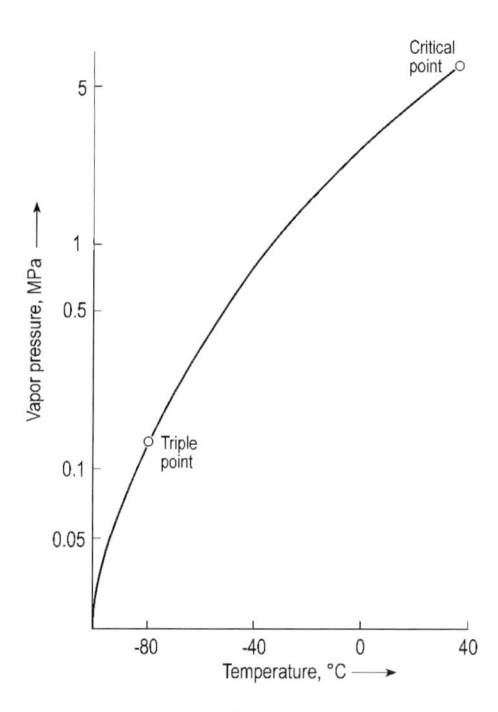

Figure 1. Vapor pressure of acetylene [7], [8]

The crystalline structure of solid acetylene changes at − 140.15 °C from a cubic to an orthorhombic phase. The heat of reaction for this phase change is 2.54 kJ/mol [9]; two different values for the enthalpy of fusion are reported in the literature [7], [8]. Figure 2 shows the density of liquid and gaseous acetylene.

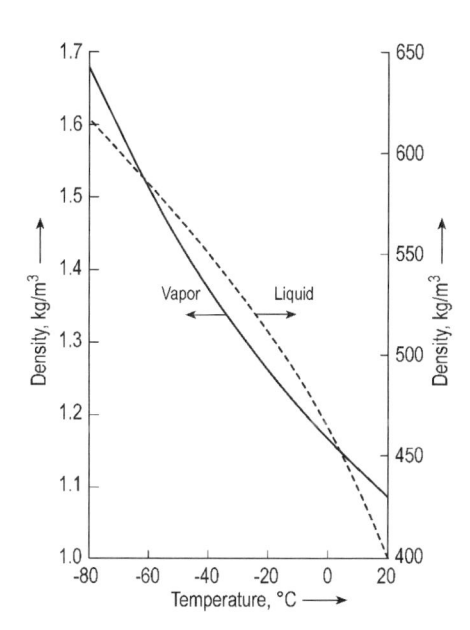

Figure 2. Density of acetylene vapor (at 1.013 bar) and liquid

Details about flame properties, decomposition, and safety measures are given in Chapter 5.

Solubility coefficients of acetylene in organic solvents are listed in Table 2 [10]. Further solubility data are available as coefficients of absorption α (20 °C, m^3 (STP) m^{-3} atm^{-1}), as solubilities (g/kg of solvent), and for different pressures (see [11]). The solubilities of acetylene at infinite dilution are shown in Figure 3 for water, methanol, DMF, and N-methyl-2-pyrrolidone (NMP) [872-50-4]. Figure 4 shows the solubility of acetylene in acetone for various partial pressures and temperatures. The heat of solution depends on the concentration of acetylene in the solvent: dissolving 0.5 kg of acetylene in 1 kg of solvent generates 293 kJ for acetone and 335 kJ for DMF. For details on the influence of water, of partial pressure, and deviations from Henry's law, see [10], [11]. The temperature dependence of the solubility of acetylene in DMF at infinite dilution is compared with those of ethylene and ethane in Figure 5 (see also [10], [12] for selectivities).

The solubility of acetylene in water at 25 °C is 0.042 mol L^{-1} bar^{-1}. Under pressure of acetylene (e.g. > 0.5 MPa at 0 °C) and at temperatures between 268 and 283 K, waxy hydrates of the composition $C_2H_2 \cdot (H_2O)_{\sim 5.8}$ are formed [10], [11]. The hydrates can block equipment; shock waves may initiate self-decomposition.

Liquid oxygen dissolves only traces of acetylene (5.5 ppm at 90 K [13]); the solubilities of ethylene and ethane in oxygen are much higher (factor of 350 and 2280, respectively). The prepurification of the process air in air separation plants with molecular sieves removes acetylene to < 1 ppb provided there is no breakthrough of carbon dioxide. This fact guarantees a safe operation of the downstream equipment [14].

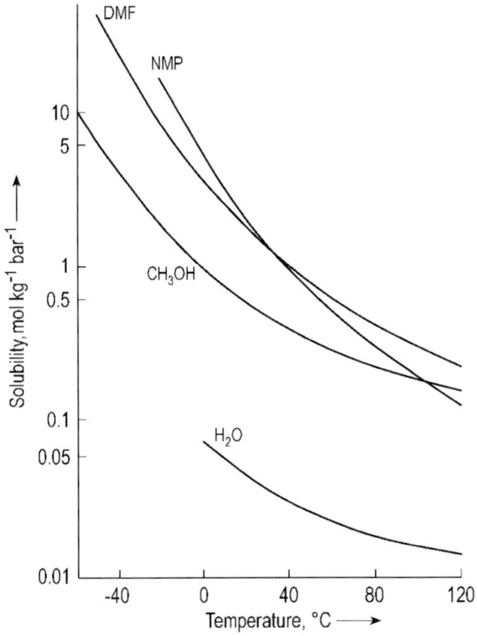

Figure 3. Solubility of acetylene in various solvents at infinite dilution

Table 2. Solubility coefficients of C_2H_2 in various solvents (in $mol\,kg^{-1}\,bar^{-1}$)

Solvent	C_2H_2 pressure, bar	$-20\,°C$	$25\,°C$
Methanol	0.98	1.979	0.569
Ethanol	0.98	0.851	0.318
n-Butanol	0.245 – 0.657		0.237
1,2-Dichlorethane	0.4-1.05	0.569	0.218
Carbon tetrachloride	0.98	0.164	0.075
n-Hexane	6.90	0.523	0.264
n-Octane	0.196 – 14.71	0.205	0.146 (0° C)
Benzene	0.98		0.225
Toluene	0.98	0.619	0.214
Xylene (tech.)	0.98	0.528	0.189
4-Methyl-1,3-dioxalan-2-one	0.98	1.137	0.350
Tri-n-butylphosphate	0-0.4	2.366	0.614
Methyl acetate	0.98	2.912	0.878
Triethylene glycol	0.98		0.205
Acetone	0.98	4.231	1.069
N-Methyl-2-pyrrolidone	0.98	5.687	1.319
N,N-Dimethylformamide	0.98	5.096	1.501
Dimethyl sulfoxide	0.98		1.001
Ammonia	0.98	7.052	2.229

Typical adsorption isotherms of acetylene are shown in Figure 6 for molecular sieves, activated carbon and silica gel at 25 °C [15]; additional information for activated carbon is summarized in [16].

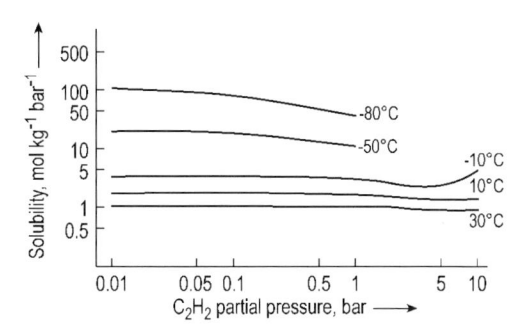

Figure 4. Solubility of acetylene in acetone [10]

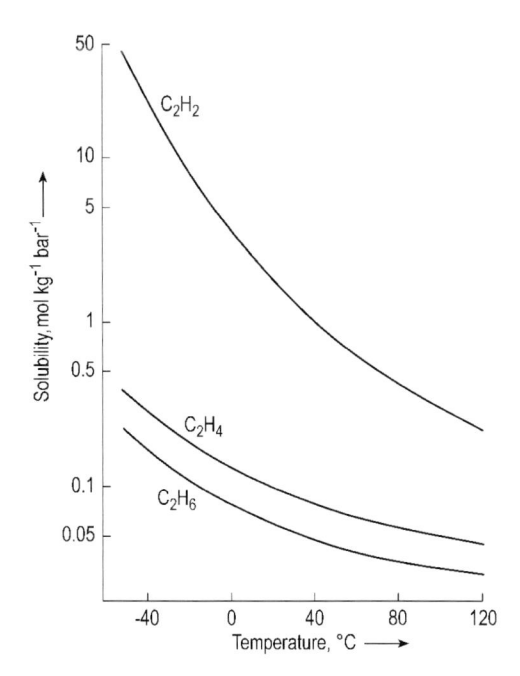

Figure 5. Solubility of C_2 hydrocarbons in DMF at infinite dilution

3. Chemical Properties

Because of its strongly unsaturated character and high positive free energy of formation, acetylene reacts readily with many elements and compounds. As a result acetylene is used as raw material for a great variety of substances. Important are addition reactions, hydrogen replacements, polymerization, and cyclization.

Acetylene is more susceptible to nucleophilic attack than, for instance, ethylene. In addition, the polarized C-H bond makes acetylene acidic (pK_a=25) [17]. Because of this acidity, acetylene is very soluble in basic solvents [18], [19], forming hydrogen bonds with them [20]. Therefore, the vapor pressures of such solutions cannot be described by Raoult's law [21].

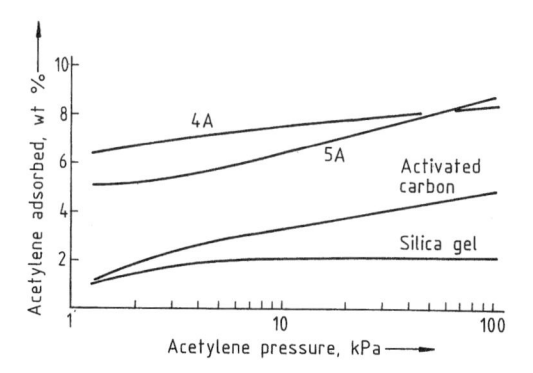

Figure 6. Adsorption isotherms for acetylene on 4A and 5A molecular sieves, activated carbon, and silica gel at 25 °C [15]

The development of the acetylene pressure reactions by W. REPPE (1892–1969), BASF Ludwigshafen (Federal Republic of Germany) [22], [23], [24] began modern acetylene chemistry. The most interesting groups of reactions are vinylation, ethynylation, carbonylation, and cyclic and linear polymerization.

3.1. Industrially Important Reactions

Vinylation Reactions and Products [25]. Vinylation is the addition of compounds with a mobile hydrogen atom, such as water, alcohols, thiols, amines, and organic and inorganic acids, to acetylene to form vinyl compounds chiefly used for polymerization.

The two types of vinylation reactions are *heterovinylation* and the less usual *C vinylation*. In the former, the hydrogen atom originates from the heteroatoms O, S, and N, whereas C vinylation occurs when the mobile hydrogen atom is directly bound to a carbon atom. Examples of C vinylation are dimerization and trimerization of acetylene, the synthesis of acrylonitrile from acetylene and hydrogen cyanide, and the addition of acetylene to unsaturated hydrocarbons with activated hydrogen atoms, such as cyclopentadiene, indene, fluorene, and anthracene.

The first industrial vinylation products were acetaldehyde, vinyl chloride, and vinyl acetate. Many other products followed.

Some examples of industrial vinylation processes are given below:

Acetaldehyde [75-07-0] (→ Acetaldehyde):

$$HC \equiv CH + H_2O \longrightarrow CH_3CHO$$

Catalyst: acidic solutions of mercury salts, such as $HgSO_4$ in H_2SO_4. Liquid-phase reaction at 92 °C.

Vinyl chloride [75-01-4] (→ Chlorinated Hydrocarbons):

$$HC \equiv CH + HCl \longrightarrow CH_2 = CHCl$$

Catalyst: $HgCl_2$ on coal. Gas-phase reaction at 150–180 °C.
Vinyl acetate [108-05-04] (→ Vinyl Esters):

$$HC\equiv CH + CH_3COOH \longrightarrow CH_2=CHOOCCH_3$$

Catalyst: cadmium, zinc, or mercury salts on coal. Gas-phase reaction at 180–200 °C.
Vinyl ethers (→ Vinyl Ethers), conjectured reaction steps:

$$ROH + KOH \xrightarrow{-H_2O} ROK \xrightarrow{C_2H_2} RO-CH=CHK$$
$$RO-CH=CHK + ROH \rightarrow RO-CH=CH_2 + ROK$$

where R is an alkyl group. Reaction temperature of 120–150 °C; pressure high enough to avoid boiling the alcohol used, e.g., 2 MPa with methanol to produce methyl vinyl ether (acetylene pressure reaction).
Vinyl phenyl ether [766-94-9], vinylation with KOH catalyst:

$$HC\equiv CH + C_6H_5OH \longrightarrow C_6H_5O-CH=CH_2$$

Vinyl sulfides, KOH catalyst:

$$HC\equiv CH + RSH \longrightarrow CH_2=CH\text{-}S\text{-}R$$

Vinyl esters of higher carboxylic acids:

$$HC\equiv CH + R\text{-}COOH \longrightarrow RCOO\text{-}CH=CH_2$$

Catalyst: zinc or cadmium salts. Liquid-phase reaction.
Vinyl amines, vinylation with zinc or cadmium compounds as catalyst:

$$R^1R^2NH + HC\equiv CH \longrightarrow R^1R^2N\text{-}CH=CH_2$$

where R^1 and R^2 are alkyl groups.
N-Vinylcarbazole [1484-13-5], vinylation of carbazole in a solvent, e.g., *N*-methylpyrrolidone, at 180 °C.
Vinylation of ammonia, complex Co and Ni salts as catalysts, reaction temperature of 95 °C:

$$4\ HC\equiv CH + 4\ NH_3 [\longrightarrow 4\ CH_2=CH\text{-}NH_2]$$
$$\longrightarrow \text{(5-ethyl-2-methylpyridine)} + 3\ NH_3$$

Vinylation of acid amides, potassium salt of the amide as catalyst:

$$HC\equiv CH + RCO\text{-}NH_2 \longrightarrow RCO\text{-}NH\text{-}CH=CH_2$$

N-Vinyl-2-pyrrolidone [88-12-0], vinylation of 2-pyrrolidone with the potassium salt of the pyrrolidone as catalyst.

Acrylonitrile [*107-13-1*], C-vinylation of HCN in aqueous hydrochloric acid with CuCl and NH$_4$Cl catalyst:

$$HC \equiv CH + HCN \longrightarrow H_2C=CH\text{-}CN$$

Ethynylation Reactions and Products [26]. Ethynylation is the addition of carbonyl compounds to acetylene with the triple bond remaining intact. REPPE found that heavy metal acetylides (see Section 3.2), especially the copper(I) acetylide of composition Cu$_2$C$_2 \cdot$ 2 H$_2$O $\cdot$ 2 C$_2$H$_2$, are suitable catalysts for the reaction of aldehydes with acetylene. Alkaline catalysts are more effective than copper acetylide for the ethynylation of ketones. The generalized reaction scheme for ethynylation is:

$$HC \equiv CH + RCOR' \longrightarrow HC \equiv C\text{-}C(OH)RR'$$

where R and R' are alkyl groups or H.

The most important products from ethynylation are propargyl alcohol and butynediol.

Propargyl alcohol, 2-propyn-1-ol [*107-19-7*] ($\rightarrow$ Alcohols, Aliphatic):

$$CH \equiv CH + HCHO \longrightarrow HC \equiv CCH_2OH$$

Catalyst: Cu$_2$C$_2 \cdot$ 2 H$_2$O $\cdot$ 2 C$_2$H$_2$.

Butynediol, 2-butyne-1,4-diol [*110-65-6*] ($\rightarrow$ Butanediols, Butenediols, and Butynediol):

$$HC \equiv CH + 2\ HCHO \longrightarrow HOCH_2C \equiv CCH_2OH$$

Catalyst: Cu$_2$C$_2 \cdot$ 2 H$_2$O $\cdot$ 2 C$_2$H$_2$.

Other examples of ethynylation are the reactions of aminoalkanol and secondary amines with acetylene:

$$HC \equiv CH + (CH_3)_2N\text{-}CH_2OH \xrightarrow{-H_2O} (CH_3)_2N\text{-}CH_2\text{-}C \equiv CH$$

$$HC \equiv CH + 2\ (CH_3)_2N\text{-}CH_2OH \xrightarrow{-2\ H_2O}$$
$$(CH_3)_2N\text{-}CH_2\text{-}C \equiv C\text{-}CH_2\text{-}N(CH_3)_2$$

$$R^1R^2NH \xrightarrow{C_2H_2} R^1R^2N\text{-}CH=CH_2 \xrightarrow{C_2H_2}$$
$$R^1R^2N\text{-}\underset{\underset{CH_3}{|}}{CH}\text{-}C \equiv CH$$

Carbonylation Reactions and Products [27]. Carbonylation is the reaction of acetylene and carbon monoxide with a compound having a mobile hydrogen atom, such as water, alcohols, thiols, or amines. These reactions are catalyzed by metal carbonyls, e.g., nickel carbonyl, Ni(CO)$_4$ [*13463-39-3*]. Instead of metal carbonyls, the halides of metals that can form carbonyls can also be used.

Acrylic acid [*79-10-7*] ($\rightarrow$ Acrylic Acid and Derivatives):

$$HC \equiv CH + CO + H_2O \longrightarrow CH_2=CH\text{-}COOH$$

The reaction of acetylene with water or alcohols and carbon monoxide using Ni(CO)$_4$ catalyst was first reported by W. Reppe [27]. If water is replaced by thiols, amines, or carboxylic acids, then thioesters of acrylic acid, acrylic amides, or carboxylic acid anhydrides are obtained.

Ethyl acrylate [140-88-5] (→ Acrylic Acid and Derivatives):

$$4\,C_2H_2 + 4\,C_2H_5OH + Ni(CO)_4 + 2\,HCl \longrightarrow 4\,CH_2\!=\!CHCOOC_2H_5 + H_2 + NiCl_2 \quad (1)$$

$$C_2H_2 + C_2H_5OH + CO \longrightarrow CH_2\!=\!CHCOOC_2H_5 \quad (2)$$

Catalyst: nickel salts. Reaction temperature: 30–50 °C. The process starts with the stoichiometric reaction (1); afterwards, most of the acrylate is formed by the catalytic reaction (2). The nickel chloride formed in the stoichiometric reaction (1) is recovered and recycled for carbonyl synthesis.

Hydroquinone [123-31-9] is formed in a suitable solvent, e.g., dioxane, at 170 °C and 70 MPa [28]. The catalyst is Fe(CO)$_5$:

$$2\,HC\!\equiv\!CH + 3\,CO + H_2O \longrightarrow \underset{\text{(hydroquinone)}}{C_6H_4(OH)_2} + CO_2$$

Hydroquinone is formed at 0–100 °C and 5–35 MPa if a ruthenium carbonyl compound is used as catalyst [29]:

$$2\,HC\!\equiv\!CH + 2\,CO + H_2 \longrightarrow C_6H_4(OH)_2$$

Bifurandiones: The reaction of acetylene and CO in the presence of octacarbonyldicobalt, (CO)$_3$Co–(CO)$_2$–Co(CO)$_3$ [10210-68-1], forms a cis–trans mixture of bifurandione. The reaction is carried out under pressure (20–100 MPa) at temperatures of about 100 °C [30]:

$$2\,HC\!\equiv\!CH + 4\,CO \longrightarrow \text{cis- and trans-bifurandione}$$

New aspects of such CO insertion reactions have been reported [31].

Cyclization and Polymerization of Acetylene. In the presence of suitable catalysts, acetylene can react with itself to form cyclic and linear polymers.

Cyclization was first observed by Berthelot, who polymerized acetylene to a mixture of aromatic compounds including benzene and naphthalene. In 1940, Reppe synthesized 1,3,5,7-cyclooctatetraene [*629-20-9*] with a 70% yield at an only slightly elevated pressure:

$$4\ HC\equiv CH \longrightarrow \text{[cyclooctatetraene]} + \text{byproducts}$$

Reaction temperature of 65–115 °C, pressure of 1.5–2.5 MPa, Ni(CN)$_2$ catalyst.

The reaction is carried out in anhydrous tetrahydrofuran. The byproducts are mostly benzene (about 15%), chain oligomers of acetylene of the empirical formulas $C_{10}H_{10}$ and $C_{12}H_{12}$, and a black insoluble mass, called niprene after the nickel catalyst.

If dicarbonylbis(triphenylphosphine)nickel [*13007-90-4*], Ni(CO)$_2$[(C$_6$H$_5$)$_3$P]$_2$, is used as catalyst, the cyclization products are benzene (88% yield) and styrene (12% yield). The reaction is carried out in benzene at 65–75 °C and 1.5 MPa [32], [33].

Linear polymerization of acetylene occurs in the presence of a copper (I) salt such as CuCl in hydrochloric acid. Reaction products are vinylacetylene, divinylacetylene, etc. [34]:

$$HC \equiv CH + HC \equiv CH \longrightarrow H_2C{=}CH{-}C\equiv CH$$

A particular polymerization product, known as *cuprene*, is formed when acetylene is heated to 225 °C in contact with copper sponge. Cuprene is chemically inert, corklike in texture, and yellow to dark brown.

Polyacetylene [35], [36] is formed with Ziegler–Natta catalysts, e.g., a mixture of triethylaluminum, Al(C$_2$H$_5$)$_3$, and titanium tetrabutoxide, Ti(*n*-OC$_4$H$_9$)$_4$, at 10^{-2} to 1 MPa:

$$n\ C_2H_2 \xrightarrow{t>100°C} \text{trans-polyacetylene}$$
$$n\ C_2H_2 \xrightarrow{t<-75°C} \text{cis-polyacetylene}$$

Polymerization can be carried out in an auxiliary inert liquid, such as an aliphatic oil or petroleum ether. The monomer can also be copolymerized in the gas phase.

Polyacetylene is a low-density sponge-like material consisting of fibrils with diameters of 20–50 nm. The ratio *cis*- to *trans*-polyacetylene depends on the reaction temperature.

Polyacetylene doped with electron acceptors (I$_2$, AsF$_5$), electron donors (Na, K), or protonic dopants (HClO$_4$, H$_2$SO$_4$) is highly conductive and has the properties of a one-dimensional metal [36].

3.2. Other Reactions; Derivatives

Metal Acetylides [37]. The hydrogen atoms of the acetylene molecule can be replaced by metal atoms (M) to yield metal acetylides. Alkali and alkaline-earth acetylides can be prepared via the metal amide in anhydrous liquid ammonia:

$$C_2H_2 + MNH_2 \longrightarrow MC_2H + NH_3$$

The direct reaction of the acetylene with a molten metal, such as sodium, or with a finely divided metal in an inert solvent, such as xylene, tetrahydrofuran, or dioxane, at a temperature of about 40 °C, is also possible:

$$2\,M + C_2H_2 \longrightarrow M_2C_2 + H_2$$

The very explosive copper acetylides, e.g., $Cu_2C_2 \cdot H_2O$, can be obtained by reaction of copper(I) salts with acetylene in liquid ammonia or by reaction of copper(II) salts with acetylene in basic solution in the presence of a reducing agent such as hydroxylamine. Copper acetylides can also form from copper oxides and other copper salts. For this reason copper plumbing should be avoided in acetylene systems.

Silver, gold, and mercury acetylides, which can be prepared in a similar manner, are also explosive.

In sharp contrast to the highly explosive $Cu_2C_2 \cdot H_2O$, the catalyst used for the synthesis of butynediol, $Cu_2C_2 \cdot 2\,H_2O \cdot 2\,C_2H_2$, is not as sensitive to shock or ignition.

Halogenation. The addition of chlorine to acetylene in the presence of $FeCl_3$ yields 1,1,2,2-tetrachloroethane [79-34-5], an intermediate in the production of the solvents 1,2-dichloroethylene [540-59-0], trichloroethylene [79-01-6], and perchloroethylene [127-18-4].

Bromine and iodine can also be added to acetylene. The addition of iodine to acetylene stops with formation of 1,2-diiodoethylene.

Hydrogenation. Acetylene can be hydrogenated, partly or completely, in the presence of Pt, Pd, or Ni catalysts, giving ethylene or ethane.

Organic Silicon Compounds [38], [39]. The addition of silanes, such as $HSiCl_3$, can be carried out in the liquid phase using platinum or platinum compounds as catalysts:

$$HC \equiv CH + HSiCl_3 \longrightarrow CH_2 = CH\text{-}SiCl_3$$

Oxidation. At ambient temperature acetylene is not attacked by oxygen; however, it can form explosive mixtures with air or oxygen (see Chap. 5). The explosions are initiated by heat or ignition. With oxidizing agents such as ozone or chromic acid, acetylene gives formic acid, carbon dioxide, and other oxidation products. The reaction of acetylene with dilute ozone yields glyoxal.

Hydrates. At temperatures below ca. 15 °C, under pressure, hydrates of the composition $C_2H_2 \cdot 6\,H_2O$ are formed (see Section 2).

Chloroacetylenes [40]. *Monochloroacetylene,* $HC\equiv CCl$, M_r 60.49, *bp* −32 to −30 °C, a gas with nauseating odor that irritates the mucous membranes, is obtained by reaction of 1,2dichloroethylene with alcoholic NaOH in the presence of $Hg(CN)_2$. It ignites in the presence of traces of oxygen. In air it explodes violently. Chloroacetylene is very poisonous.

Dichloroacetylene, $ClC\equiv CCl$, M_r 94.93, *mp* - 66 to - 64.2 °C, a colorless oil of unpleasant odor, explodes in the presence of air or on heating. It is obtained from acetylene in strongly alkaline potassium hypochlorite solution [41] or by reaction of trichloroethylene vapor with caustic alkali.

4. Production

4.1. Thermodynamic and Kinetic Aspects

The production of acetylene from hydrocarbons, e.g.,

$$2\ CH_4 \rightleftharpoons C_2H_2 + 3\ H_2 \qquad \Delta H\ (298\ K) = 376.4\ kJ/mol$$

requires very high temperatures and very short reaction times. The main reasons for the extreme conditions are the temperature dependence of the thermodynamic properties (molar enthalpy of formation, ΔH_f, and molar free energy of formation, ΔG_f) of the hydrocarbons; the position of the chemical equilibria under the reaction conditions; and the kinetics of the reaction.

Thermodynamic data relevant to the hydrocarbon – acetylene system are shown in Table 3 and Figure 7. These data show clearly that at normal temperatures acetylene is highly unstable compared to the other hydrocarbons. However, Figure 7 also shows that the free energy of acetylene decreases as temperature increases, whereas the free energies of the other hydrocarbons increase. Above about 1230 °C, acetylene is more stable than the other hydrocarbons. The temperature at which the acetylene line intersects an other line in Figure 7 is higher the shorter the chain length of the hydrocarbons. Acetylene production from methane requires higher reaction temperatures than production from heavier hydrocarbons.

The equilibrium curve for the methane reaction as a function of temperature (Fig. 8) shows that acetylene formation only becomes apparent above 1000 K (730 °C). Therefore, a very large energy input, applied at high temperature, is required.

However, even at these high temperatures acetylene is still less stable than its component elements, carbon und hydrogen (see Fig. 7). In fact, the large difference in free energy between acetylene and its component elements favors the decomposition of acetylene to carbon and hydrogen up to temperatures of about 4200 K.

$$C_2H_2 \longrightarrow 2\ C(s) + H_2\ (g) \qquad -\Delta G_f\ (298\ K) = -209.3\ kJ/mol$$

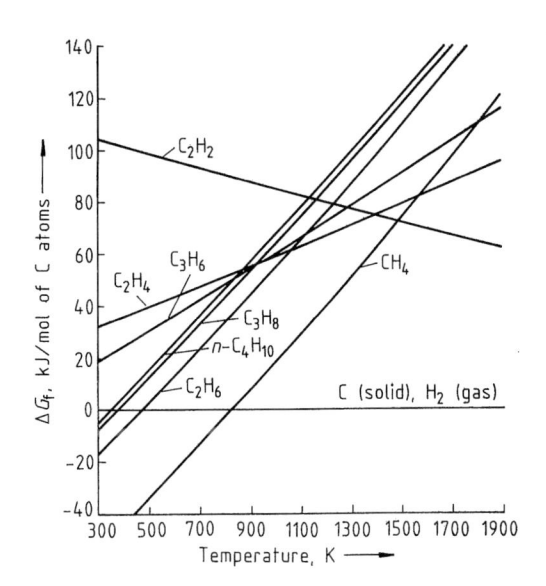

Figure 7. Gibbs free energy of formation per carbon atom of several hydrocarbons as a function of temperature

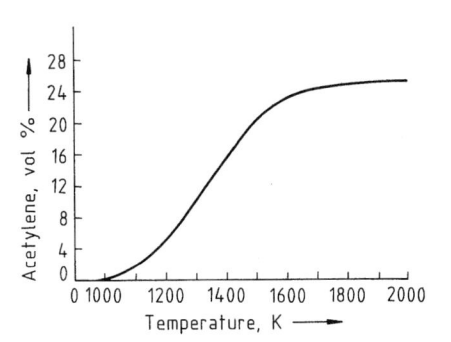

Figure 8. Equilibrium curve for the methane cracking reaction, $2\,CH_4 \rightleftharpoons C_2H_2 + 3\,H_2$

Thus cracking and recombination of the hydrocarbons and decomposition of acetylene compete. To achieve reasonable acetylene yields and to avoid the thermodynamically favorable decomposition into the elements, rapid quenching of acetylene produced in the cracking reaction is necessary. In practice, the residence time at high temperature is between 0.1 and 10 ms.

Higher temperatures also increase the rate of conversion of acetylene to byproducts. Again, the residence time must be sufficiently short to prevent this.

In the case of cracking by partial oxidation, the combustion reaction of the hydrocarbon supplies the energy necessary for the production of acetylene from the other part of the hydrocarbon feed:

$CH_4 + O_2 \longrightarrow CO + H_2 + H_2O \qquad \Delta H\,(298\,K) = -277.53\ kJ/mol$

$CO + H_2O \longrightarrow CO_2 + H_2 \qquad \Delta H\,(298\,K) = -41.19\ kJ/mol$

Table 3. Standard molar enthalpies of formation and Gibbs free energy of formation at 298 K.

	ΔH_f (kJ/mol)	ΔG_f (kJ/mol)
C (s)	0	0
H_2 (g)	0	0
CH_4 (g)	− 74.81	− 50.82
C_2H_2 (g)	+226.90	+209.30
C_2H_4 (g)	+ 52.30	+ 68.15
C_2H_6 (g)	− 84.64	− 32.90
C_3H_6 (g)	+ 20.43	+ 62.75
C_3H_8 (g)	−103.90	− 23.48
n-C_4H_{10} (g)	−126.11	− 17.10

From these reaction enthalpies, the amount of oxygen needed to produce the high reaction temperature can be calculated. Therefore, in addition to the short residence time, the correct methane : oxygen ratio, which also determines the reaction temperature, is essential to obtain good acetylene yields.

4.2. Partial Combustion Processes

In this group of processes, part of the feed is burnt to reach the reaction temperature and supply the heat of reaction. The necessary energy is produced where it is needed. Almost all carbon-containing raw materials can be used as feedstocks: methane, ethane, natural gas liquids (NGL), liquefied petroleum gas (LPG), naphtha, vacuum gas oil, residues, and even coal or coke. Natural gas is especially suitable because it is available in many parts of the world and because its only other uses are for heating and for the production of synthesis gas. Only under the conditions of acetylene synthesis can methane be transformed into another hydrocarbon in a single process step, and this is the essential reason for using the thermodynamically unfavorable acetylene synthesis.

The partial combustion processes for light hydrocarbons, from methane to naphtha, all follow similar schemes. The feed and a certain amount of oxygen are preheated separately and introduced into a burner. There they pass through a mixing zone and a burner block into the reaction zone, where they are ignited. On leaving the reaction zone the product mixture is cooled rapidly, either by water or oil. Cooling by water is easier, and more common, but it is thermally less efficient than cooling by oil. Alternatively, the gases can be cooled with light hydrocarbon liquids, which leads to additional acetylene and ethylene formation between 1500 and 800 °C. These processes are usually called *two-step processes*.

Burner design is very important for all partial combustion processes. The residence time of the gas in the reaction zone must be very short, on the order of a few milliseconds, and it should be as uniform as possible for all parts of the gas. Flow velocity within the reaction zone is fixed within narrow limits by the requirements of high yield and the avoidance of preignition, flame separation from the burner block, and coke depositions. A survey of the processes operating according to these principles is given in [9], [42]. Only the BASF process is described here in detail, because it is the most widely used process for the partial combustion of natural gas.

The *submerged flame process*, SFP, was developed by BASF with the aim of producing acetylene from crude oil or its heavy fractions, and thus to be independent of the more expensive refined oil products used in olefin chemistry. One unit of this kind was built in Italy, but it became uneconomic and was shut down after a year of operation [43]. Nevertheless, the process is described in some detail below because of its simple cracking section, because of the simultaneous formation of acetylene and ethylene, and because of its high thermal efficiency and its high degree of carbon conversion (perhaps of even greater importance in the future).

The *partial combustion carbide process*, also developed by BASF, uses coke, oxygen, and lime as feed. It was developed in the 1950s to reestablish the competitive position of carbide in the face of the new acetylene processes on a petrochemical – natural gas basis. Some attention is given here to the basics of this process, although it has never gone beyond the pilot-plant stage. When petrochemical feedstocks become scarce, this process may have a place in a future coalbased chemistry because it has a higher degree of carbon conversion and a higher thermal efficiency than the electric carbide process.

All these acetylene processes based on partial combustion yield a number of by-products, such as hydrogen and/or carbon monoxide, which may cause problems if acetylene is the only product desired. Within a complex chemical plant, however, these may be converted to synthesis gas, pure hydrogen, and pure CO and can actually improve the economics of acetylene production.

4.2.1. BASF Process (Sachsse-Bartholome)

The BASF process for the production of acetylene from natural gas has been known since 1950 [44]. Worldwide, some 13 plants used this process in 1983, a total capacity of about 400 000 t/a. All use a water quench, except the plant in Ludwigshafen (Federal Republic of Germany) operated with an oil quench [45].

The basic idea of partial combustion involves a flame reaction on a premixed feed of hydrocarbon and oxygen. In this way the rate of hydrocarbon conversion is made independent of the gas-mixing rate, which is governed by diffusion. Only then can the residence time in the reaction zone be made much smaller than the average decay time of acetylene. The separate preheating of the reactants to the highest temperature possible before introduction into the burner reduces the consumption of oxygen and the hydrocarbon within the burner. It also causes a higher flame propagation speed and therefore a higher mass flow within the acetylene burner.

The smallest, but most important, part of a partial oxidation acetylene plant is the burner, Figure 9. Its design is nearly identical in the two process variants (i.e., oil and water quench).

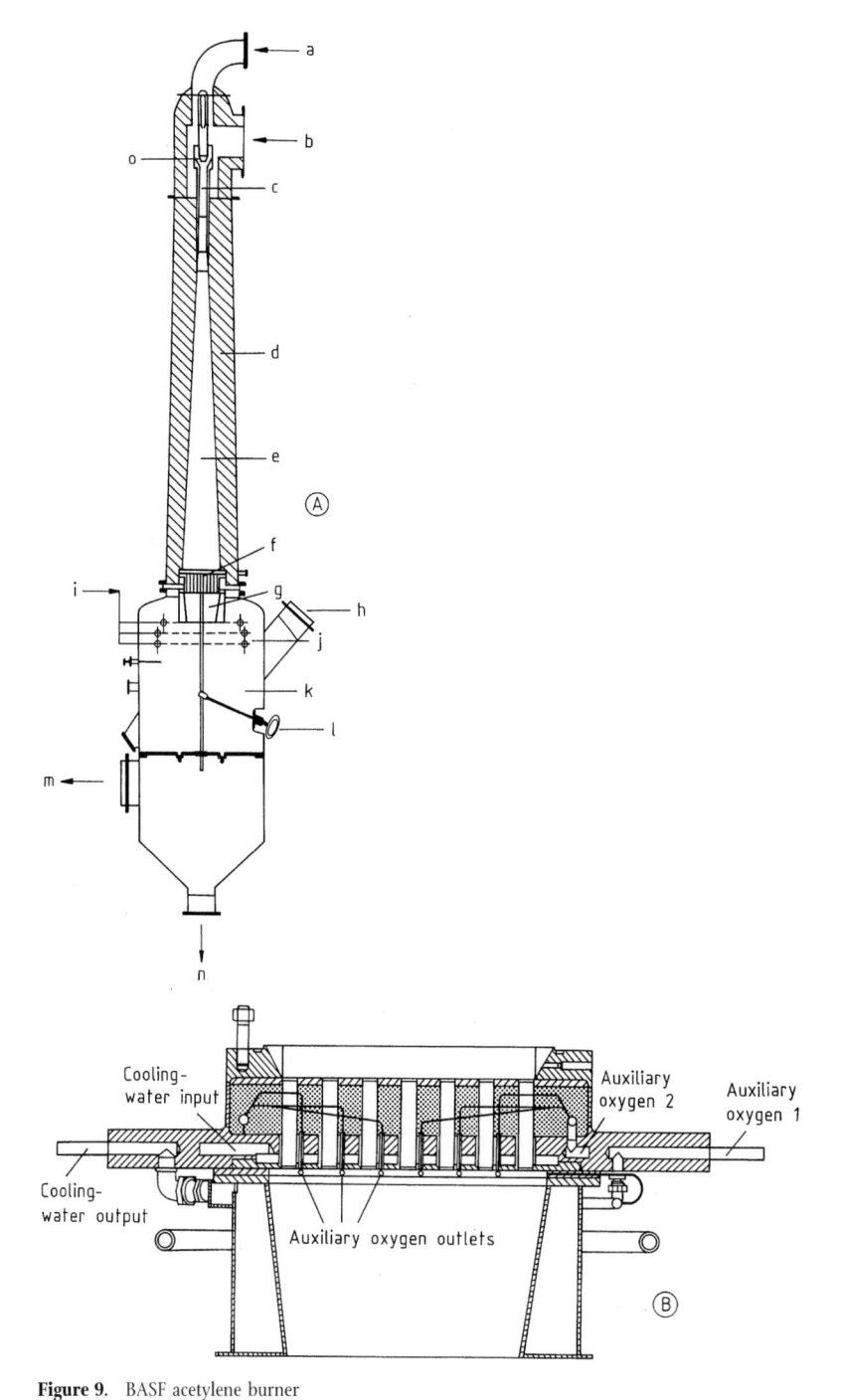

Figure 9. BASF acetylene burner
A) The burner: a) Oxygen: b) Hydrocarbon; c) Mixer; d) Concrete lining; e) Diffuser; f) Burner block; g) Reaction chamber; h) Rupture disk; i) Quench-medium inlet; j) Quench rings; k) Quench chamber; l) Manual scraper; m) Cracked-gas outlet; n) Quench-medium outlet. B) The burner block

At the top of the burner, the preheated reactants, (600 °C in the case of methane) must be mixed (c) so rapidly that there are no domains with a high oxygen concentration. Such domains cause preignition before the reactants are introduced into the reaction zone (g). In fact, the reaction mixture ignites after an induction time depending on the hydrocarbon used as feed and on the preheat temperature, on the order of a few tenths of a second. The maximum preheat temperature is lower for higher hydrocarbons than for methane. Backmixing of the gas between the mixing and the reaction zones is avoided by the diffuser (e), a tube which connects the mixing zone and the burner block (f). Because of its smooth surface and the small opening angle the reaction feed is decelerated gently and backmixing does not occur. The burner block (f) consists of a water-cooled steel plate with a large number of small channels. The flow velocity through these channels is substantially higher than the flame propagation speed, so that the flame below the burner block cannot backfire into the diffuser. The lower side of the burner block has small openings between the channels through which additional oxygen is fed into the reaction mixture. At these openings small flames form and initiate the flame reaction. The strong turbulence below the burner block stabilizes the flame.

Under unfavorable conditions the flame may appear above the burner block. In this case the oxygen feed must be shut off immediately and replaced by nitrogen. This extinguishes the preignition before it can cause any damage to the equipment. Such preignitions can result from a momentary shift in the oxygen : hydrocarbon ratio or the entrainment of small particles of pyrophoric iron formed from rust in the preheaters.

As mentioned above, the hot gas leaves the reaction chamber within a few milliseconds and passes through sprays of water or oil, which cool the gas almost instantaneously, to about 80 °C in the case of water or 200–250 °C in the case of oil. The quench system consists of a set of nozzles that are fed by three annular tubes below the reaction chamber.

The concentrations of the major constituents of the cracked gas depend on the oxygen : hydrocarbon ratio in the feed as shown in Figure 10. As the oxygen supply is increased, the acetylene concentration increases until it passes through a smooth maximum. At the same time there is an increase in the volume of the cracked gas. Thus maximum acetylene production is attained when a little more oxygen is used than the amount required for maximum acetylene concentration in the cracked gas. This is clear from the consumption of natural gas per ton of acetylene produced and the reduction in unconverted methane. When the oxygen : hydrocarbon ratio is too low, the reaction time is insufficient for complete conversion of oxygen, and the cracked gas contains free oxygen. Free oxygen can be tolerated only up to a certain concentration. When the oxygen : hydrocarbon ratio is too high, the increased velocity of flame propagation exceeds the flow velocity in the channels of the burner block, leading to preignitions.

Coke deposits in the reaction chamber have to be removed from time to time with a manual or an automatic scraper. Normally, a burner produces 25 t of acetylene per day from natural gas and 30 t per day from liquid feedstocks.

Acetylene Water Quench Process (AWP), Soot Removal (Figure 11). After quenching with water the cracked gas leaves the burner (b) at 80 – 90 °C. A certain amount of soot is formed in the reaction chamber in spite of the very short reaction time. When natural gas is used as a feedstock, the soot is 50 kg per ton of acetylene,

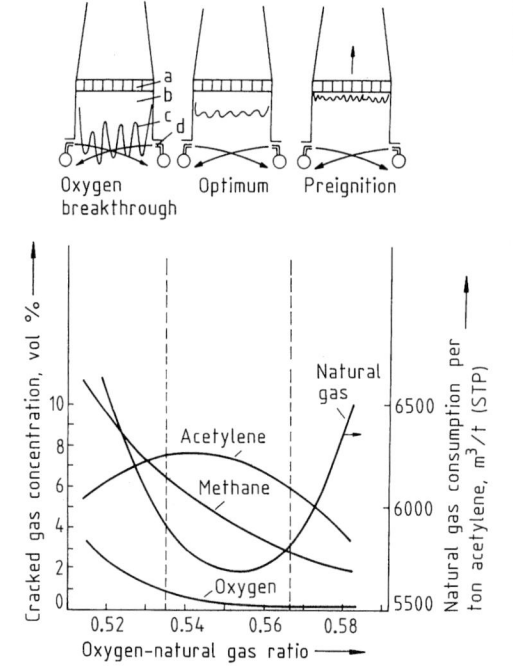

Figure 10. Burner characteristics
a) Burner block; b) Reaction chamber; c) Flame front; d) Quench-medium inlet

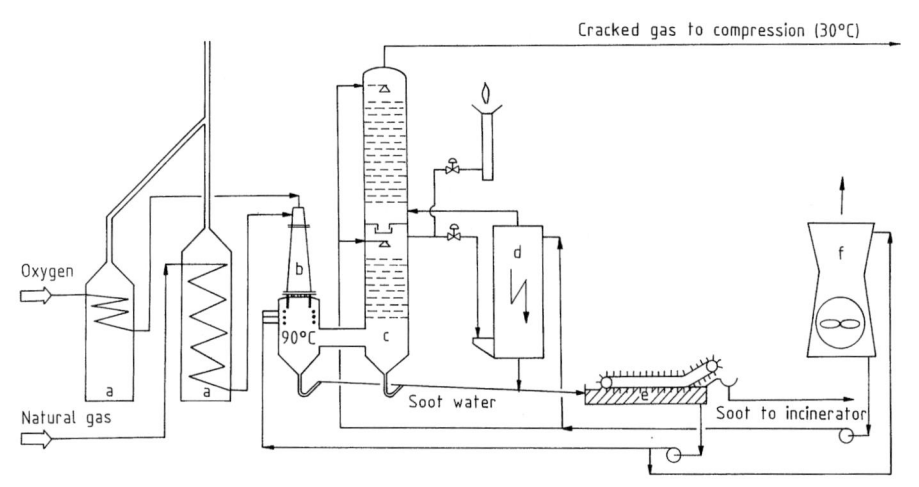

Figure 11. Acetylene water quench process (AWP)
a) Preheaters; b) Acetylene burner; c) Cooling column; d) Electrofilter; e) Soot decanter; f) Cooling tower

with LPG feedstock it is 250 kg, and with naphtha it is 350 kg. The soot is partly removed from the gas by the quench, then by washing with recirculated water in a cooling column (c), and by passing the gas through an electrofilter (d). After cooling and soot removal, the gas has a pressure slightly above atmospheric, a temperature of

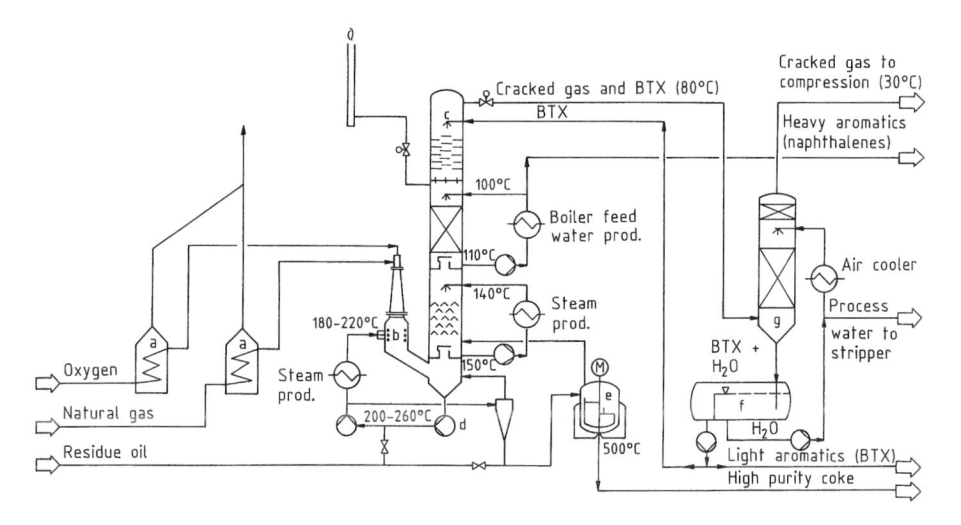

Figure 12. Acetylene oil quench process (AOP)
a) Preheaters; b) Acetylene burner; c) Burner column; d) Mill pump; e) Coker; f) Decanter; g) Final cooler

about 30 °C, and a soot content of about 1 mg/m^3. The water effluents from the quench system, the cooling column, and the electrofilter carry the washed-out soot. Some gas remains attached to the soot, causing it to float when the soot-containing water flows slowly through basin decanters (e). The upper soot layer, which contains 4–8 wt% of carbon, depending on the feedstock, is scraped off the water surface and incinerated.

Acetylene Oil Quench Process (AOP), Soot Removal (Figure 12). In this process the cracked gas is quenched with oil sprays and leaves the burner at 200–250 °C. The oil absorbs the heat from the gas and then passes through waste heat boilers before returning to the quench. The sensible heat of the cracked gas represents more than 15% of the heating value of the feedstock. The pressure of the generated steam depends on the process configuration and can reach 15 bar (1.5 MPa).

Unlike the water quench process, where the scraped coke deposits sink to the bottom of the quench chamber and are easily removed, in the oil quench the coke deposits do not settle immediately. In order to prevent plugs in the quench nozzles a mill pump (d) is installed immediately underneath the burner column. The coke and soot content in the quench circuit is kept near 25% by sending a fraction of the coke-containing oil to externally heated, stirred kettles (coker (e)). In the kettles the volatile matter evaporates very quickly, leading to fluidization of the coke bed. The vapor is returned to the burner column, while the soot is agglomerated. A fine-grained coke is withdrawn from the bottom of the coker.

Because of the cracking losses in the quench a certain amount of quench oil has to be added continuously to the process. This makeup oil is at least 0.15 to 0.3 t per ton of acetylene, depending on the stability of the oil used. When residual oil from steam crackers is used, it can be desirable to add

Table 4. BASF acetylene oil quench process, cracked gas composition (vol%)

Component *	Raw material (ΔH, kJ/mol)		
	Methane (400)	LPG (325)	Naphtha (230)
H_2	56.5	46.4	42.7
CH_4	5.2	5.0	4.9
C_2H_4	0.3	0.4	0.5
C_2H_2	7.5	8.2	8.8
C_{3+} **	0.5	0.6	0.7
CO	25.8	35.0	37.9
CO_2	3.2	3.4	3.5
O_2	0.2	0.2	0.2
Inerts		balance	

* Dry gas, water, and aromatic compounds condensed out;
** Hydrocarbons with three or more carbon atoms

up to 1 t of oil per ton of acetylene, because the excess oil is partially converted to light aromatic hydrocarbons.

The cracked gas leaving the quench is cooled in a burner column (c), where there are additional oil circuits for the production of 3-bar steam and for boiler feedwater preheat. At the top of the column a small amount of a low-boiling oil (BTX = benzene, toluene, and xylene) is added to prevent deposit-forming aromatics (mainly naphthalene) from passing downstream into other parts of the plant. The cracked gas, which has to be compressed before separation, is cooled further (g) by water. At this stage most of the BTX condenses and is separated from the water in a large decanter (f).

Table 4 shows the cracked gas compositions for the BASF acetylene oil quench process when natural gas, liquid petroleum gas (LPG), or naphtha is used as feedstock. The water quench process gives very similar compositions. The relative amounts of hydrogen and carbon monoxide formed depend on the hydrogen : carbon ratio of the feedstock used. Even when naphtha is used, almost no ethylene forms. This is because the reaction takes place above 1200 °C where the formation of ethylene is thermodynamically impossible. Only a prequench with additional naphtha or LPG produces additional acetylene and ethylene at intermediate temperatures, as in the case of two-step processes. The higher hydrocarbons require a somewhat lower reaction temperature than methane and have a less endothermic heat of reaction: oxygen consumption per ton of acetylene is lower for the higher hydrocarbons in spite of the lower preheating temperature.

Comparison of Oil Quench and Water Quench Processes. The advantage of the oil quench process is obvious: the heat recovery in the form of steam makes the overall thermal efficiency in relation to primary energy input rather high. If the thermal efficiency for the production of electricity is 33%, over 70% of the net heating value of the overall primary energy input is recovered in the form of products and steam. A comparison between the oil quench and water quench (see Table 6) shows that the oil quench requires a net heating value input of 300–330 GJ per ton of acetylene, of which 82 GJ (27–25%) is lost, whereas the water quench requires a 288 GJ input, of which 113 GJ (39%) is lost.

Acetylene Recovery. Liquid acetylene is a dangerous product, even at low temperatures. Separation of the cracked gas by cryogenic processes such as those used in olefin production is clearly ruled out. One exception to this rule is the acetylene recovery unit of the submerged flame process (Section 4.2.3) [46], in which all hydrocarbons except methane are condensed at −165 °C. Otherwise, acetylene is recovered by *selective* absorption into a solvent. This procedure is economical only when the cracked gas is compressed. The upper limit for the pressure is determined by the danger of explosions, and as a rule the partial pressure of acetylene should be kept below 1.4 bar (0.14 MPa).

The solubility of acetylene in the solvents used is between 15 and 35 m^3 (STP) per m^3 of solvent under process conditions. The dissolved gas is recovered by depressurizing the solvent and by vapor stripping at higher temperatures. All solvents used commercially, *N*-methylpyrrolidone (NMP), methanol, ammonia, and dimethylformamide (DMF), are miscible with water. They are recovered from the gas streams leaving the plant by water scrubbing and distillation.

The kinetics of acetylene formation always lead to the formation of higher homologues of acetylene as byproducts [47], mainly diacetylene, but also methylacetylene, vinylacetylene, and others. These compounds polymerize very easily and must be removed from the cracked gas as soon as possible. Because they are much more soluble in the solvents than acetylene, scrubbing the cracked gas with a small amount of solvent before it enters the acetylene recovery stages is sufficient.

Absorption Section (Fig. 13). Acetylene recovery is illustrated here by the BASF process. *N*-Methylpyrrolidone is used to separate the cracked gas into three streams:

1) Higher homologues of acetylene and aromatics, the most soluble part of the cracked gas. (This is a small stream of gas, which is diluted with crude synthesis gas for safety reasons and is used as fuel.)
2) Product acetylene, less soluble than the higher acetylenes, but much more soluble than the remainder of the gas
3) Crude synthesis gas (off-gas), mainly hydrogen and carbon monoxide

In the prescrubber (b) the cracked gas is brought into contact with a small amount of solvent for removal of nearly all the aromatic compounds and C_4 and higher acetylenes except vinylacetylene. This is done after the compression of the gas if screw compressors are used but before compression if turbo compressors are used because turbo compressors cannot tolerate deposits on their rotors. In the main scrubber (d) the gas is brought into contact with a much larger amount of *N*-methylpyrrolidone (NMP), which dissolves all the acetylene, the remaining homologues, and some carbon dioxide. Crude synthesis gas (off-gas) leaves at the top of the column. The NMP solution is degassed in several steps in which the pressure is reduced and the temperature increased. The stripper (e) operates at pressures and temperatures slightly above ambient. In this tower, the solution is put in contact with a counter-current gas stream from the subsequent degassing step (f). This leads to the evolution of carbon

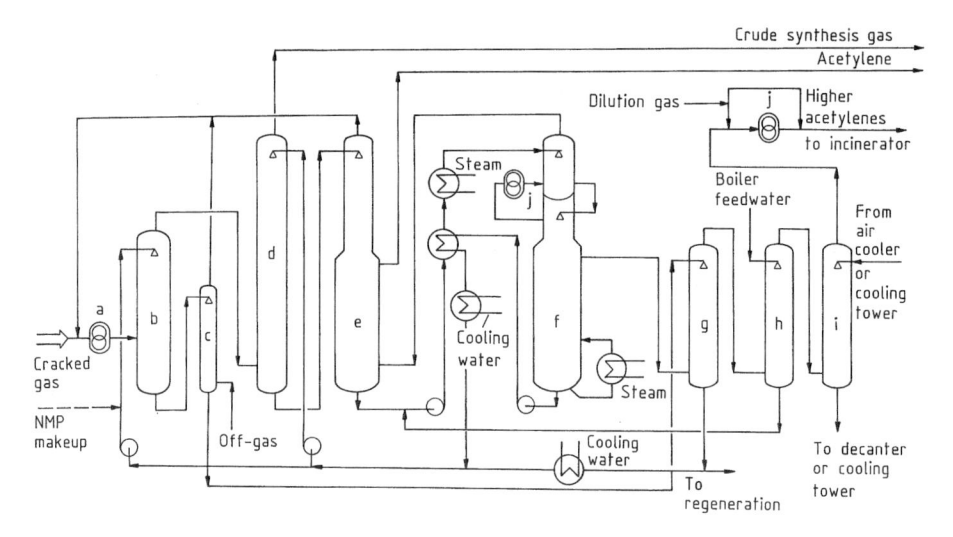

Figure 13. BASF acetylene process — *N*-methylpyrrolidone absorption section
a) Compressor; b) Prescrubber; c) Acetylene stripper; d) Main scrubber; e) Stripper; f) Vacuum column; g) Vacuum stripper; h) Side column; i) Condenser; j) Vacuum pumps

dioxide, the least soluble of the dissolved gases, at the top of the stripper. The carbon dioxide is recycled to the suction side of the compression and thereby is shifted into the crude synthesis gas. The acetylene product is withdrawn as a side stream from the stripper. The *N*-methylpyrrolidone solution is then completely degassed (f) in two further steps at 110–120 °C, first at atmospheric, then at reduced pressure. Vinylacetylene, methylacetylene, and excess process water are withdrawn as bleed streams from the vacuum column (f). The water content of the solvent is controlled by the reboiling rate in the vacuum column. At the bottom of the vacuum column, degassing is completed, and the solvent is cooled and returned to the main scrubber (d).

The small amount of solvent from the prescrubber (b) is stripped with crude synthesis gas for recovery of the dissolved acetylene, the overhead gas being recycled to the suction side of the compressor. The solvent is then degassed completely in the vacuum stripper (g), a column which also accepts the bleed stream from the vacuum column (f) containing the excess process water together with some higher acetylenes. The overhead vapor of the vacuum stripper contains the higher acetylenes, water, and some NMP vapor. In a side column (h) the NMP is recovered by scrubbing with a small amount of water, which is recycled to the main solvent stream. The gas is cooled (i) by direct contact with water from a cooling circuit to condense most of the water vapor. The higher acetylenes are diluted with crude synthesis gas before they enter and after they leave the vacuum pump (j). The diluted higher acetylenes, which are now at a pressure slightly above atmospheric, can be used as fuel gas, e.g., for soot incineration.

In order to minimize the polymer content of the solvent, about 2% of the circulating flow is withdrawn continuously from the vacuum stripper circuit and distilled under reduced pressure, leaving the polymers as a practically dry cake for disposal.

The acetylene product from the process as described above has a purity of about 98.4%, the remainder consisting mainly of propadiene, methylacetylene, and nitrogen.

Table 5. Purity of the acetylene from the BASF process

Component	Crude acetylene, vol%	Purified acetylene, vol%
Acetylene	ca. 98.42	99.70
Propadiene	0.43	0.016
Propyne	0.75	traces
Vinylacetylene	0.05	0
1,3-Butadiene	0.05	0
Pentanes	0.01	0.01
Carbon dioxide	ca. 0.10	0
Nitrogen	ca. 0.30	0.30

For most applications the purity is increased to 99.7% by scrubbing with sulfuric acid and sodium hydroxide solutions. Table 5 compares the compositions of crude and purified acetylene. Table 6 compares the consumption and product yields per ton of acetylene for the oil quench process with those for the water quench process.

4.2.2. Other Partial Combustion Processes

The main features of the BASF process described in detail above are common to all partial oxidation processes. Therefore only the differences between the BASF acetylene burner and burners used in the *Montecatini* and the *SBA processes* [42], [48] are described. These two processes have also attained some importance. The details of the acetylene recovery process depend on the properties of the solvent, but here too the basic principles are the same for all processes.

Montecatini Process. The Montecatini burner [49] has the same main components as the BASF burner: mixing unit, gas distributor, reaction chamber, and quench. The essential difference is the pressure for acetylene synthesis, which can be as high as several bar. This saves compression energy, improves heat recovery from the quench water, which is obtained at 125 °C, and is claimed to make soot removal easier because the cracked gas is scrubbed with water above 100 °C. Although it is well known [9] that acetylene decomposition is accelerated under pressure at high temperatures (> 1000 °C), the acetylene yield is comparable to that obtained at atmospheric pressure because of the short residence time in the reactor. Methanol is used at cryogenic temperatures for acetylene recovery. The main steps of the gas separation are absorption of higher acetylenes and of aromatics, absorption of acetylene, stripping of coabsorbed impurities, and desorption of acetylene.

SBA Process (of the Société Belge de l'Azote). The SBA burner [50] has the same main components as the other processes. However, it has a telescope-like reaction chamber and a device for shifting the quench up and down. Thus it is possible to adjust the length of the reaction zone for optimum residence time at any throughput. The walls of the reaction chamber are sprayed with demineralized water to prevent coke

Table 6. BASF acetylene process, consumption and product yields per ton of acetylene

Consumption and product yields	Oil quench		Water quench	
Feed and energy requirements				
Natural gas, 36000 kJ/m^3(STP) (LHV)*	5833 m^3	= 210 GJ	5694 m^3	= 205 GJ
Oxygen, 0.55 kWh/m^3(STP)**	3400 m^3	= 20.4 GJ	3400 m^3	= 20.4 GJ
Fuel gas		= 12.0 GJ		= 18.0 GJ
Residue oil minimum (surplus)	0.3 (1.0) t	= 12.0 (40.0) GJ		
Sulfuric acid	160 kg		160 kg	
Sodium hydroxide	5 kg		5 kg	
N-Methylpyrrolidone	5 kg		5 kg	
Electric energy**	3200 kWh	= 34.9 GJ	3100 kWh	= 33.8 GJ
Steam, 4 bar	5.0 t	= 11.7 GJ	4.5 t	= 10.5 GJ
Energy input		301.0 (329.0) GJ		287.7 GJ
Product yields				
Acetylene, 48650 kJ/kg (LHV)	1.0 t	= 48.6 GJ	1.0 t	= 48.6 GJ
Crude sythesis gas, 12100 kJ/m^3(STP) (LHV)	10600 m^3	= 128.3 GJ	10150 m^3	= 122.8 GJ
Coke (with residue surplus), 35500 kJ/kg	0.3 (0.46) t	= 10.7 (16.3) GJ	–	
BTX (with residue surplus), 40250 kJ/kg (LHV)	0.05 (0.12) t	= 2.0 (4.8) GJ	–	
Naphthalenes (with residue surplus), 38770 kJ/kg (LHV)	0.0 (0.41) t	= – (15.9) GJ	–	
Steam (up to 15 bar)	13.0 (14.0) t	= 30.3 (32.6) GJ	1.5 t	= 3.5 GJ
Energy output		219.9 (246.5) GJ		174.9 GJ
Thermal efficiency		73.0 (74.9)%		60.8%
Energy losses, absolute per ton acetylene		81.1 (82.5) GJ		112.8 GJ

* If the natural gas contains inerts and higher hydrocarbons, the required input will remain approximately the same on a heating value basis (LHV = low heating value), but the cracked gas analyses and the crude synthesis gas analyses will differ slightly.
** Thermal efficiency of electricity production is assumed to be 33%.

deposits. This eliminates the need to scrape the reaction chamber periodically. Acetylene recovery is carried out with several scrubbing liquids — kerosene, aqueous ammonia, caustic soda, and liquids ammonia, each with its own circuit. After soot is separated from the gas in an electrofilter, higher hydrocarbons are absorbed in kerosene or gas oil. Carbon dioxide is scrubbed in two steps, first with aqueous ammonia and then with caustic soda solution. The acetylene product is absorbed into anhydrous ammonia and must be scrubbed with water after desorption. All the ammonia–water mixtures are separated in a common distillation unit. This recovery scheme leads to exact separation of the various cracked gas components.

Additional Remarks. The Montecatini and SBA processes can also be operated with two-stage burners. A prequench with light hydrocarbons cools the cracked gas to about

800 °C. After a residence time at this intermediate temperature the gas is cooled down with water. In this way the heat content of the hot gases is used for further cracking of hydrocarbons to yield extra acetylene and olefins. The presence of additional components in the cracked gas requires more process steps in the gas separation units.

4.2.3. Submerged Flame Process

The submerged flame process (SFP) of BASF attracted considerable interest up to 1973 as a partial combustion process for the production of acetylene, ethylene, C_3 and C_4 hydrocarbons, and synthesis gas from feedstock of crude oil and residues, such as Bunker C oil and vacuum residue [45], [46]. Although it was abandoned at the end of 1973, recently the need to make the most economic use of raw materials has renewed interest in this process [51].

Oxygen compressed to 16 bar (1.6 MPa) feeds a flame that is submerged in the oil. The oil surrounding the flame is partially burnt to obtain the necessary reaction temperature and also acts as the quenching medium. This process differs from the partial oxidation processes using natural gas and lighter hydrocarbons in five main respects:

1) Crude oil can be gasified without the formation of residues, and the process can be operated under certain conditions with heavy fuel oil.
2) All the soot formed is consumed when crude oil feedstock is used, eliminating all the problems associated with the storage, disposal, or utilization of acetylene soot.
3) The heat of reaction is removed by steam generation at 8 bar (0.8 MPa).
4) The process is operated at 9 bar (0.9 MPa) so that the oxygen is the only compressed stream. The cracked gas is formed at a pressure sufficient for economic separation.
5) The design of the cracking unit is greatly simplified because the reaction feed, fuel, and quenching medium are identical.

The process is described in detail in the literature cited; therefore, only general overviews of the cracking unit (Fig. 14) and the separation unit (Fig. 15) are shown here. The capacities of a submerged flame burner for acetylene and ethylene are 1 t/h and 1.15 t/h, respectively. To produce these quantities, 5000 m^3 (STP) of oxygen and 8–10 t of oil are required per hour. The cracked gas shows the following average composition (vol%, the components grouped as streams leaving the separation unit):

Main products
   Acetylene   6.2
   Ethylene   6.5
Crude synthesis gas
   Carbon monoxide   42.0
   Hydrogen   29.0
   Methane   4.0
   Inerts   0.6

Other hydrocarbons	
Ethane	0.5
Propane	0.1
Propene	1.2
Propadiene, propyne	0.7
1,3-Butadiene	0.5
Other C_4 and C_{5+}* hydrocarbons	1.5
Remainder	
Carbon dioxide	7.0
Hydrogen sulfide	0.05 – 0.5
Carbon oxide sulfide	0.03 – 0.3

* C_{5+}, five or more carbons

Unlike all other processes the submerged flame process uses low temperatures (–165 °C) to separate the off-gas, consisting of carbon monoxide, hydrogen, and methane, from the C_2 and higher hydrocarbons. On account of the acetylene in the condensed phase, extensive decomposition tests have been carried out. Whereas the cracking unit (Fig. 14) and the amine scrubbing unit have been tested by Soc. Ital. Serie Acetica Sintetica, Milan, on a commercial scale, the remaining purification units (Fig. 15) have not. However, the experience obtained with a pilot plant indicates that major difficulties are not to be expected.

The submerged flame process may become competitive because of its ability to use crude oil and especially residues and because of its low losses on the primary energy input.

4.2.4. Partial Combustion Carbide Process

Calcium carbide production from lime and coal requires a large high-temperature heat input (see Section 4.3.4). In the thermal process some of the coal must be burnt to attain the necessary reaction temperature and supply the heat of reaction. The thermal carbide process was developed by BASF [9], [52] from 1950 to 1958 to eliminate the input of electrical energy necessary in the classic carbide process. Starting in 1954, a large pilot plant, with a nominal carbide capacity of 70 t/d, was operated, but in 1958 the more economical petrochemical acetylene production halted further development. However, because of the enormous price rises in petrochemical feedstocks, worldwide efforts to develop coal chemistry are now being renewed. Carbide production is just one way of converting coal chemically; other methods include pyrolysis, hydrogenation, and gasification. The question arises as to the conditions under which a thermal carbide process using oxygen can compete with the electric carbide process. The biggest drawback of carbide production in a shaft furnace (Fig. 16) compared to the electric carbide process is the lack of commercial-scale operational experience. Specific disadvantages are greater susceptibility to disruption because of plugging of the furnace feed, more stringent specifications for the raw materials, more handling of solids, and the large amount of byproduct. There are two main advantages:

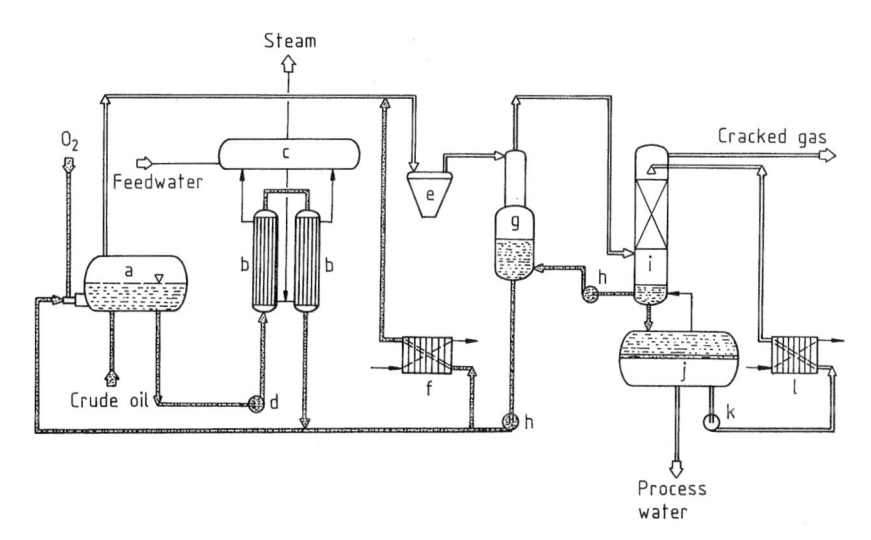

Figure 14. Submerged flame process (SFP) — cracking unit
a) Reactor; b) Oil cooler; c) Steam generator; d) Oil recycle pump; e) Scrubber; f) Naphtha cooler; g) Naphtha separator; h) Naphtha pump; i) Spray cooler; j) Separating vessel; k) Recycle-water pump; l) Recycle-water cooler

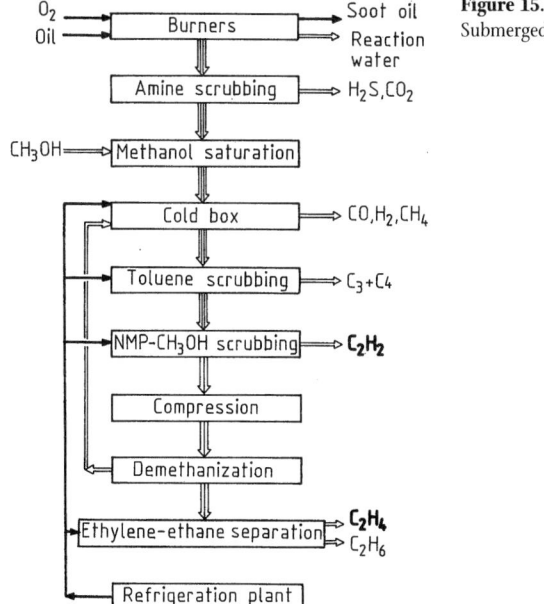

Figure 15. Submerged flame process — purification unit

1) A thermal efficiency of about 50 % versus about 30 % for the electrothermal process if the thermal efficiency of electricity production is 33 %
2) Carbon monoxide production, which is desirable because carbon monoxide can be converted to synthesis gas by the water-gas shift reaction

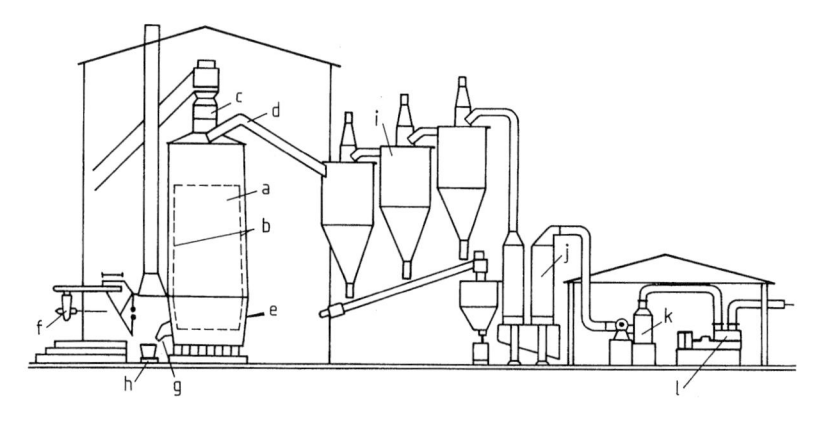

Figure 16. Partial combustion carbide process
a) Carbide furnace; b) Refractory brick lining; c) Charging hopper; d) Gas outlet; e) Oxygen jet; f) Tapping burner; g) Tapping chute; h) Bogey; i) Cyclone; j) Washing column; k) Desintegrator; l) Compressor

Table 7. Partial combustion carbide process, consumption and product yields per ton acetylene

Raw materials	
Coke, dry (88% C)	5700 kg
Lime (92% CaO)	3140 kg
Oxygen (98%) 3560 m^3 (STP)	5090 kg
Total consumption	13930 kg
Products	
Carbide (80.5%)	2850 kg ≙ 1000 kg acetylene
Carbon monoxide 7980 m^3 (STP)	9975 kg
(CO 95.5, H$_2$ 2.0, N$_2$ 2.0, CO$_2$ 0.5 vol%)	
Dust	900 kg
Losses	205 kg
Total products	13930 kg

If the carbon monoxide is converted to synthesis gas and the electrical energy is produced from fossil fuels, production costs are about one third lower for the thermal process than for the electrical process [5] based on the pilot-plant consumption data (Table 7).

4.3. Electrothermic Processes

Because calcium carbide is produced electrothermally, the production of acetylene from this material also is discussed in this group of processes (Section 4.3.4).

Electrothermic processes have the following advantages over partial oxidation:

— The energy requirement for the formation of acetylene can be made independent of the hydrocarbons used as feedstock.
— Hydrocarbon consumption can be reduced by 50%.

— Provided that electrical energy is available under favorable conditions (nuclear power, hydroelectric power, cheap coal) and/or the availability of hydrocarbons is limited, electrothermic processes are more economical.

In the case of acetylene formation, the electric-arc process offers optimal conditions for the endothermic reaction at high temperatures.

The development of the electric-arc process for cracking light hydrocarbons to acetylene began in 1925 in Germany. The acetylene was to be used as feedstock for butadiene production. In 1940, the first commercial plant was put on stream at Chemische Werke Hüls in Marl, Germany. The Hüls process has been improved since then, and the capacity raised to its current 120 000 t/a, but it is still based on the original principles [53].

A small electric-arc plant is still operated at Borcesti, Rumania, whereas Du Pont, in the United States, ran its electric-arc process only from 1963 to 1968, with a capacity of 25 000 t/a.

Feedstock for electric-arc processes may be gaseous or liquid hydrocarbons or even solids such as coal. The design of the arc furnace and the purification section for the cracked products have to be adapted to the different feedstocks. For gaseous or gasified hydrocarbons the classical one-step process is used: the arc burns directly in the gas being cracked. For liquid and solid feeds, a one- or two-step process may be used. In the two-step process hydrogen is first heated in the arc furnace, and then liquid or solid feed is injected into the hydrogen plasma [54]. Figure 17 shows both types of arc furnaces. Because of hydrogen formation during the cracking reaction, the arc burns in a hydrogen atmosphere in both processes. The conductivity and the high rate of ion – electron recombination for hydrogen mean that arcs above a certain length cannot be operated with alternating current at normal frequency and high voltage. All commercial plants therefore run on direct current.

4.3.1. Production from Gaseous and/or Gasified Hydrocarbons (Hüls Arc Process)

The plant for the Hüls arc process includes the arc furnace section itself (Fig. 17 A), which is operated at a pressure of 1.2 bar, and a low and high pressure purification system.

Arc Furnace. A cathode, a vortex chamber, and an anode make up the arc furnace. Cathode and anode are water-jacketed tubes of carbon steel 0.8 m and 1.5 m long, respectively, and with inner diameters of 150 and 100 mm, respectively. The arc burns between cathode and anode with a length of about 1.2 m and with a current of 1200 A. The cathode is connected to the high-voltage side of the rectifier (7.1 kV) and electrically isolated from the other parts of the furnace. Between cathode and anode is the vortex chamber. The gas is injected into it tangentially at a specific velocity to stabilize the arc

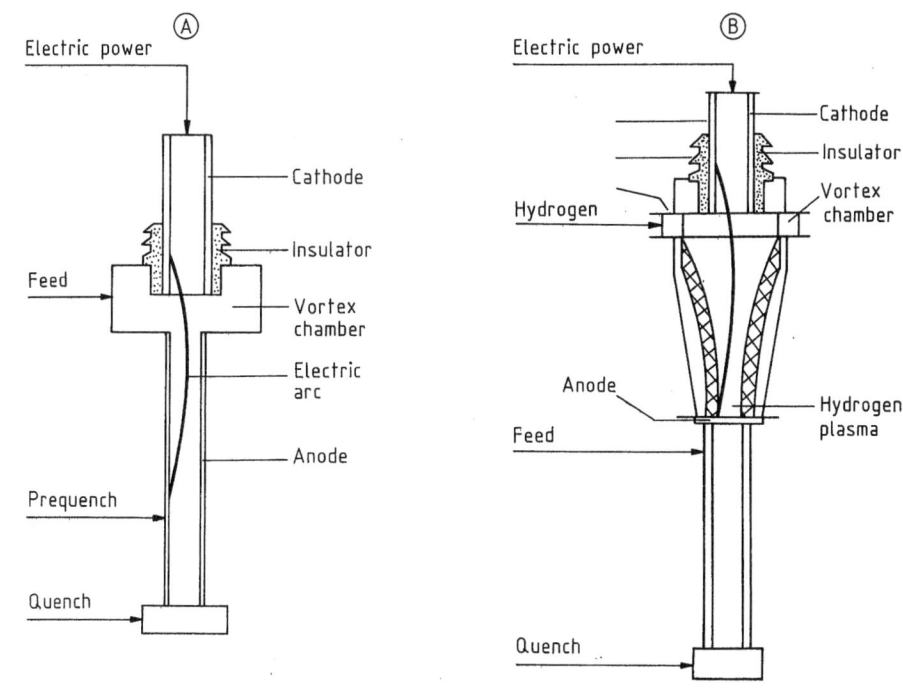

Figure 17. Hüls electric-arc furnaces for gaseous, liquid, and solid feed
A) One-step process; B) Two-step process

by creating a vortex. The arc burns in the dead zone, and the striking points of the arc on the electrodes are forced into a rapid rotation so that they only burn for fractions of a millisecond at one point, which gives the electrodes a lifetime up to 1000 h. Temperatures reach 20 000 °C in the center of the arc. Because of the tangential flow of the gas, the arc is surrounded by a sharply decreasing coaxial temperature field, and the temperatures at the wall of the electrode are only 600 °C. Thermal losses are therefore limited to less than 10 % of the electrical power input of 8.5 MW.

The residence time of the gas in the arc furnace is a few milliseconds. In this interval, the hydrocarbons are cracked, mainly into acetylene, ethylene, hydrogen, and soot. At the end of the arc furnace, the gases are still at a temperature of about 1800 °C. The high heat content of this gas can be exploited for additional ethylene production by means of a prequench with liquid hydrocarbons. This lowers the temperature to about 1200 °C. Because acetylene rapidly decomposes into soot and hydrogen at these temperatures, the gases must be quenched immediately with water to about 200 °C, i.e., a quench rate of 10^6 °C/s must be achieved.

The *specific energy requirement* (SER) and the acetylene yield depend on the geometry and dimensions of the furnace and electrodes, the velocity distribution of the gas, and the kind of hydrocarbon to be cracked. Once the furnace has been designed, only the hydrocarbons can be varied.

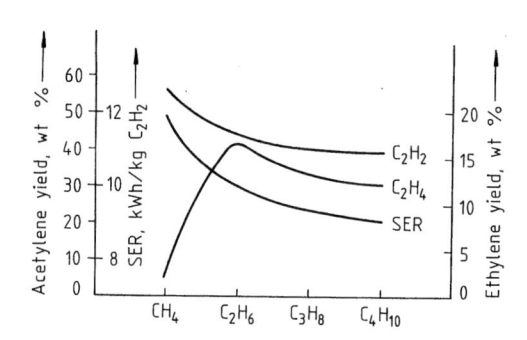

Figure 18. Acetylene yield, ethylene yield, and energy consumption for various hydrocarbons in the Hüls arc process

Process Without Prequench. Figure 18 shows acetylene and ethylene yields and the specific energy requirement (SER) of various saturated hydrocarbons under constant conditions without prequench. Methane shows the highest SER and acetylene yield, but the lowest ethylene yield. As the chain length is increased, both acetylene yield and SER decline, corresponding to the declining heat of acetylene formation from the various hydrocarbons.

Normally, pure hydrocarbons are not available. The results obtained from mixtures of hydrocarbons can be expressed as a function of the carbon number, which is the number of moles of carbon atoms bound in hydrocarbons per mole of the gaseous mixture. Figure 19 shows specific amounts of acetylene, ethylene, and hydrogen formed and of hydrocarbon consumed as a function of carbon number. This function enables the Hüls process to be optimized within certain limits, for example, for hydrogen output in relation to acetylene production.

Process with Prequench. Cracking in the prequench section is essentially an ultrasevere steam cracking process. The kind and amount of hydrocarbons used for the prequench can be varied. Figure 20 shows the specific product yield for different prequench rates for feeding methane to the arc furnace and propane to the prequench. Acetylene and hydrogen yields are unaffected, whereas ethylene shows a slight maximum and declines when the temperature is not sufficient at a given residence time. Propene shows a steady increase, and the $C_3:C_2$ ratio is below 0.25. The relative ethylene yield from various hydrocarbons is as follows: ethane 100, propane 75, n-butane 72, isobutane 24, 1-butene 53.

Oil quench. Because the gas temperature of the furnace gas after prequench is on the order of 1200 °C, an oil quench system has been developed to regain about 80% of the sensible heat content of the furnace gas as steam by heat exchange. The soot–oil mixture formed can be upgraded to a sulfur- and ash-free high-grade petroleum coke. Figure 21 shows the newly designed Hüls system with oil quench.

The Purification System. The process of purification depends on the type of the quench system. In the case of water quenching, 80% of the carbon black is removed by cyclones as dry carbon black, the remaining 20% as soot in water-operated spray

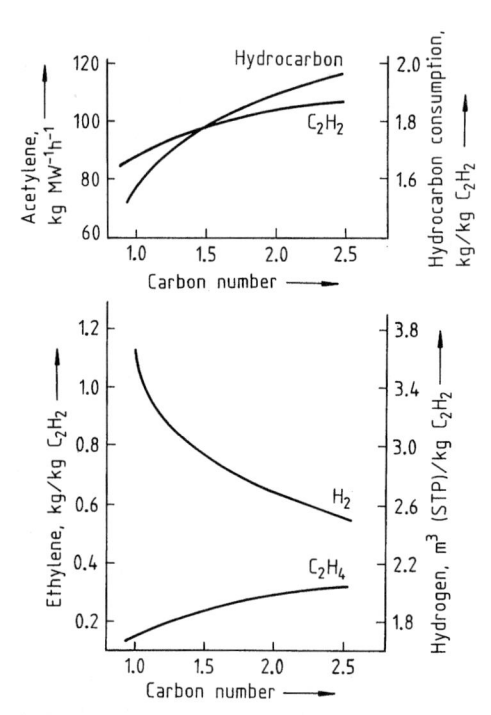

Figure 19. Specific values for acetylene and hydrogen formation

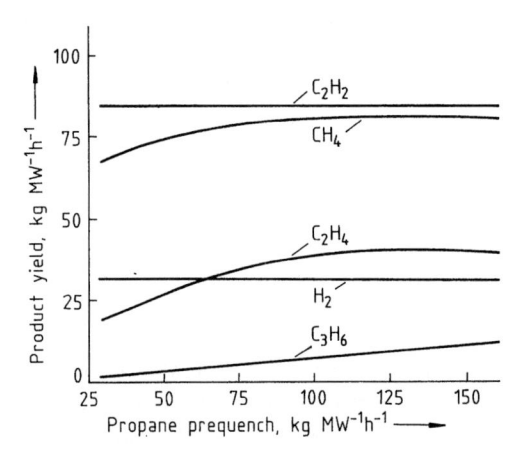

Figure 20. Specific product yield for different prequench rates

towers. In a combined oil–water scrubbing system, aromatic compounds are removed and benzene, toluene, and xylene (BTX) are recovered in a distillation process.

Figure 22 shows the principle separation and purification steps for the furnace gas. The gas leaves the first three purification sections with a carbon black content of 3 mg/m^3 and is compressed by four-stage reciprocating compressors to 19 bar (1.9 MPa). The gas is washed in towers with water in a countercurrent flow. At the bottom of the tower, the water is saturated with acetylene, whereas the overhead gas contains less than 0.05 vol% acetylene. The acetylene–water solution is decompressed in

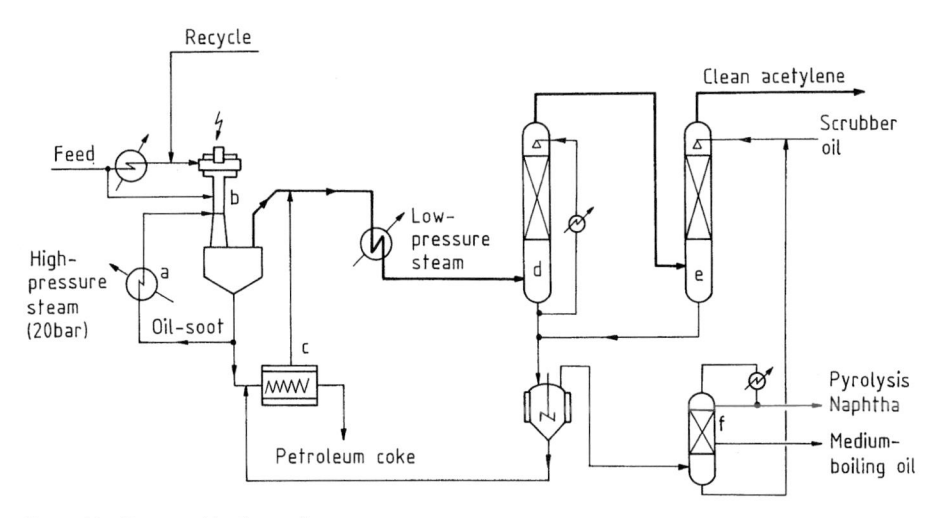

Figure 21. Process with oil quench system
a) Heat recovery; b) Arc furnace; c) Oil recovery; d) Separation of medium-boiling compounds; e) Separation of low-boiling compounds; f) Oil regeneration

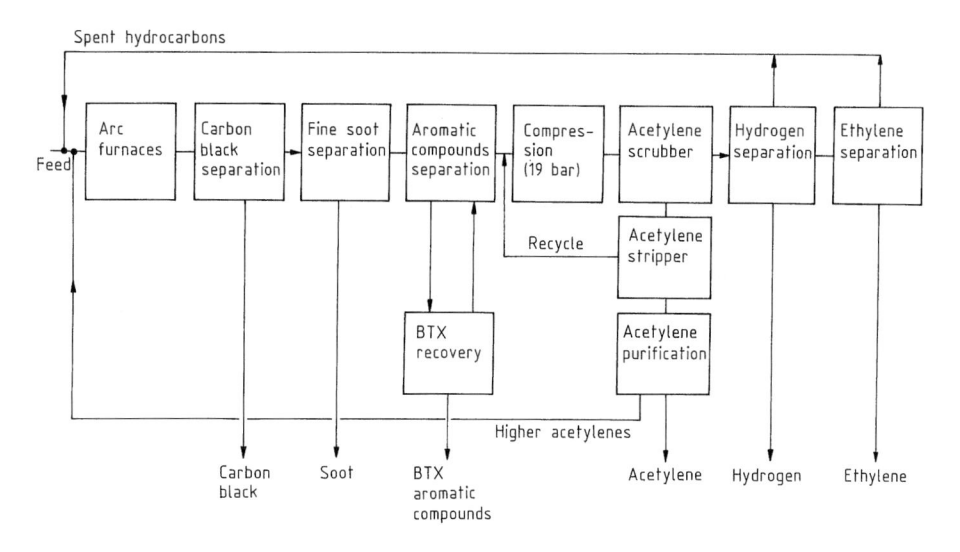

Figure 22. Principal separation and purification steps for the furnace gas of the Hüls arc process

four stages. Gas from the first decompression stage returns to the compressor to improve selectivity. The last two stages operate at 0.2 and 0.05 bar (20 and 5 kPa). The gas still contains about 10 vol% of higher acetylenes, which are removed by a cryogenic process. The higher acetylenes are liquefied, diluted with flux oil, stripped, and returned to the arc furnace together with spent hydrocarbon. Today, a more selective solvent such as N-methylpyrrolidone or dimethylformamide is preferred to the water wash. Linde and Hüls have designed an appropriate purification system. Hydrogen and ethylene are separated by well-known technology, such as the cryogenic process or pressure-swing adsorption.

Table 8. Typical analysis of feed and cracked gas

	Feed gas, vol%	Cracked gas, vol%
C_2H_2	0.4	15.5
C_3H_4	1.4	0.4
C_4H_2	1.2	0.3
C_4H_4	1.7	0.4
C_2H_4	0.8	6.9
C_3H_6	3.6	1.0
Allene	0.4	0.2
C_4H_8	1.0	0.2
C_4H_6	0.9	0.2
C_5H_6	0.6	0.2
C_6H_6	0.5	0.5
CH_4	64.6	13.8
C_2H_6	7.5	0.4
C_3H_8	3.6	0.3
C_4H_{10}	4.6	1.0
C_5H_{12}	0.5	0.1
H_2	4.5	57.6
CO	0.5	0.6
O_2	0.1	0.0
N_2	1.6	0.4

Process Data. Hüls operates its plant with a mixture of natural gas, refinery gas, and liquefied petroleum gas. The carbon number varies between one and two. Table 8 shows a typical analysis of feed gas and cracked gas.

The Hüls plant has 19 arc furnaces, the number operated depending on the electricity supply. The arc furnaces can be started up and shut down immediately. Two large gas holders provide a storage volume of 350 000 m^3 so that the purification section operates on permanent load and there is a dependable supply of products, even if the arc furnace section is operated at higher or lower load.

The plant has an annual capacity of 120 000 t acetylene, 50 000 t ethylene, 400×10^6 m^3 (STP) hydrogen, 54 000 t carbon black and soot, and 9600 t aromatic compounds. The energy consumption is 1.5×10^6 MW h/a.

Specific data for consumption of hydrocarbons and energy and the production of byproducts per ton of acetylene produced are as follows:

Hydrocarbons to the arc furnace	1.8 t
Hydrocarbons for prequench	0.7 t
Energy for the arc furnace	9800 kW h
Energy for gas purification	2500 kW h
Ethylene	0.42 t
Hydrogen	3300 m^3 (STP)
Carbon black and soot	0.45 t
Aromatics	0.08 t
Residue	0.12 t
Heating gas	0.12 t

4.3.2. Production from Liquid Hydrocarbons (Plasma Arc Process)

Two different plasma furnaces, each with the appropriate reactor for the cracking of liquid hydrocarbons, were developed by Hoechst and Chemische Werke Hüls in close coorperation. Both units were tested on an industrial scale at a power level of 8–10 MW [55]. However, neither process has actually come into use for acetylene production on account of the economics.

The scheme of the plasma generator used by Hüls is shown in Figure 17 B. The unit consists of three parts: the arc furnace, the reactor, and the quench system. The arc burns over a length of 1.6 m at 7 kV d.c. and 1.2 kA, resulting in a power input of 8.5 MW. It is stabilized by hydrogen injected tangentially through the vortex chamber. The thermal efficiency of the furnace is ca. 88% of the electrical power input. The hydrogen plasma jet passing through the anode nozzle has an energy density of 3.5 kW h/m^3 (STP), corresponding to an average temperature of 3500 K. The liquid hydrocarbons (e.g., crude oil) to be cracked are injected into the cylindrical reactor to achieve good mixing with the plasma jet and to avoid the formation of carbonaceous deposits on the wall. Within several milliseconds the hydrocarbons are heated and cracked to acetylene, ethylene, hydrogen, soot, and other byproducts before the mixture is quenched with oil to 300 °C. The acetylene ratio can be adjusted by varying the residence time. By operation of an oil quench with the high-boiling residue of the crude oil, 80% of the sensible heat content of the cracked gas can be recovered as steam. The soot is taken up by the quench oil and is removed from the system as an oil–soot dispersion having 20% soot concentration. The unconverted vaporized fractions of the oil are condensed in oil scrubbers at a lower temperature, simultaneously cleaning the gas of the aromatic components and fine soot. These oil fractions are recycled to the reactor and the quench system, respectively.

Tests were carried out with a variety of hydrocarbons from propane to naphtha, but mainly with crude oil and residue oils. The cracking results depend on the chemical nature of the feed. Consumption figures and yields for various feedstocks are given in [55]. For high-boiling petroleum fractions, the acetylene and ethylene yields increase with the content of low-boiling components in the feed (see Fig. 23). Consumption and byproduct yields per ton of acetylene for a Libyan crude oil are summarized below:

Consumption
    Crude oil consumed   3.5 t
    Power (d.c.)   10500 kW h
Byproducts
    Ethylene   0.46 t
    Hydrogen (99.9%)   1100 m^3 (STP)
    Fuel gas   0.74 t
    Soot–oil mixture (20% soot)   1.2 t

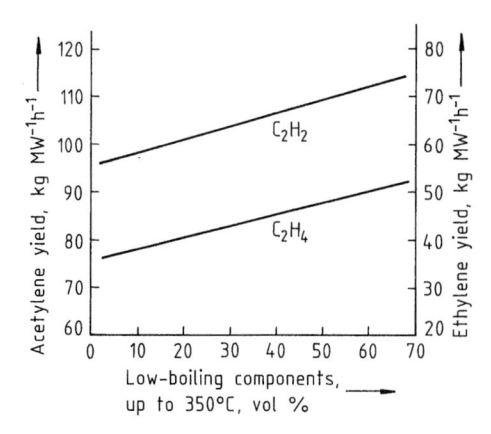

Figure 23. Acetylene and ethylene yields as a function of the low-boiling components

Table 9. Consumption and yield per ton of acetylene for the Hoechst arc process and naphtha feed for different reactor designs

	Low ethylene yield	High ethylene yield
Consumption		
Naphtha	1.92 t	2.50 t
Quench oil	0.53 t	0.63 t
Energy (2-phase a.c.)	9300 kW h	10500 kW h
Byproducts		
Ethylene	0.5 t	0.95 t
Hydrogen	1450 m^3 (STP)	1500 m^3 (STP)
Soot–oil mixture (20% soot)	0.75 t	1.00 t

Hoechst used a high-intensity three-phase a.c. arc furnace at 1.4 kV and 4.2 kA, giving a power input of 10 MW [55]. The thermal efficiency was 90%. Because of the high amperage the graphite electrodes had to be replenished continually. The generator was lined with graphite. Different reactor designs for ethylene:acetylene ratios of 0.5 and 1.0 were developed by varying the mixing intensity of the hydrogen plasma jet with the liquid hydrocarbon. The tests were carried out with naphtha feed (see Table 9). The cracked gas was quenched with residue oil, in a manner similar to that described in the Hüls process.

The acetylene concentration in the Hüls process and the Hoechst process was ca. 14 vol% so that in principle the same acetylene separation process can be used as described above for the arc process.

4.3.3. Production from Coal (Arc Coal Process)

Numerous laboratory tests for the conversion of coal to acetylene using the arc or plasma processes have been carried out since the early 1960s [56]. The results can be summarized as follows:

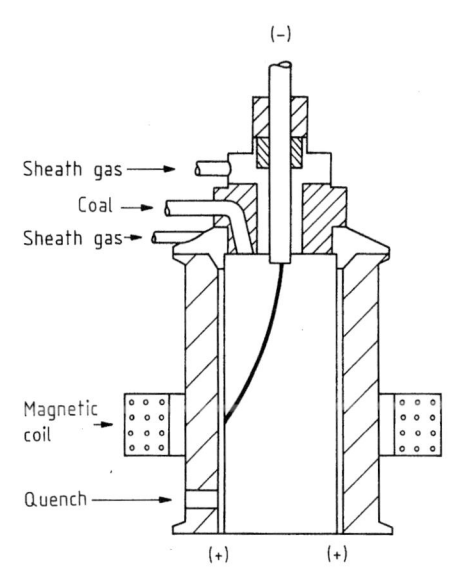

Figure 24. Principal scheme of the AVCO plasma furnace for the pyrolysis of coal

— Acetylene yields up to 30% can be obtained.
— Because of the rapid heating of the coal in the plasma jet, a higher total gas yield can be achieved than is indicated by the volatiles of the coal measured under standard conditions.
— Hydrogen (instead of argon) plasma gas considerably increases the acetylene yield.

Recently, the AVCO Corp. in the United States [57] and Chemische Werke Hüls in Germany [58] brought pilot plants on stream for the technical development of the process. The AVCO arc furnace (Fig. 24) consists of a water-cooled tungsten-tip cathode and a water-cooled anode. The arc is stabilized by a magnetic field surrounding the anode, forcing the anode striking point of the arc to rotate rapidly and so avoiding burnthrough. The dried and finely ground coal is injected by means of a hydrogen gas flow around the cathode. Additional gas without coal is introduced around the cathode and at the anode as a sheath. On passing the arc zone the coal particles are heated up rapidly. The volatiles are released and are cracked to acetylene and byproducts, leaving a residue of fine coke particles covered with soot. After a residence time of some milliseconds the gas–coke mixture is quenched rapidly with water or gases. The use of a prequench system similar to that of the Hüls arc process was also tested. The system pressure can be varied between 0.2 and 1.0 bar (20 and 100 kPa).

Hüls' pilot plant uses the same plasma furnace as for the crude oil cracking, but with 500 kW of power. The dried and ground coal is injected into the plasma jet, and the coal is cracked to acetylene and byproducts in the reactor. The reactor effluent can be prequenched with hydrocarbons for ethylene production or is directly quenched with water or oil. Char and higher boiling components are separated by cyclones and scrubbers, respectively. The problem in the reactor design is to achieve thorough

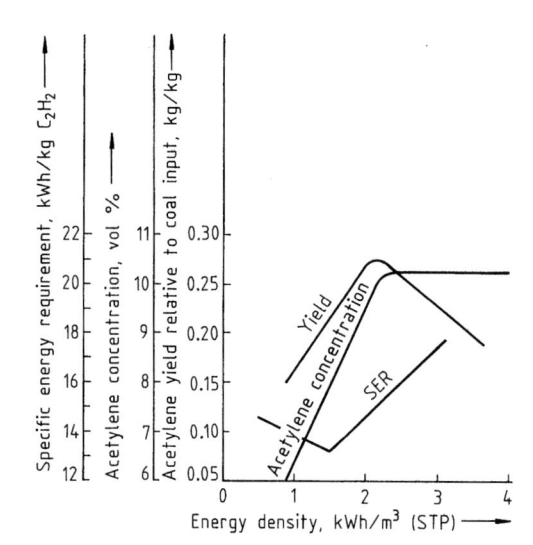

Figure 25. Effect of the energy density of the plasma on the cracking of coal (AVCO)

and rapid mixing of the coal with the plasma jet and to avoid forming carbonaceous deposits on the wall. Smaller amounts of deposits can be removed by periodic wash cycles with water. Operation times of 2.5 h by AVCO and 5 h by Hüls have been reported.

Experiments published by Hüls and AVCO show that at the optimal residence time the energy density of the plasma jet, the specific power, and the pressure all greatly affect the acetylene yield (Fig. 25 and Fig. 26). Other parameters affecting the yield are the amounts of volatiles in the coal and the particle size. The lowest figures for the specific energy consumption published by AVCO are of the order of 27–37% based on water-free coal.

In addition to acetylene, the exit gas contains considerable amounts of CO, depending on the oxygen content of the coal. Because nitrogen and sulfur are present in the coal, other byproducts are HCN, CS_2, COS, and mercaptans. The gas separation system is therefore designed accordingly [59]. Depending on the hydrogen content of the coal, the process is either self-sufficient in hydrogen or has a slight surplus. The total gas yield of the coal based on a volatile content in the coal of 33% is up to 50%. Thus 50% of the coal remains as char. Tests with a view to using this char in the rubber industry have been unsuccessful so far. Thus the char can be used only for gasification or as a fuel.

In all the processes under development, the production of ethylene from coal requires several process steps (Fig. 27), resulting in a high capital demand for a production plant. In contrast, the acetylene production from coal arc pyrolysis is straightforward, leading to lower investment costs. Demonstration units on a higher power level are therefore scheduled by both AVCO and Hüls.

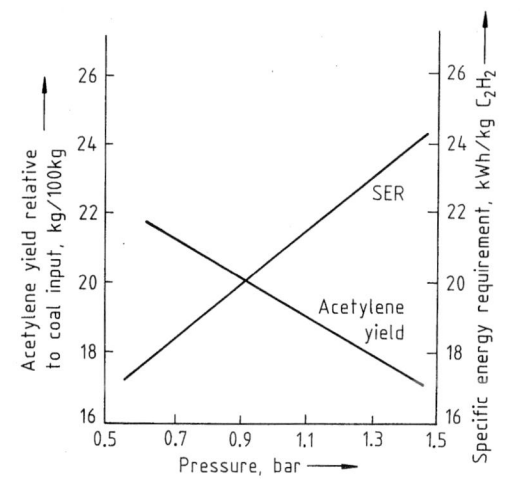

Figure 26. Acetylene yield and specific energy requirement as a function of pressure (Hüls)

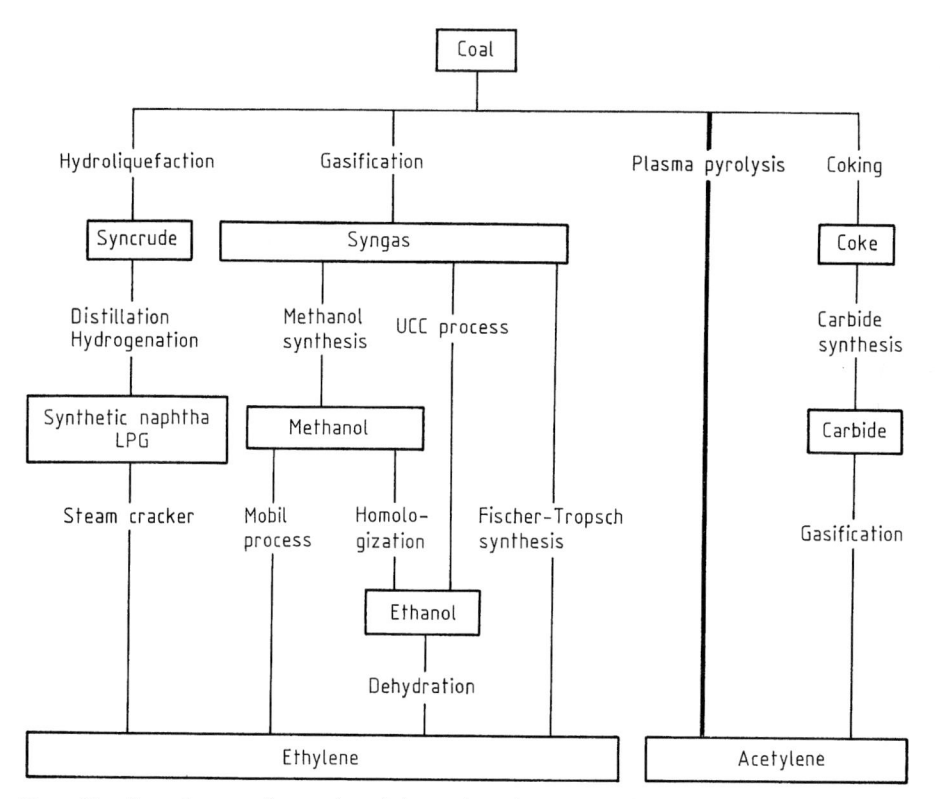

Figure 27. Alternative routes from coal to ethylene and acetylene

4.3.4. Production from Calcium Carbide

At present, the generation of acetylene from calcium carbide is of primary importance for welding and for the production of carbon for batteries. The particular raw-material situation and the use of special processes are two common reasons for continuing to use acetylene generated from carbide in the chemical industry.

The reaction of calcium carbide and water to form acetylene and calcium hydroxide is highly exothermic:

$$CaC_2 + 2\,H_2O \longrightarrow C_2H_2 + Ca(OH)_2 \qquad \Delta H = -129\ \text{kJ/mol}$$

The acetylene generator used for commercial production must therefore be designed to allow dissipation of the heat of reaction. In the event of inadequate heat dissipation, for example, when gasification proceeds with insufficient water, the carbide may become red-hot. Under certain circumstances (including increasing pressure), this may cause the thermodynamically unstable acetylene to decompose into carbon and hydrogen. (For safety precautions see Chap. 5.) Carbide for the production of acetylene is used in the following grain sizes (mm): 2–4, 4–7, 7–15, 15–25, 25–50, 50–80. This classification is virtually identical in most countries: DIN 53922 (Federal Republic of Germany); BS 642:1965 (United Kingdom); JIS K 1901–1978 (Japan); Federal Specification 0–6-101 b/GEN CHG NOT 3 (United States). In addition, grain size 0–3 is used for the dry generation of acetylene.

Pure calcium carbide has a yield number of 372.66. This means that the gasification of 1 kg of carbide yields 372.66 L acetylene at 15 °C and 1013 mbar (101.3 kPa). Commercially available carbide has a yield number of 260–300.

A distinction is made between two groups of acetylene generators (with continuous rates of production greater than 10 m^3 acetylene per hour): the wet type and the dry type.

In *wet generators*, the acetylene is converted with a large water excess. In most cases, a lime slurry containing 10–20 wt % calcium hydroxide is obtained. The heat of reaction increases the temperature of the generator water and is removed from the reactor with the lime slurry.

In *dry generators*, the water mixed with the carbide is just sufficient for chemical reaction and for dissipating the heat of reaction. The calcium hydroxide is obtained in the form of a dry, easily pourable powder having a residual moisture content of 1–6%. The heat of reaction is dissipated by evaporation of part of the generator water.

Generators are classified according to their working pressure as either low or medium pressure. This classification is governed by the regulations concerning acetylene plants and calcium carbide storage facilities (*Acetylenverordnung*) [60] issued by Deutscher Acetylenausschuß (German acetylene committee) and the associated technical rules for acetylene plants and calcium carbide storage facilities TRAC (*Technische Regeln für Acetylenanlagen und Calciumcarbidlager*) [61]. The *Acetylenverordnung* and *TRAC* constitute a comprehensive set of rules for handling acetylene. Recommendations in other countries (e.g., United States) deal only with some aspects, such as safety precautions [62].

Low-pressure generators are designed for a maximum allowable working pressure of 0.2 bar. They must be rated for an internal pressure of at least 1 bar. Lower pressure ratings are possible if proof is given in each particular case that the generator can withstand the expected stress (maximum working pressure, water filling, agitator, etc.; *TRAC 201*).

Medium-pressure generators have a maximum allowable working pressure of 1.5 bar. They must be rated for an internal pressure of 24 bar. A design pressure of 5 bar suffices whenever the generators are equipped with rupture disks of a defined size and specified response pressure (3 – 4.5 bar, *TRAC 201*).

4.3.4.1. Wet Generators

Wet generators are used primarily for the production of small amounts of acetylene, e.g., for welding purposes. Wet generators work by one of three different principles [63]:

1) The *carbide-to-water principle*, where the carbide is mixed with a large excess of water at a rate corresponding to the gas withdrawal rate. Most generators today work by this principle.
2) The *water-to-carbide principle* (drawer type generators), where water is added at a controlled rate to the carbide, which is held in a replaceable container (drawer).
3) The *contact principle* (basket generators), where the carbide, which is held in a basket, is immersed into the generator water. This type is designed so that the water drifts off the basket as a result of the gas pressure at low gas withdrawal rates and, conversely, returns to the basket when gas withdrawal rates increase.

Medium-Pressure Generators. The Messer Griesheim MF 1009 is a typical carbide-to-water generator (Fig. 28).

The carbide skip (a) is filled with carbide of the 4 – 7 grain size. The skip is connected by gas-tight gates to the hopper (b) and is purged of air by nitrogen or acetylene. The carbide drops into the hopper (b) and is fed continuously by the feeding system (c) to the gasification chamber (d). The gasification chamber contains water up to a level defined by the generator capacity and is equipped with an agitator (e) for whirling the lime slurry. The heat evolved in gasification heats the generator water. For continuous operation the water temperature must not exceed 90 °C; therefore, fresh water is admitted continuously to the gasification chamber. If the defined water level is exceeded, the slurry valve (f), controlled by a float, opens, allowing the excess water and the lime slurry to be discharged from the generator. The acetylene generated collects above the water and is withdrawn. The feeding system (c) is controlled by the gas pressure, i.e., the rate at which carbide drops into the gasification chamber varies directly with the rate of gas withdrawal.

The carbide stock in the hopper (b) is sufficient for about one hour, but the skip (a) can be refilled with carbide and replaced on the hopper so that continuous operation is possible. The wet generator described has a continuous hourly output of 75 m^3 of acetylene. The skip holds 1000 kg of carbide.

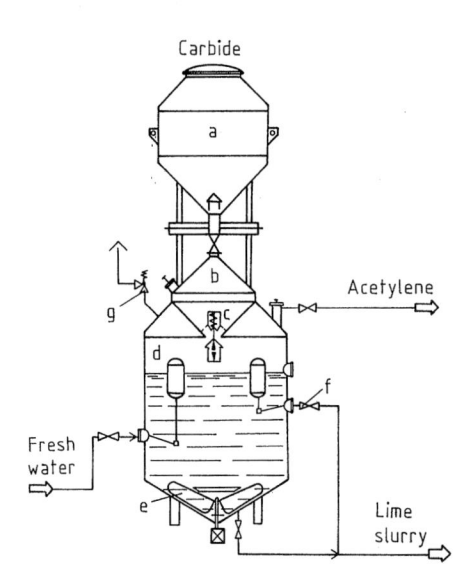

Figure 28. Medium-pressure wet generator
a) Carbide skip; b) Hopper; c) Feeding system; d) Gasification chamber; e) Agitator; f) Slurry valve, g) Safety device

Low-Pressure Generators. The working principle of the low-pressure carbide-to-water generators is very similar to that of the medium-pressure carbide-to-water generator described above. In most cases, a downstream acetylene holder, normally of the floating gas bell type, is provided. In contrast to the medium-pressure generator, in which the carbide feed rate is controlled by the acetylene gas pressure, the feed rate in the low-pressure generator is controlled by the position of the bell in the acetylene holder, i.e., by the gas quantity.

Products. The acetylene generated in the wet generator can be used for welding, often without further purification. In certain cases, coke- or gravel-filled purifiers or a wet scrubber are connected downstream from the generator for separating solid or liquid particles. Before it is fed to a synthesis unit, the acetylene must be purified chemically (see Section 4.3.4.3).

The lime slurry formed is fed into pits. Here, the calcium hydroxide settles in the form of a lime dough containing 35–75 wt% water (wet lime, carbide lime dough). This dough is used as carbide lime.

4.3.4.2. Dry Generators

Dry generators are mainly used for the production of large quantities of acetylene for chemical synthesis.

Compared to the wet generator, the primary advantage of the dry generator is that the dry calcium hydroxide formed as a byproduct can be used in other processes more easily, more cheaply, and in a more diversified way than the lime slurry obtained in the wet generator [64]. Moreover, lime recycling into the carbide production process is only possible with dry calcium hydroxide.

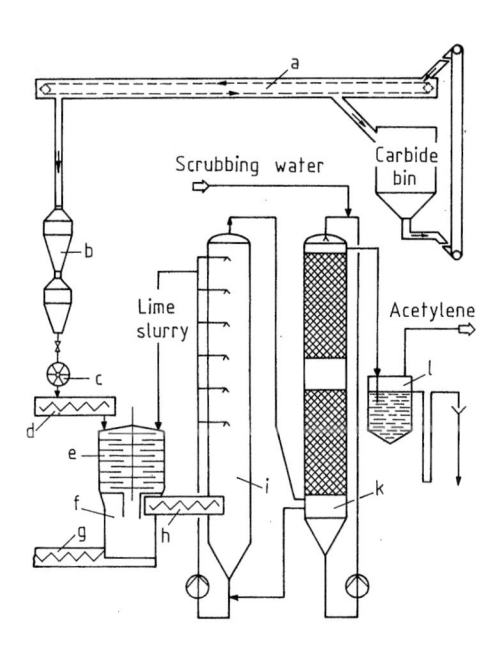

Figure 29. Knapsack dry generator
a) Chain conveyor; b) Feed bin; c) Star wheel; d) Carbide feed screw; e) Generator; f) Lime lock-hopper; g) Lime discharge screw; h) Lime scraper; i) First scrubbing tower; k) Second scrubbing tower; l) Dip seal

A high gasification rate and the elimination of the risk of overheating were originally the most important criteria for the design of dry generators. Early designs of dry generators worked by continuous renewal of the reaction surfaces of coarse carbide with the aid of rotating drums, blades, vibrating screens, and similar equipment.

Typical examples are the early generator of Shawinigan Chemicals [65] and the Piesteritz dry generator [66].

Although a number of factors affect the gasification rate of carbide, e.g., density, porosity, and crystalline structure, above all it is the specific surface that affects the carbide gasification rate the most. Hence, dry generators work with finely ground carbide (0–3 mm), which gasifies in a fraction of the time needed for coarse carbide. The result is a high space–time yield.

A typical application of this principle is the large-scale *Knapsack dry generator*, which was developed at the Knapsack works of Hoechst. This type of generator is used worldwide and is described in more detail below (Fig. 29).

Carbide of the grain size 0–3 falls from the chain conveyor (a) into the subdivided feed bin (b). The chain conveyor is loaded with material from the carbide bin. Because of the recirculating stream of carbide the feed bin (b) is full at all times. The carbide layer in the feed bin (b) acts as a gas seal between the generator and the carbide conveying system. The carbide is fed to the generator (e) via the star wheel (c) and the carbide feed screw (d). The largest generator of this kind built to date has a diameter of 3.5 m and an overall height of approximately 8.0 m. The generator has up to 13 circular trays. These are so designed as to leave alternate annular gaps on the shell side and at the central agitator shaft. The agitator shaft moves stirrer paddles across the trays.

The carbide first reaches the uppermost tray where the generator water is also admitted. The reaction mixture consisting of carbide, water, and calcium hydroxide is pushed by the stirrer paddles towards the outer edge, drops on to the second tray, returns towards the center, etc. When it reaches the last tray, the carbide has been fully gasified. The calcium hydroxide, which still contains up to 6 % water, drops into the lime lockhopper (f). Here, a lime layer two meters deep serves as the gas closure between generator and lime conveying system. The lime is withdrawn continuously.

The gas leaving the generator through the lime scraper consists of 25 % acetylene and ca. 75 % water vapor. The water vapor is the result of dissipating the major portion of the reaction heat. Depending on the generator load, up to several hundred kilograms of lime hydrate dust are carried along with the acetylene. The lime scraper (h) retains the major portion of this dust and returns it to the generator. The remainder is sent together with the gas into the first scrubbing tower (i). Here, lime slurry is sprayed into the hot acetylene gas (ca. 90 °C) to scrub out the lime dust; part of the water vapor condenses because of the simultaneous cooling. In the second scrubbing tower (k), the acetylene is sprayed with atomized water to cool the gas below 40 °C; additional water vapor condenses here. Any ammonia still present in the gas is also removed.

The acetylene leaves the generator via the dip seal (l). It still contains certain impurities in the form of sulfur and phosphorus compounds.

The Knapsack dry generator is suitable for a carbide throughput of 15 t/h, corresponding to an acetylene quantity of 3750 m^3/h. During this process, about 17.5 t of calcium hydroxide per hour are obtained. The pressure in this low-pressure generator amounts to approximately 1.15 bar (115 kPa).

The dry generator of Shawinigan Chemicals, Montreal [67], also processes finely ground carbide and has a variety of applications. It consists of several superimposed troughs. Carbide and water are fed into the uppermost trough. The reacting mixture, which is constantly kept in motion by blades, flows over a weir onto the trough below, etc. At the uppermost trough, water is admitted at such a rate that carbide-free calcium hydroxide can be withdrawn at the lowermost trough. The generated acetylene is purified in two scrubbing towers and cooled.

The calcium hydroxide formed (carbide lime) has a wide range of applications, e.g., in the building industry (for preparing mortar, cement, etc.), in the chemical industry (for neutralization and for recycling to the carbide furnace), in agriculture (as fertilizer), and for water purification and waste water treatment [64].

4.3.4.3. Acetylene Purification

During the gasification of carbide with water, gaseous compounds become mixed with the acetylene, and these must be removed because they have a harmful effect on the downstream chemical synthesis processes. The impurities are mainly sulfur and phosphorus compounds. They can be removed by one of the following purification processes.

In the first process, *dilute chlorinated water* is used as the oxidizing agent. The chlorine concentration of the water is limited to 1.5 g/L to prevent the formation of unstable chlorine compounds, which present an explosion hazard. The chlorine scrubbing step is followed by a caustic soda scrubber to remove the hydrogen chloride formed during the oxidation process. The disadvantage of this purification process is that considerable quantities of scrubbing water are produced.

The second process uses 98% *sulfuric acid* as the oxidizing agent [68]. Because very small quantities of sulfuric acid are admitted, it is difficult to dissipate the heats of absorption and reaction. Heating the acetylene results in increased formation and settling of polymerization products in the purification stage. For this reason the gas requires additional cooling in the event that the acetylene contains appreciable quantities of impurities. Moreover, it is recommended that a second scrubbing tower be kept on standby if a high onstream factor is desired (e.g., 91% ≙ 8000 h/a operating time).

The sulfuric acid scrubber is followed by a caustic soda scrubber, in which the sulfur dioxide formed during oxidation is removed.

The main advantage of this purification method is that virtually no waste water is obtained. The small amount of polluted, highly concentrated sulfuric acid can be used, for example, in fertilizer plants.

These two purification processes yield the following acetylene purities (by volume):

Acetylene	> 99.5%
Sulfur, as H_2S	< 10 ppm
Phosphorus, as H_3P	< 10 ppm

As a result of the extremely good sorption properties of the concentrated sulfuric acid, very pure acetylene can be expected.

4.4. Other Cracking Processes

4.4.1. Thermal Cracking By Heat Carriers

Well-known processes using heat carriers, such as the Wulff and Hoechst high-temperature pyrolysis (HTP) processes, are no longer used because they require refined petrochemical feedstocks such as naphtha and liquid petroleum gas. The Wulff process uses refractory material as the heat carrier, whereas the Hoechst HTP process uses hot combustion gases.

Newer processes, which are able to convert crude and heavy distillates into olefins and considerable amounts of acetylene, are still in the pilot-plant stage. These processes include the advanced cracking reactor process developed by Kureha, Chiyoda, and Union Carbide, using high-temperature superheated steam, and Dow's partial combustion cracking process, using hot combustion gases produced from oxygen and fuel oil as the heat carrier.

Wulff Process [69], [70, p. 58]. This process is based on indirect heat transfer, an approach fundamentally different from the partial-oxidation and electric-arc processes. The hydrocarbon feed is cracked in refractory ovens previously heated by combustion gas. After cracking, the products are quenched outside the reactor. Soot formation is a serious problem because the feed cannot be heated as rapidly as in the partial-oxidation or arc processes. This problem can be diminished by using a feed with a higher hydrogen:carbon ratio. However, methane is not suitable because of the high tempera-

Figure 30. High-temperature pyrolysis (HTP) process

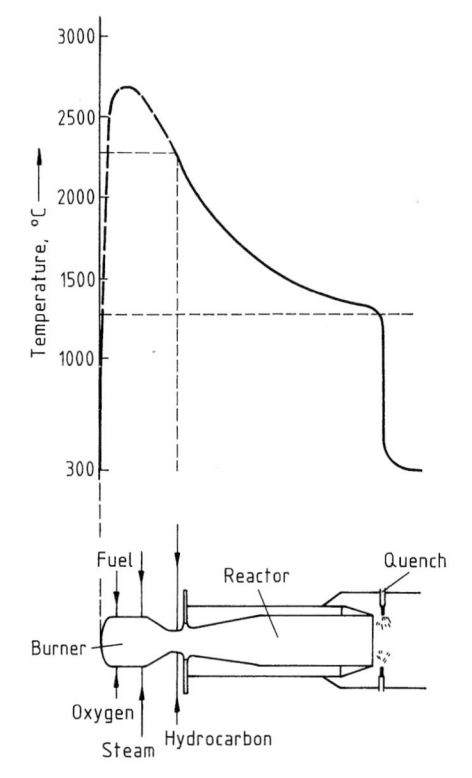

ture and high heat of reaction required, resulting in a low conversion rate. Thus the best feed for the Wulff process is ethane or propane.

Hoechst High-Temperature Pyrolysis (HTP) Process (Fig. 30) [1], [70, p. 55], [71]. This is a two-stage process. In the first stage, heat is produced in the burner by the combustion of residual cracked gas from the acetylene recovery section (CO, H_2, CH_4) with oxygen. Immediately after combustion, the temperature is about 2700°C; this is moderated to about 2300°C by the injection of steam before the reactor is entered. In the second stage, the feedstock naphtha is injected, and the adiabatic cracking reaction takes place. A final temperature of about 1300°C is reached: This determines the cracked gas composition. By varying the feed rate of naphtha the acetylene – ethylene ratio can be altered from 30 : 70 to 70 : 30. However, thermodynamic and economic considerations show that the optimum ratio is 40 : 60.

After a reaction time of a few milliseconds the cracked gas is quenched to approximately 250 °C by the injection of cracked oil from the process. The oil absorbs heat from the cracked gas and is passed through waste heat boilers, raising the steam pressure. No soot is formed in this process, even when crude oil is used as feed because of the high steam content of the carrier gas.

After the oil crisis of 1973 the process became uneconomical in spite of its high thermal efficiency, and in 1976 it was shut down after 15 years of operation. However, one unit is still running in Czechoslovakia.

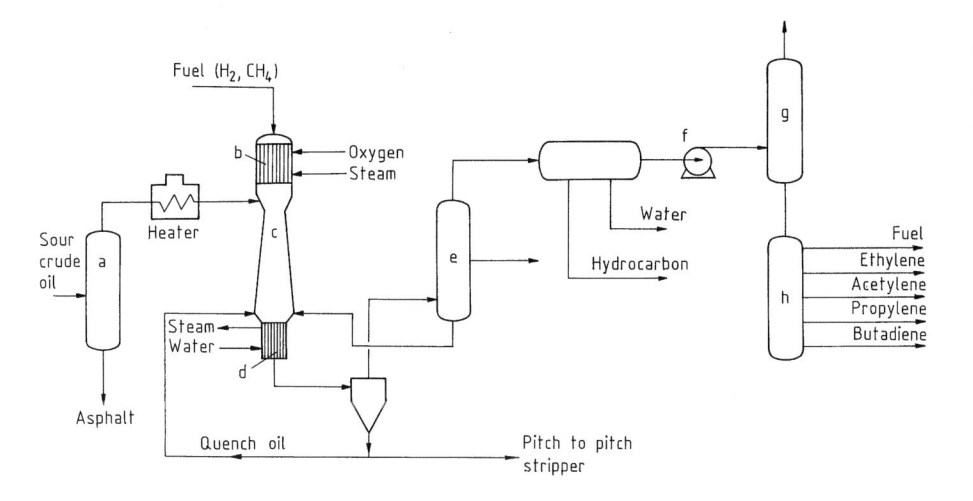

Figure 31. Advanced cracking reactor process (ACR)
a) Crude distillation column; b) Burner; c) Advanced cracking reactor; d) Ozaki quench cooler; e) Oil gasoline fractionator; f) Compressor; g) Acid gas removal column; h) Gas separator

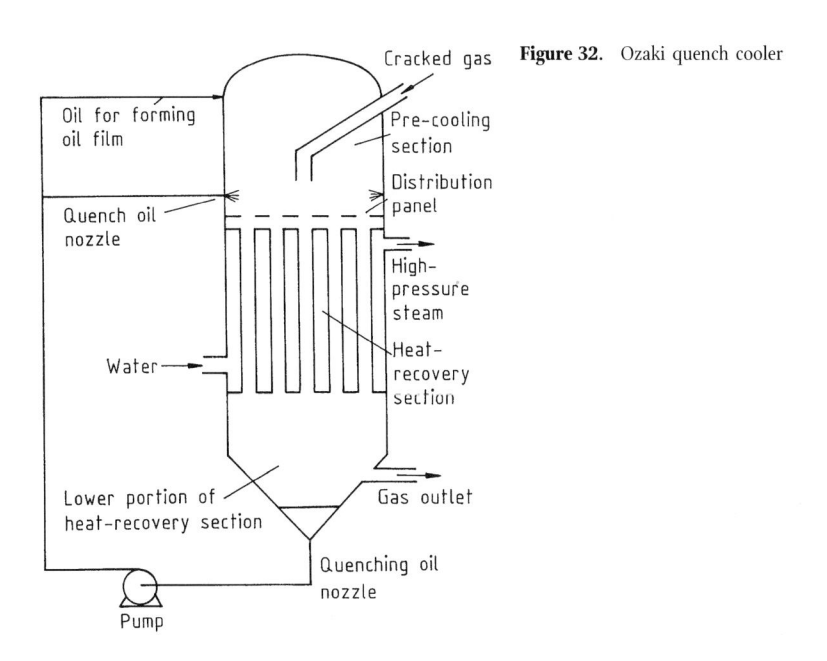

Figure 32. Ozaki quench cooler

Kureha, Chiyoda, Union Carbide Advanced Cracking Reactor (ACR) Process

[72]. To avoid dependence on oil refineries or gas processors for the supply of feedstocks, processes for directly cracking crude oil have been developed by various companies for the production of olefins. Some of these processes operate at reaction temperatures intermediate between those of the usual crack processes for olefins and

165

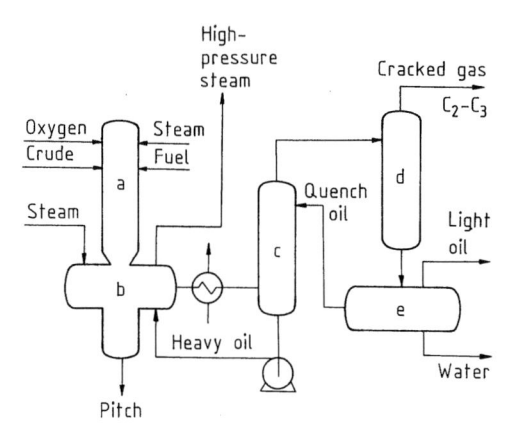

Figure 33. Dow partial combustion cracking process (PCC)
a) Reactor; b) Quench boiler; c) Quench column; d) Stripper; e) Decanter

those for acetylene. The ACR process (Fig. 31) uses a multi-port burner to produce a heat carrier gas of 2000 °C by the combustion of $H_2 - CH_4$ mixtures with oxygen in the presence of steam preheated to 800 °C.

The oil to be cracked is introduced through nozzles into the stream of carrier gas and passes into an advanced cracking reactor, where the reaction takes place adiabatically at 5 bar (0.5 MPa). The initial temperature is 1600 °C; the final temperature at the exit of the reactor is 700 – 900 °C after a residence time of 10 – 30 ms. The cracked gas is quenched by oil in an Ozaki quench cooler (Fig. 32), where steam production up to 120 bar (12 MPa) is possible. This particular boiler design was developed for a high heat transfer rate without coke formation on the exchanger surfaces. Yields reported for Arabian light crude oil are 11.2 wt % hydrogen and methane, 40.7 wt % olefins, and 4.2 wt % acetylene. The acetylene yield is about ten times higher than in usual olefin processes.

Dow Partial Combustion Cracking (PCC) Process [72]. The basic idea of this process is to reduce coking and soot formation considerably when heavy feeds are cracked and when hydrogen is present in the reaction mixture. The PCC process (Fig. 33), which accepts crude oil and heavy residue as feedstock, attains a high partial pressure of hydrogen in the reaction zone by recycling the quench oil (produced in the process) to the burner where it is partially oxidized to yield synthesis gas. Thus there is no need to find a use for the quench oil as in the case of the ACR process. Starting from residual oil boiling above 343 °C, yields are given as 12.4 wt % methane, about 38 wt % alkenes, and 2.5 wt % acetylene. This is seven to eight times more acetylene than that obtained from a steam cracker, but less than the acetylene yield of the ACR process, because of a residence time in the reaction zone which is three to ten times longer.

4.4.2. Acetylene as a Byproduct of Steam Cracking

In a steam cracker saturated hydrocarbons are converted to olefinic products such as ethylene and propylene. Besides these desired components, acetylene and many other products are formed in the cracking process (Fig. 36). The concentration of acetylene depends on the type of feed, the residence time, and temperature (cracking severity: expressed as conversion or propene/ethylene ratio P/E). Typical data are given in Table 10. The acetylene concentration in the off-gas from the furnace varries between

Table 10. Yields of unsaturated components (wt %) in raw gas from steam cracking

Feedstock	Cracking severity	Acetylene	Propyne	Propadiene
Ethane	65 % convers.	0.4–0.50	0.04	0.02
LPG	90 % convers.	0.65–1.20	0.63	0.35
Full-range naphtha	P/E: 0.4	0.9–1.05	0.81	0.54
Full-range naphtha	P/E: 0.53	0.5–0.70	0.68	0.50
Full-range naphtha	P/E: 0.65	0.25–0.42	0.46	0.38
Atmospheric gas oil	P/E: 0.55	0.40	0.34	0.29
Hydrocracker residue	P/E: 0.55	0.50	0.36	0.31

0.25 and 1.2 wt %; the corresponding content of acetylene in the C_2 fraction is about 0.4–2.5 wt %. An ethylene plant producing 400 000 t/a ethylene produces 4500–11 000 t/a acetylene. The acetylene is removed by catalytic selective hydrogenation or solvent extraction.

Acetylene Hydrogenation. Most ethylene plants are equipped with a hydrogenation unit. Acetylene is converted selectively to ethylene on a Pd-doped catalyst. Typical process conditions are temperatures of about 40–120 °C, pressures of 15–40 bar and space velocities of 1000–120 000 $kg\,L^{-1}\,h^{-1}$. Depending on the type of feed and the plant, there are two process options:

1) Front-end hydrogenation (C_2 stream containing H_2, CO, methane, C_2H_2, C_2H_4 and C_2H_6)
2) Tail-end hydrogenation (pure C_2 stream containing C_2H_2, C_2H_4, and C_2H_6; separate addition of an equimolar amount of hydrogen)

General aspects of the process and the catalyst requirements are reviewed in [74], [75] (see also → Ethylene).

Acetylene Recovery. Acetylene is extracted from the C_2 fraction of the steam cracker. The solvent must fulfil the following criteria:

- Melting point lower than the dew point of the feed gas
- High solubility of acetylene at a temperature near the dew point of the C_2 fraction
- High acetylene selectivity ([10], [12])

Acetylene

– High chemical and thermal stability
– No foaming tendency due to traces of hydrocarbons
– Low toxicity
– Low vapor pressure at the operating temperature

After testing many solvents, including DMF, *N*-methyl-2-pyrrolidone (NMP), and acetone, the most suitable solvent for such a process proved to be DMF. The solubility of acetylene as a function of temperature is shown in Figure 5.

The process for the recovery of high-purity acetylene is shown in Figure 34 [76]. The

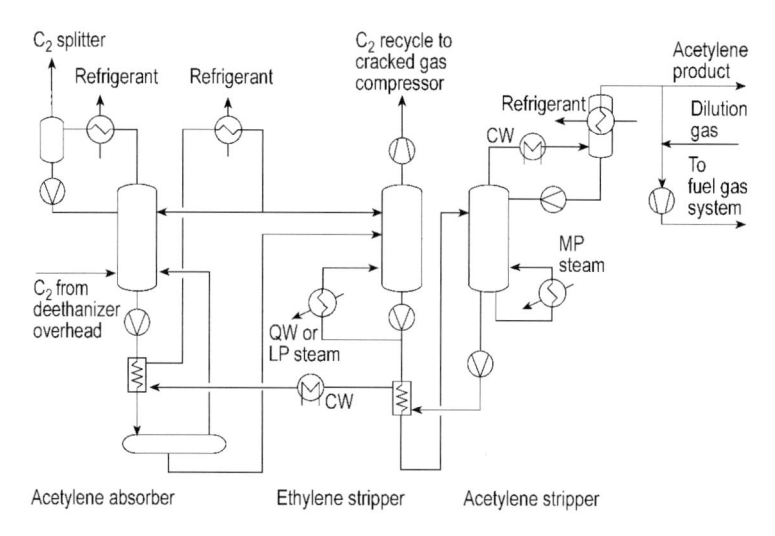

Figure 34. Acetylene recovery process [76]

gaseous C_2 mixture, consisting of ethylene, ethane, and acetylene, is fed to the acetylene absorber; the gas stream is contacted with counterflowing lean DMF at a pressure of 0.8 – 3.0 MPa. The process is suitable for the full pressure range prevailing in any of the known ethylene processes. The entire acetylene and some of the ethylene and ethane are dissolved by the solvent. Entrainment of DMF at the top of the column is avoided by a reflux stream. The purified C_2 fraction, containing < 1 ppm of acetylene, is fed to the C_2 splitter. The rich solvent stream is sent to the ethylene stripper, which operates slightly above atmospheric pressure. Ethylene and ethane are stripped off and recycled to the first stage of the gas compressor for the cracked gas. Any acetylene entrained with the overhead gas is recovered by washing with cold solvent at the top of the stripper. In the acetylene stripper, pure acetylene is isolated from the top of the column. After cooling and heat recovery, the acetylene-free solvent is recycled to the absorber and ethylene stripper. The acetylene product has a purity of > 99.8 % and a DMF content of less than 50 ppm and is available at 10 kPa and ambient temperature.

The material balance and the utilities consumption of an acetylene recovery unit are listed in Tables 11 and 12. At present, more than 112 000 t/a of petrochemical acetylene from twelve olefin plants worldwide is recovered by this technology; three other

Table 11. Material balance (mol %) for an acetylene recovery process operating on the C_2 fraction from a plant producing 400 000 t/a C_2H_4 (Linde)

	Gas to absorber	Purified gas	Recycle gas	Product C_2H_2
Methane	trace	trace		
Acetylene	2	1 ppm	4	99.8
Ethylene	82	83.5	85.7	0.2
Ethane	16	16.5	10	trace
C_3	trace	trace	0.3	trace
DMF		1 ppm	trace	trace
Temperature, K	252	249	255	258
Pressure, MPa	2 ·	1.98	0.11	0.12
Flow rate, kmol/h	2186	2126	17.5	52.5

Table 12. Consumption of utilities for an acetylene recovery process operating on the C_2 fraction from a plant producing 400 000 t/a C_2H_4 (Linde)

DMF, kg/h	1.3
Heating steam, t/h	3.9
Cooling water, m³/h	100
Electrical energy, kW	125
Refrigerant, GJ/h	6.3
Quench water, GJ/h	3.1
Plot area, m × m	15 × 40

facilities are under construction (1998). Figure 35 shows an industrial plant with a design capacity of 14 400 t/a of high-purity acetylene.

A material balance for ethylene plant outputs including acetylene extraction or hydrogenation is shown in Figure 36 [73]. The economic evaluation shows that petrochemical acetylene remains attractive even if the price of ethylene is doubled. It is economical to retrofit acetylene absorption in an existing olefin plant equipped with a catalytic hydrogenation.

A similar process is available for propyne (see Chap. 7)

Figure 35. Acetylene recovery plant (name plate capacity: 14 400 t/a of high-purity acetylene)

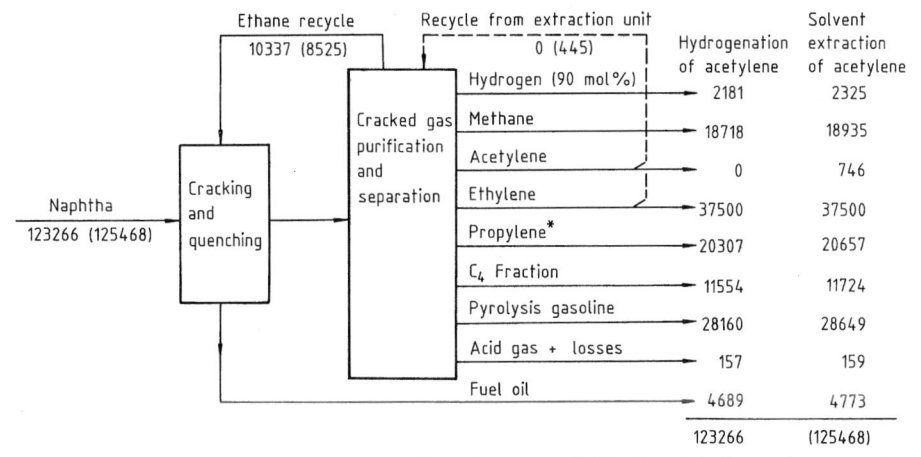

Figure 36. Material balances of a 300 000 t/a ethylene plant equipped with either C_2 hydrogenation or acetylene extraction (all rates in kg/h, the numbers in parentheses are for the solvent extraction process) [73]
* Chemical grade

5. Safety Precautions, Transportation, and Storage

General literature is given in [11], [78], [79].

5.1. General Safety Factors and Safety Measures

Decomposition and Combustion. Acetylene is thermodynamically unstable under normal conditions. Decomposition into carbon and hydrogen can achieve temperatures of about 3100 °C, but due to formation of other products, the temperature reached adiabatically is 2800–2900 °C. The decomposition can be initiated by heat of reaction, by contact with a hot body, by an electrostatic spark, by compression heating, or by a shock wave. The decomposition induced by heating the wall of the container or pipe is very sensitive to the pressure, the size and shape of the container or the diameter of the pipe, the material of the container, and traces of impurities or other components. Solid particles such as rust, charcoal, alumina, and silica can lower the ignition temperature compared to clean steel pipe.

Decomposition gives rise to different scenarios:

– Working range I (deflagration): a flame produced by decomposition or combustion and propagates at a velocity below the velocity of sound into the unconverted gas (pressure rises simultaneously in front of and behind the flame front)

171

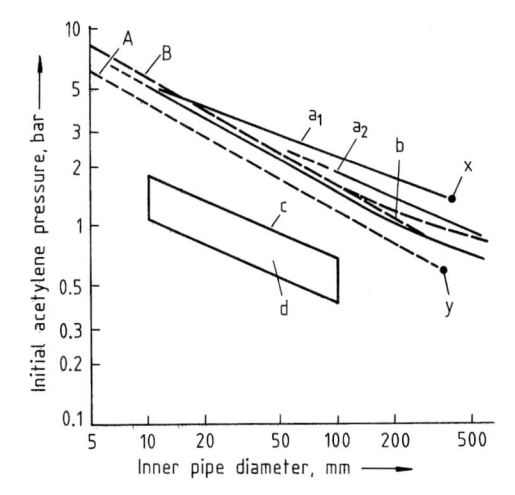

Figure 37. Detonability limits of acetylene [81], [86]
A) Deflagration limit; B) Detonation limit
Detonation limits: a) Thermal ignition in a plain pipe (a_1 melting wire, 20–80 J; a_2 detonator cup, ≈2400 J); b) Thermal ignition plus orifice; c) Ignition by chemical reaction in a shock wave; d) Range of possible quasi-detonation depending on ignition energy of shock wave; x) and y) Limiting ignition pressure for thermal ignition with melting Pt wire and with detonator cup, respectively

Table 13. Stability pressure (bar) of acetylene and acetylene mixtures for two methods of ignition and pressure increase

Mixture	Ignition method		p_{ex}/p_a
	Reppe	BAM	(BAM)
100% C_2H_2	1.4	0.8	
90% C_2H_2/10% N_2	1.8	1.0	8.6
90% C_2H_2/10% CH_4	2.1	1.0	7.2
90% C_2H_2/10% H_2	1.6	0.9	6.9
50% C_2H_2/50% N_2	9.0	3.6	6.3
50% C_2H_2/50% CH_4	14.7	12.9	
50% C_2H_2/50% H_2	4.7	2.5	5.5

* p_{ex}/p_a: ratio of maximum pressure to pressure before ignition.

- Working range III (detonation): the flame propagates at ultrasonic velocity into the unconverted gas (shock wave between low pressure in the unconverted gas and high pressure in the converted gas)
- Working range II (intermediate between I and III): often the propagation velocity of a deflagration is not constant (it increases with increasing density, temperature, and turbulence), and therefore a change from deflagration to detonation is observed

As consequence, design criteria for piping and other components are proposed for the different working ranges and depend on the diameter of the pipe. Limit lines for deflagration and detonation are given in [80] on the basis of the work of SARGENT [81]. An extended Sargent diagram is shown in Figure 37.

The limits are influenced by the method of ignition (e.g., melting wire or a detonator cap). Changing the method of ignition from a melting Pt wire (Reppe) to the exploding wire ignitor (ignition energy ca. 70 J) used by BAM resulted in lower stability pressures for acetylene mixtures [82] and pure acetylene [83] (Table 13).

Mixtures of acetylene with methane have higher stability pressures than those with nitrogen or hydrogen. Further information on the effect of additional gases on acetylene

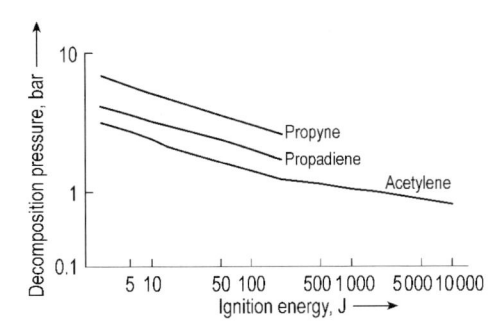

Figure 38. Decomposition pressure versus ignition energy for unsaturated hydrocarbons

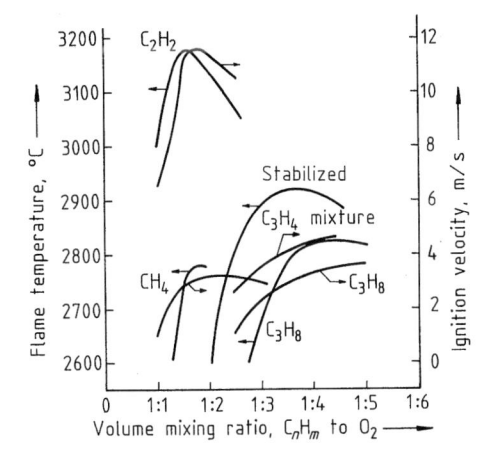

Figure 39. Flame temperatures and ignition velocities of acetylene–oxygen mixtures and mixtures of other hydrocarbons with oxygen

decomposition is given in [84]. The dependence of stability pressure on the energy of the ignition source has led to ongoing discussion about its relevance for industrial design and operations [85].

Figure 38 shows the decomposition pressure of acetylene, propyne, and propadiene as function of ignition energy. This relationship is the basis for the safe design of processes for the recovery of acetylenic components (see Section 4.4.2 and Chap. 7). Additional investigations have been published on the dependence of deflagration pressure on the flow in pipes [86] and the decomposition of high-pressure acetylene in branched piping [87]. Solid acetylene is not critical with regard to decomposition, provided it is the only material involved [9]. In liquid oxygen, solid acetylene can readily ignite on mechanical impact and react violently [88]. Recommendations for equipment used in gas welding and cutting technology, such as rubber hoses, safety devices, and flame arresters, are given in [89], [90], [91].

Combustion of Acetylene in Oxygen (Air). The reaction of acetylene and oxygen at 25 °C and 1 bar to form water and CO_2 generates 1255.6 kJ/mol. Temperatures of around 3100 °C can be reached. Figure 39 shows flame temperatures and flame front velocities for mixtures of oxygen with hydrocarbons [92].

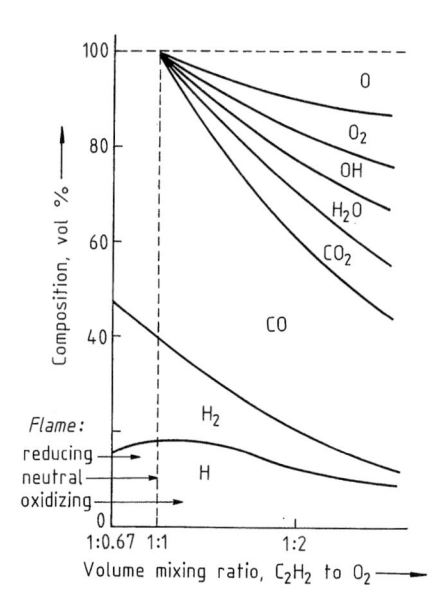

Figure 40. Chemical composition of an oxygen–acetylene flame at its tip versus mixing ratio

Table 14. Fundamental safety data for acetylene–air and acetylene–oxygen mixtures

	Air	Oxygen
Lower flammability limit, vol %	2.5	2.4
Upper flammability limit, vol %	82	93
Flame temperature*, K	2863	3343
Flame front velocity, m/s	1.46	7.6
Increase of pressure (deflagration)	11	50
Detonation velocity, m/s	2300	2900

* Stoichometric mixture.

Acetylene allows the highest temperatures and flame front velocities to be attained. The maximum temperature is very sensitive to the mixing ratio, which also determines whether a reducing, neutral, or oxidizing flame exists (Fig. 40).

Fundamental safety data for acetylene–air and acetylene–oxygen mixtures are listed in Table 14. At atmospheric pressure and 25 °C mixtures of 2.4–93.0 vol % acetylene in oxygen are explosive; the possibility of self-decomposition at high acetylene must also be taken into account.

Handling of Acetylene. For pure acetylene the prescribed safety instructions, for example, the *Technische Regeln für Acetylenanlagen und Calciumcarbidlager* (Technical regulations for acetylene plants and calcium carbide depots), *TRAC*, [93], have to be strictly followed. However, it is not possible to formulate general safety instructions for the great variety of chemical processes with acetylene as reaction component under diverse reaction conditions.

Both handling acetylene and experiments with it necessitate critical examination of sources of possible danger. The literature cited can only serve as an aid to decisions on precautions. The safety regulations mentioned above have been determined in experiments with well-defined apparatus dimensions (length, diameter, geometry). For other dimensions they can only serve as an indication of explosive behavior and should not be considered as rigid limits. The development of economical chemical processes involving acetylene at elevated pressures or under other hazardous conditions calls for decomposition tests for the crucial stages where decomposition could occur. This must be done in close cooperation with official testing institutions, such as the Bundesanstalt für Materialprüfung (BAM) in the Federal Republic of Germany.

In general, the following rules should be observed in handling acetylene:

— Temperature and pressure must be selected so as to avoid liquefaction of acetylene.

— Reactions of acetylene in solvents or with liq-uid reaction components must be carried out at such acetylene concentrations that explosive decomposition of the acetylene in the liquid phase cannot occur. In many cases this condition is fulfilled at an acetylene loading below 100 m^3 (STP) per cubic meter of solvent. Higher loadings are only permitted if additional precautions are taken, such as filling the volumes containing the liquid with steel packings. The formation of a separate gas phase has to be avoided.

— The technical rules (such as *TRAC* [93]) are valid for pure gaseous acetylene. If an inert gas, such as nitrogen, is added to the acetylene, higher acetylene partial pressures are permitted.

— In the design of apparatus the partial pressure of acetylene should be selected so that the minimum distance to the decomposition limit is about 20%. The apparatus should be designed to withstand pressures (1) 12-fold the initial pressure for pure acetylene systems or (2) the initial pressure plus 12-fold the acetylene partial pressure for mixtures and solutions.

— Formation of hydrates (see Chap. 2) under pressure must be avoided because this leads to obstructions in the apparatus and pipelines. The melting point of these compounds is in the range 0–13 °C; therefore, pressurized acetylene containing water has to be kept above 15 °C.

In addition to the measures for building construction, electrical installations, fire protection, purging, and leak detection, acetylene plants and distribution systems are provided with flame traps and flashback and release valves and locks [94], [95]. Flame traps consist either of tubes immersed into water-filled cylinders (wet trap) or cylinders filled with a packing of high surface area to decelerate the decomposition. A wet arrester, which is used for an acetylene distributing line, is shown in Figure 41. Suitable materials for dry-trap packings are sinter metals, ceramic beads (e.g., Raschig rings), bundles of small tubes, and corrugated metal foils [96], [97].

Tapping points for acetylene distribution units which meet the German *TRAC* rules include a nonreturn valve to avoid intrusion of air from downstream, a sinter-metal

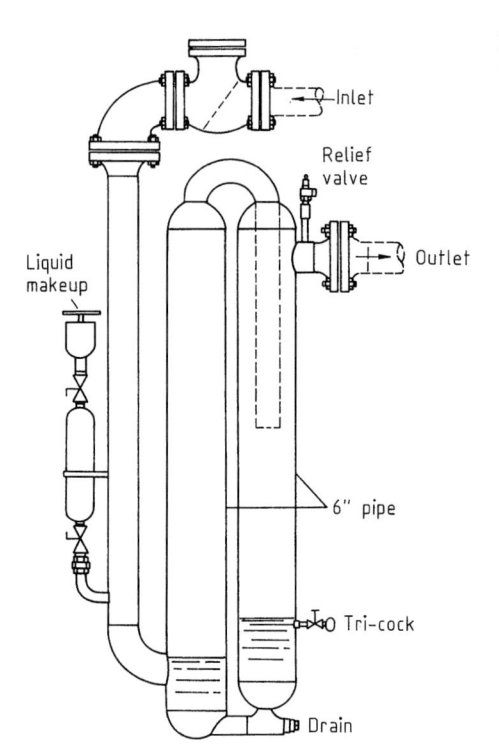

Figure 41. Hydraulic flame trap for acetylene lines (Union Carbide), [98]

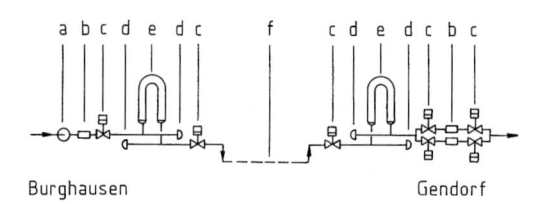

Figure 42. The 8-km acetylene pipeline from the Marathon refinery, Burghausen, to Farbwerke Hoechst, Gendorf, Federal Republic of Germany [99]
a) Compressor; b) Control points; c) Automatic quick-closing valves; d) Rupture disks to atmosphere; e) Flame traps; f) Pipeline

flame trap, and a thermo- or pressure-sensitive spring lock. The last closes if a flame is stopped by the trap but still burns outside of the flame trap. Detailed information is given in [94].

Transportation in Pipelines. Acetylene is occasionally transported in pipelines. Figure 42 shows the safety components of an acetylene pipeline between Burghausen and Gendorf, Federal Republic of Germany. The pipeline was operated until 1976 without incident. Its length is 8 km, and the pipes are 300 mm in diameter. Design pressure was 100 bar, although operation pressure was only 2 bar at the inlet and 1.25 bar at the outlet. The pipeline was provided with rupture disks, which open to atmosphere in case of decomposition. Quick closing valves are initiated simultaneously to protect both upstream and downstream equipment. At each end, part of the pipeline is filled with tube bundles to stop propagation of any acetylene decomposition. The flame traps consist of 600-mm-diameter U's filled with Raschig rings.

A report [98] is available on an acetylene-decomposition event in a pipeline system, which demonstrates the need for safety measures. Instead of transportation of pure acetylene, pipeline transportation of acetylene solutions in acetone was proposed as safer [99]. In the United States, transportation of acetylene solutions in liquid ammonia was considered for existing ammonia distribution systems [100].

Hazardous Acetylene Traces in Low-Temperature Processes. Acetylene is the most dangerous component in gas mixtures processed in low-temperature plants. In air separators, for example, acetylene can be suspended in liquid oxygen as a solid or as a segregated liquid phase that is quite unstable and tends to uncontrollable and violent decomposition. (The solubility of acetylene in liquid oxygen is low, see Chap. 2) Therefore much attention must be given to checking for and removing acetylene in low-temperature separation plants.

Normally air contains some acetylene, up to 0.3 mL/m^3. In industrial areas, especially in the proximity of petrochemical plants, higher concentrations (up to 1 mL/m^3) can occur. Without any measures, acetylene in the feed air of an air separator would be enriched in the cold section of the unit.

In modern air separators alternating molecular sieve adsorbers are used. The adsorbers obey the following breakthrough sequence:

$$CH_4 - C_2H_6 \genfrac{}{}{0pt}{}{\nearrow C_2H_4 \searrow}{\searrow C_3H_8 \nearrow} C_2H_2 - C_3H_6 - C_4 - H_2O$$

To avoid acetylene breakthrough the adsorber is operated for sufficient CO_2 removal and is regenerated when the CO_2 concentration at the adsorber outlet starts to increase. Further details are given in [101]–[105].

Another possible way to remove acetylene and other combustible air contaminants is catalytic oxidation prior to the separation [103], [106]. This method, being expensive, is rarely used.

All air separators are provided with routine analysis systems for acetylene. Routine analysis concentrates on the liquid oxygen of the main condenser [107].

The removal of acetylene in cracked gas separation is treated in Section 4.4.2. The processing of other acetylene-containing gases, e.g., coke-oven gas, by low-temperature separation is described in [108].

5.2. Acetylene Storage in Cylinders

Because of its tendency to deflagrate or to detonate, acetylene cannot be compressed and stored in gas cylinders like other gases.

For desensitizing, acetylene stored in gas cylinders is dissolved in a solvent in which the acetylene is very soluble. This solvent is dispersed in a porous solid that completely

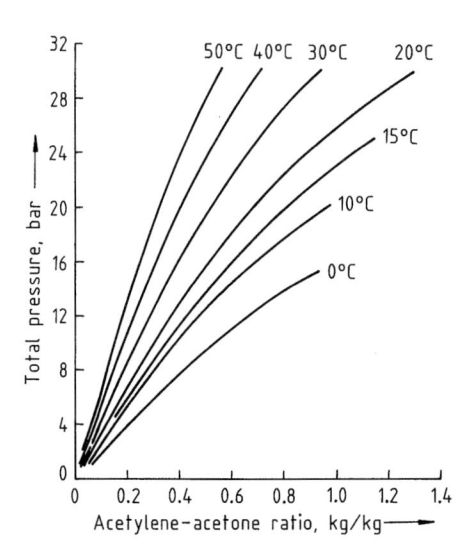

Figure 43. Total pressure of acetylene solution in acetone as a function of acetylene : acetone ratio and temperature [9]

fills the gas cylinder. As well as giving better solvent distribution the porous material arrests any local acetylene decomposition induced, for instance, by flashback.

Acetone and dimethylformamide are the preferred solvents for acetylene in cylinders. An advantage of dimethylformamide is its lower vapor pressure, resulting in lower solvent losses during acetylene discharge. A disadvantage is its higher toxicity. The total pressure of an acetone-containing acetylene cylinder depends on the acetylene : acetone ratio and on temperature as is shown in Figure 43. Deviations from the plotted curves resulting in higher pressures are caused by the porous filling of the gas cylinder, which absorbs acetone, changing the effective acetylene : acetone ratio [109].

Impurities in the acetylene decrease the dissolving capacity of the acetone. Figure 44 shows the effect of moisture on acetylene solubility. As a result, acetylene produced from calcium carbide and water has to be dried.

Calcium carbide-based acetylene contains further impurities that have to be scrubbed out to avoid decreased solubility in acetone. Examples are divinyl sulfide and phosphine: 1 wt% divinyl sulfide in the acetone reduces the acetylene solubility from 35 g/kg to 31 g/kg at 20 °C and an acetylene pressure of 0.1 MPa. Further values, also for phosphine, are given in [110], [111]. The impurities have to be scrubbed out to residual concentrations of 0.5 g of phosphorus and 0.1 g of sulfur per m^3 of acetylene.

During acetylene production from calcium carbide, disperse calcium hydroxide (0.1–1.0 µm) is produced. This contaminates the product gas. The calcium hydroxide present in acetylene filled into acetone-containing gas cylinders catalyzes aldol condensation of the solvent and reduces the solubility for acetylene:

$$2\,(CH_3)_2CO \longrightarrow (CH_3)_2COHCH_2COCH_3 \qquad \text{diacetone alcohol}$$

Therefore, the solids content of calcium carbide-based acetylene for filling acetone-containing gas cylinders must be kept below 0.1 mg/m^3 [110], [112].

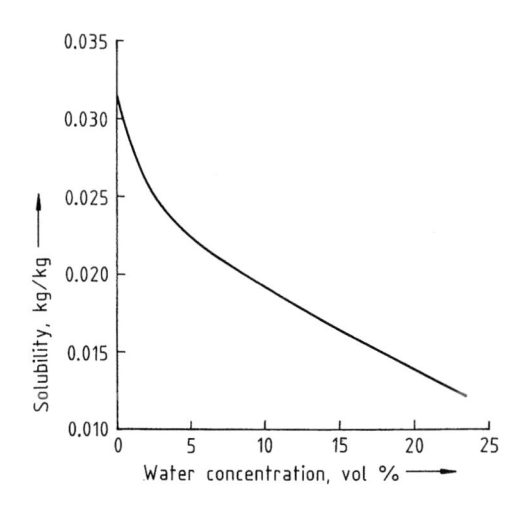

Figure 44. Solubility of acetylene in water-containing acetone at 25 °C and $p_{C_2H_2} = 1$ bar (0.1 MPa) [110]

Table 15. Permitted acetylene and acetone filling of seamless gas cylinders (satisfying German standards and safety rules)

Gas cylinder			Acetone filling, kg			Acetylene filling, kg
Vol. of gas cylinder, L	Outer diameter, mm	Length, mm	Minimum	Maximum General	Exceptional	
3	140	300	0.789	0.8625	0.9375	0.4725
5	140	460	1.315	1.4375	1.5625	0.7875
10	140	850	2.630	2.8750	3.125	1.575
20	204	810	5.260	5.750	6.250	3.150
27	204	1040	7.101	7.7625	8.4375	4.2525
40	204	500	10.520	11.50	12.50	6.30
40	229	1210	10.520	11.50	12.50	6.30

The gas cylinders have to be filled with definite amounts of acetylene and solvent. Commercial-grade, seamless gas cylinders which meet specified standards (in the Federal Republic of Germany, DIN 4664) may be filled with the amounts listed in Table 15.

The amounts are fixed by regulations for handling pressurized gases [114]. Corresponding regulations in the United States have been issued by the Department of Transportation [115].

To determine maximum acetylene filling of gas cylinders, extensive ignition, impact, and heating tests have been worked out [116], [117].

The porous material in the acetylene gas cylinders must satisfy the following requirements: no interaction with the cylinder material, acetylene, or acetone and suitable mechanical properties, such as sufficient impact resistance. Suitable materials include pumiceous compounds, silica, charcoal, asbestos fiber, and alkaline carbonates. The porosity of these materials varies between 70 and 80% [110], [118]. Modern monolithic

Table 16. Examples of approved silica-based porous materials in German acetylene cylinders

Material		Approved filling			
Type	Origin[a]	Acetylene[b], kg	Acetone[c], kg	Acetylene: acetone, kg/kg	Maximum pressure[d], bar
Linde M1	Linde, Munich	8.0	12.7	0.63	19
AGA 2	AGA, Hamburg	8.0	12.4	0.645	19
SIAD 2	SIAD, Sabbio	8.0	12.4	0.645	19

[a] Approval only when porous filling is prepared at place of origin;
[b] Maximum;
[c] Desired value;
[d] Gage, at 15 °C;

Figure 45. Porous silica material Linde M1 for acetylene cylinders (magnification 1:10 000)

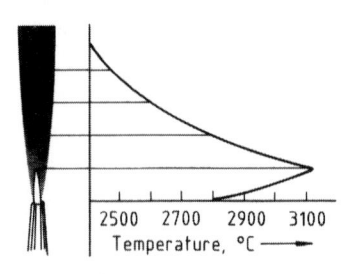

Figure 46. Temperature profile of an oxyacetylene flame

materials are made preferably from silica, lime, and glass fiber. The mixtures are suspended in water to obtain a pasty material which is filled into the gas cylinders. The material is hardened at about 200 °C and subsequently dried and activated at 350–400 °C. A porosity of about 90% is obtained.

Any porous material to be used for acetylene cylinders has to be examined and approved by competent authorities. The examination includes the determination of the maximum acetylene and solvent filling, the maximum filling pressure, and ignition and impact testing. Table 16 lists three porous materials approved for use in Germany. Figure 45 shows a photograph of "Linde M1" magnified 1 : 10 000, clearly revealing the porous structure of such materials.

Methods for examining the materials have been standardized by CPI (Commission Permanente Internat. de l'Acétylène, de la Soudure Autogène et des Industries qui s'y rattachent, Paris) and ISO (International Organization for Standardization) [118].

Discharging acetylene from a gas cylinder leads to acetone losses because the partial pressure of acetone at 15 °C ranges from 0.14 bar at 15 bar total pressure to 0.18 bar at 1 bar total pressure. Solvent loss has to be replaced when an acetylene cylinder is reloaded.

6. Uses and Economic Aspects

6.1. Use in Metal Processing

Acetylene has many applications in the processing of metals and other materials. This is because of the high flame temperature and propagation velocity resulting in high energy densities and rapid heat transfer to the piece being worked. The properties of an oxyacetylene flame given here supplement those in Section 5.1.

The temperature profile of an oxyacetylene flame consists of a hotter primary flame and a scattered secondary flame. The highest flame temperature is at the tip of the primary flame (Fig. 46). For material processing the primary flame is the more important.

The heating efficiency of the primary flame is the product of the volume-based heat released by the primary flame and the propagation velocity. This is plotted in Figure 47 for the oxyacetylene flame and some other flames. The heat transferred in welding is generated by radiation, convection, and thermal conduction (see Table 17). The heat transfer is promoted by a high temperature gradient between flame and workpiece.

Oxidizing, neutral, or reducing (carburizing) flames can be obtained by varying the oxygen:acetylene ratio (Fig. 40). For steel, alumina, and copper welding usually neutral or slightly reducing flames are used, whereas oxidizing flames are preferred for brass welding, cutting, pickling, and surface hardening [120]. Acetylene is burned with oxygen in single torches or in bundles of torches, the chief components of which are the connections for acetylene and oxygen, regulating valves, a mixing chamber (usually

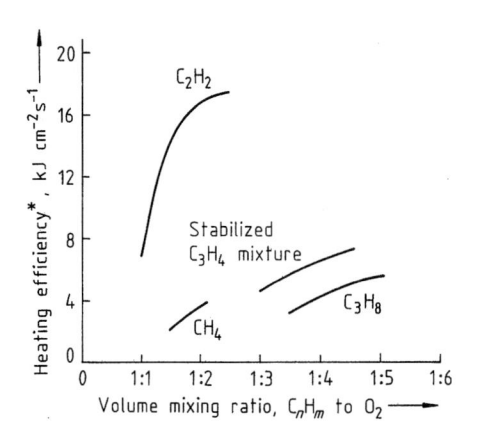

Figure 47. Heating efficiency of acetylene – oxygen mixtures and mixtures of other hydrocarbons with oxygen
* based on area of primary flame cone

Table 17. Heat transfer in welding

Gas temperature K	Welding temperature, K					
	800		1200		1600	
	Q_S	Q_K	Q_S	Q_K	Q_S	Q_K
1000	2.3	13.4				
2000	3.8	55.4	3.4	37.0	2.3	18.5
3000	4.0	83.2	3.8	68.0	3.6	50.4

Q_S, heat transfer by radiation (kJ cm^{-2} h^{-1});
Q_K, heat transfer by convection (kJ cm^{-2} h^{-1}) at a gas velocity of 50 m/s

of the injection type), a flashback protection element, and a nozzle adapted to the specific applications [121], [122].

Oxyacetylene flames are used in welding, cutting, brazing, soldering, surfacing, flame spraying, heating, hardening, straightening, cleaning, pickling, rust removal, and decarbonizing.

Acetylene – air flames are occasionally used for tin brazing, hot air welding of thermoplasts, glassworking, and paint removal [121], although the convenience and safety of fuels such as propane or butane has displaced acetylene in those applications. Soft and hard soldering, flame hardening, and flame tempering are important applications for the softer acetylene – air flame. For acetylene – air mixtures, self-aspirating Bunsen-type and acetylene – compressed air burners are used.

The different uses of oxyacetylene and acetylene – air flames in metal working, the procedures, and the equipment are comprehensivelydescribed in [9] and [123]; other sources of information for oxyacetylene flame properties in welding are [119], [120], and [124].

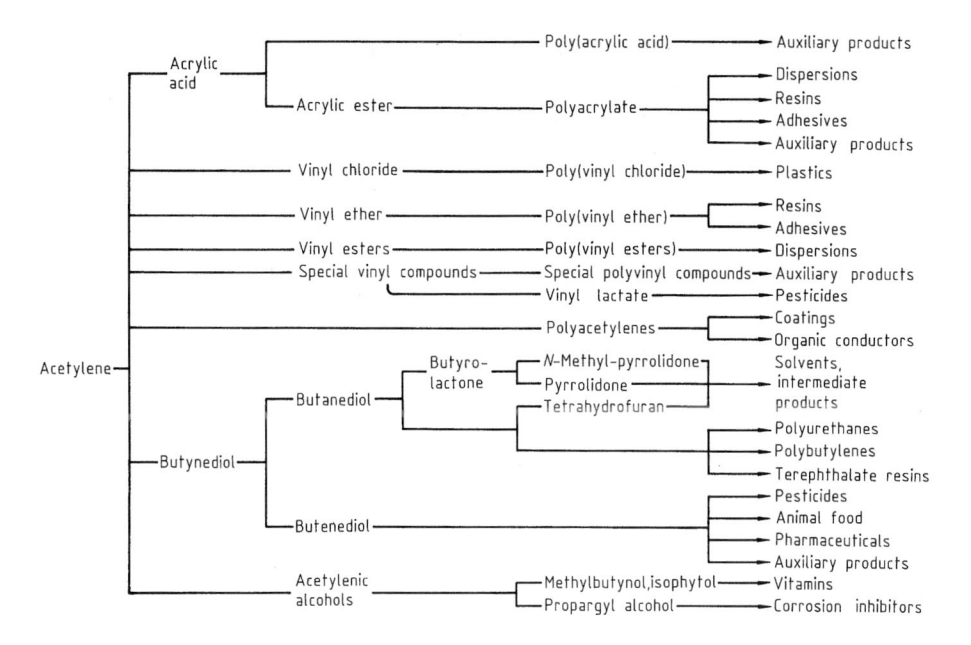

Figure 48. Acetylene as a starting material for industrial products

6.2. Use as Raw Material in Chemical Industry

Because of the diversity of acetylene chemistry (see Section 3.1), acetylene has been used as a starting material for a great variety of industrially important products. These are summarized, together with their applications, in Figure 48.

Between 1960 and 1970, when worldwide acetylene production peaked, most of the products listed in Figure 48 were produced via acetylene. During the last 15 years the competition between acetylene and the olefins (see Section 6.3) has resulted in substitution of ethylene and propene for acetylene, especially in the production of acetaldehyde and acrylonitrile. At present, acetylene is used mainly for the production of vinyl chloride, vinyl acetate, and other vinyl esters; acrylic acid; acetylene black; and acetylenic chemicals such as 1,4-butynediol and acetylenic alcohols. For the acetylenic chemicals the acetylene route is either the only commercial production process available or the predominant process. Vinyl chloride, vinyl acetate, and acrylic acid, formerly the main products from acetylene, are produced today mainly from ethylene and propene [3].

6.3. Competitive Position of Acetylene as Chemical Feedstock

Today, acetylene plays an important role only in the production of the acetylenic chemicals. The fact that acetylene production has not decreased further seems to indicate that the competition from the olefins is no longer as strong as it was. The main reason for this is that European olefin chemistry depends on refinery products, which have become more expensive than natural gas, the main feedstock for acetylene today. Another contributing factor is that acetylene is produced only in old plants, which have low capital costs.

In addition, process improvements, such as an increase in thermal efficiency and optimum use of byproducts by other plants, can make acetylene more competitive. The position of acetylene in chemical industry may be advanced because of the variety of valuable products to which it can be converted in high yields with known technology. Acetylene must compete with ethylene for the production of vinyl chloride and vinyl acetate, and for the production of acrylic acid and its esters it must compete with propene. The Stanford Research Institute [3] has investigated this question in detail. The results, which take into account both capital investment (25 % return) and specific consumption figures for the various processes, show that the prices at which the alternate routes are competitive can be expressed by the following equations:

for vinyl chloride $A = 1.10\ E + 0.42$
for vinyl acetate $A = 1.23\ E + 0.40$
for acrylic acid $A = 1.74\ P + 0.23$

where A is the acetylene price, E is the ethylene price, and P is the propene price, all in \$/kg. For example, if ethylene costs 0.65 \$/kg, acetylene can cost 1.15 \$/kg for the production of vinyl chloride and 1.21 \$/kg for the production of vinyl acetate. If propene costs 0.49 \$/kg, acrylic acid can be profitably produced from acetylene if it costs 1.08 \$/kg or less.

Acetylene prices of 1.08 to 1.21 \$/kg or less can be reached in a new plant only with optimal integration of energy and byproducts within an integrated chemical plant. Figure 49 shows the flow sheet of a 100 000 t/a gas-based partial oxidation acetylene plant in a chemical complex. The required oxygen facility and methanol plant based on acetylene synthesis gas (off-gas) are included. The main products of the complex are acetylene and methanol. The acetylene process is operated with improved quenching technology, allowing a high proportion of energy to be regained in the form of steam. The aromatic residue oil from a steam cracker is converted into high-purity coke and light aromatics.

A production cost estimate based on power, feedstock, and product prices roughly corresponding to market prices in 1982 shows that low production costs for acetylene are possible under the conditions described, in spite of the relatively high natural gas price of 5.5 \$/10^6 BTU (0.021 \$/kW h), which in heating value terms comes very close

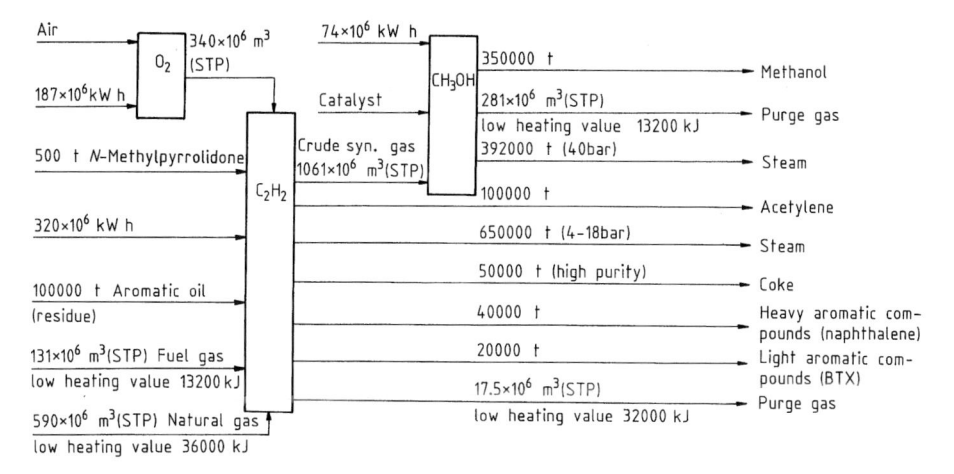

Figure 49. Acetylene – methanol plant

to that of crude oil (0.022 $/kW h). The difference in production costs between acetylene from natural gas and ethylene from naphtha (price 333 $/t, corresponding to 0.028 $/kW h) is so slight that acetylene from new plants can once again compete with ethylene for certain syntheses, provided that there is a difference between the costs of natural gas and naphtha. This was nearly 0.007 $/kW h in 1982 in Germany. A similar calculation for the acetylene production by the Hüls arc process based on an ethylene price of 600 $/t and a hydrogen price 40 % above the heating value results in the same acetylene product value at 0.038 $/kW h for electrical power. Lower electrical energy costs favor the arc process.

7. Propyne

Propyne [74-99-7], methylacetylene, is obtained in cracking processes mostly as a byproduct together with its isomer propadiene [463-49-0], allene. Typical concentrations depend on the feedstock and the cracking conditions (Table 10) and vary between 0.3 and 0.8 wt %. The corresponding figures for propadiene are 0.3 – 0.55 wt %.

Pure propyne is a colorless, nontoxic, flammable gas. The important physical properties of propyne are listed in Table 18 [7], [8], [125]. Vapor pressure curves of propyne and propadiene including melting and critical points are shown in Figure 50 [125]. Table 19 summarizes solubility coefficients of propyne for various solvents [10]; further data are available in [11]. The solubility of propyne in various solvents is shown in Figure 51; the solubilities of the C_3 hydrocarbons in DMF are plotted in Figure 52 for infinite dilution.

The equilibrium between the two isomers of C_3H_4 is reached in the presence of catalysts (for example $Al_2O_3/_3$, SiO_2, actived carbon, and γ-Al_2O_3/Na_2CO_3) [126], [127].

Table 18. Physical properties of propyne

Molecular mass	40.065
Critical temperature	402.39 K (129.24 °C)
Critical pressure	5.626 MPa
Critical volume	0.1635 m^3/kmol
Melting point	170.45 K (−102.7 °C)
Dipole moment	2.61 × 10^{-30} C·m
Density (liquid)	638.92 kg/m^3 (273.15 K)
Normal boiling point	249.97 K (−23.18 °C)
Enthalpy of vaporization (273.15 K)	20.765 kJ/mol (273.15 K)
Molar volume	0.05962 m^3/kmol (249.91 K)
Enthalpy of formation	185.5 ± 1.0 kJ/mol (298.15 K)
Gibbs free energy of formation	193.8 ± 1.0 kJ/mol (298.15 K)
Entropy of formation	248.4 J mol^{-1} K^{-1} (298.15 K)
Enthalpy of combustion	1938.943 kJ/mol (298.15 K)
Heat capacity constant pressure	59.842 J mol^{-1} K^{-1} (273.15 K)
Viscosity (liquid)	1.7500 × 10^{-4} Pa·s (273.15 K)
Viscosity (gaseous)	8.3300 × 10^{-6} Pa·s (293.15 K)
Thermal conductivity (liquid) *	0.14560 W m^{-1} K^{-1} (233.45 K)
Thermal conductivity (gas) *	0.014310 W m^{-1} K^{-1} (273.15 K)
Surface tension *	1.47 × 10^{-2} N/m (273.15 K)

* Predicted or estimated.

Table 19. Solubility coefficients of propyne in various solvents (mol kg^{-1} bar^{-1})

Solvent	C$_3$H$_4$ pressure, bar	−20 °C	25 °C
Methanol	0.1	9.099	1.865
1,2-Dichlorethane	0.098−0.196	3.276 (0 °C)	1.546
Carbon tetrachloride	0.25	4.732	0.842
n-Octane	0.098	3.412	
Toluene	0.25/1.0	8.644	2.047
Xylene (technical).	0.49/0.98	9.782	1.683
4-Methyl-1,3-dioxalan-2-one	0.49/1.0	8.644	1.183
Triethylene glycol	1.0		0.400 (30 °C)
Acetone	0.6/1.0	35.03	4.186
N-Methyl-2-pyrrolidone	≤ 0.78	6.597 (0 °C)	2.502
DMF	≤ 0.15	12.512	3.003
Water	1.0		0.071
Ammonia	≤ 0.04	20.018	4.436

At 270 °C the equilibrium mixture contains 82 % propyne, and at 5 °C, 91.1 % propyne. For calculated data, see [128]. This equilibrium is important for the industrial propyne recovery process.

The decomposition pressure as a function of the ignition energy for propyne (and propadiene) is plotted in Figure 38. The lower and upper flammability limits of propyne in air are about 2.3 and 16.8 vol %. For propadiene only ranges are available: 1.7−2.5 and 12−17 vol % [129].

Production. Propyne and propadiene can be recovered from cracked gas by solvent extraction. The process is outlined in Figure 53 [73], [130] and is similar to the recovery

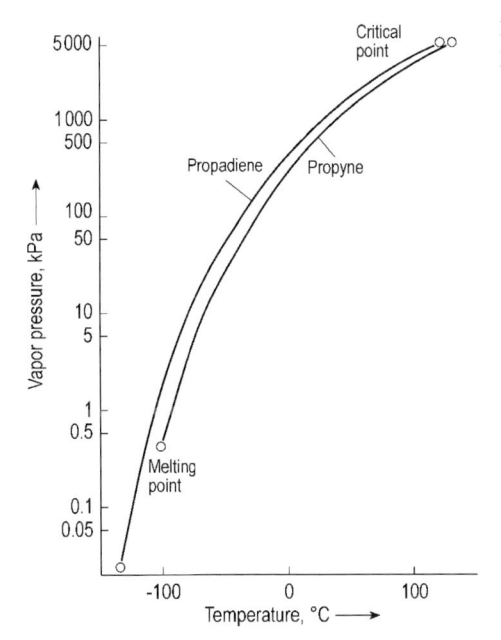

Figure 50. Vapor pressure curves of propyne and propadiene [125]

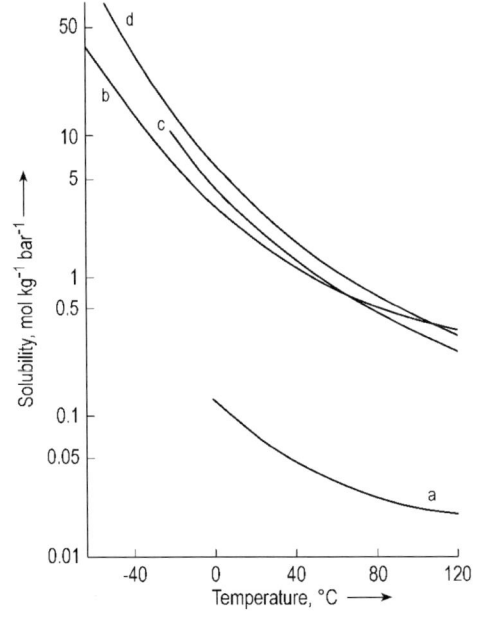

Figure 51. Solubility of propyne in various solvents at infinite dilution
a) Water; b) Methanol; c) N-methylpyrrolidone; d) DMF

of acetylene (see Chap. 4.4.2) An important step is the catalytic isomerization of propadiene to propyne in the liquid phase on a catalyst [127].

The bottom product from propene/propane splitter is routed to depropanizer II for removal of traces of C_{4+} hydrocarbons originating from the feed and the isomerization

Figure 52. Solubility of the C_3 hydrocarbons in DMF at infinite dilution
a) Propyne; b) Propadiene; c) Propene; d) Propane

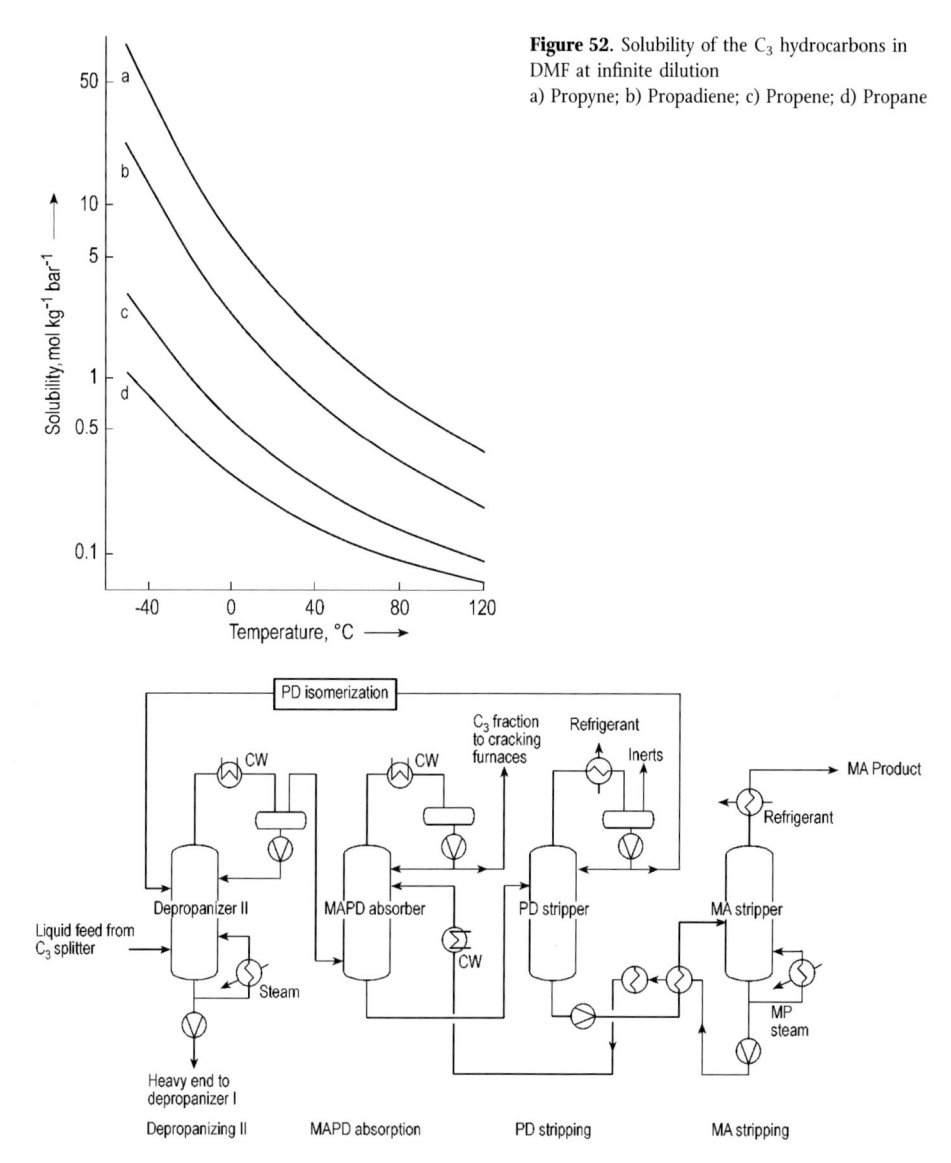

Figure 53. Propyne recovery process (Linde, Shell [76])
MA: methylacetylene; PD: propadiene

reactor effluent. The overhead is sent to the absorption column. The propane (and some propene) is routed back to the cracking furnace. Propyne (MA) and propadiene (PD) are stripped off in the corresponding columns. The propadiene fraction is converted to propyne in the isomerization reactor and recycled to the feed. Traces of solvent in the pure propyne are removed by cooling. The product can be sent directly to methyl methacrylate (MMA) unit, where MMA is manufactured by reaction of propyne with carbon monoxide and an alcohol in the presence of a Pd-based carboxylation catalyst [127], [131].

Usually propyne and propadiene are undesireable byproducts of the steam cracking process and they are removed by selective hydrogenation. The reaction is carried out in the liquid or gas phase. Typical conditions are pressures of 1.5 – 5 MPa and inlet temperatures of 20 – 100 °C. The vapor-phase processes are declining in importance; liquid-phase (or composite phase) processes have some advantages:

- Operation at lower temperatures (higher selectivity)
- Lower operating costs (no vaporization of the feed and recondensation of the product)
- Regeneration of reactor in situ (removal of polymers)
- Lower frequency of catalyst regeneration
- For high concentration of C_3H_4 in the feed, two reactors are often sufficient; the gas-phase process requires more reactors
- Control of conversion by injection of hydrogen (no excess hydrogen) reduces investment cost.
- Control of temperature by use of vaporization minimizes risk of runaway

The selectivity in case of liquid-phase hydrogenation to propylene is about 60 – 70 % and depends on the concentration of C_3H_4 in the C_3 feed. For better control of reaction heat, some product is normally recycled to the feed. The cycle time of the catalyst is very sensitive to contaminants (and to byproduct formation in gas-phase processes), especially in the first reactor. A reactivation procedure is necessary when the catalyst loses its activity.

Stabilized Propyne – Propadiene Mixtures. As a replacement for acetylene, stabilized propyne – propadiene mixtures are available commercially. Trade names are Tetrene or MAPP gas. These mixtures are stabilized by propane, propene, and/or butane and are used for metal cutting, welding, hardening, and brazing. The flame properties are closer to those of propane – propene mixtures. Therefore stabilized C_3H_4 mixtures have not yet won a large-scale market. Further information is available in [132].

8. Toxicology and Occupational Health

Pure acetylene is a simple asphyxiant. When generated from calcium carbide, acetylene is frequently contaminated with arsine, hydrogen sulfide, or phosphine, and exposure to this impure acetylene has often resulted in serious consequences. Today, commercial acetylene no longer contains these impurities and is therefore less harmful [133].

The lowest published lethal concentration for rats is 9 vol% [134]. Dogs are less sensitive: 80 vol% acetylene in the air is necessary to produce a narcosis accompanied by an increased blood pressure and a decreased pulse frequency (stimulation of vasomotor and vagus centers) [135]. In humans, the inhalation of air containing 10 vol% acetylene has a slight intoxicating effect, marked intoxication occurs at 20 vol%, incoordination at 30 vol%, and unconsciousness within 5 min on exposure to 35 vol%. Inhaling 35 vol.% for 5–10 min or 10 vol% for 30–60 min is lethal. Symptoms of intoxication are excitement, coma, cyanosis, weak and irregular pulse, and memory failure [136]–[138].

There is no evidence that *repeated exposure* to tolerable levels of acetylene has effects deleterious to health [139]. Inhalation of air with 33 vol% of acetylene by humans led to unconciousness within 6 min, but when the experiment was repeated within the week the susceptibility to acetylene decreased: 9 min were required on the second exposure and more than 33 min on the third exposure to produce unconsciousness [135].

Acetylene does not irritate the mucous membranes [133]. Neither threshold limit value (TLV) nor a MAK has been established. The standard air concentration allowed for OSHA and NIOSH is 2500 ppm [137].

9. References

[1] *Acetylene or Ethylene as Feedstocks for the Chemical Industry*, Conference proceedings Dechema and SCI, Frankfurt/Main, 27–29 March 1968, SCI, London 1968.E. Schenk: "Acetylen oder Äthylen als Rohstoffe der chemischen Industrie," DECHEMA, Frankfurt/Main 1968.

[2] O. Horn, *Erdöl Kohle Erdgas Petrochem.* **26** (1973) no. 3, 129.

[3] Yen-Chen Yen, *Acetylene*, Report Nr. 16, Supplement A, Stanford Research Inst., Menlo Park, Calif., Nov. 1981, and Information of the VCI, Frankfurt/Main.

[4] A. Stratton: *Energy and Feedstocks in the Chemical Industry*, Ellis Horwood, Chichester, England, 1983, p. 191.

[5] J. Schulze, M. Homann, "Die mögliche Stellung des Acetylens in der zukünftigen Kohlechemie," *Erdöl Kohle Erdgas Petrochem.* **36** (1983) no. 5, 224.

[6] D. A. Plattner, Y. Li, K. N. Houk: "Modern Computational and Theoretical Aspects of Acetylene Chemistry", Chapter 1.2. in P. J. Stang, F. Diederichs (eds.): *Modern Acetylene Chemistry*, VCH Verlagsgesellschaft, Weinheim 1995.

[7] DIPPR Database, STN International, Design Institute for Physical Property Data c/o DIPPR Project Staff, Pennsylvania State University 167 Fenske Lab, University Park, Pa 16802 U.S.A American Institute of Chemical Engineers, 3345 E. 47th Street, New York, NY 10017, U.S.A., basis February 1998.

[8] Beilstein Database, STN International, Beilstein Informationssysteme Carl-Bosch-Haus, Varrentrappstr. 40-42, D-60486 Frankfurt am Main, Germany, basis February 1998.

[9] W. Wiechmann, *Amts- und Mitteilungsblatt der Bundesanstalt für Materialforschung und -prüfung (BAM)* (1987) 3, 505.

[10] A. Kruis: "Gleichgewicht der Absorption von Gasen in Flüssigkeiten," in H. Hausen (ed.): *Landolt-Börnstein Zahlenwerte und Funktionen,* **4.** Teil, Bestandteil c, Springer Verlag, Berlin – Heidelberg – New York 1976.

[11] S. A. Miller: *Acetylene, Its Properties, Manufacture and Uses,* Ernest Benn, London 1965.

[12] F. Rottmayr, H. Reimann, U. Lorber, *Linde Ber. Tech. Wiss.* **30** (1971) 3.

[13] H. Schmidt, D. Forney: *Oxygen Technology Survey,* vol. **9,** NASA SP-3090, ASRDI, 1975.

[14] E. Lassmann, *Linde Reports on Science and Technology* **57** (1996) 16.

[15] UCC isotherm data sheet No. 44, Union Carbide Corp., New York 1980.

[16] D. P. Velenuela, A. L. Myers: *Adsorption Equilibrium Data Handbook,* Department of Chem. Engineering, University of Pennsylvania, Philadelphia, Prentice Hall, Englewood Cliffs, NJ 1989.

[17] T. F. Rutledge: *Acetylenic Compounds – Preparation and Substitution Reactions,* Reinhold Publ. Co., New York 1968.

[18] A. C. Mc Kinnis, *Ind. Eng. Chem.* **27** (1962) 2928.

[19] S. A. Miller: *Acetylene, Its Properties, Manufacture and Uses,* Ernest Benn, London 1965.

[20] R. C. West, C. S. Kraihanzel, *J. Am. Chem. Soc.* **83** (1961) 765.

[21] H. J. Copley, C. E. Holley Jr., *J. Am. Chem. Soc.* **61** (1939) 1599.

[22] W. Reppe: *Neue Entwicklungen auf dem Gebiet der Chemie des Acetylens und Kohlenoxids,* Springer Verlag, Berlin-Göttingen-Heidelberg 1949.

[23] W. Reppe: *Chemie und Technik der Acetylen-Druckreaktionen,* 2nd ed., Verlag Chemie, Weinheim 1952.

[24] N. von Kutepow in: *Ullmann,* 4th ed., vol. **7,** p. 44.

[25] W. Reppe, *Justus Liebigs Ann. Chem.* **601** (1956) 81.

[26] W. Reppe et al., *Justus Liebigs Ann. Chem.* **569** (1955) 1.

[27] W. Reppe et al., *Justus Liebigs Ann. Chem.* **582** (1953) 1.

[28] W. Reppe, N. von Kutepow, A. Magin, *Angew. Chem.* **81** (1969) 717.

[29] Lonza, DE 1251329, 1964 (P. Pino, G. Braca, G. Sbrana).

[30] BASF, DE-AS 1071077, 1965 (W. Reppe, A. Magin).Du Pont, DE 1054086, 1955 (J. C. Sauer). J. C. Sauer, R. D. Cramer, V. A. Engelhardt, T. A.Ford, E. H. Holmquist, B. W. Howk, *J. Am. Chem. Soc.* **81** (1959) 3677. G. Albanesi, M. Toraglieri, *Chim. Ind. (Milan)* **41** (1959) 189.

[31] G. Palyi, G. Varadi, J. T. Horvath, *J. Mol. Catal.* **13** (1981) 61.

[32] W. Reppe et al., *Justus Liebigs Ann. Chem.* **560** (1948) 1.

[33] G. Schröder: *Cyclooctatetraen,* Verlag Chemie, Weinheim 1965.

[34] J. A. Nieuwland, W. S. Calcott, F. B. Downing, A. S.Carter, *J. Am. Chem. Soc.* **53** (1931) 4197.

[35] J. C. W. Chien, *Polym. News* **6** (1979) 52. D. Bloor, *New Sci.* **93** (1982) no. 1295, 577. H. Shirakawa, *Kotai Butsuri* **16** (1981) no. 7, 402; *Chem. Abstr.* **96** (1982) 172568. A. J. Heger, A. G. Macdiarmid, *Int. J. Quantum Chem. Quantum Chem. Symp.* **15** (1981) 243. H. Shirakawa, *Kagaku (Kyoto)* **37** (1982) no. 3, 181; *Chem. Abstr.* **96** (1982) 153398. H. Shirakawa, *Kagaku, Zokan (Kyoto)* 1980, no. 87, 165; *Chem. Abstr.* **94** (1981) 139105 x.

[36] G. Wegner, *Angew. Chem.* **93** (1981) 352.

[37] T. Mole, J. R. Suertes, *Chem. Ind. (London)* 1963, 1727.

[38] P. N. Rylander, "Platinum Metal Catalysts in Organosilicon Chemistry," *Engelhard Ind. Tech. Bull.* **10** (1970) no. 4, 130.

[39] J. L. Speier, *Adv. Organomet. Chem.* **17** (1979) 407.

[40] E. Ott et al., *Ber. Dtsch. Chem. Ges.* **76** (1943) 80 – 91.

[41] I. G. Farbenind., DE 495787, 1927.

[42] *Chem. Week* **98** (1966) Apr. 16, 90.

[43] *Eur. Chem. News* **25** (1974) Feb. 1, 5.

[44] H. Sachsse, *Chem. Ing. Tech.* **26** (1954) 245. E. Bartholome, *Chem Ing. Tech.* **26** (1954) 253. T. P. Forbath, B. I. Gaffney, *Pet. Refiner* **33** (1954) 160.

[45] H. Friz, *Chem. Ing. Tech.* **40** (1968) 999.

[46] K. G. Baur, K. Taglieber, *Chem. Ing. Tech.* **47** (1975) 385.

[47] M. J. Zundel: *Problèmes de Fabrication d'Acetylène et d'Ethylène à partir d'Hydrocarbures*, Societe de Chimie Industrielle, Paris 1962.

[48] *Hydrocarbon Process. Pet. Refiner* **44** (1965) no. 11, 163.

[49] Montecatini, GB 1000480, 1962.

[50] J. L. Petton et al., *Pet. Refiner* **37** (1958) 180.

[51] *Hydrocarbon Process.* **54** (1975) no. 11, 104. Stanford Research Report no. 109, September 1976. L. Verde, R. Riccardi, S. Moreno, *Hydrocarbon Process.* **57** (1978) no. 1, 159.

[52] G. Hamprecht, M. Gettert: "Sauerstoff-Thermisches Calziumcarbid," Festschrift für Carl Wurster, BASF, Ludwigshafen 1960, p. 43.

[53] H. Gladisch, *Hydrocarbon Process. Pet. Refiner* **41** (1962) 159–164.

[54] H. Gladisch, *Chem. Ing. Tech* **61** (1969) 204–208.

[55] K. Gehrmann, H. Schmidt, *Water Air Conserv. Pet. Ind. 1971*, 379.

[56] R. L. Bond, W. R. Ladner, G. I. McCommet, *Fuel* **45** (1966) 381–395. R. E. Gannon, V. J. Krukonis, Th. Schoenberg, *Ind. Eng. Chem. Prod. Res. Dev.* **9** (1970) 343–347. R. E. Gannon, S. K. Ubhayakar, *Fuel* **56** (1977) 281. AVCO Corp., US 179144, 1980 (Ch. Kim). D. Bittner, H. Baumann, C. Peuckert, J. Klein, H. Jüntgen, *Erdöl Kohle Erdgas Petrochem.* **34** (1981) no. 6, 237–242. R. Müller, C. Peuckert, *Int. Symp. on Plasma Chem.*, *5th*, Edinburgh 1981, p. 197–202.

[57] *Eur. Chem. News* **36** (1981) Feb. 2, 19.

[58] R. Müller, *World Hydrogen Energy Conf., Conf. Proc., 4th*, Pasadena 1982, p. 885–900.

[59] GAF, US 219756, 1980 (M. Katz, F. Carluccio).

[60] *Verordnung über Acetylenanlagen und Calcium-carbidlager* (Acetylenverordnung – Acet V) vom 27. Februar 1980, Carl Heymanns Verlag, Köln 1980.

[61] *TRAC 201 – Acetylenentwickler*, Carl Heymanns Verlag, Köln.

[62] Acetylene Transmission for Chemical Synthesis (Recommended Minimum Safe Practices for Piping Systems) International Acetylene Association, New York 1980.

[63] C. Hase, W. Reitze: *Fachkunde des Autogenschweißens*, 7th ed., Girardet, Essen 1965.

[64] *Carbidkalk, Hinweise für seine Verwendung*, Fachbuchreihe Schweißtechnik, 9th ed., Deutscher Verlag für Schweißtechnik (DVS), Düsseldorf 1968, p. 16.

[65] Shawinigan Chemicals, US 1872741, 1931 (R. S. Jane).

[66] Bayerische Stickstoff-Werke, DE 714 323, 1938 (R. Wendlandt, R. Neubner).

[67] Shawinigan Chemicals, US 1343185, 1942 (A. C. Holm, E. Poirier).

[68] Linde, DE 906 005, 1951 (F. Rottmayr). Linde, DE 2 549 399, 1975 (E. Laßmann).

[69] *Hydrocarbon Process.* **46** (1967) no. 11, 139. T. Wett, *Oil Gas J.* **70** (1972) 101–110.

[70] *Ullmann*, 4th ed., vol. **7**.

[71] H. K. Kamptner, *Erdöl Kohle Erdgas Petrochem.* **16** (1963) 547. H. K. Kamptner, W. R. Krause, H. P. Schilken, *Hydrocarbon Process. Pet. Refiner* **45** (1966) no. 4, 187. K. Lissa, *Chem. Anlagen Verfahren* 1970, no. 6, 83.

[72] Y. C. Hu, *Hydrocarbon Process.* **61** (1982) no. 11, 109.

[73] D. Sohns, *Linde Reports Science and Technology* **30** (1979) 21.

[74] F. Mey, H. D. Neubauer, R. Schubert, *Petrochemicals and Gas Processing Petroleum Technology quaterly* (1997) Autumn, 119.

[75]   K. J. Sasaki, *Petrochemicals and Gas Processing Petroleum Technology quaterly* (1997) Autumn, 113.
[76]   P. Cl. Haehn, Dr. E. Haidegger, Dr. N. Schödel, *Hydrocarbon Engineering* (1997) January/February, 41.
[77]   Linde München folder C/3.3.e/93
[78]   G. Marcks, *Schadenprisma* **3** (1982) 37. D. Lietze, *Amts- und Mitteilungsblatt der Bundesanstalt für Materialprüfung (BAM)* **16** (1986) 1, 23.
[79]   W. Reppe, *Chem. Ing. Tech.* **22** (1950) 273.
[80]   CEN/TC 121/SC 7/WG3 German Proposal: *Pipelines in Acetylene Systems*, November 1996, NAS-DIN Deutsches Institut für Normung e. V., D 10772 Berlin or *"Empfehlungen für Acetylenleitungen auf der Grundlage von Arbeitsbereichen"*, Industrial Gases Committee IGC Document 9/78/D, F 75880 Paris, CEDEX 18
[81]   H. B. Sargent, *Chem. Eng.* **64** (1957) 250.
[82]   Th. Schendler, H.-P. Schulze, *Chem.-Ing.-Tech.* **62** (1990) 1, 41. W. Reppe: *Chemie und Technik der Acetylen-Druck-Reaktionen*, Verlag Chemie, Weinheim 1951.
[83]   D. Lietze, H. Pinkofsky, T. Schendler, H.-P. Schulze, *Chem.-Ing.-Tech.* **61** (1989) 9, 736. H. Große-Wortmann, N. Kalkert, H.-G. Schecker, *Chem.-Ing.-Tech.* **53** (1981) 461.
[84]   A. Williams, D. B. Smith, *Chem. Rev.* **70** (1970) 267. B. A. Ivanov, S. M. Kogarko, *Int. Chem. Eng.* **4** (1964) 4, 670. C. M. Detz, *Combust. Flame* **34** (1979) 187. M. A. Glikin et al., *Sov. Chem. Ind. (Engl. Transl.)* **7** (1975) 1373. A. Baumeier, D. Conrad, S. Dietlen, W. Pezold, T. Schendler und H.-P. Schulze, *Chem.-Ing.-Tech.* **64** (1992) 3, 260.
[85]   R. Grätz, M. Wagenknecht, *vfdb, Zeitschrift für Forschung, Technik und Management im Brandschutz* (1994) 3, 103.
[86]   D. Lietze, *Chem. Ing.-Tech.* **63** (1991) 11, 1148.
[87]   D. Lietze, *Chem.-Ing.-Tech.* **62** (1990) 3, 238.
[88]   E. Karwat, *Chem Eng. Progr.* **54** (1958) 10, 96.
[89]   D. Lietze, *Journal of Hazardous Materials.* **54** (1997) 227.
[90]   D. Lietze, *J. Loss Prev. Process Ind.* **8** (1995) 6, 319.
[91]   D. Lietze, *J. Loss. Process Ind.* **8** (1995) 6, 325.
[92]   A. D. Hewitt: Technology of Oxy-Fuel Gas Processes, Welding and Metal Fabrication, Part 2, 1972 (November) 382. H. Meinass, H. Manhard, J. Schlander, L. Fruhstorfer, *Linde Reports on Science and Technology* **24** (1976) 32. L. Kögel, *Linde Reports on Science and Technology* **32** (1981) 36.
[93]   *Technische Regeln für Acetylenanlagen und Calciumcarbidlager, TRAC 207, 208,* Deutscher Acetylenausschuß, Bundesanstalt für Arbeitsschutz und Unfallforschung, Dortmund 1969.
[94]   K. H. Roch, *Amts Mitteilungsbl. Bundesanst. Materialprüf. Berlin* **12** (1982) 283.
[95]   K. H. Roch, *Schweissen + Schneiden* **25** (1973) 94.
[96]   D. Lietze, *Amts Mitteilungsbl. Bundesanst. Materialprüf. Berlin* **2** (1972) 9.
[97]   D. Lietze, *Berufsgenosssenschaft* 1976, 435.
[98]   M. E. Sutherland, M. W. Wegert, *Chem. Eng. Prog.* **69** (1973) no. 4, 48.
[99]   C. Isting, *Erdöl Kohle Erdgas Petrochem.* **23** (1970) 29.
[100] *Chem. Eng. (N.Y.)* **76** (1969) no. 1, 89.
[101] E. Karwat, *Chem. Eng. Prog.* **53** (1957) no. 4, 27.E. Karwat, *Linde Ber. Tech. Wiss.* **13** (1962) 12; "Safety in Air and Ammonia Plants," *Chem Eng. Prog. Tech. Man.* **5** (1963) 43. G. Klein in: "Luftzerlegungsanlagen," Linde-Arbeitstagung, München 1975.
[102] L. W. Coleman: "Safety in Air and Ammonia Plants," *Chem. Eng. Prog. Tech. Man.* **4** (1962) 26.
[103] E. Karwat, *Chem. Eng. Prog.* **57** (1961) no. 4, 5.

[104] F. G. Kerry, *Chem. Eng. Prog.* **52** (1956) no. 11, 3.
[105] J. Reyhing, *Linde Rep. Sci. Technol.* **36** (1983) 14.
[106] Engelhard Ind., DE-AS 1283805, 1968 (J. G. Cohn A. J. Haley, Jr.).
[107] G. Klein, *Linde Ber. Tech. Wiss.* **17** (1964) 24. Cryogenics Safety Manual, Part II, British Cryogenics Council, London 1970. H. H. Hofmaier: "Safety in Air and Ammonia plants", Chem Eng. Prog. Tech. Man. 5 (1963) 22.
[108] E. Karwat, *Chem. Eng. Prog. Tech. Man.* **2** (1960) A-18.
[109] G. Drewes, M. Ermscher, *Chem. Tech. (Leipzig)* **35** (1983) 57.
[110] K. H. Möller, C. Stöber, K. Schulze, *Arbeitsschutz* **23** (1972) no. 1, 18.
[111] P. Hölemann, R. Hasselmann, *Forschungsberichte des Landes Nordrhein-Westfalen*, no. 765, Westdeutscher Verlag, Köln-Opladen 1959.
[112] P. Hölemann, *Forschungsberichte des Landes Nordrhein-Westfalen*, no. 888 and 1151, Westdeutscher Verlag, Köln-Opladen 1960 and 1963.
[113] K. H. Möller, *Arbeitsschutz* **22** (1971) no. 1, 6.
[114] "Technische Grundsätze für ortsbewegliche Druckgasbehälter," Ziffer 29 and 31, *Arbeitsschutz* **21** (1970) no. 3."Allgem. Verwaltungsvorschrift zu 14, 17–19 der Verordnung über ortsbewegliche Behälter und über Füllanlagen für Druckgase," 20. 6. 1968.
[115] Department of Transportation, Office of Hazardous Materials Regulations, Code of Federal Regulations 49 (.
[116] K. H. Möller, *Arbeitsschutz* **23** (1972) no. 2, 30.
[117] International Organization for Standardization – ISO/TC 58/WG 1.
[118] K. H. Möller, *Berufsgenossenschaft* 1972, 375, 422.
[119] H. Springmann, *Linde Rep. Sci. Technol.* **34** (1982) 54.
[120] E. Zorn, *Mitt. BEFA* **14** (1963) no. 6, 2.
[121] *Safety in the Production and Use of Acetylene*, Commission permanente internationale de l'acetylène, Paris 1968.
[122] A. D. Hewitt: *Welding and Metal Fabrication*, part 1, Oct. 1972, 347; part 2, Nov. 1972, 382; part 3, Dec. 1972, 416, IPC Science and Tech. Press, Guildford, England.
[123] F. Houldcroft: *Welding Processes*, Oxford University Press, London 1975. J. Ruge: *Handbuch der Schweißtechnik*, Springer Verlag, Berlin 1980. Bibliographies, American Welding Soc., New York (appears annually).
[124] H. Weiler, *Schweißen + Schneiden* **26** (1974)220. L. Kögel, *Linde Ber. Tech. Wiss.* **48** (1980) 36.
[125] Engineering Sciences Data Unit, "London Vapour Pressures and critical points of liquids", Part 2C: alkadienes and alkynes, Number 86001 (1986)
[126] J. F. Cordes, H. Günzler, *Chem. Ber.* **92** (1959) 1055. J. F. Cordes, H. Günzler, *Z. Naturforschung* **15b** (1960) 682. C. P. Khulbe, R. S. Mann, *Prep.-Can. Symp. Catal.*, 5th (1977) 384. P. Kos, I. Kiricsi, K. Varga, P. Fejes, *Acta Phys. et. Chem. Szeged* **33** (1987) 109. F.-D. Zeiseler, G. Zimmermann, *Journal für prakt. Chemie* **319** (1977) 4, 655.
[127] Shell Internationale, EP 0 392 601, 1990 (M. J. Doyle, J. Van Gogh, J. Van Ravenswaay Claasen)
[128] D. A. Frank-Kamenetzki, V. G. Markovich, *Acta Physicochim, (U.R.S.S.)* **17** (1942) 308. H. Zeise (ed.): *Thermodynamik*, vol. **III/1**, Verlag Hirzel, Leipzig 1954.
[129] T. Redeker, G. Schön, *Sicherheitstechnische Kennzahlen brennbarer Gase und Dämpfe, 6. Nachtrag* Deutscher Eichverlag GmbH, Braunschweig 1990.
[130] Linde AG, US 4 701 190, 1986 (P. C. Haehn), EP 0 224 748, 1986 (P. C. Haehn).
[131] J. Happel, S. Umemura, Y. Sakakibara, H. Blanck, S. Kunichika, *Ind. Eng. Chem. Process Des. Develop.* **14** (1975) 1, 44. E. Drent, P. Arnoldy, P. H. M. Budzelaar, *Journal of Organometallic*

Chemistry, **455** (1993) 247. Shell Internationale, EP 0 539 628, 1993 (J. Hengeveld, P. B. de Blank).

[132] R. F. Huston, C. A. Barrios, R. A. Holleman, *J. Chem. Eng. Data* **15** (1970) 168.R. D. Green, *Met. Prog.* **108** (1975) 2, 71.A. Farwer, *Gas Aktuell* **1982** (Nov.) 24.

[133] L. T. Fairhall: *Industrial Toxicology,* 2nd ed., The Williams and Wilkins Co., Baltimore 1957, p. 270.

[134] The International Technical Information Institute: *Toxic and Hazardous Industrial Chemicals Safety Manual,* Tokyo 1975, p. 10.

[135] T. Sollmann: *A Manual of Pharmacology and its Application to Therapeutics,* 7th ed., W. B. Saunders Co., Philadelphia-London 1949, p. 662.

[136] F. Flury, F. Zernik: *Schädliche Gase,* Springer Verlag, Berlin 1931, p. 270.

[137] NIOSH, U.S. Dept. of Health and Human Services: Registry of Toxic Effects of Chemical Substances, vol. 1, Washington, DC, 1980, p. 70.

[138] E. R. Plunkett: *Handbook of Industrial Toxicology,* Heyden, Barberton, Ohio, 1976, p. 8.

[139] F. A. Patty: *Industrial Hygiene and Toxicology,* 2nd ed., Interscience Publ., New York 1962, p. 1205.

Acridine

GERD COLLIN, DECHEMA e.V., Frankfurt/Main, Federal Republic of Germany

HARTMUT HÖKE, Weinheim, Federal Republic of Germany

1.	**Physical Properties**	197	4.	**Uses**	198
2.	**Chemical Properties**	197	5.	**Toxicology**	198
3.	**Production**	197	6.	**References**	198

Acridine [260-94-6], $C_{13}H_9N$, was discovered in coal-tar anthracene oil by C. GRAEBE and H. CARO in 1870, and identified as dibenzopyridine by C. RIEDEL, A. BERNTHSEN, and F. BENDER in 1883.

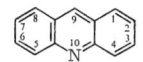

1. Physical Properties

M_r 179.22, mp 110 °C, bp 345 °C (at 1013 hPa), ϱ at 20 °C 1.1005 g/cm^3. Colorless needles or prisms; volatile in steam; slightly soluble in boiling water, readily soluble in organic solvents such as alcohol, ether, carbon disulfide, and benzene; solutions show blue fluorescence.

2. Chemical Properties

Acridine is a weak tertiary base. With strong acids, it forms crystalline, yellow salts that decompose readily in boiling water. Acridine is stable when heated with concentrated alkali or hydrochloric acid. Nitration with nitric acid yields mainly 2- and 4-nitroacridine and small amounts of dinitroacridines. Reduction gives acridane (9,10-dihydroacridine [92-81-9]) preferentially. Oxidation results in the formation of acridinic acid (quinoline-2,3-dicarboxylic acid [643-38-9]) or acridone, depending on the oxidizing agent used. Acridine forms quaternary acridinium salts with alkyl or aryl halides and sulfates.

3. Production

Acridine can be separated as sodium acridone sulfonate from the coal-tar anthracene oil boiling between 300 and 360 °C ($\rightarrow$ Tar and Pitch) by extraction with dilute sulfuric acid [1] or with aqueous sodium bisulfite. The free base is obtained by decomposition of the salt with caustic soda [2]. Acridine can be synthesized by reduction of acridone or 9-chloroacridine.

Acridone [*578-95-0*], pale yellow needles, *mp* 354 °C, is obtained by the cyclization of diphenylamine-2-carboxylic acid (phenylanthranilic acid), which is synthesized from aniline and 2-chlorobenzoic acid. *9-Chloroacridine* [*1207-69-8*], almost colorless crystals, *mp* 120 °C, is obtained by reacting phosphoryl chloride with diphenylamine-2-carboxylic acid.

Acridone

4. Uses

Acridine as such has no commercial significance. In the patent literature it is proposed as an antioxidant, as a polymerization and corrosion inhibitor, as an additive to peroxidic vulcanizing agents for ethylene–propylene rubber, as a thermal stabilizer for polyolefins, and also as an occult blood test reagent, used in paper strips for blood tests. The acridine dyes are made from other starting materials.

5. Toxicology

Animal studies have shown that acridine is slightly toxic: the LD_{50} value is about 2000 mg/kg (rat, oral) [3], [4]. Industrial observations show that exposure to acridine dust or vapor causes strong irritations of the skin and mucous membranes combined with sneezing, with itching or even inflammation of the skin. Some acridine derivatives cause sensitization of the skin, especially on exposure to light. Exposure limits at the workplace (MAK, TLV) are not specified. Resorptive intoxication is unknown.

6. References

General References

Beilstein **20**, 459; **20 (2)**, 171.
A. Adrien: *The Acridines*, 2nd ed., St. Martin's Press, New York 1966.
R. M. Acheson (ed.): *Acridines. Chemistry of Heterocyclic Compounds*, 2nd ed., vol. **9**, J. Wiley & Sons, New York 1973.
N. Campbell in S. Coffey, M. F. Ansell (eds.): *Rodd's Chemistry of Carbon Compounds*, 2nd ed., vol. **4 (G)**, Elsevier, Amsterdam 1978,pp. 1–82.

Specific References

[1] Rütgerswerke, DE 688335, 1938 (N. Hviid).
[2] H. J. V. Winkler: *Der Steinkohlenteer und seine Aufarbeitung*, Verlag Glückauf, Essen 1951, p. 191.
[3] W. S. Spector: *Handbook of Toxicology*, vol. **1**, W. B. Saunders Co., Philadelphia – London 1956, p. 12.
[4] V. B. Kapitul'skii, *Klin. Patog. Profil. Profzabol. Khim. Etiol. Predpr. Tsvetn. Chern. Metall* 1969, no. 2, 179–183.

Acrolein and Methacrolein

DIETRICH ARNTZ, Degussa Corporation, Mobile, Alabama, United States

MATHIAS HÖPP, Degussa AG, Zweigniederlassung Wolfgang, Hanau, Federal Republic of Germany

SYLVIA JACOBI, Degussa AG, Zweigniederlassung Wolfgang, Hanau, Federal Republic of Germany

JÖRG SAUER, Degussa AG, Zweigniederlassung Wolfgang, Hanau, Federal Republic of Germany

TAKASHI OHARA, Nippon Shokubai Kagaku Kogyo Co., Ltd., Osaka, Japan

TAKAHISA SATO, Nippon Shokubai Kagaku Kogyo Co., Ltd., Osaka, Japan

NOBORU SHIMIZU, Nippon Shokubai Kagaku Kogyo Co., Ltd., Osaka, Japan

GÜNTER PRESCHER, Degussa AG, Zweigniederlassung Wolfgang, Hanau, Federal Republic of Germany

HELMUT SCHWIND, Degussa AG, Zweigniederlassung Wolfgang, Hanau, Federal Republic of Germany

OTTO WEIBERG, Degussa AG, Zweigniederlassung Wolfgang, Hanau, Federal Republic of Germany

1.	Introduction	199	4.	Quality and Analysis	211
2.	Properties	200	5.	Handling, Storage, and Transportation	212
2.1.	Physical Properties	200	6.	Uses and Production Data	213
2.2.	Chemical Properties	200	7.	Toxicology and Ecotoxicology	215
3.	Production	207	7.1.	Toxicology	215
3.1.	Acrolein by Propene Oxidation	207	7.2.	Ecotoxicology	217
3.2.	Methacrolein	210	8.	References	218

1. Introduction

Acrolein [107-02-8], propenal, acrylaldehyde, $CH_2=CH\text{-}CHO$, the simplest unsaturated aldehyde, is a colorless, volatile, toxic, and lacrimatory liquid with a powerful odor.

The commercial production of acrolein by heterogeneously catalyzed gas-phase condensation of acetaldehyde and formaldehyde was established by Degussa in 1942. Today, acrolein is produced on a large commercial scale by heterogeneously catalyzed gas-phase oxidation of propene.

Acrolein is an important intermediate for numerous substances (see Chap. 6). The main use of commercial, isolated acrolein is currently the production of D,L-methionine, an essential amino acid used as an animal feed supplement. In the production of acrylic acid, acrolein is not isolated from the gas-phase reaction mixture but is oxidized further on a heterogeneous catalyst.

Several review articles [1], [2] (see also [73]) and a monograph [3] describe the preparation, reactions, and uses of acrolein.

Methacrolein [*78-85-3*], 2-methylpropenal, α-methylacrolein, $CH_2=C(CH_3)\text{-}CHO$, is a colorless, volatile, toxic, and lacrimatory liquid with a piercing odor. It is an intermediate in one of the processes for the production of methyl methacrylate (see Section 3.2).

2. Properties

2.1. Physical Properties

The important physical properties of acrolein and methacrolein are compiled in Table 1. The solubility of acrolein in water is limited. It is soluble in many organic solvents, such as alcohols, ethers, and aliphatic or aromatic hydrocarbons. Methacrolein has properties similar to acrolein but is less soluble in water.

2.2. Chemical Properties

Acrolein is an extremely reactive chemical because of its conjugated vinyl and aldehyde groups. It undergoes reactions characteristic of both an unsaturated compound and an aldehyde. The conjugation between the carbon–carbon double bond and the carbonyl group increases the reactivity of both groups, which can react either together or separately. Highly exothermic polymerization can occur spontaneously (see Chap. 5).

Diels–Alder Reaction. Acrolein reacts as both a diene and a dienophile. Thus two molecules of acrolein can form a cyclic dimer, 3,4-dihydro-2*H*-pyran-2-carboxaldehyde [*100-73-2*] [4]:

The dimer is formed by the uncatalyzed, thermal reaction of acrolein at approximately 190 °C (1 h for 75 % acrolein conversion) together with polymeric side products. Sufficient stabilization with hydroquinone or complexing compounds, such as polyvalent organic acids [6], is necessary for high yields. This "thermal dimer" is a clear liquid with an unpleasant odor: bp 151.3 °C at 101.3 kPa, d_{20}^{20} 1.0775. Acid hydrolysis of the dimer yields 2-hydroxyadipaldehyde [*141-31-1*], which can be hydrogenated to form 1,2,6-hexanetriol [*106-69-4*].

Table 1. Physical properties of acrolein and methacrolein

	Acrolein	Methacrolein
M_r	56.06	70.09
bp (101.3 kPa), °C	52.69	68.4
(1.33 kPa), °C	−36	−25
mp, °C	−86.95	−81.0
Relative density, d_{20}^{20}	0.8427	0.8474
Refractive index, n_D^{20}	1.4013	1.4169
Vapor pressure (20 °C), kPa	29.3	16.1
Viscosity (20 °C), mPa · s	0.35	0.49
Solubility (20 °C), g/kg		
in water	260	50
water in	73	36
Critical temperature, °C	233	257
Critical pressure, MPa	5.07	4.36
Critical volume, mL/mol	189	
Heat of vaporization (101.3 kPa), kJ/mol	28.2	29.0
Heat of combustion (25 °C), kJ/mol	1632	2299
Heat of formation (gas, 25 °C), kJ/mol	−74.5	−70.8
Heat of polymerization, kJ/mol	71−80	
Specific heat capacity, kJ mol^{-1}K^{-1}		
c_p (liquid) (17−44 °C)	0.120	
c_p (gas) (27 °C)	0.067	
Flash point, open cup, °C	−18	−15
closed cup, °C	−26	
Flammability limits in air, vol%		
upper	31	
lower	2.8	6.0
Autoignition temperature in air, °C	234	280

Properties

Acrolein behaves as a 1,3-diene in reactions with dienophiles in which the electron density of the carbon-carbon double bond is increased by electron-releasing substituents. Vinyl ethers [3], [5], [7], [8] and vinylamines [9] react readily with acrolein to form dihydropyrans. The reaction of methyl vinyl ether and acrolein to form 3,4-dihydro-2-methoxy-2H-pyran [4454-05-1] is a commercially important example. At a reaction temperature of 160−190 °C, reported yields are 80−90% [5].

Acid hydrolysis of this product leads to glutaraldehyde [111-30-8] (for uses, see p. 213).

The electron-deficient vinyl group of acrolein reacts readily with conjugated dienes, such as butadiene or substituted butadienes, forming derivatives of 3-cyclohexene-1-carbaldehyde [7], [10]:

3-Cyclohexene-1-carbaldehyde (1,2,3,6-tetrahydrobenzaldehyde) [*100-50-5*] is formed at 100–150 °C in ca. 80–90% yield and is a valuable intermediate for various commercial products. With pentaerythritol a cyclic acetal is formed, which can be used as a stabilizer against ozone degradation in natural or synthetic rubber [11]:

The carbaldehyde also is used for the synthesis of cycloaliphatic epoxides, such as the following:

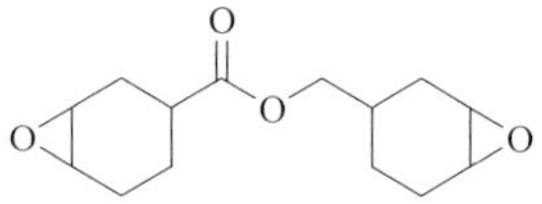

Addition to the Carbon–Carbon Double Bond. The β-carbon atom of acrolein, which is polarized by the carbonyl group, behaves as an electrophile. Therefore nucleophilic reagents, such as alcohols, thiols, water, amines, active methylene compounds, and inorganic and organic acids, add to the carbon–carbon double bond of acrolein in the presence of acidic or basic catalysts. These reactions must be carried out under carefully controlled conditions in order to minimize undesirable side reactions [3].

$CH_2=CHCHO$		
	+ ROH → $ROCH_2CH_2CHO$	[12], [14]
	+ H_2O → $HOCH_2CH_2CHO$	
	+ H_2O + H_2 → $HOCH_2CH_2CH_2OH$	
	+ CH_3COOH → $CH_3COOCH_2CH_2CHO$	[20]
	+ HCl → $ClCH_2CH_2CHO$	
	+ HCl + ROH → $ClCH_2CH_2CH(OR)_2$	
	+ Cl_2 → $ClCH_2CHClCHO$ → $CH_2=CCl\text{-}CHO$	
	+ CH_3SH → $CH_3SCH_2CH_2CHO$	
	+ $CH_2(COOR)_2$ * → $(ROOC)_2CHCH_2CH_2CHO$	[3], [21]

* or other compounds with an active methylene group

The addition of water under mild acidic conditions gives 3-hydroxypropionaldehyde [*2134-29-4*] with high selectivity. Buffer solutions with a pH of 4–5 [13], [15] or weak acidic ion-exchange resins [16] are preferentially used as catalysts. Further hydrogenation of the aqueous solutions gives 1,3-propanediol [*504-63-2*] [17]. Direct oxidation of aqueous solutions of 3-hydroxypropionaldehyde at pH 3 with precious metal catalysts produces 3-hydroxypropionic acid [18]. If the oxidation is conducted at above pH 7, the malonate anion is formed in high yield [19].

Acrolein reacts rapidly with hydrogen chloride or hydrogen bromide to form 3-chloropropionaldehyde [*19434-65-2*] [22], [23] or 3-bromopropionaldehyde, but these products easily polymerize, e.g., to trimers and tetramers, in the presence of acids. The preferred procedure for making 3-halopropionaldehyde acetals is therefore simultaneous hydrogen halide addition and acetalization. Acetal yields are about 90% [23], [24].

Chlorine and bromine add to acrolein in dilute aqueous solution to give 2,3-dihalopropionaldehydes with about 85 % yields; these products can be dehydrohalogenated to form the 2-haloacroleins [25].

These 2-haloacroleins are considered to be potent mutagens [26]. Further halogenation of the 2-haloacroleins provides the 2,2,3-trihalopropionaldehydes, which are valuable intermediates in the synthesis of folic acid [27].

The addition of hydrogen sulfide to two equivalents of acrolein followed by an aldol reaction forms 3-formyl-5,6-dihydrothiopyran [30058-79-8] [28].

$$2\ H_2C{=}CHO + H_2S \longrightarrow \text{(3-formyl-5,6-dihydrothiopyran)} + H_2O$$

The base-catalyzed addition of methanethiol to form 3-(methylthio)propionaldehyde [3268-49-3] is the commercially most important reaction used in the synthesis of the essential amino acid D,L-methionine (→ Amino Acids).

Reactions of the Aldehyde Group. The selectivity of the acid-catalyzed *acetalization* strongly depends both on the nature of the alcohol used and on the catalysts [3]. The most important side reactions yield the corresponding 3-alkoxypropionaldehydes and 3-alkoxypropionaldehyde acetals.

Usually cyclic acetals are formed much more easily, especially with branched diols such as 2-methyl-1,3-propanediol, than the acyclic acetals of lower alcohols [29], [30]. For the preparation of various cyclic acetals, several continuous processes are known [31]. High yields (more than 90 %) of the dimethyl and diethyl acetals can be realized by means of special processes in combination with extraction [32].

Acrolein acetals are valuable intermediates. They have gained interest in the last years as safer forms for the transport of acrolein, especially 2-vinyl-1,3-dioxolane [3984-22-3]. The acetals can be cleaved back into acrolein easily in aqueous acidic solutions [33]. These solutions are of particular interest for the use of acrolein as an aquatic herbicide and as a hydrogen sulfide scavenger in oil-field waters.

One of the most interesting cyclic acetals is the spiro diacetal, diallylidene pentaerythritol [78-19-3], prepared by condensing pentaerythritol with two moles of acrolein. It can be polymerized with vinyl or other monomers to form spirane resins [34].

$$2\ H_2C{=}CHO + \text{C(CH}_2\text{OH)}_4 \xrightarrow{H^+} \text{spiro diacetal} + H_2O$$

The hydroformylation of acrolein acetals, especially cyclic acetals, leads to the monoacetals of 1,4-butanedial. Subsequent hydrolysis and hydrogenation of this hydroformylation product produces 1,4-butanediol [29], [35].

$$H_2C\!=\!CH\!-\!\overset{O\!-\!CH_2}{\underset{O\!-\!CH_2}{\diagdown}}(CH_2)_n \xrightarrow[Rh]{H_2/CO} OHC\!-\!CH_2\!-\!CH\overset{O\!-\!CH_2}{\underset{O\!-\!CH_2}{\diagdown}}(CH_2)_n$$

$$\xrightarrow[H_2O]{H^+} OHC\!-\!CH_2\!-\!CH_2\!-\!CHO \xrightarrow{H_2} HO\!-\!(CH_2)_4\!-\!OH$$

A synthesis of D,L-tryptophan from 1,4-butanedial has been described [36]

Acrolein diacetate [*869-29-4*] 2-propene-1,1-diol, diacetate, is easily prepared in 90% yield by the acid-catalyzed reaction of acetic anhydride with acrolein [3]:

$$H_2C\!=\!CH\!-\!CHO + (CH_3CO)_2O \xrightarrow{H^+} H_2C\!=\!CH\!-\!CH(OCOCH_3)_2$$

Hydrogen cyanide addition in the presence of mild alkaline catalysts gives acrolein cyanohydrin [*5809-59-6*] in yields of more than 90% [3], [37]. Acrolein cyanohydrin is a vesicant; it decomposes vigorously in the presence of alkali and is stabilized by traces of acid. Its reaction with acetic anhydride yields acrolein cyanohydrin acetate, which is a valuable intermediate in the synthesis of pharmaceutically and biologically active substances [38] – [40].

Simultaneous Reaction of the Aldehyde and Vinyl Groups. Acrolein has long been recognized as an intermediate in the Skraup synthesis of quinolines from aromatic amines and glycerol [3]. However, little commercial use has been made of acrolein or methacrolein as starting materials for this reaction.

$$R^1\!-\!C_6H_4\!-\!NH_2 + H_2C\!=\!CR^2\!-\!CHO \longrightarrow \text{(substituted quinoline)} + H_2O$$

The heterogeneously catalyzed gas-phase condensation of acrolein and ammonia over multicomponent Al_2O_3 or $SiO_2\!-\!Al_2O_3$ catalysts gives 45% 3-methylpyridine [*108-99-6*] with 20 – 25% pyridine [*110-86-1*] as a byproduct [3], [41] – [43].

$$2\ H_2C\!=\!CH\!-\!CHO + NH_3 \xrightarrow[\text{cat.}]{400\text{-}500\ °C} \text{3-methylpyridine} + 2\ H_2O$$

This reaction, usually carried out in a fluidized-bed reactor, has been generalized to a universal synthesis for substituted pyridines [43] (→ Pyridine and Pyridine Derivatives).

Isocyanuric acid reacts with acrolein to form an adduct useful as a cross-linking agent [44].

The bifunctional reactivity of acrolein also has been used in various syntheses of heterocycles. Examples include the reactions with phenols, yielding substituted chromenes or chromanes [45], and the reaction with 2-aminophenols to produce 8-quinolinols [46]. The reaction of acrolein with excess phenol under acidic conditions leads to a mixture of polyphenols with 1,1,3-tris(4-hydroxyphenyl)propane [4137-11-5] as the main product [47].

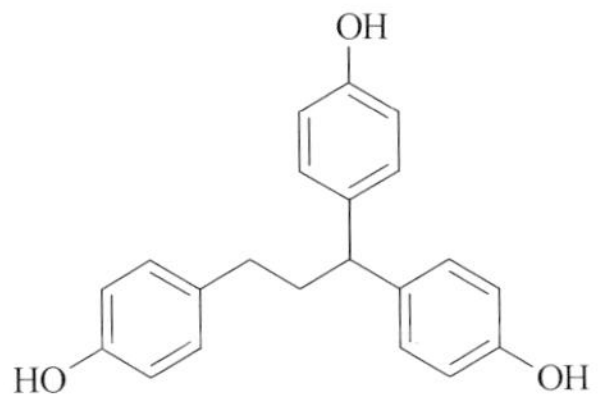

Further reaction with epichlorohydrin leads to glycidyl polyphenols which are useful as epoxy hardeners [48]. Reaction with glycidyl methacrylate followed by polymerization gives heat-resistant materials with high impact resistance [47].

The reaction of acrolein with urea leads to (hexahydro-2-oxo-4-pyrimidinyl urea) [31036-27-8], which has been proposed as a fertilizer [3], [49]:

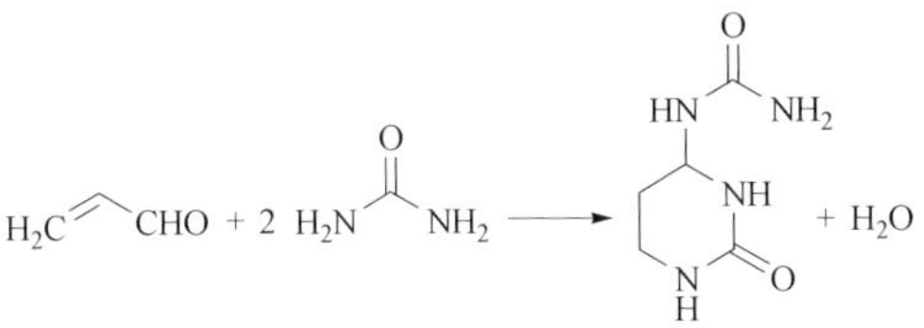

The condensation of acrolein with hydrazine and substituted hydrazines gives pyrazolines in 80% yield [50]:

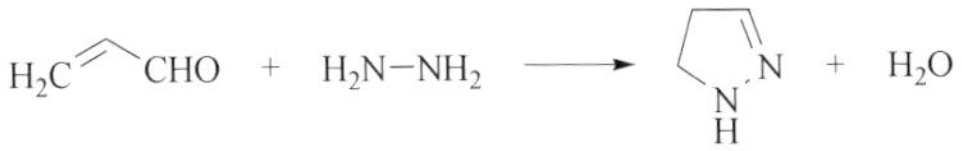

Acrolein reacts with formaldehyde under alkaline conditions to give pentaerythritol [115-77-5] [51]. Yields are lower than in the commercial synthesis of pentaerythritol via acetaldehyde and formaldehyde.

Sodium bisulfite reacts with acrolein to form a stable adduct, the disodium salt of 1-hydroxypropane-1,3-disulfonic acid [35850-94-3]. This reaction can be conveniently applied for deodorization of spilled acrolein.

Reduction. The selective reduction of acrolein to allyl alcohol [107-18-6] in high yields (90%) is accomplished by hydrogen transfer from secondary alcohols, such as 2-propanol, in a Meerwein–Ponndorf reaction in the liquid phase [3] or from ethanol or 2-propanol in the gas phase over magnesium oxide catalysts [52]–[54] or rare earth oxide catalysts [55] at 300–500 °C.

Moreover, the aldehyde group is selectively reduced by metal hydrides, such as sodium borohydride or lithium aluminum hydride [56]. The catalytic hydrogenation to allyl alcohol [52], [57] is either less selective or selective only at very low conversions.

Oxidation. Glycidaldehyde [*765-34-4*] is easily prepared in 80–90% yield by reaction with aqueous hydrogen peroxide at pH 8–8.5 [58].

Hydration of the oxirane group gives an aqueous solution of D,L-glyceraldehyde [*56-82-6*] [59].

Acrolein is oxidized commercially with molecular oxygen to form acrylic acid (→ Acrylic Acid and Derivatives).

Polymerization. Acrolein readily polymerizes on heating, on exposure to light, or by the action of various initiators. The properties of the acrolein polymer depend on the polymerization conditions, such as the type of initiator, solvent, and reaction temperature [60].

The functional groups of acrolein (vinyl and aldehyde) can polymerize either separately or together. Radical-initiated polymerization takes place exclusively at the vinyl group, leading to polymers with aldehyde groups **1**, which very easily yield tetrahydropyran structures **2**. Ionic polymerization gives rise mostly to polymers with vinyl groups **3** along with such polymers as **1** and **4**.

Radical polymers are infusible and insoluble in common solvents, whereas the products of anionic initiation are soluble in many organic solvents. In addition to homopolymers, copolymers with most vinylic and acrylic comonomers have been obtained. Furthermore, the carbonyl groups of polyacrolein (**1**) can be modified chemically under mild conditions. Most of the polyacrolein derivatives are soluble in organic solvents or even in water.

These polymers have recently gained interest because of their biocidal activity [61], [62]. The biocidally active component is acrolein, which is slowly released from these polymers [62].

3. Production

3.1. Acrolein by Propene Oxidation

The first process for acrolein production was commercialized by Degussa in 1942. It was based on the vapor-phase condensation of acetaldehyde and formaldehyde, catalyzed by sodium silicate on silica supports at 300–320 °C [63]. This method prevailed until 1959, when Shell began producing acrolein by the vapor-phase oxidation of propene over a cuprous oxide catalyst. The catalyst performance in this process was very poor. In 1957 Standard Oil of Ohio (Sohio) discovered the bismuth molybdate catalyst system, which yielded a fairly good selectivity but still a low propene conversion. Since then catalyst performance has significantly improved.

$$CH_2=CH\text{-}CH_3 + O_2 \longrightarrow CH_2=CH\text{-}CHO + H_2O \qquad \Delta H = -340.8 \text{ kJ/mol}$$

Byproducts of this reaction are acrylic acid and carbon oxides in addition to minor products such as acetaldehyde, acetic acid, formaldehyde, and polyacrolein.

Catalysts for the Oxidation of Propene. The discovery that propene could be oxidized rather selectively to acrolein over copper(I) oxide [64] marked the beginning of the current process of catalytic alkene oxidation to aldehydes over metal oxide catalysts [65]. However, low conversion of propene (20%) per pass [66], significant recycle of unreacted propene, and low acrolein selectivities were reported for this catalyst. The propene oxidation gathered impetus with the discovery of the bismuth molybdate – bismuth phosphomolybdate system in 1957 by Sohio [67], [68]. Bismuth and molybdenum oxides also are essential components of commercial catalysts. However, reasonable selectivity (max. 72%) of acrolein is obtained only at low propene conversions (57%), using $Bi_9PMo_{12}O_{52}$ on an SiO_2 support.

Modern catalysts are multicomponent metal oxide systems. Knapsack [69] first proposed a three- or four-component catalyst containing MoBiFe oxides. Later, Nippon Kayaku [70] developed a MoBiFeCoNiP oxides catalyst, Degussa [71] the MoBiFeCoNiPSm oxides system, and Nippon Shokubai [72] the MoBiFeCoWKSi oxides system. The multicomponent catalyst system has been modified frequently [65], [73], [74]. The catalysts operate by a catalyst-reduction cycle (selective product formation), and a catalyst-reoxidation cycle (lattice-oxygen regeneration).

Table 2 lists various catalysts together with the corresponding reaction temperatures and acrolein yields according to the patent specifications, which usually refer to short-

Table 2. Propene oxidation catalysts for acrolein production

Catalyst composition (neglecting oxygen)	Reaction temperature, °C	Conversion, %	One-pass yield, mol%			Reference
			Acrolein	Acrylic acid	Total	
$Mo_{12}BiFeNi_{10.5}P$	300	98.0	71.0	19.4	90.4	[75]
$Mo_{12}BiFe_2Co_3NiP_2K_{0.2}$	305	96.0	88.0	3.0	91.0	[76]
$Mo_{10}BiFeCo_4W_2K_{0.06}Si_{1.35}$	320	97.0	90.2	6.0	96.2	[72]
$Mo_{10}BiFeCo_4W_2Mg_{0.06}Si_{1.35}$	320	98.0	85.5	8.8	94.3	[77]
$Mo_{12}Bi_{0.83}Fe_{0.33}Ni_{7.5}CrSn_{0.5}$	326	93.6	75.9	11.8	87.7	[78]
$(Bi_2W_2O_9)_{0.5}Mo_{12}Co_5Fe_{2.5}Si_{1.6}K_{0.05}$	338	97.9			96.8	[79]
$Mo_{10}W_2BiFeCo_4K_{0.06}Si_{1.5}$	315	98.0	85.7	10.0	93.8	[80]
$Mo_{12}BiFe_2Co_7Cs_{0.05}Si$	310	97.7	83.8	8.3	91.8	[81]
$Mo_{12}Bi_5Co_2Ni_3Fe_{0.3}Na_{0.35}Ca_{0.1}B_2K_{0.8}Si_{12}$	310	99.1	90.8	4.5	95.3	[82]
$Mo_{12}Fe_{1.2}CoNi_4Bi_{0.5}P_{0.8}K_{0.025}Sm_{0.1}Si_{30}$	331	94.2	81.8	7.0	88.8	[83]

time results in small, optimized laboratory reactors. However, acrolein yields depend not only on the chemical compositions of these catalysts, but also on their physical properties, such as shape, porosity, pore-size distribution, and specific surface area, as well as on the reaction conditions.

At present the maximum acrolein yield at high propene conversions (90–95%) using commercial catalysts is approximately 80%, with acrylic acid yields of 5–10%. In commercial plants the catalysts are employed at reaction temperatures of 300–400 °C, contact times of 1.5–3.5 s, and propene concentrations of 5–10 vol% of the feed gas at inlet pressures of 150–250 kPa. The catalysts have lifetimes of up to ten years, after which they generally have to be replaced because yields decrease or there is a larger pressure drop in the reactor.

Production Processes. Production processes are described in [66], [84]–[86]. A simplified flow sheet for the production of acrolein by propene oxidation is shown in Figure 1.

Propene is mixed with air and steam in a molar ratio of approximately 1:8:(2–6); steam can be replaced by inert gas, e.g., the off-gas from the absorber. The inlet gas mixture is fed to a multitubular fixed-bed reactor (a) which is cooled by a recirculating molten salt bath.

The reactor is usually operated at 300–400 °C and inlet pressures of 150–250 kPa. The conversion rate of propene is approximately 95% so that there is no need to recover and recycle unreacted propene. The reactor effluent is quenched at the exit to prevent subsequent reactions of acrolein [90]. The reaction gas is then scrubbed with water or water/solvent mixtures in a first column (b) to remove acrylic acid, polymeric compounds, and traces of acetic acid [91]. Byproduct acrylic acid can be recovered from the bottoms [85] and purified; acrylic acid usually forms in 5 to 10 mol% yield based on propene.

The gas is then passed to an absorber (c) where an aqueous solution of acrolein is obtained by absorbing the gas in cold water. Part of the off-gas from the absorber can be used as inert gas for the reactor because it contains only noncondensable compo-

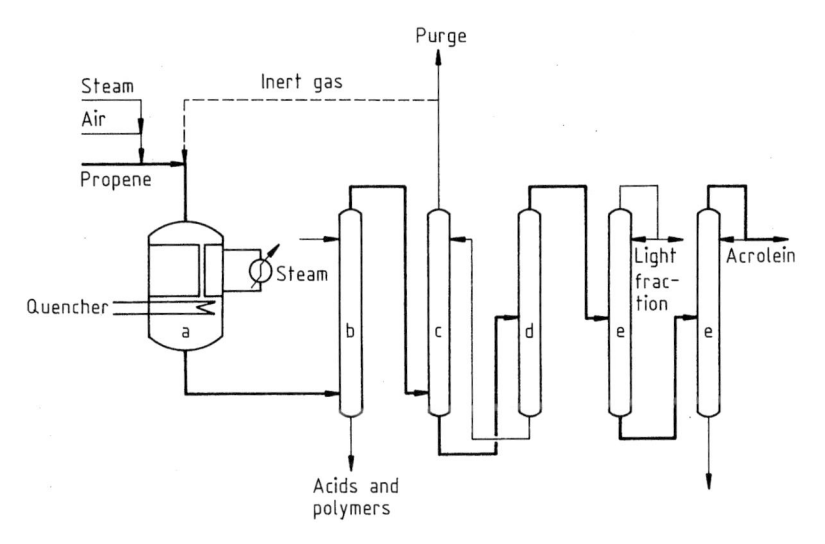

Figure 1. Acrolein production by propene oxidation
a) Oxidation reactor; b) Scrubber; c) Absorber; d) Desorber; e) Fractionators

nents, such as unreacted propene, carbon oxides, oxygen, and nitrogen. The rest is purged as waste gas after it passes through a combustion system.

The aqueous solution of acrolein is sent to a desorption column (d), where it is stripped to give crude acrolein; the bottom stream from this column is cooled and reused as an absorbent. The crude acrolein is distilled to remove low-boiling byproducts, such as acetaldehyde, and heavy ends; acrolein is then obtained as a 96% pure product that contains only traces of acetaldehyde. Sometimes crude acrolein is used directly. To minimize polymerization, the whole system is stabilized by hydroquinone or a similar agent. Pipelines and apparatus are constructed preferably of stainless steel.

New Developments. During the last decade the attempt to convert alkanes or natural feedstocks to functionalized organic intermediates has encouraged the reevaluation of alternative routes to acrolein. The development of catalysts for the selective catalytic oxidation of propane to acrolein has not led to a commercial process for the production of acrolein [87]. The direct combination of propane dehydrogenation and selective oxidation of propene to acrolein [88] has also not yet been commercialized. For the production of acrolein from natural feedstocks, a catalytic dehydration of glycerol [89] could be an interesting route to acrolein.

3.2. Methacrolein

Methacrolein was produced by Union Carbide in the 1950s and early 1960s by 2-methylpropene (isobutene) oxidation over a copper(I) oxide catalyst. Today, several companies produce methacrolein commercially as an organic intermediate for limited applications. However, large quantities of methacrolein are produced as an intermediate in newly developed processes for the production of methyl methacrylate or methacrylic acid. These processes use either C_4 feedstocks (2-methylpropene or *tert*-butyl alcohol) or the C_2 hydrocarbon ethylene as raw material.

Vapor-Phase Oxidation of 2-Methylpropene or *tert*-Butyl Alcohol. Catalysts for the oxidation of 2-methylpropene or *tert*-butyl alcohol are multicomponent metal oxide systems similar to those used to make acrolein. Most of the catalysts for the oxidation of propene have been claimed in the literature to be suitable for the oxidation of 2-methylpropene or *tert*-butyl alcohol. However, the yields were generally lower than for acrolein. Yields of 75–85% methacrolein and 1–5% methacrylic acid for 90–98% conversions in short-time laboratory experiments have been reported [92]. For high selectivity, the process generally requires greater dilution of the feed gas with steam or inert gas than is needed for propene oxidation. Impurities, such as butadiene, often oligomerize on the oxidation catalyst causing serious deterioration of its activity.

Several plants for producing methyl methacrylate via methacrolein have been constructed in Japan and Korea. In 1983, Nippon Shokubai brought on stream a 15 000 t/a demonstration plant based on the oxidation of 2-methylpropene [93], and later in the same year Mitsubishi Rayon started a 40 000 t/a commercial plant based on the oxidation of *tert*-butyl alcohol [94]. These were the first plants using the new technology. Process conditions are similar to those for propene oxidation [95].

In 1996 nearly half of the Japanese methyl methacrylate capacity of 570 000 t/a [96] was produced with processes based on 2-methylpropene or *tert*-butyl alcohol. This corresponds to a methacrolein capacity of 250 000 t/a.

Production from C_2 and C_1 Feedstocks. BASF has developed a process for the production of methacrolein from C_2 and C_1 feedstocks. Ethylene is hydroformylated to propanal, which can be condensed with formaldehyde in a Mannich type reaction to give methacrolein:

$$CH_2=CH_2 + CO + H_2 \longrightarrow CH_3CH_2CHO$$

$$CH_3CH_2CHO + CH_2O \longrightarrow CH_2=C(CH_3)CHO + H_2O$$

The Mannich condensation can be carried out in aqueous solution in the presence of a dialkylammonium salt. Methacrolein can be distilled from the resulting solution in a yield of 95%; the aqueous solution can be recycled [97]. Since 1989 BASF has used this process as one reaction step in their production process for methyl methacrylate starting from ethylene (capacity 36 000 t/a).

Alternatively, heterogeneouslycatalyzed vapor-phase condensation processes have been developed [63], [98]. The catalyst is Al_2O_3 or basic SiO_2. At reaction temperatures of 300 – 350 °C yields of approximately 90 % can be reached.

Dehydrogenation of Isobutyraldehyde. Methacrolein can be prepared by vapor-phase oxidative dehydrogenation of isobutyraldehyde using a catalyst with heteropoly anions containing molybdenum and phosphorus as the main components [99]. Yields are approximately 80 % methacrolein and 7 % methacrylic acid. However, as long as isobutyraldehyde is available only from propene as a byproduct of the oxo synthesis, this route will not be a major process for methacrolein.

4. Quality and Analysis

Quality. Typical specifications (wt %) guaranteed by producers of acrolein are as follows:

Purity of acrolein 95 – 97
Water content < ca. 3
Hydroquinone < 0.1 or 0.2
Acetaldehyde < 0.3

Some producers specify the acetaldehyde limit as < 2 % and indicate propene oxide contents of up to 1.5 %.

Analysis. Impurities include water, acetaldehyde, and, depending on the process, small amounts of propionaldehyde, acetone, propene oxide, methanol, and traces of allyl alcohol and ethanol. The contents of acrolein, water, and acetaldehyde can be determined by gas chromatography with a thermal conductivity detector. The Karl Fischer method is not applicable for determining the water in acrolein. Hydroquinone is determined colorimetrically using Millon's reagent to form a yellow complex [3].

5. Handling, Storage, and Transportation

Acrolein is classified as a very toxic, flammable liquid. The liquid vaporizes easily and the vapors are readily flammable in air at between 2.8 and 31.0 vol%. Acrolein vapors are twice as heavy as air. The flash point, −26 °C, is very low. Preferred firefighting agents are foam, powder (not alkaline), spray water, and carbon dioxide. The high toxicity and volatility of acrolein necessitate the use of a respirator in case of fire.

Acrolein polymerizes easily and exothermally; therefore it is stabilized with 0.1 or 0.2% hydroquinone against radical-initiated polymerization, which can be catalyzed by light, air, heat, or peroxides. Acrolein should be stored and transported in the dark under a blanket of nitrogen at temperatures below 20 °C, and it should be used within three months. The hydroquinone content should be determined (see Chap. 4) and if it is below 0.05%, hydroquinone must be added to bring it up to 0.1%.

Highly exothermic ionic polymerization is catalyzed by alkaline compounds, such as caustic soda, ammonia, and amines, or by mineral acids, such as concentrated sulfuric acid. These chemicals initiate polymerization at an explosive rate. Because even a trace amount of contaminant can initiate polymerization, equipment for acrolein handling must be cleaned thoroughly before use. Common inhibitors such as hydroquinone are *absolutely not effective* in preventing these ionic polymerization reactions. Depending upon the amount of ionic contaminants, the temperature rise may be slow enough that the injection of an emergency buffer solution (84% acetic acid, 8% hydroquinone, 8% anhydrous sodium acetate) suffices to control further reaction. Addition of water to stored acrolein must be avoided completely. The acrolein-containing water layer is particularly prone to polymerization. Acrolein vapors polymerize upon condensation.

Because acrolein is toxic, it is not allowed in wastewaters. It is also strongly lacrimatory, and contact with skin and eyes must be strictly avoided because exposure can result in severe injury. But it can be handled safely under controlled conditions in properly designed equipment [100].

Regulations Governing Transport. Acrolein is classified as a dangerous, flammable, and poisonous substance in various international and regional regulations. For the transportation of inhibited acrolein, the following regulations are mandatory:

International sea transport (IMDG Code): Class 6.1, UN no. 1092, PG I. International air transport (IATA-DGR): forbidden. European road (ADR) and rail (RID) transport: Classes 6.1, 8a). Proper shipping name: Acrolein, inhibited. National regulations: United States (CFR 49): § 172.101 Toxic liquid, flammable. Germany (GGVS): Special permission required for road transport of 1000 kg or more if transported in tanks with a capacity exceeding 3000 L.

For the transportation of inhibited methacrolein the following regulations are mandatory:

International sea transport (IMDG Code): Class 3.2, UN-No. 2396, PG. II. International air transport (IATA-DGR): Class 3, UN no. 2396, PG II. European road (ADR) and rail (RID) transport: Class 3, 17b). Proper shipping name: Methacrylaldehyde, inhibited. National regulations: United States (CFR 49): § 172.101 Flammable liquid, toxic.

6. Uses and Production Data

Methacrolein is an intermediate in two processes for the production of methyl methacrylate: the oxidation of 2-methylpropene or *tert*-butyl alcohol (see Section 3.2) and the Mannich reaction of propionaldehyde with formaldehyde (see Section 3.2). It has also found limited commercial application in the synthesis of flavors and fragrances [101].

Acrolein is used commercially as a very effective broad-spectrum biocide in very low concentrations of approximately 10 ppm [102], [103]. For example, it is applied to control the growth of aquatic weeds in irrigation waterways [104] or of algae and mollusks in recirculating water systems [103], [105]. Of particular importance is the use of acrolein as a biocide in oil-field brines; it increases the efficiency of oil-field water flooding and is useful in brine disposal operations. Furthermore, it is used in oil-field waters to scavenge malodorous hydrogen sulfide completely [106].

Mainly, acrolein is used as an intermediate for the following products:

D,L-Methionine. Most of the acrolein currently produced in the world is used for the production of D,L-methionine [59-51-8] (→ Amino Acids), which is used as an animal feed supplement. Acrolein is also used to make the methionine hydroxy analogue ($CH_3SCH_2CH_2CH(OH)COOH$) [583-91-5], either as an 88% aqueous solution [107], [108] or the calcium salt [109]. Different values for the bioefficacy of this acid have been published, but it is about 22–28% lower than that of D,L-methionine on an equimolar basis.

Acrylic acid is another commercially important product derived from acrolein (→ Acrylic Acid and Derivatives). It is used to make acrylates.

Glutaraldehyde, glutardialdehyde (for preparation from acrolein, see p. 200), is supplied industrially in the form of a 25% or a 50% aqueous solution; the anhydrous compound is unstable. This dialdehyde is used mainly for leather tanning [110]. Applications also include use as a biocide in oil recovery operations [111] and as a disinfectant and chemical sterilizer for hospital equipment [112].

Pyridines. Only Daicel is currently using acrolein in the commercial synthesis of 3-methylpyridine (see p. 204). Other substituted pyridines can be prepared from acrolein and substituted acroleins [43], [113] (→ Pyridine and Pyridine Derivatives).

Tetrahydrobenzaldehyde (see p. 200) is another interesting product, e.g., for the synthesis of pharmaceuticals, fungicides [114], and fragrances [115].

Flavors and Fragrances. A broad variety of compounds, synthesized mainly by Diels–Alder reactions of acrolein or methacrolein (see Section p. 200), are described as flavors and fragrances. Commercially interesting substances are, for example, lyral (**5**) [*31906-04-4*] [116], [117], myrac aldehyde (**6**) [*80450-04-0*] [117], and 5-norbornene-2-carbaldehyde (**7**) [118].

Herbicides. Acrolein cyanohydrin acetate is produced as an intermediate for the production of phosphinotrycin, a nonselective herbicide. Phosphinotrycin is very effective in crop protection together with genetically modified phosphinotrycin-resistant crops [119].

Allyl Alcohol and Glycerol. In a process developed by Shell in 1959, allyl alcohol was synthesized from acrolein by hydrogen transfer from 2-propanol in the vapor phase. The allyl alcohol was then reacted with hydrogen peroxide to form glycerol. However, this process was closed down in 1980.

1,3-Propanediol (→ Propanediols) is produced commercially by Degussa from acrolein. An alternative route via hydroformylation of ethylene oxide and subsequent hydrogenation of the intermediate 3-hydroxypropionaldehyde is claimed by Shell. A new large market for 1,3-propanediol could be in polyester coatings and in the production of poly(trimethylene terephtalate), a new material for the production of high-quality carpet fibers. 1,3-Propanediol-di-*p*-aminobenzoate is used as a curative for special polyurethane resins [120]. 1,3-Propanediol has been suggested for use in making highly resistant lubricants [121].

D,L-Glyceraldehyde (see p. 206), which is commercially available as a 40% aqueous solution, has been proposed as a water-soluble hardener for leather tanning and related applications, as well as for various syntheses.

Acrolein Polymers. Acrolein itself has seldom been used as a monomer for commercial polymerization because of the difficulty in orienting its mode of polymerization and the likelihood of complex cross-linking leading to insoluble products. However, a poly(aldehyde carboxylic acid), produced by oxidative copolymerization of acrolein and acrylic acid [122], is applied industrially as a sequestering agent [123]. Furthermore, many applications of acrolein polymers have been proposed, including those in textile treatment [124], reinforcement of paper [125], and photography [126]. The water-soluble polymeric condensation product of acrolein and formaldehyde in 40% aqueous solution is used as a biocide, e.g., as an algicide in recirculating cooling waters [127].

Production Data. The current production capacities (t/a) of acrolein in the western world is estimated as follows:

Degussa	110 000
Union Carbide	72 000
Atochem	25 000
Sumitomo Chem.	15 000
Wolshski Orgsynthes	10 000
Daicel	9000
Ohita Chem.	4500

In the production of D,L-methionine by Rhône-Poulenc in France, approximately 30 000 t/a acrolein are produced and consumed. Moreover, all acrylic acid processes involving vapor-phase propene oxidation include the intermediate production and consumption of acrolein.

7. Toxicology and Ecotoxicology

7.1. Toxicology

Acrolein is severely irritating to the skin and the mucous membranes. Its vapor causes strong eye and nasal irritation. Direct contact of liquid acrolein with the eye or skin results in severe burns. Acrolein is very toxic after inhalation and oral exposure and toxic after skin contact.

Acute toxicity data are:

LD_{50} (rat, oral)	< 11 up to 46 mg/kg
LD_{50} (rabbit, skin, neat acrolein)	562 mg/kg
LD_{50} (rabbit, skin, 20% aqueous acrolein solution)	335 mg/kg
LC_{50} (rat, inhalation, 4 h)	0.02 mg/L

Even very dilute solutions of acrolein are strongly irritating to skin and mucous membranes [128]. Severe damage to the eye was observed in rabbits with a 1% acrolein solution in glycol [129]. Due to the high reactivity of the acrolein molecule it is primarily bound locally to the application site. After absorption in the gastrointestinal tract, the main pathway of biotransformation is conjugation with glutathione and subsequent oxidation or reduction of the aldehyde group [128]. Acrolein was reported to be nonsensitizing in the guinea pig maximization test [128].

In subacute to chronic inhalation studies conducted in various species, irritation and inflammation with hyper- and metaplastic changes in the respiratory tract were the primary observed effects. At higher exposure concentrations, additional inflammatory changes in the liver and kidney were observed. After exposure for 13 weeks, (6 h/d, 5 d per week), a concentration of 0.9 mg/m^3 was proven to be the NOAEL level in hamsters and rabbits [130], while exposure for 8 to 13 weeks (6 h/d, 5 d per week) at a concentration of 0.9 mg/m^3 caused slight damage to the nasal mucosa and the lung epithelium in some strains of rat [130] – [132]. After continuous exposure of rats (24 h/d, 7 d per week) for 61 d, the NOAEL was 0.15 mg/m^3 [128].

Chronic oral administration of acrolein to rats, mice and dogs resulted in reduced body weight gain, increased mortality and changes in clinicochemical parameters at high dosages. Pathological organ changes, however, were not observed. The NOAEL for rats and dogs was 0.5 mg/kg and 2.0 mg/kg for mice [128].

In vitro acrolein reacts with nucleic acids and inhibits their synthesis. On the basis of the high cytotoxicity as well as the high reactivity of acrolein, there are difficulties in testing the genotoxic potential. From the available studies it may be concluded that acrolein seems to show genotoxic properties predominantely at high, cytotoxic concentrations in in vitro test systems (bacteria, yeasts, mammalian cell cultures), while no mutagenic properties were noted in in vivo studies in mammalian animals.

Acrolein is noncarcinogenic after oral administration to rats and mice. After inhalation exposure of hamsters for 52/81 weeks no increase in tumor incidence was observed. In a dermal initiation/promotion study in mice (18 weeks) there was no indication of a possible cocarcinogenic effect of acrolein [128].

Acrolein is classified by IARC in group 3 (inadequate evidence for carcinogenicity in humans, with inadequate evidence in experimental animals and no information in humans) [133].

Acrolein did not show any reproductive or developmental toxicity after oral or inhalative exposure in multigeneration experiments in rats [128]. No developmental toxicity was observed in doses that were not maternally toxic after oral and i.v. administration in rabbits [128].

In humans exposed to acrolein vapors for 5 min, an increased frequency of eyelid closure was observed at 0.69 mg/m^3 and a decrease in respiratory frequency at 1.4 mg/m^3 [134]. The odor threshold was reported to be between 0.05 and 0.8 mg/m^3. At 0.23 mg/m^3, 50% of the test persons were able to detect the odor [128]. Thus the odor threshold seems to be close to the threshold for eye irritation.

Occupational exposure limits are 0.1 ppm (0.25 mg/m^3) in Germany (8 h TWA and STEL) [135], and 0.1 ppm (0.23 mg/m^3, 8 h TWA) and 0.3 ppm (0.7 mg/m^3, STEL) in the United Kingdom [136] and the United States [137].

Methacrolein is of moderate toxicity after oral, dermal, and inhalative administration to experimental animals.

Acute toxicity data are:

LD$_{50}$ (rat, oral)	140 mg/kg
LD$_{50}$ (rabbit, skin)	111 mg/kg
LC$_{50}$ (rat, inhalation, 4 h)	560 mg/m^3

The effects after inhalative exposure are characterized by irritation of the respiratory tract [138].

Methacrolein is corrosive to rabbit skin and severely irritating to rabbit eyes in concentrations of 1–5%. Therefore undiluted methacrolein is expected to be corrosive to the eye [138].

After subacute (15 d, 6 h/d, 5 d per week) and subchronic (90 d, 6 h/d, 5 d per week) inhalation in rats, the only adverse effect observed was severe irritation of the upper, middle, and lower respiratory tract. The NOAEL was 13 mg/m^3 (14 d study) and 14 mg/m^3 (90 d study) respectively. In the 90 d study, signs of reversibility were observed within the four week recovery period [138].

Methacrolein showed conflicting results in mutagenicity tests in microorganisms [138], which may be related to bacteriotoxic effects. In an in vitro chromosomal aberration assay in Chinese hamster V79 cells, methacrolein caused structural chromosomal aberrations with and without metabolic activation [138].

No embryotoxic or teratogenic effects were observed in maternally nontoxic doses in an inhalation study for developmental toxicity in rats [138].

7.2. Ecotoxicology

Acrolein is of moderate to high toxicity towards aquatic organisms and microorganisms. Toxicity does not seem to increase considerably with exposure time. Ranges of acute toxicity on species of three trophic levels including microorganisms are as follows: In fish LC$_{50}$ (24–96 h) values ranged from 0.014 to 2.5 mg/L. In amphibia (tadpoles) an LC$_{50}$ (96 h) of 0.007 mg/L was reported, Daphnia showed EC$_{50}$ (24–48 h) values between 0.022 and 0.093 mg/L. In green algae the EC$_{50}$ (1–25 h) was 0.69–1.8 mg/L. Microorganisms showed IC$_{10}$ (16–48 h) values of 0.04–1.7 mg/L [128].

Acrolein in a concentration of 1.4 mg/m^3 damaged the leaves of higher plants after an exposure period of 3 h in a fumigation chamber. No toxic effects occurred after 9 h exposure at a concentration of 0.233 mg/m^3 [128].

In a three-generation chronic toxicity study (64 d) in *Daphnia magna*), the MATC (maximal acceptable toxicant concentration) in the second and third generation for reproductive toxic effects ranged from 0.0169 to 0.0336 mg/L [128].

Biodegradation of acrolein only takes place with adapted microorganisms [128]. Based on the low octanol/water partition coefficient ($\log P_{ow}$ = – 1.1 – 1.02 [1]) and its high reactivity, bioaccumulation of acrolein is not expected.

Methacrolein. No data on fish, daphnia, or algae toxicity are available. In a respiration inhibition test with activated sludge, the toxicity threshold concentration of methacrolein was 0.05 mmol/L. Biodegradation, measured as BOD, was about 35% after 10 d with unadapted microorganisms and > 60% after 8 d with adapted microorganisms. This indicates ready biodegradability only after acclimatization [139]. Due to its low octanol/water partition coefficient ($\log P_{ow}$ = 0.59 [140]) bioaccumulation of methacrolein is not expected.

8. References

[1] W. Weigert, H. Haschke, *Chem. Ztg.* **98** (1974) 2, 61.
[2] A. Isard, A. Lakodey, F. Weiss, *Chim. Ind. Genie Chim.* **103** (1970) 11, 1341.
[3] C. W. Smith: *Acrolein*, J. Wiley & Sons, New York-London 1962.
[4] G. Jenner, H. Abdi-Oskoni, J. Rimmelin, *Bull. Soc. Chim. Fr.* 1977, 341, 983.
[5] Distillers, GB 698 736, 1953 (R. G. Hall, B. K. Howe). Union Carbide, GB 739 128, 1955 (R. I. Hoaglin, R. G. Kelso).
[6] Shell Oil, US 3 159 651, 1962 (G. F. Johnson, L. C. Teague).
[7] D. P. Schirmann, F. Weiss, G. Bonnard, *Bull. Soc. Chim. Fr.* 1968, 3326.
[8] F. Y. Kasumov, S. K. Kyazimov, R. A. Sultanov, *Epoksidnye Monomery Epoksidnye Smoly* 1975, 150; *Chem. Abstr.* **85** (1975) 62907 q.
[9] G. Opitz, I. Loschmann, *Angew. Chem.* **72** (1960) 523. G. Opitz, H. Hoffmann, *Liebigs Ann. Chem.* **684** (1965) 79.
[10] T. Mukaiyama, S. Aizawa, T. Yamagushi, *Bull. Chem. Soc. Jpn.* **40** (1967) 2641. L. Givaudan et Cie, EP 12 224, 1978.S. Watanabe, S. Suga, H. Tsuruta, T. Sato, *J. Appl. Chem. Biotechnol.* **27** (1977) 423. P. V. Alston, R. M. Ottenbridge, *J. Org. Chem.* **40** (1975) 322.
[11] Bayer, EP 73 374, 1981, EP 67 353, 1981 (M. Blazejak, D. Grotkopp, J. Haydn), DE 2 548 911, 1975 (E. Roos, T. Kempermann, W. Redetzky).
[12] Toa Gosei Chem. Ind., JP 7404204, 1974 (H. Ito, H. Inone, K. Kimura, M. Sato).
[13] Du Pont, US 3 536 763, 1970 (H. S. Elenterio, T. A. Koch). Shell, GB 1 185 615, 1969 (E. T. Lutz).
[14] Degussa, US 5 284 979, 1992 (T. Haas, G. Boehme, D. Arntz).Du Pont, US 3 536 763, 1970 (H. S. Elenterio, T. A. Koch). Shell, GB 1 185 615, 1969 (E. T. Lutz).
[15] R. Hall, E. S. Stern, *J. Chem. Soc.* 1950, 490.
[16] Degussa, EP 487 903, 1992 (D. Arntz, N. Wiegand).

[17] Degussa, EP 572 812, 1993 (D. Arntz, T. Haas, N. Wiegand). Degussa, EP 535 565, 1993 (D. Arntz, T. Haas, A. Schaefer-Sindlinger). Ruhrchemie, DE 2 054 601, 1972 (W. Rottig, H. Tummes, B. Cornils).

[18] Degussa, DE-A 19 629 371, 1996 (T. Haas, M. Meier, C. Broßmer, D. Arntz, A. Freund).

[19] Degussa, DE-A 19 629 372, 1996 (T. Haas, M. Meier, C. Broßmer, D. Arntz, A. Freund).

[20] Mitsubishi Rayon, JP 7307089, 1973 (Y. Takayami, Y. Nakoyama, M. Asao, Y. Tokumichi). Mitsubishi Chem. Ind., JP 4043931, 1970; *Chem. Abstr.* **81,** 120208 t.

[21] G. V. Krishtal, V. V. Kulganek, V. F. Kucherov, L. A. Yanooskaya, *Synthesis* 1979, 2, 107.

[22] Degussa, DE 2 115 327, 1972 (H. Wagner, K. Udluft).

[23] *Org. Synth.* **2** (1963) 137.

[24] Y. Sato, Sh. Sugasawa, *Proc. Jpn. Acad. Ser. B.* **56** (1980) 573.

[25] A. Berlande, *Bull. Soc. Chim. Fr.* **37** (1925) 4, 1385.

[26] J. D. Rosen, Y. Segall, J. E. Casida, *Mutation Research* **78** (1980) 113.

[27] Takeda Chemical Ind., JP 59 225 155, 1984.

[28] Degussa, DE 1 919 504, 1969 (H. Wagner, K. Udluft). BASF, DE 3 427 404, 1984 (M. Sauerwald, T. Dockner, W. Rohr, G. Reissenweber).

[29] Du Pont, US 4 024 159, 1974 (C. C. Cumbo, K. K. Bhatia).

[30] Celanese, US 4 003 918, 1975 (O. R. Hughes).

[31] Degussa, US 5 688 973, 1995D. Arntz et al. Degussa, US 5 216 179, 1991 (M. Hoepp, D. Arntz, S. Bartsch, A. Schaefer-Sindlinger, W. Boeck). Du Pont, 4 108 869, 1977 (H. B. Copelin).

[32] Degussa, DE 3 403 426, 1984 (G. Prescher, J. Andrade, D. Arntz).

[33] Betz Laboratories, US 4 851 583, 1988 (E. Bockowski, C. R. McDaniel). Degussa, US 5 696 052, 1995 (P. Werle, M. Trageser, O. Helmling, H. Jakob).

[34] J. I. Mateo, R. Vallve, *Rev. Plast. Mod.* **198** (1972) 24, 921. Nippon Kayaku JP Kokai 74 101 310, 74 101 311, 1973; Chem. Abstr. 82, (1974) 72534 t and 111579 h. E. Takiyama, T. Hanyuda, *Nippon Setchaku Kyokaishi* **13** (1977) 9, 330. E. Takiyama, *Plast. Age* **21** (1975) 9, 93.

[35] Mitsubishi Petrochemical, JP-Kokai 06 305 998, 1993. Mitsubishi Petrochemical, JP-Kokai 06 107 586, 1992.

[36] Degussa, DE 3 043 259, DE 3 043 252, 1980 (A. Kleemann, M. Samson).

[37] Nippon Kayaku, JP 7118733, 1971 (G. Yamaguchi, H. Yoshikawa). Asahi Chemical Industries, JP 7405917, 1974 (C. Shibuya, S. Onchi, S. Hayashi).

[38] Hoechst, EP 19 227, 1979 (R. Mündnich, M. Finke, W. Rupp, K. Dehmer), EP 19750, 1979 (M. Finke, R. Mündnich), DE 3 047 024, 1980 (M. Finke, H. Erpenbach).

[39] R. Palm, H. Ohse, H. Cherdron, *Angew. Chem.* **78** (1966) 1093.

[40] McIntosh, *Can. J. Chem.* **55** (1977) 4200.

[41] H. J. Vebel, K. K. Moll, M. Muehlstaedt, *J. Prakt. Chem.* **312** (1971) 849.

[42] S. Doi, I. Hayashi, G. Amano, Y. Hachihama, *Kogyo Kagaku Zasshi.* **63** (1960) 828. Ube Ind., JP 7622, 1982 (T. Kawato, K. Koga, M. Kuniyoshi).

[43] H. Beschke, H. Friedrich, *Chem. Ztg.* **101** (1977) 377. Degussa, DE 2 639 701, 1976, DE 2 712 694, 1977, and DE 2 639 702, 1976 (H. Beschke, H. Friedrich), DE 2 819 196, 1978 (H. Beschke, A. Kleemann).

[44] Monsanto, EP 55 090 and 55 091, 1982 (S. M.Cohen, T. R. Le Blanc).

[45] Hoffmann La Roche, US 4 003 919, 1972 (J. W. Scott, D. R. Parrisch, G. Saucy). G. Sartori, G. Casiraghi, L. Bolzoni, G. Casnati, *J. Org. Chem.* **44** (1979) 803.

[46] Yuki Gosei Kogyo, JP 7312745, 1970 (S. Wagatsuma, I. Yamashita, Y. Kinoshita); *Chem. Abstr.* **79** (1979) 53191 a.

[47] BASF, DE-A 3 520 113, 1985 (B. Czauderna, J. Nieberle, R. Peter, D. Nissen).

[48] Ciba-Geigy, EP 6535, 1980 (F. Lohse, F. Gutekunst, R. Schmid, A. Schmitter).
[49] Asahi Chemical Industries, JP 57983, 1973 (K. Tanaka, S. Furuhashi, S. Yokoyama).
[50] Bayer, EP 48373, 1980 (U. Heinemann et al.). Z. Brzozowski, E. Pormanacka-Jankowska, S. Angielski, *Acta Pol. Pharm.* **34** (1977) 3, 279; *Chem. Abstr.* **88** (1978) 120742 g.
[51] Montecatini Edison, DE 1947419, 1968 (I. Dahli et al.).
[52] Celanese, JP-Kokai 18506, 1978, US 4072727, 1978 (T. H. Vanderspurt). Stamicarbon, US 3686333, 1972 (C. J. Duyrerman).
[53] Shell Development Co., US 2767221, 1953 (A. B. Seaver, H. de V. Finch).
[54] A. Shigaki, *Chem. Econ. Eng. Rev.* **2** (1970) 8, 47.
[55] Mitsubishi Petrochemicals, EP 582277, 1993 (Y. Watanabe, M. Kurashige).
[56] M. R. Johnson, B. Rickborn, *J. Org. Chem.* **35** (1970) 1041.
[57] H. S. Broadbent, G. C. Campbell, W. B. Bartley, J. H. Johnsons, *J. Org. Chem.* **24** (1959) 1847. Standard Oil (Indiana), US 4292452, 1978 (R. J. Lee, D. H. Meyer, D. M. Senneke). P. N. Rylander, D. R. Stelle, *Tetrahedron Lett.* **20** (1969) 1579. Celanese, US 4020116, 1976, DE 2734811, 1976, US 4127508, 1978, JP-Kokai 18506, 1978 (T. H. Vanderspurt). Y. Nagase, H. Hittori, K. Tanabe, *Chem. Lett.* **1983**, 1615.
[58] G. B. Payne, *J. Am. Chem. Soc.* **80** (1958) 6461, **81** (1959) 4501.
[59] Shell Oil, US 2941006, 1957 (Ch. R. Greene).
[60] R. C. Schulz: "Acrolein Polymers" in *Encyclopedia of Polymer Science and Engineering,* vol. **1** John Wiley, New York 1985.
[61] Biopolymers Ltd., PCT/WO 8804671, 1988 (G. J. H. Melrose, C. M. Kleppe, J. W. Langley, J. M. Stewart).
[62] Degussa, EP 97101161, 1997 (P. Werle, H.-P. Krimmer, M. Trageser, F.-R. Kunz).
[63] H. Schulz, H. Wagner, *Angew. Chem.* **62** (1950) 105.
[64] Shell Development Co., US 2451485, 1948 (G. W. Hearne, M. L. Adams).
[65] D. J. Hucknall: *Selective Oxidations of Hydrocarbons,* Academic Press, London-New York 1974.
[66] L. F. Hatch, S. Matar, *Hydrocarbon Process.* **57** (1978) 6, 149.
[67] Standard Oil, US 2941007, 1960 (J. L. Callahan, R. W. Foreman, F. Veatch).
[68] C. R. Adams, *Chem. Ind.* (*London*) **52** (1970) 1644. J. L. Callaghan, R. K. Grasselli, E. C. Milberger, H. A. Strecker, *Ind. Eng. Chem. Prod. Res. Dev.* **9** (1970) 134. Sohio, US 2904580, 1959 (J. D. Idol).
[69] Knapsack, US 3171859, 1965 (K. Sennewald, W. Vogt, K. Gehrmann, S. Schafer).
[70] Nippon Kayaku, US 3454630, 1969 (G. Yamaguchi, S. Takenaka).
[71] Degussa, DE 1792424, 1968 (E. Koberstein et al.).
[72] Nippon Shokubai Kagaku, US 3825600, 1974 (T. Ohara, M. Ueshima, I. Yanagisawa).
[73] Surveys on bismuth molybdate catalyst developments: G. W. Keulks: "Selective Oxidation of Propylene", in G. W. Smith (ed.): Catalysis in Organic Syntheses, Academic Press, New York 1977. R. Higgins, P. Hayden, *Catalysis* (London) 1 (1977) 168. T. Ohara, *Shokubai* **19** (1977) 3, 157; *Chem. Abstr.* **87** (1977) 151636. I. Matsuura, *Shokubai* **21** (1979) 6, 409. H. Offermanns, G. Prescher in
Houben-Weyl: *Methoden der Organischen Chemie,* vol. **E3**: Aldehyde, p. 231, Georg Thieme Verlag, Stuttgart 1983.
R. K. Grasselli, J. D. Burrington in W. R. Moser (ed.): *Catalysis of Organic Reactions* (Chem. Ind. Ser., vol. **5**), Marcel Dekker, New York 1981.
[74] Y. Moro-Oka, W. Ueda, *Adv. Catal.* **40** (1994) 233.
[75] Nippon Kayaku, US 3454630, 1969 (G. Yamaguchi, S. Takenaka).
[76] Nippon Kayaku, US 3778386, 1973 (S. Takenaka, Y. Kido, T. Shimabara, M. Ogawa).

[77] Nippon Shokubai Kagaku, JP 42242, 1972 (T. Ohara et al.), US 3833649.
[78] Toa Gosei Chemical Co., FR 2028164 (H. Ito, S. Nakamura, T. Nakano).
[79] BASF, EP 0575 897, 1993 (H. P. Neumann, H. Martan, H. Petersen, W. Doerflinger).
[80] Nippon Shokubai Kagaku, EP 0450 596, 1991 (T. Kawajiri, H. Hironaka, S. Uchida, Y. Aoki).
[81] Sumitomo Chemical Co., EP 0630879, 1994 (Y. Nagaoka, Y. Nomura, K. Nagai).
[82] Mitsubishi Petrochemical Co., EP 0239071, 1992 (K. Sarumaru, E. Yamamoto, T. Saito).
[83] Degussa, EP 0417 722, 1990 (W. Böck, D. Arntz, G. Prescher, W. Burkhardt).
[84] G. E. Schaal, *Hydrocarbon Process.* **52** (1973) 9, 218.
[85] W. M. Weigert, *Chem. Eng.* **80** (1973) 15, 68.
[86] PCUK, Inf. Chim., Spec. Exp. 97 (1977/78).
[87] M. M. Bettahar, G. Costentin, L. Savary, J. C. Lavalley, *Appl. Catal. A* **145** (1996) 1–48.
[88] BASF, EP 0731 077, 1996 (W. Hefner, O. Machhammer, H. P. Neumann, A. Tenten, W. Ruppel, H. Vogel).
[89] Degussa, DE 4 238 493, 1992 (A. Neher, T. Haas, D. Arntz, H. Klenk, W. Girke).
[90] Degussa, DE 1 910 795, 1969 (H. Hillenbrandt, E. Liebetanz, Th. Lüssling, E. Noll, K. Simon).
[91] Degussa, DE 2 263 496, 1972 (E. Noll, H. Schaefer, H. Schmid, W. Weigert).
[92] Asahi Kasei Kogyo, DE 2 941 341 (A. Aoshima, Y. Kohoku, R. Mitsui, T. Yamaguchi). BASF, DE 2 909 597, 1979 (H. Engelbach, H. Krabetz, G. Duembgen, C.-H. Willersinn, W. Breitelschmidt). Celanese, DE 2 943 707, 1979, DE 2 943 704, 1979 (T. H. Vanderspurt). Rohm and Haas, US 4 151 117, 1979 (F. W. Schlaefer). Mitsubishi Rayon, JP-Kokai 8195135, 1980, US 4 219 670; *Chem. Abstr.* **95** (1981) 203334.
[93] N. Shimizu, *Petrotech (Tokyo)* **6** (1983) 9, 778.
[94] *Chem. Econ. Eng. Rev.* **15** (1983) 4, 49.
[95] T. Hatsuike, H. Matsuzawa, *Hydrocarbon Process.* **58** (1979) 2, 105.Y. Oda et al., *Hydrocarbon Process.* **54** (1975) 10, 115.
[96] *Chemical Week*, November 12, 1997, p. 36.
[97] BASF, EP 58 927, 1982 (F. Merger, H. J. Foerster), EP 92 097, 1982 (G. Duembgen, G. Fouguet, R. Krabetz, E. Lucas, F. Merger, F. Nees).M. Tramontini, *Synthesis* 1973, 703. Celanese, US 2 848 499, 1956 (A. F. Mac Lean, B. G.Frenz). Ruhrchemie, DE 2 855 504, 1978 (W. Bernhagen et al.).
[98] S. Malinowski, H. Jedrzejawska, S. Basinski, S. Benbenek, *Chim. Ind. (Paris)* **85** (1961) 6, 885.
[99] Mitsubishi Chemical Ind., JP-Kokai 138499, 1977, US 4 146 574.
[100] Degussa(company publication): *Acrolein*, Frankfurt Manufacturing Chemists Assoc.: *Acrolein*, Chemical Safety Data Sheet, SD-85, 1961. G. Hommel: *Handbuch der gefährlichen Güter*, 3rd ed., vol. **1**, Merkblatt 218, Springer-Verlag, Berlin 1980. Kühn-Birett: *Merkblätter gefährliche Arbeitsstoffe*, Verlag moderne Industrie, München 1981.
[101] U. Harcher, *Chem. Ztg.* **99** (1975) 54. Rhône Poulenc,US 3 023 247, 1960, GB 850360, 1960. International Flavors and Fragrances, US 3 185 629, 1965.
[102] Shell, US 2 959 476, 1969 (J. van Overbeek).
[103] C. v. Oppel, A. Schiffers, *Energie* **26** (1974) 173.
[104] J. L. Brady, C. L. Kissel: "Acrolein in Irrigation Waterways," Proc. Western Aquatic Plant Management Society, 1st Symposium, Denver, Colo. 1982.
[105] J. M. Donokue, A. J. Piluso, J. R. Schieber, *Mater. Prot.* **5** (1966) 22.
[106] Magna Corp., US 4 215 147, 4 215 148, 1978 (C. L. Kissel, F. F. Caserio).
[107] Monsanto, US 4 353 924, 1979 (J. W. Beher, D. L. Mansfield, D. J. Weinkauff).
[108] Monsanto, *Chem. Week* **127** (Jul. 16, 1980) 3, 37.
[109] Du Pont, US 4 335 257, 1980 (E. W. Cummins, S. I. Gleich, R. M. Vigilant).

[110] Union Carbide, US 2 941 859, 1960 (M. L. Fein, E. M. Filachione). BASF, DE 2 215 948, 1972 (H. Erdmann, F. F.Miller).
[111] Union Carbide, US 4 244 876, 1978 (G. H. Warrer, L. F. Theiling, M. G. Freid).
[112] Schülke & Mayr, DE 3 032 794, 1980 (W. Münzenmaier et al.). S. Sankara Subramanian, *Sci. Res. News* **1** (1978) 42.
[113] A. Kleemann, *Chem. Ztg.* **191** (1979) 389.
[114] BASF, DE 3 121 349, 1981 (W. Graulich, W. Himmele, Ch. Martin, E.-H. Pommer, H. Siegel).
[115] H. Bolens, J. Heydel, *Chem. Ztg.* **97** (1973) 8. Fritzsche Dodge and Olcott, US 4 287 100, 1978 (K. Kulka, T. Zazulka, J. M. Yurecko).
[116] International Flavors & Fragrances, US 4 007 137, 1977 (J. M. Sanders, W. L. Schreiber, J. B. Hall).
[117] International Flavors and Fragrances, DE 2 643 062, 1976 (J. M. Sanders et al.).
[118] Takasago Perf. Co., DE 2 833 283, 1977 (T. Kobayashi, H. Tsuruta, T. Yoshida).
[119] Hoechst, EP 275 957, 1988, (E. Strauch, W. Arnold, R. Alijah, W. Wohlleben, A. Pühler, P. Eckes, G. Donn, E. Uhlmann, F. Hein, F. Wengenmayer).
[120] Polaroid, US 3 932 360, 1974 (L. D. Cerankowski, N. Mattucci, R. C. Baron). R. C. Baron, R. E. Brooks, K. C. Frisch: *Trimethylene Glycol p-Aminobenzoate,* Can. Ureth. Manufacturers Assoc., April 1978.
[121] Chevron Research, US 3 819 521 (M. J. Sims).
[122] Degussa, DE 1 071 339, 1959 (K. H. Rink).
[123] H. Haschke, G. Morlock, P. Kuzel, *Chem. Ztg.* **96** (1972) 199.
[124] M. M. Ishanov, U. A. Azizov, M. Nigmankhodzhaeva, *J. Polym. Sci. Polym. Chem. Ed.* **9** (1971) 1013.
[125] American Cyanamid, US 3 819 555, 1974 (E. D.Kaufmann).
[126] GAF, US 3 595 663, 1971 (S. Emmi), US 3 615 623, 1971 (N. D. Field, D. I. Randall, J. D. Fitzpatrick).
[127] Degussa, DE 3 205 487, 1982 (K. H. Rink, W. Merk).
[128] Beratergremium für umweltrelevante Altstoffe der Gesellschaft deutscher Chemiker: *Acrolein (2-Propenal), BUA Stoffbericht Nr. 157,* S. Hirzel Wiss. Verlagsges., 1995.
[129] T. B. Albin in C. W. Smith (ed.): *Acrolein* Hüthig, Heidelberg, 1975.
[130] V. J. Feron, A. Kruysse, H. P. Til, H. R. Immel, *Toxicology* **9** (1978) 47–57.
[131] R. S. Kuztmann, E. A. Popenoe, M. Schmaeler, R. T. Drew, *Toxicology* **34** (1985) 139–151.
[132] R. S. Kutzmann, R. W. Wehner, S. B. Haber, *Toxicology* **31** (1984) 53–65.
[133] International Agency For Research on Cancer, World Health Organisation, IARC Monographs on the Evaluation of Carcinogenic Risks to Humans, vol. **63**, Dry Cleaning, some Chlorinated Solvents and other Industrial Chemicals, WHO, 1995, pp. 337–372.
[134] A. Weber-Tschopp, T. Fischer, R. Gierer, E. Grandjean, *Z. Arbeitswiss.* **32** (1977) 166–171.
[135] TRGS 900 Grenzwerte in der Luft am Arbeitsplatz, BArbBl. Nr. 10 (1996) 88, Nr. 4 (1997) 42, Nr. 11 (1997) 27.
[136] Health and Safety Executive (HSE), EH 40/97 Occupational Exposure Limits 1997.
[137] ACGIH, American Conference of Governmental Industrial Hygienists, Threshold Limit Values for Chemical Substances and Physical Agents, Biological Exposure Indices, 1997.
[138] Berufsgenossenschaft der chemischen Industrie Toxikologische Bewertung Methacrolein, Nr. 108, 1995.
[139] V. T. Stack Jr., *Ind. Eng. Chem.* **49** (1957) 913–917.
[140] M. D. Barrat, *Toxicol. in Vitro* **10** (1996) 247–236.

Acrylic Acid and Derivatives

Takashi Ohara, Nippon Shokubai Kagaku Kogyo Co., Ltd., Osaka, Japan.
Takahisa Sato, Nippon Shokubai Kagaku Kogyo Co., Ltd., Osaka, Japan.
Noboru Shimizu, Nippon Shokubai Kagaku Kogyo Co., Ltd., Osaka, Japan.
Günter Prescher, Degussa AG, Zweigniederlassung Wolfgang, Hanau, Federal Republic of Germany
Helmut Schwind, Degussa AG, Zweigniederlassung Wolfgang, Hanau, Federal Republic of Germany
Otto Weiberg, Degussa AG, Zweigniederlassung Wolfgang, Hanau, Federal Republic of Germany
Klaus Marten, Sichel-Werke GmbH, Hannover, Federal Republic of Germany (Chap. 2)

1.	Acrylic Acid and Esters	223	1.6.	Uses	238
1.1.	Physical Properties	224	1.7.	Some Special Acrylates	238
1.2.	Chemical Properties	224	1.8.	Acrylic acid, economic aspects	239
1.3.	Production	229	1.9.	Toxicology and Occupational Health	240
1.3.1.	Propene Oxidation	231			
1.3.2.	Esterification	234	2.	Cyanoacrylates	241
1.4.	Quality Specifications and Analysis	236	3.	Acrylamide	243
1.5.	Storage and Transportation	237	4.	References	244

1. Acrylic Acid and Esters

Acrylic acid [79-10-7], 2-propenoic acid, CH_2=CHCOOH, and its esters CH_2=CHCOOR, which are also known as acrylates, are flammable, volatile, mildly toxic, colorless liquids. Hydroquinone or its monomethyl ether is usually added to commercial preparations to inhibit polymerization. Until recently, acrylic acid and acrylates were produced industrially via a variety of routes such as acrylonitrile hydrolysis and the modified Reppe process (see Section 1.3). However, remarkable progress on the catalytic oxidation of propene to acrylic acid via acrolein has led to almost complete replacement of these earlier processes.

Esters such as methyl, ethyl, *n*-butyl, and 2-ethylhexyl acrylates, as well as acrylic acid, are in worldwide use, primarily for polymers. Other esters, including multifunctional acrylates, are produced for special applications.

Chemically, *acrylamide* (see Chap. 3) is a derivative of acrylic acid but the amide is produced by hydration of acrylonitrile instead of by amidation of the acid.

1.1. Physical Properties

Acrylic Acid is a clear, colorless liquid, bp 141.0 °C (101.3 kPa), mp 13.5 °C; it forms crystalline needles in the solid state. Other important physical constants are listed below [1]–[6]:

M_r	72.06
Refractive index	n_D^{20} 1.4224, n_D^{25} 1.4185
Density	1.060 (10 °C), 1.040 (30 °C), 1.018 (50 °C) g/cm^3
Viscosity at 25 °C	1.149 mPa · s
Critical temperature	380 °C
Critical pressure	5.06 MPa
Heat of vaporization at 101.3 kPa	45.6 kJ/mol
Heat of combustion	1376 kJ/mol
Heat of melting at 13 °C	11.1 kJ/mol
Heat of neutralization	58.2 kJ/mol
Heat of polymerization	77.5 kJ/mol
Dissociation constant at 25 °C	5.510^{-5}; pK_a= 4.26

Vapor pressure as function of temperature:

t, °C	0	20	40	60	100	120	141
p, kPa	0.31	1.03	2.93	7.2	33.2	63.3	101.3

Acrylic acid is highly miscible with water, alcohols, esters, and many other organic solvents. Figure 1 gives the density of the aqueous solution as a function of water content. Table 1 shows the freezing points of various acetic acid–acrylic acid and water–acrylic acid solutions.

Derivatives. Table 2 lists physical properties of representative derivatives other than esters, Table 3 those of five commercial acrylates, and Table 4 those of other acrylates including some diesters.

1.2. Chemical Properties

Acrylic acid and its esters undergo reactions characteristic of both unsaturated compounds and aliphatic carboxylic acids or esters. The high reactivity of these compounds stems from the two unsaturated centers situated in a conjugated position. The β carbon atom, polarized by the carbonyl group, behaves as an electrophile; this favors the addition of a large variety of nucleophiles and active hydrogen compounds to the vinyl group. Moreover, the carbon-carbon double bond undergoes radical-initiated addition reactions, Diels-Alder reactions with dienes, and polymerization reactions.

The carboxyl function is subject to the displacement reactions typical of aliphatic acids and esters, such as esterification and transesterification.

Table 1. Freezing points of acrylic acid mixtures: A with acetic acid, B with water

System A, wt% acetic acid	Freezing point,°C	System B, wt% water	Freezing point, °C [5]
0	13.5	0	13.5
10	7.5	5	5.5
20	0.7	10	1.0
40	-14.1	20	-5.5
50	-23.5	30	-10.3
50.2	-23.8	37	-12.5
60	-13.4	40	-12.0
80	3.7	60	-8.0
100	16.6	80	-4.0
		100	0

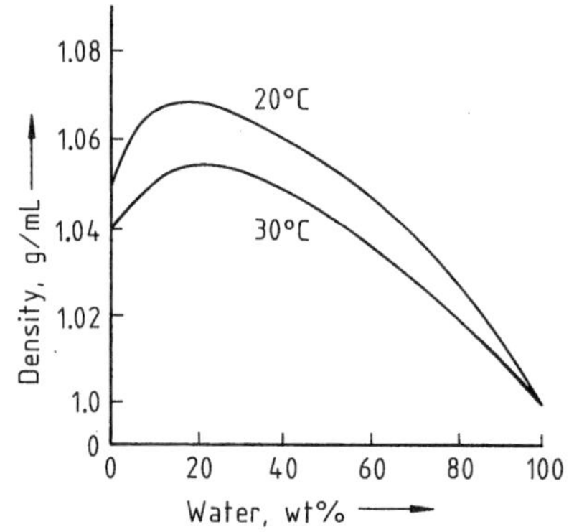

Figure 1. Relationship between density of aqueous acrylic acid solution and water content

Table 2. Physical properties of acrylic acid derivatives

	Acrylic anhydride	Acryloyl chloride	Acrylamide
CAS registry number	[2051-76-5]	[814-68-6]	[79-06-1]
Structural formula	$(CH_2=CHCO)_2O$	$CH_2=CHCOCl$	$CH_2=CHCONH_2$
Molecular formula	$C_6H_6O_3$	C_3H_3ClO	C_3H_5NO
M_r	126.11	90.51	71.08
mp, °C			84.5
bp, °C/p in kPa	38/0.27	75/101	125/16.6
Density, g/cm^3		1.113 (20 °C)	1.122 (30 °C)
Refractive index, n_D^{20}	1.4487	1.4337	

Joint reactions of the vinyl and carboxyl functions, especially with bifunctional reagents, often constitute convenient routes to polycyclic and heterocyclic substances.

Acrylic acid and its esters polymerize very easily. The polymerization is catalyzed by heat, light, and peroxides and inhibited by stabilizers, such as the monomethyl ether of hydroquinone or hydroquinone itself. These phenolic inhibitors are effective only in the presence of oxygen. The highly exothermic, spontaneous polymerization of acrylic acid is extremely violent.

Table 3. Physical properties of the most important acrylates

Property	Methyl acrylate	Ethyl acrylate	n-Butyl acrylate	Isobutyl acrylate	2-Ethylhexyl acrylate
CAS registry number	[96-33-3]	[140-88-5]	[141-32-2]	[106-63-8]	[103-11-7]
Molecular formula	$C_4H_6O_2$	$C_5H_8O_2$	$C_7H_{12}O_2$	$C_7H_{12}O_2$	$C_{11}H_{20}O_2$
M_r	86.09	100.12	128.17	128.17	184.28
mp, °C	-76	-72	-64.6	-61	-90
bp at 101.3 kPa, °C	80.3	99.4	147.4	138	216
Specific heat (l), kJ mol^{-1} K^{-1}	0.48	0.47	0.46	0.46	0.46
Solubility at 25 °C					
in water (g/100 g)	5	1.5	0.2	0.2	0.01
of water in ester (g/100 g)	2.5	1.5	0.7	0.6	0.15
Azeotropes					
with water, bp, °C	71	81.1	94.5		
water content, wt%	7.2	15	40		
with methanol, bp, °C	62.5	64.5			
methanol content, wt%	54	84.4			
with ethanol, bp, °C	73.5	77.5			
ethanol content, wt%	42.4	72.5			
with n-butanol, bp, °C			119		
n-butanol content, wt%			89		
Heat of vaporization at bp, kJ/mol	33.2	34.8	36.5	38.1	47.0
Heat of polymerization, kJ/kg	84.7	77.9	77.3		60.1
Vapor pressure, kPa					
at 0 °C	4.2	1.2	0.14		
at 20 °C	9.3	3.9	0.44		
at 50 °C	35.9	17.3	2.82		0.16
at 100 °C			21.9		2.1
at 150 °C					14.6
Refractive index, n_D^{20}	1.4040	1.4068	1.4190	1.4150	1.4365
Relative density, d_4^{20}	0.9535		0.8998		0.8852
d_{20}^{20}	0.9565	0.9231	0.9015	0.890	0.8869
Viscosity, mPa s					
at 20 °C	0.53	0.69	0.90	0.78	1.7
at 25 °C	0.49	0.55	0.81		1.54
at 40 °C		0.50	0.70		1.2
Autoignition temperature, °C	393	355	267	340	230
Flammability range in air, vol%	2.8–25	1.8–saturated	1.5–9.9	1.9–8.0	0.6–1.8
Flash point					
closed cup, °C	-3	9	41	33	87
open cup, °C	-2	19	47		92

In this section are listed typical examples of reactions other than polymerization, which is discussed in Section 1.6. Several review articles and monographs [1]–[7] describe the rich chemistry of acrylates and acrylic acid.

Addition Reactions. Acrylic acid and acrylates combine readily with substances, such as hydrogen, hydrogen halides and hydrogen cyanide, that customarily add to olefins [8]:

Table 4. Physical properties of acrylic esters and diesters

Ester	CAS registry number	Molecular formula	M_r	bp, °C/p, kPa	Refractive index, n_D^{20}	Relative density, d_4^{20}
n-Propyl	[925-60-0]	$C_6H_{10}O_2$	114.15	44/5.3	1.4130	0.9078
n-Pentyl	[2998-23-4]	$C_8H_{14}O_2$	142.20	48/0.9	1.4240	0.8920
n-Hexyl	[2499-95-8]	$C_9H_{16}O_2$	156.23	40/0.2	1.4280	0.8882
n-Heptyl	[2499-58-3]	$C_{10}H_{18}O_2$	170.25	57/0.1	1.4311	0.8846
Isopropyl	[689-12-3]	$C_6H_{10}O_2$	114.15	52/14	1.4060	0.8932
sec-Butyl	[2998-08-5]	$C_7H_{12}O_2$	128.17	60/6.7	1.4140	0.8914
tert-Butyl	[1663-39-4]	$C_7H_{12}O_2$	128.17	120/101.3	1.408	0.879
Allyl	[999-55-3]	$C_6H_8O_2$	112.13	47/5.3	1.4320	0.9441
2-Hydroxyethyl	[818-61-1]	$C_5H_8O_3$	116.12	74/0.7	1.4505	1.1038 (25 °C)
2-Hydroxypropyl	[999-61-1]	$C_6H_{10}O_3$	130.14	77/0.7	1.4443	1.5036
Ethylene glycol diester	[2274-11-5]	$C_8H_{10}O_4$	170.17	70/0.1	1.4529	
1,2-Propanediol diester	[25151-33-1]	$C_9H_{12}O_4$	184.19	60/0.04	1.4470	
1.4-Butanediol diester	[31442-13-4]	$C_{10}H_{14}O_4$	198.22	83/0.1	1.4538	

$$H_2C=CHCOOR + HX \rightarrow H_2CX-CH_2COOR$$

where R=H, alkyl, or aryl, and X=H, halogen, or CN.

Michael additions of organic substances take place in the presence of basic catalysts, such as tertiary amines, quaternary ammonium salts, and alkali alkoxides:

where X = $-C(NO_2)R'R''$, $-CH(COOR')_2$,

$$-\underset{\underset{COR''}{COOR'}}{CH}, \quad -\underset{\underset{COOR'}{CN}}{CH}, \quad \text{or} \quad -\underset{\underset{C_6H_5}{CN}}{CH}$$

and R',R'' = alkyl or aryl.

Ammonia and amines are sufficiently basic to react without a catalyst: where X=-NH_2, -NHR' [9], [10], -NR'R'' [11], heterocycles [12]–[14], -NR'COR'', or -NHNR$_2$ [15]. The addition of only one molecule of NH_3 (for addition of two, see below) can be achieved with an aqueous solution of ammonia and ammonium carbonate [16].

The addition of aromatic amines or amides and tert-alkyl primary amines is more effectively promoted by acids. Amines may attack both the vinyl and carboxyl functions, but the products of such reactions decompose to give N-substituted amides.

Alcohols [17], phenols, hydrogen sulfide [14], [18], and thiols [19], [20] also add under basic conditions. Hydrogen sulfide in the presence of sulfur and ammonium polysulfide or amine catalysts gives polythiodipropionic acids and esters [19], [21]:

$$CH_2=CH-COOR \xrightarrow[S, H_2S]{(NH_4)_2S_x} S_x(CH_2CH_2COOR)_2$$

where R=H, alkyl, or aryl.

Other examples of HX additions to acrylic acid and acrylates are:

where X = $-O_2SR'$, $-SO_3Na$, $-OP(OR')_2$,

$-\overset{O}{\underset{}{C}}-R'$ [22],[23], pyrrolyl, or furyl (24)

and R' = alkyl or aryl. [22] – [24]

Additions of aromatic hydrocarbons are promoted more efficiently by Lewis acids [25].

If further acidic hydrogen atoms are available in the addition product, a second (and third) molecule of acrylic acid or ester adds. This is the case in the reaction of acrylic acid or ester with H_2S, NH_3, RNH_2, and pyrrole.

Other examples of addition reactions are the following:

$H_2C=CH-COOR$

$+ Hal_2 \longrightarrow CH_2-CH-COOR$ [26]
                          $\;\;\;\;\;\;|\;\;\;\;\;\;|$
                          $\;\;\;\;\;\;Hal\;\;Hal$

$+ HOCl \longrightarrow CH_2-CH-COOR$ [3]
                       $\;\;\;\;\;\;|\;\;\;\;\;\;|$
                       $\;\;\;\;\;\;OH\;\;Cl$

$+ Hg(OCOCH_3)_2 + CH_3OH \longrightarrow$
$\;CH_2-CH-COOR + CH_3COOH$ [3]
$\;|\;\;\;\;\;\;\;|$
$\;OCH_3\;HgOCOCH_3$

$+ C_6H_5N_2^+Cl^- \longrightarrow CH_2-CH-COOR + N_2$ [3]
$\;|\;\;\;\;\;\;|$
$\;C_6H_5\;Cl$

$+ CH_2=C(CH_3)_2 \longrightarrow$
$\;\;\;\;\;\;\;\;CH_2=C-CH_2CH_2CH_2COOR$ [3], [27]
$\;\;\;\;\;\;\;\;\;\;\;\;\;\;|$
$\;\;\;\;\;\;\;\;\;\;\;\;\;\;CH_3$

where R = H, alkyl, or aryl. [3], [26], [27]

Heterocyclic substances often can be formed by subsequent reaction of the carboxyl function, especially with bifunctional nucleophiles [23], [28]:

$H_2C=CH-COOR + (NH_2)_2C=O \longrightarrow$ (cyclic urea product) + ROH

$CH_2=CH-COOCH_3$ + phenol $\longrightarrow$ chromanone + CH_3OH [29]

where R = H, alkyl, or aryl.

Substituted ring compounds are formed readily by Diels-Alder reactions [30], [31]:

$CH_2=CH-COOCH_3$ + furan $\longrightarrow$ bicyclic adduct with COOCH$_3$

Acrylates also undergo cobalt- or rhodium-catalyzed hydroformylation reactions [32], [33]:

$$2\,CH_2=CH-COOR + 2\,CO + 2\,H_2 \xrightarrow{catalyst} CH_2(CHO)-CH_2COOR + CH_3CH(CHO)-COOR$$

where R = alkyl or aryl.

At elevated temperature or on longer storage acrylic acid dimerizes:

$$2\,CH_2{=}CH{-}COOH \longrightarrow CH_2{=}CH{-}COOCH_2\,CH_2COOH$$

In the presence of catalysts such as tributylphosphine, acrylates can also dimerize to give 2-methyleneglutarates [34]:

$$2\,CH_2=CH-COOR \xrightarrow{catal.} ROOC-\underset{\underset{CH_2}{\|}}{C}-CH_2CH_2-COOR$$

where R = alkyl or aryl.

Reactions of the Carboxyl Group. Acrylic acid is converted readily into its corresponding salts, into acrylic anhydride by reaction with acetic anhydride, or into acryloyl chloride by reaction with benzoyl or thionyl chloride. The esterification of acrylic acid and transesterification of acrylic esters are economically the most important reactions (see Section 1.3.2).

Some other examples are:

$$CH_2=CHCOOH$$

$$+\; \overset{O}{\overset{|}{CH_2}}\!-\!CH_2 \longrightarrow CH_2=CHCOOCH_2CH_2OH$$

$$+\; HOCH_2CH_2SO_3Na \longrightarrow CH_2=CHCOOCH_2CH_2SO_3Na + H_2O \quad [35]$$

$$+\; CH_2=CR'R'' \xrightarrow{H^+} CH_2=CHCOOCR'R''CH_3 \quad [36]$$

$$+\; ClCH_2CH_2OCH=CH_2 \xrightarrow{\gamma\text{-collidine}} CH_2=CHCOOCH_2CH_2OCH=CH_2 + HCl \quad [37]$$

where R', R'' = alkyl or aryl [35] – [37].

1.3. Production

Currently most of the commercial acrylic acid is produced from propene, which is also the raw material for the production of acrolein. In the past, acrylic acid and its esters were produced by various processes some of which are summarized here (see [5], [6], [38], [39]) and are still in use to a small extent.

Processes Based on Acetylene (→ Acetylene). The stoichiometric synthesis of acrylic acid and its esters from acetylene proceeds at atmospheric pressure and at 40 °C in the presence of acid and nickel carbonyl:

$$4\ C_2H_2 + 4\ ROH + 2\ HCl + Ni(CO)_4 \longrightarrow 4\ CH_2=CHCOOR + NiCl_2 + H_2$$

where R=H, alkyl, or aryl.

The reaction was discovered by W. Reppe in 1939. Röhm & Haas and Toa Gosei Chemical have used this method as well as the modified, non-stoichiometric Reppe process, but both have been abandoned because of the difficulties in handling the toxic and corrosive nickel carbonyl.

High-Pressure Reppe Process. The process employed by BASF and Badische Corp. proceeds at approximately 14 MPa and 200 °C with a nickel bromide–copper(II) bromide catalyst:

$$C_2H_2 + CO + H_2O \xrightarrow{\text{catal.}} CH_2=CHCOOH$$

However, the safety and pollution control problems with nickel carbonyl (formed in the process) and the high cost of acetylene are disadvantages of this process. It has largely been replaced by the direct oxidation of propene although BASF still produces part of its acrylic acid by this process.

Acrylonitrile Hydrolysis (→ Acrylonitrile). This method is economically unattractive because of the low yield based on propene and the large quantities of NH_4HSO_4 waste. The process has been abandoned by Ugine Kuhlmann, Mitsubishi Petrochemical, and Mitsubishi Rayon. However, it is still on stream at Asahi Chemical.

Ketene Process [6], [40]. Acetic acid or acetone is pyrolyzed to ketene in this process which has long been abandoned by Celanese and B. F. Goodrich. The many steps and toxicity of β-propiolactone are major disadvantages.

$$CH_3COOH \xrightarrow{-H_2O} CH_2=C=O \xrightarrow{+CH_2O} \begin{array}{c} CH_2-C=O \\ | \quad\quad | \\ CH_2-O \end{array} \xrightarrow[H_2SO_4]{ROH} CH_2=CHCOOR + H_2O$$

where R = H or alkyl.

R = H or alkyl.

Ethylene Cyanohydrin Process. Ethylene cyanohydrin is generated by addition of hydrogen cyanide to ethylene oxide. The product then is hydrolyzed to acrylic acid using sulfuric acid. This process was used by Union Carbide and Röhm & Haas, but has been abandoned because of problems in dealing with HCN and the NH_4HSO_4 waste.

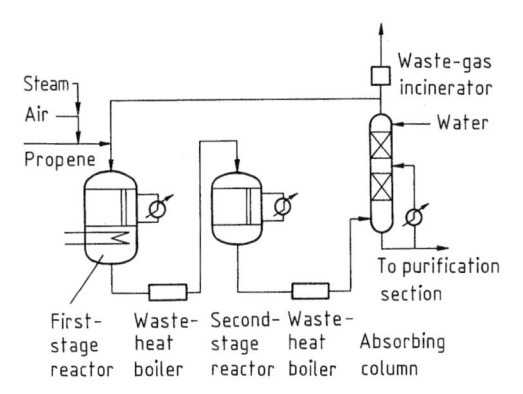

Figure 2. Schematic diagram of acrylic acid production (oxidation section)

1.3.1. Propene Oxidation

Propene oxidation involves heterogeneous catalytic oxidation of propene in the vapor phase with air and steam to give acrylic acid. Generally the product leaving the reactor is absorbed in water, extracted with an appropriate solvent, and then distilled to give technical grade glacial acrylic acid.

Oxidation Catalysts. Research on catalysts for propene oxidation to acrylic acid began in the latter half of the 1950 s. The two methods for the heterogeneously catalyzed gas-phase oxidation of propene are single-step and two-step processes:

Single-step process:

$$CH_2=CHCH_3 + 3/2\, O_2 \longrightarrow CH_2=CHCOOH + H_2O \quad \Delta H = -594.9 \text{ kJ/mol}$$

Two-step process:

$$CH_2=CHCH_3 + O_2 \longrightarrow CH=CH_2CHO + H_2O \quad \Delta H = -340.8 \text{ kJ/mol}$$
$$CH_2=CHCHO + 1/2\, O_2 \longrightarrow CH_2=CHCOOH \quad \Delta H = -254.1 \text{ kJ/mol}$$

Many patents have been issued in both cases. The yield in the *single-step process* is at best approximately 50 – 60 % [38], [41] – [43]. Another drawback is limited lifetime of the catalyst, which is a multicomponent system composed of polyvalent oxides with molybdenum oxide as the main component and tellurium oxide as the promoter. The life of the catalyst is short because of the tendency of tellurium oxide to sublime.

The *two-step reaction* (Fig. 2) requires different reaction conditions and different catalysts to produce optimum conversion and selectivity in each step. Recent research has focused on this process, in which the oxidation of propene to acrolein and the oxidation of acrolein to acrylic acid employ separate catalysts. The steps are operated at different temperatures to permit high overall efficiency.

First-stage catalysts are acrolein-selective propene-oxidation catalysts. The total yield of acrolein and acrylic acid is more than 85 % (→ Acrolein and Methacrolein).

The early second-stage catalysts [43] for acrolein oxidation to acrylic acid were based mainly on cobalt-molybdenum oxides [44]. They had fairly low activity even at high

Table 5. Catalyst for the second step of acrylic acid production

Catalyst composition (support) neglecting oxygen	Reaction temperature, °C	Acrolein conversion, %	One-pass yield of acrylic acid, mol%	References
$Mo_{12}V_{1.9}Al_{1.0}Cu_{2.2}$ (Al sponge)	300	100	97.5	[46]
$Mo_{12}V_3W_{1.2}$ (SiO$_2$)	240	98.0	87.0	[47]
$Mo_{12}V_3W_{1.2}Mn_3$	255	99.0	93.0	[48]
$Mo_{12}V_2W_2Fe_3$	230	99.0	91.0	[49]
$Mo_{12}V_3W_{1.2}Cu_1Sb_6$	272	99.0	91.0	[50]
$Mo_{12}V_{4.6}Cu_{2.2}W_{2.4}Cr_{0.6}$ (Al$_2$O$_3$)	220	100.0	98.0	[51]
$Mo_{12}V_2(Li_2SO_4)_2$	300	99.8	92.4	[52]
$Mo_{12}V_{4.8}Cu_{2.2}W_{2.4}Sr_{0.5}$ (Al$_2$O$_3$)	255	100.0	97.5	[53]
$Mo_{12}V_{2.4}Cu_{0.24}$ (SiC)	290	99.5	94.8	[54]
$Mo_{12}V_3W_{1.2}Ce_3$	288	100	96.1	[55]
$Mo_{12}V_{4.7}W_{1.1}Cu_{6.3}$	260	99.0	96.0	[56]

reaction temperatures and gave yields of less than 70 mol%. Almost all the catalysts currently used are composed of molybdenum and vanadium oxides. In 1959, Distillers first proposed a molybdenum-vanadium catalyst system in which the atomic ratio of molybdenum to vanadium was one-to-one [45]. The maximum yield obtained was 30% at about 400 °C. Since then further studies have shown that only a relatively small amount of vanadium is required. In addition, other elements and carriers have been shown to increase the activity and yield. They have been used for the preparation of multi-component metal-oxide catalysts that contain one or more of the elements copper, arsenic, uranium, aluminum, tungsten, silver, manganese, germanium, gold, barium, calcium, strontium, boron, tin, cobalt, iron, or nickel in addition to molybdenum and vanadium. Supported on an aluminum sponge, the catalyst described in [46] shows good activity and yield. Table 5 lists patented acrolein oxidation catalysts that have relatively high activities and yields. All of these catalysts are metal oxides.

Process Conditions. The conditions in the first step correspond to conditions in the acrolein synthesis (see → Acrolein and Methacrolein). The catalysts currently used in the second step require reaction temperatures from 200 to 300 °C and contact times from 1 to 3 s. They give almost 100% conversion of acrolein and yields of acrylic acid greater than 90%.

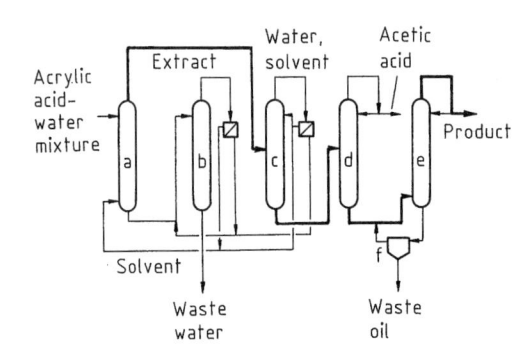

Figure 3. Schematic diagram of acid recovery and purification section
a) Extraction column; b) Raffinate-stripping column; c) Solvent-separation column; d) Light-ends cut column; e) Product column; f) Decomposition evaporator

Acid Recovery and Purification. The effluent gas from the second-stage multi-tube reactor in Figure 2 is cooled to about 200 °C and then fed to the absorbing column to be scrubbed with water. Because the effluent gas contains a large amount of steam, acrylic acid usually is obtained as an aqueous solution of 20 to 70 wt% [57]. Alternatively, the acid may be absorbed by an organic solvent such as biphenyl, diphenyl ether, or a carboxylic ester with a boiling point higher than 160 °C [58]. Then the steam in the reaction gas does not condense in the absorbing column, but is discharged with other gases from the column top. This method reduces energy consumption in the subsequent purification step, but it also increases the loss of acrylic acid and solvent from the column top.

After the absorption in water, the acrylic acid is purified by extraction with an organic solvent and then distillation. Various solvents can be used for the extraction. The first group (light solvents) includes those with boiling points lower than acrylic acid, such as ethyl acetate, butyl acetate, ethyl acrylate, and 2-butanone, as well as combinations of these [59]. The second group (heavy solvents) has boiling points higher than acrylic acid (e.g., *tert*-butyl phosphate, isophorone, and aromatic hydrocarbons [60]). Mixtures of these light and heavy solvents form a third group [61], [62].

Figure 3 represents the separation and purification process using a *light extraction solvent*. The aqueous acrylic acid from the absorbing column is introduced into the extraction column (a) countercurrent to an organic solvent. The solvent must have a high distribution coefficient for acrylic acid and low solubility in water, and it must form an azeotrope containing a high percentage of water. The extract from the top of the extraction column goes to the solvent-separation column (c), where the solvent and water are distilled overhead and the solvent is separated and recycled to the extraction column.

The bottom stream from the extraction column and the water from the overhead of the solvent-separation column are sent to the raffinate-stripping column (b), where a small amount of solvent is recovered by distillation. The waste water from the raffinate-stripping column is biologically treated or incinerated. The bottom fraction from the solvent-separation column is fed to the light-ends cut column (d), where acetic acid is distilled off and, if desired, recovered. The crude acrylic acid from the bottom of the light-ends cut column is sent to the product column (e), where acrylic acid of high purity is obtained overhead. The material from the bottom of the product column containing acrylic acid dimer is fed to the evaporator (f), where the dimer is decomposed to the monomer. The evaporator residue, composed of acrylic acid oligomers, polymers, and inhibitors, is withdrawn and burned as waste oil.

Because acrylic acid is readily polymerized, distillation columns are operated with an inhibitor, such as hydroquinone or hydroquinone monomethyl ether, in the presence of oxygen, and at reduced pressure to lower the distillation temperature. The purity of acrylic acid produced by this process usually exceeds 99.5 wt%, and the purified yield is about 98%.

In a *heavy-solvent extraction process* the solvent is not distilled and therefore the energy consumption is less than in a light-solvent process. However, other problems exist, such as the loss of solvent by decomposition and the inferior quality of the product. In the processes using mixtures of light and heavy solvents, the purification system is complex. The light-solvent process is therefore the most suitable for a commercial plant.

Other purification methods also have been reported. In one of these, the acrylic acid is first oligomerized in aqueous solution in the presence of a catalyst such as sulfuric or phosphoric acid. Next the water is distilled, and finally the residual oligomer is decomposed at 120 to 200 °C to obtain acrylic acid [63]. In another process, the acrylic acid is extracted from the aqueous solution with butyl acrylate or octanol. It is then directly esterified with an alcohol to form an acrylate without isolation of the acrylic acid [64], [65]. These processes have not yet been used commercially, probably because of high energy consumption or problems with the product quality.

1.3.2. Esterification

Although acrylic acid can be esterified in the vapor phase [66], [67], the liquid phase esterification is industrially more important. Two types of acid catalyst are used: a strong acid, such as sulfuric acid or *p*-toluenesulfonic acid [68], or a solid acid, such as a cation-exchange resin [69]. Although sulfuric acid is superior to ion-exchange resins, its use causes problems in waste disposal. In general, cation-exchange resins are favored for esterification using such alcohols as methanol and ethanol, whereas sulfuric acid is favored for higher alcohols having slower rates of esterification (e.g., pentanols and octanols). Liquid phase reaction of acrylic acid with ethylene in the presence of sulfuric acid does not seem economically feasible for producing ethyl acrylate [70], because of the large quantities of sulfuric acid that are needed.

Lower Alkyl Acrylates (see Fig. 4). Acrylic acid and a small excess (10–30%) of an alcohol are fed into the fixed-bed reactor (a) which is packed with a cation-exchange resin and operated at a temperature of 60 to 80 °C. The reaction liquid then goes to the ester stripper (b) where the desired ester, water, and unreacted alcohol are removed overhead using part of the bottoms from the light-ends column (e) as reflux. The bottom liquid from b contains unreacted acid and is recycled to the reactor. Part of the recycled liquid is fed into the bottom stripper (c), where high-boiling materials, such as inhibitors, impurities, and polymers, are removed to prevent their accumulation in the reaction system.

The acid-free mixture of ester and alcohol distilled from the ester stripper (b) is fed into the extraction column (d), where the alcohol is extracted with water fed from the top of the column. The raffinate from the top of the column goes into the light-ends cut column (e), where light-ends such as water, acetate, and alcohol are separated overhead.

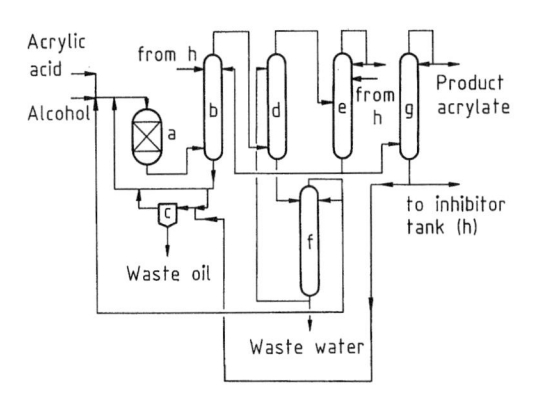

Figure 4. Esterification – lower alkyl acrylate process
a) Esterification reactor; b) Ester stripper; c) Bottom stripper; d) Extraction column; e) Light-ends cut column; f) Alcohol-recovery column; g) Product column

The extract from the bottom of the extraction column is fed into the alcohol recovery column (f), where the alcohol is recovered for reuse in the reaction. Part of the bottom liquid is reused as extracting water; the rest is taken out as waste, concentrated, and either treated biologically or incinerated.

Crude ester from the bottom of the light-ends column is distilled in the product column (g) to obtain acrylate of high purity. The bottom liquid from the product column is recycled (via the inhibitor tank) to the ester stripper (b) and light-ends cut column (e) to be reused as an inhibitor. However, a part of it is sent to the bottom stripper (c) to recover ester and separate high-boiling materials such as polymers.

Polymerization inhibitors, such as hydroquinone or phenothiazine, are added to each column. The light-ends cut column and the product column are operated at reduced pressure to permit lower distillation temperatures.

This process for making alkyl acrylates is quite economical because only a small excess of alcohol is applied and the inhibitor is reused; this leads to low energy and inhibitor consumptions. The yield reaches 95% and 97% based on acrylic acid and on alcohol, respectively. The purity of the product exceeds 99.5 wt%.

Higher Alkyl Acrylates (see Fig. 5). The esterification reaction is preferably carried out batchwise in the presence of an organic solvent as entrainer and sulfuric acid as catalyst. The water formed is separated through the top of the azeotropic-distillation column (b). The reaction conditions are: atmospheric pressure, temperature 85–95°C, reaction time 3–5 h, molar ratio (alcohol to acid) 1.0–1.1.

After completion of the esterification, the reaction liquid is cooled to 60°C and then transferred to tank (c) where the sulfuric acid is neutralized with alkali. The oil and water layers are separated and stored in tanks d and e, respectively. The oil layer is fed into the solvent-recovery column (f) and subsequently into the alcohol-recovery column (g) for distillation. The solvent and alcohol are recovered overhead and reused in the reaction.

The crude ester obtained from the bottom of the alcohol-recovery column is fed into the product column (h) where purified acrylic ester is obtained by distillation. The bottom liquid mainly is recycled to the reactor and the alcohol-recovery column to be reused as supplementary inhibitor. However, part of the liquid is fed into the bottom stripper (i) to recover valuable materials that are resupplied to the

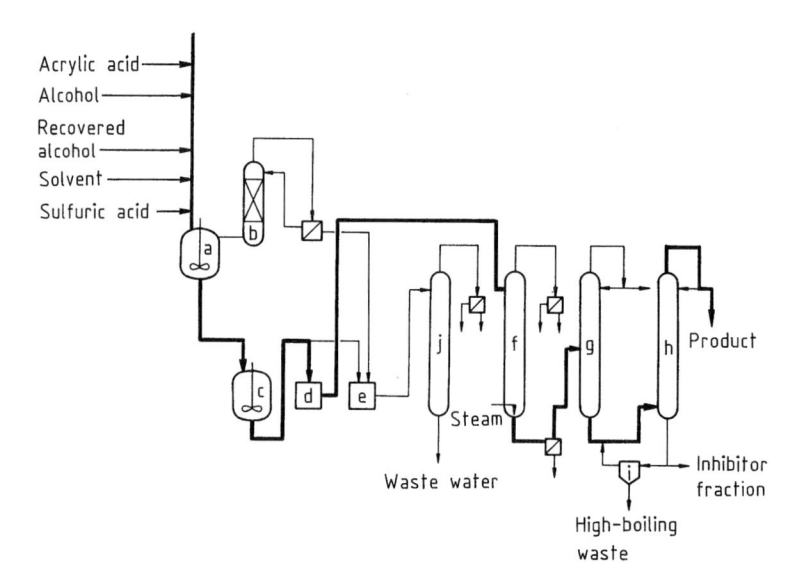

Figure 5. Esterification – higher alkyl acrylate process
a) Esterification reactor; b) Azeotropic-distillation column; c) Neutralization tank; d) Oil-layer tank; e) Water-layer tank; f) Solvent-recovery column; g) Alcohol-recovery column; h) Product column; i) Bottom stripper; j) Organic stripper

product column. High-boiling waste composed of polymers, inhibitors, and other impurities is taken out of the bottom stripper and incinerated.

The water from the water-layer tank (e) is fed into the organic stripper (j) together with the water layer from the top and the bottom of the solvent-recovery column (f). The oil layer obtained from the top of the solvent-recovery column (f) is recycled into the oil-layer tank. The waste obtained from the bottom of the organic stripper (j) is either treated biologically or incinerated after concentration.

As in the lower alkyl acrylate process, hydroquinone, its monomethyl ether, or phenothiazine is added to each column, and the alcohol-recovery and product columns are operated at reduced pressure.

The yield reaches 95% and 96% based on acrylic acid and on alcohol, respectively. The purity of the product exceeds 99.5 wt%.

1.4. Quality Specifications and Analysis

Production control requires monitoring the propene and oxygen concentrations in the gas phase of the oxidation. These are checked periodically to maintain optimum reaction conditions and avoid entering the range of flammability. Propene is determined by GLC with flame ionization detection, oxygen by a magnetic meter [71].

The purity of acrylic acid and its esters depends on the production method employed. Table 6 shows current quality standards for some of these products. The purity

Table 6. Quality specifications of acrylic acid and esters

		Acrylic acid 99%	Acrylic acid 80%, aq.	Methyl acrylate	Ethyl acrylate	Butyl acrylate	Octyl acrylate
Purity[b]	wt%, min	99.0	80.0	99.0	99.0	99.0	99.0
Acid[c]	wt%, max	–	–	0.005	0.005	0.005	0.005
Water	wt%, max	0.20	–	0.05	0.05	0.05	0.05
Color	APHA, max	20	20	20	20	20	20
Inhibitor[d]	ppm	200	200	15±5	15±5	15±5	15±5

[a] Authorized by the Japanese Acrylic Acid and Esters Industrial Association; [b] determined by GLC; [c] as acrylic acid; [d] as hydroquinone monomethyl ether.

of acrylic acid and its esters is commonly determined from the percentage of impurities measured by GLC with a flame ionization detector. Occasionally the purity of acrylic acid is determined by titration with a base.

In addition to the purity, the polymerization characteristics are important because acrylic acid and acrylates are used chiefly to make polymers. Polymerization characteristics are determined by examining polymerization patterns such as induction period and temperature elevation under fixed polymerization conditions (temperature, concentration, catalyst, etc.) The degree of polymerization of highly purified acrylic acid is greatly decreased in the presence of trace amounts of heavy metals such as copper, or of aldehydes such as acrolein and furfural.

1.5. Storage and Transportation

Acrylic acid and its esters are usually stabilized with inhibitors such as phenothiazine, hydroquinone, or hydroquinone monomethyl ether. Because phenolic inhibitors are only effective in the presence of oxygen, the monomers must be stored under air (usually normal air for acrylic acid and air with reduced oxygen concentration for esters). The safe handling of these products requires the use of proper protective equipment such as rubber gloves and vapor-proof goggles and masks.

Acrylic Acid. Acrylic acid normally contains 50 to 500 ppm of an inhibitor to prevent polymerization. Because of its relatively high corrosiveness, it should be stored in equipment made of or lined with glass, polyethylene, polypropylene, or stainless steel. In addition, it should be kept at 15 to 30 °C and away from direct sunlight. Freezing should be avoided because it tends to localize the inhibitor. If acrylic acid should freeze, however, it should be melted by using a warm water or air bath below 30 °C. Agitation of the acrylic acid during the melting is recommended to avoid any localized heating. Acrylic acid often is used as an 80% aqueous solution which has a freezing point of –3 to –5 °C.

Acrylic Esters. In general, a lower level of inhibitor is required for acrylic esters than for the acid, although the range is still 50 to 500 ppm. The esters are less corrosive than the acid and thus can be stored in equipment made of or lined with carbon steel or phenolic resin, in addition to glass, polyethylene, and polypropylene. Grades of acrylates containing little or no inhibitor are available. These products should be carefully stored at temperatures of 0 to 10 °C. Methyl and ethyl acrylates have very low flash points and form explosive gas mixtures in air, even at room temperature. Thus, even though oxygen is an effective inhibitor, the oxygen concentration in large storage tanks is usually kept at 6 to 8 vol% to prevent the formation of a flammable mixture.

1.6. Uses

Acrylic Acid. The primary use of acrylic acid is as an intermediate in the production of acrylates. Polymers of the acid and its sodium salts are used increasingly in flocculants and dispersants with the polymeric sodium salts having more industrial importance.

Acrylic Esters. Acrylic esters are used exclusively for the production of polymers. The polymers are used mainly for coatings, paints, adhesives, and binders for leather, paper, and textiles.

About 80% of the methyl ester produced is used as a copolymer component of acrylic fibers. The ethyl ester is used for both solvent- and water-based paints, and in textiles as a binder in nonwoven fabrics and flocking. It generally is used in areas where more rigidity is required than can be obtained with the butyl ester. The butyl ester is growing in use, mainly in water-based paints and adhesives. The 2-ethylhexyl ester is used for almost the same purposes as the butyl ester, with a large demand for it in stick-on labels and in the caulking of building materials.

1.7. Some Special Acrylates

Esters with Polyhydric Alcohols. Representative *multifunctional acrylates* are trimethylolpropane triacrylate, pentaerythritol tri- or tetraacrylate, 1,4-butanediol diacrylate, 1,6-hexanediol diacrylate, and poly(ethylene glycol) diacrylate ($n = 2-14$). They usually are produced by direct esterification of acrylic acid with the corresponding polyhydric alcohol in the presence of an entrainer and an acid catalyst, such as sulfuric acid or *p*-toluenesulfonic acid. Because these esters have high boiling points, they cannot be purified by ordinary distillation. Instead, the reaction mixture is neutralized, the entrainer removed, and the product washed with water [3], [72].

The esters are used as cross-linking agents and modifiers in rubber and synthetic resins, in adhesives, and as active diluents in photosensitive resins. They are also

Table 7. Estimated regional production capacities for acrylic acid and its esters in the western world (t/a; Dec. 1982)

	Acid	Esters
United States	430 000	700 500
Western Europe	282 000	380 000
Japan	117 500	182 000
Others	10 000	15 000
Total	839 500	1 277 500

applied in the coating and ink industries because they can be cured with ultraviolet light [73], [74] or electron-beam radiation [75]. Proper protection is required when handling these eye and skin irritants.

2-Hydroxyalkyl Acrylates. Two industrially important multifunctional esters are 2-hydroxyethyl acrylate and 2-hydroxypropyl acrylate. These are produced by liquid-phase esterification of acrylic acid with ethylene oxide or propylene oxide in the presence of a Lewis acid catalyst, such as a chromium [76] or ruthenium [77] compound, or the iron salt of an organic acid [78]. Because this reaction readily produces di(alkylene glycol) monoacrylates and alkylene glycol diacrylates as byproducts, a highly efficient catalyst is required. Although vapor-phase catalytic synthesis using magnesium oxide has recently been proposed [79], the liquid-phase esterification is the one currently used. These esters are used especially as cross-linking agents in heat-cured paints, adhesives, textile preparations, etc. They are toxic and lacrimatory, cause blistering of the skin, and may give rise to long-term sensitivity. Inhalation of the vapor causes nose, eye, and throat irritation.

Other Derivatives. Halogenated derivatives such as 2-chloroacrylic acid [80], [81], 2,3-dibromopropyl acrylate [82], tetrafluoropropyl acrylate, and octafluoropentyl acrylate, have potential uses as fine chemicals. Dialkylaminoethyl acrylates are produced by transesterifying methyl acrylate with the corresponding amino alcohol.

1.8. Acrylic acid, economic aspects

Western world production capacities of acrylic acid and its esters as of late 1982 were 839 500 t/a and 1 277 500 t/a, respectively (Table 7). These figures may be a little greater than the real capacities because plants using acrylonitrile hydrolysis and the modified Reppe process probably are operated only as a supplement to the propene oxidation facilities now predominant. Plants based on other processes, such as the ketene and cyanohydrin methods (see Section 1.3), were shut down during the past decade because of inefficiency.

In fact it appears that the propene oxidation route will continue to be the most economical process for quite some time. Announced additional capacities of 280 000

Table 8. Estimated distribution of end uses of acrylic esters (% of total)

	United States	Western Europe	Japan
Surface coatings	42	35	34
Textiles	23	18	16
Acrylic fibers	6	7	14
Adhesives	5	15	20
Others	24	25	16

and 340 000 t/a for the acid and esters, respectively, will all be based on propene oxidation.

Consumption of acrylic acid for uses other than as an intermediate in ester production ranges from 5 to 9 % of the total, although demand for and consumption of both acid and esters varies from region to region. Table 8 gives estimated end-use percentages in three regions. Surface coatings provide the largest market for the esters in all three regions.

1.9. Toxicology and Occupational Health

Acrylic Acid. Acrylic acid is moderately toxic and very corrosive [83]. Ingestion may cause severe gastrointestinal burns. The vapor is an irritant to the eyes and respiratory tract and skin contact may cause burns. Physiological response data are:

LD_{50} 340 mg/kg (rat, oral)
LC_{50} 3600 mg/m^3 (rat, inhalation, 5 L, 4 h)
LD_{50} 280 mg/kg (rabbit, skin)

The TLV on a time weighted average (TWA) is 10 ppm or 30 mg/m^3.

Acrylic Esters. Acrylic esters are of moderately acute toxicity, which decreases with an increase in the number of carbon atoms in the alkyl group (Table 9). Liquid methyl and ethyl acrylates severely irritate the skin and mucous membranes and are corrosive to the eyes, whereas the butyl and 2-ethylhexyl acrylates have less severe effects. Methyl and ethyl acrylate vapors are very lacrimatory, extremely irritating to the respiratory tract, and are corrosive to the eyes, causing corneal injury. The lacrimatory effect of the butyl and 2-ethylhexyl esters is weak, but their vapors may cause dizziness, headache, nausea, and vomiting.

Methyl and ethyl acrylates can be absorbed through the skin in toxic amounts, and overexposure to the vapor can result in fatal pulmonary edema. However, their noticeable odors and irritating effects reduce the likelihood of significant exposure.

Table 9. Physiological response data and exposure levels of some acrylates

	Methyl acrylate	Ethyl acrylate	Butyl acrylate	2-Ethylhexyl acrylate
LD_{50} (rat, oral), mg/kg	300	1020	3730	5660
LCLo (rat, inhalation, 4 h), mg/m^3	3500	4000	5500	
LD_{50} (rabbit, dermal), mg/kg	1243	1950	2000	8480
TLV (TWA)	10 ppm, 35 mg/m^3	5 ppm, $\approx$ 20 mg/m^3	10 ppm, 55 mg/m^3	
MAK	10 ppm, 35 mg/m^3	25 ppm, 100 mg/m^3		

2. Cyanoacrylates

The monofunctional 2-cyanoacrylates, $CH_2=C(CN)-COOR$, have been known for many years. Technical developments, based on the original patents of 1949 [84], led in 1954 to the first viable production process [85], [86]. Since then these compounds have achieved a considerable growth rate on the market. According to a recent study [87], this will result in a worldwide requirement of 2500 t/a by 1988. The 2-cyanoacrylates are utilized almost exclusively as adhesives. The methyl, ethyl, butyl, allyl, and methoxyethyl esters are available with different setting characteristics and rheological properties, depending on the requirements of the application.

Physical Properties. Pure 2-cyanoacrylates are clear, colorless liquids at room temperature and have a characteristic odor. Their physical properties are listed in Table 10.

Chemical Properties. The vinyl structure of the 2-cyanoacrylates makes them liable to spontaneous polymerization. This is desirable for their use as reactive adhesives, but causes significant difficulties in synthesizing and especially in purifying them.

$$n\ CH_2=C\begin{subarray}{c}CN\\COOR\end{subarray} \longrightarrow -CH_2\underset{COOR}{\overset{CN}{C}}\left[CH_2\underset{COOR}{\overset{CN}{C}}\right]_{n-1}\cdots$$

Because the chain propagation reaction can be initiated by either an ionic or a radical mechanism, the rate of polymerization depends on temperature, humidity, light, and the presence of polymerization accelerators, such as peroxides and bases. The 2-cyanoacrylates are, however, adequately stable if stored under cool conditions in the presence of suitable inhibitors. In addition to polymerization, they undergo the other reactions typical of vinyl compounds (e.g., addition).

Production. Many different processes can be used to manufacture 2-cyanoacrylates. The important step in most of the published syntheses is the classical Knoevenagel reaction using formaldehyde and a cyanoacetic ester. In this first step oligomeric

Table 10. Physical properties of industrially important 2-cyanoacrylates

Property	Methyl	Ethyl	Butyl	Allyl	Methoxyethyl
CAS registry number	[137-05-3]	[7085-85-0]	[6606-65-1]	[7324-02-9]	[27816-23-5]
Molecular formula	$C_5H_5NO_2$	$C_6H_7NO_2$	$C_8H_{11}NO_2$	$C_7H_7NO_2$	$C_7H_9NO_3$
M_r	111.10	125.13	153.18	137.14	155.15
bp, °C/p, kPa	48–49/ 0.33–0.36	54–56/ 0.34–0.40	53–56/ 0.27–0.33	115/2.53	80–82/0.13
Viscosity, mPa s	2.20	1.86	2.08	6.4	–
Heat of polymerization in isobutyronitrile at 25 °C, kJ/mol	57.7	58.2	67.8	–	–
Refractive index, n_D^{25}	1.4406	1.4349	1.4291	1.4426 (20 °C)	–
Density at 20 °C, g/cm^3	1.1044	1.0501	1.0009	1.0578	
Vapor pressure at 25 °C, kPa	<0.27	<0.27	<0.27	–	–

cyanoacrylates, water, and other byproducts are formed. The reaction can be catalyzed by bases (e.g., amines) [85], [86], alkali metal tetraborates, metal carbonyls [88], or phase-transfer catalysts [89]. The raw condensation product from the first reaction is depolymerized thermally (either continuously or discontinuously) [90] and purified by distillation, chromatography on alumina or silica gel [91], crystallization in polyethylene columns [92], or treatment with zinc chloride [93].

Alternatively, 2-cyanoacrylates can be manufactured by the ethoxycarbonylation of cyanoacetylene in the presence of nickel carbonyl [94].

Uses. 2-Cyanoacrylate-based formulations have been a valuable component of modern adhesive technology for many years. Their ability to join the most dissimilar materials quickly, firmly, and durably has insured a wide field of applications. They are used in the electrical and electronics industries as well as many areas of mechanical engineering, such as automobile, ship, and aircraft construction. In additon to these purely technical applications, the 2-cyanoacrylates are used in medicine to close wounds.

Special properties to meet specific requirements are obtained by the use of additives or a manufacturer's proprietary process. Such parameters as hardening time [95]–[97], viscosity [98], [99], relative strength [100], resistance to hydrolysis [101], [102], heat resistance [103], [104], flexibility of the mature adhesive [105], [106], and shelf life [107]–[109] commonly are varied.

Toxicology and Occupational Health. Both the MAK and the TLV exposure limits for 2-cyanoacrylates have been set at 2 ppm. Adequate ventilation of the work place is therefore necessary if these substances are in continual use. Further information can be obtained from the relevant documentation provided by suppliers. Although many years of practical use have not yet led to any significant damage to health, the toxicology of 2-cyanoacrylates requires further investigation [110], [111].

Figure 6. Catalytic hydration of acrylonitrile

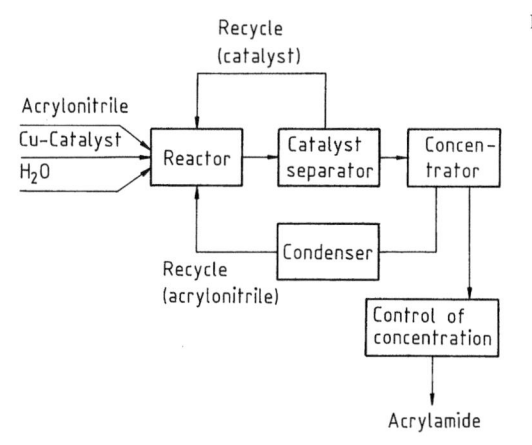

3. Acrylamide

Properties. Acrylamide [*79-06-1*], 2-propenamide, $CH_2=CHCONH_2$, M_r 71.08, mp 84.5 °C, usually forms white crystalline platelets. It is very toxic.

Acrylamide is most conveniently handled as an aqueous solution because the dry powder forms a fine, toxic dust. It is soluble in water, alcohols, and acetone but insoluble in benzene and heptane.

Acrylamide and polyacrylamide can be converted into various products by the usual addition reactions of the double bond and especially by reactions of the amido group:

$$RCONH_2 + HCHO \longrightarrow RCONHCH_2OH$$
$$+ 2\,HCHO \longrightarrow RCON(CH_2OH)_2$$
$$2\,RCONH_2 + HCHO \longrightarrow (RCONH)_2CH_2$$
$$RCONH_2 + HCHO + NaHSO_3 \longrightarrow$$
$$RCONH-CH_2SO_3Na + H_2O$$
$$+ HCHO + R'R''NH \longrightarrow$$
$$RCONH-CH_2NHR'R'' + H_2O$$
$$+ NaOCl + 2\,NaOH \longrightarrow$$
$$RNH_2 + Na_2CO_3 + NaCl + H_2O$$
$$+ H_2N-NR'R'' \longrightarrow RCONH-NR'R'' + NH_3$$

N-Alkyl derivatives of acrylamide are prepared by the reaction of acryloyl chloride with the corresponding amine, by the dehydrochlorination of 2-chloropropionamide, or by the amination of acrylic acid or esters [112].

Production. Acrylamide was first produced on a large scale, by American Cyanamid, in 1954. It is currently produced from acrylonitrile either by homogeneous sulfuric acid hydration or by heterogeneous catalytic hydration.

Sulfuric Acid Process. One mole of Acrylonitrile (→ Acrylonitrile) is added to a solution of 1 mol sulfuric acid and 1 mol of water at 60 °C. The mixture is slowly heated to 80 °C, kept for an hour at this temperature, and then cooled to 40 °C. Ammonium sulfate precipitates on neutralization with ammonia and is filtered out.

The acrylamide, which crystallizes on cooling the mother liquor to below 10 °C, is purified by recrystallization from benzene.

Catalytic Hydration Process. Figure 6 shows a flow diagram of this process. A 50 wt% solution of acrylonitrile in water is slurried with a catalyst such as Raney copper and kept at 120 °C for 2.5 h. Conversion of acrylonitrile is greater than 50% and selectivity to the amide is nearly 100%. This process conveniently makes an aqueous solution of 30 to 50 wt% acrylamide because of the large amount of water required in the reaction.

Biological Process. A patented biocatalyst enables the hydration of acrylonitrile at room temperature and pressure in a process developed by Nitto Chemical Industry Co. (Tokyo). The catalyst is an enzyme derived from one of several bacterial genera [113].

Uses. Acrylamide is made mainly into water-soluble polymers and copolymers used in flocculants, papermaking aids, thickening agents, surface coatings, and enhanced oil recovery [112], [114].

Toxicology and Occupational Health. Acrylamide is highly toxic. Its toxicity to humans and animals has been studied extensively [115], [116]. It is readily absorbed through the skin and mucous membranes and quickly accumulates on chronic exposure, causing symptoms of common neuritis. Repeated exposure may increase sensitivity. The MAK and TLV are 0.3 mg/m^3(TWA, skin) and 0.6 mg/m^3 (STEL, skin).

Subjective Symptoms of Exposure. Generally, fatigue and weight loss are observed after serious exposure to acrylamide. The skin develops depigmentation, erythema, and peeling of the palms and soles. Nausea, abdominal pain, and anorexia are observed. Sweating is increased. The peripheral nervous system reacts with numbness, paresthesia, myalgia, weakness of the hands and legs, speech disorder, and urinary incontinence. Effects on the central nervous system result in gait disorder, tremor, and somnolence.

Objective symptoms are hyporeflexia, hypoethesia, muscle atrophy, disorder of vibration sense, abnormal electromyogram, abnormal gait, and abnormal electroencephalogram. Blood, urine, and cerebrospinal fluid, as well as liver functions, are in their normal ranges.

4. References

[1] E. C. Leonard: *Vinyl and Diene Monomer,* part 1, Wiley-Interscience, New York 1970, p. 148.
[2] E. Ohmori: *Acrylic Acid and its Polymers,* vol. **I** and **II**, Shokodo, Tokyo 1973 and 1975.
[3] E. H. Riddle: *Monomeric Acrylic Esters,* Reinhold, New York 1954.
[4] H. Rauch-Puntigam, T. Völker: *Acryl- und Methacrylverbindungen,* Springer Verlag, Berlin 1967.
[5] M. Sittig: *Vinyl Monomers and Polymers,* Noyes Dev. Corp., Park Ridge, N.J., 1966.
[6] L. S. Luskin, *High Polym.* **24** (1970) 105.
[7] Sh. Suzuki, *Yuki Gosei Kagaku Kyokaishi* **28** (1970) no. 12, 1272.
[8] Rohm & Haas, US 2647923, 1953 (C. Burton). *Org. Synth. Coll.* **3** (1955), 576.

[9] R. Madhav, C. A. Snyder, P. L. Southwick, *J. Heterocycl. Chem.* **17** (1980) no. 6, 1231.V. P. Nekoroshkov, G. L. Kamalov, N. M. Vladyka, *Khim. Promst. Ser. Reakt. Osobo Chist. Veshchestva* 1980, no. 6, 17; *Chem. Abstr.* **95**, 42519 t.

[10] K. Thiele, K. Posselt, H. Offermanns, *Arzneim. Forsch.* **28** (1978) no. 11, 2047.

[11] Union Carbide, US 4012445, 1972, US 4011223, 1972 (D. C. Priest, M. R. Sander, D. J. Trecker).

[12] P. Brun, A. Tenaglia, B. Waegell, *J. Organomet. Chem.* **194** (1980) C 39.

[13] R. M. Acheson, P. W. Poulter, *J. Chem. Soc.* 1960, no. 2, 2138.

[14] A. Etienne, G. Louchambon, P. Givaudean, *C. R. Acad. Sci. Ser. C* **289** (1979) no. 9, 263.

[15] A. Le Berre, C. Porte, *Bull. Soc. Chim. Fr.* 1978, 602. B. B. Snyder, R. S. E. Coun, S. Sealfon, *J. Org. Chem.* **44** (1979) 218.

[16] Tokyo Fine Chemical, JP 31137, 1980, US 3846489, 1974 (H. Uesugi, T. Takeda).

[17] C. E. Rehberg, *J. Am. Chem. Soc.* **68** (1946) 544, **69** (1947) 2966, **72** (1950) 2205.

[18] BASF, DE 2759162, 1977 (H. Distler, K. Schneider, R. Widder, G. Paulus, H. Schoeppl).

[19] Witco Chemical Corp., US 4052440, 1973 (S. Gladstone, S. R. Rao, C. J. Rosshirt).

[20] C. D. Hurd, L. L. Gershbein, *J. Am. Chem. Soc.* **69** (1947) 2328.

[21] Rohm & Haas, US 3769315, 1973 (R. L. Keener, H. Raterink).

[22] Soda Sangyo, JP-Kokai 34720, 1978 (Y. Nakatomi, S. Izukawa, T. Ohtsuka); *Chem. Abstr.* **89**, 108175 n.

[23] L. Novak, G. Baan, J. Marosfalvi, C. Szantay, *Chem. Ber.* **113** (1980) 2939–2949.

[24] Inst. of Phys. and Chem. Res., JP-Kokai 57784, 1980, DE 3007592 (K. Koh, H. Yamazaki).

[25] Inst. of Phys. and Chem. Res., JP-Kokai 115829, 1980, DE 3007592 (K. Koh, H. Yamazaki). Universal Oil Products, US 3935234, 1976 (P. H.Reichenbacher, T. M. Forsythe, A. K. Sparks, T. Symon).

[26] N. D. Lee, D. Faulkner, H. C. Highet, A. R.Philpotts, W. Thain, *J. Appl. Chem.* **3** (1953) 481.

[27] C. J. Albisetti, N. G.Fisher, M. J. Hogsed, R. M. Joyce, *J. Am. Chem. Soc.* **78** (1956) 2637, 2640.

[28] Smithkline Corp., US 4005208, 1975 (P. E. Bender, B. Loev).

[29] V. B. Reddy, *Chem. Ind.* (London) 1976, 414.

[30] P. De Strong, N. E. Lowmaster, *Synth. Commun.* **13** (1983) no. 7, 537.

[31] J. G. Martin, R. Hill, *Chem. Rev.* **61** (1961) 537.

[32] BASF, BE 585725 , 1978.

[33] F. Falbe (ed.): *New Syntheses with Carbon Monoxide,* Springer Verlag, Berlin 1980, p. 465.

[34] M. Hidai, A. Misono, *Aspects Homogeneous Catal.* **2** (1974) 159.

[35] Mitsubishi Rayon, JP-Kokai 30958, 1981 (Y. Nakai, A. Yanagase).

[36] Sumitomo Chemical, JP-Kokai 72937, 1982 (S. Shimizu, K. Nakai); *Chem. Abstr.* **97**, 128256 h. Rohm & Haas, JP-Kokai 9219, 1979 (R. J. Piccolini). Sumitomo Chemical, JP-Kokai 65815, 1978 (T. Matsuda, T. Motohashi).

[37] Nippon Oil Seal Ind., JP-Kokai 39533, 1980 (H. Idemitsu, M. Kishida, T. Nishikubo, S. Ukai); *Chem. Abstr.* **78**, 158964 n.

[38] F. T. Maler, W. Bayer, *Encycl. Chem Process. Des.* **1** (1976) 401.

[39] D. J. Hadley, E. M. Evans: *Propylene and its Industrial Derivatives,* J. Wiley & Sons, New York 1973, p. 416–497.

[40] Celanese Corp., US 2820058, 1958 (O. V. Luke, M. O. Robeson, W. E. Taylor). Celanese Corp., US 3069433, 1962 (K. A. Dunn). B. F. Goodrich Co., US 2 356 459, 1944, US 2361036, 1944 (E. F. King), US 3002017, 1961 (N. Wearsch, A. J. De Paola).

[41] B. F. Goodrich Co., US 3392196, 1964. Nippon Shokubai, US 3475488, 1969 (N. Kurata, T. Ohara, K. Oda).

[42] S. Sakuyama, T. Ohara, N. Shimizu, K. Kubota, *Chemtech.* **3** (1973) no. 6, 350.
[43] D. J. Hucknall: *Selective Oxidations of Hydrocarbons,* Academic Press, London 1974, p. 52.
[44] Distillers, JP 20838, 1963, GB 915799, 1963 (D. J. Hadley, R. H. Jenkins).
[45] Distillers, GB 903034, 1962 (J. R. Bethell, D. J.Hadley, E. J. Gasson, R. F. Neale).
[46] Rikagaku Kenkyusho, JP 26287, 1969, US 3567772, 1971 (M. Yanagita, M. Kitahara).
[47] Nippon Kayaku, BE 698273, 1967 (S. Takenaka).
[48] Celanese, GB 1267189, 1972, US 3644509, 1972 (G. C. Allen).
[49] BASF, FR 2032915, 1970, US 3845120, 1974 (R. Krabetz, H. Engelback).
[50] Degussa, JP-Kokai 8360, 1972, DE-OS 2055155, 1972 (J. Hensel).
[51] Nippon Shokubai, JP 11371, 1974, DE 2152037 1972 (M. Wada, I. Yanagisawa, M. Ninomiya, T. Ohara).
[52] Sumitomo Chemical, JP-Kokai 31923, 1972 (T. Shiraishi, S. Kishiwada, S. Shimizu, S. Honmaru, Y. Nagaoka, K. Jinpo); *Chem. Abstr.* **78,** 30464 g.
[53] Nippon Shokubai, JP-Kokai 117419, 1974, DE 2413206 1974 (M. Wada, I. Yanagisawa, N. Ninomiya, T. Ohara).
[54] Mitsubishi Petrochem., JP 169, 1974, DE 2164905 1972 (Y. Kadowaki, T. Koshikawa).
[55] Sohio, JP-Kokai 83280, 1975, DE 2448804, 1975 (S. R. Dolhyi, E. C. Milberger).
[56] Società Italiana: JP 28889, 1978, DE 2456100, 1975 (N. Ferlazzo).
[57] Knapsack, US 3717675, 1966 (K. Sennewald, A. Hauser, K. Gehrmann, H.-K. Steil, W. Lork).Celanese, US 4156633, 1979, JP-Kokai 59813, 1976 (T. Horlenko et al.). Nippon Shokubai, JP-Kokai 95217, 1975 (N. Shimizu); *Chem. Abstr.* **83,** 206785 g.
[58] BASF, DE 2241714, 1974, JP 21010, 1981 (D. Gerd et al.). Rhône-Poulenc, JP 11896, 1982, US 4219389, 1977, FR 7718136, 1977 (G. Biola Y. Komorn, G. Schneider).
[59] Knapsack, US 3553261, 1971 (K. Sennewald et al.). Asahi Kasei, JP 18967, 1971, DE 1 950750 1970 (M. Honda et al.).
[60] Rhône-Poulenc, JP 11896, 1982, US 4219389, 1980 (G. Biola et al.). Knapsack, JP 32843, 1974, DE 1965014, 1971 (K. Sennewald et al.).
[61] Sumitomo Chemical, GB 1427 223, 1972, US 3968153, 1976 (T. Ohrui et al.).
[62] BASF, US 3962074, 1976, JP-Kokai 18 412, 1975, DE 2323328, 1974 (W. K. Schropp et al.).
[63] Rohm & Haas, DE-OS 1168609, 1967, FR 1573704, 1967.
[64] Celanese, GB 1182809, 1966, FR 1544368, 1966 (E. G. Schiedling, J. R. Muratorio).
[65] Nippon Geon, DE-OS 2035228, 1969 (R. Sato et al.).
[66] Toa Gosei Chemical, US 3639461, 1972 (H. Ito et al.).
[67] Asahi Kasei, JP 6131, 1978 (N. Kominami et al.).
[68] Sumitomo Chemical, US 3875212, 1975 (T. Ohrui et al.).
[69] Nippon Shokubai, JP 37 404, 1972 (K. Kubota et al.), *Chem. Abstr.* **78,** 16746 u.
[70] Mitsubishi Rayon, JP 35 883, 1972 (T. Kita et al.); *Chem. Abstr.* **78,** 85047 d.
[71] L. S. Luskin in F. D. Snell, C. L. Hilton (eds.): *Encyclopedia of Industrial Chemical Analysis,* vol. **4,** J. Wiley & Sons, New York 1967, p. 181.
[72] Sumitomo Chemical, JP-Kokai 70215, 1979 (T. Matsuda, H. Takamatsu); *Chem. Abstr.* **91,** 192843 a.
[73] C. B. Rybny, C. A. Defazio, I. K. Shahide, J. C. Trebellas, J. A. Vona, *J. Paint Technol.* **46** (1974) no. 569, 60.
[74] S. E. Young, *Prog. Org. Coat.* **4** (1976) 225–249.
[75] C. Bluestein, *Adhes. Age* **25** (1982) no. 12, 19.
[76] Nippon Shokubai, JP 300, 1982 (S. Yoshida, M. Daigo, S. Matsumoto, N. Shimizu). Dow, BR Pedido PI 7800250, 1979 (A. E.Gurgiolo).

[77]   Union Carbide, US 4223160, 1980 (L. G. Hess).
[78]   Nippon Shokubai, JP-Kokai 161349, 1981 (N. Shimizu, M. Daigo, S. Matsumoto).
[79]   Nippon Kayaku, JP 61259, 1982 (H. Hayami); *Chem. Abstr.* **87**, 53823 r.
[80]   Knapsack, JP-Kokai 56917, 1974 (E. Auer, W. Vogt, K. Gehrmann).
[81]   E. M. Movsumsade, M. G. Mamerov, I. A. Shikhuev, V. B. Papin, *Khim. Promst. (Moscow)* **4** (1980) 204.
[82]   Sankyo, EP 57058, 1982. (Y. Morisawa, K. Konishi, M. Kataoka).
[83]   *Registry of Toxic Effects of Chemical Substances,* U.S. Department of Health, Education, and Welfare, Public Health Service Center for Disease Control, NIOSH, Cincinnati, Ohio 1980. *Patty's Industrial Hygiene and Toxicology,* 3rd ed., vol **2 A,** Wiley-Interscience, New York 1981, p. 2294,
[84]   B. F. Goodrich, US 2467927, 1949, US 2467926, 1949 (A. E. Ardis).
[85]   Eastman Kodak, US 2721858, 1955 (F. B. Joyner, G. F. Hawkins).
[86]   Eastman Kodak, US 2794788, 1957 (H. W. Coover Jr., N. Shearer Jr.).
[87]   H. S. Holappa et al., *Adhes. Age* **22** (1979) 43.
[88]   V. Winkovic, DE-OS 2738285, 1979 (Viktor Winkovic).
[89]   Matsumoto, DE-OS 2944085, 1979 (S. Iwakichi et al.).
[90]   Schering, DE-AS 2027502, 1970 (W. Imöhl, P. Borner).
[91]   Toa Gosei, JP 7897036, 1978 (K. Kimura, K. Sakabe, A. Motegi, H. Tatemichi); *Chem. Abstr.* **89**, 216475 a.
[92]   Kores, FR 2143023, 1973 (L. Kamlander).
[93]   Toa Gosei, JP 78128689, 1978 (K. Kimura, E. Osowa, H. Tatemichi); *Chem. Abstr.* **90**, 104657 a.
[94]   Matsumoto, JP 7435608, 1974 (S. Iwakichi, E. Kiyoshi); *Chem. Abstr.* **82**, 125805 r.
[95]   Eastman Kodak, DE-OS 2223026, 1972, US 3 728 375, 1973 (H. W. Coover, J. M. McIntire).
[96]   Toa Gosei, DE-OS 2 816 836, 1978 (M. Akira et al.).
[97]   Toaka, JP 79123147, 1979 (K. Ohashi, S. Kusayama).
[98]   Eastman Kodak, US 3654239, 1972 (J. M. McIntire, H.Wicker, Jr.).
[99]   A. Kawabata, JP-Kokai 79043247, 1979 (A. Kawabata); *Chem. Abstr.* **91**, 40425 c.
[100] Eastman Kodak, FR 2010590, 1969.
[101] Eastman Kodak, US 33 54 128, 1967 (T. H. Wicker).
[102] Toa Gosei, JP 7 776 344, 1977 (K. Ito, A. Yamada, K. Kimura).
[103] Schering, DE-AS 2349799, 1973 (B. Brinkman, W. Imöhl).
[104] Loctite, DE-AS 2201547, 1971.
[105] Loctite, DE-OS 2049 744, 1970 (D. J. O'Sullivan, B. J. Bolger).
[106] Toa Gosei, JP 7780336, 1977 (K. Kimura, A. Yamada); *Chem. Abstr.* **87**, 185547 w.
[107] Loctite, DE-AS 2042 334, 1970 (D. J. O'Sullivan, B. J. Bolger).
[108] Schering, DE-OS 2128985, 1971 (W. Imöhl, P. Borner).
[109] Toa Gosei, JP 7313 334, 1969 (I. Kenji, K. Sayoshi, K. Kishichiro, Z. Hiroshi); *Chem. Abstr. 80,* 74347 s.
[110] R. A. W. Lehmann, G. J. Hayes, F. Leonhard: "Toxicity of alkyl 2-cyanoacrylates, I: Peripheral Nerve," *Arch. Surg. (Chicago)* **93** (1966) 441–446, "II: Bacterial growth," *Arch. Surg. (Chicago)* **93** (1966) 447–450.
[111] K. C. Pani, G. Gladieux, R. K. Kulkarni, G. Brandes, F. Leonhard: "The degradation of n-butyl alpha cyanoacrylate tissue adhesive," *Surgery (St. Louis)* **63** (1968) 481–489.
[112] Röhm, GB 2 102 426, JP-Kokai 26849, 1983, DE 3130508, 1982 (S. Besecke, G. Schroeder).
[113] *Chem. Eng.* **91** (1984) no. 8, 10. *Eur. Chem. News* **42** (1984) no. 1134, 19.

[114] Jefferson Chemical Co., US 3 878 247, 1975 (P. H. Moss, R. M. Gipson).Lubrizol Corp., US 3917594, 1975 (D. J. Hoke).Texaco, US 4031138, 1976 (E. C. Nieh, P. H. Moss). BASF, DE-OS 2730094, 1977 (H. Pohlemann, H. Naarmann).Chem. Fabrik Stockhausen, EP 13416, 1979 (B. Gossens, E. Barthell, K. Dahmen, E. Küster).
[115] K. Hashimoto, *Sangyo Igaku* **22** (1980) 233.
[116] P. S. Spencer, H. H. Schaumburg, *EHP Environ. Health Perspect.* **11** (1975) 129–133. D. D. McCollister, F. Oyen, V. K. Rowe, *Toxicol. Appl. Pharmacol.* **6** (1964) 172–181.

Acrylonitrile

PATRICK W. LANGVARDT, The Dow Chemical Co., Midland, Michigan 48640, USA

1.	Introduction	249	6.	Storage and Transportation	256
2.	Physical Properties	250	7.	Uses	256
3.	Chemical Properties	252	8.	Economic Aspects	257
4.	Production	252	9.	Toxicology and Occupational Health	258
5.	Quality Specifications and Chemical Analysis	255	10.	References	259

1. Introduction

In 1893, the French chemist, Ch. MOUREAU, first prepared acrylonitrile [107-13-1], acrylic acid nitrile, propene nitrile, vinyl cyanide, $CH_2=CH-C\equiv N$, M_r 53.06, by dehydrating either acrylamide or ethylene cyanohydrin with phosphorus pentoxide [1]. However, no significant technical or commercial applications were discovered for acrylonitrile until the late 1930s.

Shortly before the Second World War, I. G. Farbenindustrie introduced a synthetic rubber, Buna N, based on a copolymer of butadiene and acrylonitrile. This synthetic rubber was highly resistant to swelling in gasoline, oils, and other nonpolar solvents. At about the same time research began in the United States on similar copolymers termed GR-A, NBR, or nitrile rubber. Projects concerning acrylonitrile-containing polymers received special support during the Second World War because of obvious strategic importance, thus establishing acrylonitrile as a monomer with commercial significance.

Since that time the dramatic increase in the demand for acrylonitrile has been attributed not to nitrile rubber but largely to acrylic fibers, first introduced commercially in 1950 by Du Pont under the trademark Orlon. In addition, acrylonitrile is used in resins, thermoplastics, elastomers, and as an intermediate for organic synthesis, most notably for producing adiponitrile and acrylamide.

2. Physical Properties [2]–[7]

Acrylonitrile is a colorless liquid with a slightly pungent odor. Its physical properties are listed below.

Boiling point:

p, mbar	1013	666.5	333.2	133.3	66.7
bp, °C	77.3	64.7	45.5	23.6	8.7

Freezing point	-83.55 ± 0.05 °C
Density (20 °C)	0.8060 g/cm^3
Viscosity (25 °C)	0.34 mPa · s
Refractive index (n_D^{25})	1.3888
Dielectric constant (33.5 MHz)	38
Dipole moment	
(liquid)	3.51 D
(vapor)	3.88 D
Molar refractivity (D line)	15.67
Surface tension (25 °C)	26.63 mN/m
Vapor density (theoretical; air = 1)	1.83
Critical pressure	3.54 MPa (35.4 bar)
Critical temperature	246 °C
Critical volume	3.798 cm^3/g

Acrylonitrile is miscible with numerous organic solvents, including acetone, benzene, carbon tetrachloride, diethyl ether, ethyl acetate, ethylene cyanohydrin, petroleum ether, toluene, some kerosenes, and methanol. Azeotropes between acrylonitrile and several of these solvents are described in Table 1. The water solubility of acrylonitrile at several temperatures is shown in Table 2. The liquid–liquid equilibria of sulfuric acid–water–acrylonitrile systems have also been studied [8], as have vapor–liquid equilibria for systems containing various combinations of acetonitrile, acrolein, hydrogen cyanide, water, and acrylonitrile [9]–[13].

The thermodynamic data for acrylonitrile are summarized in Table 3. Additional flammability data have been reported for ternary mixtures of acrylonitrile or acetonitrile with air and nitrogen [14], as have sorption isotherms of acrylonitrile on an acrylonitrile–styrene copolymer [15].

Acrylonitrile has been characterized by electron diffraction [16], microwave [17], infrared, Raman, and ultraviolet absorption spectroscopy [18], and mass spectrometry [19].

Table 1. Azeotropes of acrylonitrile

Azeotrope	bp, °C	Acrylonitrile concentration, wt%
Benzene	73.3	47
Isopropyl alcohol	71.7	56
Methanol	61.4	39
Carbon tetrachloride	66.2	21
Water	71	88
Chlorotrimethylsilane	57	7
Tetrachlorosilane	51.2	89

Table 2. Solubilities of acrylonitrile in water

t, °C	Mass fractions, %	
	Acrylonitrile in water	Water in acrylonitrile
0	7.15	2.10
10	7.17	2.55
20	7.30	3.08
30	7.51	3.82
40	7.90	4.85
50	8.41	6.15
60	9.10	7.65
70	9.90	9.21
80	11.10	10.95

Table 3. Thermodynamic properties of acrylonitrile

Ignition temperature	481 °C
Flash point (open cup)	- 5 °C
Explosive limits (air, 25 °C)	$3.05-17.0 \pm 0.5$ vol%
Heat of combustion (l, 25 °C)	- 1761.89 kJ/mol
Heat of vaporization (25 °C)	32.65 kJ/mol
Heat of polymerization	72.4 ± 2.1 kJ/mol
Molar heat capacity (l)	2.09 kJ kg^{-1} K^{-1}
Molar heat capacity (g, 50 °C, 101.3 kPa)	1.204 kJ kg^{-1} K^{-1}
Molar heat of fusion	6.641 kJ/mol
Entropy S (g, 25 °C, 101.3 kPa)	274.06 kJ mol^{-1} K^{-1}
Free energy of formation ΔG^0(g, 25 °C)	195.31 kJ/mol
Enthalpy of formation ΔH^0 (g, 25 °C)	184.93 kJ/mol
Enthalpy of formation ΔH^0 (l, 25 °C)	150.21 kJ/mol

3. Chemical Properties [2], [3], [18], [20]

Acrylonitrile is a very reactive compound. The double bond in the acrylonitrile molecule is activated by conjugation with the polar nitrile group and will react in a variety of ways. Acrylonitrile can undergo spontaneous, exothermic polymerization and so must be inhibited for storage. The homo- and copolymerization of acrylonitrile take place rapidly in the presence of radiation, anionic initiators, or free-radical sources, such as peroxides or diazo compounds. The reaction involves charge transfer complexes between various monomers [21] and can be produced in the vapor, liquid, or solid phase, in solution, and in dual-phase systems. Only the latter two methods have had industrial impact.

Other notable reactions of the double bond of acrylonitrile include Diels-Alder reactions, hydrogenation, cyanoethylation, hydrodimerization, and hydroformylation [18], [20].

The most important reactions of the nitrile moiety of acrylonitrile are hydrolysis and alcoholysis. Acrylonitrile can be hydrolyzed partially to acrylamide or completely to acrylic acid, depending on the concentration of the acid used [22], [23]. For years the first step in the commercial production of acrylamide was the partial hydrolysis with sulfuric acid to acrylamide sulfate; however, now acrylonitrile is converted directly to acrylamide using various copper-based catalysts [24]. Hydrolysis with hydrochloric acid leads to hydrochlorination of the double bond as well, forming 3-chloropropionamide or 3-chloropropionic acid [25]. Although base-catalyzed hydrolysis of acrylonitrile is possible [26], it can lead to undesired reactions of the double bond.

Acrylic esters can be produced from acrylonitrile and primary alcohols in the presence of sulfuric acid. This reaction has been used commercially to produce methyl acrylate.

Other sulfuric acid-catalyzed reactions of acrylonitrile include those with olefins or tertiary alcohols to yield N-substituted acrylamides and with formaldehyde to form N,N'-methylenebisacrylamide or 1,3,5-triacrylhexahydro-s-triazine [17, p. 22].

4. Production [20, Chap. 1–7], [27], [28]

Today nearly all acrylonitrile is produced by ammoxidation of propene [*115-07-1*]. Although the first report of the preparation of acrylonitrile from propene occurred in a patent by the Allied Chemical and Dye Corporation in 1947 [29], it was a decade later when Standard Oil of Ohio (Sohio) developed the first commercially viable catalyst for this process [30]. Today, all of the United States capacity and approximately 90% of the world capacity for acrylonitrile is based on the Sohio process.

Sohio Process [31], [32]. In the Sohio process (Fig. 1) propene, oxygen (as air), and ammonia are catalytically converted directly to acrylonitrile using a fluidized-bed

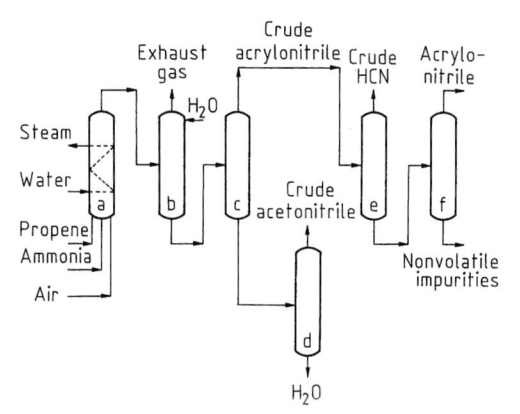

Figure 1. Simplified diagram of the Sohio acrylonitrile process
a) Fluidized-bed reactor; b) Absorber column; c) Extractive distillation column; d) Acetonitrile stripping column; e) Lights fractionation column; f) Product column

reactor operated at temperatures of 400–500 °C and gauge pressures of 30–200 kPa (0.3–2 bar):

$$2\ CH_2=CH\text{-}CH_3 + 2\ NH_3 + 3\ O_2 \longrightarrow 2\ CH_2=CH\text{-}C\equiv N + 6\ H_2O$$

Complex mechanisms for this reaction have been described [33]–[35].

Approximately stoichometric amounts of the starting materials are passed through the reactor with residence times of a few seconds. The process is highly selective, requiring no recycling to produce high acrylonitrile yields of approximately 0.8–0.9 kg from 1 kg of propene. Acetonitrile [75-05-8] (0.02–0.11 kg) and hydrogen cyanide [74-90-8] (HCN, 0.15–0.20 kg) are the principal byproducts (from 1 kg propene). The heat of reaction can be recovered as high-pressure steam.

The catalyst used in the early Sohio process was a bismuth–phosphomolybdate combination. Since that time there has been a continuous search for alternatives and for superior performance, resulting in patents by more than 30 companies. Sohio introduced Catalyst 21 (antimony–uranium) in 1967, Catalyst 41 (ferrobismuth–phosphomolybdate) in 1972, and Catalyst 49 (undisclosed) in 1978. All of these changes were aimed at improved efficiency and reduction in byproducts.

The reactor effluent is cooled and scrubbed with water in a countercurrent absorber. The off-gas, consisting primarily of nitrogen, is vented. The reaction products remain in the aqueous phase. Acetonitrile is removed by extractive distillation. Crude acrylonitrile and hydrogen cyanide are distilled overhead while water and acetonitrile are removed from the bottom of the columns. In subsequent distillations, hydrogen cyanide is separated from wet acrylonitrile, the water content of the product is then reduced, and finally, nonvolatile impurities are removed.

The major byproducts of this process, hydrogen cyanide and acetonitrile, normally are incinerated as their supply often exceeds demand. Unused ammonia can be recovered as ammonium sulfate and then disposed of, but it commonly is vented to the atmosphere. Aqueous wastes containing cyanides, sulfates, and various organic byproducts must be disposed of by incineration or deepwell injection or be pretreated for subsequent biological waste treatment.

Four companies, American Cyanamid, Du Pont, Monsanto, and Sohio (formerly Vistron Corp.) produce acrylonitrile in the United States via the Sohio process, with a total operating capacity of 9.5×10^5 t/a in 1980.

Other Ammoxidation Processes [36]. Others who manufacture acrylonitrile by ammoxidation of propene include Distillers/Ugine, Societa Nazionale Metandotti (SNAM), Montedison-UOP, and Chemie Linz. All use different catalyst systems and, with the exception of Montedison-UOP, fixed-bed reactors.

Production from Ethylene Cyanohydrin. Germany (I.G. Farben, Leverkusen) [37] and the United States (American Cyanamid) first produced acrylonitrile on an industrial scale in the early 1940s. These processes were based on the catalytic dehydration of ethylene cyanohydrin [38], [39]. Ethylene cyanohydrin was produced from ethylene oxide and aqueous hydrocyanic acid at 60 °C in the presence of a basic catalyst. The intermediate was then dehydrated in the liquid phase at 200 °C in the presence of magnesium carbonate and alkaline or alkaline earth salts of formic acid.

$$HO\text{-}CH_2\text{-}CH_2\text{-}C \equiv N \longrightarrow CH_2\text{=}CH\text{-}C \equiv N + H_2O$$

An advantage of this process was that it generated few impurities; however, it was not economically competitive. American Cyanamid and Union Carbide closed plants based on this technology in the mid-1960s.

Production from Acetylene and Hydrocyanic Acid. Before the development of the propene ammoxidation process, a major industrial route to acrylonitrile involved the catalytic addition of hydrocyanic acid to acetylene [27].

$$H\text{-}C \equiv C\text{-}H + HCN \longrightarrow CH_2\text{=}CH\text{-}CN$$

Although a vapor-phase reaction has been reported, the commerical reaction usually was carried out at 80 °C in dilute hydrochloric acid containing cuprous chloride. Unreacted acetylene was recycled. The yield from this reaction was good; however, the raw materials were relatively expensive, some undesirable impurities, divinylacetylene and methyl vinyl ketone, were difficult to remove, and the catalyst required frequent regeneration. Du Pont, American Cyan-amid, and Monsanto employed this process until about 1970.

Other Routes to Acrylonitrile. Several other routes to acrylonitrile include nitrosation of propene [40], [41], ammonation of propionaldehyde [42], dehydrogenation of propionitrile [43], [44], and the reaction of hydrogen cyanide and acetaldehyde [45]. None of these processes has achieved the commercial status of those described previously.

Future Processes. Several other chemicals have been studied as possible alternative precursors to acrylonitrile. Ethylene, propane, and butane react with ammonia at high temperatures (750 – 1000 °C) to yield acrylonitrile [46]. Monsanto, Power Gas, and ICI [47] have developed catalytic ammoxidation processes based on propane. Propane is of particular interest because of a cost advantage over propene. However, this price difference is not likely to be great enough in the near future to dictate change. High conversions to acrylonitrile also have been obtained on a laboratory scale from ethylene, hydrogen cyanide, and oxygen using a palladium-based catalyst [48].

See Chapter 9 for special precautions and regulations concerning the production of acrylonitrile.

5. Quality Specifications and Chemical Analysis

Below is a list of typical sales specifications.

Appearance	Clear, free of suspended matter
Color, Pt – Co scale	10 max.
Refractive index (25 °C)	1.3880 – 1.3895
Acetone	300 ppm max.
Acetonitrile	500 ppm max.
Acidity (as acetic acid)	20 ppm max.
Aldehydes (as acetaldehyde)	50 ppm max.
Hydrogen cyanide	5 ppm max.
Soluble iron	0.2 ppm max.
Nonvolatile matter	100 ppm max.
Peroxides (as H_2O_2)	0.2 ppm max.
Water	0.25 – 0.45 wt %
Inhibitor (hydroquinone monomethyl ether)	35 – 50 ppm

Although customers will commonly request additional quality specifications for specific applications, these specifications often are critical to the properties of the final product and therefore are classified as proprietary information. Specifications that have been reported include acrylonitrile 99% min.; divinylacetylene 5 ppm max.; methyl vinyl ketone 300 ppm max.; pH value 6.0 – 7.5; distillation range 74.5 – 78.5 °C at 101.3 kPa [3], [49].

A variety of standard, analytical techniques are used for the chemical analysis of acrylonitrile [50] – [52].

6. Storage and Transportation [6, Chap. 2 and 3]

Acrylonitrile may be fatal if absorbed through the skin and can be harmful if inhaled or ingested (Chap. 9). In addition, acrylonitrile also is a flammable liquid and its vapors can form explosive mixtures with air under ambient conditions (Chap. 2).

The toxicity, flammability, and vapor pressure of acrylonitrile dictate that it be stored in closed systems.

Storage vessels and piping for use at ambient temperature and pressure may be constructed from carbon steel. Stainless steel is recommended for more severe conditions. Tanks should be electrically grounded and equipped with scrubbers or vent condensers to prevent vapor leaks to the atmosphere.

Other storage considerations include preserving product quality and minimizing the potential for polymerization.

Acrylonitrile is transported in tank cars, barges, steel drums, and via pipeline. International transportation of acrylonitrile is governed by the International Maritime Dangerous Goods (IMDG) code published by the Intergovernmental Maritime Consultative Organization (IMO): IMDG code no. 3105, class 3.1, UN no. 1093. In the United States, acrylonitrile is classified as a flammable liquid and as a poison, and its transportation is governed by the U.S. Department of Transportation (DOT) Safety Act, title CFR 172.101 et seq. The DOT freight classification is RQ/ Acrylonitrile/ Flammable Liquid/ UN 1093/ Poison. Transportation in Europe is regulated by RID, ADR, and ADRN: class 6.1, no. 2a (from 1985: class 3, no. 11 a), RN 601, 2601, and 6601 resp. Blue Book (UK): flammable liquid, IMDG E 3022.

7. Uses [28], [53]

Acrylic textile fibers are by far the largest end-use product for acrylonitrile (Table 4). Acrylic fibers find use primarily in wearing apparel and in home furnishings such as carpets and draperies. In 1981, acrylic fibers accounted for 53% of the acrylonitrile used in the United States, for 76% in Western Europe, and for 67% in Japan.

The production of acrylonitrile–butadiene–styrene (ABS) and styrene–acrylonitrile (SAN) resins consumes the second largest quantity of acrylonitrile. The former resins are produced by grafting acrylonitrile and styrene onto polybutadiene or a styrene–butadiene copolymer. These products are used to fabricate components for automotive and recreational vehicles, pipe fittings, and appliances. The SAN resins are styrene–acrylonitrile copolymers containing approximately 25–30 wt% of acrylonitrile. The superior clarity of SAN resin allows it to be used in automobile instrument panels, for instrument lenses, and for houseware items. In 1981 ABS and SAN resins accounted for 17% of United States acrylonitrile use, 9.5% in Western Europe, and 17% in Japan.

Table 4. End-uses of acrylonitrile in 1981 (thousand tons) *

	United States	Japan	FR of Germany	Italy	United Kingdom	France
Acrylic fibers	315	356	216	209	104	74
ABS/SAN resins	104	89	22	12	13	9
Adiponitrile	98	27	**	**	**	**
Acrylamide	28	28	**	**	**	**
Nitrile rubber	23	15	12	4	2	8
Other	30	10	12	3	76	3
Total	598	525	262	228	195	94

* Reproduced with permission of the World Petrochemicals Program, SRI International [53].
** Included in "other", along with acrylic acid/acrylates.

In the past decade, the production of the chemical intermediates adiponitrile and acrylamide have surpassed nitrile rubbers as end-use products of acrylonitrile in the United States and Japan. Adiponitrile is further converted to hexamethylenediamine (HMDA), used to manufacture nylon 66 ($\rightarrow$ Hexamethylenediamine). Acrylamide is used to produce water-soluble polymers or copolymers used for paper manufacturing, waste treatment, mining applications, and enhanced oil recovery.

Nitrile rubbers, the original driving force behind acrylonitrile production, have taken a less significant place as end-use products. They find industrial applications in areas where their oil resistance and low-temperature flexibility are important, such as in the fabrication of seals (O-rings), fuel hoses, and oil well equipment.

In the future, acrylonitrile may be used increasingly for such things as high-nitrile barrier resins [54], high-resilient polyurethane foams, production of specialty chemical intermediates, and graft copolymers with starch.

8. Economic Aspects [28], [53], [55]

Table 5 summarizes 1981 acrylonitrile supply and demand information for selected countries. The total world capacity for acrylonitrile production in January 1983 was just under 4×10^6 t/a. The 1983 United States capacity was down to just over 1×10^6 t/a. Capacities for the United States, Western Europe, and Japan are not expected to increase significantly through 1983. A worldwide surplus of acrylonitrile is predicted for some time as several developing countries plan to build plants.

The demand for acrylonitrile in the United States, Western Europe, and Japan increased in the 1970s by an average of about 6% per year. However, growth in the next decade is expected to fall to about 2% per year in the United States and 1% in Western Europe and Japan. This outlook reflects predicted slow growth in acrylic fibers and ABS resins.

Table 5. Acrylonitrile supply and demand in 1981 (thousand tons) *

Country	Year-end capacity	Production	Imports	Exports	Consumption
Mexico	74	54	20	-	72
United States	1131	906	Neg**	291	598
Brazil	60	57	-	35	24
France	90	56	53	16	94
FR of Germany	370	250	37	26	262
Italy	230	119	97	-	228
United Kingdom	395	225	45	75	195
Spain	75	59	44	-	100
Japan	733	477	71	40	515
South Korea	77	52	76	-	159
Taiwan	132	103	Neg**	9	93

* Reproduced with permission of the World Petrochemicals Program, SRI International [53].
** Neg = Negligible.

9. Toxicology and Occupational Health

The effects of human exposure to acrylonitrile have been a matter of public health concern and speculation for some time, but a good under-standing of the toxic effects of acrylonitrile has just evolved in the last decade. Because so vast a number of studies have been directed at understanding the toxicity of acrylonitrile, only a general overview can be given here.

Acrylonitrile is toxic to laboratory animals, regardless of the route of exposure. Acrylonitrile exerts its toxic action by two simultaneous mechanisms: inhibition of the activity of cyto chrome oxidase by liberation of cyanide, and the inhibition of sulfhydryl-dependent enzymes of intermediary metabolism by cyanoethylation of sulfhydryl groups. Furthermore, there is some evidence that, contrary to past belief, cyanide plays little role in acrylonitrile lethal effects. Coadministration of certain aromatic compounds has been reported to increase the lethal effects of acrylonitrile [56]. A wide range of acute LD_{50} values has been found for different laboratory animals and for different routes of administration. Mice (25–50 mg/kg) [57]–[59] are more sensitive to acrylonitrile than are rats (78–150 mg/kg) [58], [60] and guinea pigs (56 mg/kg) [61].

The consequences of human acrylonitrile exposure depend both on the route and the degree of exposure. Acrylonitrile may cause death by ingestion, inhalation of vapor, or absorption of the liquid through the skin. Nonfatal intoxication of people working with acrylonitrile has been reported in several instances [57], [60], [62], [63]. Acrylonitrile poisoning results in toxic symptoms characteristic of the cyanide ion. Sequentially, one experiences irritation of eyes and nose, limb weakness, labored breathing, dizziness and impaired judgement, nausea, collapse, irregular breathing, and convulsions, possibly followed by cardiac arrest. Direct skin contact with acrylonitrile can cause severe skin irritation [64] and, in some cases, allergic dermatitis [65], [66]. Despite its large-scale use, no fatal accidental poisonings from acrylonitrile are known in industry, although several deaths have been reported following applications of fumigants containing acrylonitrile [67]. A description of protective clothing that should be worn when handling this compound is given in reference [68].

Chronic effects potentially can occur after prolonged, excessive exposure to acrylonitrile. Complaints of headache, weakness, fatigue, nausea, nosebleeds, and insomnia came from Japanese workers manufacturing acrylonitrile [69]. Others exposed to 5–20 ppm of acrylonitrile were found to have abnormal liver functions [70]. Skin irritation and allergic dermatitis also have been observed in workers after chronic exposure to acrylonitrile [71].

A number of long-term studies with laboratory animals have added significantly to the understanding of acrylonitrile toxicity [72]–[76], particularly in relation to carcinogenicity. Ingestion or inhalation of acrylonitrile has caused tumors of the central nervous system and zymbal gland in rats. Acrylonitrile also appears to be mutagenic in certain bacterial [77], [78] and mammalian [79] test systems.

The results of laboratory experiments and an epidemiology study that suggested above average cancer levels among workers at a Du Pont textile plant [80], prompted OSHA to regulate acrylonitrile as a carcinogen, and its use in the United States must be in strict conformance to standards set forth in the Federal Register [64, p. 45 809–45 819]. These regulations set the permissible exposure limit (PEL) to acrylonitrile at 2 ppm as an 8-h time-weighted-average (TWA) concentration, with a ceiling level of 10 ppm for any 15-min period. In addition, the standard established an action level of 1 ppm (8-h TWA) and included requirements for employee training, medical surveillance, record keeping, and analytical procedures for monitoring employee exposure (appendix D of the standard). Among other things, the standard requires the employer to provide protective clothing and equipment, including respirators, and to establish regulated areas where acrylonitrile concentrations may exceed the permissible limits.

Legal actions have been taken in the United States against certain applications of acrylonitrile. In 1977, the U. S. Federal Drug Administration declared acrylonitrile to be an indirect food additive and banned use of beverage containers made from acrylonitrile. In other food-packaging applications limits were established for allowable residual monomer concentrations.

The German MAK commission classifies acrylonitrile in group III A 2 for compounds presenting a carcinogenic risk for humans [81].

10. References

[1] Ch. Moureau, *Ann. Chim. Phys.* **2** (1894) 187–191.
[2] K. Stueben, *High Polym.* **24** (1970) no. 1, 1–80.
[3] M. V. Norris in F. D. Snell, C. L. Hilton (eds.): *Encyclopedia of Industrial Chemical Analysis,* vol. **4,** Wiley-Interscience, New York 1967, p. 368–681.
[4] H. L. Finke, J. Messerly, S. S. Todd, Am. Petrol. Inst. Res. Project no. 62 (1971).

[5] J. J. Jaspers, *J. Phys. Chem. Ref. Data* **1** (1972) 889.
[6] Monsanto Chemical Intermediates Co. (ed.): *Acrylonitrile Handling and Storage*, Publ. no. 162, St. Louis, Mo., 1982, Chap. 1.
[7] D. R. Stull, E. F. Westrum, Jr., G. C. Sinke: *The Chemical Thermodynamics of Organic Compounds*, J. Wiley & Sons, New York 1969, p. 474.
[8] G. A. Chubarov, S. M. Danov, V. I. Logutov, *Zh. Prikl. Khim. (Leningrad)* **55** (1982) no. 5, 1032–1034; *Chem. Abstr.* **97** (1982) 29274 f.
[9] N. M. Sokolov, *Massoobmennye Protsessy Khim. Tekhnol.* **4** (1969) 62.
[10] A. C. Zawisza, S. Glowska, *Bull. Acad. Pol. Sci., Ser. Sci. Chim.* **17** (1969) 373–379.
[11] N. M. Sokolov, *Rev. Chem.* **20** (1969) 169–172.
[12] N. M. Sokolov, N. N. Sevryugova, N. M. Zhavoronkor, *Theor. Osn. Khim. Tekhnol.* **3** (1969) 449–453.
[13] N. M. Sokolov, *Proc. Int. Symp. Distill.* **3** (1969) 110–117.
[14] S. DeMicheli, V. Tartari, *J. Chem. Eng. Data* **27** (1982) 273–275.
[15] A. Orr, J. Miltz, S. G. Gilbert, *Eur. Polym. J.* **17** (1981) 1149–1153.
[16] T. Fukuyama, K. Kuchitsu, *J. Mol. Struct.* **5** (1970) 131–135.
[17] M. C. L. Gerry, K. Yamada, G. Winnewisser, *J. Phys. Chem. Ref. Data* **8** (1979) 107–123.
[18] American Cyanamid Co.: *The Chemistry of Acrylonitrile*, 1st ed., New York 1951, p. 14–15.
[19] U.S. National Bureau of Standards: *EPA/NIH Mass Spectral Data Base*, vol. **1**, Washington, D.C., 1978, p. 5.
[20] M. A. Dalin, I. K. Kolchin, B. R. Serebryakov: *Acrylonitrile*, Technomic, Westport, Conn., 1971, Chap. 8.
[21] J. Vialle, *J. Macromol. Sci. Chem.* **5** (1971) 1229–1240.
[22] Standard Oil, US 3546289, 1967 (O. A. Kiikka).
[23] Dow Chemical, DE-OS 2001904, 1970.
[24] Dow Chemical, US 3597481, 1971; US 3631104, 1971; US 3758578, 1973 (B. A. Tefertiller, C. E. Habermann).
[25] R. A. Barnes, E. R. Kraft, L. Gordon, *J. Am. Chem. Soc.* **71** (1949) 3523–3528.
[26] Y. Mamiya, S. Matui, S. Kanbara, *J. Soc. Chem. Ind. Japan* **44** (1941) 125–126 B; *Chem. Abstr.* **38** (1944) 3481.
[27] D. J. Hadley, E. G. Hancock (ed.): *Propylene and Its Industrial Derivatives*, Halsted Press, New York 1973, Chap. 11.
[28] S. A. Cogswell: *Acrylonitrile*, Chemical Economics Handbook Marketing Research Report, Stanford Research Institute International, Menlo Park, Calif., 1981, 607.5032A- 607.5033O.
[29] Allied, US 2481826, 1947 (J. N. Cosby).
[30] Standard Oil of Ohio, US 2904580, 1957 (J. D. Idol, Jr.).
[31] F. Veatch, J. L. Callahan, J. D. Idol, E. C. Milberger, *Chem. Eng. Progr.* **56** (1960) 65–67.
[32] F. Veatch, J. L. Callahan, J. D. Idol, E. C. Milberger, *Hydrocarbon Process.* **41** (1962) 187–190.
[33] M. Cathala, *Bull. Soc. Chim. Fr.* **11** (1970) 4114–4119.
[34] C. R. Adams, T. J. Jennings, *J. Catal.* **2** (1963) 63–68; **3** (1964) 549–558.
[35] W. M. H. Sachtler, N. H. DeBoer, *Proc. 3rd. Int. Congr. Catal.* **1** (1965) 252–255.
[36] R. B. Stobaugh, S. G. Clark, G. D. Camirand, *Hydrocarbon Process.* **50** (1971) 109–120.
[37] *Chem. Eng. News* **23** (1946) 1841–1845.
[38] American Cyanamid, US 2690452, 1954 (E. L. Carpenter).
[39] Stamicarbon, US 2729670, 1956 (P. H. DeBruin).
[40] Du Pont, US 2736739, 1956 (D. C. England, G. V. Nock).
[41] Du Pont, US 3184415, 1965 (E. B. Huntley, J. M. Kruse, J. W. Way).

[42] Phillips Petroleum, US 2412437, 1946 (C. R. Wagner).
[43] Du Pont, US 2553482, 1951 (N. Brown).
[44] Rohm & Haas, US 2385552, 1945 (L. R. V. Spence, F. O. Haas).
[45] K. Sennewald, *Proc. 5th. World Petroleum Congr.* **4** (1960) 217–227.
[46] M. Sittig: *Acrylonitrile,* Chemical Process Monograph no. 14, Noyes Development Corp., Park Ridge, N.J., 1965.
[47] P. Townsend, *Appl. Polym. Symp.* **25** (1974) 311–319.
[48] J. Perkowski, *Przem. Chem.* **51** (1972) 17–21.
[49] G. Caporali, U. Sansoni (eds.): *Acrylonitrile,* in Enciclopedia della Chimica, vol. **1,** Scientifiche, Firenze, Italy, 1971.
[50] American Society for Testing and Materials: *Standard Test Method for Water Using Karl Fischer Reagent,* Annual Book of ASTM Standards, Philadelphia, Pa., 1982, Part 30, E-203, 815–824.
[51] American Society for Testing and Materials: *Standard Test Method for Trace Amounts of Peroxides in Organic Solvents,* Annual Book of ASTM Standards, Philadelphia, Pa., 1982, Part 30, E-299, 909–912.
[52] Monsanto: *Acrylonitrile,* Monsanto Data Sheet, Publ. no. 155-A, St. Louis, Mo., May 5, 1981.
[53] *Acrylonitrile,* World Petrochemicals Program-Propylene, vol. **1,** SRI International, Jan. 1983, 206.3–206.10.
[54] *Chem. Week* **132** (1983) 12–13.
[55] Acrylonitrile, *Chem. Mkt. Rep.*, 1983, Jan. 10, p. 50.
[56] I. Gut, J. Kopecky, J. Nerudova, *G. Ital. Med. Lav.* **3** (1981) 131–136.
[57] H. Zeller, H. T. Hofmann, A. M. Thiess, H. Hey, *Zentralbl. Arbeitsmed. Arbeitsschutz* **19** (1969) 226–238.
[58] V. Benes, V. Cerna, *J. Hyg. Epidemiol. Microbiol. Immunol.***3** (1959) 106–116.
[59] H. Yoshikawa, *Igaku to Seibutsugaku,* **77** (1968) no. 1, 1–4.
[60] R. H. Wilson, G. V. Hough, W. E. McCormick, *Ind. Med.* **17** (1948) 199–207.
[61] V. Jedlicka, A. Pasek, J. Gola, *J. Hyg. Epidemiol. Microbiol. Immunol.* **1** (1958) 116–125.
[62] E. Sartorelli, *Med. Lav.* **57** (1966) 184–187.
[63] International Agency For Research on Cancer: *Some Monomers, Plastics and Synthetic Elastomers and Acrolein,* IARC Monograph, vol. 19, (1979) 73–113.
[64] H. C. Dudley, P. A. Neal, *J. Ind. Hyg. Toxicol.* **24** (1942) no. 2, 27–36.
[65] B. R. Balda, *Hautarzt* **26** (1975) 599–601.
[66] K. Hasimoto, T. Kobayasi, *Quart. J. Labor. Res.* **9** (1961) 21–24.
[67] *Occupational Exposure to Acrylonitrile,* vol. **43,** U.S. Federal Register, no. 192, 45815 (1978).
[68] J. H. Davis, J. E. Davies, A. Raffonelli, G. Reich in W. D. Deichmann (ed.): *Pesticides and the Environment,* vol. **2,** Intercontinental Medical Book Corp., New York 1973, p. 547–556.
[69] H. Sakurai, M. Onodera, T. Utsunomiya, H. Minakuchi, H. Iwai, H. Matyumura, *Br. J. Ind. Med.* **35** (1978) no. 3, 219–225.
[70] H. Sakurai, M. Kusumoto, *J. Sci. Labour (Tokyo)* **48** (1972) 273–282.
[71] M. Spassovski, *Environ. Health Perspect.* **17** (1976) 199–202.
[72] J. F. Quast, C. G. Humiston, B. A. Schwetz, L. A. Frauson, C. E. Wade, J. M. Norris: *A Six-Month Oral Toxicity Study Incorporating Acrylonitrile in the Drinking Water of Pure-Bred Beagle Dogs,* Toxicology Research Laboratory, Dow Chemical, Midland, Mich., prepared for the Chemical Manufacturers Association, Washington, D.C., 1975.
[73] J. F. Quast, R. M. Enriguez, C. E. Wade, C. C. Humiston, B. A. Schwetz: *Toxicity of Drinking Water Containing Acrylonitrile in Rats: Results After 12 Months,* Toxicology Research Laboratory,

Dow Chemical, Midland, Mich., prepared for the Chemical Manufacturers Association, Washington, D.C., 1977.
[74] C. Maltoni, A. Ciliberti, V. DiMaio, *Med. Lav.* **68** (1977) 401–411.
[75] J. F. Quast, C. E. Wade, C. G. Humiston, R. M. Careon, E. A. Hermann, C. N. Park, B. A. Schwetz: *A Two-Year Toxicity and Oncogenicity Study with Acrylonitrile Incorporated in the Drinking Water of Rats,* Toxicology Research Laboratory, Dow Chemical, Midland, Mich., prepared for the Chemical Manufacturers Association, Washington, D.C., 1980.
[76] J. F. Quast, D. J. Schuetz, M. F. Balmer, T. S.Gushow, C. N. Park, M. J. McKenna: *A Two-Year Toxicity and Oncogenicity Study with Acrylonitrile Following Inhalation Exposure of Rats,* Toxicology Research Laboratory, Dow Chemical, Midland, Mich., prepared for the Chemical Manufacturers Association, Washington, D.C., 1980.
[77] J. McCann, N. W. Spingarn, J. Kobori, B. N. Ames, *Proc. Natl. Acad. Sci. U.S.A.* **72** (1975) 979–983.
[78] R. E. McMahon, J. C. Cline, C. Z. Thomson, *Cancer Res.* **39** (1979) 682–693.
[79] R. A. Parent, B. C. Castro, *J. Natl. Cancer Inst.* **62** (1979) 1025–1029.
[80] M. T. O'Berg, *J. Occup. Med.* **22** (1980) 245–252.
[81] Deutsche Forschungsgemeinschaft (ed.): *Maximale Arbeitsplatzkonzentrationen (MAK) 1983,* Verlag Chemie, Weinheim 1983.

Adipic Acid

DARWIN D. DAVIS, E. I. Du Pont de Nemours & Co., Victoria, Texas 77901, United States

1.	Introduction	263	7.	Storage and Transportation	269
2.	Physical Properties	263	8.	Derivatives	270
3.	Chemical Properties	264	8.1.	Adiponitrile	270
4.	Production	265	8.2.	Other Derivatives	271
4.1.	Nitric Acid Oxidation of Cyclohexanol	265	9.	Uses	272
			10.	Economic Aspects	273
4.2.	Other Routes	268	11.	Toxicology and Occupational Health	274
5.	Byproducts	268			
6.	Quality Specifications	269	12.	References	274

1. Introduction

Adipic acid, hexanedioic acid, 1,4–butanedicarboxylic acid, $C_6H_{10}O_4$, M_r 146.14, $HOOCCH_2CH_2CH_2CH_2COOH$ [124-04-9], is the most significant commercially of all the aliphatic dicarboxylic acids. Appearing in nature in only minor amounts, it is synthesized on a very large scale worldwide. Its primary use is in the production of nylon 66 polyamide, discovered in the early 1930s by W. H. CAROTHERS of Du Pont.

Manufacture of this polymer has grown to become one of the dominant processes in the synthetic fibers industry. The historical development of adipic acid production was reviewed recently [5].

2. Physical Properties [6]

Adipic acid is isolated as colorless, odorless crystals having an acidic taste, d_4^{25} 1.360, mp 152.1 ± 0.3 °C, and specific gravity 1.085 (at 170 °C).

Boiling point

p, kPa	101.3	13.3	2.67	0.67	0.1333
bp, °C	337.5	265	222	191	159.5

It is very soluble in methanol and ethanol, soluble in water and acetone, and very slightly soluble in cyclohexane and benzene. Adipic acid crystallizes as monoclinic prisms from water, ethyl acetate, or acetone–petroleum ether. Solubility in water increases rapidly with temperature:

Solubility in 100 g H_2O

t, °C	15	40	60	80	100
Adipic acid, g	1.42	4.5	18.2	73	290

Values in the literature vary widely at the higher temperatures.

The pH of a 0.1% solution is 3.2 (at 25 °C), of a saturated solution it is 2.7 (at 25 °C). The dissociation constant (at 18 °C) is k_1 4.6×10^{-5}, k_2 3.6×10^{-6}. Specific heat of the liquid is 2.253 kJ kg^{-1}K^{-1}, and of the vapor (at 300 °C) is 1.680 kJ kg^{-1} K^{-1}; heat of fusion is 115 kJ/kg; heat of vaporization is 549 kJ/kg; heat of solution in water is −214 kJ/kg (at 10–20 °C), −241 kJ/kg (at 90–100 °C); melt viscosity is 4.54 MPa · s (at 160 °C), 2.64 MPa · s (at 193 °C).

The bulk density of the crystalline solid is 600–700 kg/m^3 depending on particle size; the closed cup flash point is 196 °C, and the Cleveland open cup flash point is 210 °C; the autoignition temperature is 420 °C; the dust cloud ignition temperature is 550 °C; the minimum explosive concentration (dust in air) is 0.035 kg/m^3, and the minimum cloud ignition energy is 6.0×10^{-2} J; the maximum rate of pressure rise is 18.6 MPa/s.

3. Chemical Properties

Adipic acid is stable in air under ordinary conditions, but heating the molten material leads to some decarboxylation above 230–250 °C. The reaction is markedly catalyzed by several metals, including calcium [7] and barium [8]. The product in either case is cyclopentanone [*120-92-3*], *bp* 131 °C. The tendency for loss of water to produce cyclic anhydrides is much less pronounced with adipic acid than with glutaric and succinic acids [9]. Adipic acid readily reacts at one or both of the carboxyl groups to produce esters, amides, salts, etc. (Chap. 8). The acid is quite stable to several oxidizing materials, as evidenced by its synthesis in their presence. However, nitric acid attacks it autocatalytically above 180 °C, producing carbon dioxide, water, and nitrogen oxides.

4. Production

Early processes for production of adipic acid used a two-step, air oxidation of cyclohexane [110-82-7]; however, now essentially all production is derived from the nitric acid oxidation of a mixture of cyclohexanone [108-94-1] – cyclohexanol [108-93-0], or ketone–alcohol (KA) oil.

Differences among the processes arise mostly in the KA manufacturing step. Historically the six-carbon, saturated ring was obtained by hydrogenation of either benzene or phenol. The cyclohexane from the former subsequently is air oxidized to the KA mixture. In recent years, there has been a shift almost exclusively to the lower cost cyclohexane-based processes [10]. However, the patent literature indicates active development of alternate processes in view of the uncertainty of petrochemical costs and supply. (For a discussion of KA production, → Cyclohexanol and Cyclohexanone).

4.1. Nitric Acid Oxidation of Cyclohexanol

Reaction Mechanism. The second step of the conventional process, developed by Du Pont in the late 1940s, involves the nitric acid oxidation of cyclohexanol, cyclohexanone, or a mixture of both [11], [12]. Adipic acid is obtained in greater than 90% yield. Nitrogen oxides, carbon dioxide, and some lower dicarboxylic acids are the major byproducts, as well as oxidation products arising from impurities in the KA intermediate.

The chemical mechanism was discussed originally in 1956 [13], and later in great detail [14], [15]. The latter reports included kinetic and reactor design considerations. Results of related studies, especially on the later stages of the reaction, were published at about the same time [16]–[18]. A summary of the findings of these investigators is given in Figure 1.

Cyclohexanol (**1**) is oxidized to cyclohexanone (**2**), accompanied by the generation of nitrous acid. The ketone then reacts by one of three possible pathways leading to the formation of adipic acid (**8**). The major fraction of the reaction occurs via nitrosation to produce 2-nitrosocyclohexanone (**3**), then by further reaction with nitric acid to form the 2-nitro-2-nitrosoketone (**6**). Hydrolytic cleavage of this intermediate gives 6-nitro-6-hydroximinohexanoic acid, or "nitrolic acid" (**9**). This breaks down further to give adipic acid and nitrous oxide, the main nonrecoverable nitric acid reduction product. Typically 2.0 mol of nitric acid are converted to nitrous oxide for each mole of adipic acid produced. Nitration predominates at higher temperature, and the route via such intermediates as the dinitroketone (**4**) becomes significant.

Another path proposed by the early investigators involves the intermediate formation of the 1,2-diketone (**5**) or its dimer. Conversion of this material to adipic acid in

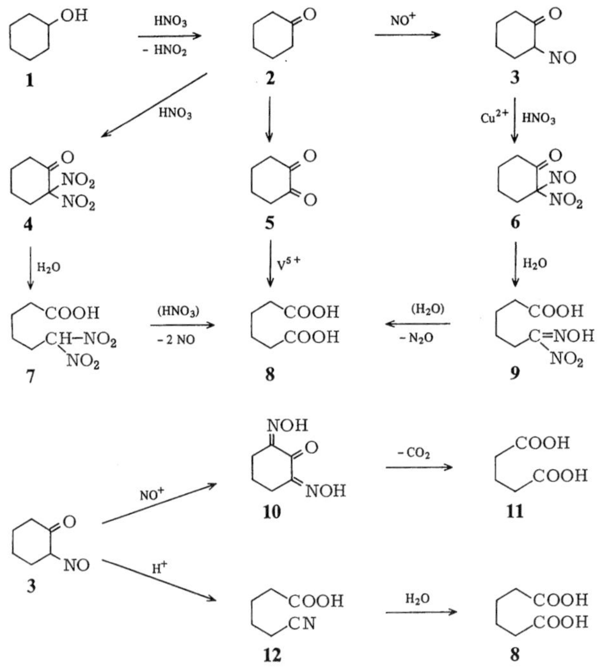

Figure 1. Reaction paths in nitric acid oxidation of cyclohexanol

good yield requires the use of a vanadium catalyst, and the effect of vanadium on the overall yield suggests a significant contribution by this path. The intermediate nitrosoketone (**3**) can undergo two important side reactions. Multiple nitrosation leads to intermediate **10**, which loses CO_2 to produce glutaric acid (**11**) or succinic acid from subsequent reaction with nitric acid. Copper shot is added to the nitric acid to inhibit these reactions. In systems containing a relatively high steady-state concentration of the nitrosoketone (**3**) or its tautomer, the oximinoketone, a Beckmann-type rearrangement leads to 5-cyanopentanoic acid (**12**) in minor amounts. This material slowly hydrolyzes to adipic acid.

Commercial Nitric Acid Oxidation Processes. The basic technology for carrying out the nitric acid oxidation of cyclohexanone-cyclohexanol (KA) remains similar to that described in the early patent literature. Recent developments have centered on improvement in byproduct removal and catalyst and nitric acid recovery. Because of the corrosive nature of nitric acid, plants are constructed of stainless steel (type 304L), or of titanium in areas of most severe exposure. The block flow diagram in Figure 2 shows a typical layout for a commercial nitric acid oxidation process [19], [5]. The reaction is carried out in a continuously circulated loop of nitric acid mother liquor passing through the entire system.

The reactor (a), controlled at 60–80 °C and 0.1–0.4 MPa, is charged with the recycled nitric acid stream, the KA feed material, and makeup acid containing 50–60% nitric acid and copper–vanadium catalyst [20], [21]. Residence time in (a)

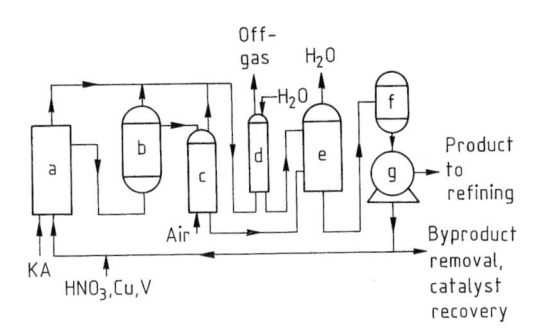

Figure 2. Flow diagram of a process for nitric acid oxidation of cyclohexanone-cyclohexanol a) Reactor; b) Optional cleanup reactor; c) NO$_x$ bleacher; d) Nitric acid absorber; e) Concentrator; f) Crystallizer; g) Filter or centrifuge

is of the order of a few minutes. The effluent may also pass through a second reactor (b) at elevated temperature (115 °C) [22]. The reaction is very exothermic (6280 kJ/kg), and normal heat-exchange surfaces tend to frost, leading to loss of temperature control. Imperial Chemical Industries has patented several reactor systems for removing the heat of reaction and minimizing energy use in the process [23]–[28]. An excess of recycled nitric acid stream over the KA feed stream of at least 3:1 up to >300:1 is maintained to control the reaction and improve the yield [21].

The product stream is passed through a bleacher column (c) where excess dissolved nitrogen oxides are removed with air and sent to (d), where the oxides are reabsorbed and recovered as nitric acid. The off-gas from (d) has been used to aid in initiation of reaction at low oxidation temperature by passing it into the KA feed stream prior to oxidation [29]–[31]. Removal of NO$_x$ from the off-gas by scrubbing with KA also has been described [32]. Water produced in the oxidation is then removed in a concentrator column (e) usually operated under vacuum. The concentrated product stream is either recycled back to the reactor with a portion diverted to product recovery or passed to product recovery prior to recycle of the filtrate. Adipic acid is removed from the product stream by crystallization (f) and subsequent filtration or centrifugation (g) [33]–[35]. The mother liquor stream, or a portion thereof, which contains high concentrations of glutaric and succinic acid byproducts, is processed further to recover the copper and vanadium catalysts and to remove the byproducts. Ion exchange is used most often for this purpose [36].

Recent improvements of the conventional process have been described [37], especially in connection with schemes to separate and recover the dibasic acid byproducts [38]–[41]. The crude adipic acid is refined to varying degrees, depending upon the end use, but usually is recrystallized from water. Removal of impurities by refluxing in 60% nitric acid containing dissolved vanadium metal recently was claimed to produce high quality product [42].

4.2. Other Routes

In addition to the two-step air-nitric acid process for adipic acid synthesis from cyclohexane, several alternative routes have been investigated. One-step oxidations of cyclohexane with nitric acid [43], [44], nitrogen dioxide [45], or air have been described. Major developers of the latter process include Gulf Research and Development [46]–[48] and Asahi Chemical Ind. [49], [50], among others [51]. For example, cyclohexane is oxidized in one step to adipic acid (70–75% yield) in the presence of a cobalt acetate catalyst and acetic acid solvent [49]. Several recent BASF patents describe the synthesis of adipic acid via hydrocarboxylation [52] and the esters via stepwise carboalkoxylation of butadiene [53], [54], [55], usually with cobalt carbonyl and a nitrogen base. Related work was done at Texaco Development Corp. [56]. Hydrocarboxylation in the presence of a metal amalgam also was reported to give the free acid [57].

Research at Monsanto on the palladium halide catalyzed dicarbonylation of 1,4–disubstituted–2–butenes was reported in early 1984 [58]. The preferred process produces adipic acid from 1,4–dimethoxy–2–butene and palladium chloride at 100 °C after the resulting unsaturated dimethyl ester is hydrogenated and hydrolyzed.

Adipic acid can be produced from cyclohexene by ozonolysis [59], or by addition of a carboxylic acid and nitric acid oxidation of the resulting carboxylate ester [60]. Formation of adipic acid derivatives by electrolytic coupling of acrylates is described in [61].

5. Byproducts

Major byproducts of the nitric acid oxidation of KA include glutaric acid [*110-94-1*] and succinic acid [*110-15-6*], with minor amounts of pentanoic and hexanoic acids being formed. In commercial systems the nitric acid reaction medium contains high concentrations of glutaric and succinic acids, resulting from recycling the mother liquor after crystallization of adipic acid. A portion of this stream is diverted and processed to remove the byproduct acids and recover nitric acid and catalyst (Section 4.1.). Although the byproducts were discarded for many years, more recently they have been recovered (mostly as the esters) for a variety of uses.

Following removal of copper and vanadium by ion exchange and distillation of nitric acid and water, methanol often is added to convert the acids to their methyl esters. Then the esters are distilled to give a mixture or the individual esters [62]–[64]. Sometimes the acids are removed by distillation to produce a mixture of the acids and anhydrides, especially glutaric anhydride [*108-55-4*] and succinic anhydride [*108-30-5*] [65]–[69]. Separation of the individual acids by crystallization and by extraction with organic solvents has been described [70], [71]. A combination process involving a series of evaporations, crystallizations, and distillation was described recently [72].

Other means for separating the byproduct acids include addition of inorganic salts [73], a C_1-C_6 primary alkylamine [74], or urea [75], and extraction by a ketone solvent [76].

6. Quality Specifications

Commercial adipic acid is one of the purest chemicals manufactured on a large scale because of the stringent requirements of its major user, the synthetic fibers industry. Some of the material has been approved as a food additive by the U.S. Food and Drug Administration. Because essentially all adipic acid producers use a nitric acid oxidation process, impurity types are similar. Purity is affected mostly by variations in synthesis of the ketone–alcohol intermediate and in the extent of refining. Some typical specifications for the major quality parameters of food-grade adipic acid are: color, APHA equivalence (Hazen) 10 max., water 0.2% max., ash on ignition 10 ppm max., iron 1.0 ppm max., adipic acid content 99.6% min. [77].

Procedures for analysis of food-grade acid have been described in the literature [78]. General methods for water (Karl Fischer), color in methanol solutions (APHA), iron, and other metallic impurities in commerical acid also have been summarized [79]. Resin-grade acid frequently has limits for succinic (≈ 50 ppm) and caproic (≈ 30 ppm) acids, and for hydrocarbon oils (≈ 15 ppm). Carboxylic acids may be determined by gas chromatography of the esters or liquid chromatographic separation of the free acids [80]. Total nitrogen also may be determined by chemical reduction and distillation of ammonia from an alkaline solution. Hydrocarbon oil may be determined by IR analysis of a halocarbon extract of a solution of the salt.

7. Storage and Transportation

Adipic acid is conveyed pneumatically or mechanically from the drying equipment to the storage or shipping container. These may be aluminum or stainless steel railroad hopper cars, trucks, paper bags, or drums. Principal hazards in handling adipic acid are the danger of dust explosion (Chap. 2), and skin or mucous membrane irritation on exposure to the dust (Chap. 11).

Particle size control and flow characteristics also are important factors due to the tendency of acid containing excessive fines to cake during storage.

8. Derivatives

8.1. Adiponitrile

By far the most important derivative of adipic acid is hexanedinitrile, adiponitrile, 1,4–dicyanobutane, [*111-69-3*], M_r 108.14, bp 298 – 300 °C (at 101.3 kPa), 154 °C (at 1.3 kPa), fp 2.4 °C, n_D^{25} 1.4370, d_4^{25} 0.9599, an intermediate in the manufacture of the other major nylon 66 component, 1,6-hexanediamine (→ Hexamethylenediamine). The original production process involved conversion of the acid to the dinitrile by either liquid [81] or vapor phase dehydration [82] of the ammonium salt in the presence of phosphoric acid or boron-phosphorus catalysts. Several plants using this process have closed in recent years. Also shut down recently was the Celanese process for diamine based on ammonolysis of 1,6-hexanediol, which in turn was made by hydrogenation of adipic acid [83]. In 1948 Du Pont also introduced and for several years operated a process based on furfural [84].

The newer adiponitrile processes are based on butadiene and propene. From 1951 to 1983 Du Pont operated a butadiene chlorination process [85]. The intermediate 1,4-dichloro-2-butene was converted to 3-hexenedinitrile with sodium cyanide and then hydrogenated to adiponitrile. In 1965 Monsanto (United States) introduced a new process involving the electrolytic coupling of acrylonitrile [86]. This process or a variation has been used also in the United Kingdom and Japan. Du Pont began the direct hydrocyanation of butadiene in 1972 [87], and now the bulk of Du Pont's United States production and a joint venture with Rhône-Poulenc in France use this route. The process consists of a two-step hydrocyanation, catalyzed by nickel(0) phosphite complexes, and promoted by certain Lewis acids [88]–[90]. The mixture of isomeric pentenenitriles and methylbutenenitriles produced in the first step is isomerized to predominately 3- and 4-pentenenitrile [91]–[93]. Subsequent anti-Markovnikov addition of hydrogen cyanide to the pentenenitrile produces adiponitrile.

Other potential routes currently being investigated include one by ICI, in which acrylonitrile is catalytically dimerized over an organophosphinite or phosphonite to predominately 3-hexenedinitrile [94]–[96]. This material is hydrogenated to adiponitrile by standard methods. Another ICI development involves the hydrocyanation of butadiene over a copper halide catalyst to 3-pentenenitrile [97], followed by disproportionation to dicyanobutenes and butenes. Alternatively, dicyanobutene is produced directly by reacting butadiene, hydrogen cyanide, and oxygen in the presence of a copper-containing catalyst [98]. Finally, another dimerization route to adiponitrile involves the conversion of acrylonitrile to 2-methyleneglutaronitrile in the presence of zinc or cobalt complexes and a Lewis base [99]. The dimer is then hydrocyanated to 1,2,4-butanetricarbonitrile and dehydrocyanated to 3-hexenedinitrile [100].

Table 1. Boiling points of adipic acid esters

Ester		p, kPa	bp, °C
Monomethyl	[627-91-8]	1.3	158
Dimethyl	[627-93-0]	1.7	115
Monoethyl	[626-86-8]	0.9	160
Diethyl	[141-28-6]	1.7	127
Di-n-propyl	[106-19-4]	1.5	151
Di-n-butyl	[105-99-7]	1.3	165
Di-2-ethylhexyl	[103-23-1]	0.67	214
Di-n-nonyl	[151-32-6]	0.67	230
Di-n-decyl	[105-97-5]	0.67	244

8.2. Other Derivatives

Salts. Adipic acid forms alkali metal or ammonium salts that are water soluble and alkaline-earth metal salts that are only moderately soluble. Their solubilities in 100 g of water are: diammonium salt [3385-41-9] 40 g (14 °C), disodium salt [7486-38-6] 59 g of hemihydrate (14 °C), dipotassium salt [19147-16-1] 65 g (15 °C), calcium salt [22322-28-7] 4 g of monohydrate [18850-78-7] (13 °C), 1 g of anhydrous salt (100 °C), barium salt [60178-85-0] 12 g (12 °C), 7 g (100 °C). By far the most common salt is poly(1,6-hexanediammoniumhexanedioate), produced by interaction with 1,6-hexanediamine. This water-soluble salt, the precursor to nylon 66, is shipped readily or stored prior to its polyamidation by removal of water. The chemistry of this step has been reviewed [101].

Esters, Polyesters. Conversion to the ester constitutes the largest non-polyamide use of the acid. Esters made from long-chain alcohols are used as plasticizers and lubricants, whereas those from short-chain alcohols are used primarily as solvents. Monomethyl adipate, along with the diester, can be produced by refluxing adipic acid and methanol in the presence of an acid catalyst. Electrolysis of the salt of this monoester (Kolbe synthesis) produces dimethyl sebacate, another polyamide precursor. Boiling points of several esters are shown in Table 1. Melting points above 0 °C include the monomethyl ester 9 °C, the dimethyl ester 10.3 °C, and the monoethyl ester 29 °C. The esters dissolve readily in most organic solvents. Di-2-ethylhexyladipate is the most widely used adipate plasticizer. Other simple adipate plasticizers include the n-octyl, n-decyl, isodecyl, and isooctyl esters. More complex polymeric plasticizers, prepared from glycols, accounted for roughly half of adipic acid consumption for plasticizers in 1979. Low molecular mass polyester polyols having hydroxyl end groups are used as ingredients with polyisocyanates to produce polyurethane resins. Polyurethanes accounted for almost 4% of the total United States adipic acid consumption in 1979.

Anhydrides. The usual form of the anhydride produced on dehydrating adipic acid is the linear, polymeric form [2035-75-8]. Distillation of the polymeric anhydride is said to

Table 2. Adipic acid consumption, 10^3 t/a

	United States		Western Europe		Japan	
	1979	1981	1979	1981	1979	1981
Nylon 66 fiber	659	490	542	432	40	36
Nylon 66 resin	70	64	78	*	14	*
Plasticizers	27	20	77	16	11	12
Polyurethane resins	29	26	70	16	*	*
Miscellaneous	14	14	8	82	20	22

* data unavailable

produce the monomeric cyclic form, but this is very unstable and reverts readily to the linear, polymeric anhydride.

Amide. The diamide, $C_6H_{12}N_2O_2$ [628-94-4], mp 228 °C, is practically insoluble in cold water. It can be prepared from the dimethyl ester by treatment with concentrated ammonium hydroxide, or by heating the diammonium salt of adipic acid in a stream of ammonia. Other substituted amides can be prepared from amines by the usual synthetic methods.

9. Uses

More than 87% of worldwide adipic acid consumption is for manufacture of nylon 66 fibers and resins. Table 2 summarizes consumption in three major regions in the world. Some of the acid is used captively to produce adiponitrile, the other major nylon 66 intermediate.

Large quantities are converted to the esters for use in plasticizers, in lubricants, and in a variety of polyurethane resins.

The monomeric esters are important plasticizers for poly(vinyl chloride) and other resins, whereas polymeric esters are used when unusually high plasticizer levels are required. Polyurethane resins employing adipic acid are produced from polyisocyanates and polyester polyols (adipates). These are used in specialty foams, lacquers, adhesives, and surface coatings, and in spandex fibers for stretchwear. The plasticizer and polyurethane markets consume a slightly larger share of production in Western Europe than in the United States because of the greater competition of nylon 66 in the European textile fiber market.

Adipic acid is added to gelatins and jams as an acidulant and to other foods as a buffering or neutralizing agent. It is also added to modify properties of unsaturated polyesters for use in reinforced plastics and alkyd coatings. Polyamide-epichlorohydrin resins employing adipic acid are used to increase the wet strength of paper products. Other miscellaneous applications are in the insecticide, adhesives, tanning and dyeing, and textiles industries. The acid recently has been used as a buffer in flue gas desulfurization treatment in power plants [104].

Table 3. Worldwide adipic acid capacity as of January 1, 1983

Region	Capacity*	Major producer*
North America	905	Du Pont (619), Monsanto (272), Allied
Western Europe	841	Rhône-Poulenc (230), BASF (200), Imperial Chemical Industries (300), UCB-Ftal sa, Bayer, Chemische Werke Hüls, Montedison, Rhodiatoce**
Far East	70	Asahi (50), Honshu Chemicals, Kanto Denka Kogyo, Sumitomo Chemicals, Ube Industries
Other	142	

* Capacity, 10^3 t/a
** Now wholly owned by Montedison.

10. Economic Aspects

Capacities. Total worldwide annual capacity for adipic acid was 2.2×10^6 t/a in January 1980. United States capacity was 8.7×10^5 t/a, or 39% of the total, whereas Western Europe accounted for 48%, distributed mostly among the United Kingdom, France, and the Federal Republic of Germany. Imports and exports have not been a significant factor. In 1979 United States exports were only 1.2% of United States consumption [105]. Worldwide the total was 2%. From 1971 to 1979 United States consumption grew at an average annual rate of 3.8% but declined over the next 3 years. Growth rate for 1982–1987 was estimated at 4.5–6.0% per year [106]. Regional capacities are shown in Table 3, along with annual capacities for the major producing companies.

Production. Adipic acid production is dominated by nylon 66 fiber and resin manufacturers, hence the economic picture for the acid is strongly influenced by the markets for their materials. Fewer than 10% of United States production is sold on the merchant market, essentially for nonnylon uses. The ratio is higher in Western Europe and Japan. Synthesis of adiponitrile from the acid, once significant, now consumes only a minor fraction of total production. Consumption for nonnylon uses grew at an average annual rate of 3.0% from 1968 to 1979. Production costs closely paralleled raw material prices, which rose sharply beginning in 1973 reflecting the price of crude oil. As of early 1983 crude oil price cuts again were bringing down prices for petrochemicals [108]. Because of projected slow growth in the nylon 66 market, the supply/demand picture is expected to remain relatively constant for the foreseeable future [109]. Several marginal units were shut down in the late 1970s and early 1980s. Shipments of all types of synthetic fiber were down in 1982, with United States shipments of nylon fiber off 13% vs. 1981 [110].

Table 4. Toxicity data for adipic acid derivatives

Derivative	Oral LD_{50} (rat) mg/kg	Inhalation LC_{50} (rat, 4 h), mg/m^3	Other LD_{50}, mg/kg
Adiponitrile	300	1710	50 (scu[a], guinea pig)
Di-2-ethylhexyladipate	9110	–	900 (ivn[b], rat)
Dimethyl adipate	–	–	1809 (ipr[c], rat)
Adipamide	500	–	
Magnesium adipate	–	–	180 (ivn[b] mouse)

[a] subcutaneous,
[b] intraveneous,
[c] intraperitoneal

11. Toxicology and Occupational Health

Adipic acid is a minor irritant of low oral toxicity. The lowest published lethal dose (LDL_0) is 3600 mg/kg (rat, oral), LD_{50} 275 mg/kg (rat or mouse, intraperitoneal), LD_{50} 1900 mg/kg (mouse, oral) [111]. Some delayed body weight increases and changes in certain enzymes and in urea and chloride content in the blood were observed in chronic feeding tests [112]. No teratogenic activity was detected in studies with pregnant mice [113]. In metabolism studies with rats fed [^{14}C]-labeled adipic acid, both unchanged acid and normal metabolic products were detected in the urine [114], [115]. Aquatic toxicity is LC_{50} 88 mg/L (t =96 h).

Exposure of the mucous membranes (eyes, respiratory tract) produces irritation; prolonged exposure to the skin can be drying or irritating. In case of spills or leaks, personnel should be protected from inhalation or excessive skin contact. Dusting should be controlled, and static sparks should be avoided. Water may be used to flush the area.

Although no TLV or MAK has been established, the airborne exposure should be less than that of an organic nuisance dust: ACGIH (1979) 8 h TWA 10 mg/m^3 (total dust) and 8 h TWA 5 mg/m^3 (respirable dust) (OSHA TLV is 15 mg/m^3 for total dust). Toxicity data for representative types of adipic acid derivatives are shown in Table 4.

12. References

General References

[1] S. A. Cogswell: "Organic Chemicals A-B," in *Chemical Economics Handbook*, SRI International, Menlo Park, Calif., 1983, 608.5031A-608.5033F.
[2] M. Sittig: *Dibasic Acids and Anhydrides*, Noyes Development Corp, Park Ridge, N.J., 1966, p. 35–50.
[3] *Kirk-Othmer*, 3rd ed., vol. **1**, p. 510–531.
[4] *Ullmann*, 4th ed., vol. **7**, p. 106–113.

Specific References

[5] V. Luedeke in J. McKetta, W. Cunningham (eds.): *Encyclopedia of Chemical Processing and Design*, vol. **2,** Marcel Dekker Inc., New York 1977, p. 128–146.
[6] *Ullmann* 4th ed. vol. **7,** p. 106.
[7] W. Hentzchel, J. Wislicenus, *Liebigs Ann. Chem.* **275** (1983) 312.
[8] G. Vavon, A. Apchie, *Bull. Soc. Chim. Fr.* **43** (1928) 667.
[9] J. W. Hill, *J. Am. Chem. Soc.* **52** (1930) 4110.
[10] S. A. Cogswell: "Organic Chemicals A-B," in *Chemical Economics Handbook*, SRI International, Menlo Park, Calif., 1983, 608.5032A.
[11] Du Pont, US 2 557 282, 1951 (C. Hamblett, A. MacAlevy).
[12] Du Pont, US 2 703 331, 1953 (M. Goldbeck, F. Johnson).
[13] H. Godt, J. Quinn, *J. Am. Chem. Soc.* **78** (1956) 1461–1464.
[14] D. van Asselt, W. van Krevelen, *Recl. Trav. Chim. Pays Bas* **82** (1963) 51–56, 429–437, 438–449.
[15] D. van Asselt, W. van Krevelen, *Chem. Eng. Sci.* **18** (1963) 471–483.
[16] I. Y. Lubyanitskii, R. Minati, M. Furman, *Russ. J. Phys. Chem. (Engl. Transl.)* **32** (1962) 294–297.
[17] I. Y. Lubyanitskii, *Zh. Obshch. Khim.* **36** (1962) 3431
[18] I. Y. Lubyanitskii, *Zh. Prikl. Khim. (Leningrad)* **36** (1963) 819–823.
[19] *Ullmann*, 4th ed., vol. **7,** p. 107.
[20] BASF, US 3 564 051, 1971 (E. Haarer, G. Wenner).
[21] BASF, GB 1 092 603, 1969 (G. Riegelbauer, A. Wegerich, A. Kuerzinger, E. Haarer).
[22] Du Pont, US 3 359 308, 1967 (O. Sampson).
[23] ICI, US 3 754 024, 1973 (F. Foster, P. Hay).
[24] ICI, US 3 950 410, 1976 (J. Lopez–Merono).
[25] ICI, US 3 997 601, 1976 (P. Langley).
[26] ICI, GB 1 366 082, 1974 (J. Lopez–Merono).
[27] ICI, DE 2 435 387, 1973 (P. Langley).
[28] F. Hearfield, *Chem. Eng. (London)* 1980 no. 361, 625–627.
[29] Y. A. Lubyanitskii, SU 433 784, 1971.
[30] El Paso Products Co., US 3 673 245, 1972 (S. Mims).
[31] Chem. Werke Hüls, US 3 761 517, 1973 (H. Rohl, W. Eversmann, P. Hegenberg, G. Hellemanns).
[32] ICI, GB 1 510 397, 1976 (F. Foster, N. Hutchinson, D. Potter).
[33] Du Pont, US 2 713 067, 1955 (C. Hamblett).
[34] Vickers-Zimmer, US 3 476 804, 1969 (F. Bende, H. Vollinger, K. Pohl).
[35] Vickers-Zimmer, US 3 476 805, 1969 (H. Vollinger, K. Pohl, F. Bende).
[36] Monsanto, US 3 186 952, 1965 (D. Brubaker, D. Danly).
[37] Celanese, US 3 965 164, 1976 (J. Blay).
[38] Celanese, US 3 983 208, 1976 (J. Blay).
[39] ICI, GB 1 470 169, 1977 (B. Darlow, R. Chase, J. Peters).
[40] ICI, GB 1 480 480, 1977 (A. Bowman).
[41] Asahi, US 3 673 068, 1972 (M. Seko, A. Yomiyama, T. Miyake, H. Iwashita).
[42] BASF, DE 2 624 472, 1977.
[43] Du Pont, US 3 306 932, 1967 (D. Davis).
[44] Monsanto, US 3 654 355, 1972 (W. Mueller, C. Campbell, J. Hicks).

[45] Kogai Boshi Chosa, JP-Kokai 105 416, 1978 (W. Ando, I. Nakaoka).
[46] Gulf R & D, US 3 231 608, 1966.
[47] Gulf R & D, US 4 032 569, 1977 (A. Onopchenko, J. Schulz).
[48] Gulf R & D, US 4 263 453, 1981 (J. Schulz, A. Onopchenko).
[49] K. Tanaka, *Chem. Technol.* **4** (1974), no. 9, 555.
[50] Asahi, JP-Kokai 100 022, 1974 (K. Tanaka, S. Handa).
[51] Honshu Chem., JP-Kokai 33 891, 1979 (T. Abe, H. Kasamatsu, Y. Ayabe, Y. Isoda).
[52] BASF, US 3 876 695, 1975 (N. von Kutepow).
[53] BASF, US 4 169 956, 1979 US 4 171 451, 1979 (R. Kummer).
[54] BASF, US 4 258 203, 1981 (R. Platz, R. Kummer, H. Schneider).
[55] BASF, US 4 259 520, 1981 (R. Kummer, H. Schneider, F. Weiss).
[56] Texaco Dev. Corp., US 4 172 087, 1979 (J. Knifton).
[57] Montecatini Edison, US 3 686 299, 1972 (G. Carraro).
[58] *Chem. Eng. News* **62** (1984) no. 18, 28 – 29.
[59] Dainippon Ink. & Chem., JP-Kokai 32 245, 1982.
[60] Toray Ind., GB 1 402 480, 1975.
[61] Monsanto, GB 1 447 772, 1976 (C. Campbell, D. Danly, W. Mueller).
[62] El Paso Products Co., US 4 316 775, 1982 (W. Nash).
[63] El Paso Products Co., DE 3 043 051, 1982 (N. Cywinski).
[64] Du Pont, US 3 991 100, 1976 (S. Hochberg).
[65] ICI, US 4 191 616, 1980 (B. Baker).
[66] Allied Chem., FR 1 347 525, 1963 (J. Benfield, R. Belden).
[67] ICI, US 3 511 757, 1970 (W. Costain, B. Terry).
[68] BASF, US 3 564 051, 1971 (E. Haarer, G. Wenner).
[69] Du Pont, CA 707 340, 1965.
[70] Du Pont, US 3 338 959, 1967 (C. Sciance, L. Scott).
[71] Monsanto, US 3 329 712, 1967 (D. Danly, G. Whitesell).
[72] Monsanto, US 4 254 283, 1981 (G. Mock).
[73] Asahi, JP-Kokai 115 314, 1979 (J. Nishikido, A. Tomura, Y. Fukuoka).
[74] BASF, DE 3 002 256, 1981 (W. Rebofka, G. Heilen, W. Klink).
[75] Asahi, US 4 146 730, 1979 (J. Nishikido).
[76] Veba-Chemie, DE 2 309 423, 1974 (H. Heumann, W. Hilt, H. Liebing, M. Schweppe).
[77] E. I. Du Pont de Nemours Co., *Adipic Acid Product Bulletin E-18722,* Wilmington, Del., 1983, p. 1.
[78] National Academy of Sciences: *Food Chemicals Codex,* 2nd ed., National Academy Press, Washington, D.C., 1972, p. 21 – 22.
[79] R. Keller in F. Snell, C. Hilton (eds.): *Encyclopedia of Industrial Chemical Analysis,* vol. **4,** Wiley-Interscience, New York 1967, p. 408 – 423.
[80] R. Schwarzenbach, *J. Chromatogr.* **251** (1982) 339 – 358.
[81] Rhodiatoce, US 3 299 116, 1967 (R. Romani, M. Ferri).
[82] Du Pont, US 2 200 734, 1940. Monsanto, US 3 574 700, 1968 (R. Somich).
[83] Celanese, FR 1 509 288, 1968 (P. Volpe, W. Humphrey).
[84] J. Hardy in H. Simonds, J. Church (eds.): *The Encyclopedia of Basic Materials for Plastics,* Reinhold Publ. Co., New York 1967, p. 293.
[85] Du Pont, US 2 680 761, 1952 (R. Halliwell). Du Pont, US 2 518 608, 1947 (M. Farlow).
[86] M. M. Baizer, D. E. Danly, *Chem. Technol.* **10** (1980) no. 10, 161 – 164, 302 – 311.
[87] *Eur. Chem News* **23** (1973) no. 2, 17.

[88]   Du Pont, US 3 496 217, 1970 (W. Drinkard, R. Kassal).
[89]   Du Pont, US 3 496 218, 1970 (W. Drinkard).
[90]   Du Pont, US 3 766 237, 1973 (W. Drinkard).
[91]   Du Pont, US 3 526 654, 1970 (G. Hildebrand).
[92]   Du Pont, US 3 536 748, 1970 (W. Drinkard, R. Lindsey).
[93]   Du Pont, US 3 542 847, 1970 (W. Drinkard, R. Lindsey).
[94]   ICI, US 4 138 428, 1979 (J. Jennings, P. Hogan, L. Kelly).
[95]   ICI, US 4 316 857, 1982 (A. Gilbert).
[96]   ICI, US 4 059 542, 1977 (J. Jennings, L. Kelly).
[97]   ICI, US 4 210 558, 1980 (G. Crooks). ICI, US 4 088 672, 1978 (D. Waddan).
[98]   ICI, GB 2 077 260, 1981 (D. Waddan).
[99]   Halcon, US 3 954 831, 1976 (O. Onsager).
[100] Halcon, US 3 795 694, 1974 (O. Onsager).
[101] M. I. Kohan: *Nylon Plastics,* J. Wiley & Sons, New York 1973, p. 14–82.
[102] S. A. Cogswell: "Organic Chemicals A-B," in *Chemical Economics Handbook*, SRI International, Menlo Park, Calif., 1983, 608.5032 J,V,W.
[103] K. L. Ring, R. T. Gerry: "Organic Chemicals A-B," in *Chemical Economics Handbook*, SRI International, Menlo Park, Calif., 1980, 608.5032 H, 608.5033 K,L.
[104] *Chem. Eng.* **87** (1980) no. 3, 60.
[105] U.S. Dept. of Commerce, Bureau of Census: *U.S. Exports, Schedule B by Commodity,* U.S. Government Printing Office, Washington, D.C., 1980, EM 546.
[106] S. A. Cogswell: "Organic Chemicals A-B," in *Chemical Economics Handbook*, SRI International, Menlo Park, Calif., 1983, 608.5031D.
[107] S. A. Cogswell: "Organic Chemicals A-B," in *Chemical Economics Handbook*, SRI International, Menlo Park, Calif., 1983, 608.5032 C-F.
[108] *Chem. Week* **132** (1983) no. 9, 8.
[109] "Global Strategies for Man-Made Fibers in the 80's" *A. D. Little Report* (1981), 16.
[110] *Chem. Week* **132** (1983) no. 10, 20.
[111] J. M. Nielsen (ed.): *Material Safety Data Sheets,* vol. **1,** General Electric Co., Schenectady, N. Y., 1979 . no. 400.
[112] M. Krapotkina, *Gig. Tr. Prof. Zabol.* **5** (1981) 46–47.
[113] Food and Drug Research Labs Inc. (ed.): "Teratogenic Evaluation of FDA-71-50 (Adipic Acid)" *NTIS No. PB221802,* East Orange, N.J., 1972, p. 1.
[114] I. Rusoff, *Toxicol. Appl. Pharmacol.* **2** (1960) 316–330.
[115] D. Guest, G. Katz, B. Astill in G. Clayton, F. Clayton (eds.): *Patty's Industrial Hygiene and Toxicology,* 3rd ed., vol. **2C,** Wiley-Interscience, New York 1982, p. 4945.
[116] R. Tatken, R. Lewis, Sr. (eds.): *Registry of Toxic Effects of Chemical Substances,* vol. **1,** 1981–82 ed., U.S. Department of Health and Human Resources, U.S. Government Printing Office, Washington, D.C., 1983, p. 286–7.

Alcohols, Aliphatic

Individual keywords: → *Alcohols, Polyhydric;* → *Butanols;* → *Cyclohexanol and Cyclohexanone;* → *Ethanol;* → *2-Ethylhexanol ;* → *Fatty Alcohols;* → *Methanol;* → *Pentanols;* → *Propanols ; for Allyl Alcohol* → *Allyl Compounds*

JÜRGEN FALBE, Henkel KGaA, Düsseldorf, Federal Republic of Germany (Chap. 2–4)

HELMUT BAHRMANN, Ruhrchemie AG, Oberhausen, Federal Republic of Germany (Chap. 2–4)

WOLFGANG LIPPS, Naturwissenschaftlich-technische Akademie Prof. Dr. Grübler, Isny, Federal Republic of Germany (Chap. 2–4)

DIETER MAYER, Centre International de Toxicologie, Évreux, France (Chap. 5)

1.	Introduction	280	2.3.13. Other Processes	292
2.	Saturated Alcohols	280	2.4. Individual Alcohols	293
2.1.	Physical Properties	280	2.4.1. C_6 Alcohols	293
			2.4.2. C_7 Alcohols,	295
2.2.	Chemical Properties	283	2.4.3. C_8 Alcohols	298
2.3.	Production	284	2.4.4. C_9 Alcohols	300
2.3.1.	Synthesis from Carbon Monoxide and Hydrogen	285	2.4.5. C_{10} Alcohols	301
2.3.2.	Oxo Synthesis	285	2.4.6. Mixtures of Linear C_{12}–C_{18} Alcohols (Detergent Alcohols)	302
2.3.3.	Hydrogenation of Aldehydes, Carboxylic Acids, and Esters	286	2.4.7. C_{13}–C_{18} Isoalcohols	302
2.3.4.	Aldol Condensation of Lower Aldehydes and Hydrogenation of the Alkenals	287	2.5. Economic Aspects	303
			2.6. Quality Specifications	307
2.3.5.	Oxidation of Trialkylaluminum Compounds	287	2.7. Storage and Transportation	308
			3. Unsaturated Alcohols	308
2.3.6.	Oxidation of Saturated Hydrocarbons	288	4. Alkoxides	311
			4.1. Properties	311
2.3.7.	Hydration of Olefins	290	4.2. Preparation	312
2.3.8.	Homologation of Alcohols	290	4.3. Uses	314
2.3.9.	Reppe Process	291	5. Toxicology	314
2.3.10.	Hydrocarboxymethylation	291	6. References	316
2.3.11.	Fermentation	292		
2.3.12.	Guerbet Alcohols	292		

1. Introduction

Industrially, the most important alcohols are methanol, ethanol, 1-propanol, 1-butanol, 2-methyl-1-propanol (isobutyl alcohol), the plasticizer alcohols (C_6-C_{11}), and the fatty alcohols ($C_{12}-C_{18}$), used for detergents. They are prepared mainly from synthesis gas alone (methanol), from olefins via the oxo synthesis, or by the Ziegler process.

Apart from the applications mentioned above, alcohols are used as solvents and diluents for paints (mainly C_1-C_6 alcohols) [4], as intermediates in the manufacture of esters and a whole range of organic compounds, as flotation agents, as lubricants, and in recent times increasingly as fuel or fuel additives, e.g., methanol, ethanol, *tert*-butyl alcohol.

For industrial purposes, isomeric mixtures often are preferred because the pure alcohols are too expensive. Moreover, mixtures of alcohols with differing numbers of carbon atoms can be advantageous for certain purposes. Therefore, the amounts of alcohol mixtures available on the market are similar to the quantities of the pure, individual alcohols.

2. Saturated Alcohols

2.1. Physical Properties

Some physical properties of saturated C_1-C_{26} alcohols are given in Table 1. Up to C_{10}, the straight-chain alcohols are colorless liquids at room temperature with characteristic odors. The higher alcohols are solid, waxy substances. Because of hydrogen bonding, the boiling points are considerably higher for alcohols, especially lower-molecular-mass ones, than for the corresponding hydrocarbons. With increasing molecular mass, however, the influence of the hydroxyl group becomes less. For example, the degree of association at room temperature is 3.17 for methanol but only 1.12 for decanol.

Methanol, ethanol, and the propanols are completely miscible with water. With increasing molecular mass, the solubility decreases significantly. For example, the solubility of 1-butanol in water is 7.7 %, and that of decanol less than 0.01 % [5]. The data for some azeotropic mixtures with water are compiled in Table 2.

The temperature dependence of vapor pressure, heat of vaporization, specific heat, density, viscosity, surface tension, and thermal conductivity is given in [6] for methanol, ethanol, propanol, and butanol. For nomograms to determine the thermodynamic data of straight-chain alcohols, see [7]. The physical data of alcohols employed as solvents are summarized in [8], [9].

Table 1. Physical properties of saturated alcohols

IUPAC Name		Other common name	Formula	M_r	bp, °C	mp, °C	n_D^{20}	d_4^{20}
Methanol	[67-56-1]	methyl alcohol	CH_3OH	32.04	64.7	−97.8	1.3285	0.7910
Ethanol	[64-17-5]	ethyl alcohol	CH_3CH_2OH	46.07	78.32	−114.5	1.3614	0.7893
1-Propanol	[71-23-8]	propyl alcohol	$CH_3CH_2CH_2OH$	60.10	97.2	−126.2	1.3859	0.8035
2-Propanol	[67-63-0]	isopropyl alcohol	$CH_3CH(OH)CH_3$	60.10	82.4	−87.8	1.3771	0.7850
1-Butanol	[71-36-3]	butylalcohol	$CH_3(CH_2)_2CH_2OH$	74.12	117.7	−89.3	1.3991	0.8098
2-Methyl-1-propanol	[78-83-1]	isobutyl alcohol	$(CH_3)_2CHCH_2OH$	74.12	107.9	−107.9	1.3959	0.8027
2-Butanol	[78-92-2]	sec-butyl alcohol	$CH_3CH_2CH(OH)CH_3$	74.12	99.5	−114.7	1.3972	0.8065
2-Methyl-2-propanol	[75-65-0]	tert-butyl alcohol	$(CH_3)_3COH$	74.12	82.55	25.6	1.3841	0.7867
1-Pentanol	[71-41-0]	amyl alcohol	$CH_3(CH_2)_3CH_2OH$	88.15	137.8	−78.5	1.4100	0.8150
2-Pentanol	[6032-29-7]	sec-amyl alcohol	$CH_3(CH_2)_2CH(OH)CH_3$	88.15	119.3		1.4053	0.8090
3-Pentanol	[584-02-1]	–	$CH_3CH_2CH(OH)CH_2CH_3$	88.15	115.6	−75	1.4098	0.8218
2-Methyl-1-butanol	[137-32-6]	–	$CH_3CH_2CH(CH_3)CH_2OH$	88.15	128	<−70	1.4098	0.816
3-Methyl-1-butanol	[123-51-3]	isopentyl alcohol	$CH_3CH(CH_3)CH_2CH_2OH$	88.15	131.4	−117.2	1.4078	0.812
2-Methyl-2-butanol	[75-85-4]	tert-amyl alcohol	$CH_3CH_2C(OH)(CH_3)CH_3$	88.15	101.8	−11.9	1.4052	0.809
3-Methyl-2-butanol	[598-75-4]	–	$CH_3CH(CH_3)CH(OH)CH_3$	88.15	112.5		1.4095	0.819
2,2-Dimethyl-1-propanol	[75-84-3]	neopentyl alcohol	$CH_3C(CH_3)_2CH_2OH$	88.15	113.4	53.0		0.812
1-Hexanol	[111-27-3]	hexyl alcohol	$CH_3(CH_2)_4CH_2OH$	102.18	157.1	−44.6	1.4178	0.8136
2-Methyl-1-pentanol	[105-30-6]	2-methylpentyl alcohol	$CH_3(CH_2)_2CH(CH_3)CH_2OH$	102.18	148		1.4190	0.8254
4-Methyl-1-pentanol	[626-89-1]	–	$(CH_3)_2CH(CH_2)_2CH_2OH$	102.18	152.3		1.4134^a	0.8110^a
4-Methyl-2-pentanol	[108-11-2]	methylamyl alcohol	$(CH_3)_2CHCH_2CH(OH)CH_3$	102.18	131.8	−90	1.4113	0.8066
2-Ethyl-1-butanol	[97-95-0]	2-ethylbutyl alcohol	$CH_3CH_2CH(C_2H_5)CH_2OH$	102.18	146.5	−114	1.4224	0.8348
1-Heptanol	[111-70-6]	heptyl alcohol	$CH_3(CH_2)_5CH_2OH$	116.20	176	−35	1.4233	0.8221
2-Heptanol	[543-49-7]	–	$CH_3(CH_2)_4CH(OH)CH_3$	116.20	158–160		1.4213	0.8173
3-Heptanol	[589-82-2]	–	$CH_3(CH_2)_3CH(OH)CH_2CH_3$	116.20	156.2	−70	1.4222	0.8210
4-Heptanol	[589-55-9]	–	$CH_3(CH_2)_2CH(OH)(CH_2)_2CH_3$	116.20	156	−41.5	1.4199	0.8183
2,4-Dimethyl-3-pentanol	[600-36-2]	–	$CH_3-CH(CH_3)-CH(OH)-CH(CH_3)-CH_3$	116.20	140		1.4246	0.8294

Saturated Alcohols

Table 1. (continued)

IUPAC Name		Other common name	Formula	M_r	bp, °C	mp, °C	n_D^{20}	d_4^{20}
1-Octanol	[111-87-5]	octyl alcohol	$CH_3(CH_2)_6CH_2OH$	130.23	195.15	−16.3	1.4300	0.827
2-Octanol	[123-96-6]	capryl alcohol	$CH_3(CH_2)_5CH(OH)CH_3$	130.23	179	−38.6	1.4260	0.8205
2-Ethyl-1-hexanol	[104-76-7]	2-ethylhexyl alcohol	$CH_3(CH_2)_3CH(C_2H_5)CH_2OH$	130.23	184.7	−75	1.4315	0.8329
3,5-Dimethyl-1-hexanol	[13501-73-0]	–	$(CH_3)_2CHCH_2CH(CH_3)CH_2CH_2OH$	130.23	182.5		1.4250	0.8297
2,2,4-Trimethyl-1-pentanol	[123-44-4]	–	$(CH_3)_2CHCH_2C(CH_3)_2CH_2OH$	130.23	168	−70	1.4300	0.839
1-Nonanol	[143-08-8]	nonyl alcohol	$CH_3(CH_2)_7CH_2OH$	144.26	213.5	−5	1.4323	0.8271
5-Nonanol	[623-93-8]	–	$CH_3(CH_2)_3CH(OH)(CH_2)_3CH_3$	144.26	193–194	−36	1.4299	0.8356
3,5-Dimethyl-4-heptanol	[19549-79-2]	–	$C_2H_5CH(CH_3)CH(OH)CH(CH_3)C_2H_5$	144.26	171		1.4330[b]	0.836[b]
2,6-Dimethyl-4-heptanol	[108-82-7]	diisobutyl carbinol	$CH_3CH(CH_3)CH_2CH(OH)CH_2CH(CH_3)CH_3$	144.26	178	−65	1.4231	0.8121
3,5,5-Trimethyl-1-hexanol	[3452-97-9]	–	$CH_3C(CH_3)_2CH_2CH(CH_3)CH_2CH_2OH$	144.26	194	−70	1.4300[a]	0.8236[a]
1-Decanol	[112-30-1]	decyl alcohol	$CH_3(CH_2)_8CH_2OH$	158.29	232.9	6.4	1.4359	0.8320
1-Undecanol	[112-42-5]	undecyl alcohol	$CH_3(CH_2)_9CH_2OH$	172.31	245	14.3	1.4392	0.8298
1-Dodecanol	[112-53-8]	lauryl alcohol	$CH_3(CH_2)_{10}CH_2OH$	186.34	259	23.8	1.4428	0.8306[a]
2,6,8-Trimethyl-4-nonanol	[123-17-1]	–	$(CH_3)_2CHCH_2CH(CH_3)CH_2CH(OH)CH_2CH(CH_3)_2$	186.34	225	−60	1.4345	0.8193
1-Tridecanol	[112-70-9]	tridecyl alcohol	$CH_3(CH_2)_{11}CH_2OH$	200.33	276	30.6	1.4475	0.8454
1-Tetradecanol	[112-72-1]	myristyl alcohol	$CH_3(CH_2)_{12}CH_2OH$	214.39	170–173[d]	38	1.4358[c]	0.8165
1-Pentadecanol	[629-76-5]	pentadecyl alcohol	$CH_3(CH_2)_{13}CH_2OH$	228.42	170[e]	44		0.8215[c]
1-Hexadecanol	[124-29-8]	cetyl alcohol	$CH_3(CH_2)_{14}CH_2OH$	242.45	177[e]	49	1.4392[f]	0.8157[f]
1-Heptadecanol	[1454-85-9]	margaryl alcohol	$CH_3(CH_2)_{15}CH_2OH$	256.48	191[e]	54		
1-Octadecanol	[112-92-5]	stearyl alcohol	$CH_3(CH_2)_{16}CH_2OH$	270.50	210[g]	57.6–58.0		0.8124
1-Nonadecanol	[1454-84-8]	–	$CH_3(CH_2)_{17}CH_2OH$	284.53		62		
1-Eicosanol	[629-96-9]	eicosanyl alcohol	$CH_3(CH_2)_{18}CH_2OH$	298.56	251[e]	66		
1-Hexacosanol	[506-52-5]	ceryl alcohol	$CH_3(CH_2)_{24}CH_2OH$	382.72	305[h]	79.5		

[a] 25 °C; [b] 18 °C; [c] 50 °C; [d] 26.7 mbar; [e] 13.3 mbar; [f] 60 °C; [g] 20 mbar; [h] 26.7 mbar.

2.2. Chemical Properties

Reactions of alcohols can be characterized by cleavage of the O-H bond or the C-O bond either homolytically or ionically. The chemical properties of greatest industrial importance are as follows.

Oxidation and Dehydrogenation. Under normal conditions alcohols are stable. Oxidation with chemical agents, such as chromic acid or permanganate, leads to a variety of products depending on the nature of the alcohol. Primary alcohols are oxidized first to aldehydes and then to carboxylic acids; secondary alcohols are oxidized to ketones. Catalytic oxidation or dehydrogenation of primary and secondary alcohols on copper, silver, iron, molybdenum, etc., catalysts lead to the formation of aldehydes and ketones. In the Oppenauer oxidation a secondary alcohol is dehydrogenated by an excess of a ketone, e.g., acetone, in the presence of aluminum isopropoxide:

$$R^1-CHOH-R^2 + CH_3-CO-CH_3 \underset{}{\overset{Al(OC_3H_7)_3}{\rightleftarrows}} R^1-CO-R^2 + CH_3-CHOH-CH_3$$

Reduction. With hydrogen iodide or zinc and hydrochloric acid, alcohols are converted to hydrocarbons. Catalytic hydrogenolysis is especially successful with benzyl alcohols.

Dehydration. Water can be split from alcohols by heating them in the presence of strong acid or by passing them over aluminum oxide, silicic acid, or synthetic zeolites. As a rule, not only those products that result from a β-elimination are formed, but also compounds with an isomerized double bond. The isomerization can be suppressed by the addition of amines. Tertiary alcohols can be dehydrated more easily than secondary or primary alcohols.

Under less severe conditions symmetric ethers are formed from alcohols in an intermolecular reaction:

$$2\ ROH \longrightarrow ROR + H_2O$$

The same compounds that accelerate the intramolecular dehydration are suitable as catalysts.

Alcohols as Alkylation Reagents. Alcohols react with ammonia and amines to form *N*-alkyl or *N,N*-dialkylamines. Aromatic hydrocarbons are alkylated by alcohols in the presence of Friedel-Crafts catalysts.

Esterification. In the presence of acid catalysts, alcohols react with organic or inorganic acids, acid chlorides, or anhydrides to form esters ($\rightarrow$ Esters, Organic).

Addition Reactions. Alcohols add to aldehydes and ketones to form acetals ($\rightarrow$ Aldehydes, Aliphatic and Araliphatic). Alkylpolyglycol ethers are obtained with alkylene oxides. The addition to acetylene gives vinyl ether; addition to olefins yields mixed ethers.

For further reactions of alcohols see [10].

Table 2. Azeotropic mixtures with water [9]

Alcohol	bp of alcohol, °C	bp of azeotrope, °C	Water in azeotrope, wt%
Ethanol	78.32	78.174	4.0
1-Propanol	97.2	87.7	28.3
2-Propanol	82.4	80.3	12.6
1-Butanol	117.7	92.3	37.0
Isobutyl alcohol	107.9	89.9	33.2
sec-Butyl alcohol	99.4	87.5	27.3
tert-Butyl alcohol	82.5	79.9	11.76
1-Pentanol	138.0	95.8	54.4
2-Pentanol	119.3	91.7	36.5
3-Pentanol	115.4	91.7	36.0
2-Methyl-1-butanol	128.0	93.8	41.5
3-Methyl-1-butanol	132.0	95.15	49.6
2-Methyl-2-butanol	101.8	87.35	27.5
3-Methyl-2-butanol	112.5	91.0	33
1-Hexanol	157.1	97.8	75
1-Heptanol	176.2	98.7	83
1-Octanol	195.15	99.4	90
2-Ethylhexanol	184.7	99.1	80.0

2.3. Production

The following processes have been realized on an industrial scale:

1) Synthesis from carbon monoxide and hydrogen (C_1)
2) Oxo synthesis (mostly combined with hydrogenation of the initially formed aldehydes; $C_3 - C_{20}$)
3) Hydrogenation of aldehydes, carboxylic acids, or esters
4) Aldol condensation of lower aldehydes and hydrogenation of the alkenals ($C_3 \rightarrow C_6$, $C_4 \rightarrow C_8$, $C_8 \rightarrow C_{16}$)
5) Oxidation of trialkylaluminum compounds (Ziegler process)
6) Oxidation of saturated hydrocarbons
7) Hydration of olefins ($C_2 - C_4$)
8) Homologation of alcohols
9) Hydrocarbonylation by the Reppe process
10) Hydrocarboxymethylation
11) Fermentation processes ($C_2 - C_5$)
12) Guerbet process

The most important industrial processes are the methanol synthesis, with an annual production of ca. 12×10^6 t, and the oxo synthesis, with ca. 4.5×10^6 t. However, the hydration of ethylene and propene to ethanol and 2-propanol, and the oxidation of trialkylaluminum compounds (Alfol process or Ziegler process) also have achieved considerable commercial significance. Fermentation, especially for the production of ethanol, has become important again in certain regions because of the increased price of oil.

2.3.1. Synthesis from Carbon Monoxide and Hydrogen

On a commercial scale only methanol is prepared from synthesis gas (→ Methanol). Synol, isobutylol, oxyl, and similar processes led to a mixture of oxygen-containing compounds with alcohols as the main components [11], [12]. These methods are no longer used in the Western World. Newer developments also produce oxygen-containing compounds [13] or alcohol mixtures [14].

In the IFP (Inst. Français du Pétrole) process for higher alcohols [14], highly activated catalysts are used under low-pressure methanol synthesis conditions. The catalyst contains mixed oxides of copper and cobalt plus at least one other metal (Al, Ce, Cr, Fe, La, Mn, Pr, Nd, Y, or Zn) and at least one Group I or II metal compound. The higher alcohol content can be varied from 20–50 wt% by changing the catalyst composition. A number of byproducts, such as hydrocarbons, esters, and ketones, also are produced.

However, none of these newer processes are used in industry.

Small amounts of higher alcohols are formed as byproducts of the Fischer-Tropsch synthesis [15].

2.3.2. Oxo Synthesis

Alcohols in the range C_3–C_{20} can be prepared by the oxo synthesis, in which olefins react with synthesis gas to form aldehydes, which in turn are hydrogenated.

One particular version of the oxo synthesis is the *Shell process*; the strong hydrogenating activity of the catalyst, $HCo(CO)_3PR_3$, leads to the direct hydrogenation in the oxo reactor of the initially formed aldehyde [16]:

$$R-CH=CH_2 + CO + 2H_2 \xrightarrow{cat.} R-CH_2CH_2CH_2OH$$

The process was first applied commercially on a propene basis in 1963 and used for the manufacture of 1-butanol and 2-ethylhexanol [17]. In 1965 it was developed further to produce detergent alcohols. The process currently is used in the United States and various other countries primarily to make higher alcohols.

Planned or actual expansions of the Shell process capacity in the United States (previously 250 000 t/a) by 46 000 t/a, of surfactant alcohols in the Federal Republic of Germany by 56 000 t/a, and of Mitsubishi's in Japan by 70 000 t/a [18], will increase world capacity to some 470 000 t/a.

The Shell process has the advantage that olefins with an internal double bond can be hydroformylated, because under the reaction conditions, isomerization of the double bond takes place, and α-olefins are formed. For example, ω-olefin fractions obtained from ethylene by the "SHOP process" of Shell [19] can be converted with synthesis gas to alcohols. The alcohol mixtures formed consist of up to 80% linear compounds and are used in the plasticizer and detergent fields.

2.3.3. Hydrogenation of Aldehydes, Carboxylic Acids, and Esters

Aldehydes can be hydrogenated in the presence of homogeneous or heterogeneous catalysts [20], [21]. Homogeneous systems are advantageous only if sulfur-containing starting materials, which would poison the heterogeneous catalysts, are used or if the hydrogen for the hydrogenation contains carbon monoxide.

Generally, however, heterogeneous catalysts are preferred. These are effective both in the gas phase at temperatures of 90–180 °C and pressures of 25 bar and in the liquid phase at 80–220 °C and pressures up to 300 bar. The hydrogenation temperature applied in industrial processes represents a compromise between the best possible energy utilization and high catalyst lifetimes.

For continuous processes, catalysts in fixed-bed systems are favored. The aldehyde, either as vapor strongly diluted with excess hydrogen [22] or as liquid together with hydrogen [23], is fed through the high-pressure pipe containing the catalyst bed. To remove the heat generated by the reaction, the hydrogen is circulated through a heat exchanger. The especially important hydrogenation of 2-ethyl-2-hexenal usually is carried out in a single step on a nickel-containing catalyst [24]. Similar one-step processes have been suggested [25]. However, two-step processes are also common, whereby the main hydrogenation takes place in the gas phase (e.g., on copper-containing catalysts), and the second in a liquid phase or trickle-bed process (e.g., on nickel-containing catalysts).

The catalysts usually are supported on aluminum oxide or silica gel. In addition to nickel and copper, also zinc, chromium, and combinations of these metals have been used successfully as catalysts [26].

For the manufacture of fatty alcohols, the corresponding carboxylic acid esters are hydrogenated (for details → Fatty Alcohols). Starting materials are natural fats and oils, which are first transesterified to the methyl esters and then reduced to the alcohols, either with sodium (Bouveault-Blanc reduction) or by catalytic hydrogenation. The reduction with sodium enables the preparation of unsaturated fatty alcohols from the esters of unsaturated fatty acids.

The hydrogenation of fatty acids and fatty acid esters requires more drastic conditions than the aldehyde hydrogenation. The process operates continuously or discontinuously and makes use of copper–chromium oxide catalysts (Adkins catalysts), which are either suspended or placed in a fixed bed. Reaction temperatures of 240–300 °C and pressures of 200–300 bar are normal. By modifying the catalysts (e.g., by addition of cadmium) unsaturated fatty acid esters can be converted directly to unsaturated fatty alcohols [27].

2.3.4. Aldol Condensation of Lower Aldehydes and Hydrogenation of the Alkenals

In industry, the only source of aldehydes for the aldol condensation is the oxo synthesis [21]. After the isoaldehydes and byproducts are removed, the condensation is catalyzed by acids or bases. Because the reactivity of each aldehyde depends on the chain length and the degree of branching, the reaction conditions are adapted according to the individual compound. The alkenals, formed from the aldols by elimination of water, are hydrogenated over heterogeneous catalysts. As a rule the same catalysts are used as for the hydrogenation of oxo aldehydes. 2-Ethylhexanol, 2-methylpentanol, and limited amounts of highly branched, isomeric C_{16} and C_{18} alcohols are prepared by this method. The so-called aldox process, in which the aldehyde mixture formed in the oxo synthesis is subjected to aldol condensation in the oxo reactor, has not become established because of the presence of mixed aldols in the final products; however, some mixed aldols of acetaldehyde with higher aldehydes are important.

2.3.5. Oxidation of Trialkylaluminum Compounds

Ethylene can add to triethylaluminum to form a mixture of trialkylaluminum compounds of higher molecular mass [28]. These products can be oxidized with air to the corresponding aluminum alkoxides, which are then hydrolyzed to a mixture of linear primary alcohols with the same number of carbon atoms as the alkyl groups of the trialkylaluminum components [1]–[3]:

$$-\overset{|}{Al}-C_2H_5 + x\,C_2H_4 \longrightarrow -\overset{|}{Al}-(CH_2CH_2)_x-C_2H_5$$
$$\xrightarrow{O_2} -\overset{|}{Al}-O-(CH_2CH_2)_x-C_2H_5$$
$$\xrightarrow{H_2O} HO-(CH_2CH_2)_x-C_2H_5 + -\overset{|}{Al}-OH$$

On the basis of this reaction, called the Ziegler process, two commercial processes have been developed, one by Conoco, in operation in the United States since 1962 and in the Federal Republic of Germany by Condea Chemie (Conoco and Deutsche Texaco) since 1964; and the other by the Ethyl Corp., in operation since 1965. The principal differences between the two processes are the chain-length distribution and the linearity of the alcohols produced, as well as the technical process characteristics for the generation and control of this distribution.

Table 3. Composition of the alcohol mixtures from the Ziegler process

C no.	Conoco	Ethyl Corp.
6	9.6	1.4
8	16.9	3.2
10	20.7	7.7
12	19.4	34.5
14	15.1	26.3
16	9.8	16.7
18	5.3	8.9
20	3.2	1.3

Alfol Alcohol Process (Conoco Process). The chain-growth reaction is carried out at a temperature as low as possible in order to prevent displacement reactions that lead to the formation of olefins. The chain-length distribution corresponds to a Poisson curve. The resulting alcohols are practically 100% linear. A broad range of alcohols ($C_2 - C_{28}$) is typical for the Alfol alcohol process. Although the process can be varied to either increase or decrease the chain lengths of the manufactured alcohols, the distribution pattern remains the same.

Process with Controlled Linear Chain Growth. The Ethyl Corp. has successfully developed a process which forms predominantly C_{12} and C_{14} alcohols. The alcohols are up to 95% linear.

The product distributions obtained in the Conoco and the Ethyl Corp. processes are compared in Table 3.

2.3.6. Oxidation of Saturated Hydrocarbons

Bashkirov Oxidation. The oxidation of aliphatic hydrocarbons with air in the presence of boric acid gives boric acid esters in high yield. These are hydrolyzed in a second step to secondary alcohols in which the hydroxyl groups are distributed statistically along the molecular chain [29]–[31].

The reaction passes through an intermediate secondary hydroperoxide, for the reaction mechanism see [31], [32]. Normally a mixture of *n*-hydrocarbons with chain lengths between 10 and 16 is used as feedstock. The oxidation is carried out in the liquid phase at 150–170 °C in the presence of 4–5 wt% metaboric acid. A nitrogen-oxygen mixture (ca. 3.5% O_2) is used at normal or slightly elevated pressure. In newer plants amines are added as co-catalysts [31]. The resulting metaboric acid esters or boroxines are resistant to oxidation and thermally stable. To obtain economically acceptable selectivities of 80–85%, the level of conversion must be held below 20%.

Starting material and oxidation byproducts are removed by flash evaporation and are cleaned in alkaline and water scrubbers; the *n*-hydrocarbons are recycled [30]. The metaboric acid esters at the bottom of the flash column are hydrolyzed by the addition of small amounts of water at 80–100 °C. Impurities (mostly carbonyl compounds) are removed by alkaline and water washes. After fractional distillation, alcohols are obtained with a purity greater than 98%. Processing concludes with "hydrofinishing" (hydrogenation over heterogeneous nickel catalysts) to remove colored and odorous substances.

The orthoboric acid in the aqueous solution is converted to metaboric acid by dehydration. Recovery also is possible by crystallization. In more recent process variants the boric acid concentrate is mixed with the n-hydrocarbons and the mixture dehydrated, preventing agglomerations and stoppages. The dried slurry is returned to the oxidation reactor.

In the 1950s BASHKIROV developed the original German work into a commercial process [28], [33]. The first plant began production in Shebekino/Belgorod, USSR, in 1959 [34]. Plants are operating today in the Soviet Union [34] and in Japan [31]. The Union Carbide (UCC) plant in the United States was closed in 1977 [31].

The secondary alcohols produced in this process usually are converted to alkylphenol ethoxides and used as detergents. The costly recirculation of hydrocarbons and boric acid as well as disadvantages in the application of secondary alcohols [35], however, have prevented this process from achieving any great significance in the Western World.

A further important use of boric acid-catalyzed oxidation is the preparation of cyclohexanol and cyclohexanone from cyclohexane [36] (intermediate products in the manufacture of caprolactam, adipic acid, and phenol), and of cyclododecanol (intermediate product in the nylon 12 synthesis) and cyclododecanone from cyclododecane [37]; → Cyclohexanol and Cyclohexanone.

Oxidation with Alkyl Hydroperoxides. Alcohols also can be obtained by hydroxylation of alkanes with alkyl hydroperoxides [38]. Iron porphyrins are especially useful as catalysts.

$$RH + R'OOH \xrightarrow{cat.} ROH + R'OH$$

Alcohols from Fatty Acids Produced by Hydrocarbon Oxidation. The oxidation of hydrocarbons with air in the presence of manganese catalysts leads to a complex mixture of reaction products [39]. The initially formed alcohols are oxidized further to ketones and acids, → Fatty Acids. At present the process is operated mainly in the Soviet Union. The production of fatty acids in this way has roughly the same order of magnitude as that of fatty acids from native raw materials in the United States [3].

Processing of the reaction mixture is difficult, because not all byproducts can be removed. Therefore, these fatty acids do not reach the same standards of quality as natural fatty acids.

Alcohols are formed from fatty acids by esterification of the fractionated raw acids (mostly in the $C_{10}-C_{15}$ range) with methanol or butanol and subsequent hydrogenation. With linear hydrocarbons as starting materials, linear alcohols can be produced; however, depending on the catalyst, the specific hydrocarbon, and the reaction conditions, the alcohol mixture contains 5–15% branched-chain alcohols. Owing to the impurities present in the fatty acids, the alcohols prepared from them contain odorous substances, so their range of applications is limited.

According to estimates [3], about 10% of these synthetic fatty acids are processed to alcohols. The annual production in Eastern Block countries is presumed to be at least 50 000 t/a.

2.3.7. Hydration of Olefins

A common method for the production of lower alcohols is the hydration of alkenes. In accordance with Markovnikov's rule, secondary and tertiary alcohols are formed (except in the case of ethylene).

$$R-CH=CH_2 + H_2O \xrightarrow{H^+} R-CHOH-CH_3$$

The rate of this reaction is determined by the stability of the intermediate carbenium ion (tertiary > secondary > primary). Therefore the hydration of isobutene proceeds at room temperature in the presence of low H^+ ion concentrations owing to the relative stability of the intermediate tertiary carbenium ion. The hydration of ethylene, in contrast, requires elevated temperatures and pressures [10].

Industrially two variants of the hydration reaction are used. In the so-called *indirect process*, the liquid-phase reaction takes place in two steps. In the first, the olefin reacts with sulfuric acid to form mono- and dialkylsulfates, which, after dilution with water, are hydrolyzed to the alcohol. However, in order to recycle the sulfuric acid, a costly reconcentration is necessary.

In the *direct process*, hydration occurs in the gas phase. Because the reaction is exothermic and is accompanied by a reduction in volume (2 mol of reactant form 1 mol of product), the alcohol formation is favored by high pressure and low temperature. Because the conversion is incomplete, a costly gas recycle is necessary. Phosphoric acid-containing materials, e.g., celite, are efficient catalysts. Recently, however, increased use has been made of ion exchangers.

The primary use of direct hydration is for the preparation of ethanol from ethylene (→ Ethanol) and of isopropyl alcohol from propene (→ Propanols). It is also important in the production of 2-butanol from a mixture of 1-butene and 2-butene (raffinate II) and of *tert*-butyl alcohol from isobutene.

2.3.8. Homologation of Alcohols

Homologation is the reaction of alcohols with synthesis gas in the presence of complex, multicomponent catalyst systems. Depending on reaction conditions the products are aldehydes or alcohols containing one CH_2 group more than the starting materials [40]–[43]:

$$RCH_2OH + CO + H_2 \xrightarrow{cat.} RCH_2CHO + H_2O$$
$$RCH_2OH + CO + 2H_2 \xrightarrow{cat.} RCH_2CH_2OH + H_2O$$

Although the reaction was originally conceived for the synthesis of ethanol from methanol [44], [45], the scope has been extended to include the production of homologous aldehydes (acetaldehyde from methanol), carboxylic acids (propionic acid from acetic acid), carboxylic acid esters (ethyl acetate from methyl acetate), as well as the synthesis of styrene (via the homologation of benzyl alcohol to 2-phenylethanol with subsequent dehydration) [28], [31].

So far, homologation has not been used industrially because conversion and selectivity, despite considerable advances [28], are still insufficient and because there are problems with recycling the complex homologation catalysts.

2.3.9. Reppe Process

The Reppe hydrocarbonylation of olefins with carbon monoxide and water and using ammonium salts of tetracarbonyldihydrido iron as catalyst leads to alcohols [46]:

$$C_3H_6 + 3CO + 2H_2O \xrightarrow{cat.} C_4H_9OH + 2CO_2$$

As in the oxo synthesis, branched-chain products also are formed (molar ratio of linear to branched-chain alcohols is ca. 9:1).

Propene reacts at 90–110 °C and 5–20 bar to form butanols with yields of 90%. Approximately 4% of the propene is hydrogenated to propane. The conversion of higher olefins requires more extreme conditions.

The process cannot compete with hydroformylation. The only plant using it to manufacture butanol from propene (Japan Butanol Co., capacity 30 000 t/a) closed several years ago [47].

2.3.10. Hydrocarboxymethylation

Hydrocarboxymethylation is a variant of the Reppe process in which higher olefins react with carbon monoxide and methanol in the presence of a cobalt-pyridine catalyst. The products are esters of carboxylic acids containing one more carbon atom in the parent chain than the olefin feedstock [48]. The esters can be hydrogenated to the alcohols. For economic reasons – at the present time the products can be prepared more cheaply from natural raw materials – the process has not as yet achieved any significance.

2.3.11. Fermentation

Fermentation, probably the oldest process for the manufacture of ethanol (→ Ethanol), is still practiced on a large scale [49]. The butanol-acetone fermentation of carbohydrate raw materials is no longer of any importance [50]. On a small scale, pentanols are recovered from fusel oils [10], [51].

2.3.12. Guerbet Alcohols

In the Guerbet process saturated primary alcohols are dimerized to α-branched primary alcohols [52]. Normally the reaction is carried out by refluxing the alcohol in the presence of an alkaline condensation agent and a hydrogenation-dehydrogenation catalyst, e.g.:

$$2\, n\text{-}C_6H_{13}OH \xrightarrow[\text{Raney Ni}]{\text{NaOH}} C_6H_{13}\overset{\overset{\displaystyle C_4H_9}{\displaystyle |}}{C}H-CH_2OH + H_2O$$

The water and small amounts of hydrogen produced in the reaction are removed continuously. For the mechanism of the reaction see [53], [54]. If heating is carried out for long periods, trimeric α-branched primary alcohols also are formed.

The yield of dimeric alcohols is about 80 %. It can be increased by recycling the residues and portionwise addition of fresh catalyst [55]. Metallic sodium, as well as a number of other substances have been proposed as condensation agents [10]. For industrial purposes alkali metal hydroxides are preferred.

Alcohols having chains shorter than cetyl alcohol (C_{16}) preferably are dimerized at higher pressure [56]. Using this process, short-chain alcohols can be converted into $C_{10}-C_{20}$ alcohols. Examples include 2-hexyldecanol [57] and 2-octyldodecanol [58], which find application in cosmetics as oily components with favorable solvent properties.

Because alcohols with the typical α-branching [59] are prepared more easily by other methods, e.g., 2-ethyl-1-hexanol by hydroformylation of propene to give butanal and subsequent aldol condensation, the Guerbet reaction has not become established as a large-scale industrial process.

2.3.13. Other Processes

Of commercial interest is the epoxidation of linear α-olefins and subsequent hydrogenating cleavage to plasticizer (C_6-C_{10}) or detergent ($C_{10}-C_{16}$) alcohols [60]. For hydrogenating cleavage a number of catalysts can be applied [61]. The selectivity towards primary alcohols can be influenced by the choice of catalyst, the hydrogenation conditions, and the use of solvents [10], [62].

The hydrolysis of carboxylic acid esters [63] is also of some importance. For a comprehensive review of other processes see [10].

2.4. Individual Alcohols

The following sections describe individual alcohols and special alcohol mixtures, classified according to increasing number of carbon atoms. In each case the specific method of manufacture, the main producers, and trade names, as well as the necessary basic feedstocks are mentioned.

The C_1-C_5 alcohols are described under individual keywords.

2.4.1. C_6 Alcohols

For physical properties see Table 1.

1-Hexanol [*111-27-3*], $CH_3(CH_2)_4CH_2OH$, is prepared according to the Ziegler process (see Section 2.3.5) from ethylene (Alfol 6, Condea; Epal 6, Ethyl Corp.) or is made from natural products derived from coconut or palm oils (Lorol C 6, Henkel). It is used as a solvent, as a basic material for the perfume industry, and for the production of plasticizers (in this case usually as a mixture with higher *n*-alcohols). Nitrates of 1hexanol are recommended as cetane number improvers. Commercial specifications are given in Table 4.

2-Methyl-1-pentanol [*105-30-6*] is prepared by aldol condensation of propionaldehyde and subsequent hydrogenation of the intermediate 2-methyl-2-pentenal, (see Section 2.3.4). It is used as a solvent.

$CH_3(CH_2)_2CH(CH_3)CH_2OH$

Commercial specifications are given in Table 4. Manufacturers are, e.g., Ruhrchemie and UCC [69], [70].

4-Methyl-2-pentanol [*108-11-2*] is a byproduct of the synthesis of methylisobutylketone [*108-10-1*] (→ Ketones).

$(CH_3)_2CHCH_2CH(OH)CH_3$

The alcohol is sold by UCC, for example, under the name Methylamylalcohol [69]. It is used as a solvent in the paint industry, as a brake fluid, as a flotation aid [71], and as a fungicide [72].

Table 4. Typical specifications of commercial plasticizer alcohols

	Alcohol, wt%	Color, Hazen no.[a]	Density[b], g/cm³	Boiling range[c], °C	Hydroxyl value[d], mg KOH/g	Acidity[e], mg KOH/g	Moisture[f], wt%	Carbonyl number[g]
1-Hexanol	98	<10	0.819	150–170	540–555	<0.05	<0.5	–
2-Methyl-1-pentanol	≥98	5	0.820	148	–	0.1	<0.3	≈0.5
2-Ethyl-1-butanol	≥98	<10	0.831–0.834	156–150	–	<0.2	<0.2	<0.3
C₆ Alcohol mixture	≥98	10	0.819–0.821	151–159	543	0.003	0.2	0.2
Isooctanol	99	10	0.830–0.834	184–191	428	0.001	0.1	0.15
Isodecanol	99	<10	0.835–0.841	215–225	350	0.05	<0.1	0.3
Isotridecanol	99	10	0.843–0.848	250–266	285	0.05	<0.1	0.4

[a] DIN 53409; ASTM D 1209–69
[b] DIN 51757; ASTM D 1298–67
[c] DIN 53171; ASTM D 1078–70
[d] DIN 53240; ASTM D 1957–63
[e] DIN 53402; ASTM D 1613–66
[f] Karl Fischer reagent
[g] Oximation

2-Ethyl-1-butanol [97-95-0] is prepared by aldol condensation (→ Aldehydes, Aliphatic and Araliphatic) of 1-butanal and acetaldehyde and subsequent hydrogenation.

$(C_2H_5)_2CHCH_2OH$

It is a neutral, colorless, pleasant-smelling liquid and is employed as a solvent and flow improver for paints and varnishes, as a component in the manufacture of penetrating oils and corrosion inhibitors, as a cleaning agent for printed circuits, as an extracting agent for metal ions, and as the alcohol component in the manufacture of plasticizer phthalates for special uses, of nitric acid esters as cetane number improvers, and of phosphoric acid esters as plasticizer auxiliaries.

Manufacturers are, e.g., Ruhrchemie and UCC. Commercial specifications are given in Table 4.

C_6 Alcohol Mixtures. These are prepared by the hydroformylation of 1-pentenes and subsequent hydrogenation. Esso/Enjay offers, for example, an isomeric mixture under the name Hexanol that consists of 44% 1-hexanol, 53% 2-methylpentanol, and 3% 2-ethylbutanol [64]. The corresponding plasticizer, DHP (dihexylphthalate), has found application only in special cases. Commercial specifications are given in Table 4.

Hexanols have good solvent properties for fats and oils. Because of their lower volatility and higher viscosity, they are superior to pentanols in several applications, e.g., as flow improvers.

C_6–C_{10} Alcohol Mixtures. Condea [65] and the Ethyl Corp. [73] offer mixtures of C_6, C_8, and C_{10} alcohols under the trade names Alfol 610 and Epal 610, respectively. The 100% linear alcohols are prepared by the Ziegler process (see Section 2.3.5). For the composition and typical data of alcohol mixtures with various C numbers, see Table 5.

The phthalates are suitable as plasticizers for coatings, artificial leather, poly(vinyl chloride) films and sheets, and anticorrosive pastes. The esters of aliphatic acids, also plasticizers, provide cable coverings, which are extremely resistant to cold, films, and artificial leather. They also find application in food wrapping.

Mixtures of 1-hexanol and 1-octanol [111-87-5] serve as frothing agents (bubbling promoters) in flotation, e.g., of coal [77]. Large amounts are added to aqueous drilling muds to prevent frothing during drilling for oil and gas.

2.4.2. C_7 Alcohols,

For physical properties see Table 1.

1-Heptanol [111-70-6] in its pure form has very little commercial value. It can be prepared from 1-hexene by oxo synthesis. Isomeric mixtures of various heptanols are of greater significance.

Alcohols, Aliphatic

Table 5. Composition and typical data of alcohol mixtures with various carbon numbers

Manufacturer	Trade name	Chain-length distribution, wt% alcohol	Boiling range, °C	Density (20 °C), g/cm³	n_D^{20}	Max. water content, wt%	n-Alcohol content, wt% (approx.)
Monsanto	Oxo alcohol 7911	C_7 30–34, C_9 35–43, C_{11} 27–31	178–288	0.826–0.832	1.420–1.440	0.01	65–70
	Oxo alcohol 7900	C_7 41–49, C_9 51–59, C_{11} max. 1	n.a.*	0.825–0.831	n.a.	0.01	
	Oxo alcohol 1100	C_{11} approx. 99	243–274	0.831–0.837	1.439	0.01	
ICI	Alphanol 79	$C_7/C_8/C_9$ mixture	177–202	0.835	1.431	0.02	n.a.
Condea (Conoco)	Alfol 610	C_6 20, C_8 35, C_{10} 44	165–235	0.826	n.a.	0.3	100
	Alfol 810	C_8 43, C_{10} 55	195–240	0.827	n.a.	0.2	100
Shell	Dobanol 91	C_9/C_{11} mixture	225–248	0.835	n.a.	0.1	80
	Linevol 79	C_7 44, C_8 36, C_9 20	183–214	0.827–0.833	1.430	0.1	80
	Linevol 911	C_9 20, C_{10} 50, C_{11} 30	228–248	0.833–0.839	1.439	0.1	80
Ethyl Corp.	Epal 610	C_6 17, C_8 36, C_{10} 47	183–242	0.825	1.428	0.08	100
	Epal 810	C_6 1, C_8 45, C_{10} 54	205–240	0.827	1.431	0.02	100
Henkel	Lorol 810	C_6 0–5, C_8 50–66, C_{10} 30–40, C_{12} 3–8	n.a.	approx. 0.82	n.a.	<0.5	100
Ugine Kuhlmann	Linopol 7–11	C_7 33–36, C_9 38–44, C_{11} 21–28	183–245	0.827–0.830	n.a.	<0.1	70
	Acropol 91	C_9 50–53, C_{11} 47–50	215–248	0.830–0.834	n.a.	<0.1	70

* not available.

C$_7$ Alcohol Mixtures. Heptanol mixtures are prepared by the hydroformylation of isohexene, the dimerization product of propene. They have been used by Chisso and Nissan in Japan and recently also by Exxon for the manufacture of the phthalate plasticizer DIHP (diisoheptyl phthalate) or Jayflex 77, a fast-dissolving plasticizer with good low-temperature properties that is said to be particularly suited for poly(vinyl chloride) flooring and for use in plastisols [78].

C$_7$–C$_{11}$ Alcohol Mixtures. From a linear C$_6$/C$_8$/C$_{10}$ olefin stream of the Ethyl Corp., Monsanto produces the corresponding alcohols with 7, 9, and 11 carbon atoms via oxo synthesis. Depending on the reaction conditions, the content of iso compounds is 30–35%. In addition to C$_7$/C$_9$/C$_{11}$ mixtures, C$_7$/C$_9$ and C$_{11}$ fractions also are available [74]. In the United States, practically the whole production is processed further to phthalates or adipates, used as plasticizers (trade name Santicizer).

Shell also produces mixtures of C$_7$, C$_9$, and C$_{11}$ alcohols that are sold under the trade names Dobanal 91 (Europe) and Neodol 91 (United States, Canada). They are prepared by the Shell oxo process from linear olefins with internal double bonds and are ca. 80% linear; corresponding product distributions are obtained by using α-olefins. The C$_7$/C$_9$ and C$_9$/C$_{11}$ fractions are sold under the trade names Linevol 79 and Linevol 911 [75].

Imperial Chemical Industries (ICI) offers C$_7$/C$_8$/C$_9$ alcohol mixtures under the trade name Alphanol 79 [66]. Linopol 7–11, a mixture of C$_7$/C$_9$/C$_{11}$ primary and α-alkyl-branched alcohols with ca. 70% *n*-isomer content, is produced by Kuhlmann [76].

Typical data for these mixtures as given in the various companies' data sheets are presented in Table 5. The mixtures are colorless liquids, miscible with most organic solvents but not with water.

Uses. Approximately 70–80% are processed to plasticizers, mostly phthalates, which to a certain extent replace the plasticizer DOP, dioctyl phthalate, di-(2-ethylhexyl)phthalate, made from phthalic anhydride and 2-ethylhexanol. However, DOP still commands the major share of the market.

These esters of dicarboxylic acids are used almost exclusively as plasticizers for PVC. The linear esters exhibit somewhat better low-temperature properties, greater resistance toward oxidation, and lower volatility than DOP. For example, in the United States the plasticizers Santicizer 711 (phthalate) and Santicizer 97 (adipate), manufactured by Monsanto from C$_7$/C$_9$/C$_{11}$ alcohol mixtures, have practically displaced DOP in automobile interiors because, owing to the lower volatility, no fogging of the windows takes place.

Further applications of the alcohols depend on their solvent properties. For alcohols with an odd number of carbon atoms, better solvent and wetting properties are claimed than for those with an even number [9], [68].

The alcohol mixtures are used as solvents or solubilizers in the paint and printing ink sector, as components in textile auxiliaries and pesticides, for hormone extraction, and in the surfactant field as foam boosters or antifrothing agents.

2.4.3. C$_8$ Alcohols

For physical properties see Table 1.

2-Ethyl-1-hexanol [*104-76-7*] is the most important C$_8$ alcohol ($\rightarrow$ 2-Ethylhexanol).

CH$_3$(CH$_2$)$_3$CH(C$_2$H$_5$)CH$_2$OH

1-Octanol [*111-87-5*], CH$_3$(CH$_2$)$_6$CH$_2$OH, capryl alcohol. Its esters are widespread in nature; e.g., they occur in grapefruits, oranges, or green tea. 1-Octanol is manufactured by the Alfol process and from natural products. It is sold under the trade names Alfol 8, by Condea, and Lorol C 8, by Henkel. The alcohol is used, for example, in the perfume industry.

2-Octanol [*123-96-6*], CH$_3$(CH$_2$)$_5$CH(OH)CH$_3$, is obtained by the alkaline hydrolysis of castor oil. It is used as a solvent in the paint industry, as a wetting agent in the textile industry, and as a component of brake fluids.

C$_8$ Alcohol Mixtures (Isooctyl Alcohol). Butenes and propene are codimerized to heptenes which, in turn, are hydroformylated to mixtures of isomeric C$_8$ aldehydes. These are hydrogenated to yield mixtures of primary alcohols. In addition to 3,4-, 3,5-, and 4,5-dimethyl1-hexanol and 3- and 5-methyl-1-heptanol, such mixtures contain varying amounts of heptanols and nonanols. They are colorless, slightly viscous, neutral liquids and have a mild fragrance. They are practically immiscible in water (0.06 wt% at 20 °C), but absorb up to 3.5 wt% water at 20 °C. They are miscible with most organic solvents. The mixtures are available under the names isooctanol or isooctyl alcohol from Ruhrchemie, Kuhlmann, ICI, Esso/Enjay, and UCC. The specifications of a typical commercial product are summarized in Table 4.

Isooctyl alcohol is an excellent solvent for many organic compounds, such as fats, oils, and waxes, as well as various rubber formulations and resins. Isooctyl alcohol is especially suitable for baking enamels because, owing to its low vaporization tendency, cratering and blistering do not occur.

Further applications are as a foam suppressant, an extracting agent for metal salts (Mo, Re, Co) and mineral acids (e.g., phosphoric acid, boric acid from aqueous solutions), an emulsifier and stabilizer in oil emulsions, a modifier for PVC pastes, a polishing agent for acrylic ester polymers, and as a bath additive in galvanization.

Derivatives. As with all higher oxo alcohols, the principal use of isooctyl alcohol is as an esterification component in the preparation of plasticizers, mainly diisooctyl phthalate (DIOP), but also of esters of adipic, sebacic, azelaic, and trimellitic acids. Triisooctyl trimellitate, Jayflex 80-TM, is now available from Exxon [78]. As a highly permanent monomeric plasticizer with low viscosity and low specific mass, it competes with polymeric plasticizers.

Various derivatives of isooctyl alcohol, for example, ethers or acetates are suitable as solvents for polymers. If metallic soaps are added, these products can be used as lubricating greases. Some esters are suitable as lubricants or hydraulic fluids.

For the preparation of surfactants, ethoxides or sulfuric acid esters of isooctyl alcohol are recommended. Organophosphorus derivatives, such as metal salts of diisooctyldithiophosphoric acid, are reported to be suitable as additives for lubricants, fuels, and hydraulic fluids.

In agriculture, the isooctyl esters of 2,4-dichloro- and 2,4,5-trichlorophenoxyacetic acid are used as nonvolatile herbicides.

C_8–C_{12} Alcohol Mixtures. Henkel supplies C_8/C_{10} and C_{10}/C_{18} alcohol fractions based on natural fats and oils under the names Lorol 810 and Lorol 818 [67]. The C_8/C_{10} fraction contains 3–8% C_{12} alcohols (Table 5); Lorol 818 contains a wide range of C_8–C_{18} alcohols. Because alcohols from coconut and palm oils are available in only limited quantities, however, by far the greater portion of the alcohols in this range is derived from petrochemical sources. For example, Condea produces Alfol 810 and Alfol 1012 (C_8/C_{10} and C_{10}/C_{12} alcohol mixtures) from ethylene by the Ziegler process [65]. Similar products from the same feedstock are the Epal C_8/C_{10} and C_{10}/C_{12} fractions offered by the Ethyl Corp. [73].

The C_8/C_{10} and C_{10}/C_{12} alcohol mixtures, produced from natural products and by petrochemical methods, are both practically 100% linear.

Mixtures of 1-octanol and 1-decanol are used by the tobacco industry as growth inhibitors [79].

Derivatives. Esters of aliphatic acids and of trimellitic acid with C_8–C_{10} alcohols are suitable as plasticizers for weather-resistant films and artificial leather as well as for special poly(vinyl chloride) applications, such as thermally stable cable coverings.

Upon addition of 3–6 mol ethylene oxide to C_8–C_{10} alcohols, ethoxides are obtained that possess some advantages over longer-chain detergents owing to their excellent solubility in water, improved viscosity, and especially good wettability. They are used as frothing agents in oil drilling, as components in gentle and low-temperature washing liquids, in technical cleansers, and also in the textile industry.

By sulfatation of the ethoxides the corresponding ether sulfates are produced, which possess high stability toward electrolytes and alkaline-earth metal ions. They are used as industrial foaming agents, e.g., for plaster and mortars, and for oil drilling. The sodium salts of alcohol sulfates are used to a slight extent as wetting agents with high stability towards electrolytes.

Primary alkylphosphates are prepared from the alcohols and phosphorus pentoxide. After neutralization with amines or alkali hydroxides, they are used as water-soluble anionic surfactants with high stability toward hydrolysis and as emulsifiers. Other derivatives of these alcohols, especially in the C_8–C_{10} range, include the corresponding amines. They are used for the solvent extraction of heavy metals, such as molybdenum, vanadium, uranium, and thorium, from acidic aqueous solutions. Quaternary ammonium salts with C_8–C_{12} alkyl groups are added to such cleaning agents as bactericides and fungicides.

Table 6. Composition of various octenes (in wt%)

Trade name	DIB[a]	Codibutylen[a]	Dimersol[b]	Octene[c]
Butene combination	i + i	i + n	n + n	n + n
Methylheptenes	–	8	52	4
Dimethylhexenes	7	25	40	90
Trimethylpentenes	93	65	2	2

[a] Manufacturer, e.g., Erdölchemie;
[b] IFP/Nissan;
[c] UOP.

Ether amines are obtained by cyanoethylation with acrylonitrile and subsequent hydrogenation (Sherex, Henkel). They are used as selective collectors for the flotation of silicates in low-quality iron ores.

2.4.4. C$_9$ Alcohols

For physical properties see Table 1.

C$_9$ Alcohol Mixtures. Nonanol mixtures are typical plasticizer alcohol components for the preparation of diisononylphthalate (DINP). Mixtures of C$_9$ alcohols are prepared by oxo synthesis from dimeric 1-butene and 2-butene (the mixture also is known as raffinate II) or from dimeric isobutene. Because of increased demand for isobutene for the production of methyl *tert*-butyl ether (MTBE), starting materials are now the relatively cheap mixed dimers of *n*-butenes and isobutene (codibutylenes) and dimers of *n*-butene in particular.

Typical commercial olefins of the latter variety are Dimersol from the IFP process and octenes from the UOP process. Because of the differences in the manufacturing methods, these olefins have considerably different compositions, see Table 6. For specifications of the various alcohols obtained from them, see Table 7. The nonanol from diisobutene is the only one consisting primarily of 3,5,5-trimethyl-1-hexanol; the other alcohols are more or less complex alkyl-branched mixtures. Esso's isononyl alcohol, for example, contains 80% dimethyl-1-heptanols and 20% trimethyl-1-hexanol.

Derivatives. Isononyl alcohol is a useful high-boiling solvent for fats, oils, waxes, and resins. It is frequently used as a solvent in the paint industry. In nitrocellulose-based paints, it acts as a latent solvent; the addition of small amounts facilitates the blending with hydrocarbons. In paints based on alkyd resins, shellacs, or urea-formaldehyde resins the viscosity and the flow properties are improved; in baking enamels cratering and blistering are prevented.

As a rule, the mixtures immediately are processed further to plasticizers (for phthalates and esters of other dicarboxylic acids.

Isononyl esters of phosphorous, thiophosphorous, and dithiophosphoric acid are employed as additives for high-quality lubricants. The polyacrylates and polymethacrylates of isononyl alcohol are recommended for lowering the pour point temperatures

Table 7. Specifications of various isononanols

Olefin basis Tradename of the alcohol	DIB Nonanol[a]	Codibutylene	Dimersol Oxocol-900	UOP octene
Boiling range (101.3 kPa), °C	193–202	201.7–204.4	204.5–207.1	201.2–205.1
Density d_4^{20}	0.827–0.831	0.843	0.835–0.837	0.834
Refractive index n_D^{20}		1.4390	1.4365–1.4369	1.4365
Color no., Hazen	$\leq 10^{\,a}$	–	5	–
Carbonyl no., mg KOH/g	$< 0.01^{\,b}$	0.1	0.1	0.1
Acid no., mg KOH/g	$\leq 0.001^{\,c}$	0.02	0.03	0.02
Water content, %	≤ 0.10	–	–	0.01

[a] Method ISO R 1843;
[b] Method ISO R 1847;
[c] Method ISO R 1846.

of lubricating oils. They also are used as sealing liquids and hydraulic fluids, e.g., for vacuum and diffusion pumps.

Isononyl alcohol is the first in the series of oxo alcohols used for the industrial synthesis of surfactants. Neutralized isononyl sulfates or phosphates are wetting agents in the textile industry. Sodium diisononyl succinate is suitable as a wetting agent and as an emulsifier for various oils.

2,6-Dimethyl-4-heptanol [108-82-7] can be prepared by aldol condensation of acetone and subsequent hydrogenation.

$[(CH_3)_2CHCH_2]_2CHOH$

The alcohol is sold by UCC under the name diisobutylcarbinol. It is used as a reaction medium for the preparation of hydrogen peroxide and as an effective defoaming agent, in this case also in its ester form. The phosphates are extraction solvents for uranium and rare metals.

2.4.5. C$_{10}$ Alcohols

For physical properties see Table 1.

1-Decanol [112-30-1], $CH_3(CH_2)_8CH_2OH$, is prepared by a number of producers, e.g., Henkel (Lorol C 10, Lorol C 10–98), by the catalytic reduction at high pressure of coconut oil or coconut fatty acids or esters. Greater, but still not significant quantities are manufactured by the Ziegler process. Companies offering the product include Conoco/Condea and the Ethyl Corp. The perfume industry uses 1-decanol as a raw material for detergents and as a defoaming agent.

C$_{10}$ Alcohol Mixtures. Relatively complex, multi-branched mixtures of primary alcohols are manufactured almost exclusively by the hydroformylation of tripropylene. The main components of the mixture are isomeric trimethylheptanols and 3,5-dimethyloctanol. Isodecanol is sold by Hoechst, Kuhlmann, Esso, UCC, and ICI among others (Table 4).

The C_{10} alcohol mixtures are processed mainly to phthalate plasticizers (DIDP). Because of their low volatility and good low-temperature properties they are gaining steadily in importance for this application. Following esterification with dicarboxylic acids, the C_{10} oxo alcohol mixtures yield synthetic ester oils with outstanding properties [1]. The C_{10} alcohol mixtures are also starting materials for a wide range of derivatives, which are employed as textile auxiliaries, stabilizers, flotation agents, and antifoaming agents, and are used in detergents, pharmaceuticals, cosmetics, and pesticides (esters with 2,4-dichloro- and 2,4,5-trichlorophenoxyacetic acid are herbicides).

2.4.6. Mixtures of Linear C_{12}–C_{18} Alcohols (Detergent Alcohols)

For more details → Fatty Alcohols.

n-Alcohols with carbon chain lengths of 12 to 18, in former times prepared solely from natural products and consequently known as fatty alcohols, are now also manufactured from ethylene by the Ziegler process and by the Shell version of the oxo process (*n*-alcohol content ca. 80%). These alcohols are important intermediates for a large number of chemical products, but over 95% of them are used in detergents.

To a lesser extent they are used directly as wetting, emulsifying, and foaming agents. For most applications, they are modified by insertion of a hydrophilic group, e.g., by ethoxylation with ethylene oxide to ethoxides, $RO(CH_2CH_2O)_nH$, by sulfatation of the ethoxides to the ether sulfates, e.g., $RO(CH_2CH_2O)_nCH_2CH_2OSO_3Na$, and by direct sulfatation to alcohol sulfates, $ROSO_3Na$.

The products prepared by esterification of the alcohols with acrylic or methacrylic acid and subsequent polymerization are employed mainly in lubricating oils as viscosity index improvers. The C_{12}–C_{15} region is preferred, but C_{16}–C_{18} alcohols also are used.

2.4.7. C_{13}–C_{18} Isoalcohols

Isotridecyl alcohol [27458-92-0] is prepared from tetrapropylene by oxo synthesis. The product isolated is actually a complex mixture of primary alcohols with an average of 13 carbon atoms.

Isotridecyl alcohol is a colorless, clear, slightly viscous, neutral liquid with a mild odor. It is manufactured by Hoechst, Kuhlmann, ICI, Esso, and UCC, and consists principally of tetramethyl-1-nonanols. A compilation of the standard specifications is given in Table 4.

Because of its extremely low volatility, the phthalate has special applications as a plasticizer in wire insulation based on PVC. The alcohol also is used as a surfactant raw material, as an antifrothing agent, and as a solvent where low volatility is required, e.g., for preventing cratering and blistering during the drying of paint surfaces.

Isohexadecyl alcohol [*36311-34-9*] is prepared by aldol condensation of isooctylaldehyde and subsequent hydrogenation of the isohexadecenal. A complex mixture of primary alcohols of 2,2-dialkyl-1-ethanols is obtained, wherein the alkyl groups consist of methyl-branched C_6 and C_8 units. The alcohol is manufactured by Esso and Shell.

Isohexadecyl alcohol is used in lubricants, e.g., as the esters of higher fatty acids, dicarboxylic acids, orthophosphoric acid, or unsaturated polybasic acids. Hexadecyl alcohol is also used as a component of detergents and wetting agents, textile aids, softeners, cosmetics, evaporation preventers, and foam breakers. In pure form it is applied as a flotation and extraction agent.

Isooctadecyl alcohol [*27458-93-1*] is a highly branched, primary C_{18} alcohol with 5,7,7-trimethyl-2-(1,3,3'-trimethylbutyl)–1-octanol [*36400-98-3*] as the main component. It is prepared by aldol condensation of isononylaldehyde and subsequent hydrogenation of the unsaturated aldehyde intermediate. It is a highly viscous, neutral, odorless liquid that is practically insoluble in and hydrophobic to water. The alcohol is miscibile with most organic liquids in all proportions.

The alcohol is available from Hoechst. As a highly branched alcohol it is suitable for many applications requiring a low pour point. It is used, therefore, for the manufacture of synthetic lubricants and hydraulic fluids. Because of the low vapor pressure, it is often applied to water surfaces to prevent evaporation (by formation of monomolecular layers).

Isostearyl alcohol is available from Mitsubishi Chem. under the name Diadol 18 G. It is a slightly-branched alcohol with a definite structure (2-heptylundecanol) and is prepared either by the aldol condensation of nonanal or by the hydroformylation of a C_{17}-vinylidene olefin. It is used as the waxy component for high-quality cosmetics.

2.5. Economic Aspects

The economic significance of alcohols can be seen clearly from the production figures (in 10^6 t, 1980):

Methanol	12.0
C_6-C_{11} Alcohols	2.0
Ethanol	≈ 4.0
$C_{12}-C_{15}$ Alcohols	0.6
C_3-C_7 Alcohols	2.0
$C_{16}-C_{18}$ Alcohols	0.4

Table 8. World capacity of C_6–C_{18} alcohols, 1980 (10^3 t)

	C_6–C_{11} Alcohols	C_{12}–C_{15} Alcohols	C_{16}–C_{18} Alcohols	Total	%
World capacity	2000	615	410	3025	100
Petrochemical basis	1980	510	164	2654	88
Natural sources	20	105	246	371	12

Table 9. Plasticizer alcohols, consumption, production, and capacities in Western Europe, United States, and Japan, 1980 (10^3 t)

	Western Europe	United States	Japan	Total
Consumption				
2-Ethylhexanol	425	193	297	915
Other C_6–C_{11} alcohols	215	259	55	529
	640	452	352	1444
Production				
2-Ethylhexanol	602	167	250	1019
Other C_6–C_{11} alcohols	305	311	57	673
	907	478	307	1692
Capacity				
2-Ethylhexanol	808	215	333	1356
Other C_6–C_{11} alcohols	435	379	83	897
	1243	594	416	2253

Crude oil and natural gas still predominate as raw materials for their manufacture. However, the oil crisis in the 1970s induced a number of countries to develop coal and natural materials as a substitute. For example, production of ethanol by fermentation or from C_{16}–C_{18} alcohols from fats and oils should increase worldwide.

As a result of the development of new competitive gasification processes, coal is expected to be used increasingly as raw material for the production of synthesis gas from which methanol and C_3–C_{15} alcohols can be made [80]. Table 8 gives a survey of the basic raw materials for higher alcohols. About 90% of these alcohols are produced from petrochemicals.

About two thirds of the higher alcohols can be classified as plasticizers (C_6–C_{11}) and one third as detergents (C_{12}–C_{18}).

Plasticizer Alcohols (C_6–C_{11} and C_{13}-Branched Alcohols). Table 9 shows the 1980 consumption, actual production, and production capacities of plasticizer alcohols in the major Western industrial countries. About 75% of the available production capacity was being utilized; 85% of the alcohols produced were consumed in the producer countries, and about 15% were exported, especially to East Asia. With a near 60% share of the market, 2-ethylhexanol is the most important plasticizer alcohol. Insufficient production capacity (USA) and an unfavorable raw material supply (Japan) have led to shortages in these countries. Therefore the other C_6–C_{11} alcohols have

Table 10. Production of plasticizer alcohols in the United States, Europe, and Japan, 1980 (10^3 t)

	Process	United States	Western Europe	Japan	Total	%
2-Ethylhexanol	oxo synthesis/aldol cond.	167	602	250	1019	60
Branched C_6–C_{13}	oxo synthesis	186	240	41	467	28
Linear C_7–C_{11}	oxo synthesis	73	25	11	109	6
Linear C_6–C_{10}	Ziegler process	52	30	–	82	5
Linear C_6–C_{10}	from natural sources	–	≈10	5	≈15	1
		478	907	307	1692	100

Table 11. Production of plasticizer alcohols, 1980 (10^3 t), apart from 2-ethylhexanol

	United States	Europe	Japan	Total	%
Hexyl alcohol (non-Ziegler)	7	–	–	7	1
Heptyl alcohol	–	–	28	28	4
Isooctyl alcohol	32	105	–	137	22
Isononyl alcohol	66	50	3	119	19
Isodecyl alcohol	59	75	6	140	22
Linear C_6–C_{11} alcohol	125	25	16	166	26
Tridecyl alcohol	23	10	3	36	6
Total	312	265	56	633	100

gained in significance, particularly in the United States. An outline of the C_6–C_{11} alcohol producers in 1980 with capacity and process type is given in [81].

Table 10 contains a list of the plasticizer alcohols produced in Western Europe, the United States, and Japan in 1980, classified according to process and raw material source. Table 11 shows the relative market shares of various plasticizer alcohols and alcohol mixtures.

With plasticizer alcohols the term "linear" has a different meaning; alcohols from the oxo synthesis are customarily termed linear even if, depending on the process variant, they contain 20–50% α-alkyl-branched isomers mixed with the *n*-alcohol components. In the strict sense of the word, therefore, only the alcohols from the Ziegler process (5%) and on the basis of natural raw materials (1%) can be termed "linear."

Future Aspects. The further development of plasticizer alcohols and the phthalates produced from them is closely connected with the development of PVC, as up to 50% phthalate is mixed with the polymer. In the Western industrial countries relatively small growth rates are expected.

The following key developments will have an influence on the relative market shares of the different alcohols [1], [82]:

1) The changes in the costs of raw materials. This favors *n*-butene (dimersol) as a feedstock to the detriment of ethylene.

2) The shrinking market for linear phthalates and the trend toward "softer" vinyls favors DIDP.
3) The trend toward plasticizers of higher quality than DOP (fogging problem in automobile furnishings and vinyl roofing) will lead to a shift from 2-ethylhexanol and isooctyl alcohol toward higher alcohols.
4) The trend toward a reduction of the proces-sing temperature in the manufacture of PVC products and toward the use of extra fillers.
5) More stringent environmental regulations during processing (volatility of plasticizers) is promoting the trend toward less volatile plasticizers.

Detergent Alcohols (C_{12}–C_{18}). The following figures show the relative worldwide feedstocks for C_{12}–C_{18} alcohols [18]:

	C_{12}–C_{15}	C_{16}–C_{18}
Petrochemical (%)	83	40
Natural (%)	17	60

Because most of the natural oils and fats (tallow) consist almost entirely of C_{16}–C_{18} fatty acids, which are available in large quantities [18], [83], [84], their significance as raw materials will increase in the years to come, a trend that also will be promoted by the price increase of natural and petrochemical raw materials [80]. On the other hand, the possibilities of using natural raw materials in the C_{12}–C_{15} range are limited because the corresponding lauric oils only make up about 5–7% of world oil and fat production (1980: 3.3×10^6 t) [85] and are mostly used in the food industry. Therefore this range is produced primarily from petrochemical feedstocks.

In 1980, when the total production capacity for surfactant alcohols in the Western World was about 911 000 t [86], only 516 000 t actually were produced. For producers, process type, and installed capacities in the Western World, see [86]. The relative production figures % for the different processes are:

Oxo process	41%
Ziegler process	21%
Natural basis	35%
Paraffin oxidation	3%

Consumption figures for the surfactant alcohols and their derivatives are given in Table 12.

Future Aspects. A large growth in the production of surfactant alcohols is expected in the next few years [87] owing to the increasing demand for non-ionic detergents, which exhibit good washing properties even in low-phosphate detergent formulations (environmental regulations). In view of the limited supplies of C_{12}–C_{15} raw materials and the uncertainty in the price development of lauric oils [88], there will be a continued increase of the quantities in this range produced on a petrochemical basis. The natural

Table 12. Consumption of detergent alcohols, 1980 (10^3 t)

	United States	Western Europe	Japan
Alcohol sulfates	72	31	9
Alcohol ethoxides	63	53	21
Alcohol ether sulfates	64	52	25
Alcohol glyceryl ether Sulfonates	6	–	–
Fatty amine oxides	1	–	–
Methacrylate derivatives	–	18	
Use as free alcohol	18	21	23
Other derivatives	39		
Total	263	175	78

raw materials otherwise will increase their share in the production of C_{15}–C_{18} alcohols; for corresponding data see [26].

2.6. Quality Specifications

The quality control specifications for alcohols are given by DIN [89] or ASTM [90]; analytical procedures are compiled by the Deutsche Gesellschaft für Fettwissenschaft [91] and by Henkel [3]; see also [2].

The most important analytical procedure by far is *gas chromatography*. Chain-length distribution, the proportion of unsaturated, branched, or secondary alcohols, as well as small amounts of esters and hydrocarbons can be determined using suitable columns and conditions.

An important quality criterion for alcohols is the *carbonyl number*, which indicates the degree of impurity caused by carbonyl compounds. It is determined by titration of the hydrochloric acid released by reaction of the carbonyl groups with excess $NH_2OH \cdot HCl$. Its numerical value expresses the mass of CO (mg) in the form of carbonyl groups, that are present in 1 g of product.

The *hydroxyl value* (HV=mg of KOH equivalent to the hydroxyl content of 1 g of alcohol) measures the hydroxyl groups and reflects both the molecular mass and the purity of the sample.

The *saponification number* is the number of mg KOH that is required to neutralize the acids and saponify the esters contained in 1 g of substance. The *iodine value* (g of iodine consumed by 100 g of alcohol) is a measure of the content of unsaturated compounds. The *water content* is determined by the Karl Fischer method [3]. The *color* of the alcohol is usually given in Hazen units (the mass of platinum (mg) present in a solution of K_2PtCl_6 and $CoCl_2 \cdot 6 H_2O$ in a mass ratio of 1:0.8025 in 1000 ml aqueous HCl that has the same color as the alcohol). Hazen numbers of 5 to 10 should not be exceeded.

A number of other tests relating to smell, melting point, boiling range, and color stability also serve as indications of quality and storage stability. One special test is, e.g., the color determination after treatment with concentrated sulfuric acid or after saponification with phthalic acid in the presence of sulfuric acid.

The recognition of the *source* of alcohols and alcohol mixtures (natural, Ziegler, oxo, paraffin oxidation, etc.) is summarized in [92]. A further valuable aid is ^{13}C NMR spectroscopy [93]. Using this technique, it is even possible to identify alcohol fractions from several different feedstock sources [3].

The *heavy-metal content* of alcohols should not exceed 1×10^{-6} g/g because of catalytic effects (reduced storage stability).

How pure an alcohol should be and what impurities it may contain usually depend on the further processing requirements and subsequent application. The specifications usually are agreed upon individually by the manufacturer and customer; therefore, alcohols of differing qualities are obtainable from the same producer.

2.7. Storage and Transportation

Short-chain liquid alcohols can be stored in containers made of noncorroding unalloyed carbon steel or stainless steel. Aluminum containers also are suitable, but not for long-term storage of lower alcohols. For higher alcohols, containers of pure aluminum and frequently of Al-Mg-Mn alloy (DIN 1725/1745) are used. Stainless steel is necessary only for the storage of extremely water-free alcohols. In aluminum vessels, corrosion is to be expected because of the formation of aluminum alkoxides. The storage temperature should be as low as possible.

In order to avoid oxidation, an inert-gas unit should be installed to enable flushing with pure nitrogen, etc. Water content of ca. 0.1% has a stabilizing effect; for extremely water-free alcohols the danger of autoxidation is very great.

Liquid alcohols are transported in painted, corrugated iron vessels, in road tankers, or in tank wagons made of normal steel, aluminum, or stainless steel. Solid products, such as the higher alcohols, are sold as flakes contained in multilayer polyethylene-lined paper bags.

Transportation is controlled by a number of general instructions and legal regulations for volatile and flammable substances.

The alcohols are classified on the basis of their specific properties, such as flash point, boiling point, etc.

3. Unsaturated Alcohols

Compared with saturated alcohols, unsaturated ones are of minor importance. In the United States the consumption of allyl alcohol, the most important unsaturated alcohol (→ Allyl Compounds), amounted to 60 000 t in 1979. The proportion attributed to the other unsaturated alcohols, which are prepared in processes based either on acetylene or on natural feedstocks, is even smaller.

The unsaturated alcohols are generally colorless, pungent-smelling liquids and even 9-*trans*-octadecenol has a melting point of only 37 °C. The solubility in polar solvents, especially of the lower members, is high.

Some physical properties of unsaturated alcohols are given in Table 13. The accumulation of several functional groups in the molecule leads to a high reactivity.

Table 13. Physical properties of unsaturated alcohols

Name		Formula	M_r	mp, °C	bp, °C	d_4^{20}	n_D^{20}
Allyl alcohol	[107-18-6]	$CH_2=CH-CH_2OH$	58.08	−129	96.95	0.8520	1.4134
Crotyl alcohol, cis	[4088-60-2]	$CH_3-CH=CHOH$	72.11	−90.15	123.6	0.8662	1.4342
trans	[504-61-0]		72.11		121.2	0.8521	1.4288
Propargyl alcohol	[107-19-7]	$CH\equiv C-CH_2OH$	56.06	−51 to −48	114−115	0.9478	1.4306
3-Buten-2-ol	[598-32-3]	$CH_2=CH-CH(OH)-CH_3$	72.11	<−100	97.4	0.8413	1.4127
2-Butyn-1-ol	[764-01-2]	$CH_3-C\equiv C-CH_2OH$	70.09	−2.2	142.7−142.9	0.9373	1.4550
3-Butyn-2-ol	[2028-63-9]	$CH\equiv C-CH(OH)-CH_3$	70.09		107	0.8858	1.4265
10-Undecen-1-ol	[112-43-6]	$CH_2=CH-(CH_2)_8-CH_2OH$	170.30	−2	250	0.8495	1.4506
9-Octadecen-1-ol		$CH_3(CH_2)_7CH=CH(CH_2)_7CH_2OH$					
cis, oleyl alcohol	[143-28-2]		268.49	5.5−7.5	333−335	0.8491	1.4607
trans, elaidyl alcohol	[506-42-3]		268.49	35	333	0.8338	1.4522
1-Hexyn-3-ol	[105-31-7]	$CH\equiv C-CH(OH)(CH_2)_2CH_3$	98.08	−80	142	0.882	1.4350
4-Ethyl-1-octyn-3-ol	[5877-42-9]	$CH\equiv C-CH(OH)-CH(C_2H_5)-(CH_2)_3CH_3$	154.14	−45	197.2	0.873	1.4502
2-Methyl-3-butyn-2-ol	[115-19-5]	$CH\equiv C-C(OH)(CH_3)-CH_3$	84.06	2.6	103.6	0.8672	1.4211
3-Methyl-1-pentyn-3-ol	[77-75-8]	$CH\equiv C-C(CH_3)(OH)-CH_2CH_3$	98.08	−30.6	121.4	0.8721	1.4318

Therefore, they are capable of undergoing a large number of chemical reactions and frequently serve as intermediates in the chemical industry.

Crotyl alcohol [6617-91-5], 2-buten-1-ol, $CH_3CH=CHCH_2OH$, is made by aldol condensation of acetaldehyde and subsequent hydrogenation. It is of little industrial importance since butanol has been manufactured via the oxo synthesis rather than from crotonaldehyde.

Propargyl Alcohol [107-19-7], 2-propyn-1-ol, $HC\equiv C-CH_2OH$, is formed as a by-product ($\sim 5\%$) in the butynediol synthesis from acetylene and formaldehyde [94] ($\rightarrow$ Butanediols, Butenediol, and Butynediol):

$$HC\equiv CH + H-\overset{O}{\underset{\|}{C}}-H \xrightarrow{CuC_2} HC\equiv C-CH_2OH$$

Copper acetylide serves as catalyst. The reaction can be controlled in such a way that either propargyl alcohol (high acetylene pressure, low formaldehyde concentration) or 2-butyne-1,4-diol (excess formaldehyde) is formed preferentially.

Main manufacturers using this process are BASF, GAF, Du Pont, and GAF/Chemische Werke Hüls. Total capacity for butanediol from butynediol has increased since 1975 from 70 000 t/a to over 280 000 t/a, so that the total capacity of propargyl alcohol is now ca. 14 000 t/a.

The alcohol is an excellent rust inhibitor, superior in this respect to allyl alcohol. It retains its rust-inhibiting properties at high temperatures and therefore is used for oil drilling.

Methylbutynol, 2-methyl-3-butyn-2-ol, [115-19-5], $HC\equiv C-C(CH_3)_2OH$, is formed by the ethynylation of acetone in base. Since 1977 it has been used as an intermediate for the manufacture of isoprene. Further applications are the preparation of vitamin A and other products and use in metal pickling and plating operations.

Oleyl Alcohol, *cis*-9-octadecen-1-ol, [143-28-2], $CH_3(CH_2)_7CH=CH(CH_2)_7CH_2OH$, is the most important natural-based unsaturated alcohol. All commercial fatty alcohols derived from natural sources contain oleyl alcohol to a greater or lesser extent. It is prepared by catalytic hydrogenation under pressure of the methyl esters of unsaturated fatty acids; mixed catalysts containing zinc are employed, and the reaction conditions correspond to those for the production of saturated fatty alcohols [95]. The methyl esters of the fatty acids are prepared from beef tallow and olive oil.

Oleyl alcohol is sold by Henkel under the trade name HD-Ocenol. Other manufacturers are New Japan Chemical Co., Ashland Chem. Co., and Kedzierzyn, Poland. Total capacity amounts to ca. 35 000 t/a. The alcohol is used principally in the detergent field, where the double bond offers possibilities for applications not covered by the saturated fatty alcohols.

4. Alkoxides

Metal alkoxides can be formally regarded as salts of very weak acids (J. LIEBIG, 1837).

The alkoxides of alkali metals, magnesium, aluminum, titanium, zirconium, and antimony are commercially the most important ones, e.g., as catalysts in condensation and transesterification reactions, as reducing agents, and as paint additives.

4.1. Properties

The properties of the alkoxides depend on the position of the metal in the periodic table and on the character of the alcohol component. The alkoxides of the highly electropositive metals are solid, nonvolatile, very basic compounds of ionic nature. Polymeric, covalent compounds with a low volatility are formed by the higher multivalent elements; the alkoxides of the lighter transition elements are distillable, monomeric, covalent liquids. The alkali and alkaline-earth metal alkoxides are soluble in organic solvents to only a limited extent owing to their polar nature. Alkali-metal alkoxides crystallize from alcohol solution mostly with 1–3 mol alcohol (crystal alcohol), which can be removed by raising the temperature or by azeotropic distillation with benzene or toluene [96]. Heating to above 200 °C generally leads to decomposition [96]. The lower alkoxides of aluminum (with the exception of methoxide) and titanium can be distilled in the vacuum without decomposition. The physical properties of metal alkoxides have been discussed in detail, cf. [96]–[98].

Alkoxides are very reactive toward compounds having acidic hydrogen atoms, for example, with water:

$$ROM + H_2O \rightleftharpoons ROH + MOH$$

Therefore, alkoxides are stable only in the absence of water. With highly electropositive metals the reaction is reversible so that, e.g., the sodium ethoxide can be obtained from sodium hydroxide and ethanol. With the covalent multivalent alkoxides, however, this reaction is irreversible. The hydrolysis of titanium and aluminum alkoxides has been examined thoroughly [98]. With other alcohols transalcoholization occurs until equilibrium is established:

$$ROM + R'OH \rightleftharpoons R'OM + ROH$$

In this way the alkoxides of higher alcohols can be obtained from lower alkoxides. Reaction with phenol leads to the phenoxides:

$$M(OR)_n + n\ C_6H_5OH \longrightarrow M(OC_6H_5)_n + n\ ROH$$

By reaction with carboxylic acids, the metal salts of the carboxylic acids and the free alcohols are formed:

$$M(OR)_n + n\,R'COOH \longrightarrow M(R'COO)_n + n\,ROH$$

Reaction of trialkyl silanols with alkoxides gives the corresponding trialkylsiloxy metal derivatives [99]:

$$M(OR)_n + n\,R'_3SiOH \longrightarrow M(OSiR'_3)_n + n\,ROH$$

β-Dicarbonyl compounds form chelate complexes with alkoxides.

The alkoxides react with carbonyl compounds in different ways depending on their basicity. With alkali-metal alkoxides enolization or condensation reactions occur. The Tishchenko reaction with aluminum alcoholates leads to esters by disproportionation [100]:

$$2\,RCHO \xrightarrow{Al(OR)_3} RCOOCH_2R$$

In the Meerwein-Ponndorf-Verley reaction, ketones are reduced with alcohols in the presence of aluminum triisopropoxide:

$$R\text{-}CO\text{-}R + R'_2CHOH \rightleftharpoons R_2CHOH + R'\text{-}CO\text{-}R'$$

Transesterification reactions of esters with alkoxides also are known [101]:

$$M(OR)_n + n\,R''COOR' \rightleftharpoons M(OR')_n + n\,R''COOR$$

Other alkoxides catalyze transesterification reactions; e.g., benzoic acid butyl ester and ethanol are obtained from benzoic acid ethyl ester and butanol in the presence of titanium isopropoxide [546-68-9] [102]. A detailed description of reactions and applications of alkali and alkaline-earth metal alkoxides has been published [103].

4.2. Preparation

Alkoxides may be obtained by the reaction of alcohols with metals, metal hydroxides, metal halides, or alkoxides of other alcohols; the method depends on the metal and the alcohol component. A summary of the procedures can be found in [96].

Preparation from Metals and Alcohols.

$$M + n\,ROH \longrightarrow M(OR)_n + n/2\,H_2$$

This reaction is the simplest method for the preparation of alkoxides; it is limited to the alkali metals and to magnesium and aluminum. The alkali metals are suspended in an inert solvent [104] or are used as an amalgam [105]. The amalgam is brought into contact with the alcohol in a packed column according to the countercurrent principle. A graphite or Fe-Cr-C alloy contact electrode that cannot be amalgamated is used to speed up the reaction [106]. The longer the alcohol chain, the less vigorous the reaction. Primary alcohols react fastest, tertiary slowest. Magnesium and aluminum must be activated by the addition of catalysts, e.g., mercury chloride, HCl, or iodine [107]. Other catalysts, especially for magnesium, are *p*-toluenesulfonic acid and orthoformic acid esters [108]. By heating the alcohols in a column filled with aluminum turnings, the corresponding aluminum alkoxides can be produced without the aid of a catalyst [109].

Reaction of Alcohols with Metal Hydroxides. The equilibrium reaction of alcohols with metal hydroxides can be driven to completion by azeotropic distillation of the water formed with benzene, toluene, or the alcohol itself [110].

$$n \text{ ROH} + \text{M(OH)}_n \rightleftharpoons \text{M(OR)}_n + n \text{ H}_2\text{O}$$

The method is suitable only for the highly electropositive elements; lithium hydroxide does not react [111].

Reaction of Alcohols with Metal Halides. For the alkoxides of the less electropositive elements the corresponding metal chlorides are reacted with the alcohols:

$$\text{MCl}_n + n \text{ ROH} \rightleftharpoons \text{M(OR)}_n + n \text{ HCl}$$

The process is primarily suitable for the production of alkoxides of multivalent metals, such as titanium, germanium, and antimony. Mixtures with chloroalkoxides always are obtained. Bases, in particular ammonia, are used to shift the equilibrium toward the formation of pure alkoxides:

$$\text{TiCl}_4 + 2 \text{ ROH} \rightleftharpoons \text{TiCl}_2(\text{OR})_2 + 2 \text{ HCl}$$
$$\text{TiCl}_2(\text{OR})_2 + 2 \text{ ROH} + 2 \text{ NH}_3 \rightleftharpoons \text{Ti(OR)}_4 + 2 \text{ NH}_4\text{Cl}$$

Reaction of Metal Alkoxides with Metal Halides and Other Metal Compounds. In particular the alkoxides of multivalent metals can be prepared from their halides (MX_n) by reaction with other alkoxides (preferably of monovalent metals):

$$\text{MX}_n + n \text{ MOR} \rightleftharpoons \text{M(OR)}_n + n \text{ MX}$$

Nearly all metal alkoxides can be manufactured by this process. Sometimes other metal compounds are employed instead of halides; see [103] for details.

Reaction with Other Alkoxides. The alkoxides of higher or multivalent alcohols also can be prepared by alcohol transfer:

$$n \text{ R'OH} + \text{M(OR)}_n \rightleftharpoons (\text{R'O})_n\text{M} + n \text{ ROH}$$

The more volatile alcohol distills from the reaction mixture [112]. However, when a large excess of the lower alcohol is used, a lower alkoxide can be obtained from a higher one.

4.3. Uses

Alkoxides of alkali metals are used predominantly as condensation agents in organic synthesis and to introduce the alkoxy group into other compounds. The methoxides and ethoxides of sodium and potassium and the potassium *tert*-butoxide have attained particular significance. For the use of sodium ethoxides, see [103], [113], [114]. Magnesium alkoxides are used as raw materials for driers in paints and varnishes and as components of highly active Ziegler catalysts [115]. Aluminum isopropoxide and aluminum *sec*-butoxide are used commercially as specific reducing agents for various organic compounds (Meerwein-Ponndorf-Verley reaction, see Section 4.1). They also find application in paints to improve resistance to chemicals [116], in textiles to produce impregnating agents, and in cosmetics as components of antiperspirants.

Titanium alkoxides are used in the manufacture of corrosion-proof, high-temperature paints, as esterification catalysts, for the water proofing of textiles, as bonding agents in adhesives, and as catalyst components for the polymerization of olefins.

Besides the solid alkoxides, solutions in the corresponding alcohols are on the market. Alkoxides are very easily hydrolyzed by water, extremely easily in the case of the short-chain alkoxides. Contact with atmospheric moisture must be avoided because spontaneous ignition may occur. Alkoxide fires must be extinguished with dry sand or powder but never with water.

Solid alkoxides are sold mostly in polyethylene bags packed in drums or in corrugated barrels fitted with special lids. Alkoxide solutions, e.g., the 25% solution of sodium methoxide in methanol, however, also are transported in tankers. During unloading, a dry, inert-gas atmosphere must be provided.

Compared with alcohols, the alkoxides are of minor commercial importance. The 1982 world production is estimated at 10 000 –12 000 t.

5. Toxicology

1-Hexanol. The oral LD_{50} is 4.87 g/kg for rats [117] and 4.0 g/kg for mice [118]. The maximum nonlethal period of time that rats can inhale a saturated atmosphere is 8 h [119].

1-Hexanol is a severe irritant to the rabbit eye [119]. It is moderately irritating to the skin of rabbits after 24-h skin contact [120]. The dermal LD_{50} is 2.53 g/kg, an indication of considerable dermal absorption [119]. 1-Hexanol is not a human skin irritant according to the Epstein test [121]. Also, it does not act as a skin sensitizer [122].

2-Methyl-1-pentanol. The oral LD_{50} for rats is reported to be 1.41 g/kg. For rabbits the dermal LD_{50} is 3.56 g/kg. In the open-patch test, 2-methyl-1-pentanol caused mild skin irritation in rabbits. In humans, inhalation of 50 ppm irritates mucous membranes [123].

4-Methyl-2-pentanol. The oral LD_{50} for rats is 2.59 g/kg [119]. Symptoms of acute intoxication are anesthesia, gastrointestinal irritation, and congestion of the mesenteric blood vessels [119].

In the rabbit eye, 4-methyl-2-pentanol was found to be a mild irritant [119]; after prolonged exposure corneal lesions can occur [124]. The skin irritation is reported to be mild [79], but absorption is considerable. Repeated exposure causes severe drying of the skin due to degreasing [125].

Mice that had inhaled concentrations of 4600 ppm for 1 min showed irritated respiratory passages, drowsiness, and anesthesia; after 10 min, death occurred occasionally, and after 15 min, 8 out of 10 mice had died [125]. For humans, eye irritation is reported after an exposure to 50 ppm for 15 min; more than 50 ppm causes irritation of nose and throat. Exposure to 25 ppm does not cause symptoms to appear [126]; therefore, the TLV and MAK values are set at 25 ppm.

2-Octanol. The oral LD_{50} is 3.2 g/kg for rats [127] and 4.0 g/kg for mice [118]. Rabbit eyes develop a pronounced redness on contact with 2-octanol [118]. In guinea pigs, 2-octanol is a slight skin irritant [128]. Rats were exposed for 4.5 months to a maximum inhalation concentration of 56.3 ppm for 2 h/day, 6 days/week. Hematological changes (anemia) and changes in liver metabolism, kidneys, and myocard were observed [128].

C_9 Alcohol Mixtures. Oral LD_{50} for rats range from 2.98 g/kg [129] to 6.4 g/kg [130]. Percutaneous LD_{50} fall in a similar range [131]. Inhalation of 21.7 mg/L for 4 h causes death in rats [129]. Eye contact causes conjunctivitis and keratitis in rabbits. Inhalation of 33, 99, and 136 ppm by rabbits for 2 months caused degenerations of glial cells in the cerebral cortex and subcortex [132].

C_{10} Alcohol Mixtures (Decyl Alcohols). LD_{50} values differ considerably because of the variable composition of the mixtures. After oral administration to rats the values range from 4.72 g/kg [129] to 25.6 g/kg [130]. Dermal LD_{50} for rabbits are reported to be between 3.56 mL/kg and 18.8 mL/kg [133]. The inhalative LD_{50} for mice is 525 ppm [132].

Eye contact in rabbits causes severe irritative conjunctivitis with corneal injuries [129]. Skin contact in guinea pigs [130] and rabbits [129] for 24 h causes moderate to severe skin irritation.

In a 60-week skin painting study in mice, *n*-decanol showed tumor promoting activities in the presence of dimethylbenzanthracene initiator; however, these concentrations of *n*-decanol alone irritated the skin considerably [134].

6. References

General References

[1] E. J. Wickson: "Monohydric Alcohols," *ACS Symp. Ser.* 1981, no. 159 (March 25–26, 1980).
[2] J. A. Monick: *Alcohols, Their Chemistry, Properties and Manufacture*, Reinhold Publ. Co., New York 1968.
[3] *Fettalkohole, Rohstoffe, Verfahren, Verwendung*, 2nd ed., Henkel, Düsseldorf 1982.

Specific References

[4] J. Falbe, B. Cornils, *Fortschr. Chem. Forsch.* **11**(1968) 101.
[5] L. H. Horsley: *Azeotropic Data*, Am. Chem. Soc., Washington, D.C., 1952.
[6] R. W. Gallant, *Hydrocarbon Process.* **45** (1966) no. 10, 171.
[7] A. M. P. Tans, *Hydrocarbon Process.* **48** (1969) no. 5, 168.
[8] I. Mellan: *Industrial Solvents Handbook*, Noyes Data Corp., Park Ridge 1970, p. 123.
[9] *Lösemittel Hoechst, ein Handbuch für Laboratorium und Betrieb*, 5th ed., Hoechst, Frankfurt 1974.
[10] B. Cornils in J. Falbe (ed.): *Methodicum Chimicum*, vol. **5**, Thieme Verlag, Stuttgart 1975.
[11] *Ullmann*, 3rd. ed., vol. **9**, p. 740, vol. **12**, p. 410.
[12] B. Cornils, W. Rottig in J. Falbe (ed.): *Chemierohstoffe aus Kohle*, Thieme Verlag, Stuttgart 1977, p. 323.
[13] Union Carbide, US 4 162 262, 1976 (P. C. Ellgen, M. Bhasin). Hoechst, DE 2 925 571, 1979 (H. Hachenberg, F. Wunder, E. Leopold, H.-J. Schmidt).
[14] Institut Français du Pétrole, US 4 122 110, 1978 (A. Sugier, E. Freund). *Eur. Chem. News* 1983, no. 12, 25.
[15] M. E. Dry, *CHEMTECH* **12** (1982) 744.
[16] A. Lundeen, R. Poe, *Encycl. Chem. Process. Des.* 1977, no. 2, 465.
[17] R. E. Vincent: "Higher Linear Oxo Alcohols Manufacture," paper presented at ACS Ind. and Eng. Chem. Div. Symposium, Houston, Texas, March 25, 1980.
[18] P. Hofmann, *Fette Seifen Anstrichm.* **85** (1983) no. 3, 127.
[19] M. Sherwood, *Chem. Ind. (London)* 1982, 994.
[20] C. L. Thomas: *Catalytic Processes and Proven Catalysts*, Academic Press, New York-London 1970.
[21] J. Falbe: *New Syntheses with Carbon Monoxide,* Springer Verlag, Berlin-Heidelberg-New York 1980.
[22] Shell, US 3 239 569, 1969 (L. H. Slaugh, P. Hill, R. D. Mullineaux).
[23] Chem. Werke Hüls, DE 1 643 856, 1968 (M. Reich).
[24] BASF, DE 1 269 605, 1966 (K. Adam, E. Haarer).
[25] BASF, DE 1 277 232, 1967 (H. Corr, E. Haarer, H. Hoffmann, S. Winderl). Shell, US 3 278 612, 1966 (C. R. Greene).
[26] *Hydrocarbon Process.* **60** (1981) no. 11, 165. National Distillers, US 3 288 866, 1966.
[27] H. Bortsch, H. Reinheckel, K. Haage, *Fette Seifen Anstrichm.* **71** (1969) 357, 785, 851.
[28] K. Ziegler et al., *Justus Liebigs Ann. Chem.* **629** (1960) 1.
[29] *Ullmann*, 4th ed., vol. **7**, p. 211, vol. **11**, p. 436.
[30] N. J. Stevens, J. R. Livingston, *Chem. Eng. Prog.* **64** (1968) no. 7, 61.
[31] *Hydrocarbon Process.* **60** (1981) no. 11, 171; **57** (1978) no. 1, 145.
[32] R. Landan, D. Brown, J. L. Russel, *Prepr. Div. Pet. Chem. Am. Chem. Soc.* **15** (1970) 63–67.

[33] K. Lindner, R. Roland in K. Lindner (ed.): *Tenside, Textilhilfsmittel, Waschrohstoffe,* 2nd ed., vol. **1,** Wissenschaftliche Verlagsgesellschaft, Stuttgart 1964, p. 344.B. W. Werdelmann, *Fette Seifen Anstrichm.* **76** (1974) 1.

[34] I. M. Towbin, D. M. Boljanowskii, *Maslo. Zhir. Promst.* **32** (1966) 29.

[35] F. Püschel, *Tenside* **3** (1966) 71.

[36] F. Püschel, *Tenside Deterg.* **11** (1971) 147.

[37] F. Broich, H. Grasemann, *Erdöl Kohle Erdgas Petrochem.* **18** (1965) 250.

[38] D. Mansuy, J. F. Bartoli, J. C. Chottard, M. Lange, *Angew. Chem.* **92** (1980) 938; *Angew. Chem. Int. Ed. Engl.* **19** (1980) 909.

[39] D. Osteroth, *Seifen Öle Fette Wachse* **95** (1969) no. 19, 659.

[40] H. Bahrmann, W. Lipps, B. Cornils, *Chem. Ztg.* **106** (1982) 249.

[41] M. E. Fakey, R. A. Head, *Appl. Catal.* **5** (1983) 3.

[42] M. Röper, H. Loevenich, J. Korff, *J. Mol. Catal.* **8** (1982) no. 17, 315.

[43] H. Bahrmann, B. Cornils, *Chem. Ztg.* **104** (1980) 39.

[44] BASF, DE 843 876, 867 849, 875 346, 1941 bis 1943 (G. Wietzel, A. Scheuermann).

[45] J. Wender, M. Orchin, et al., *J. Am. Chem. Soc.* **71** (1949) 4160; **73** (1951) 2656; Science (Washington D.C.) **113** (1951) 206.

[46] H. Pichler: *Advances in Catalysis,* vol. **4,** Academic Press, New York 1952.

[47] *CEER Chem. Econ. Eng. Rev.* **9** (1976) 58.

[48] J. F. Knifton, *J. Am. Oil. Chem. Soc.* **55** (1978) 496. P. Hofmann, W. H. E. Müller, *Hydrocarbon Process.* **60** (1981) no. 10, 151. P. Hofmann, *Chem. Ztg.* **105** (1981) 311. P. Hofmann, *Fette Seifen Anstrichm.* **85** (1983) no. 3, 126.

[49] K. Esser, U. Schmidt, *Process Biochem.* **17** (1982) 46. B. Maiorella, C. R. Wilke, *Adv. Biochem. Eng.* **20** (1981) 43.J. Tremolieres: *Alcohols and Derivatives,* vol. **2,** Pergamon Press, Oxford-London 1970, p. 81.

[50] J. J. H. Hastings, *Econ. Microbiol.* **2** (1978) 31. W. L. Faith, D. B. Keyes, R. L. Clark: *Industrial Chemicals,* J. Wiley & Sons, New York 1965, p. 181.

[51] *Ullmann,* 4th ed., vol. **7,** p. 531.

[52] M. Guerbet, *C. R. Hebd. Seances Acad. Sci.* **128** (1899) no. 51, 1002. S. Veibel, J. I. Nielsen, *Tetrahedron* **23** (1967) 1723. H. Krauch, W. Kunz: *Reaktionen der organischen Chemie,* 4th ed., Hüthig Verlag, Heidelberg 1961, p. 271.

[53] E. Supp, *Hydrocarbon Process.* **60** (1981) no. 3, 71.

[54] H. Machemer, *Angew. Chem.* **64** (1952) 213.

[55] Henkel, DE 2 703 746, 1978 (R. Karl).

[56] C. Weizmann, E. Bergmann, M. Salzbacher, *J. Org. Chem.* **15** (1950) 54. R. E. Miller, G. E. Bennett, *Ind. Eng. Chem.* **63** (1961) 33.

[57] *Seifen Öle Fette Wachse* **97** (1971) 790.

[58] *Technical Service Report: Organische Produkte CPI,* Henkel, Düsseldorf 1980, p. 28.

[59] W. Straßberger, *Fette Seifen Anstrichm.* **71** (1969) 215.

[60] W. Stein, H. Rutzen, Paper presented at DGF-Tagung in Mainz (1967). Mainz (1967). Henkel Referate **4** (1969) 18.

[61] M. S. Newmann, G. Underwood, M. Renoll, *J. Am. Chem. Soc.* **71** (1949) 3362.

[62] Henkel, DE 1 139 477, 1963 (W. Stein, H. Rutzen).

[63] E. K. Euranto in S. Patai (ed.): *Chemistry of Carboxylic Acid Esters,* Interscience, New York 1969, p. 508.

[64] *Technical Service Report on Alcohols,* Esso Chemie, Hamburg 1983.

[65] *Technical Service Report on Alfol-Alcohols,* Condea Chemie, Brunsbüttel 1983.

[66] *Technical Service Report on Alcohols*, Deutsche ICI, Frankfurt/Main 1983.
[67] *Technical Service Report on Fatty Alcohols*, Henkel, Düsseldorf 1983.
[68] *Technical Service Reports on Oxoalcohols:* n-Propanol, n- and iso-Butanol, Amylalcohol, Isooctanol, 2-Ethylhexanol, Isononanol, Isodecanol, Isotridecylalcohol, Isooctadecylalcohol, Isohexadecylalcohol, 2-Ethylbutanol, Hoechst, Frankfurt/Main 1974.
[69] *Technical Service Report on Alcohols*, Union Carbide Corp., New York 1983.
[70] *Oxo Synthesis Products from Ruhrchemie*, 2nd ed., Hoechst, Frankfurt/Main 1971.
[71] G. I. Mathier, R. W. Bruce, *Can. Min. J.* **95** (1974) no. 6, 75.
[72] Cotton Inc., US 3 778 509, 1973 (H. L. Lewis, E. L. Frick, R. T. Burchill).
[73] *Technical Service Report on Alcohols*, Ethyl Internat., Baton Rouge 1976.
[74] *Technical Service Report on Oxo Alcohols*, Monsanto (FRG), Düsseldorf 1983.
[75] *Shell-Industrie-Chemikalien*, Deutsche Shell Chemie, Frankfurt/Main 1978.
[76] *Technical Service Report on Alcohols*, Ugine Kuhlmann, Paris-Cedex 1983.
[77] H. Schubert: *Aufbereitung fester mineralischer Rohstoffe*, vol. **2**, VEB-Deutscher Verlag für Grundstoffindustrie, Leipzig 1967.
[78] *Mod. Plast.* **9** (1981) 47.
[79] Emery, US 3 438 765, 1969.
[80] F. Asinger, *Erdöl Kohle Erdgas Petrochem.* **36** (1983) no. 1, 28.
[81] *Chemical Economic Handbook:* "Oxo Chemicals," SRI International, Menlo Park, Calif., 1981.
[82] *Chem. Ind. (Düsseldorf)* **34** (1983) 3.
[83] M. Buchold, *Chem. Eng.* **90** (Feb. 21, 1983) 42.
[84] R. Leysen, *Chem. Ind. (London)* 1982, 428.
[85] *Seifen Öle Fette Wachse* **107** (1981) no. 6, 140.
[86] *Surfactants and Detergent Raw Materials*, Process Economics Program, SRI International, Menlo Park, Calif., August 1981.
[87] D. E. Haupt, P. B. Schwin, *J. Am. Oil Chem. Soc.* **55** (1978) 28. P. L. Layman, *Chem. Eng. News* **60** (1982) no. 1, 13.
[88] W. Stein, *Fette Seifen Anstrichm.* **84** (1982) no. 2, 45.
[89] *Mineral- und Brennstoffnormen*, DIN-Taschenbuch 32, Beuth-Vertrieb, Berlin-Köln-Frankfurt 1972.
[90] *Annual Book ASTM Standards*, part 15, 23, 24, 25, 27, 29 (1978).
[91] *Deutsche Einheitsmethoden zur Untersuchung von Fetten, Fettprodukten und verwandten Stoffen*, Wissenschaftliche Verlagsgesellschaft, Stuttgart 1950 – 1979.
[92] *Ullmann*, 4th ed., vol. **11**, p. 439.
[93] E. Breitmeier, W. Voelter: *^{13}C-NMR-Spectroscopy*, Verlag Chemie, Weinheim 1974, p. 138.
[94] GAF, GB 968 928, 1962 (M. E. Chiddix, O. F.Hecht). BASF, DE 1 284 964, 1967 (D. Neubauer, N. v. Kutepow). Cumberland Chemical Corp., US 3 257 465, 1966 (M. W. Leeds, H. L. Komarowski).
[95] Henkel, US 3 729 520, 1973 (H. Rutzen, W. Rittmeister).
[96] Houben-Weyl: *Methoden der organischen Chemie, Sauerstoffverbindungen I*, vol. **6**, part 2, Thieme Verlag, Stuttgart 1963, p. 5 – 70.
[97] O. C. Dermer, *Chem. Rev.* **14** (1934) 408.
[98] D. C. Bradley, F. A. Cotton, *Prog. Inorg. Chem.* **2** (1960) 303.
[99] D. C. Bradley, J. M. Thomas, *J. Chem. Soc.* 1959, 3404.
[100] H. Krauch, W. Kunz: *Reaktionen der organischen Chemie*, 3rd ed., Hüthig Verlag, Heidelberg 1966, p. 592.
[101] R. C. Mehrota, *J. Am. Chem. Soc.* **76** (1954) 2266.

[102] Du Pont, US 2 822 348, 1958 (J. H. Haslam).
[103] *Alkali and Alkali Earth Metal Alcoholates-Reactions and Applications,* Dynamit Nobel, Troisdorf-Oberlar.
[104] Ethyl Corp., FR 1 070 601, 1952 (T. P. Whaley).
[105] Wacker Chemie, DE 928 467, 1953 (A. Gerber, O. Leschhorn, H. Müller, J. Rambausek).
[106] J. Rambausek).Mathieson Alkali Works, US 2 069 403, 1936 (G. L. Cunningham).
[107] Armour & Co., GB 840 499, 1960 (S. O. Grimsdick).
[108] Dynamit Nobel, BE 741 121, 1968 (M. M. A. Lenz, O. Bleh, E. Termin).
[109] Stauffer-Chemical, GB 954 892, 1960 (W. E. Smith, A. R. Anderson).
[110] Chemische Werke Hüls, DE 968 903, 1952 (A. Coenen).
[111] Dow Chemical, US 2 796 443, 1956 (R. H. Meyer, A. K. Johnson).
[112] Continental Oil, GB 874 585, 1960.
[113] G. D. Byrkit, E. C. Soule, *Chem. Eng. News* **22** (1944) 1903.
[114] M. Sittig: *Sodium, Its Manufacture, Properties and Uses,* Reinhold Publ., New York 1956, p. 260.
[115] Hoechst, DE 2 021 831, 1970 (B. Dietrich).
[116] F. Schlenker, *Kunststoffe* **47** (1957) 7.
[117] E. Bar, F. Griepentrog, *Med. Ernähr.* **8** (1967) 244.
[118] G. N. Zaeva, V. J. Fedorova, *Toksikol. Nov. Prom. Khim. Veshchestu.* **5** (1963) 51.
[119] H. F. Smyth, C. P. Carpenter, C. S. Weil, *AMA Arch. Ind. Hyg. Occup. Med.* **4** (1951) 119.
[120] D. L. Opdyke, *J. Food Cosmet. Toxicol.* **13** (1975) 695.
[121] A. M. Kligman, *J. Invest. Dermatol.* **47** (1966) 393.
[122] W. L. Epstein, *Report to RIFM,* August 27, 1974.
[123] *Registry of Toxic Effects of Chemical Substances,* vol. **II,** U.S. Dept. of Health and Human Services, 1979. 1960. Center for Disease Control, NIOSH, Cincinnati, Ohio 45226, Sept. 1980, p. 205.
[124] Shell Chemical Corporation, Ind. Hyg. Bull. 1957.
[125] W. A. McOmic, H. H. Anderson, *Univ. Calif. Berkeley Publ. Pharmacol.* **2** (1949) 217.
[126] L. Silverman, H. F. Schulte, M. W. First, *J. Ind. Hyg. Toxicol.* **28** (1946) 262.
[127] D. W. Fassett in G. D. Clayton, F. E. Clayton (eds.): *Patty's Industrial Hygiene and Toxicology,* 3rd ed., vol. **2,** Wiley-Interscience, New York 1963.
[128] E. Browning: *Toxicity of Organic Solvents,* Med. Res. Council, Ind. Health Research Report no. 80, London 1937.
[129] R. A. Scala, E. G. Burtis, *Am. Ind. Hyg. Assoc. J.* **34** (1973) 493.
[130] G. D. Clayton, F. E. Clayton (eds.): *Patty's Industrial Hygiene and Toxicology,* 3rd ed., vol. **2A,** Wiley-Interscience, New York 1981, p. 4622.
[131] H. F. Smith, Jr., C. P. Carpenter, C. S. Weil, *J. Ind. Hyg. Toxicol.* **31** (1949) 60.
[132] Y. L. Egorov, L. A. Andrianov, *Uch. Zap. Mosk. Nauchno. Issled. Inst. Gig. im. F. F. Erismana* **9** (1961) 47.
[133] Y. L. Egorov, *Toksikol. Gig. Prod. Neftekhim. Neftekhim. Proizvod. Vses. Konf. Dokl 1 2nd* 1972, 98.
[134] J. Sice, *Toxicol. Appl. Pharmacol.* **9** (1966) 70.

Alcohols, Polyhydric

Individual keywords: → Ethylene Glycol, → Propanediols, → Butanediols, Butenediol, and Butynediol, → Glycerol.

PETER WERLE, Degussa AG, Hanau-Wolfgang, Federal Republic of Germany (Section 2.9, Chaps. 3, 4, 5, 6)

MARCUS MORAWIETZ, Degussa AG, Hanau-Wolfgang, Federal Republic of Germany (Chap. 2, Sections 2.1 – 2.8)

1.	General Aspects	321
2.	Diols	326
2.1.	1,5-Pentanediol	326
2.2.	1,3-Propanediols	326
2.2.1.	2,2-Dimethyl-1,3-propanediol (Neopentyl Glycol)	326
2.2.2.	Hydroxypivalic Acid Neopentyl Glycol Ester	329
2.2.3.	2-Methyl-2-propyl-1,3-propanediol and 2-Butyl-2-ethyl-1,3-propanediol	330
2.2.4.	2-*sec*-Butyl-2-methyl-1,3-propanediol	330
2.2.5.	1,3-Propanediol	330
2.2.6.	2-Methyl-1,3-propanediol	331
2.3.	1,6-Hexanediol	331
2.4.	Hexynediols	332
2.4.1.	3-Hexyne-2,5-diol	332
2.4.2.	2,5-Dimethyl-3-hexyne-2,5-diol	333
2.5.	1,10-Decanediol	333
2.6.	2,2-Bis(4-hydroxycyclohexyl)propane	333
2.7.	1,4-Bis(hydroxymethyl)cyclohexane	334
2.8.	2,2,4-Trimethyl-1,3-pentanediol	336
2.9.	Vicinal Diols by Hydroxylation of Olefins with Peracids	337
2.9.1.	1,2-Pentanediol	337
2.9.2.	Other 1,2-Diols	337
3.	Triols	338
3.1.	Trimethylolpropane	338
3.2.	Trimethylolethane	339
4.	Tetrols	340
4.1.	Ditrimethylolpropane	340
4.2.	Pentaerythritol	340
5.	Higher Polyols	342
5.1.	Dipentaerythritol	342
5.2.	Tripentaerythritol	343
6.	Toxicology	343
7.	References	344

1. General Aspects

The properties common to all polyhydric alcohols are determined by the hydroxyl groups: hydrogen bonding leads to high boiling points, high viscosity, and solvency for polar substances. Other important characteristics are the structure of the alcohols and the accessibility of the hydroxyl groups (primary, secondary, or tertiary), which influence the chemical behavior and the thermal stability of the hydroxyl functions. Alcohol groups next to a *neo* structure (e.g., pentaerythritol or neopentyl glycol) exhibit higher resistence towards elimination and degradation because of the lack of β-hydrogen atoms. Some of the important physical properties of the alcohols discussed in this article are given in Table 1.

Important reactions of polyhydric alcohols are those with isocyanates to form urethanes, with acids and acid anhydrides to form esters, and with aldehydes or ketones to form acetals or ketals.

Polyhydric alcohols, especially triols and tetrols, are normally regarded as polyols, but in the literature poly(ether alcohol(s) such as poly (ethylene glycol)s and/or poly-(ester alcohols) are sometimes also classified as polyols.

Production. Each class of polyhydric alcohol has its own basic production method and raw materials. The most important synthetic strategies are presented in Scheme 1.

The reactions can be classfied into the following general categories:

a) Epoxidation of alkenes followed by acid-catalyzed cleavage of the epoxide ring to obtain the 1,2-diol.
b) Base-catalyzed aldol addition of formaldehyde to the appropriate higher aldehyde, followed by reduction, either by Cannizzaro reaction with excess formaldehyde and base or by catalytic hydrogenation. Reduction by the Cannizzaro reaction causes the formation of large amounts of formate salt, so that several recrystallization steps are required for the separation of the salts from the products. For the catalytic hydrogenation to be worthwhile, the equilibrium of the aldol addition must lie sufficiently to the right, that is, sufficient β-hydroxyaldehyde must be formed in the first step.
c) Addition of water to α,β-unsaturated aldehydes also results in formation of β-hydroxyaldehydes. The corresponding 1,3-diol can be obtained by subsequent hydrogenation.
d) Hydroformylation of epoxides represents the third method for producing β-hydroxyaldehydes for hydrogenation to 1,3-diols. In contrast to the aldol method (b), the methods based on Michael addition of water (c) and the cleavage of epoxides (a) are limited to the generation of one 1,3-diol unit.
e) Addition of aldehydes or ketones to acetylene to give the alkynediols or the corresponding 1,4-diols by subsequent hydrogenation.
f) Catalytic hydrogenation of carboxylic acids and their esters.
g) Hydrogenation of phenols or phenol derivatives (e.g., bisphenol A).

Uses. The chief use of polyhydric alcohols is as components of saturated and unsaturated polyester resins and alkyd resins. These resins, which are known for their excellent stability, are used in a wide range of varnishes and coatings. The addition of polyisocyanates to polyetherpolyols and polyesterpolyols derived from polyhydric alcohols leads to polyurethanes, which are used for the production of elastomers and various types of foams. Other uses are as synthetic lubricants, plasticizers, dispersants, fibers, and as additives to poly(vinyl chloride). Some of the diols are used in the production of pharmaceuticals and fragrances.

Table 1. Physical properties of polyhydric alcohols

Compound	M_r Molecular formula	ϱ, g/cm³ (°C)	bp, °C (kPa)	mp, °C	Dynamic viscosity, mPa · s (°C)	Flash point, °C
1,5-Pentanediol	104.15 $C_5H_{12}O_2$	0.992 (20)	240–244 (101.3)	ca. –16	128 (20) 48 (40)	136[a]
2,2-Dimethyl-1,3-propanediol (Neopentyl glycol)	104.15 $C_5H_{12}O_2$	1.060 (20)	209–210 (101.3)	130	174 (25)	103[b] 151.6[c]
Hydroxypivalic acid neopentyl glycol ester	204.26 $C_{10}H_{20}O_4$	1.019 (20) 1.000 (85)	292 (decomp., 103.3) 163 (1)	50–51	70 (60)	161
2-Methyl-2-propyl-1,3-propanediol	132.20 $C_7H_{16}O_2$	0.913 (60)	230 (102.3) 111–113 (0.5)	56–58		130
2-Butyl-2-ethyl-1,3-propanediol	160.20 $C_9H_{20}O_2$	0.9266 (50)	132–133 (0.7)	40–43	8.5 (100)	144
2-sec-Butyl-2-methyl-1,3-propanediol	146.22 $C_8H_{18}O_2$		92–97 (0.1)	52.5 (decomp.)		
2-Methyl-1,3-propanediol	90.12 $C_4H_{10}O_2$		213 (101.3)	–91	68 (25)	> 230
1,6-Hexanediol	118.18 $C_6H_{14}O_2$	0.965 (50) 0.927 (104)	252 (101.3)	40–42	46.86 (48.6) 6.901 (104)	147[a]
3-Hexyne-2,5-diol	114.15 $C_6H_{10}O_2$	1.023 (13)	ca. 100 (0.1) 113–114 (2)	40–70	6 (90)	
2,5-Dimethyl-3-hexyne-2,5-diol	142.20 $C_8H_{14}O_2$			96–97		
1,10-Decanediol	174.29 $C_{10}H_{22}O_2$	0.89 (80)	160 (0.65)	71–73	15.8 (80)	152[a]
2,2-Bis(4-hydroxycyclohexyl)-propane[d]	240.37 $C_{15}H_{28}O_2$	0.958 (155)	235–245 (2.7)	125–163	114 (152)	152[a]
1,4-Bis(hydroxymethyl)cyclo-hexane	144.24 $C_8H_{16}O_2$	1.041 (20) 0.994 (100)	286 (100) 163.9 (1.33)	31.5[e]	218 (70) 40.2 (100)	162
2,2,4-Trimethyl-1,3-pentanediol	146.23 $C_8H_{18}O_2$	0.897	215–235 (101.3)	46–55	27 (50)	110[f]

Alcohols, Polyhydric

Table 1. (continued)

Compound	M_r Molecular formula	ϱ, g/cm^3 (°C)	bp, °C (kPa)	mp, °C	Dynamic viscosity, mPa·s (°C)	Flash point, °C
1,2-Pentanediol	104.15 $C_5H_{12}O_2$	0.980 (20)	210 (101.3)			105
2,3-Dimethyl-2,3-butanediol (Pinacol)	118.18 $C_6H_{14}O_2$	0.967 (20)	174 (101.3)	42		75
Trimethylolpropane	134.18 $C_6H_{14}O_2$	1.084 (20)	285 (101.3)	58		
Trimethylolethane	120.15 $C_5H_{12}O_2$	1.210 (20)	283 (101.3)	202		
Ditrimethylolpropane	250.34 $C_{12}H_{26}O_5$			112–114		
Pentaerythritol	136.15 $C_5H_{12}O_4$	1.396 (20)	276 (4)	260–262		240[a]
Dipentaerythritol	254.28 $C_{10}H_{22}O_7$	1.365 (20)		222		
Tripentaerythritol	372.42 $C_{15}H_{32}O_{10}$	1.300 (20)		248		

[a] Pensky-Marten (closed cup).
[b] BASF.
[c] Eastman.
[d] Isomer distribution: cis/cis 3–3.5%; cis/trans 30–35%; trans/trans 60–65%.
[e] Softening point.
[f] Cleveland (open cup).

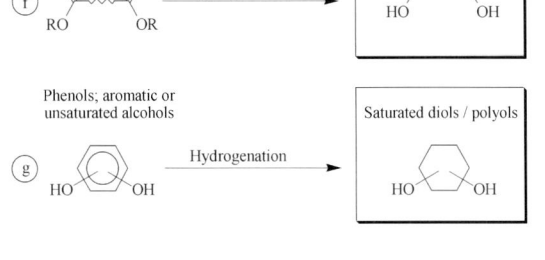

Figure 1. Synthetic routes to polyhydric alcohols

2. Diols

2.1. 1,5-Pentanediol

1,5-Pentanediol [111-29-5], pentamethylene glycol, HOCH$_2$(CH$_2$)$_3$CH$_2$OH, is a colorless liquid (see Table 1), soluble in water, alcohol, acetone, and relatively insoluble in aliphatic and aromatic hydrocarbons.

Production. 1,5-Pentanediol usually is produced by catalytic hydrogenation of glutaric acid or of its esters, although dicarboxylic acid mixtures containing glutaric acid also may be used (see Section 2.3). The crude products are purified by distillation. Manufacturers of 1,5-pentanediol are BASF and Ube Ind.; the world capacity is about 500 t/a.

Uses. 1,5-Pentanediol is the starting material for the production of various heterocyclic compounds, such as 1-methylpiperidine, which are used in the synthesis of pharmaceuticals and pesticides. The substance is also used in the production of fragrances and as a component of polyesters and polyurethanes.

2.2. 1,3-Propanediols

2.2.1. 2,2-Dimethyl-1,3-propanediol (Neopentyl Glycol)

Neopentyl glycol [126-30-7] has become an industrially important and versatile diol [1], particularly as a building block for polyesters and polyurethanes.

$$HOCH_2C(CH_3)_2CH_2OH$$

This is because of the wide availability of the starting materials, isobutyraldehyde and formaldehyde, and the extraordinary stability of its derivatives resulting from the quaternary structure and the absence of hydrogen atoms in the β-position.

Physical Properties. Neopentyl glycol (see Table 1), colorless crystals, has a phase transition at 40–42 °C. It is hygroscopic at relative humidities greater than 50 % and sublimes readily, even somewhat below the melting point. Neopentyl glycol is soluble in water, alcohols, and ketones, moderately soluble in hot aromatic solvents, such as benzene and toluene, and relatively insoluble in aliphatic and cycloaliphatic solvents.

Chemical Properties. Pure neopentyl glycol is thermally stable up to the boiling point. However, in the presence of alkali salts or bases, neopentyl glycol decomposes above 140 °C.

In addition to the reactions typical of primary alcohols, such as ester, ether, and carbamate formation, 1,3-diols give six-membered cyclic derivatives with carbonyl compounds, carbonates, phosphites, sulfites, and borates.

Production. The aldol addition of isobutyraldehyde (2-methylpropanal) and formaldehyde gives hydroxypivaldehyde (3-hydroxy-2,2-dimethylpropanal) [597-31-9], which is then reduced to neopentyl glycol:

$$\text{H}_3\text{C-CH(CH}_3\text{)-CHO} + \text{CH}_2\text{O} \rightleftharpoons \text{HO-CH}_2\text{-C(CH}_3\text{)}_2\text{-CHO}$$

$$\longrightarrow \text{HO-CH}_2\text{-C(CH}_3\text{)}_2\text{-CH}_2\text{-OH}$$

Both aldol addition and subsequent reduction are exothermic. Hydroxypivaldehyde can be reduced either by a crossed Cannizzaro reaction with equimolar amounts of formaldehyde and a base, or by catalytic hydrogenation (see Scheme 1).

Production by the Cannizzaro Reaction. In this process, aldol addition and reduction can be performed either simultaneously [2] or in two stages [3]. However, the individual reactions can be controlled more selectively in the two-stage process. Reduction of hydroxypivaldehyde with formaldehyde/sodium hydroxide solution leads to the coproduction of 0.8–0.9 t sodium formate per tonne of neopentyl glycol and to high production costs. The sodium formate has a limited commercial value.

Production by Catalytic Hydrogenation. This process usually is performed in two stages. The type of catalysts used for the aldol addition has a significant effect on the technical requirements of the hydrogenation and purification stages.

If *inorganic bases* such as potassium carbonate [4] or sodium hydroxide [5], [6] are used as catalysts, satisfactory yields (ca. 80 %) of neopentyl glycol can be obtained only with an excess of isobutyraldehyde. Side products formed by aldol addition, Cannizzaro reactions, or Tishchenko reactions are 2,2,4-trimethyl-1,3-pentanediol, neopentyl glycol isobutyrate, and hydroxypivalic acid neopentyl glycol ester (Section 2.2.2). Salts are also formed which interfere with the hydrogenation catalyst or cause decomposition during distillation. Sodium hydroxide always gives some Cannizzaro reaction, which can be partially suppressed by excess isobutyraldehyde and/or water soluble solvents such as aliphatic alcohols [6].

Hydroxypivaldehyde is separated from salts by extraction with dibutylether and hydrogenated on copper chromite catalyst between 175 and 220 °C.

In a variant of this process [7] hydrogenation and hydrogenolysis are performed stepwise at different temperatures (120–160 °C and 175–190 °C). At the higher temperature, ester byproducts also are hydrogenated to the corresponding alcohols.

Ruhrchemie [8] has avoided the interference of alkali salts, formed during the aldol addition, in the hydrogenation and in the purification stages by diluting the hydroxypivaldehyde with isobutanol, evaporating the mixture, and then hydrogenating the hydroxypivaldehyde in the gas phase at 110–150 °C over cobalt, copper, or nickel catalysts.

Tertiary amines are the state-of-the-art catalysts for the aldol addition.[9]. Hydroxypivaldehyde is formed rapidly by the reaction of isobutyraldehyde (up to 10 % excess), formaldehyde, and trialkylamine. The reaction is characterized by the complete conversion of formaldehyde, which is the main difference to the classical processes for trimethylolpropane and pentaerythritol. The excess of isobutyraldehyde is distilled and recycled together with the amine. The reaction runs very selectively and therefore most of the side products which are typical of the production based on the Cannizzaro reaction are negligible. The conditions of the hydrogenation reaction vary from 80 to 200 °C and 35–300 bar. Heterogeneous catalysts based on cobalt, copper, or nickel can be used in the fixed-bed reactor [10]. After fractional distillation a very good yield of high-purity neopentyl glycol is obtained.

Quality, Storage, and Transportation. Commercial neopentyl glycol has a purity of about 99 % (determination by GLC). Impurities may include neopentyl glycol monoformate, neopentyl glycol monoisobutyrate, 2,2,4-trimethyl-1,3-pentanediol, and the cyclic acetal from the reaction of hydroxypivaldehyde and neopentyl glycol.

During storage, higher temperatures and stacking pressures can lead to caking of the lower stack layers. As a hygroscopic material, neopentyl glycol must be stored dry.

Producers. Manufacturers are: BASF, Hoechst Celanese, Degussa-Hüls, Eastman Kodak, Koei, Mitsubishi Gas Chemical, Perstorp, and Polioli. World capacity is estimated at about 220 000 t/a.

Uses. The main use of neopentyl glycol is in saturated polyester resins for coil coatings and internal coating of tins, packing material varnishes, car varnishes, electrical insulation, and wire coatings, as well as in alkyd resins for paint varnishes, household varnishes, and self-drying industrial and ship varnishes. The alkyd resins can be used as solvent-containing, radiation-cured, water-dilutable, and powder varnishes and as high-solids paints. Neopentyl glycol is gaining importance particularly as a diol component in powder coating systems. In addition to the technological advantages of these varnishes, they are highly stable with respect to hydrolysis, to attack by alkali, and to heat; they also are weather-resistant.

Neopentyl glycol imparts the same advantages to unsaturated polyester resins, which are used for gel coats, corrosion resistant containers, and in molding materials for building components, etc. In addition, neopentyl glycol is a component of polyesterpolyols in plasticizers, its esters are synthetic lubricants, it is used as dispersant for titanium dioxide, and as starting material for polyurethanes.

2.2.2. Hydroxypivalic Acid Neopentyl Glycol Ester [11]

Hydroxypivalic acid neopentyl glycol ester, (3-Hydroxy-2,2-dimethylpropyl)-3-hydroxy-2,2-dimethylpropionate (HPN) [1115-20-4] is the monohydroxypivalate of neopentyl glycol:

$$\text{HO-CH}_2\text{-C(CH}_3)_2\text{-C(O)-O-CH}_2\text{-C(CH}_3)_2\text{-CH}_2\text{OH}$$

The compound has properties similar to neopentyl glycol but in addition specific characteristics that are responsible for its increasing use. Manufacturers are Eastman Chemical, BASF, Mitsubishi Gas, and Union Carbide. World capacity is estimated at 2000–2500 t/a.

Physical Properties. Hydroxypivalic acid neopentyl glycol ester is a colorless crystalline solid with a melting point of 50 °C (see Table 1). It is highly soluble in polar organic solvents; its solubility in water is limited to 27.4 % at 25 °C; conversely, 52.4 % water dissolves in HPN.

Chemical Properties. Pure HPN can be distilled at reduced pressure without decomposition up to about 200 °C. In the presence of alkali metal salt impurities, significant decomposition occurs even at temperatures of about 150 °C. Both hydroxyl groups undergo the usual reactions of primary alcohols. As with neopentyl glycol, the quaternary structure contributes to the excellent stability of HPN derivatives.

Production. Hydroxypivalic acid neopentyl glycol ester is produced by a Tishchenko reaction directly from hydroxypivaldehyde in the presence of a basic catalyst (e.g., aluminum oxide) [12] or classically by thermal treatment [13] (Section 2.2.1). Alternatively, it can be synthesized by esterification of neopentyl glycol with hydroxypivalic acid [14]. High-purity hydroxypivalic acid neopentyl glycol ester is obtained by subsequent distillation [15].

Uses. Hydroxypivalic acid neopentyl glycol ester is a diol modification agent in polyesters, polyurethanes, and plasticizers. As a component of polyester varnish systems, HPN has special advantages with respect to flow properties, adhesion firmness, flexibility, thermal stability, resistance to light (UV), and impact strength at low temperature. The ester is used as a diol component of powder varnishes, coil coatings (e.g., internal coating of cans), electrophoresis varnishes, and unsaturated polyester resins; it is used also in radiation-cured varnish resins, binders for water lacquers, plasticizers, and polyurethane foams.

2.2.3. 2-Methyl-2-propyl-1,3-propanediol and 2-Butyl-2-ethyl-1,3-propanediol

2-Methyl-2-propyl-1,3-propanediol [*78-26-2*] and 2-butyl-2-ethyl-1,3-propanediol [*115-84-4*] (see Table 1) are produced either by the salt-free procedure based on aldol addition followed by hydrogenation [16] or by the classical combination of hydroxymethylation and Cannizarro-type disporportionation with formaldehyde [17] (see Figure 1).

The production costs are considerably higher than for neopentyl glycol, since the synthesis of 2-methylpentanal and of 2-ethylhexanal from propanal and butanal requires two more steps: aldol condensation and hydrogenation.

Since ca. 1980 2-butyl-2-ethyl-1,3-propanediol has become a very useful diol for polyester or polyurethane coatings [18]. The world production of 2-butyl-2-ethyl-1,3-propanediol is estimated at about 10 000 t/a; the main producer is Neste.

The best known fine chemicals produced from 2-methyl-2-propyl-1,3-propanediol are the dicarbamates Meprobamate (β-methyl-β-propyltrimethylene dicarbamate [*57-53-4*]), used as a tranquilizer, and Carisoprodol (*N*-isopropyl-β-methyl-β-propyl-trimethylene dicarbamate [*78-44-4*]), used as a muscle relaxant.

2.2.4. 2-*sec*-Butyl-2-methyl-1,3-propanediol

2-*sec*-Butyl-2-methyl-1,3-propanediol (see Table 1) [*813-60-5*] can be made via 2,3-dimethylbutanal by simultaneous aldol addition and Cannizzaro reaction with formaldehyde [25]. So far it has been used only in the production of the tranquilizer Mebutamate (β-*sec*-butyl-β-methyl-trimethylene dicarbamate [*64-55-1*]).

2.2.5. 1,3-Propanediol

1,3-Propanediol [*504-63-2*] is a colorless liquid at room temperature in contrast to 2,2-dialkyl-1,3-diols (see Table 1). 1,3-Propanediol is produced commercially by Degussa starting from acrolein [*107-02-8*].

$$CH_2CHCHO + H_2O \longrightarrow HOHCH_2CH_2CHO$$
$$HOHCH_2CH_2CHO + H_2 \longrightarrow HOHCH_2CH_2CH_2OH$$

The addition of water under mild acidic conditions gives 3–hydroxypropionaldehyde [*2134-29-4*] with high selectivity (see Scheme 1). Preferentially buffer solutions with a pH 4 – 5 [19] or weak acidic ion exchange resins [20] are used as catalysts. Further hydrogenation of this aqueous solutions gives 1,3-propanediol [21]. There is an alternative route via hydroformylation of ethyleneoxide and subsequent hydrogenation of the intermediate 3-hydroxypropionaldehyde [22].

A new large market for 1,3-propanediol will be in polyester coatings and in the production of poly(trimethylene terephthalate), a new material for the production of high quality carpet fibres [23].

2.2.6. 2-Methyl-1,3-propanediol

2-Methyl-1,3-propanediol [2163-42-0] is a fairly new commercially available diol and is produced by Arco with an estimated capacity of about 7000 t/a [24]. The process is designed for the production of 1,4-butanediol [110-63-4] as an alternative to Reppe technology and the acetoxylation of butadiene.

The raw material propylene oxide [75-56-9] is isomerized to allyl alcohol [107-18-6] in the first step. Hydroformylation of allyl alcohol gives 4-hydroxybutanal (79 %) [25714-71-0] and 3-hydroxy-2-methylpropanal (11 %) [38433-80-6]. The last step comprises hydrogenation of the *n* and *iso* intermediates to the corresponding diols. 2-Methyl-1,3-propanediol is used in polyester coatings.

2.3. 1,6-Hexanediol

1,6-Hexanediol [629-11-8], hexamethylene glycol, $HOCH_2(CH_2)_4CH_2OH$, colorless crystals (see Table 1), is soluble in water and other polar solvents.

Production. 1,6-Hexanediol is produced industrially by the catalytic hydrogenation of adipic acid or of its esters. Mixtures of dicarboxylic acids and hydroxycarboxylic acids with C_6 components formed in other processes (e.g., in cyclohexane oxidation) also can be used.

The acids are hydrogenated continuously at 170–240 °C and at 15.0–30.0 MPa (150–300 bar) on a suitable catalyst either in a trickle-flow (downflow) or a bubble-flow (upflow) fixed-bed reactor. The reactor temperature is controlled by circulating part of the reactor discharge. The hydrogen required for the hydrogenation is fed together with the recycle gas through the recycle gas compressor to the reactor (see Fig. 2).

Side products of the synthesis are alcohols, ethers, diols, and esters. Pure 1,6-hexanediol is obtained by fractional distillation of the crude reactor discharge.

For the hydrogenation of dicarboxylic acids, catalysts containing cobalt, copper, or manganese are suitable [26]. For the hydrogenation of esters, catalysts such as copper chromite or copper with added zinc and barium are used as "full catalysts" or on inert carriers [27], [28]. Both acids and esters also may be hydrogenated using suspended catalysts.

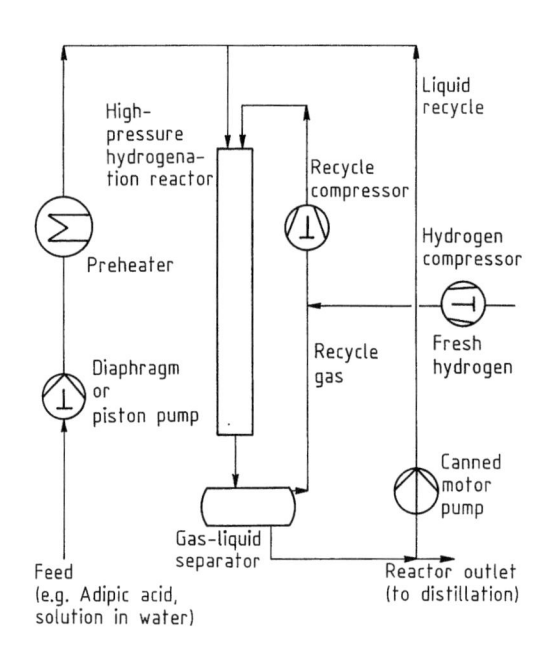

Figure 2. Hydrogenation of adipic acid or of its esters (trickle-flow fixed bed)

Quality and Analysis. The assay of the pure product is about 98%; impurities are various diols and ϵ-caprolactone as well as traces of water. The color number of the product determined photometrically according to the Pt/Co scale must not exceed 15 APHA. Above 70 °C, 1,6-hexanediol tends to turn yellow.

Storage and Transportation. 1,6-Hexanediol in the form of flakes or as a solidified melt is stored and shipped in barrels. It is transported as a melt in stainless steel containers, tank trucks, or tank cars.

Uses and Capacity. The most important use of 1,6-hexanediol is in the synthesis of polyesters and polyurethanes. In addition it is used in the production of varnishes, adhesives, pharmaceuticals, and textile auxiliaries.

The world capacity of 1,6-hexanediol is ca. 25 000 t/a; manufacturers are BASF and Ube.

2.4. Hexynediols

2.4.1. 3-Hexyne-2,5-diol

3-Hexyne-2,5-diol [3031-66-1] is a more or less solid yellow mass of crystals (see Table 1) depending on the relative content of the *meso*form. It is miscible with water and polar solvents.

$$CH_3CH(OH)-C\equiv C-CH(OH)CH_3$$

3-Hexyne-2,5-diol is produced by BASF from acetylene and acetaldehyde by the Reppe ethynylation method (→ Acetylene). 3-Hexyne-2,5-diol is used, for example, as a brightener for nickel baths (80 % aqueous solution) or as starting material for production of 2,5-hexanediol [28], but especially in the production of 4-hydroxy-2,5-dimethyl-3-(2*H*)-furanone [*3658-77-3*], trade name Furaneol [29], used as pineapple and strawberry flavor.

2.4.2. 2,5-Dimethyl-3-hexyne-2,5-diol

2,5-Dimethyl-3-hexyne-2,5-diol [*142-30-3*], $(CH_3)_2C(OH)-C\equiv C-C(OH)(CH_3)_2$, white crystals (see Table 1), is produced by addition of acetone to acetylene. It is an intermediate of minor importance in the production of 2,5-bis(*tert*-butylperoxy)-2,5-dimethyl-3-hexyne, which is used as an initiator in polymerization processes, as well as in the production of moschus fragrances, of chrysanthemumic acid, and of surfactants (surfinols) by oxalkylation (Air Products). Manufacturers are Enichim and Air Products.

2.5. 1,10-Decanediol

1,10-Decanediol [*112-47-0*] (see Table 1) is produced from sebacic acid by a process analogous to that described for 1,6-hexanediol [30] (see Section 2.3). It is an intermediate of minor importance in the production of polyesters and polyurethanes.

$$HOCH_2(CH_2)_8CH_2OH$$

Manufacturers are Ashland and Degussa-Hüls.

2.6. 2,2-Bis(4-hydroxycyclohexyl)propane

2,2-Bis(4-hydroxycyclohexyl)propane [*80-04-6*] is a crystalline, colorless, and odorless compound (see Table 1).

The conformation of the cyclohexane rings gives rise to cis-cis, cis-trans and trans-trans isomers. The industrial product is a mixture of these stereoisomers. The physical properties depend on the relative amounts of the isomers present. The trans-trans compound has the lowest energy and is formed from the other isomers at higher temperature. It has the lowest solubility in solvents such as methanol, butanol, and acetone and has the highest melting point of these isomers.

Production. 2,2-Bis(4-hydroxycyclohexyl)propane is produced by hydrogenation of bisphenol A, 2,2-bis(4-hydroxyphenyl)propane [*80-05-7*], an industrially important compound made from acetone and phenol (→ Phenol Derivatives). Bisphenol A is hydrogenated at about 25.0 MPa (250 bar) and at 200 °C on a nickel, cobalt, or ruthenium catalyst [31]–[33].

Uses. 2,2-Bis(4-hydroxycyclohexyl)propane is used mainly in synthetic materials as a bifunctional component, e.g., as a modifier in the production of saturated and unsaturated polyester resins, and of oil- and fatty-acid-containing alkyd resins. Exchange of hydroxyl for amino gives a diamine which can be used in the production of glass-clear polyamides.

2.7. 1,4-Bis(hydroxymethyl)cyclohexane

1,4-Bis(hydroxymethyl)cyclohexane [*105-08-8*] is commonly known as 1,4-cyclohexanedimethanol (CHDM), also as 1,4-dimethylolcyclohexane, or hexahydro-*p*-xylylene glycol.

The compound became known in the 1960s mainly for its use in polyester films and fibers highly resistant to hydrolysis.

Physical Properties (see also Table 1). 1,4-Cyclohexanedimethanol is colorless and has a slight odor. The softening point of the *cis* isomer is 43 °C, that of the *trans* isomer 67 °C. *Cis – trans* mixtures can be isomerized at 200 °C in the presence of alkoxides to an equilibrium mixture of 42 % *cis* and 76 % *trans*. The commercial product consists of a mixture of *cis* and *trans* isomers with a typical ratio of 1/3.

1,4-Cyclohexanedimethanol is miscible with water as well as with low molecular mass alcohols; it is very slightly soluble in hydrocarbons and ethers. Its solubility at 20 °C (g in 100 g solvent) in benzene is 1.1, in trichloromethane 5.7, in water 92.0, and in methanol 92.2.

Chemical Properties. CHDM undergoes the characteristic reactions of a diol with two primary hydroxyl functions. In contrast to 1,2- and 1,3-diols reactions of CHDM with aldehydes or ketones do not result in the formation of heterocyclic products, such as 1,3-dioxolanes or 1,3-dioxenanes.

Production. The commercially important process for producing CHDM is the hydrogenation of dimethyl terephthalate (DMT) [120-61-6] via dimethyl hexahydroterephthalate (DMHT) [94-60-0] in a methanolic solution or in the molten state [34]:

$$\underset{(DMT)}{\text{DMT}} \xrightarrow[\text{Pd}]{3 H_2} \underset{(DMHT)}{\text{DMHT}} \xrightarrow[\text{Cu/Cr}]{4 H_2} \underset{(CHDM)}{\text{CHDM}}$$

The hydrogenation plant consists of two reactors. In the first reactor, a continuous circulation of product DMHT and of DMT is maintained. The molten DMT is pumped into the inlet of this reactor, which operates at 30–48 MPa and 160–180 °C with a commercial supported Pd catalyst. Temperature control is achieved by operating at sufficiently high cross-sectional loadings of the mixture of ca. 10 % in product DMHT. This allows the radial dissipation of heat via the reactor walls and largely avoids high temperature peaks. The cooled reactor effluent, which contains only minor amounts of unreacted DMT, is divided: 8–10 parts are recycled via a pump and heat exchanger to the reactor inlet. The remaining 1–2 parts (the exact amount corresponds to the amount of fresh DMT) are fed continuously into the second reactor for the final Pd-catalyzed hydrogenation.

An advantage of this process is that DMT can be hydrogenated as ca. 10 % dilute feed without handling large amounts of liquid. The yield of DMHT is typically 97–98 %, with methyl 4-methyl-4-cyclohexanecarboxylic acid methyl ester [51181-40-9] and some 1-hydroxymethyl-4-methylcyclohexane [34885-03-5] as main byproducts.

1,4-Bis(hydroxymethyl)cyclohexane (CHDM) is formed in the second stage of the hydrogenation. The crude DMHT of stage one is used in a plant analogous to that of the first stage [40], although any conventional hydrogenation plant can be used since the heat of the ester hydrogenation is comparatively small.

The industrial processes use commercial copper chromite catalysts. The effects of catalyst, residence time, and temperature must be carefully adjusted in order to achieve a *cis/trans* ratio of 1/3 to 1/4 in the CHDM product.

A constant isomer ratio is important for further processing of CHDM in polyester formation since the crystal structures of polyesters of *cis*- and *trans*-CHDM differ and thus influence the melting range and density of polyester fibers.

In addition to the side products of the first hydrogenation stage, 4-methyloxymethyl-hydroxymethylcyclohexane and bis(4-hydroxymethylcyclohexyl) ether can be formed in the second stage of ester hydrogenation. The Eastman process (see above) [34] avoids the formation of high-boiling compounds and thus saves cumbersome separation of side products by fractionation in vacuum. The diol is purified by simply removing methanol and low-boiling compounds. The CHDM of fiber quality obtained by this method is ca. 99 % pure.

Pure CHDM may also be obtained by recycling waste poly(1,4-cyclohexylene dimethylene terephthalate) by first cleaving the polyester in the presence of low molecular mass alcohols and then subjecting the resulting mixture to a hydrogenation analogous to that described above [35]. Estimated world capacity is 55 000 t/a.

Uses. An important application of CHDM is still the production of terephthalate – polyester fibers with lower densities and higher melting points than analogous fibers using ethylene glycol. They are more stable to hydrolysis and have more favorable electrical properties than similar polyesters made from other diols [35].

Unsaturated resins based on CHDM are characterized by very good properties with respect to water absorption, thermal stability, and resistence to enviromental and chemical exposure [36]. The use of CHDM in saturated polyester resins leads to powder, waterborne, and solvent-based coatings with excellent hardness and durability [37]. 1,4-Cyclohexanedimethanol is of increasing interest for application in polyurethanes and polycarbonates [38] as well as an antifogging agent and as a sensitizing additive in silver bromide emulsions.

2.8. 2,2,4-Trimethyl-1,3-pentanediol

2,24-Trimethyl-1,3-pentanediol [*144-19-4*], TMPD glycol, is a white crystalline solid (see Table 1). It is a versatile diol with a wide range of applications including polyester resins, polyurethane foams, and lubricants.

Physical and Chemical Properties. TMPD glycol is only slightly soluble ($< 2\%$) in water and kerosene, but very soluble (up to 80 %) in alcohols. The solubility in benzene, acetone, and ethers is at most 29 %.

TMPD glycol, with one primary and one secondary hydroxyl function, shows the typical reactivity of a diol. In comparison to other diols, such as CHDM (see Section 2.7) or NPG (see Section 2.2.1), TMPD glycol has an unsymmetrical structure and shielded hydroxyl groups.

Production. TMPD glycol is produced by homo-aldol condensation of isobutyraldehyde [*78-84-2*] in the presence of an alkaline catalyst followed by hydrogenation of the resulting cyclic acetal to give TMPD glycol and isobutanol. Eastman Chemical is the only manufacturer of TMPD glycol.

Uses. TMPD glycol is mainly used as raw material in the production of unsaturated and saturated polyesters (e.g. for waterborne or high-solids coatings) and as an intermediate for polyesters used in the manufacture of polyurethane elastomers and foams [39]. These resins are very stable to hydrolysis and resistant to corrosion, as well as having other desirable features, such as low density, low viscosity, and good compatibility.

2.9. Vicinal Diols by Hydroxylation of Olefins with Peracids

Some vicinal diols are manufactured on a commercial scale by the reaction of an alkene with an organic peracid, such as performic acid, produced in situ, or peracetic acid [41]. An epoxide is formed initially; acid-catalyzed cleavage of the epoxide ring leads to a 1,2-diol or a 2-hydroxyalkyl-ester or a mixture of the two. The 2-hydroxyalkyl ester is hydrolyzed to the 1,2-diol by the subsequent treatment with a base.

2.9.1. 1,2-Pentanediol

1,2-Pentanediol [5343-92-0], 1,2-dihydroxypentane, $CH_3(CH_2)_2CH(OH)CH_2OH$, is a colorless and odorless liquid which is miscible with water and polar organic solvents (see also Table 1).

The compound is made by treating a mixture of 1-pentene and formic acid with 40 % hydrogen peroxide at 30 °C [42]. 1,2-Pentanediol has attained commercial importance as a synthetic building block for systemic fungicides [43] and as a moisturizing agent in cosmetics.

2.9.2. Other 1,2-Diols

Pinacol [76-09-5], 2,3-dimethyl-2,3-butanediol, $(CH_3)_2C(OH)C(OH)(CH_3)_2$, colorless crystals, is soluble in hot water, alcohol, and ether and slightly soluble in cold water and carbon disulfide. The hexahydrate, *mp* 45 °C, crystallizes as four-sided plates (see also Table 1).

1,2-Hexanediol [6920-22-5], (*bp* 108 °C/0.5 hPa; ϱ 0.953 g/cm^3 at 20 °C) is a colorless liquid miscible with water. It is used in printing inks and in cosmetics.

1,2-Octanediol [1117-86-8] (*bp* 140 °C/1.6 hPa) is a white waxy solid, slightly soluble in water. Its production is described in [42], [44]. It is used in printing inks and in cosmetics.

3. Triols

3.1. Trimethylolpropane

Properties. Trimethylolpropane [*77-99-6*], 2-ethyl-2-hydroxymethyl-1,3-propanediol, $CH_3CH_2C(CH_2OH)_3$, is a colorless, crystalline, trivalent alcohol (see Table 1). Trimethylolpropane is quite soluble in water and polar organic solvents. The three primary hydroxyl groups undergo the normal OH group reactions.

Production. Trimethylolpropane is made by the base-catalyzed aldol addition of butyraldehyde with formaldehyde followed by Cannizzaro reaction of the intermediate 2,2-bis(hydroxymethyl)butanal with additional formaldehyde and at least a stoichiometric quantity of base (see Scheme 1):

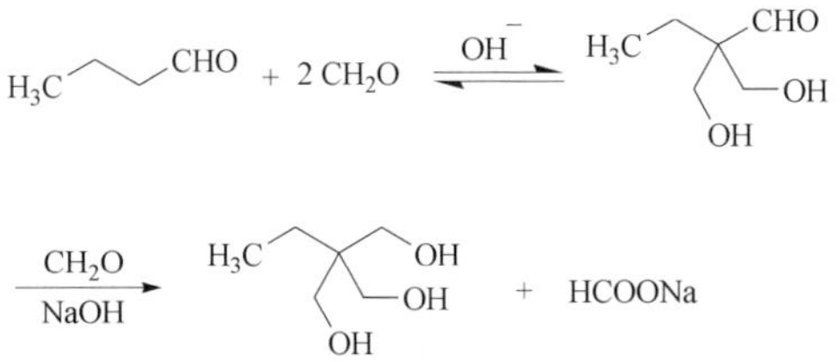

Sodium hydroxide [45], [46] or calcium hydroxide [47], [48] are used as bases in industrial-scale production. The yield in both cases is ca. 90 %. Many attempts have been made to avoid the large quantities of formate byproduct by catalytic hydrogenation of the 2,2-bis(hydroxy-methyl)butanal [49], [50]. Recently, these processes have been successful by using tertiary amines as catalyst and transforming the formed ammonium formate into methyl formate [87] or even avoiding the formation of larger amounts of formates [88].

Various procedures have been suggested for separating the trimethylolpropane from the formate. For instance, the concentrated reaction solution can be extracted with an organic solvent, which is then evaporated and the crude trimethylolpropane is purified by vacuum distillation [51], [52], [53]. In another variant [54], the aqueous reaction solution is evaporated until most of the sodium formate crystallizes and is removed by hot filtration. The liquid trimethylolpropane which remains is liberated of residual salts with an ion exchanger and then distilled. The removal of discoloring impurities is described in [55].

Cyclic diethers of the acetal type (1,3-dioxanes, formals) can be converted into TMP and methanol by metal-catalyzed hydrogenation [56].

Uses. Large quantities of trimethylolpropane and its ethoxylated derivatives are used as precursors for urethanes and polyester resins. Another important field of application is in medium-oil and short-oil alkyd resins. The resulting lacquers are characterized by excellent resistance to alkali, detergents, and water, combined with outstanding impact resistance and flexibility, as well as excellent clearness and clearness retention.

Reaction products with fatty acids (C_5-C_{10}) are components of synthetic lubricants. A strongly growing market will be the use of these blended polyol esters in chlorine-free, purely fluorocarbon based refrigerant systems. Powder and high-solids coatings represent a strong potential market for TMP, because it contributes to low viscosity. TMP acrylates are used as reactive diluents in UV-cured systems for inks and coatings, and TMP allyl ethers in unsaturated polyesters.

Economic Aspects. World production capacity for trimethylolpropane was estimated at 150 000 t in 1998, of which about 67 000 t was accounted for by Western Europe, 60 000 t by the United States, and 15 000 t by Japan. The main producers are Bayer, Perstorp, and Polioli in Western Europe; Celanese and Perstorp in the United States; and Koei and Mitsubishi Gas in Japan.

3.2. Trimethylolethane

Properties. Trimethylolethane [77-85-0], 2-hydroxymethyl-2-methyl-1,3-propanediol, $CH_3C(CH_2OH)_3$, is a colorless, crystalline substance (see Table 1), soluble in water (140 g per 100 g at 25 °C), alcohols, and acetone.

Production. Trimethylolethane is made by aldol condensation of propionaldehyde with formaldehyde, followed by reaction of the intermediate 2,2-bis(hydroxymethyl)-propanal with excess formaldehyde in the presence of sodium hydroxide or lime as basic component (see Figure 1).

$$H_3CCH_2CHO + 3 H_2CO + NaOH \longrightarrow H_3CC(CH_2OH)_3 + HCOONa$$

The resulting aqueous solution is freed from excess formaldehyde by distillation under pressure and further concentrated. The formate byproduct can be separated by extracting the trimethylolethane from the residue with organic solvents such as 2-propanol [57] or by almost complete removal of the water and dissolving the trimethylolethane in acetone or methanol.

Uses and Economic Aspects. Trimethylolethane, like trimethylolpropane, is a polyol component of short- and medium-oil alkyd resins, in which it results in a shorter drying time and, because of its shorter alkyl chain, a harder lacquer film compared with trimethylolpropane. However, its economic importance is small compared with that of pentaerythritol or trimethylolpropane. It is produced by Mallinckrodt (USA) and Mitsubishi Gas (Japan) in amounts of several thousand tonnes per annum.

4. Tetrols

4.1. Ditrimethylolpropane

Ditrimethylolpropane [*23235-61-2*] 2,2-[oxybis(methylene)]-bis(2-ethyl)-1,3-propanediol is a colorless and crystalline substance (*mp* 112–114 °C), only slightly soluble in water at ambient temperature (2.6 g/100 g at 25 °C).

$$H_3C\text{—}\overset{OH\ \ OH}{\diagdown\diagup}\text{—}O\text{—}\overset{OH\ \ OH}{\diagdown\diagup}\text{—}CH_3$$

It is a byproduct of TMP production and can be isolated from the residue of the TMP distillation step. Newer investigations show that ditrimethylolpropane can be prepared in high yield and selectivity by the reaction of TMP, formaldehyde, and 2-ethylacrolein [58]. It is used in water-thinnable alkyd resins and as a polyol component in PVC stabilizers based on Ca/Zn alkyl carboxylates. Producers are Perstorp (Sweden) and Mitsubishi Gas (Japan).

4.2. Pentaerythritol

Pentaerythritol [*115-77-5*], 2,2-bis(hydroxymethyl)-1,3-propanediol, $C(CH_2OH)_4$, was discovered by TOLLENS in 1882 when an aqueous solution of formaldehyde containing some acetaldehyde was allowed to stand with barium hydroxide [59].

Properties. Pentaerythritol is a colorless, crystalline alcohol with four primary hydroxyl groups. It crystallizes in a tetragonal system with. The tetragonal crystals convert into the cubic form at 180–190 °C. The compound sublimes on heating in vacuum without dissociation. Important physical data are listed below (see also Table 1):

Heat of combustion (p = const.)	2767 kJ/mol
Heat of formation	931 kJ/mol
Specific heat capacity at 100 °C	255 J/mol
Ignition temperature	390 °C
Solubility in water (per 100 g water)	
at 25 °C	7 g
at 97 °C	77 g

Mixtures of air and pentaerythritol dust are explosive at 490 °C and at dust concentrations of more than 30 g/m^3. The four primary hydroxyl groups undergo the normal reactions of OH groups [60]. Pentaerythritol is oxidized to tris(hydroxymethyl)acetic acid by air in the presence of platinum or palladium [61].

Production. Pentaerythritol is prepared by the reaction of acetaldehyde with formaldehyde in alkaline medium. The aldol addition results in replacement of the three α-hydrogen atoms, leading to trimethylolacetaldehyde (pentaerythrose), which is reduced to pentaerythritol by excess formaldehyde in the presence of base (sodium hydroxide or calcium hydroxide). The yield is approximately 90 %. Catalytic hydrogenation of pentaerythrose, which would avoid the formate byproduct, is not practicable. Higher homologues of pentaerythritol with the general formula

$$\left[\text{HO} \underset{\text{OH OH}}{\overset{}{\diagdown\diagup}} \text{O} \right]_n \text{H} \quad n = 2 - 4$$

are byproducts of the production of pentaerythritol. The product with $n = 2$, known as dipentaerythritol, is the primary contaminant. To minimize its formation it is necessary to increase the ratio of formaldehyde to acetaldehyde beyond the stoichiometric ratio of 4:1 [60]. Additionally, certain amounts of formaldehyde acetals are formed [62]. Both batch and continuous processes for the production of pentaerythritol have been described [63], [64].

To isolate the product, the aqueous reaction solution is first freed of excess formaldehyde and most of the water by distillation under pressure. After further concentration, usually in vacuo, the crude pentaerythritol crystallizes. The mother liquor contains the sodium formate plus residual pentaerythritol. Separation of the formate, recycling of the residual pentaerythritol, and elimination of the wastewater occur in subsequent process steps. The crude pentaerythritol is purified by recrystallization from water after treatment with activated carbon. By using lime, the process runs reversely; first calcium formate separates because of its low water solubility, the mother liquor than contains the dissolved pentaerytrithol. Residual Ca^{2+} or Na^+ ions can be removed by ion-exchange resins. The wastewater contains a certain amount of dissolved pentaerythritol, which can be recovered [65].

Quality and Analysis. Pentaerythritol is available in various grades which differ in their content of the above-mentioned byproducts including residual ash (alkali metal salts and/or alkaline earth metal salts). Higher ash content often has a detrimental effect on the color of the alkyd resin obtained from the product. Products termed pentaerythritol monograde generally contain more than 97 % pentaerythritol, whereas the technical grade consists of 80 – 95 % pentaerythritol and 5 – 15 % dipentaerythritol.

For *quality control*, the melting point, hydroxyl number, ash content, and the color or iodine number of the melt are determined. Gas chromatographic analysis after silylation with agents such as *N*-methyl-*N*-trimethylsilyltrifluoracetamide [66] gives precise information on the composition of the sample. Analysis by high-pressure liquid chromatography has also been described [67]

Uses. The major application of pentaerythritol (ca. 70 %) is the production of long-oil and medium-oil alkyd resins. The pentaerythritol structure gives the resin outstanding properties both in processing and end use. Resins prepared from pentaerythritol and rosin or tall oil are used for the modification of printing inks and nitrocellulose lacquers and for the preparation of adhesives and core binders. Solvent-free, water-thinnable alkyd resin emulsions can be prepared by using tribasic acids in the esterification process, which gives water-dispersible salts after neutralization.

Pentaerythritol esters of short- to medium-chain carboxylic acids (C_5-C_{10}) are applied as high-temperature synthetic lubricants, lubricant additives, and plasticizers. An emerging market for these esters is as refrigerant lubricants. Because of the ozone-depleting properties of CFCs they are to be replaced by chlorine-free fluorocarbons. However, since this class of compounds differs in polarity the compressor must be lubricated with oils of higher polarity [68], [69] Pentaerythritol tetranitrate is used as a detonator in fuses and, to a limited extent, as a military explosive. The tetranitrate also is used in medicine as a long-term coronary vasodilator.

Finely ground pentaerythritol in combination with metal soaps serves as a nontoxic costabilizer for poly(vinyl chloride). It also is used in flame retardant compositions (intumescent paints). Under the action of heat an insulating foam layer forms, up to 50 times the thickness of the original coating. Pentaerythritol triacrylate is used as an active diluent in radiation-cured paints. Pentaerythritol adducts with ethylene oxide are used as emulsifiers, inorganic phosphoric or phosphoric esters are used as additives in hydraulic fluids, and tetraesters with sterically hindered phenylcarboxylic acids act as a highly effective antioxidant in plastics.

Economic Aspects. In 1998, world production capacity was estimated to be 360 000 t, 71 000 t of which was in the United States, 120 000 t in Europe, 27 000 t in Japan, and 25 000 t in Canada. Smaller producers are located in many other countries. The main manufacturers are Perstorp, Hoechst-Celanese, and Degussa-Hüls. The overall pentaerythritol market is expected to grow at a rate of 1.5 – 2 %/a. The alkyd resin market should be stable, and growing use of fatty acid esters in refrigerant systems can be assumed.

5. Higher Polyols

5.1. Dipentaerythritol

Dipentaerythritol [126-58-9], 2,2-[oxybis(methylene)-bis[2-hydroxymethyl]-1,3-propanediol, is a byproduct of pentaerythritol production. Because of its low solubility in water (0.22 g/100 g H_2O at 20 °C and 10.0 g/100 g H_2O at 100 °C) it can be separated from pentaerythritol by fractional crystallization. The yield can be increased by changing the ratio of formaldehyde/acetaldehyde towards stoichiometric. Adding pentaerythritol before starting the reaction also increases the yield. An effective method is to use acrolein instead of acetaldehyde [70], [71].

In the last decade the consumption of dipentaerythritol steadily increased. The main outlets are C_5-C_{10} ester lubricants for jet engines and refrigerator systems [72], dipentahexaacrylate as cross-linking agent for UV-curing acrylates, for water-resistent intumescent paints, alkyd resins with low viscosity (high solids), and as a nonsubliming polyol component in nontoxic PVC stabilizer systems.

The total installed capacity is ca. 6000 t. Main producers are Hercules, Koei, Perstorp, Polialco, and Degussa-Hüls.

5.2. Tripentaerythritol

Tripentaerythritol [*78-24-0*], 2,2-bis{[hydroxy-2,2-bis(hydroxymethyl)propoxy]methyl} 1,3-propanediol is only slightly soluble in cold and hot water and can be isolated during the production of dipentaerythritol. It has at present only a very limited economic importance.

6. Toxicology

Table 2 summarizes the acute toxicity data of polyhydric alcohols. According to the lethal doses shown, these compounds are generally little toxic. For comparison, the LD_{50} values of the butanediols also are included in Table 2.

TMPD Glycol, 2,2,4-trimethyl-1,3-pentanediol, is classified as "slightly toxic" [78], [79], see Table 2. In rabbits, moderate eye irritation was observed. Slight to no skin irritation occured in guinea pigs, and no skin sensitization was found. Repeated skin application studies in humans gave no evidence of irritation, sensitization, photosensitization, or systemic toxic effects. In humans, TMPD glycol is rapidly excreted in the urine, partly unchanged, partly as the glucuronide and sulfate conjugates, and partly as 2,2,4-trimethyl-3-hydroxyvaleric acid.

1,4-Bis(hydroxymethyl)cyclohexane (CHDM) has a low acute toxicity (see Table 2). For metabolism of CHDM in rats, see [80].

Trimethylolpropane is practically nontoxic (see Table 2). Skin irritation was not observed.

Pentaerythritol is practically nontoxic. Health disturbances resulting from handling pentaerythritol have never been observed. High oral doses (50 g) administered to humans resulted only in slightly increased pulse rate [85]. Diarrhea was observed at high doses in animals. Pentaerythritol is readily biodegradable under aerobic conditions according to OECD Guideline No. 301 [86].

Table 2. Lethal doses of polyhydric alcohols (in mg/kg)

1,4-Butanediol [25265-75-2]	[73], [74]
LD_{50} 1500 – 1780 (rat, oral)	
1,2-Butanediol [584-03-2]	[73]
LD_{50} 16000 (rat, oral)	
1,3-Butanediol [107-88-0]	[75], [76]
LD_{50} 29600 (rat, oral)	
1,5-Pentanediol [111-29-5]	[77]
LD_{50} 5890 (rat, oral)	
1,6-Hexanediol [629-11-8]	[77]
LD_{50} 3730 (rat, oral)	
2,5-Hexanediol [2935-44-6]	[77]
LD_{50} 5000 (rat, oral)	
LD_{50} 16 000 (rabbit, dermal)	
1,2-Hexanediol [6920-22-5]	
LD_{50} > 5000 (rat, oral)	
1,2-Octanediol [1117-86-8]	
LD_{50} 2200 (rat, oral)	
1,3-Propanediol [504-63-2]	
LD_{50} 15670 (rat, oral)	
2,2,4-Trimethyl-1,3-pentanediol, (TMPD glycol) [144-19-4]	[79]
LD_{50} 2000 (rat, oral),	
LD_{50} 145 (rat, intravenous)	
Neopentyl glycol [126-30-7]	[81]
LD_{50} 6400 – 12800 (rat, oral)	
1,4-Bis(hydroxymethyl)cyclohexane (CHDM) [105-08-8]	[77], [80]
LD_0 3200 (rat, oral)	
LD_0 1600 (mouse, oral)	
Pentaerythritol [115-77-5]	
LD_0 > 5110 (rat, oral)	[83]
LD_{50} 19 000 (mouse, oral)	[84]
Dipentaerythritol [124-58-9]	
LD_{50} > 2000 (rat, oral)	
Trimethylolpropane [77-99-6]	[82]
LD_{50} 14100 (rat, oral)	

1,2-Diols. The oral LD_{50} values in rats for 1,2-hexane-diol, 1,2-octanediol, are included in Table 2. For these 1,2-diols, skin irritation was not observed in rabbits. 1,2-Octanediol caused no eye irritation in rabbits; testing in guinea pigs resulted in a slight sensitization reacion.

7. References

[1] B. Cornils, H. Feichtinger, *Chem. Ztg.* **100** (1976) 504 – 514.
[2] F. C. Whitmore, A. H. Popkin, H. I. Bernstein, J. P. Wilkins, *J. Am. Chem. Soc.* **63** (1941) 124 – 127. R. W. Shortridge, R. A. Craig, K. W. Greenlee, J. M. Derfer, C. E. Boord, *J. Am. Chem. Soc.* **70** (1948) 946 – 949. BASF, DE 1 057 083, 1957 (E. Haarer, K. Ruhl). Riedel de Haen, DE 1 020 614, 1956 (H. Müller). Bayer, DE 1 041 027, 1955 (H. Danziger, K. Haeseler, G. Schulze). Perstorp, WO 9 601 249, 1996 (L. Erlandsaon, S. Johansson). Mitsubishi Gas Chemical, JP 61 091 144, 1986 (A. Ninomiya).

[3] BASF, DE-AS 1800506, 1968 (F. Merger, W. Fuchs).
[4] Eastman Kodak, US 2811562, 1954 (H. J. Hagemeyer, Jr.), US 2895996, 1956 (H. N. Wright, Jr., H. J. Hagemeyer, Jr.), US 3939216, 1974 (R.L. Wright).
[5] Eastman Kodak, US 3340312, 1964 (R. B. Duke, M. A. Perry, H. N. Wright). Kuhlmann, FR 1230558, 1959 (F. Le Paire). Ruhrchemie, DE 2045669, 1970 (H. Tummes, G. Schiewe, B. Cornils, W. Pluta, J. Falbe). Hoechst, DE-AS 1768274, 1968 (G. Jacobson, H. Fernholz, D. Freudenberger).
[6] Eastman Kodak, DE-OS 1804984, 1968 (H. J. Hagemeyer, Jr., S. H. Johnson).
[7] Chemische Werke Hüls, EP 6460, 1978 (M. zur Hausen, M. Kaufhold, E. Lange).
[8] Ruhrchemie, DE 2054601 (= US 4094914), 1970 (W. Rottig, H. Tummes, B. Cornils, J. Weber).
[9] BASF, DE 1 957 591, 1968 (F. Merger, S. Winderl, E. Haarer, W. Fuchs). BASF, DE 1 793 512, 1968 (F. Merger, R. Platz, W. Fuchs). Ruhrchemie, DE 3 644 675, 1988 (N. Breitkopf, W. Hoefs, H. Kalbfell, F. Thoennessen, P. Lappe, H. Springer). Eastman Kodak, WO 8 903 814, 1989 (D. L. Morris). Eastman Kodak, US 4 855 515, 1989 (D. L. Morris, B. W. Palmer, T. W. McAninch). Mitsubishi Gas Chemical, EP 343475, 1989 (N. Teruyuki, F. Tomiyoshi, K. Seiji, F. Yoshimi). Hoechst, DE 3942792, 1989 (G. Daembkes, P. Lappe, H. Springer, F. Thoennessen). Aristech, US 5 146 012, 1991 (J. S. Salek, J. Pugach, C. L. Carole, L. A. Cullo). Mitsubishi Gas Chemical, EP 708073, 1994 (T. Nimomiya, T. Watanabe, A. Mori, T. Ikebe, A. Iwamoto).
[10] Mitsubishi Gas Chemical, EP 484 800, 1990 (M. Yoneoka, K. Watabe, G. Matsuda). Mitsubishi Gas Chemical, US 5 395 989, 1990 (M. Yoneoka, K. Watabe, G. Matsuda). BASF, WO 9 532 171, 1995 (M. Brudermueller, M. Irgang, M. Schmidt-Radde, F. Merger, T. Witzel, D. Kratz, E. Danz, A. Wittwer).
[11] J. Arpe, *Chem. Ztg.* **97** (1973) 53–62.
[12] Eastman Chemical, US 5 041 621, 1991 (D. Mores, G. Luce). BASF, US 5 024 772, 1991 (L. Thurman, J. Dowd, K. Fischer).
[13] G. K. Finch, *J. Org. Chem.* **25** (1960) 2219–2220. Eastman Kodak, DE 1 168 411, 1959 (G. Finch).
[14] BASF, EP 410 167, 1990 (L. R. Thurman, J. P. Dowd, K. J. Fischer).
[15] BASF, DE 1 793 512, 1968 (F. Merger, R. Pletz, W. Fuchs). BASF, DE 2234110, 1972 (F. Merger, G. Dümbgen, W. Fuchs). Eastman Kodak, WO 9 207 815, 1992 (G. E. Butler, G. C. Luce, D. L. Morris).
[16] Kyowa Yuka, JP 4 069 351, 1990 (S. Mizutani, K. Sato, K. Muto). BASF, 1 927 301, 1969 (F. Meger, R. Pletz, W. Fuchs). Eastman Kodak, US 5 146 004, 1991 (D. L. Morris, W. A. Beavers, W. E. Choate). Neste, WO 9 500 464, 1995 (K. Kulmala, K. Ankner, L. Rintala).
[17] US Industrial Chemicals, US 2 413 803, 1947 (W. Tribit). Montrose Chemical, US 2761 881, 1956 (J. Rosin). Chisso, JP 62 129 233, 1985 (K. Tani, K. Saito). Koei, JP 2 062 836, 1988 (K. Doi, T. Moriyama).
[18] Herberts, EP 688 840, 1994 (K. Bederke, F. Herrmann, H. Kerber, T. Kutzner, H. M. Schoenrock). Nippon Polyurethanes Kogyo, JP 7 113 005, 1993 (T. Morishima, S. Hama, S. Konishi). BASF, DE 44 01 544, 1994 (H. P. Rink).
[19] Degussa, US 5 284 979, 1992 (T. Haas, G. Boehme, D. Arntz). Du Pont, US 3 536 763, 1970 (H. S. Elenterio, T. A. Koch). Shell, GB 1 185 615, 1969 (E. T. Lutz).
[20] Degussa, EP 487 903, 1992 (D. Arntz, N. Wiegand).
[21] Degussa, EP 572 812, 1993 (D. Arntz, T. Haas, N. Wiegand). Degussa, EP 535 565, 1993 (D. Arntz, T. Haas, A. Schaefer-Sindlinger). Ruhrchemie, DE 2 054 601, 1972 (W. Rottig, H. Tummes, B. Cornils).

[22] Shell, WO 9 610 550, 1996 (J. P. Arhancet, T. C. Forschner, J. B. Powell, T. C. Semple, L. H. Slaugh, T. B. Thomason, P. R. Weider). Shell, WO 9745 390, 1997 (J. P. Arhancet, A. N. Matzakos, W. R. Pledger, J. B. Powell, L. H. Slaugh, P. R. Weider). Union Carbide, US 5 449 653, 1995 (J. R. Briggs, J. M. Maher, A. M. Harrison). Hoechst Celanese, US 50 535 562, 1991 (D. T. Kwoliang et al.).

[23] Du Pont, WO 97/2354 3, 1997 (J. M. Stouffer, E. N. Blanchard, K. W. Leffew).

[24] A. M. Brownstein, *Chemtech* **1991**, 506–511. Kuraray, US 4 567 305, 1986 (K. Kikuchi, K. Koga, H. Kojima, M. Matsumoto, S. Miura, M. Tamura, S. Yamashita).

[25] Ruhrchemie, DE-AS 2933919, 1979 (W. Bernhagen, J. Weber, H. Bahrmann, H. Springer).

[26] Showa Denko, EP 724 908, 1995, (K. Morikawa et al.). BASF, DE 19 500 236, 1995 (R. Dostalek et al.).

[27] H. Adkins, *Org. React.* **8** (1954) 1. Eastman Kodak, WO 8 900 886, 1989 (B. L. Gustajson, P. S. Wehner, P. L. Mercer, G. O. Nelson). Ube Industries, DE 4 021 230, 1989 (S. Furusaki, T. Matsuzaki, Y. Yamasaki). Ube Industries, WO 9 510 497, 1995 (S. Furusaki, M. Matsuda, Y. Miyamoto, Y. Shiomi). Bayer, EP 721 928, 1995 (G. Darsow, G. M. Petruck, H. J. Alpers). BASF, WO 9 731 883, 1997 (K. G. Baur, R. Fischer, R. Pinkos, F. Stein, B. Breitscheidel, H. Rust). BASF, WO 9 731 882, 1997 (K. G. Baur, R. Fischer, R. Pinkos, F. Stein, H. Rust, B. Breitscheidel).

[28] BASF, EP 295 435, 1987 (H. Mueller, H. Toussaint, J. Schossig).

[29] L. Ré, B. Maurer, G. Ohlhoff, *Helv. Chim. Acta* **56** (1973) 1882.

[30] Hüls, DE 3 843 956, 1988 (W. Fuhrmann, G. Bub, H. M. Zur). Sagami Chemical Research and Arakawa Chemicals, JP 9 132 541, 1996 (T. Fuchigami, T. Ga, N, Wakassa, K. Tawara).

[31] BASF, DE 2 132 547, 1971 (L. Schuster, W. Franzischka, H. Hoffmann, S. Winderl). Institut Français du Pétrole, DE 2 809 995, 1977 (J. Gzillard, C. Lassau). Mitsui Toatsu Chemicals, JP 7 035 300, 1970 (C. Sas, H. Hiari). Honsu Chemical, JP 61 260 034, 1985 (H. Kasamatsu, M. Okada, T. Kimoto). Shinnittetsu Kagaku, JP 6 128 182, 1992 (T. Shimizu, M. Nagano, M. Furumoto). Shin Nippon Rika, JP 6 329 569, 1992 (N. Okajima, M. Nakazawa).

[32] Institut Français du Pétrole, DE-OS 2809995, 1977 (J. Gzillard, C. Lassau).

[33] Mitsui Toatsu Chemicals, JP 7035300, 1970 (C. Sasa, H. Hirai).

[34] Tevco, US 4 301 046, 1981 (M. L. Schlossman). Henkel, DE 2 526 312, 1976 (H. Moeller, C. Gloxhuber, H. J. Thimm). Henkel, DE 2 526 675, 1976 (H. Moeller, H. Schnegelberger, C. Gloxhuber, H. J. Thimm). N. Wilke, *Chem. Ztg.* **95** (1971) 16–21. Du Pont, US 3 027 398, 1960 (W. L. Foohey). Eastman Kodak, US 3 334 149, 1964 (G. A. Akin, H. Lewis, T. F. Reid). Eastman Chemical, US 5 414 159, 1993 (P. Appleton, M. A. Wood). Eastman Chemical, US 5 406 004, 1993 (P. H. D. Eastland, J. Scarlett, M. W. M. Tuck, M. A. Wood). Eastman Chemical, US 5395 987, 1993 (C. Rathmell, R. C. Spratt, M. W. M. Tuck). Eastman Chemical, US 5 395 986, 1993 (J. Scarlett, M. A. Wood). Arakawa Chemical, JP 6 228 028, 1993 (H. Tsuji, T. Okazaki, K. Azuma). Towa Kasei, JP 61 292 146, 1992 (M. Magara, Y. Onoda, F. Yamazaki, S. Yoneda, K. Kato).

[35] Toyobo, JP 75 142 537, 1975 (T. Mizumoto, H. Kamatani). Toyobo, JP 75 130 738 1976 (T. Mizumoto, H. Kamatani). Eastman Chemical, US 5559 159, 1995 (B. J. Sublett, G. W. Connell).

[36] Eastman Kodak, US 3 668 157, 1972 (R. L. Combs, R. T. Bogan). S. Oswitch, *Reinf. Plast.* **17** (1973) 308. E. H. G. Sargent, K. A. Evans, *Plastics* **34** (1969) 721.

[37] UCB, DE 2 542 191, 1976 (G. Slickx). Hüls, DE 25 454 800, 1976 (J. Rueter, H. Scholten). Towa Kasei, JP 2 053 881, 1988 (N. Okamoto, Y. Tateno, Y. Ishii, K. Kato).

[38] Hüls, DE 2 241 413, 1974 (D. Stoye, W. Andrejewski, A. Draexler). Hüls, EP 790 266, 1996 (E. Wolf). PPG Industries, US 4 859 743, 1988 (R. R. Amborse, J. B. O'Dwyer, B. K. Johnston, D. P. Zielinski, S. Porter, W. H. Tyger). BASF, EP 262 069, 1986 (T. K. Debroy, M. Guagliardo, J. G.

Pucknat).Toyo Boseki, JP 6 073 274, 1992 (H. Konishi, K. Fukuda, M. Sakaguchi). Akzo, EP 454 219, 1990 (A. Noomen, J. W. F. L. Seetz, H. Klinkenberg).

[39] D. J. Golob, T. A. Odom, R. L. Whitson, *Polym. Mater. Sci. Eng.* **63** (1990) 826–832. Shanco Plastics & Chemicals, US 3 979 352, 1976 (J. W. Brady, F. D. Strickland, C. C. Longwith). L. Gott, *J. Coat. Technol.* **48** (1976) 52. Bayer, EP 566 953, 1992 (J. Pedain, H. Mueller, D. Mager, M. Schoenfelder). Mitsui Toatsu Chemicals, JP 03 195 717, 1989 (M. Sakai, K. Sasaoka, M. Murata, M. Masayuki).

[40] Eastman Kodak, US 3334149, 1964, DE 1202269, 1961 (G. A. Akin, H. Lewis, T. F. Reid).

[41] W. M. Weigert (ed.): *Wasserstoffperoxid und seine Derivate,* Hüthig Verlag, Heidelberg 1978.

[42] Degussa, DE 2937840, 1981 (G. Käbisch, H. Malitius, S. Raupach, R. Trübe, H. Wittmann).

[43] Janssen Pharmaceutica N.V., DE 2551560, 1975 (G. van Reet, J. Heeres, L. Wals).

[44] Degussa, Henkel, DE 2937831, 1981 (G. Käbisch, H. Malitius, S. Raupach, R. Trübe, H. Wittmann).

[45] Celanese, US 3183274, 1955 (M. Robeson).

[46] Koei Chemical, JP 21512, 1958 (M. Yamada, H. Terada).

[47] Bayer, DE 1041027, 1949 (H. Danzinger, K. Haeseler, G. Schulze).

[48] Trojan Powder, US 2468718, 1949 (J. Wyler).

[49] J. Jelinek, M. Sýkora, CS 154421, 1973.

[50] BASF, DE 2507461, 1975 (F. Merger, S. Winderl, H. Touissant).

[51] Celanese, GB 816208–9–10, 1959.

[52] Bayer, DE 1031298, 1958 (K. Bauer, H. Danzinger, G. Schulze).

[53] Eastman Kodak, US 3956406, 1974 (B. Pallmer, D. Bondurant).

[54] Perstorp, DE 1075102, 1960 (S. Sveninge).

[55] Hoechst Celanese, US 5 603 835, 1994 (H. Cheung, R. W Laurel, G. C. Seaman).

[56] Perstorp, W 09 701 523, 1996 (B. Wickberg).

[57] Celanese, US 2790837, 1954 (M. Robeson).

[58] Mitsubishi Gas Chemical, EP 799 815, 1997 (T. Ninomiya, T. Watanabe, T.Ikebe, A. Iwamoto).

[59] B. Tollens, *Ber. Dtsch. Chem. Ges.* **15** (1882) 1629.

[60] E. Berlow, R. H. Barth, J. E. Snow: *The Pentaerythritols,* Reinhold Publ., New York 1958.

[61] Mitsui Toatsu Chemicals, JP 77100415 and JP 77105119, 1977 (T. Kiyoura).

[62] P. Werle, G. Nonnenmacher, K. Kruse, *Liebigs Ann. Chem.* 1980, 938.

[63] Koei Chemical Co., JP 6818888, 1965 (Y. N. Minato, S. N. Yasuda).

[64] Hercules Powder, US 2533737, 1950 (E. Mertz).

[65] Virginia Chemicals, US 4277620, 1980 (F. Gupton, H. Ulmer).

[66] G. Giesselmann in *Methodicum Chimicum,* vol. **1,** Thieme Verlag, Stuttgart 1973, p. 254.

[67] K. Callmer, *J. Chromatogr.* **115** (1975) 397.

[68] R. Nutiu, M. Maties, M. Nutiu, *J. Synth. Lubr.* **7** (1990) 145.

[69] Henkel, EP 272 575, 1987 (K. H. Schmid, U. Ploog, A. Meffert).

[70] CS 197 741, 1983 (G. Guba, V. Macho, L. Koudelka, L. Komora).

[71] Degussa, DE 19 708 695, 1998, (M. Höpp, D. Arntz, M. Morawietz).

[72] DEA, EP 0 499 994, 1992 (H.-D. Grasshoff, V. Synek, H. Kolenz).

[73] V. K. Rowe et al.: *Patty's Industrial Hygiene and Toxicology,* 3rd ed., vol. **2C,** Wiley-Intersciene, New York 1982, p. 3874.

[74] BASF Aktiengesellschaft (unpublished results 1959–1981).

[75] L. Fischer et al., *Z. Gesamte Exp. Med.* **115** (1949) 22.

[76] A. Loeser, *Pharmazie* **4** (1949) 263.

[77] Registry of Toxic Effects of Chemical Substances, NIOSH. U.S. Dept. of Health, Education and Welfare, Public Health Service, Center for Disease Control, Cincinnati, Ohio, 1980.
[78] H. C. Hodge, J. H. Sterner, *Am. Ind. Hyg. Assoc. Q.* **10** (1949) 93–96.
[79] Eastman Kodak, Rochester, N.Y., unpublished results, Health, Safety, and Human Factors Laboratory.
[80] G. D. DiVincenzo, D. A. Ziegler, *Toxicol. Appl. Pharmacol.* **52** (1980) no. 1, 10–15.
[81] NIOSH, Information Profile on Potential Occupational Hazards: Glycols, PB 89-215776 (1982).
[82] V. V. Stankevich, *Gig. Sanit.* **32** (1967) 107.
[83] Degussa AG, unpublished results.
[84] S. J. Plitman, *Gig. Sanit.* **36** (1971) 192–196.
[85] W. Kutscher, *Z. Physiol. Chem.* **283** (1948) 268–275.
[86] E. F. King, H. A. Painter, Assessment of Biodegradability of Chemicals in Water by Manometric Respirometry, Final Report, Contract No W/81/217, Commission of the European Communities, Directorate-General, Environment, Consumer protection and Nuclear Safety, 1983.
[87] BASF, DE 19 542 036, 1995, (D. Kratz et al.)
[88] BASF, DE 19 653 093, 1996, (D. Kratz et al.)

Aldehydes, Aliphatic and Araliphatic

Individual keywords: → *Acrolein and Methacrolein,* → *Benzaldehyde,* → *Butanals,* → *Chloroacetaldehydes,* → *Crotonaldehyde and Crotonic Acid,* → *Formaldehyde,* → *Glyoxal,* → *Propanal*

CHRISTIAN KOHLPAINTNER, Celanese GmbH, Werk Ruhrchemie, Oberhausen, Federal Republic of Germany

MARKUS SCHULTE, Celanese GmbH, Werk Ruhrchemie, Oberhausen, Federal Republic of Germany

JÜRGEN FALBE, Henkel KGaA, Düsseldorf, Federal Republic of Germany

PETER LAPPE, Ruhrchemie AG, Oberhausen, Federal Republic of Germany

JÜRGEN WEBER, Ruhrchemie AG, Oberhausen, Federal Republic of Germany

1.	Introduction	350
2.	Saturated Aldehydes	351
2.1.	Physical Properties	351
2.2.	Chemical Properties	354
2.3.	Production	357
2.4.	Individual Saturated C_5–C_{13} Aldehydes	360
2.4.1.	C_5 Aldehydes	361
2.4.2.	C_6 Aldehydes	362
2.4.3.	C_7 Aldehydes	362
2.4.4.	C_8 Aldehydes	363
2.4.5.	C_9 Aldehydes	364
2.4.6.	C_{10} Aldehydes	365
2.4.7.	C_{11} Aldehydes	365
2.4.8.	C_{12} Aldehydes	366
2.4.9.	C_{13} Aldehydes	366
3.	Unsaturated Aldehydes	366
3.1.	Physical Properties	367
3.2.	Chemical Properties	367
3.3.	Production	369
3.4.	Individual Unsaturated C_5–C_{11} Aldehydes	372
3.4.1.	C_5 Alkenals	372
3.4.2.	C_6 Alkenals	372
3.4.3.	C_8 Alkenals	373
3.4.4.	C_{10} Alkenals	373
3.4.5.	C_{11} Alkenals	374
4.	Hydroxyaldehydes	374
4.1.	Properties	374
4.2.	Production	376
4.3.	Individual Hydroxyaldehydes	378
4.3.1.	3-Hydroxypropanal	378
4.3.2.	3-Hydroxybutanal (Acetaldol)	378
4.3.3.	4-Hydroxybutanal	379
4.3.4.	2,2-Dimethyl-3-hydroxypropanal (Hydroxypivaldehyde)	379
4.3.5.	2-Ethyl-3-hydroxyhexanal (Butyraldol)	380
4.3.6.	3,7-Dimethyl-7-hydroxyoctanal (Hydroxycitronellal)	380
5.	Araliphatic Aldehydes	381
5.1.	Properties	381
5.2.	Production	383
5.3.	Individual Araliphatic Aldehydes	384
5.3.1.	Phenylacetaldehyde	384
5.3.2.	4-Methylphenylacetaldehyde	385
5.3.3.	2-Phenylpropionaldehyde (Hydratropaldehyde)	385
5.3.4.	3-Phenylpropionaldehyde (Dihydrocinnamaldehyde)	386
5.3.5.	2-(4-Isobutylphenyl)propionaldehyde	387

5.3.6.	3-(4-Isopropylphenyl)-2-methyl-propanal (Cyclamenaldehyde)...	388	6.3.2.	Glutardialdehyde (1,5-Pentanedial)	392
5.3.7.	3-(4-*tert*-Butylphenyl)-2-methyl-propionaldehyde (Lilial)	388	6.3.3.	TCD Dialdehyde	393
			7.	**Acetals**	393
5.3.8.	Cinnamaldehyde (3-Phenyl-2-propenal)	389	7.1.	**Properties**...............	394
5.3.9.	α-Alkylcinnamaldehydes (2-Alkyl-3-phenyl-2-propenals)	389	7.2.	**Production**	396
			7.3.	**Uses**	398
6.	**Dialdehydes**..............	390	8.	**Quality Control**	398
6.1.	**Properties**...............	390	9.	**Analysis**	399
6.2.	**Production**	391	10.	**Storage, Transportation, and Environmental Regulations**...	400
6.3.	**Individual Dialdehydes**......	392	11.	**Economic Aspects**	401
6.3.1.	Succinaldehyde (1,4-Butanedial) .	392	12.	**References**...............	402

1. Introduction

Aldehydes are represented by the general formula RCHO, where R can be hydrogen or an aliphatic, aromatic, or heterocyclic group.

$$R-\overset{O}{\underset{H}{C}}$$

According to IUPAC nomenclature, aldehydes are identified by the ending "al." However, many of them still are called by their common names.

Because of their high reactivity, aliphatic aldehydes are widely used as intermediates in organic synthesis. Sometimes the isolation of a pure aldehyde is very difficult; in such cases, stable derivatives or oligomers are prepared from which the aldehydes can be reisolated.

Aldehydes usually must be kept from contact with air and under certain circumstances must be stabilized during distillation, subsequent storage, and transportation. This applies particularly to unsaturated aldehydes, which have a tendency to polymerize (e.g., on contact with alkali). For commerical purposes aldehydes often are protected by the addition of stabilizers and antioxidants and by a nitrogen atmosphere. When handling aldehydes, care must be taken to prevent either the liquids or their vapors from coming into contact with respiratory organs, eyes, and skin. Gloves and safety glasses are absolutely necessary.

Aldehydes are obtained mainly via the oxo synthesis, by mild oxidation (dehydrogenation) of primary alcohols, and by special olefin oxidation processes. Low concentrations of aldehydes occur naturally in essential oils of various plants. Acetaldehyde is an intermediate product of alcohol fermentation; it is formed by decarboxylation of the intermediate pyruvic acid. Aldehydes also fulfill some important biological functions, e.g., 11-*cis*-retinal in the sight process or as pyridoxal in the transamination of amino acids. Their isolation from natural substances is of commercial significance only in a few cases, e.g., in the production of longer-chain fragrance aldehydes.

2. Saturated Aldehydes

The most important saturated aliphatic aldehydes are formaldehyde, acetaldehyde, propionaldehyde, and butyraldehydes; they are used as starting materials for organic chemicals and polymers (→ Formaldehyde, etc.). Some dialdehydes, e.g., glyoxal, are also commerical products (→ Glyoxal). The higher aliphatic aldehydes are used mostly for the production of alcohols, carboxylic acids, amines, etc. The aldehydes also serve as fragrances in the perfume industry.

2.1. Physical Properties (see Table 1)

Formaldehyde, the simplest saturated aliphatic aldehyde, is a gas at room temperature. The aldehydes up to about C_{12} are liquids; the straight-chain aldehydes from C_{13} and above are solids. Because the hydrogen atom of a formyl group has less tendency to hydrogen bond than the hydrogen atom of a hydroxy group, the boiling points of the aldehydes are considerably lower than those of the corresponding alcohols. The difference in boiling points between aldehydes and alcohols with an intermediate number of carbon atoms is 20–40 °C.

Viscosity, density, and refractive index at 20 °C increase with increasing molecular mass. The lower homologs are mobile liquids; aldehydes from heptanal to undecanal have an oily consistency.

There are practically no limits to the miscibility of formaldehyde and acetaldehyde with water. However, with increasing molecular mass the solubility of aldehydes decreases very rapidly. For example, the solubility of hexanal in water is only 0.6 wt% at 20 °C. The aliphatic aldehydes are soluble in alcohols, ethers, and other common organic solvents.

Whereas the lower aldehydes have a pungent smell, the higher homologs in the C_8-C_{13} range are components of nearly all perfumes and many fragrances. Their odors become weaker with increasing molecular mass.

The characteristic properties of most aliphatic aldehydes are their ease of autoxidation and their tendency to trimerize and/or polymerize. Therefore, aldehydes are protected with an inert gas atmosphere and minimal amounts of a stabilizer are added if required [1, p. 651].

Aldehydes, Aliphatic and Araliphatic

Table 1. Physical properties of saturated aldehydes

	Name	CAS registry number	Formula	M_r	mp, °C	bp[a], °C	Density (20 °C), g/cm³	Refract. index, n_D^{20}	Surface tension (20 °C), mN/m	Viscosity (20 °C), mPa·s	Solubility in H₂O at 20 °C, wt %
C_1	Methanal (formaldehyde)	[50-00-0]	HCHO	30.03	−92	−21	0.816[b]	1.331		0.2	95
C_2	Ethanal (acetaldehyde)	[75-07-0]	CH₃CHO	44.05	−123	20	0.778	1.362		0.4	35
C_3	Propanal (propionaldehyde)	[123-38-6]	CH₃CH₂CHO	58.08	−81	48	0.797	1.380	21.8	0.45	7.6
C_4	Butanal (butyraldehyde)	[123-72-8]	CH₃(CH₂)₂CHO	72.11	−97	75	0.803	1.373	24.6	0.45	6.7
	2-Methylpropanal (isobutyraldehyde)	[78-84-2]	CH₃CH(CH₃)CHO	72.11	−66	64	0.789		23.2		
C_5	Pentanal (valeraldehyde)	[110-62-3]	CH₃(CH₂)₃CHO	86.13	−91	103	0.808	1.394	27.4	0.54	1.4
	2-Methylbutanal	[96-17-3]	CH₃CH₂CH(CH₃)CHO	86.13	<−60	92	0.804	1.390		0.55	1.7
	3-Methylbutanal (isovaleraldehyde)	[590-86-3]	CH₃CH(CH₃)CH₂CHO	86.13	<−70	92.5	0.796	1.388	23.2	0.58	1.4
	2,2-Dimethylpropanal (pivaldehyde)	[630-19-3]	(CH₃)₃CCHO	86.13	6	74	0.783	1.379		0.67	1.1
C_6	Hexanal (capraldehyde)	[66-25-1]	CH₃(CH₂)₄CHO	100.16	−56	128	0.814	1.404			0.6
	2-Methylpentanal	[123-15-9]	CH₃(CH₂)₂CH(CH₃)CHO	100.16	−100	118	0.807	1.400		0.56	0.4
	3-Methylpentanal	[15877-57-3]	CH₃CH₂CH(CH₃)CH₂CHO	100.16		38−40 (4.3)	0.826	1.407			
	4-Methylpentanal	[1119-16-0]	CH₃CH(CH₃)(CH₂)₂CHO	100.16		117	0.813	1.402			0.5
	2-Ethylbutanal	[97-96-1]	CH₃CH₂CH(C₂H₅)CHO	100.16	−89	69−72 (21.3)	0.813	1.403	25.5	0.66	0.6
	2,3-Dimethylbutanal	[2109-98-0]	CH₃CH(CH₃)CH(CH₃)CHO	100.16							
	3,3-Dimethylbutanal	[2987-16-8]	CH₃C(CH₃)₂CH₂CHO	100.16							
C_7	Heptanal (enanthaldehyde)	[111-71-7]	CH₃(CH₂)₅CHO	114.18	−42	153	0.817	1.411			
	2-Methylhexanal	[925-54-2]	CH₃(CH₂)₃CH(CH₃)CHO	114.18		141 (96.7)		1.4088−1.4100 (25 °C)			
	2,2-Dimethylpentanal	[14250-88-5]	CH₃(CH₂)₂C(CH₃)₂CHO	114.18		126−127	0.807	1.406			
	2,3-Dimethylpentanal	[32749-94-3]	CH₃CH₂CH(CH₃)CH(CH₃)CHO	114.18		141	0.845				
C_8	Octanal (caprylaldehyde)	[124-13-0]	CH₃(CH₂)₆CHO	128.22		175	0.821	1.419			0.4
	2-Methylheptanal	[16630-91-4]	CH₃(CH₂)₄CH(CH₃)CHO	128.22	−23	51 (1.5)	0.816	1.414			

Table 1. (continued)

Name	CAS registry number	Formula	M_r	mp, °C	bp^a, °C	Density (20 °C), g/cm³	Refract. index, n_D^{20}	Surface tension (20 °C), mN/m	Viscosity (20 °C), mPa·s	Solubility in H₂O at 20 °C, wt %
2-Ethylhexanal	[123-05-7]	CH₃(CH₂)₃CH(C₂H₅)CHO	128.22	<−60	(22 °C) 163	(22 °C) 0.820	1.416		1.0	<0.02
2-Propylpentanal	[18295-59-5]	CH₃(CH₂)₂CH(C₃H₇)CHO	128.22							
2-Ethyl-3-methylpentanal	[42347-74-0]	CH₃CH₂CH(CH₃)CH(C₂H₅)CHO	128.22			0.807	1.405			
C₉ Nonanal (pelargonaldehyde)	[124-19-6]	CH₃(CH₂)₇CHO	142.24		191	0.825	1.427			
	[5435-64-3]	₃C(CH₃)₂CH₂CH(CH₃)CH₂CHO	142.24	<−60	173	0.820	1.421	25.5	1.4	<0.01
C₁₀ Decanal (caprinaldehyde)	[112-31-2]	CH₃(CH₂)₈CHO	156.27	−3	217	0.826	1.428		1.8	
C₁₁ Undecanal	[112-44-7]	CH₃(CH₂)₉CHO	170.29		120−122 (2.7)		1.433			
2-Methyldecanal	[19009-56-4]	CH₃(CH₂)₇CH(CH₃)CHO	170.29	<−60	229	0.830	1.430	31.9	2.5	<0.1
C₁₂ Dodecanal (lauric aldehyde)	[112-54-9]	CH₃(CH₂)₁₀CHO	184.33	12	254	0.828	1.435		2.8	
	[110-41-8]	CH₃(CH₂)₈CH(CH₃)CHO	184.33	−25	246	0.827	1.433	32.4		<0.1
C₁₃ Tridecanal	[10486-19-8]	CH₃(CH₂)₁₁CHO	198.34	14	126 (3.1)					

^{a} at 101.3 kPa unless otherwise specified;
^{b} effective density at −19 °C

Table 2. Azeotropic mixtures of aldehydes with water

Aldehyde	bp of aldehyde, °C	bp of azeotrope, °C	wt% H_2O in azeotrope
Propanal	47.9	47.5	2.0
Butanal	74.8	68.0	9.7
2-Methylpropanal	63.3	60.1	9.6
Pentanal	103.3	83.0	19.0
3-Methylbutanal	92.5	77.0	12.0
Hexanal	128.3	91.0	31.3
2-Ethylbutanal	116.7	87.5	23.7
2-Methylpentanal	118.3	88.5	23.0
2-Ethylhexanal	163.6	96.4	51.6

Because of the differences in boiling points between aldehydes and alcohols, aldehydes prepared by dehydrogenation of the corresponding alcohols generally are separated and purified by distillation. Apart from the lower boiling, short-chain aldehydes, distillation is mostly carried out at reduced pressure [1, p. 654]. Numerous aldehydes form azeotropic mixtures with water (see Table 2) and other substances.

2.2. Chemical Properties

The polarity of the carbonyl group of aldehydes not only facilitates the typical aldehyde reactions — addition of nucleophiles, reduction, and oxidation — but it also makes the α-hydrogen atom acidic. For these reasons, aldehydes can undergo a wide variety of reactions. The major ones are compiled in Table 3 [1, p. 609].

Addition Reactions. Because of the polarity of the carbonyl double bond, aldehydes enter into a wide variety of nucleophilic addition reactions. The simplest case is the intermolecular addition of one molecule of aldehyde to another to form an aldol. These *aldols* (β-hydroxyaldehydes) generally are unstable and react further to form secondary products, i.e., diols, unsaturated aldehydes, or alcohols.

Industrially, crossed aldol condensation of an aldehyde with formaldehyde is used for the production of trimethylolpropane, pentaerythritol, and neopentyl glycol (→ Alcohols, Polyhydric). The aldol condensation of butanal for the production of the plasticizer alcohol 2-ethylhexanol is of major commercial importance (→ 2-Ethylhexanol).

The addition of alcohols and thiols to aldehydes leads to acetals and monothiohemiacetals; these are used for the protection of formyl groups in synthesis because they are resistant to alkali. Reductive amination of aldehydes with ammonia or primary amines in the presence of hydrogen or some other reducing agent gives primary or secondary amines, respectively, e.g.,

Table 3. General reactions of aliphatic aldehydes

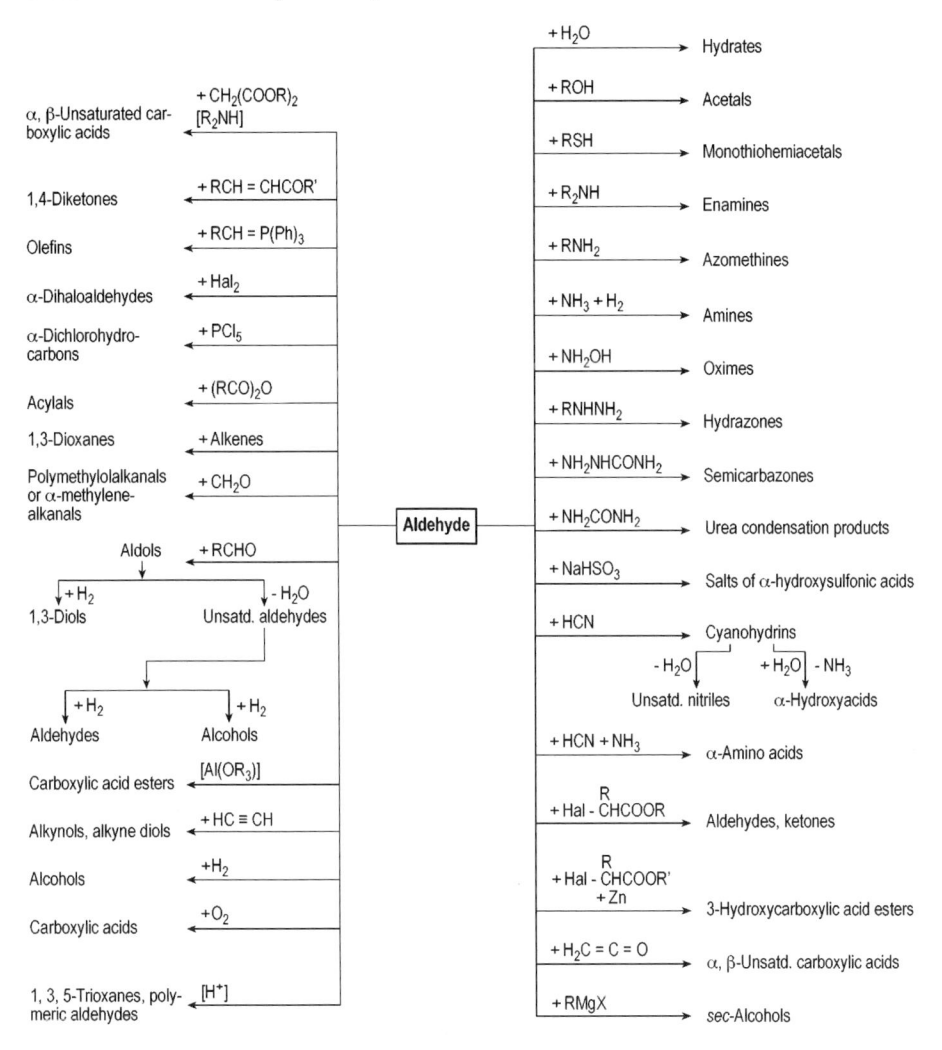

Saturated Aldehydes

Reactions on the right side of Aldehyde:

+ H_2O	Hydrates
+ ROH	Acetals
+ RSH	Monothiohemiacetals
+ R_2NH	Enamines
+ RNH_2	Azomethines
+ NH_3 + H_2	Amines
+ NH_2OH	Oximes
+ $RNHNH_2$	Hydrazones
+ $NH_2NHCONH_2$	Semicarbazones
+ NH_2CONH_2	Urea condensation products
+ $NaHSO_3$	Salts of α-hydroxysulfonic acids
+ HCN	Cyanohydrins
Cyanohydrins − H_2O	Unsatd. nitriles
Cyanohydrins + H_2O, − NH_3	α-Hydroxyacids
+ HCN + NH_3	α-Amino acids
+ Hal−CHR−COOR	Aldehydes, ketones
+ Hal−CHR−COOR' + Zn	3-Hydroxycarboxylic acid esters
+ $H_2C=C=O$	α, β-Unsatd. carboxylic acids
+ RMgX	sec-Alcohols

Reactions on the left side of Aldehyde:

Product	Reagent
α, β-Unsaturated carboxylic acids	+ $CH_2(COOR)_2$ [R_2NH]
1,4-Diketones	+ RCH=CHCOR'
Olefins	+ RCH=P(Ph)$_3$
α-Dihaloaldehydes	+ Hal_2
α-Dichlorohydrocarbons	+ PCl_5
Acylals	+ $(RCO)_2O$
1,3-Dioxanes	+ Alkenes
Polymethylolalkanals or α-methylene-alkanals	+ CH_2O
Aldols	+ RCHO
1,3-Diols	+ H_2 (from Aldols)
Unsatd. aldehydes	− H_2O (from Aldols)
Aldehydes	+ H_2 (from Unsatd. aldehydes)
Alcohols	+ H_2 (from Unsatd. aldehydes)
Carboxylic acid esters	[Al(OR)$_3$]
Alkynols, alkyne diols	+ HC≡CH
Alcohols	+ H_2
Carboxylic acids	+ O_2
1,3,5-Trioxanes, polymeric aldehydes	[H^+]

$$H_3C\text{-}CH_2\text{-}CH_2\text{-}CHO + H_3C\text{-}NH_2 \xrightarrow{-H_2O} H_3C\text{-}CH_2\text{-}CH=CH\text{-}N\text{-}CH_3 \xrightarrow{H_2} H_3C\text{-}CH_2\text{-}CH_2\text{-}CH_2\text{-}NH\text{-}CH_3$$

The reaction with secondary amines yields the corresponding enamines, which can be hydrogenated to tertiary amines.

Classically, aldehydes were identified by the preparation of crystalline derivatives, e.g., by reaction with hydrazine, substituted hydrazines (e.g., 2,4-dinitrophenylhydrazine), hydroxylamine, or semicarbazide. However, now other techniques are used (see Chap. 9).

The addition of Grignard compounds leads to secondary alcohols. The Darzens condensation of alkyl chloroacetates in the presence of a strong base yields aldehydes containing one additional methylene group, e.g.,

$$\underset{CH_3}{\overset{CHO}{|}} + \underset{Cl}{\overset{O}{\underset{\|}{C}}}\!\!-\!OR \longrightarrow \underset{H_3C}{\overset{O}{\underset{\triangle}{\diagdown}}}\!\!\!\overset{OR}{\underset{O}{\diagup}} \longrightarrow \underset{CH_3}{\overset{CHO}{\diagdown}}$$

With the alkylhaloacetates in the presence of zinc, 3-hydroxycarboxylates are obtained (Reformatsky reaction).

The addition of sodium hydrogensulfite leads to water-soluble crystalline compounds. This reaction permits the separation of α-methyl-branched aldehydes from the isomeric straight-chain aldehydes [10], [11]. The reaction of hydrogen cyanide with aldehydes is of commercial significance. The highly unstable cyanohydrins can be converted to the unsaturated nitriles by dehydration or to the α-hydroxycarboxylic acids by hydrolysis. If the reaction of hydrogen cyanide with aldehydes is carried out in the presence of ammonia, α-amino acids are obtained.

Polymerization. The formation of cyclic, 2,4,6-trialkyl-1,3,5-trioxane type trimers is catalyzed by various acids.

At higher temperature these compounds are unstable and revert to the monomer. With this process pure n-alkanals can be isolated from $n-iso$-aldehyde mixtures [12].

Mainly oligomers and polymers are formed by the lower aldehydes. Polymeric formaldehyde, referred to as paraformaldehyde, is a mixture of polyoxymethylene glycols $HO(CH_2O)_nH$, where n is 8–100. The oligomers of acetaldehyde, paraldehyde and metaldehyde are 2,4,6-trimethyl-1,3,5-trioxane and the cyclic tetramer, respectively. Polyacetaldehyde is a high-molecular mass polymer with an acetal structure.

Oxidation. A characteristic feature of most aldehydes is their great tendency to autoxidize in a radical chain reaction to the corresponding carboxylic acids. On an industrial scale the oxidation is usually carried out in the liquid phase with oxygen or air. Catalysts often are added to reduce the reaction time and lower the reaction temperature. Salts of transition metals are effective catalysts. For special purposes, hydrogen peroxide, periodic acid, nitric acid, potassium permanganate, chromium trioxide, and peroxo compounds are used as oxidizing agents [1, p. 609]. Another special method is the melting of aldehydes with alkali, e.g., [13].

$$\underset{\underset{C(CH_3)_3}{|}}{CHO} \xrightarrow[-H_2]{NaOH} \underset{\underset{C(CH_3)_3}{|}}{COONa} \xrightarrow{H_2SO_4} \underset{\underset{C(CH_3)_3}{|}}{COOH}$$

Hydrogenation (Reduction). Catalytic hydrogenation leads to primary alcohols. Both Ni and Cu catalysts have proved to be the most suitable [14]. Normally, the reaction is carried out in the liquid phase on fixed-bed catalysts at 20–200 °C and pressures of up to 30 MPa (→ Alcohols, Aliphatic). Hydrogenation in the gas phase is run continuously. Because of the good heat dissipation this is advantageous, especially with sensitive starting materials.

In addition to catalytic hydrogenation there are many other reduction processes. Complex hydrides such as lithium aluminum hydride and sodium borohydride are used most frequently as reducing agents.

An important reaction characteristic of aldehydes is the Meerwein–Ponndorf–Verley reduction with aluminum alkoxides, in particular with aluminum triisopropoxide. Because this reaction is reversible the carbonyl compound formed is distilled from the alkoxide. The process is recommended for the preparation of alcohols containing halo or nitro groups.

There are numerous other reducing agents for aldehydes; most of these are of interest only for laboratory syntheses. A few are: potassium–ammonia, trimethyl phosphite, magnesium, aluminum–mercury, trialkylboranes, and aluminum trialkyls.

The reduction of aldehydes to the corresponding saturated hydrocarbons is of no commercial significance. In laboratory syntheses it is achieved with reducing agents such as hydrazine, complex metal hydrides, lithium–ammonia, or hydrogen iodide–phosphorus.

2.3. Production

Although many aldehyde syntheses are known, only a few are used on an industrial scale. Often this is a question of the availability of feedstock.

The most important processes for the preparation of saturated aliphatic aldehydes are:

1) Hydroformylation of olefins (oxo synthesis)
2) Dehydrogenation or oxidation of primary alcohols (mainly for the production of formaldehyde from methanol)
3) Hydration of acetylene for the production of acetaldehyde
4) Oxidation of ethylene to acetaldehyde
5) Oxidation of saturated hydrocarbons (C_3, C_4) for the preparation of lower aldehydes

In addition, some special syntheses for the production of aldehydes required in the perfume industry are of industrial importance.

Oxo Synthesis. The oxo synthesis is the most important process for the production of aldehydes containing at least three carbon atoms. In this process olefins react with synthesis gas (CO, H_2) to form aldehydes with one more carbon atom than the starting material. Pure products are formed only from symmetrical or sterically hindered olefin molecules; otherwise, product mixtures of straight-chain and branched compounds are obtained:

$$R-CH=CH_2 \xrightarrow{CO/H_2} \begin{cases} R-CH_2-CH_2-CHO \\ R-CH(CH_3)-CHO \end{cases}$$

By the selection of suitable catalysts and reaction conditions the $n:iso$ ratio can be varied over a wide range.

Dehydrogenation/Oxidation of Primary Alcohols. Dehydrogenation, oxidation, and oxidative dehydrogenation proceed according to the following equations:

$RCH_2OH \rightarrow RCHO + H_2$ $\Delta H = +84$ kJ/mol for $R = CH_3$ (1)
$RCH_2OH + \frac{1}{2}O_2 \rightarrow RCHO + H_2O$ $\Delta H = -159$ kJ/mol for $R = H$ (2)
$RCH_2OH \rightarrow RCHO + H_2, H_2 + \frac{1}{2}O_2 \rightarrow H_2O$ $\Delta H = -159$ kJ/mol for $R = H$ (3)

Dehydrogenation. The endothermic dehydrogenation reaction of alcohols is carried out at atmospheric pressure and 250–400 °C, normally with Cu or Ag catalysts. The catalysts often are activated by the addition of, e.g., Zr, Co, or Cr. The advantage of this process is the simultaneous recovery of hydrogen that can be used without further purification [2, p. 282]. Catalytic dehydrogenation is an equilibrium reaction. Therefore, high temperatures and short residence times are economically advantageous.

The process still has some commercial importance in the preparation of acetaldehyde from ethanol. The gas-phase dehydrogenation is carried out at atmospheric pressure and 270–300 °C on a copper catalyst activated with cerium. Twenty-five to 50% of the ethanol is converted per throughput, with a selectivity of 90–95% acetaldehyde; ethyl acetate, ethylene, crotonaldehyde, and higher alcohols are obtained as byproducts.

Oxidation. The oxidation according to Equation (2) is carried out with an excess of air or oxygen at 350–450 °C over a catalyst containing 18–19 wt% Fe_2O_3 and 81–82 wt% MoO_3 [15]. This process is used in the production of formaldehyde ($\rightarrow$ Formaldehyde).

Oxidative Dehydrogenation. Equation (3) can be separated formally into the endothermic dehydrogenation of the alcohol and the exothermic combustion of the hydrogen formed; the overall reaction therefore can become exothermic. In the industrial process the two reactions take place simultaneously when substoichiometric quantities of oxygen or air are used. In oxidation and oxidative dehydrogenation the explosion ranges of the alcohol–air mixtures must be considered.

Oxidative dehydrogenation is the most important process for the production of aldehydes from alcohols. Silver catalysts are preferred but copper catalysts are used also.

In formaldehyde production from methanol, 75–99% conversion is achieved on silver crystals (grain size 0.2–3 mm), silver nets, or silver on Al_2O_3. Temperatures of 500–720 °C and residence times of less than 0.01 s are used (→ Formaldehyde).

In 1979 about 15% of Western Europe's acetaldehyde production was from ethanol. Silver and copper were the main catalysts employed. Ethanol conversions were 30–50% per throughput, and acetaldehyde selectivity was 85–95%. The reaction temperature ranged between 300 and 600 °C depending on the quantity of air. Ethyl acetate, formic acid, acetic acid, and carbon dioxide were byproducts.

Fragrance Aldehydes. In addition to the production of formaldehyde and acetaldehyde, dehydrogenation and oxidation also are preferred for the synthesis of fragrance aldehydes. A process developed especially for these products permits the catalytic dehydrogenation of C_5–C_{14} alcohols in the presence of hydrogen and air; it employs copper or silver catalysts, possibly combined with Zn, Cr, Cr_2O_3 [16]. Another catalyst system based on Cu/MgO, produces octanal with a selectivity of 99% and an octanol conversion of 58% at 265–330 °C [17]. Other catalyst systems described in the literature are Ag/Na_2O on supports [18], mixtures of MnO, NiO on MgO [19], or CuCl and a nitrogen-containing ligand, such as 2,2'-bipyridyl [20]. For further dehydrogenation catalysts see [2, p. 282].

Oxidation of Hydrocarbons. A process developed by Celanese for the oxidation of C_3 and C_4 alkanes [21] produces a complex reaction mixture requiring costly extraction and distillation steps. In this process propane and propane-butane mixtures are reacted in the gas phase at 425–460 °C, 0.7–0.8 MPa, and with a conversion of about 20% (with oxygen deficiency). The reaction proceeds according to a radical mechanism. In addition to acetaldehyde ($\approx 20\%$) the reaction mixture consists mainly of formaldehyde ($\approx 15\%$), methanol ($\approx 19\%$), and organic acids ($\approx 11\%$). However, the process is outdated technically. The oxidation of methane or ethane also has no practical significance.

Oxidation of Olefins. The most important industrial process for the production of acetaldehyde is the partial oxidation of ethylene in the aqueous phase in the presence of palladium and copper chlorides (Wacker-Hoechst process).

Another oxidation process for the preparation of aldehydes from olefins is the acrolein synthesis from propene (→ Acrolein and Methacrolein).

Miscellaneous Processes. The process involving addition of water to acetylene for the preparation of acetaldehyde has become more or less insignificant compared with the alternative processes based on ethylene or ethanol. In Western Europe the last plants were closed in 1980. Apart from the ready availability of the cheaper ethylene and the good selectivity of its conversion, the use of environmentally detrimental mercury sulfate as catalyst has led to the acetylene process becoming essentially obsolete.

The β-hydroxyaldehydes (aldols) and the resulting α,β-unsaturated aldehydes are of commercial significance as starting materials for the production of saturated aldehydes. Examples are the productions of butanal from crotonaldehyde (→ Butanals) and 2-ethylhexanal from 2-ethyl2-hexenal (see Section 2.4.4).

Laboratory Processes. Only a few examples of the extraordinarily large number of aldehyde syntheses commonly used on a laboratory scale are mentioned here.

The catalytic reduction of acyl chlorides (Rosenmund reduction) is one of the safest and simplest methods of converting carboxylic acids to aldehydes. Mainly Pd–C or Pd–BaSO$_4$ serve as catalysts, if necessary in the presence of a catalyst poison containing sulfur in an aromatic hydrocarbon solvent, such as benzene or toluene. The yields average between 70 and 90% [1, p. 418], [22].

Other reducing agents are complex borohydrides or aluminum hydrides, such as bis(triphenylphosphine)copper(I) tetrahydroborate, [(C$_6$H$_5$)$_3$P]$_2$CuBH$_4$ [23], and lithium tri-*tert*-butyloxyaluminum hydride [24].

Carboxylic acid esters, amides, hydrazides, and nitriles also can be reduced in many ways to the corresponding aldehydes. Most often complex metal hydrido compounds are used as reducing agents.

The oxidative cleavage of olefins with ozone via the ozonides and subsequent reductive decomposition generally take place under extremely mild reaction conditions. With this process, almost quantitive yields are attained [25].

Apart from the classic processes of oxidative glycol cleavage with lead(IV) acetate (Criegee method) [26] and with periodic acid (Malaprade method) [27], there are a number of other oxidizing agents, such as hydrogen peroxide-ruthenium(III) salts [28], manganese dioxide [29], and *N*-iodosuccinimide [30].

When Grignard compounds react with orthoformate esters, aliphatic or aromatic aldehydes are formed [31]. Allylvinyl ethers, which are easily accessible by the reaction of aldehydes with allyl alcohols, rearrange to 4-alkenals (Claisen rearrangement) when heated; the products subsequently can be hydrogenated to the saturated aldehydes [32].

Comprehensive lists of the various preparative methods are given in [1], [2, p. 203], [5, p. 943], [33].

2.4. Individual Saturated C$_5$–C$_{13}$ Aldehydes

The higher aldehydes are hydrogenated mainly to alcohols, which are used as solvents or for the production of plasticizers. In addition, they are intermediates for the production of carboxylic acids, amines, and amino acids. They also are used for the synthesis of agricultural chemicals, pharmaceuticals, disinfectants, dyes, etc. In the perfume industry they are used either as such or for the synthesis of odorants such as α-amylcinnamaldehyde.

2.4.1. C₅ Aldehydes

The most important physical properties of the four isomeric C_5 aldehydes, their CAS registry numbers, and their (low) solubilities in water are compiled in Table 1. They are colorless, flammable liquids with intense, characteristic odors, and they exhibit the typical reactions of aliphatic aldehydes. They form azeotropic mixtures with water and are freely soluble in common organic solvents.

Production. Pentanal and 2-methylbutanal are prepared by the hydroformylation of butene [34]:

$$H_3C-CH=CH_2 \xrightarrow{CO/H_2} H_3C-CH_2-CH_2-CH_2-CHO + H_3C-CH_2-CH(CH_3)-CHO$$

The ratio of the two products depends on the reaction conditions [1, p. 195].

Isomeric mixtures of pentanal and 2-methylbutanal are often commercial products. Also 2-methylbutanal can be prepared by the reaction of butanal with formaldehyde in the presence of secondary amines with subsequent hydrogenation [35].

3-Methylbutanal (isovaleraldehyde) is prepared by hydroformylation of isobutene. Small quantities ($\approx 5\%$) of 2,2-dimethylpropanal are formed also [36].

$$(CH_3)_2C=CH_2 \xrightarrow{CO/H_2} (CH_3)_2CH-CH_2-CHO + (CH_3)_3C-CHO$$

2,2-Dimethylpropanal can be obtained by isomerization of 1,1,2-trimethyloxirane at 80–170 °C in the presence of $ZnCl_2$-pumice [37].

Another way to manufacture 3-methylbutanal is by the isomerization of 3-methyl-3-butene-1-ol or of a mixture of 3-methyl-3-butene-1-ol and 3-methyl-2-butene-1-ol, prepared from isobutene and formaldehyde. The isomerization catalyst is CuO–ZnO and the reaction temperature is 200–250 °C [38].

$$(CH_3)_2C=CH_2 \xrightarrow{CO/H_2} (CH_3)_2CH-CH_2-CHO + (CH_3)_3C-CHO$$

Uses. The C_5 aldehydes are hydrogenated easily to pentanols (amyl alcohols), which are used chiefly as solvents and for the production of plasticizers, esters, and xanthogenates (→ Pentanols). Zinc diamyldithiophosphate is an additive for lubricating oils.

Further applications of the C_5 aldehydes are in the oxidation to C_5 acids (→ Carboxylic Acids, Aliphatic) and the preparation of amyl amines by aminating hydrogenation. 2-Methylbutanol (from 2-methylbutanal) can be converted to isoprene over dehydration catalysts at 300–350 °C [39] and to isoleucine.

3-Methylbutanal is a starting material for 2,3-dimethyl-2-butene, which in turn is converted to 2,3-dimethylbutane-2,3-diol (pinacol) [40] and to methyl *tert*-butyl ketone (pinacolone) [41]. Pinacolone is a valuable starting material for the synthesis of a number of pesticides.

Pharmaceuticals, such as the active substance butizide [42], are synthesized from 3-methylbutanal and the corresponding acid.

The most important manufacturers of C_5 aldehydes in Western Europe are BASF and Ruhrchemie, and in the United States, Union Carbide.

2.4.2. C_6 Aldehydes

The C_6 aldehydes are colorless, mobile liquids. Their most important physical properties and CAS registry numbers are compiled in Table 1. Hexanal occurs naturally, e.g., in lemon and orange oils.

Production. The preparation of the C_6 aldehydes, of which hexanal, 2-methylpentanal, 2-ethylbutanal, and 3-methylpentanal have medium importance, is carried out almost exclusively by hydroformylation or by aldol condensation. For example, hexanal and 3-methylpentanal are prepared by the hydroformylation of 1-pentene [43] and 2-methyl-1-butene [44], respectively. 2-Methylpentanal is obtained by aldol condensation of propanal, and 2-ethylbutanal by the reaction of butanal with acetaldehyde; in both cases the intermediate unsaturated aldehyde must be hydrogenated subsequently.

Uses. The C_6 aldehydes can easily be hydrogenated or oxidized to the corresponding alcohols or carboxylic acids.

2-Methylpentanal and 3-methylpentanal are starting materials for tranquilizers, such as meprobamate and mebutamate [44], [45]. 2-Methylpentanal also is used for the manufacture of various other products, including agrochemicals, perfumes, and catalysts for cross-linking of polyesters.

2.4.3. C_7 Aldehydes

The C_7 aldehydes are colorless, mobile liquids; all form azeotropic mixtures with water. Heptanal (enanthal) occurs naturally in ginger oil. The most important physical data of the C_7 aldehydes and the CAS registry numbers are given in Table 1.

Production. Heptanal, together with undecanoic acid, is obtained by the pyrolytic cleavage of ricinoleic acid esters [46] or by the hydroformylation of 1-hexene in the presence of modified rhodium catalysts. In the latter case, the straight-chain aldehyde is formed almost exclusively [47]. A minor byproduct is 2-methylhexanal, which can be prepared specifically by reaction of hexanal with formaldehyde followed by hydrogenation [35].

2,3-Dimethylpentanal is prepared by the hydroformylation of 3-methyl-2-pentene [48] or by the amine-catalyzed reaction of 3-methylpentanal with formaldehyde and subsequent hydrogenation [44]:

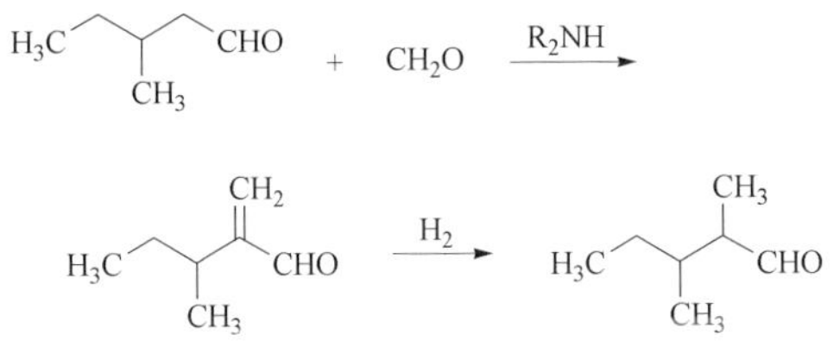

Uses. Heptanal is used principally for the preparation of α-amylcinnamaldehyde, a popular fragrance chemical especially in the field of soap perfumery. It is also a starting material for heptanoic acid, esters of which have applications in lubricants. Manufacturers of the aldehyde and/or the acid are Celanese and Monsanto in the United States and ATO Chimie in Western Europe.

2,3-Dimethylpentanal is required for the synthesis of some pharmaceuticals.

2.4.4. C$_8$ Aldehydes

The physical data and the CAS registry numbers of the most important C$_8$ aldehydes are presented in Table 1. The C$_8$ aldehydes form azeotropic mixtures with water, as do most aliphatic aldehydes, and are miscible with the common organic solvents. Octanal occurs naturally in various citrus oils. Of the C$_8$ aldehydes, octanal, isooctanal, and 2-ethylhexanal have industrial importance.

Production. Octanal (caprylaldehyde) can be obtained by hydroformylation of 1-heptene [49] or by dehydrogenation of octanol [50]. Isooctanal is used as a mixture of isomers and is prepared by hydroformylation of a heptene mixture, the product of codimerization of propene and butene.

The base-catalyzed aldol condensation of butanal leads to 2-ethyl-2-hexenal, which is converted to 2-ethylhexanal by hydrogenation:

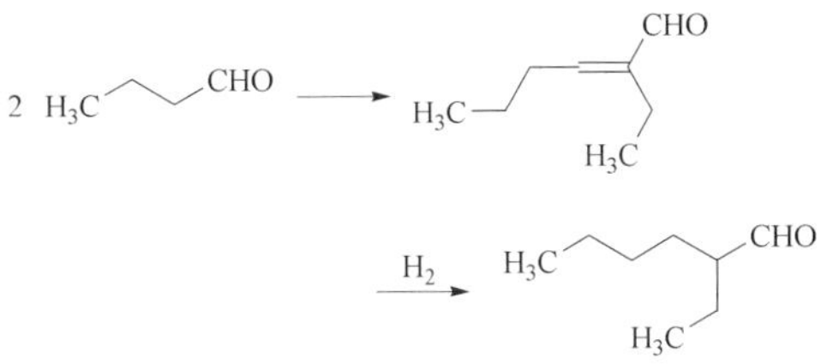

2-Propylpentanal can be prepared either by hydroformylation of 3-heptene [51] or in a multistep synthesis from pentanal [52].

Uses. Octanal is employed in the perfume industry for the preparation of synthetic citrus oils and for the synthesis of α-hexylcinnamaldehyde.

Isooctanal is used in the perfume industry and is also catalytically hydrogenated to isooctanol, which is processed further to plasticizers. The aldol reaction of isooctanal with subsequent hydrogenation leads to isohexadecanol, which can be used for the manufacture of synthetic lubricants and hydraulic fluids.

2-Ethylhexanal is used for the manufacture of 2-ethylhexanol ($\rightarrow$ 2-Ethylhexanol) and 2-ethylhexanoic acid ($\rightarrow$ Carboxylic Acids, Aliphatic). Aminating hydrogenation leads to 2-ethylhexylamines, which are important intermediates in organic syntheses. The condensation of 2-ethylhexanal with aromatic amines results in products that are used as vulcanizing agents and antioxidants for rubber.

2-Propylpentanoic acid (dipropylacetic acid) and its derivatives are prepared from 2-propylpentanal. The sodium salt is an important antiepileptic.

2.4.5. C_9 Aldehydes

Nonanal (pelargonaldehyde) occurs in several natural oils, such as cinnamon oil, lemon grass oil, citrus oil, and rose oils. For physical data and CAS registry numbers see Table 1. The most important C_9 aldehydes are the isomeric mixtures obtained by hydroformylation of commercial C_8 olefin mixtures.

Production. Nonanal is prepared by catalytic dehydrogenation of nonanol [53], by hydroformylation of 1-octene [54], or by reaction of formic acid and nonanoic acid on a titanium dioxide catalyst. Isononanal, prepared on an industrial scale by hydroformylation of commercial diisobutylene, is an isomeric mixture containing about 95 % 3,5,5-trimethylhexanal [55], [56]:

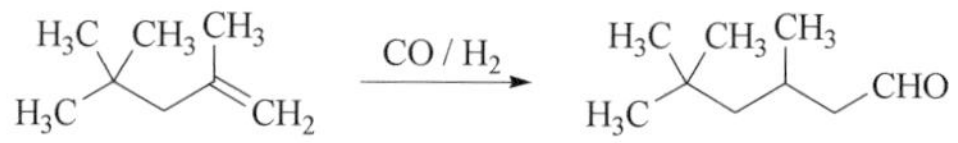

In addition to diisobutylene, other C_8 olefin mixtures are obtained by dimerization of isobutene and butene and are used as feedstock for the production of a number of isononanal mixtures.

Uses. Both nonanal and isononanal are used in the perfume industry. The various isononanals can be converted to isononanol mixtures, which serve as starting materials for the preparation of plasticizers. Upon oxidation isononanoic acid ($\rightarrow$ Carboxylic Acids, Aliphatic) is obtained.

Isononanal is an antimicrobial agent [56] and is used for the synthesis of dyestuffs.

2.4.6. C$_{10}$ Aldehydes

Some physical data and the CAS registry numbers are given in Table 1. Decanal is a colorless liquid that is immiscible with water. It has an odor similar to that of orange peel.

Production. Decanal is prepared either by hydroformylation of 1-nonene [57] or by dehydrogenation of 1-decanol on a copper catalyst [17]. 2-Methylnonanal is formed as a byproduct of the hydroformylation of 1-nonene. It can also be prepared by the reaction of nonanal and formaldehyde followed by hydrogenation. Mixtures of *iso*-C$_{10}$ aldehydes are obtained by hydroformylation of commercial C$_9$ olefins, e.g., tripropylene.

Uses. Besides the use of *iso*-C$_{10}$ aldehyde mixtures as intermediates for isodecanol and isodecanoic acid, C$_{10}$ aldehydes serve primarily as perfumes and fragrance chemicals. The main use of decanal is for citrus tones and for the manufacture of synthetic citrus oils. In addition, the C$_{10}$ aldehydes have value as intermediates in the synthesis of pharmaceuticals and in the polymer and pesticide fields.

2.4.7. C$_{11}$ Aldehydes

Of the C$_{11}$ aldehydes, undecanal and 2-methyldecanal (methyloctylacetaldehyde), possess some commercial value. When pure they are colorless liquids. They have a green, fatty character that blends well with woody and mossy perfume formulations. Undecanal occurs naturally in citrus oils and has a waxy, floral odor. The most important physical data and the CAS registry numbers are presented in Table 1. The aldehydes form azeotropic mixtures with water and are freely soluble in the common organic solvents.

Production. Undecanal is prepared by hydroformylation of decene [57], [58] or by catalytic dehydrogenation of undecanol [17]. Another method is the reduction and subsequent partial hydrogenation of the acid chloride of undecylenic acid, which in turn can be obtained by acidic cleavage of castor oil.

2-Methyldecanal is obtained together with undecanal by the hydroformylation of decene or by the reaction of decanal with formaldehyde and subsequent partial hydrogenation [35], [59]. It is formed also by the Darzens reaction of alkyl chloroacetates with 2-decanone to form glycidic esters, followed by saponification and acidulation [60].

The two isomeric aldehydes can be separated by using the bisulfite adduct [10].

Uses. Undecanal and 2-methyldecanal are used in the perfume industry. Other applications are the syntheses of pharmaceuticals, fungicides, plant growth regulators, bactericides, and disinfectants.

2.4.8. C_{12} Aldehydes

Only dodecanal (lauric aldehyde) occurs naturally in several essential oils, such as rue oil, orange peel oil, and citrus oil. Dodecanal and 2-methylundecanal, because of their pleasant odors, are among the most important toning agents in the aroma industry. The two isomeric aldehydes are colorless liquids, immiscible with water, but freely soluble in most organic solvents. Physical properties and CAS registry numbers are given in Table 1.

Production. Dodecanal is prepared by the catalytic dehydrogenation of dodecanol [61] or by the reduction of the corresponding acid chloride with sodium tetrahydridoborate [62].

2-Methylundecanal can be obtained by hydroformylation of undecene [58]. Alternatively, the mixture of undecanal and 2-methyldecanal, produced by oxo synthesis, reacts with formaldehyde, followed by partial hydrogenation and distillation [63]. 2-Methylundecanal also is formed from 2-undecanone by the Darzens reaction [60].

2.4.9. C_{13} Aldehydes

The only commercially available C_{13} aldehydes are isomeric mixtures, the composition of which depends primarily on the nature of the olefins that are hydroformylated. Physical properties and CAS registry numbers are given in Table 1.

Production. The C_{12} olefins formed by the oligomerization of butene or propene are converted to the corresponding isomeric tridecanals by hydroformylation. Ziegler α-olefins, olefins from wax cleavage, and olefins produced by metathesis (Shell's SHOP process) also are used as feedstocks.

Uses. Tridecanals are used in the perfume and aroma industries. Their catalytic reduction leads to tridecanols, which are of considerable importance as plasticizer and detergent alcohols. Salts, esters, and other derivatives of isotridecanoic acid, prepared by oxidation of the aldehyde, are of value in the paint and plastics industries.

3. Unsaturated Aldehydes

Acrolein is the most important of the unsaturated aldehydes (→ Acrolein and Methacrolein); crotonaldehyde also is manufactured industrially (→ Crotonaldehyde and Crotonic Acid). Besides these, 2-ethyl-2-hexenal (see Section 3.4.3), citral and citronellal (see Section 3.4.4), and undecylene aldehyde (see Section 3.4.5) are important unsaturated aldehydes.

3.1. Physical Properties

The most important physical data for a number of unsaturated aliphatic aldehydes are summarized in Table 4. The α,β-unsaturated aldehydes in the commercially significant C_3-C_{10} range are liquids. Their boiling points are, without exception, higher than those of the corresponding saturated aldehydes. They resemble the saturated aldehydes in that their viscosities, densities, and refractive indices increase with increasing molecular mass. The compounds are only moderately soluble in water; the solubilities decrease with increasing molecular mass. In common organic solvents, solubilities of the unsaturated aldehydes are high. The lower homologs have a pungent smell and have an irritating effect on mucous membranes. However, the higher unsaturated compounds are often used in the preparation of perfumes and aromas.

Most α,β-unsaturated aldehydes autoxidize easily and tend to polymerize, especially in the presence of base. Small amounts of stabilizers, e.g., hydroquinone, are nearly always added for these reasons.

Many α,β-unsaturated aldehydes form azeotropes with water and other substances. Data for some of these are compiled in Table 5.

Like their saturated counterparts, α,β-unsaturated aldehydes can be unequivocally characterized by IR, ^{1}H NMR, ^{13}C NMR, and by UV/visible spectroscopy.

3.2. Chemical Properties

In addition to the typical reactions of saturated aldehydes already described, the chemical properties of the α,β-unsaturated aldehydes are determined by the conjugation of the olefinic double bond with the carbonyl function. This leads to the formation of 1,2 and 1,4 adducts as a result of nucleophilic addition reactions:

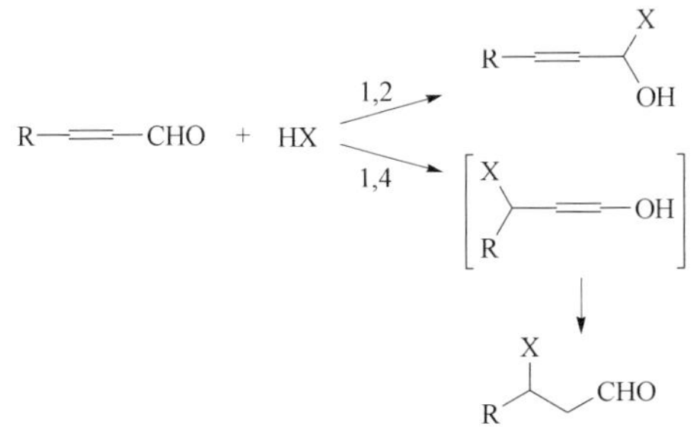

The resulting products play an important role in preparative organic chemistry. A further characteristic reaction of α,β-unsaturated aldehydes is the Diels–Alder conversion, e.g., to alkyl-substituted 2-formyl-2,3-dihydropyranes [64]:

Aldehydes, Aliphatic and Araliphatic

Table 4. Physical properties of unsaturated aldehydes

	Name	CAS registry number	Formula	M_r	mp, °C	bp*, °C	Density (20 °C), g/cm³	Refractive index, n_D^{20}	Surface tension (20 °C), mN/m	Viscosity (20 °C), mPa s	Solubility in H_2O at 20 °C, wt%
C_3	Propenal (acrolein)	[107-02-8]	CH_2=CHCHO	56.07	−86	53	0.841	1.401			20.6
C_4	trans-2-Butenal (crotonaldehyde)	[123-73-9]	CH_3CH=CHCHO	70.09	−74	102	0.852	1.437			15.5
	2-Methylpropenal (methacrolein)	[78-85-3]	CH_2=C(CH_3)CHO	70.09	−81	68	0.843	1.416		0.5	5.9
C_5	2-Methyl-2-butenal (tiglic aldehyde)	[497-03-0]	CH_3CH=C(CH_3)CHO	84.13		117 (73.8)	0.871	1.448			
	2-Pentenal	[1576-87-0]	CH_3CH_2CH=CHCHO	84.13		125	0.848	1.443			
C_6	2-Hexenal	[6728-26-3]	$CH_3(CH_2)_2$CH=CHCHO	98.15		146	0.842	1.446			
	2-Methyl-2-pentenal	[623-36-9]	CH_3CH_2CH=C(CH_3)CHO	98.15	<−60	138	0.854	1.449		0.8	1.4
	2-Isopropylpropenal	[4417-80-5]	CH_3CH(CH_3)CH(CHO)=CH_2	98.15							
	2-Ethyl-2-butenal	[19780-25-7]	CH_3CH=C(C_2H_5)CHO	98.15	<−60	135	0.857	1.440		0.7	
C_8	2-Ethyl-2-hexenal	[645-62-5]	$CH_3(CH_2)_2$CH=C(C_2H_5)CHO	126.21	<−70	175	0.852	1.453		1.28	
C_{10}	3,7-Dimethyl-6-octenal (citronellal)	[106-23-0]	$(CH_3)_2$C=CH$(CH_2)_2$CH(CH_3)CH_2CHO	154.26		207 – 208	0.851	1.448	29.26	1.93	
	3,7-Dimethyl-2,6-octadienal (citral)	[5392-40-5]	$(CH_3)_2$C=CH$(CH_2)_2$C(CH_3)=CHCHO	152.24							
	(2 E)-3,7-Dimethyl-2,6-octa-dienal (citral a, geranial)	[141-27-5]	$(CH_3)_2$C=CH$(CH_2)_2$C(CH_3)=CHCHO	152.24		229	0.889	1.490			
	(2 Z)-3,7-Dimethyl-2,6-octa-dienal (citral b, neral)	[106-26-3]	$(CH_3)_2$C=CH$(CH_2)_2$C(CH_3)=CHCHO	152.24		120 (2)	0.887	1.487			
C_{11}	10-Undecanal (undecylene aldehyde)	[112-45-8]	CH_2=CH$(CH_2)_8$CHO	168.28		103 (0.4)	0.850 (21 °C)	1.446 (21 °C)			

* at 101.3 kPa unless otherwise specified in parentheses.

Table 5. Azeotropic mixtures of unsaturated aldehydes with water

Aldehyde	bp of aldehyde, °C	bp of azeotrope, °C	wt% H$_2$O in azeotrope
Acrolein	53	52.4	2.6
Methacrolein	68	63.6	7.7
Crotonaldehyde	102.4	84	24.8
2-Ethyl-2-butenal	135.3	92.7	38
2-Methyl-2-pentenal	138.2	93.5	40
2-Ethyl-2-hexenal	176	97.6	60.9

Dienes also can enter into Diels–Alder reactions with unsaturated aldehydes:

Of special importance for the handling of α,β-unsaturated aldehydes is the previously mentioned tendency toward polymerization, which proceeds in a highly exothermic manner in the presence of base.

Each of the two functionalities present in the molecule, the olefinic double bond and the carbonyl group, can be selectively hydrogenated by a suitable choice of catalyst. For example, the olefinic double bond is hydrogenated selectively in the presence of noble-metal catalysts and the carbonyl group in the presence of modified Pt, Ru, or Os catalysts or by a Meerwein–Ponndorf–Verley reaction with aluminum alkoxides to yield the corresponding saturated aldehydes and α,β-unsaturated alcohols, respectively. The saturated alcohols are obtained by hydrogenation over nickel or copper catalysts.

3.3. Production

Although the lower, commercially important α,β-unsaturated aldehydes, such as acrolein, crotonaldehyde, or 2-ethyl-2-hexenal, are obtained exclusively by synthesis, a number of essential oils serve as raw materials for higher homologs, such as for citral or citronellal.

There are two industrial processes for the manufacture of α,β-unsaturated aldehydes:

1) Oxidation of olefins (preparation of acrolein)
2) Dehydration of aldols obtained by aldol condensation of saturated aldehydes (preparation of crotonaldehyde and 2-ethyl-2-hexenal)

In addition there are some special syntheses for the preparation of aldehydes used in the perfume industry such as dehydrogenation of unsaturated alcohols (preparation of citral from geraniol), and reduction of unsaturated acids (undecylenealdehyde from undecylenic acid).

Oxidation of Olefins. The direct oxidation of olefins is industrially important for the manufacture of acrolein from propene. The conversion proceeds at 300–480 °C over variously modified Bi–Mo oxide catalysts. The feedstock is a gaseous mixture of propene, air, and water vapor in a molar ratio of about 1:10:2. Conversion of up to 98% and acrolein yields of 78–92% are realized. Byproducts include acetaldehyde, acetic acid, and acrylic acid. Commercial acrolein has a purity of 95–97%. The stabilizer, which is added in all processing steps, is usually hydroquinone (for further details → Acrolein and Methacrolein).

Dehydration of Aldols. The β-hydroxyaldehydes (aldols), see Chapter 4 that are formed as intermediates in the aldol reaction are extremely unstable and decompose with loss of water to form α,β-unsaturated aldehydes. Depending on the reaction conditions, the aldol reaction can lead directly to the unsaturated compound. This process is used principally for the synthesis of crotonaldehyde from acetaldehyde, 2-methyl-2-pentenal from propionaldehyde, and 2-ethyl-2-hexenal from butanal.

The aldol condensation of acetaldehyde with catalytic amounts of dilute sodium hydroxide is carried out at 20–25 °C. The reaction mixture is treated with acetic acid to stop the reaction. In the subsequent first distillation, water is cleaved from the acetaldol. The selectivity towards crotonaldehyde reaches values up to 95%.

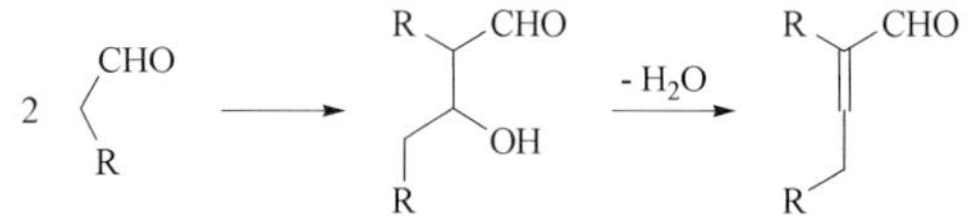

Not only can two identical aldehyde molecules undergo aldol condensation but also two different aldehyde species can do so. In the latter case a mixture of products usually results. By carefully choosing the reactants and reaction conditions, however, it is possible to obtain the desired compound as the main product. For example, acrolein was formerly prepared by the reaction of formaldehyde with acetaldehyde:

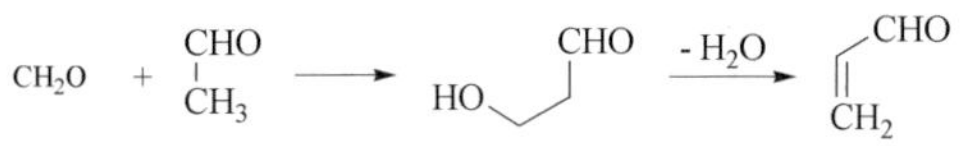

In the same way higher 2-methylenealkanals (2-alkylacroleins) can be obtained by the reaction of formaldehyde with longer chain aldehydes.

Miscellaneous Processes. For the preparation of some unsaturated aroma aldehydes, the corresponding alcohols are selectively dehydrogenated over copper, copper–zinc, or noble metal catalysts. These processes preferably are carried out under reduced pressure and find application in the manufacture of citral, citronellal, and hydroxycitronellal.

Another method of industrial interest for the synthesis of unsaturated aldehydes is the Claisen rearrangement of allyl vinyl ethers, which are formed as shown in the following reaction scheme [32]:

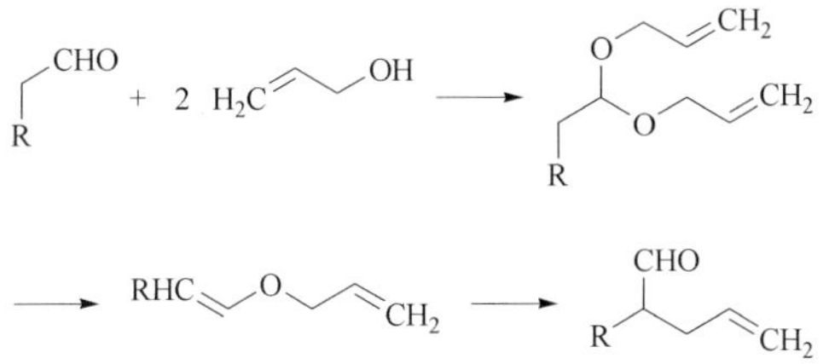

2-Alkenals are obtained by the reaction of unsaturated alkyl halides with the sodium salts of secondary nitrohydrocarbons, e.g., citral in 80% yield from 1-halo-3,7-dimethylocta-2,6-diene [65]:

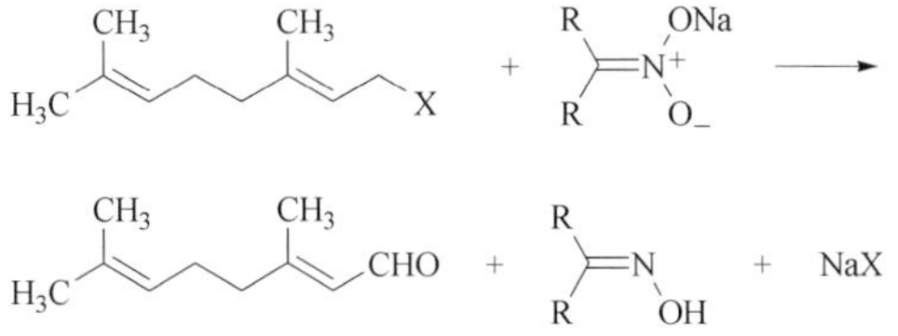

Another possibility is the treatment of acetals with vinyl ethers in the presence of boron trifluoride to give the corresponding β-alkoxyacetals. Under the influence of acids these are converted into α,β-unsaturated aldehydes [66].

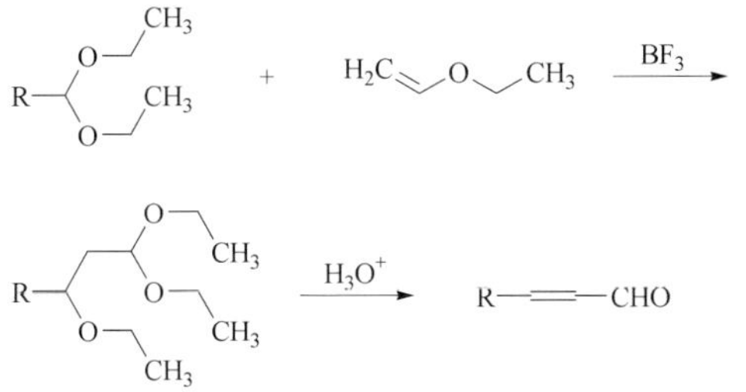

For a review of further synthetic methods, see [1], [2].

3.4. Individual Unsaturated C_5–C_{11} Aldehydes

3.4.1. C_5 Alkenals

Of the C_5 alkenals, only 2-methylcrotonaldehyde (tiglic aldehyde) has any significance. The most important physical data and CAS registry numbers of two C_5 alkenals are presented in Table 4. These compounds undergo most of the reactions characteristic of α,β-unsaturated aldehydes.

Production. Several processes for the preparation of tiglic aldehyde are described in the literature; for example, the condensation of acetaldehyde with propanal in the presence of sodium hydroxide [67]; the reaction of 3,4-epoxy-3–methylbutene with palladium acetylacetonate and triphenylphosphine [68]; the isomerization of 2-ethylacrolein [69]; or a two-step synthesis beginning with the catalytic reaction of isoprene, acetic acid, and oxygen to give 2-methyl-1,4-diacetoxy-2-butene, followed by hydrolysis catalyzed by cation-exchange resins [70].

Uses. Tiglic aldehyde is used mainly in the perfume industry. It is also a starting material for the preparation of terpenes, and the *syn*-oxime is used as a sweetener in food.

3.4.2. C_6 Alkenals

The most important physical data and CAS registry numbers for some unsaturated C_6 aldehydes are given in Table 4. The aldehydes are colorless, highly flammable liquids. A number of essential oils contain *trans*-2-hexenal. When diluted, it has an intense odor resembling that of apples.

Production. *trans*-2-Hexenal (leaf aldehyde) is prepared by reaction of butanal with vinyl ether in the presence of boron trifluoride and subsequent treatment of the reaction mixture with dilute sulfuric acid [71]:

2-Methyl-2-pentenal is formed by aldol condensation of propanal, 2-ethyl-2-butenal by aldol condensation of butanal and acetaldehyde, and 2-isopropylacrolein by reaction of 3-methylbutanal with formaldehyde [40].

Uses. The perfume and aroma industries use *trans*-2-hexenal as a constituent of floral compositions, 2-methyl-2-pentenal for the manufacture of 2-methylpentanal, and 2-isopropylacrolein as a starting material for the preparation of 2,3-dimethyl-2-butene and pinacol [40]. 2-Ethyl-2-butenal is mainly used for the synthesis of 2-ethylbutyric acid.

3.4.3. C$_8$ Alkenals

Among the many isomeric C$_8$ alkenals, *2-ethyl-2-hexenal* is of outstanding economic and technical importance. It is a clear liquid with a characteristic odor and is soluble in practically all organic solvents, although nearly insoluble in water. The most important physical data and the CAS registry number are given in Table 4.

Production. Industrially, 2-ethyl-2-hexenal is prepared by aldol condensation of butanal and subsequent dehydration. The reaction is carried out at 80–130 °C and pressures up to 1 MPa in the presence of sodium hydroxide or a basic ion-exchange resin. Many other catalysts are described in the literature [72]. The conversion is over 99 %. The organic phase usually is subjected immediately to further processing to yield either 2-ethylhexanal over a palladium catalyst, or 2-ethyl-1-hexanol over a nickel or copper catalyst [73].

Uses. The main application for 2-ethyl-2-hexenal is the manufacture of 2-ethyl-1-hexanol, currently the most important plasticizer alcohol with a production capacity (1997) of nearly 10^6 t/a in Western Europe alone and 350 000 t/a in the United States. Condensation products of 2-ethyl-2-hexenal with various aromatic amines have proved to be good vulcanizing agents and antioxidants for rubber.

2-Ethylhexanal, obtainable by selective hydrogenation, is a starting material for the synthesis of the corresponding acid and amines and for the manufacture of pharmaceuticals.

3.4.4. C$_{10}$ Alkenals

The unsaturated C$_{10}$ aldehydes are used preferentially in perfumes and fragrances. Additionally they serve as starting materials for the synthesis of a whole series of terpenoid compounds. The most important representatives are citral a and b (*cis*- and *trans*-3,7-dimethyl-2,6-octadienal) and citronellal (3,7-dimethyl-6-octenal), which occur naturally in many essential oils. They are colorless liquids with odors resembling lemons or balm. Some physical properties and the CAS registry numbers are given in Table 4.

Several industrially established syntheses exist for citral a and b and for citronellal. The isolation from essential oils is now carried out on a large scale only for citronellal.

Uses. Besides the application in the fragrance industry, citral is an intermediate in the vitamin A synthesis and in the preparation of β-ionones [74]. Its acetals, which are relatively stable toward alkali, also are used in the perfume industry. Citronellal is used in small amounts for scenting soaps and detergents. Its principal application is in the preparation of isopulegol, citronellol, and hydroxycitronellal. These compounds also have importance in the fragrance industry.

3.4.5. C_{11} Alkenals

10-Undecenal, undecylene aldehyde, the most important unsaturated C_{11} aldehyde and so far not known to occur naturally, is a colorless liquid with a stifling, flowery odor. For some physical properties and the CAS registry number see Table 4.

Undecylene aldehyde is prepared by Rosenmund reduction of undecylenoyl chloride, which in turn is obtained by cleavage of castor oil. A further possibility is the reaction of undecylenic acid with formic acid over a titanium oxide catalyst. This aldehyde is an important modulating agent in the perfume industry.

4. Hydroxyaldehydes

Of this class of compounds, some β-hydroxyaldehydes in particular are of commercial importance, primarily as intermediates for the preparation of 1,3-diols (see Sections 4.3.1, 4.3.2, and 4.3.4), and α,β-unsaturated aldehydes (see Section 3.3) and their derivatives.

The 2,2-bismethylol compounds of propanal and butanal and the trismethylol compound of acetaldehyde are intermediates in the manufacture of trimethylolethane, trimethylolpropane, and pentaerythritol, respectively (→ Alcohols, Polyhydric).

4.1. Properties

The boiling points of the β-hydroxyaldehydes (aldols) are well above those of the corresponding saturated compounds because of hydrogen bonding. The aldols obtained from lower aldehydes can be vacuum distilled without decomposition only if the salts, present from the condensation reaction, have been completely removed. They decompose upon heating at normal pressure, however [75].

Table 6. Physical properties of hydroxyaldehydes

Name	CAS registry number	Formula	M_r	mp, °C	bp (p), °C (kPa)	Density (20 °C), g/cm^3	Refractive index, n_D^{20}
2-Hydroxyethanal (glycol aldehyde)	[141-46-8]	HOCH$_2$CHO	60.05	97		1.366 (100 °C)	1.477 (19 °C)
3-Hydroxypropanal	[2134-29-4]	HOCH$_2$CH$_2$CHO	74.05		38 (0.03)		
3-Hydroxybutanal (acetaldol)	[107-89-1]	CH$_3$CH(OH)CH$_2$CHO	88.10		77 (3.0)	1.105	1.4238
4-Hydroxybutanal	[25714-71-0]	HOCH$_2$(CH$_2$)$_2$CHO	88.10		70–72 (1.5)	1.089	1.4450
2,2-Dimethyl-3-hydroxypropanal (hydroxypivaldehyde)	[597-31-9]	HOCH$_2$C(CH$_3$)$_2$CHO	102.14	ca. 70	ca. 141 (101.3)		
2-Ethyl-3-hydroxyhexanal (butyraldol)	[496-03-7]	CH$_3$CH$_2$CH$_2$CH(OH)CH(C$_2$H$_5$)CHO	144.22		98–101 (1.3)	0.9397	1.4426
3,7-Dimethyl-7-hydroxyoctanal (hydroxycitronellal)	[107-75-5]	(CH$_3$)$_2$C(OH)(CH$_2$)$_3$CH(CH$_3$)CH$_2$CHO	172.27		85–87 (0.13)	0.922	1.4488

The lower aldols are easily miscible with water and common organic solvents. Some physical data are compiled in Table 6.

For qualitative and quantitative analysis of hydroxyaldehydes, both the hydroxyl and the carbonyl groups can be exploited for the preparation of derivatives.

The chemical behavior of the hydroxyaldehydes is influenced primarily by the ease with which water is cleaved, leading to the formation of α,β-unsaturated aldehydes. This reaction, which is referred to as crotonization, is dependent on the structure of the starting material and on the reaction conditions. It is catalyzed by acids and bases.

The catalytic hydrogenation of β-hydroxyaldehydes leads to the formation of 1,3-diols. Important examples are the preparation of propane-1,3-diol from β-hydroxypropanal, 2-ethylhexane-1,3-diol from butyraldol, as well as the previously mentioned preparation of trimethylolethane, trimethylolpropane, and pentaerythritol. The last compound is also formed by reduction with formaldehyde (crossed Cannizzaro reaction).

Monomeric β-hydroxyaldehydes generally are not stable. They react with aldehydes to form so-called aldoxanes (**1**), dimerize to paraldols (**2**), or polymerize.

4.2. Production

Aldol Condensation. The aldol condensation, first mentioned in the literature by A. WURTZ in 1872, can be described in the case of a base-catalyzed reaction by the following mechanism [75]:

The aldol condensation is a reversible reaction. For the catalytic influence of amines and kinetic studies, cf. [76]; for a detailed description of the aldol reaction, cf. [77].

Aldol condensation, which is only possible for aldehydes with at least one α-hydrogen atom, can be catalyzed not only by bases, as shown above, but also by acids. In general, aldols obtained from aldehydes with more than one α-hydrogen atom can be isolated only at low temperature, because they readily lose water to form α,β-unsaturated aldehydes:

$$R-CH_2-\underset{R}{CH}-\underset{OH}{CH}-C\underset{H}{\overset{O}{\diagup}} \xrightarrow{-H_2O} R-CH_2-CH=\underset{R}{C}-C\underset{H}{\overset{O}{\diagup}}$$

If two different aldehydes with α-hydrogen atoms are subjected to aldol addition, all four possible aldol species generally are formed in varying amounts.

Of more importance for industrial purposes are reactions in which only one reactant possesses α-hydrogen atoms; in this case the second reactant is frequently formaldehyde. For the preparation of mono- or polymethylolalkanals by reaction of formaldehyde with alkanals, cf. [75, p. 89] and Section 4.3.4.

These exothermic reactions are mostly carried out in the liquid phase, and aqueous solutions of sodium hydroxide or alkali carbonates usually serve as catalysts. Other catalyst systems have been described, e.g., zinc- or magnesium-containing zeolites [78], alkali hydroxides in combination with phase-transfer catalysts [79], tertiary amines [80], and basic ion exchange resins [81].

Typical byproducts of aldol condensation, apart from the dimeric aldols, aldoxanes, and α,β-unsaturated aldehydes mentioned earlier, are cyclic acetals, Tishchenko esters, etc. This is especially the case for reactions with formaldehyde, in which a reduction of the aldol by a crossed Cannizzaro reaction can occur [82].

G. WITTIG, discovered a modification of the classic aldol reaction that involves the treatment of carbonyl compounds with metalated imines. In this case α,β-unsaturated aldehydes are practically the only products. The value of this procedure has been proved in the field of natural products [83], [84].

Miscellaneous Processes. The addition of water to α,β-unsaturated aldehydes yields β-hydroxyaldehydes. This reaction is used industrially for the preparation of 1,3-propanediol (see Section 4.3.1).

3-Hydroxypropanal also can be obtained by the hydroformylation of ethylene oxide [85]. 4-Hydroxybutanal is obtained by hydroformylation of allyl alcohol [86]. A review of the hydroformylation of unsaturated alcohols to give hydroxyaldehydes has been published [1, p. 212].

If 2,3-dihydro-1,4-pyran is hydrolized, 5-hydroxypentanal is formed; however, it exists in equilibrium with its cyclic hemiacetal [87]. For example, 2-phenyl-2,3-dihydropyran, obtained by Diels–Alder synthesis from acrolein and styrene, can be cleaved by dilute sulfuric acid to give 5-phenyl-5-hydroxypentanal [88].

4.3. Individual Hydroxyaldehydes

4.3.1. 3-Hydroxypropanal

Some physical properties of 3-hydroxypropanal and its CAS registry number are given in Table 6. This compound is the intermediate in the preparation of 1,3-propanediol. It is extremely unstable and normally is not isolated as a pure substance. It can be distilled in high vacuum and characterized as its 2,4-dinitrophenylhydrazone (*mp* 132.5 – 133 °C).

Production. The addition of water to acrolein is currently the most important process for the production of 3-hydroxypropanal. A ca. 20 % aqueous solution of acrolein is reacted in the presence of a weakly acidic catalyst or an ion exchange resin [89], [90]. The solution can be either hydrogenated directly to 1,3-propanediol or, more advantageously, first subjected to an extraction with isobutanol (→ Propanediols).

Uses. 3-Hydroxypropanal is used exclusively for the production of 1,3-propanediol.

4.3.2. 3-Hydroxybutanal (Acetaldol)

Some physical properties of 3-hydroxybutanal (acetaldol) and the CAS registry number are given in Table 6. This compound is the lowest molecular mass aldol that can be prepared by aldol condensation. Pure monomeric acetaldol boils at 59 – 60 °C (1300 – 1500 Pa). Higher distillation temperatures lead to crotonaldehyde by dehydration and to acetaldehyde by reversal of the condensation. Polymerization of the aldol during its preparation from acetaldehyde can be prevented by addition of hydroquinone or pyrogallol. When pure, acetaldol polymerizes easily to paraldol; stabilization can be achieved by addition of small amounts of water or acetaldehyde.

Production. In the presence of dilute sodium or potassium hydroxide solutions, acetaldehyde is converted to the aldol over a period of several hours in a water-cooled flow tube at 20 – 25 °C. In order to avoid side and secondary reactions, which would result in appreciable resin formation, the reaction is stopped by adding phosphoric or acetic acid when acetaldehyde conversion has reached 60 %. After the unconverted acetaldehyde is separated, a bottom product is obtained that, apart from small amounts of byproducts such as crotonaldehyde, consists of 72 – 73 wt % acetaldol and ca. 18 wt % water [91].

Uses. Acetaldol is used primarily for the preparation of crotonaldehyde. The dehydration takes place during distillation upon addition of some acetic acid. Acetaldol can also be hydrogenated to 1,3-butanediol. The secondary products, butyraldehyde, butanol, and 1,3-butadiene, which at one time were prepared from acetaldol, are now manufactured almost exclusively by other processes.

Acetaldol is used also for the manufacture of substituted quinaldines which, in turn, are valuable starting materials for dyes and pigments [92].

4.3.3. 4-Hydroxybutanal

Some physical properties of 4-hydroxybutanal and its CAS registry number are given in Table 6. Because no conjugated system can be formed, this aldehyde does not tend to split off water, but forms its cyclic hemiacetal, 2-hydroxytetrahydrofuran. Only 4 % of the free hydroxyaldehyde exists in the equilibrium with the hemiacetal. For chemical characterization the compound is transformed into the 2,4-dinitrophenylhydrazone (mp 120 – 122 °C).

Production. Allyl alcohol, obtained from propylene oxide, is converted to 4-hydroxybutanal by rhodium-catalyzed hydroformylation. The formation of 3-hydroxy-2-methylpropanal can be minimized by using a large excess of phosphine ligands [86].

Uses. 4-Hydroxybutanal is used for the synthesis of tetrahydrofuran and 1,4-butanediol. This process was commercialized by Arco in 1990.

4.3.4. 2,2-Dimethyl-3-hydroxypropanal (Hydroxypivaldehyde)

2,2-Dimethyl-3-hydroxypropanal (hydroxypivaldehyde) is characterized by the aldehyde group, the primary alcohol group, and the neopentyl configuration.

$$\text{HO}\underset{}{\overset{H_3C\ \ CH_3}{\diagup\!\!\!\diagdown}}\text{CHO}$$

Although the functional groups provide possibilities for a whole range of reactions, the neopentyl structure is responsible for a noticeable stability toward hydrolysis, heat, and light in many derivatives.

Hydroxypivaldehyde is a colorless liquid that can be distilled in vacuum. Within a short time it dimerizes to a substituted 1,3-dioxane; this reaction is reversible [82], [93]. For some physical properties and the CAS registry number see Table 6.

Production. Hydroxypivaldehyde is prepared by the reaction at ca. 50 °C of isobutyraldehyde and a 30 – 37 % aqueous solution of formaldehyde. The catalysts used are normally aqueous solutions of alkali or alkaline-earth hydroxides [94], but the use of alkali carbonates [95], alcoholic alkali hydroxides [96], tertiary amines [80], and basic ion exchangers [81] also has been described. Because the two starting materials are inexpensive and readily available, other processes for the manufacture of hydroxypivaldehyde have been unable to achieve any industrial importance.

Uses. The most important use of hydroxypivaldehyde is the manufacture of neopentyl glycol, which can be prepared either by a crossed Cannizzaro reaction with formaldehyde or by catalytic hydrogenation (→ Alcohols, Polyhydric). Further products with hydroxypivaldehyde as starting material (cf. [82], [93]) are hydroxypivalic acid neopentyl glycol ester, hydroxypivalic acid, and pantolactone (α-hydroxy-β,β-dimethyl-γ-butyrolactone), which is used in the preparation of pharmaceuticals and vitamins [97].

4.3.5. 2-Ethyl-3-hydroxyhexanal (Butyraldol)

Some physical properties and the CAS registry number are given in Table 6.

As is the case with most aldols, 2-ethyl-3-hydroxyhexanal is not isolated as a pure compound but is converted directly under the prevailing reaction conditions to 2-ethyl-2-hexenal. The aldol can be characterized as the 2,4-dinitrophenylhydrazone derivative (*mp* 172 °C).

Production. 2-Ethyl-3-hydroxyhexanal is prepared primarily by aldol condensation of butanal at 30 °C in the presence of an aqueous sodium hydroxide solution and a phase-transfer catalyst; the reaction is stopped by addition of acetic acid [98].

Other catalysts described in the literature include zeolites containing magnesium or zinc [78] and amines, e.g., tri-, tetra-, or hexamethylenediamine [76].

For the preparation of 2-ethyl-2-hexenal without isolation of the aldol see Chapter 3.4.3.

Uses. 2-Ethyl-3-hydroxyhexanal is used for the preparation of 2-ethyl-2-hexenal, 2-ethylhexanol, 2-ethylhexanal, 2-ethylhexanoic acid, and 2-ethyl-1,3-hexanediol.

4.3.6. 3,7-Dimethyl-7-hydroxyoctanal (Hydroxycitronellal)

Some physical properties and the CAS registry number are given in Table 6.

Hydroxycitronellal is a colorless liquid of low viscosity. So far it has not been found in nature. Its odor is reminiscent of linden flowers and lillies of the valley.

Hydroxycitronellal is used in numerous perfumes. The acetals also occasionally find application as perfumes because of their higher stability toward soap.

5. Araliphatic Aldehydes

Araliphatic aldehydes are compounds with the general formula $ArCH_{2x}CHO$ ($x \geq 1$) where Ar is some aromatic group, usually phenyl.

Although numerous araliphatic aldehydes occur naturally as components of essential oils, they are usually synthesized.

Araliphatic aldehydes resemble their aliphatic counterparts in that they readily undergo autoxidation and polymerization. Therefore, when stored, air should be excluded, and suitable stabilizers should be added.

Many araliphatic aldehydes and secondary products, such as the corresponding acetals, alcohols, and esters, are established compounds in the perfume industry. They are also starting materials for pharmaceuticals, agricultural chemicals, plasticizers, etc.

5.1. Properties

Physical Properties (see also Table 7). The industrially important araliphatic aldehydes are colorless to pale yellow liquids. They can be distilled, preferably under vacuum, without any appreciable decomposition. All have boiling points above 190 °C at atmospheric pressure, and the densities range between 0.95 and 1.05 g/cm^3. With the exception of phenylacetaldehyde which is slightly soluble in water, the higher araliphatic aldehydes are practically insoluble. The compounds are miscible with alcohols, ethers, and other common organic solvents.

Chemical Properties. Apart from their ease of oxidation, the reactivity of the araliphatic aldehydes is determined by the polarity of the carbonyl group, the resulting acidity of the α-hydrogen atoms (if available), and the aromatic substituents present in the molecule. These lead to reactions typical of aromatic systems in addition to the reactions of the aldehyde group (see Table 3).

In many cases the aldehyde functionality must be protected before carrying out aromatic electrophilic substitution. However, normally the substituent is already present in the aromatic system before the aldehyde is synthesized.

More important than reactions of the aryl ring are those of the aldehyde group, especially the formation of acetals, which are used in perfumes because of their comparative stability.

In addition to the saturated araliphatic aldehydes, some compounds with olefinic structure, e.g., cinnamaldehyde and α-amylcinnamaldehyde, are important as fragrances. If the aliphatic double bond is conjugated with the carbonyl group, the possibility of 1,4-addition exists (see Section 3.2).

Aldehydes, Aliphatic and Araliphatic

Table 7. Physical properties of araliphatic aldehydes

Name	CAS registry number	Formula	M_r	mp, °C	bp*, °C	Density (20 °C), g/cm³	Refractive index, n_D^{20}
2-Phenylacetaldehyde	[122-78-1]	$C_6H_5CH_2CHO$	120.16		195	1.027	1.526
4-Methylphenylacetaldehyde	[104-09-6]	$CH_3C_6H_4CH_2CHO$	134.18		221–222	1.005	1.526
3-Phenylpropanal (dihydrocinnamaldehyde)	[104-53-0]	$C_6H_5CH_2CH_2CHO$	134.18		223 (74.5)	1.019	1.527
2-Phenylpropanal (hydratropic aldehyde)	[93-53-8]	$C_6H_5CH(CH_3)CHO$	134.18		202–205	1.009	1.518
2-(4-Isobutylphenyl)-propanal	[51407-46-6]	$(CH_3)_2CHCH_2C_6H_4CH(CH_3)CHO$	190.29				
3-(4-Isopropylphenyl)-2-methyl-propanal (cyclamenaldehyde)	[103-95-7]	$(CH_3)_2CHC_6H_4CH_2CH(CH_3)CHO$	190.29		108–108.5 (0.3)	0.950	1.507
3-(4-tert-Butylphenyl)-2-methylpropanal (lilial)	[80-54-6]	$(CH_3)_3C$-C_6H_4-$CH_2CH(CH_3)CHO$	204.31		126–127 (0.8)	0.939	1.505
Cinnamaldehyde (3-phenyl-2-propenal)	[104-55-2]	$C_6H_5CH=CHCHO$	132.17	−8	253	1.050	1.620
α-Amylcinnamaldehyde	[122-40-7]	$C_6H_5CH=C(C_5H_{11})CHO$	202.3		174–175 (2.0)	0.971	1.538
α-Hexylcinnamaldehyde	[101-86-0]	$C_6H_5CH=C(C_6H_{13})CHO$	216.32		174–176 (2.0)	0.950	1.527 (25 °C)
α-Methylcinnamaldehyde	[101-39-3]	$C_6H_5CH=C(CH_3)CHO$	146.19		150 (10.0)	1.041	1.606 (17 °C)

[a] at 101.3 kPa unless otherwise specified

5.2. Production

Most araliphatic aldehydes occur naturally, often widespread, but usually in low concentrations. Therefore, their isolation from natural sources is of relatively little industrial value.

An extraordinarily large number of methods is available for the preparation of araliphatic aldehydes, but only a few are industrially important.

Aldol condensation with subsequent hydrogenation if necessary (Eq. 4):

Hydroformylation of styrenes (Eq. 5):

Isomerization of phenylated oxiranes (Eq. 6):

Friedel-Crafts reactions (Eq. 7):

Darzens glycidic ester synthesis (Eq. 8):

Production of the individual araliphatic aldehydes using these processes is described in the following section.

5.3. Individual Araliphatic Aldehydes

5.3.1. Phenylacetaldehyde

Phenylacetaldehyde, a constituent of numerous essential oils, is a colorless liquid with an odor resembling that of hyacinths and narcissi. Some physical properties and the CAS registry number are given in Table 7.

This extremely reactive compound is highly susceptible to oxidation and polymerization reactions. Alcohols are often added, for example, in perfumery, to stabilize the aldehyde by hemiacetal formation. For a comprehensive review, see [99].

Production. The most important method industrially is the isomerization of styrene oxide (Eq. 6). Ion-exchange resins [100], catalysis by chromium trioxide–tungsten trioxide on graphite [101] or silicon dioxide–aluminum trioxide [102], and thermolysis are recommended for the rearrangement.

Phenylacetaldehyde can also be made by the direct oxidation of styrene in the presence of palladium salts and copper(II) chloride in aqueous solutions of glycol ethers [104].

Other possibilities include the catalytic dehydrogenation of 2-phenylethanol on silver or gold catalysts [105], [106], the hydroformylation of benzyl halides in the presence of dicobalt octacarbonyl and sodium carbonate in acetonitrile [107], and the Darzens glycide ester synthesis from benzaldehyde and alkyl chloroacetates (Eq. 8) [108].

Uses. Phenylacetaldehyde is used in the perfume industry to obtain hyacinth and rose nuances. It also imparts these aromas to tea, tobacco, and coffee. Further applications include the preparation of pharmaceuticals, insecticides, acaricides, disinfectants, and the use as a rate-controlling additive in the polymerization of polyesters with other monomers.

The reaction of phenylacetaldehyde with ammonia and hydrogen cyanide (Strecker reaction) leads to the formation of phenylalanine, an intermediate for the sweetener aspartame. Phenylacetic acid can be obtained by oxidation of the aldehyde.

5.3.2. 4-Methylphenylacetaldehyde

Some physical properties and the CAS registry number are given in Table 7.

4-Methylphenylacetaldehyde has been identified as a constituent of maize oil. The colorless liquid resembles the corresponding unsubstituted aldehyde in that it is extremely reactive and must be stabilized. The compound has a strong floral odor.

Production. The methods of preparation correspond closely to those of phenylacetaldehyde. Thus, the compound can be prepared by glycidic ester synthesis from 4-methylbenzaldehyde and chloroacetate ester [108], by hydroformylation of 4-methylbenzylhalide with $Co_2(CO)_8Na_2CO_3$ in acetonitrile [107], or by oxidation of 4-methylstyrene [109].

Uses. 4-Methylphenylacetaldehyde is used in perfumes, e.g., in soap, in tobacco fragrances, in the preparation of pesticides, etc.

5.3.3. 2-Phenylpropionaldehyde (Hydratropaldehyde)

For some physical properties and the CAS registry number see Table 7.

2-Phenylpropionaldehyde (hydratropaldehyde) has long been established in the fragrance industry. It occurs in many natural products and has a strong fruity smell, reminiscent of hyacinths. Surprisingly, 2-phenylpropanal is much less sensitive to oxidation and polymerization than 2-phenylacetaldehyde. For a review on 2-phenylpropanal and the corresponding alcohol, see [110].

Production. In the last few years a series of new syntheses for 2-phenylpropionaldehyde has been described. However, only a few have achieved industrial significance. The most important is the rhodium-catalyzed hydroformylation of styrene.

In this process [HCo(CO)$_4$] is unsuitable as a catalyst because then styrene is largely hydrogenated [110], [111]. The use of chiral rhodium complexes leads to the formation of optically active aldehydes when prochiral olefins, such as styrene, are used. Asymmetric hydroformylation, however, has so far found no industrial application because of insufficient purity of the enantiomers [112].

Other synthetic methods are the gas-phase oxidation of α-methylstyrene on Cu, Bi–Mo–P–SiO$_2$ [113], or Pd salt catalysts [114], the rearrangement of epoxidized α-methylstyrene [100]–[103], and the dehydrogenation of 2-phenyl-1-propanol on Ag catalysts at reaction temperatures of ca. 600 °C [105], [106].

Uses. 2-Phenylpropanal and its hydrogenation product, 2-phenylpropanol, are used in perfumes. The aldehyde also is a valuable starting material for a series of pharmaceuticals and pesticides, and it serves in the plastics field as a stabilizer, catalyst, and hardener.

5.3.4. 3-Phenylpropionaldehyde (Dihydrocinnamaldehyde)

Some physical properties and the CAS registry number are given in Table 7.

3-Phenylpropionaldehyde resembles most araliphatic aldehydes in that it is only storable to a limited extent. It is a colorless liquid with a pronounced cinnamon aroma. The aldehyde occurs naturally in Ceylon cinnamon oil. For a review, see [110].

Production. The most common preparative process for 3-phenylpropanal is the partial hydrogenation of cinnamaldehyde [115].

The 3-phenylpropanal that is formed during the hydroformylation of styrene can be separated from the isomeric 2-phenylpropanal as the bisulfite adduct [10]. The Rosenmund reduction of dihydrocinnamoyl chloride in the presence of Pd/BaSO$_4$ gives good yields of the aldehyde [116]. The reaction of 2-propen-1-ol and phenylmercury chloride with CuCl$_2$ and LiPdCl$_4$ catalysts in methanol, and the reaction of phenyl bromide and allyl alcohol in the presence of PdCl$_2$ and NaHCO$_3$ in nonpolar solvents also lead to the formation of 3-phenylpropanal [117]. A review of further synthetic methods is given in [110].

Uses. The uses of 3-phenylpropanal correspond in general to those of 2-phenylpropanal.

5.3.5. 2-(4-Isobutylphenyl)propionaldehyde

Some physical properties and the CAS registry number are given in Table 7.

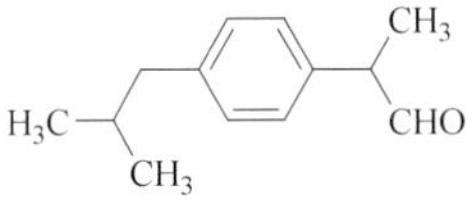

2-(4-Isobutylphenyl)propionaldehyde is a colorless, mobile liquid with a pleasant odor.

Production. Numerous methods are known for the synthesis of 2-(4-isobutylphenyl)propionaldehyde. It can be prepared by reaction of *p*-isobutylacetophenone with methyl chloroacetate using sodium methoxide as catalyst, followed by reaction of the glycidic ester with BF_3 to give 2-hydroxy-3-(4-isobutylphenyl)-3-butenoic acid ester, which is subsequently treated with mineral acid [118]:

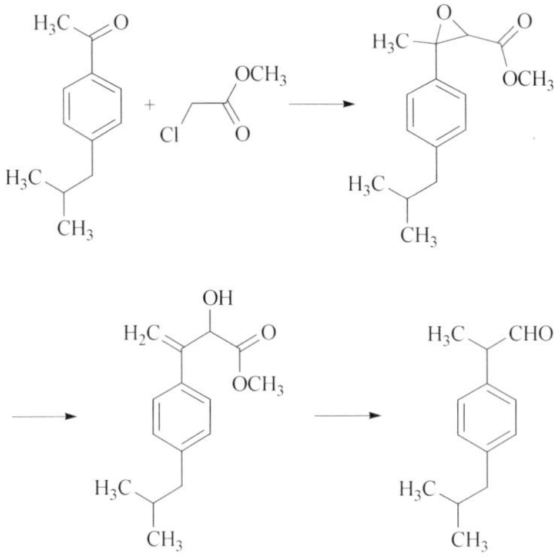

A variant of the process is the hydrolysis of the intermediately formed glycidic ester with alkali to give the corresponding salt of glycidic acid, which is then decarboxylated [119].

2-(4-Isobutylphenyl)propanal also is formed by isomerization of 2-(4-isobutylphenyl)-2-methyloxirane on Al_2O_3-SiO_2 [120] or anhydrous $ZnCl_2$ [121], by reaction of 1-(4-isobutylphenyl)-1-chloroethane with dimethylformamide in the presence of Li or Na in tetrahydrofuran [122], and in good yields by Rh- or Co-catalyzed hydroformylation of *p*-isobutylstyrene [123].

Uses. Unlike many araliphatic aldehydes 2-(4-isobutylphenyl)propionaldehyde has no significance in the perfume industry. It is used exclusively for the preparation of the antirheumatic agent Ibuprofen, 2-(4-isobutylphenyl)propionic acid.

5.3.6. 3-(4-Isopropylphenyl)-2-methyl-propanal (Cyclamenaldehyde)

Cyclamenaldehyde, 2-methyl-3-(4-isopropylphenyl)propionaldehyde, is one of the fragrance and aroma substances of above average importance. The normal commercial racemate is a colorless to pale yellow liquid. The aldehyde is reviewed in [124].

Some physical properties and the CAS registry number are given in Table 7.

In addition to its use in numerous perfume compositions, cyclamenaldehyde is an intermediate for the preparation of fungicides and fungistatic substances [125].

5.3.7. 3-(4-*tert*-Butylphenyl)-2-methylpropionaldehyde (Lilial)

3-(4-*tert*-Butylphenyl)-2-methylpropanal, a colorless to pale yellow liquid, as yet has not been found in nature. It has a higher stability than the homologous cyclamenaldehyde and therefore is used as scent in soaps. Some physical data and the CAS registry number are summarized in Table 7. A monograph on the compound has been published [126]. More than 1000 t/a of the aldehyde is produced.

Production. The aldehyde is produced industrially almost solely by aldol condensation of 4-*tert*-butylbenzaldehyde and propionaldehyde to give 4-*tert*-butyl-α-methylcinnamaldehyde, which can be hydrogenated selectively on noble metal catalysts such as Pd, Rh, Pd–Pr$_2$O$_3$ on Al$_2$O$_3$ [127], [128], or on modified cobalt catalysts [129], [130]. In a recently developed process, the aldol condensation and the hydrogenation are carried out in one step in the presence of a hydrogenation catalyst [131].

The Friedel–Crafts reaction of 4-*tert*-butylbenzene with methacrolein or methacrolein diacetate (Eq. 7, in p. 383) proceeds in an analogous manner to the preparation of cyclamenaldehyde [132].

Further possibilities are the Rh-catalyzed hydroformylation of 1-(4-*tert*-butylphenyl)-1-methoxypropene and subsequent partial hydrogenation [133], the palladium salt catalyzed reaction of 4-*tert*-butylphenylhalide with methallylalcohol [134], and the dehydrogenation of 3-(4-*tert*-butylphenyl)-2-methylpropanol on silver catalysts [105], [106].

Uses. In addition to its applications in the perfume and aroma industry, lilial is used mainly for the synthesis of substituted 3-(4-*tert*-butylphenyl)-2-methylpropylamines, a new class of substances with fungicidal properties [125]. These compounds are effective against mildew in barley and wheat.

5.3.8. Cinnamaldehyde (3-Phenyl-2-propenal)

Cinnamaldehyde, the simplest unsaturated araliphatic aldehyde, is produced in considerable quantities. It has outstanding significance as a fragrance. The *trans* isomer is the major product of industrial manufacturing processes and is the predominant naturally occurring one. Some physical properties and the CAS registry number are given in Table 7.

For the *synthesis* only the base-catalyzed condensation of benzaldehyde with acetaldehyde has been adopted on an industrial scale.

Uses. In addition to its application as a fragrance, it is used for the preparation of corrosion inhibitors, as a polymerization inhibitor for conjugated dienes, and for the coating of metals [145].

Partial hydrogenation or oxidation of cinnamaldehyde give 3-phenylpropanal or cinnamic acid, respectively.

5.3.9. α-Alkylcinnamaldehydes (2-Alkyl-3-phenyl-2-propenals)

The most important of the α-alkylcinnamaldehydes are α-amyl-, α-hexyl-, and α-methylcinnamaldehyde.

$R = CH_3, C_5H_{11},$ or C_6H_{13}

Some physical properties and the CAS registry numbers are given in Table 7. The aldehydes are synthesized, analogously to cinnamaldehyde, by aldol condensation. In addition to the considerable use for these compounds in the perfume industry, they are starting materials for their hydrogenated derivatives, and for the preparation of pharmaceuticals and agrochemicals.

Table 8. Physical properties of aliphatic and alicyclic dialdehydes

	Name	CAS registry number	Formula	M_r	mp, °C	bp**, °C (p)	Density (20 °C), g/cm^3	Refractive index n_D^{20}
C$_2$	Ethanedial (glyoxal)	[107-22-2]	OHCCHO	58.04	15	50.4	1.14	1.383
C$_3$	Propanedial (Malondialdehyde, 3-hydroxyacrolein)	[542-78-9]	OHCCH$_2$CHO	72.06	72–74			
C$_4$	Butanedial (succinedialdehyde)	[638-37-9]	OHC(CH$_2$)$_2$CHO	86.09		58 (9)	1.064	1.426
C$_5$	Pentandial (glutardialdehyde)	[111-30-8]	OHC(CH$_2$)$_3$CHO	100.12	< –14	188 (decomp.)		
C$_6$	Hexanedial (adipinaldehyde)	[1072-21-5]	OHC(CH$_2$)$_4$CHO	114.14	–8	92–94 (9)	1.003	1.435
C$_{12}$	TCD dialdehyde*	[25896-97-3]	OHCC$_{10}$H$_{14}$CHO	192.25	–33	ca. 288	1.14	

* 3(4),8(9)-Bis(formyl)tricyclo[5.2.1.0$^{2.6}$]decane. ** at 101.3 kPa unless otherwise specified in parentheses.

6. Dialdehydes

Dialdehydes, compounds bearing two formyl moieties, are of significant commercial interest. Dialdehydes are used as disinfectants, corrosion inhibitors in oil recovery, leather tanning agents, hydrophobizing agents for paper, wallpaper and textiles, biological fixatives, and cross-linkers in thermosetting resins [135]. Glyoxal (ethanedial) is the simplest possible and commercially most important dialdehyde (→ Glyoxal). Because of their bifunctionality dialdehydes are potential intermediates for the synthesis of diols, dicarboxylic acids and diamines, and as monomers for the synthesis of polyesters, polyamides, and polyurethanes. Because of their reactivity, dialdehydes are often converted directly to such more stable derivatives.

6.1. Properties

Physical data of some dialdehydes are compiled in Table 8. Glyoxal has a melting point of 15 °C and boils at 50.4 °C. Propanedial exists as the isomeric 3-hydroxyacrolein. Therefore it is a solid with a melting point of 72–74 °C. The higher α,ω-dialdehydes are high boiling liquids. The boiling points of the isomeric branched dialdehydes are lower than those of the linear α,ω-dialdehydes.

Dialdehydes have an intrinsic tendency to undergo polymerization. For long chain α,ω-dialdehydes cyclopolymerization reactions are known [136]. Anhydrous glyoxal polymerizes readily to complex polymeric structures. The monomer can be recovered from this polymer by heating in the presence of phosphorus pentoxide. Similar to formaldehyde, glyoxal reacts with water in a highly exothermic reaction to yield a mixture of linear and cyclic oligomers. In aqueous solution, higher dialdehydes form cyclic hemiacetals.

6.2. Production

Although many syntheses of dialdehydes are described in the patent literature, only a few are of commercial interest, mainly due to lack of selectivity. The most important processes for the preparation of saturated aliphatic and cycloaliphatic dialdehydes are:

- Oxidation of ethylene glycol (→ Glyoxal)
- Oxidation of acetaldehyde (→ Glyoxal)
- Hydroformylation of dienes
- Oxidative ring opening of cycloalkenes
- Addition of methyl vinyl ether to acrolein (glutardialdehyde)

Hydroformylation of dienes yields different products depending on the structure of the diene and the catalyst applied [1], [3]. Conjugated dienes such as 1,3-butadiene yield saturated monoaldehydes or monoalcohols if unmodified cobalt or rhodium catalysts are used. Initially the single addition of CO/H_2 yields the unsaturated monoaldehyde. The hydroformylation of the unsaturated aldehyde than competes with hydrogenation and isomerization of the double bond. Patent literature describes the bis-hydroformylation of 1,3-butadiene to 1,6-hexanedialdehyde with special modified rhodium catalysts, but commercial realization remains questionable [137].

Nonconjugated dienes tend to isomerize to the thermodynamically more stable α,β-unsaturated aldehydes after single addition of CO/H_2 [3]. Therefore, especially nonconjugated dienes with widely separated (> C_6), preferably terminal double bonds are useful for synthesis of dialdehydes by hydroformylation.

Oxidative Ring Opening of Cycloalkenes. In the presence of catalysts, e.g. tungstic acid, cycloalkenes react with hydrogen peroxide via formation of the corresponding epoxides to give linear α,ω-dialdehydes [138].

$$\text{(CH}_2\text{)}_n \xrightarrow[\text{cat.}]{H_2O_2} \text{O}\underset{}{\diagdown}\text{(CH}_2\text{)}_n \xrightarrow[\text{cat.}]{H_2O_2} \underset{\text{OHC}}{\overset{\text{OHC}}{\diagdown}}\text{(CH}_2\text{)}_n$$

In some cases the addition of boron compounds facilitates the reaction. Oxidation of cycloalkenes in the gas phase has also been described [139].

Miscellaneous Processes. Similar to the formation of monoaldehydes by dehydrogenation, the dehydrogenation of diols yields dialdehydes. Other routes are the ozonization of cycloalkenes and the dimerization of unsaturated aldehydes [3], [140]. Hydroformylation of unsaturated acetals gives the corresponding dialdehydes.

6.3. Individual Dialdehydes

6.3.1. Succinaldehyde (1,4-Butanedial)

Some physical properties of succinaldehyde and its CAS registry number are given in Table 8. To avoid spontaneous polymerization of anhydrous succinaldehyde, it is best stored as a cyclic acetal, e.g., 2,5 dimethoxytetrahydrofuran. Aqueous solutions consist mainly of the cyclic hemiacetal 2,5-dihydroxytetrahydrofuran.

Production. Succinaldehyde is prepared by various routes, e.g., reaction of chlorine with tetrahydrofuran and subsequent acid hydrolysis. Hydroformylation of acrolein acetals (mostly cyclic) yields a mixture of 2- and 3-(1,3-dioxan-2-yl)propanal. Hydrolysis of the latter forms 1,4-butanedial [141].

Due to the similar chemical properties, succinaldehyde has been replaced extensively by glutardialdehyde, diminishing its commercial importance significantly.

Uses. Succinaldehyde is used as disinfectant, biological fixative, and cross-linker in thermosetting resins.

6.3.2. Glutardialdehyde (1,5-Pentanedial)

Some physical properties of glutardialdehyde and its CAS registry number are compiled in Table 8. Glutardialdehyde is the most important dialdehyde besides glyoxal. Due to its high reactivity glutardialdehyde is sold as a 25 or 50 wt% aqueous solution or as its bis(sodium bisulfite) adduct.

Production. Glutardialdehyde is produced by gas-phase oxidation of cyclopentene or by Diels–Alder reaction of acrolein with methyl vinyl ether [139], [142]. The latter yields 3,4-dihydro-2-methoxy-2H-pyran [4454-05-1], acidic hydrolysis of which leads to glutardialdehyde.

In both cases the dialdehyde is recovered from the reaction mixture by multistage extraction with water.

Uses. Glutardialdehyde is used mainly for leather tanning, as a disinfectant and sterilizer, as a biocide in oil recovery, and in the paper and textile industries to improve wet strength and dimensional stability of fibers.

6.3.3. TCD Dialdehyde

Some physical properties and the CAS registry number of TCD dialdehyde (isomeric mixture), 3(4),8(9)-bis(formyl)tricyclo[5.2.1.0$^{2.6}$]decane, are given in Table 8. Like other dialdehydes TCD dialdehyde is not very stable and has a significant tendency for polymerization. Distillation under reduced pressure is possible. In practice the dialdehyde is not stored but converted immediately to more stable derivatives.

Production. TCD dialdehyde is produced in up to 95 % yield by hydroformylation of dicyclopentadiene on rhodium catalysts. To achieve bis-hydroformylation, a high pressure of synthesis gas is required (150–250 bar) [143].

If unmodified cobalt catalysts are used the dialdehyde is mostly hydrogenated to TCD diol, 3(4),8(9)-bis(hydroxymethyl)tricyclo[5.2.1.0 $^{2.6}$]decane, and TCD monoaldehyde monoalcohol. At low synthesis gas pressures, mainly the monoaldehyde is formed.

Uses. TCD dialdehyde is used for the production of the corresponding diamine, dicarboxylic acid, and diol. These compounds are used as monomers for production of poylesters, polyamides, and polyurethanes. Also some applications in the field of flavors and fragrances or dental materials are known [144].

7. Acetals

Acetals can be represented by the general formula

$$R-\overset{OR'}{\underset{OR'}{\big\langle}}$$

where R can be either hydrogen or some aliphatic, aromatic, or heterocyclic group. In contrast to the hydrates (R′ = H), which normally cannot be isolated, the acetals are generally stable compounds.

Cyclic acetals can be prepared by reaction of aldehydes with diols; 1,2- and 1,3-diols are the most common.

1,3-Dioxolanes

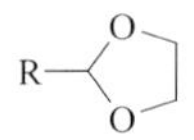

1,3-Dioxanes

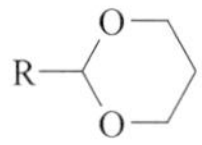

For the lower acetal homologs, trivial names are frequently used, e.g., methylal for dimethoxymethane. Because of their stability toward alkali, acetals are used as soap perfume oils. The scents of the acetals are somewhat different and less distinct than those of the corresponding aldehydes.

Acetals are found in wine and other alcoholic drinks. Piperonal, a raw material for aroma compositions, and numerous alkaloids such as piperine, a main alkaloid of black pepper, or berberine, which is found mainly in barberry roots, possess a methylenedioxy group.

7.1. Properties

Physical Properties. The lower acetals are colorless, mobile liquids, with boiling points distinctly higher than those of the corresponding aldehydes. Their solubilities in water decrease rapidly with increasing molecular mass. They have good solubilities in alcohols, ethers, and other common solvents. The most important physical data and the CAS registry numbers of some acetals are presented in Table 9.

As a rule acetals can be purified by distillation. The distillation residue must be handled with extreme care because, as a result of the ether structure of the acetals, it can contain explosive peroxides. For the same reason care should be taken when storing acetals.

Chemical Properties. Acetals are extremely stable toward base, but are hydrolyzed easily by acid. Because of these properties, the acetals provide interesting possibilities for synthesis, especially in carbohydrate chemistry.

In contrast to the aliphatic alkanals, their acetals are extremely resistant to most reducing and oxidizing agents. This is only true to a limited extent for the acetals of aromatic aldehydes, however.

Among the factors that influence the rate of hydrolysis in acids is the structure of the acetal; all the cyclic acetals are more stable than the open-chain species. Difficulties arise during hydrolysis when the aldehyde group is very reactive; secondary reactions, such as aldol condensation, occur.

Table 9. Physical properties of some acetals

Name	CAS registry number	Formula	M_r	mp, °C	bp*, °C	Density (20 °C), g/cm³	Refractive index, n_D^{20}
Formaldehyde dimethyl acetal (methylal)	[109-87-5]	$CH_2(OCH_3)_2$	76.09	−104.8	45.5	0.8593	1.3513
Formaldehyde diethyl acetal (ethylal)	[462-95-3]	$CH_2(OCH_2CH_3)_2$	104.15	−66.5	89	0.8319	1.3748
Acetaldehyde dimethyl acetal (1,1-dimethoxyethane)	[534-15-6]	$CH_3CH(OCH_3)_2$	90.12	−113.2	64.5	0.8502	1.3668
Acetaldehyde diethyl acetal (1,1-diethoxyethane)	[105-57-7]	$CH_3CH(OCH_2CH_3)_2$	118.18		103.2	0.8314	1.3834
Propionaldehyde dimethyl acetal (1,1-dimethoxypropane)	[4744-10-9]	$CH_3CH_2CH(OCH_3)_2$	101.15		84–86	0.8649	
Propionaldehyde diethyl acetal (1,1-diethoxypropane)	[4744-08-5]	$CH_3CH_2CH(OCH_2CH_3)_2$	132.21		124	0.8232	1.3924
Butyraldehyde dimethyl acetal (1,1-dimethoxybutane)	[4461-87-4]	$CH_3CH_2CH_2CH(OCH_3)_2$	118.18		112–113		1.3888
Butyraldehyde diethyl acetal (1,1-diethoxybutane)	[3658-95-5]	$CH_3CH_2CH_2CH(OC_2H_5)_2$	146.22		145–145.5	0.8320	1.3970
Pentanal diethyl acetal (1,1-diethoxypentane)	[3658-79-5]	$CH_3(CH_2)_3CH(OCH_2CH_3)_2$	160.25		162.5–163.5	0.8304	1.4021
Hexanal diethyl acetal (1,1-diethoxyhexane)	[3658-93-3]	$CH_3(CH_2)_4CH(OCH_2CH_3)_2$	174.28		155–158		
1,3-Dioxolane	[646-06-0]	$CH_2\!\!\begin{array}{c}O-CH_2\\|\\O-CH_2\end{array}$	74.08	−95	78	1.0600	1.3974
1,3-Dioxane	[505-22-6]	$CH_2\!\!\begin{array}{c}O-CH_2\\\ \ \ \ \ CH_2\\O-CH_2\end{array}$	88.12	−42	105 (100.6)	1.0342	1.4165
2-Methyl-1,3-dioxolane	[497-26-7]	$CH_3-CH\!\!\begin{array}{c}O-CH_2\\|\\O-CH_2\end{array}$	88.12		81–82	0.9811	1.4035
2-Methyl-1,3-dioxane	[626-68-6]	$CH_3-CH\!\!\begin{array}{c}O-CH_2\\\ \ \ \ \ CH_2\\O-CH_2\end{array}$	102.13		108.5–109.5 (99.3)	0.9701	1.4139

* at 101.3 kPa unless otherwise specified

Acetals can be transacetalized by other carbonyl compounds or alcohols:

$$R-CH(OCH_3)_2 + 2\,R'OH \xrightleftharpoons{H^+} R-CH(OR')_2 + 2\,CH_3OH$$

$$R-CH(OCH_3)_2 + R'COR' \xrightleftharpoons{H^+} R'_2C(OCH_3)_2 + R-CHO$$

The thermal cleavage of an acetal leads to a vinyl ether and an alcohol:

$$R-CH(OR')_2 \longrightarrow R=\!=\!OR' + R'OH$$

If R' is an allyl group, Claisen rearrangement to the corresponding unsaturated aldehyde takes place during the thermal cleavage (see p. 370).

The cleavage of 4,4-dimethyl-1,3-dioxane, a cyclic acetal, prepared by the Prins reaction of isobutene with formaldehyde (see below), is of industrial importance for the manufacture of isoprene. The dioxane is split at 240–400 °C on H_3PO_4/C or $Ca_3(PO_4)_2$ catalysts in the presence of water:

$$\text{(4,4-dimethyl-1,3-dioxane)} \longrightarrow H_2C=C(CH_3)-CH=CH_2 + CH_2O + H_2O$$

Both BF_3 or $ZnCl_2$ catalyze the reactions of acetals with vinyl ethers to give β-alkoxyacetals, which are converted to the corresponding α,β-unsaturated aldehydes upon treatment with acetic acid:

$$R-CH(OCH_3)_2 + H_2C=CH-OCH_3 \xrightarrow{BF_3}$$

$$R-CH(OCH_3)-CH_2-CH(OCH_3) \xrightarrow{CH_3COOH} R-CH=CH-CHO$$

7.2. Production

The primary process for the preparation of acetals is the reaction of an aldehyde with an alcohol. As a rule, a unstable hemiacetal is formed as an intermediate:

$$R-CHO + R'OH \rightleftharpoons R-CH(OH)(OR')$$

$$R-\underset{OR'}{\overset{OH}{\diagdown}} + R'OH \rightleftharpoons R-\underset{OR'}{\overset{OR'}{\diagdown}} + H_2O$$

Because the reaction is an equilibrium process, the water must be removed by azeotropic distillation or by the addition of water-adsorbing agents, e.g., molecular sieves, in order to improve the yield. Cyclic acetals are generally obtained in better yield than open-chain acetals. As catalysts, anhydrous sulfuric acid or *p*-toluene sulfonic acid are normally used, although other inorganic acids, oxalic or adipic acid, or ion-exchange resins or molecular sieves also can be employed.

During the acetalization of saturated aliphatic aldehydes that contain an α-hydrogen atom, 1-alkenylethers can be formed, if the reaction temperature is too high:

$$\underset{R}{\diagup}\underset{OR'}{\overset{OR'}{\diagdown}} \rightleftharpoons R=\!\!=\!\!-OR' + R'OH$$

In accordance with a process developed by L. CLAISEN acetals are formed in good yields from most aldehydes by treating them with orthoformates, mainly the methyl and ethyl esters. Strong acids are normally used as catalysts:

$$R-CHO + H-\underset{OCH_3}{\overset{OCH_3}{\diagdown}}OCH_3 \xrightarrow{H^+} R-\underset{OCH_3}{\overset{OCH_3}{\diagdown}} + H-\underset{OCH_3}{\overset{O}{\diagdown}}$$

Acetals can also be obtained by reaction of acetylene with alcohols, by addition of alcohol to vinyl ethers, and by treatment of geminal dihalides with alkoxides. Aldehydes and oxiranes form 1,3-dioxolanes, and pyrocatechols and dichloromethane form 1,3-benzodioxoles.

Of industrial importance is the acid-catalyzed reaction of olefins with formaldehyde to give 1,3-dioxanes (Prins reaction):

$$\underset{H_3C}{\overset{H_3C}{\diagdown}}\!\!=\!\!CH_2 + 2\,CH_2O \xrightarrow{H^+} \underset{H_3C}{\overset{H_3C}{\diagdown}}\!\!\diagup\!\!\overset{O}{\diagdown}\!\!\diagup$$

The reaction is carried out at 55–75 °C with an excess of formaldehyde.

The reaction of olefins with alcohols under hydroformylation conditions using, for example, phase-transfer catalysts also leads to the formation of acetals [146].

Acetaldehyde dimethyl acetal is obtained at 200 °C from methanol and synthesis gas under catalytic conditions [147]. The methanol conversion is about 60 % with a selectivity of 80–85 %. Byproducts are mainly acetaldehyde and ethyl acetate.

Vinyl esters can be converted to acetals by reaction with alcohols in the presence of palladium chloride catalyst:

For further methods of synthesis and a review of the reactions of acetals, see [1, p. 615], [148].

7.3. Uses

Acetals frequently are used in preparative chemistry to protect aldehyde functions. The acetals of lower aldehydes, e.g., formaldehyde dimethyl acetal (methylal), serve as solvents for cellulose and cellulose derivatives.

Although acetals have a somewhat weaker fragrance than the corresponding aldehydes, they are used for preparing soap perfume oils, especially because of their stability toward alkali. This applies both for open-chain as well as for cyclic acetals [149].

In polymer chemistry acetals are used for the preparation of epoxy resins [150] and aqueous dispersions [151]. The acetals of polymeric alcohols, such as polyvinyl alcohol, are employed as interlayers in safety glasses. Acetals are also used for the synthesis of pharmaceuticals, insecticides [152], fertilizers [153], and as constituents of sun lotions [154].

Substituted 1,3-dioxanes are intermediates in the synthesis of conjugated dienes. The IFP (Institut Français du Pétrole) process for the production of isoprene is the most important example [155].

8. Quality Control

The actual method used for the determination of quality primarily depends on the further processing and the special properties of the individual products. In general, high purity of the aldehydes is required.

The composition is analyzed by gas chromatography (cf. Chap. 9). In addition, the following characteristic data usually are determined:

1) Density at 20 °C (DIN 51757, ASTM D 1298)
2) Refractive index n_D^{20} (DIN 51423, Bl. 2, ASTM D 1747)
3) Acid value (DIN 53402, ASTM D 1613)
4) Hazen color index (APHA) (DIN ISO 6271, ASTM D 1209)
5) Boiling range at 101.3 kPa (DIN ISO 51777, ASTM D 1364)
6) Water content by the Karl Fischer method (methanol-free reagents and solvents)

The following data also can be useful:

1) Carbonyl value (DIN 53 173, ASTM D 2192)
2) Hydroxy value (DIN 53 240, ASTM E 222)
3) Viscosity at 20 °C (DIN 51 757, ASTM D 4052)
4) Iodine number (determination by the Kaufmann method DGF C-V116; for unsaturated aldehydes)

Safety data, such as TLV and MAK values, acute and chronic toxicology data, flash points (DIN 51755, ASTM D 56 or DIN 51758, ASTM D 93), ignition temperatures (DIN 51794, ASTM D 2155), vapor pressures, and solubilities in water, are important under certain circumstances, e.g., for classification and handling of the products during storage and transportation.

9. Analysis

Aldehydes are mainly analyzed by chromatographic and spectroscopic techniques; chemical detection methods also are used.

The most important qualitative and quantitative analytical procedure in industry is *gas chromatography*. The optimum procedure, i.e., filling of the column, carrier gas, detector, etc., depends on the nature of the sample and the individual requirements.

Today around 80 % of GC analyses of aldehydes are performed on capillary columns. Packed columns are also still used. Columns with a length of 1–3 m and an inner diameter of 2–4 mm, and capillary columns (25–100 m) with various polar stationary liquids can be used. The most frequently employed support materials are calcinated diatomaceous earths.

The amount of sample that is injected is 0.1–0.5 µL for capillary columns with flow separation and 1–3 µL for packed columns. In some cases it is advisable to convert the products into derivatives before analysis. Carrier gases which have proven suitable are N_2, He, and H_2. If hydrogen is used, particular care must be taken that both the injection system and the column are free from catalyst deposits; otherwise the aldehydes easily could be hydrogenated. For compounds with particularly low thermal stability, such as the aldols, the sample is added directly into the separating column.

The 2,4-dinitrophenylhydrazones generally are used in *liquid chromatography*. UV detectors serve to identify the constituents of the sample.

For the *chemical identification* of aldehydes, crystalline derivatives of low solubility are used, e.g., 2,4-dinitrophenylhydrazones, semicarbazones, oximes, sodium hydrogen sulfite adducts, and dimedon condensation products. For qualitative detection the reducing properties of the aldehydes (e.g., reaction with Fehling's solution or Tollen's silver nitrate solution) and numerous color reactions can be used. The reducing properties of the aldehydes are not specific, however.

Aldehydes can be detected qualitatively and identified by IR, ^{1}H NMR, and ^{13}C NMR spectroscopy. In IR spectroscopy saturated aldehydes can be characterized by the intensive CO stretching vibration in the range 1740–1720 cm^{-1}, and α,β-unsaturated compounds by the absorption at 1705–1685 cm^{-1}. The absorption is shifted to lower frequencies for conjugated, unsaturated compounds.

In the ^{1}H NMR spectrum the formyl H is found at 9.4–9.8 ppm in saturated aliphatic aldehydes; the coupling constant is 0–3 Hz for the H-C-CHO system. The corresponding values for unsaturated compounds are 9.3 – 10.2 ppm; the coupling constants for the system R–CH=CH–CHO are strongly dependent on whether the compound has E or Z configuration and which substituents are present.

For further details of the analysis of aldehydes see [156].

10. Storage, Transportation, and Environmental Regulations

For storage and transportion, containers of stainless steel normally are used. Vessels lined with polyethylene or other coatings are also suitable. For aldehydes that enter the market as solutions, such as formaldehyde, glyoxal, or glutaraldehyde, aluminum vessels or containers of standard steel should not be used, because the acids formed by autoxidation are corrosive, and the corrosion products can cause discoloration of the aldehyde.

Aldehydes are normally stored under a nitrogen atmosphere. Antioxidants and stabilizers are added to prevent autoxidation.

The formation of cyclic 2,4,6-trialkyl-1,3,5-trioxane trimers is catalyzed by the presence of strong protic and Lewis acids. The trimerization tendency of aldehydes occasionally increases at lower temperatures, so that some compounds cannot be stored over long periods at temperatures under 20 °C without stabilization [1, p. 651].

Some α,β-unsaturated aldehydes, such as acrolein, methacrolein, and crotonaldehyde, can react violently if handled incorrectly in the presence of air or oxygen or some other, in particular, alkaline compound.

Because of their health hazards, saturated and unsaturated aliphatic aldehydes, especially the lower members, should be handled only if special safety precautions are taken. Necessary measures can be found in the regulations on dangerous materials and in the directions on accident prevention, guidelines, leaflets, etc., published by the confederation of trade associations.

For the economically important aldehydes numerous transport regulations and legal directions exist (IMO, IATA, RID, ADR, etc.). The classification of a special aldehyde depends on its specific properties such as flash point, boiling point, solubility in water, toxicity, and ignition temperature.

Because the aldehydes sometimes have a very unpleasant and intense odor, special environmental measures are necessary during manufacture, storage, and dispatch. The aldehyde-containing off-gas from the production normally is drawn off centrally and burned. The wastewaters are treated chemically and biologically.

Table 10. Oxo capacities by region (in 10^6 t/a)

	1984	1994
Western Europe	2.5	2.4
United States	1.4	2.4
Japan	0.5	0.8
Others	0.6	1.6
Total	5.0	7.2

Table 11. Consumption of oxo chemicals by region (in 10^3 t/a)

	Western Europe		United States		Japan	
	1993	1998	1993	1998	1993	1998
Propionaldehyde	11	12	161	183	1	1
n-Butyraldehyde	1224	1274	1055	1178	572	622
Isobutyraldehyde	133	128	234	263	65	72
Valeraldehydes	11	12	32	35		
C_6-C_{13} plasticizer alcohols**	430	460	396	430	93	105
Detergent alcohols**	150	174	193	215	47	53
C_7-C_9 oxo fatty acids	11	12	37	44		
Branched-chain oxo acids	15	19	1	1		
Neo acids	50	61	16	19		
Total	2046	2162	2125	2368	778	853

*Excluding 2-ethylhexanol. ** Based on equivalent amounts of aldehyde.

11. Economic Aspects

The most important process for the manufactoring of aldehydes with more than two carbon atoms is the oxo synthesis. In 1994 total worldwide capacity was 7.2×10^6 t/a with further capacities of $(0.5-1.0) \times 10^6$ t/a planned or under construction. Table 10 lists oxo capacities by region for 1984 and 1994.

The most important secondary product of butyraldehyde is 2-ethylhexanol, with worldwide capacities of 2.6×10^6 t/a in 1996. The consumption of various oxo chemicals in Western Europe, the United States, and Japan is compiled in Table 11.

For the most important aldehydes used in the perfume industry, few production data are available. Estimated production for the most important fragrance aldehydes are as follows: cinnamaldehyde 1500 t/a, lilial 1500 t/a, hydroxycitronellal 1500 t/a, α-hexylcinnamaldehyde 1000 t/a, α-amylcinnamaldehyde 800 t/a, citral 500 t/a, citronellal 400 t/a, and cyclamenaldehyde 350 t/a.

12. References

General References

[1] J. Falbe (ed.): "Aldehyde," in *Houben-Weyl: Methoden der Organischen Chemie,* vol. **E3,** Thieme Verlag, Stuttgart-New York 1983.
[2] J. Falbe (ed.): "C-O-Verbindungen," in *Methodicum Chimicum,* vol. **5,** Thieme Verlag, Stuttgart-New York 1975.
[3] J. Falbe (ed.): *New Syntheses with Carbon Monoxide,* Springer Verlag, Berlin-Heidelberg-New York 1980.
[4] J. Falbe (ed.): *Carbon Monoxide in Organic Synthesis,* Springer Verlag, Berlin-Heidelberg-New York 1970.
[5] R. Brettle in: *Comprehensive Organic Chemistry,* vol. **1,** chap. 5.1, Pergamon Press, Oxford 1979.
[6] C. A. Buehler, D. E. Pearson: *Survey of Organic Syntheses,* vol. **2,** J. Wiley & Sons, New York-London-Sydney-Toronto 1977.
[7] S. Patai: *The Chemistry of the Carbonyl Group,* Interscience Publ., London-New York-Sydney 1966.
[8] S. R. Sandler, W. Karo: *Organic Functional Group Preparations,* vol. **12,** Academic Press, New York-London 1968.

Specific References

[9] L. H. Horsley: "Azeotropic Data III," *Adv. Chem. Ser.* **116** (1973), ACS, Washington.
[10] Ruhrchemie, DE 2459152, 1974 (L. Bexten, H. Noeske, H. Tummes, B. Cornils).
[11] BASF, DE 960187, 1952 (W. Hagen).
[12] BASF, DE-OS 2218305, 1972 (R. Kummer, H. J. Nienburg, G. Butz).
[13] BASF, DE 1022575, 1958 (H. J. Nienburg, H. Böhm).
[14] *Katalysatoren Hoechst,* Firmenbroschüre, Hoechst-Ruhrchemie 1979.
[15] G. Fagherazi, N. J. Pernicone, *J. Catal.* **16** (1970) 321.
[16] H. W. Knol, *Manuf. Chem. Aerosol News* **37** (1966) no. 2, 42–45.
[17] Stamicarbon, EP 37149, 1980 (T. F. De Graaf, H. J. Delahaye).
[18] Moscow Inst. of Fine Chem. Technol., SU 789491, 1978 (M. L. Kaliya et al.).
[19] Phillips Petroleum, US 4304943, 1979 (G. Bjornson).
[20] C. Jallabert, H. Riviere, *Tetrahedron* **36** (1980) no. 9, 1191–1194.
[21] Celanese, US 2128908, 1934 (J. E. Bludworth). L. F. Albright, *Chem. Eng.* **74** (1967) no. 17, 165.
[22] E. Mosettig, R. Mozingo, *Org. React. (N.Y.)* **4** (1948) 362.K. Harsanyi et al., *J. Med. Chem.* **7** (1964) 623. K. Balenovic et al., *J. Org. Chem.* **21** (1956) 115. Shell Int. Res., EP 5280, 1978.
[23] G. W. Fleet, P. J. Harding, *Tetrahedron Lett.* 1979, 975.
[24] H. C. Brown, B. C. Subba Rao, *J. Am. Chem. Soc.* **80** (1958) 5377.
[25] N. Müller, W. Hoffmann, *Synthesis* 1975, 781. J. J. Pappas et al., *Tetrahedron Lett.* 1966, 4273.
[26] R. Criegee, L. Kraft, B. Bank, *Justus Liebigs Ann. Chem.* **507** (1933) 159.
[27] L. Malaprade, *C. R. Acad. Sci.* **186** (1928) 382.
[28] Mitsui Petrochem. Ind., JP 80102528, 1979 *(Chem. Abstr.* **94** (1981) 46774).
[29] I. M. Goldman, *J. Org. Chem.* **34** (1969) 1979. A. J. Fatiadi, *Synthesis* 1976, 65.
[30] T. R. Beebe, P. Hii, P. Reinking, *J. Org. Chem.* **46** (1981) 1927.

[31] H. Stetter, E. Reske, *Chem. Ber.* **103** (1970) 643. *Vogel's Textbook of Practical Organic Chemistry,* 4th ed., Longman, London-New York 1978, p. 422.

[32] G. B. Bennett, *Synthesis* **1977**, 589. S. J. Rhoads, N. R. Raulins, *Org. React.* (N.Y.) **22** (1975) 1. Ruhrchemie, DE 2844635, 1978 (J. Weber, W. Bernhagen, H. Springer).

[33] C. A. Buehler, D. E. Pearson: *Survey of Organic Syntheses,* vol. **2,** J. Wiley & Sons, New York-London-Sydney-Toronto 1977, p. 480. S. R. Sandler, W. Karo:
Organic Functional Group Preparations, vol. **12,** Academic Press, New York-London 1968, p. 145.

[34] Chem. Verwertungs-Ges. Oberhausen, DE 953605, 1956 (G. Schiller). Conoco, US 4183825, 1978 (J. T. Carlock). Exxon Res. & Eng. Co., EP 24088, 1980 (H. L. Mitchell). Davy McKee Oil Chem., EP 16285, 1979 (T. F. Shevels, N. Harris). Agency of Ind. Sci. Technol., DE 2909041, 1978 (I. Ogata, Y. Kawaba, M. Tanaka, T. Hayashi).

[35] Ruhrchemie, DE 2855505, 1978 (J. Weber, W. Bernhagen, H. Springer).

[36] W. J. Scheidmeir, *Chem. Ztg.* **96** (1972) no. 7, 383. S. S. Kagna, A. G. Trifel, V. J. Gankin et al., GB 1524775, 1977. J. E. Knap, N. R. Cox, W. R. Privette, *Chem. Eng. Prog.* **62** (1966) no. 4, 74–78.

[37] Yaroslavl. Polytechnic Inst., SU 761450, 1978 (A. V. Irodov et al.).

[38] Kuraray Chem. Co., JP 8002632, 1978, JP 7973713, 1977, JP 79109909, 1978 (F. Yamamoto et al.).

[39] Erdölchemie, DE 2163396, 1971 (H. Fischer, G. Schnuchel). Int. Synth. Rubber Co., GB 2063297, 1979 (D. G. Timms).

[40] Ruhrchemie, DE 2917779, 1979 (J. Weber, W. Bernhagen, H. Springer).

[41] Bayer, EP 90246, 1982 (G. Rauleder, H. Waldmann).

[42] Ciba-Geigy, GB 861367, 1959.

[43] Union Carbide, EP 28378, 1979 (E. Billig, D. L.Bunning). C. U. Pittmann, G. M. Wilemon, *J. Org. Chem.* **46** (1981) no. 9, 1901–1905.

[44] Ruhrchemie, DE 2933919, 1979 (W. Bernhagen, J. Weber, H. Bahrmann, H. Springer).

[45] Carter-Wallace, DE-OS 2939105, 1979. BASF, DE-OS 1957301, 1969 (F. Merger, R. Platz, W. Fuchs).

[46] Soc. Organico, FR 952985, 1947 (P. Franck, C. Gregory, M. Genas, O. Kostelitz).

[47] Celanese, US 4201714, 1977, US 4201728, 1977 (O. R. Hughes).

[48] Ruhrchemie, BE 865728, 1977 (J. Weber, H. Springer).

[49] Shell Int. Res., GB 2068377, 1980 (P. Wilhelmus, N. M. van Leeuwen, C. F. Roobeek).

[50] Continental Oil, US 4097535, 1977 (K. Yang, K. L. Motz, J. D. Reedy).

[51] Ruhrchemie, DE 2844638, 1978 (J. Weber, V. Falk, C. Kniep).

[52] Ruhrchemie, DE-OS 2 844 636, 1978 (J. Weber, W. Bernhagen, H. Springer).

[53] Kinki University, JP 75130708, 1974 (Y. Matsubara, Y. Fujiwara, C. Hata).

[54] Air Products & Chem., US 4230641, 1977 (C. M. Bartish). Union Carbide, US 4148830, 1976 (R. L. Pruett, J. A. Smith).

[55] BASF, DE 2604545, 1976 (K. Schwirten, W. Disteldorf, W. Eisfeld, R. Kummer). Ruhrchemie, DE 2737633, 1977 (L. Bexten, B. Cornils, H. D. Hahn, H. Tummes).

[56] Henkel, DE 2914187, 1979 (J. Hagen, R. Lehmann, K. Bansemir).

[57] Johnson-Matthey, BE 890210, 1980 (M. J. H. Russel, B. A. Murrer).

[58] Rhône-Poulenc, FR 2473504, 1979 (J. Jenck).

[59] Ruhrchemie, EP 7 609, 1978 (H. Bahrmann, B. Cornils, G. Diekhaus, W. Kascha, J. Weber).

[60] M. S. Newman, B. J. Magerlein, *Org. React.* (N.Y.) 5 (1949) 413.

[61] Givaudan, US 4154762, 1973 (I. Huang, L. M.Polinski, K. K. Rao).

[62] J. H. Babler, B. J. Invergo, *Tetrahedron Lett.* 1981, 11–14.

[63] Ruhrchemie, DE-OS 2855506, 1980 (H. Springer, J. Weber).
[64] Shell, US 2479283, 1949, US 2514688, 1950 (R. R. Whetstone). W. S. Hillman, *J. Am. Chem. Soc.* **71** (1949) 324.
[65] J. N. Nazarov et al., *Zh. Obshch. Khim.* **29** (1959) 3965; *J. Gen. Chem. USSR (Engl. Transl.)* **29** (1959) 3925.
[66] R. I. Hoaglin, D. H. Hirsh, *J. Am. Chem. Soc.* **71** (1949) 3468. O. Isler et al., *Helv. Chim. Acta* **39** (1956) 249.
[67] K. Bernhauer, I. Skudrzyk, *J. Prakt. Chem.* **155** (1940) 310.
[68] Y. Nakatani, M. Sugiyama, C. Honbo, *Agric. Biol. Chem.* **39** (1975) 2431.
[69] F. G. Fischer, K. Löwenberg, *Justus Liebigs Ann. Chem.* **494** (1932) 273.
[70] BASF, DE-OS 2815539, 1978, DE-OS 2847069, EP 4622, 1979 (R. Fischer, H. Weitz).
[71] Union Carbide, US 2628257, 1953 (R. I. Hoaglin, D. H. Hirsh). GAF, US 2543312, 1948 (J. W. Copenhaver).
[72] Eastman-Kodak, US 4316990, 1980 (D. L. Morris).
[73] Ruhrchemie: "Oxo/2-Ethylhexanol Process," *Hydrocarbon Process.* 1981, 165. B. Cornils, A. Mullen, *Hydrocarbon Process.* 1980, 93.
[74] BASF, EP 44440, 1981 (L. Janitschke, W. Hoffmann).
[75] O. Bayer (ed.): "Aldehyde", in *Houben-Weyl: Methoden der organischen Chemie*, vol. **VII/1**, Thieme Verlag, Stuttgart 1954, p. 76. R. P. Bell, *J. Chem. Soc. (London)*, **1937**, 1637; **1958**, 1691; **1960**, 2983.
[76] L. P. Koshechkina, A. A. Yasnikov, *Ukr. Khim. Zh.* (Russ. Ed.) 40 (1974) no. 9, 948–957.
[77] A. T. Nielsen, W. J. Houlihan, *Org. React. (N.Y.)* **16** (1968) 1.
[78] Esso, US 3729515, 1967 (N. L. Cull, E. M. Gladrow, R. B. Mason, G. P. Hammer).
[79] Union Carbide, US 4215076, 1976 (M. L. Deem, K. C. Stueben).
[80] Celanese, GB 783458, 1957. BASF, DE-OS 2000699, 1971 (F. Merger, W. Fuchs, S. Winderl, E. Haarer).
[81] BASF, DE-AS 1235883, 1965 (R. Platz). Celanese, GB 783458, 1957.
[82] B. Cornils, H. Feichtinger, *Chem. Ztg.* **100** (1979) 504–514.
[83] G. Wittig, H. Reiff, *Angew. Chem.* **80** (1968) 8; *Angew. Chem., Int. Ed. Engl.* **7** (1968) 7. BASF, DE 1199252, 1963 (G. Wittig, H. Pommer, W. Stilz).
[84] G. Wittig et al., *Angew. Chem.* **75** (1963) 978.
[85] Y. Takegami et al., *Bull. Chem. Soc. Jpn.* **37** (1964) 672. Du Pont, US 3687981, 1972 (F. R. Lawrence, R. H. Sullivan).
[86] Shell, GB 2 282 137, 1995 (E. Drent, W. W. Jager). Arco Chemical Technology, US 5 233 093, 1993 (R. Pitchai et al.). Arco Chemical Technology, US 5 166 370, 1993 (F. J. Liotta et al.). Kuraray Co., Daicel Chemical Ind., EP 129 802, 1985 (M. Matsumoto et al.). Daicel Chemical Ind., JP 57 206 629, 1983. *Chem. Ind. (Düsseldorf)* **32** (1980) 275. M. Tamura, S. Kumano, *Chem. Econ. Eng. Ref.* **12** (1980) no. 9, 32.
[87] L. E. Schniepp, H. H. Geller, *J. Am. Chem. Soc.* **68** (1946) 1646. G. F. Woods, H. Sanders, *J. Am. Chem. Soc.* **68** (1946) 2111.
[88] C. W. Smith et al., *J. Am. Chem. Soc.* **73** (1951) 5274.
[89] R. Hall, E. S. Stern, *J. Chem. Soc.* 1950, 490. Du Pont, US 3536763, 1970 (H. S. Eleuterio, T. A. Koch).
[90] Ruhrchemie, DE 2054601, 1970 (W. Rottig, H. Tummes, B. Cornils, J. Weber).
[91] Chisso Corp., JP 78039407, 1970 (T. Nakamura, M. Takano, K. Fukatsu). Celanese, DE 2204070, 1971 (R. H. Prinz). Melle-Bezons, FR 2058532, 1969.

[92] Teijin Chem., JP 79036278, 1977 (T. Harada, O. Magoshi). Teijin Chem., JP 79011139, 1977 (T. Harada, K. Nakagawa).

[93] H. J. Arpe, *Chem. Ztg.* **97** (1973) 53. J. F. Jelinek, *Sb. Pr. Vyzk. Chem. Vyuziti Uhli Dehtu Ropy* **14** (1976) 99–113, see Chem. Abstr. 85 (1976) 176767. E. Santoro, M. Chiavarini, *J. Chem. Soc. Perkin Trans. 2*, 1978, no. 3, 189–192.

[94] J. Bathory, O. Repasy, *Petrochemia* **16** (1976) no. 4–5, 94–97. Ruhrchemie, DE-OS 2045669, 1970 (H. Tummes, G. Schiewe, B. Cornils, W. Pluta, J. Falbe).

[95] Eastman-Kodak, US 2811562, 1957 (H. J. Hagemeyer).

[96] Hoechst, DE-OS 1768274, 1968 (G. Jacobson, H. Fernholz, D. Freudenberger).

[97] BASF, DE 2758883, 1977 (M. Distler, W. Goetze). Soc. Chim. des Charbonnages, BE 843827, 1975 (P. Couderc, S. Hilmoine).

[98] Union Carbide, US 4215076, 1976 (M. L. Deem, K. C. Stueben).

[99] A. Mueller, *Seifen Öle Fette Wachse* **102** (1976) no. 1, 11–12.

[100] Toyo Soda, JP 8218643, 1980 (*Chem. Abstr.* **97** (1982) 38670).

[101] PCUK, FR 2338920, 1976 (J. C. Volta, J. M. Cognion).

[102] Ajinomoto, JP 242278, 1971 (Y. Matsuzawa, T. Yamashita, S. Ninagawa).

[103] Cosden Technology, DE-OS 2 501 341, 1974 (J. Watson).

[104] Kuraray Chem. Co., JP 8035063, 1978 (Y. Tokito, N. Yoshimura, M. Tamura). J. Vojtko et al., *Zb. Pr. Chemickotechnol. Fak. SVST* 1977, 195–200 (Chem. Abstr. 89 (1978) 146285).

[105] BASF, EP 4881, 1978 (W. Sauer, W. Fliege, C. Dudeck, N. Petri).

[106] Givaudan, FR 2231650, 1973 (I. Huang, L. M. Polinski, K. K. Rao).

[107] Sumitomo, EP 34430, 1980 (T. Takano, G. Suzukamo, M. Ishino, K. Ikimi).

[108] A. Knorr, A. Weissenborn, E. Laage, US 1899340, 1933.

[109] A. Lethbridge et al., *J. Chem. Soc. Perkin Trans. 1* 1973, no. 1, 35–38.

[110] B. Cornils, R. Payer, *Chem. Ztg.* **98** (1974) 596–606.

[111] H. Siegel, W. Himmele, *Angew. Chem.* **92** (1980) 182–187; *Angew. Chem. Int. Ed. Engl.* **19** (1980) 178. BASF, DE-OS 2132414, 1971 (W. Himmele, H. Siegel, W. Aquila, F. J. Mueller). Ethyl Corp., US 3907847, 1970 (K. A. Keblys).

[112] J. M. Brown, *Chem. Ind. (London)* **1982**, no. 10, 737. Agency of Ind. Sci. Technol., JP 7757108, 1975 (I. Ogata, M. Tanaka, Y. Ikeda, T. Hayashi). L. Marko in [1], p. 224

[113] I. L. Belostotskaya, G. A. Khmeleva, *Termokatal. Metody Pererab. Uglevodorodnogo Syr'ya* 1969, 221–224.

[114] W. Hafner et al., *Chem. Ber.* **95** (1962) 1575.

[115] Heraeus, DE-OS 2613645, 1976 (R. Siepmann, E. Hopf). Engelhard Industries, US 3372199, 1964 (P. N. Rylander, N. Himmelstein). D. V. Sokol'skii et al., *Zh. Org. Khim.* **13** (1977) no. 1, 77–80.

[116] Seibi Chem. Co., JP 6820 447, 1964 (S. Abe, K. Sato, T. Asami, T. Amakasu, T. Hakura). A. W. Burgstahler, L. O. Weigel, C. G. Shaefer, *Synthesis* 1976, no. 11, 767–768.

[117] Hercules, US 3658917, 1965 (R. F. Heck). J. B. Melpolder, R. F. Heck, *J. Org. Chem.* **41** (1976) 265. Givaudan, DE-OS 2627112, 1975 (A. J. Chalk, S. A. Magennis).

[118] Nisshin Flour Milling, DE 2404158, 1974 (K. Kogure, K. Nakagawa). Kanebo, JP 76101949, 1975 (Y. Izuka, Y. Sawa, T. Kawashima, S. Miura).

[119] Boots Pure Drug. Co., FR 1545270, 1966.

[120] Ohta Pharmaceutical Co., JP 81154428, 1981. Ota Seiyaku Co., JP 7831637, 1976 (T. Tsukama et al.).

[121] Kohjin Co., JP 7746034, 1973, JP 7626835, 1974 (S. Yoshimura, S. Takahashi, M. Ichino, T. Nakamura).

[122] Daito Koeki Co., JP 7912336, 1977 (T. Shimazaki, Y. Otsuka, H. Kondo). Daito Koeki Co., JP 7882740, 1976 (Y. Fujimori).

[123] Nado Kenkyusho, JP 7797930, 1976 (M. Arakawa). Fujimoto Pharmaceutical Co., JP 7762233, 1975 (M. Fujimoto, S. Nakayama).

[124] W. Behrends, L. M. van der Linde, *Perfum. Essent. Oil Rec.* **58** (1967) no. 6, 372–378.

[125] W. Himmele, E. H. Pommer, *Angew. Chem.* **92** (1980) 176–181; *Angew. Chem. Int. Ed. Engl.* **19** (1980) 184. BASF, DE-OS 2727482, 1977 (W. Himmele, E. H. Pommer, N. Goetz, B. Zeeh); DE-OS 2921221, 1979 (W. Himmele, W. Heberle, F. W. Kohlmann, W. Wesenberg). Hoffmann-LaRoche, DE-OS 2752135, 1976 (A. Pfiffner, K. Bohnen); EP 5541, 1978 (K. Bohnen, A. Pfiffner).

[126] D. L. J. Opdyke, *Food Cosmet. Toxicol.* **16** (1978) suppl. 1, 659.

[127] Universal Oil Prod., FR 1496304, 1966.D. V. Sokol'skii et al., *Zh. Prikl. Khim.* **49** (1976) no. 2, 407–411. BASF, DE-OS 2 832 699, 1978 (G. Heilen, A. Nissen, W. Koernig, M. Horner, W. Fliege, G. Boettger). A. M. Pak et al., *Zh. Org. Khim.* **17** (1981) no. 6, 1176–1180.

[128] Universal Oil Prod., GB 1086447, 1966 (A. Friedman, J. Levy).

[129] K. Kogami, J. Kumanotani, *Bull. Chem. Soc. Jpn.* **46** (1973) no. 11, 3562–3565.

[130] Hasegawa Co., JP 7250095, 1968, JP 7301379, 1968 (J. Kumanotani et al.).

[131] BASF, DE-OS 3105446, 1981, EP 58326, 1982 (W. Gramlich, G. Heilen, H. J. Mercker, H. Siegel).

[132] Rhône-Poulenc, DE 1145161, 1963 (I. Scriabine). Rhône-Poulenc, US 3023247, 1960. Givaudan, EP 43526, 1981 (R. Valentine, H. A. Brandman).

[133] Givaudan, EP 52775, 1980 (Y. Crameri, P. A. Ochsner, P. Schudel).

[134] A. J. Chalk, S. A. Magennis, *J. Org. Chem.* **41** (1976) no. 7, 1206.

[135] Q. Bone, K. P. Ryan, *Histochem. J.* **4** (1972) 331. Union Carbide, US 2 941 859, 1960 (M. L. Fein, E. M. Filachione). BASF, DE 2 215 948, 1972 (H. Erdmann, F. F. Miller). Union Carbide, US 4 244 876, 1978 (G. H. Warrer, L. F. Theiling, M. G. Freid). Schülke & Mayr, DE 3 032 794, 1980 (W. Münzenmaier et al.). S. S. Subramanian, *Sci. Res. News* **1** (1976) 42.

[136] J. Furukawa, J. Nishimura, *J. Polym. Sci. Polym. Symp.* **56** (1976) 437.

[137] Union Carbide, WO 9 740 002, 1997 (D. L. Packett et al.). Union Carbide, WO 9 740 001, 1997 (J. B. Briggs et al.). Union Carbide, WO 9 739 998, 1997 (A. S. Guram et al.). DSM, DuPont, WO 9 733 854, 1997 (P. M. Burke et al.).Union Carbide, US 5 312 996, 1995 (D. L. Packett).

[138] D. Jingfa, X. Xinhua, C. Haiying, J. Anren, *Tetrahedron* **48** (1992) 3503. Tonen Sekiyu Kagaku, JP 01 190 647, 1990 (H. Inagaki et al.). Toa Nenryo Kogyo, JP 62 029 546, 1987 (H. Furukawa). USSR, NL 7 312 323, 1975 (A. T. Menyailo et al.). Bayer, DE 2 252 719, 1974 (H. Waldmann, W. Schwerdtel, W. Swodenk). Bayer, DE 2 252 674, 1974 (H. Waldmann, W. Schwerdtel, W. Swodenk). Bayer, DE 2 201 455, 1973 (H. Waldmann, W. Schwerdtel, W. Swodenk).

[139] Bayer, DE 2 329 586, 1975 (L. Imre, A. Klein, K. Wedemeyer). Bayer, DE 2 201 411, 1973 (L. Irme, K. Wedemeyer).

[140] Mitsui Toatsu Chemicals, JP 57 203 024, 1983. Mitsubishi Petrochemical Co., JP 05 310 631, 1994 (J. Ookago, N. Sumya, S. Ichikawa). Mitsubishi Petrochemical Co., EP 537 625, 1993 (Y. Ohgomori, S. Ichikawa, Y. Shuji, T. Yoneyama, N. Sumitani).

[141] M. Tamura, S. Kumano, *Chem. Econ. Eng. Rev.* **12** (1980) 32. J. Maeda, R. Yoshida, *Bull Chem. Soc. Jpn.* **41** (1968) 2969. BASF, DE 2 401 553, 1975 (R. Kummer).

[142] Daicel Chemicals, JP 08 040 968, 1996 (M. Nishioka, M. Saito). Union Carbide, GB 739 128, 1955 (R. I. Hoaglin, R. G. Kelso). Distillers, GB 698 736, 1953 (R. G. Hall, B. K. Howe).

[143] Ruhrchemie, DE 1 618 384, 1972 (J. Falbe). B. Cornils, R. Payer, H. Tummes, J. Weber, *Eur. Chem. News* **27** (1975) no. 712, 36.
[144] ESPE, DE 3 902 417, 1991(O. Gasser, R. Guggenberger, K. Ellrich). ESPE, US 4 131 729, 1979 (W. Schmitt, R. Purrmann, P. Jochum, W. D. Zahler).
[145] Philips Gloeilampenfabr., NL 6501841, 1965. M & T Chemicals, GB 1168225, 1967.
[146] Ethyl Corp., US 4209643, 1978 (K. H. Shin).
[147] Union Rhein. Braunkohlen Kraftstoff, BE 882655, 1979 (J. Korff, M. Fremery, J. Zimmermann). Union Rhein. Braunkohlen Kraftstoff, BE 890964, 1980. Mitsubishi Gas Chem. Ind., GB 2079746, 1980.
[148] O. Bayer (ed.): *"Aldehyde"*, see [75] p. 413. H. Meerwein (ed.): "Acetale," in Houben-Weyl: Methoden der organischen Chemie, vol. VI/3, Thieme Verlag, Stuttgart 1965, p. 203. F. A. J. Meskens, *Synthesis* 1981, 501–522. A. J. Showler, P. A. Darley, *Chem. Rev.* **67** (1967) 427.
[149] Henkel, EP 39029, 1980 (H. Upadek, K. Bruns). Henkel, DE 2648109, 1976 (K. Bruns et al.). Firmenich, NL 7415594, 1973 (G. Ohloff, W. K.Giersch). Naarden Int., NL 7305488, 1973.
[150] Shell Int. Res., BE 816134, 1973.
[151] BASF, DE 2509739, 1975 (P. Engler et al.).
[152] BASF, DE 2636 278, 1976 (H. Zinke-Allmang, W. Scheidmeir).
[153] Akademie Wissenschaften der DDR, DD 142044, 1979 (J. Freiberg, H. Seeboth, R. Goersch, H. Bergner).
[154] Henkel, DE 2533048, 1975 (H. Moeller, J. Conrad, C. Gloxhuber, H. J. Thimm).
[155] Inst. Français du Pétrole, US 2997480, 1959.
[156] L. F. Tietze: "Nachweismethoden und quantitative Bestimmung von Aldehyden," in [1], p. 673. E. Pretsch, T. Clerc, J. Seibl, W. Simon: *Tabellen zur Strukturaufklärung organischer Verbindungen*, Springer Verlag, Heidelberg 1976. D. H. Williams, I. Fleming: *Spektroskopische Methoden in der organischen Chemie*, Thieme Verlag, Stuttgart 1971. E. Heuser: "Analytik der Carbonylgruppen," in Houben-Weyl: Methoden der organischen Chemie, vol. 2, Thieme Verlag, Stuttgart 1953, p. 434–467. W. Büchler, A. Becker, A. Walter in: *Methodicum Chimicum*, vol. **1/1**, Thieme Verlag, Stuttgart 1973, p. 216.

Allyl Compounds

LUDGER KRÄHLING, Deutsche Solvay-Werke GmbH, Rheinberg, Federal Republic of Germany (Chaps. 1, 6)

JÜRGEN KREY, Deutsche Solvay-Werke GmbH, Rheinberg, Federal Republic of Germany (Chaps. 1, 6)

GERALD JAKOBSON, Deutsche Solvay-Werke GmbH, Rheinberg, Federal Republic of Germany (Chaps. 1, 6)

JOHANN GROLIG, Bayer AG, Leverkusen, Federal Republic of Germany (Chaps. 2–5)

LEOPOLD MIKSCHE, Bayer AG, Leverkusen, Federal Republic of Germany (Chap. 6)

1.	Allyl Chloride	410
1.1.	Physical Properties	410
1.2.	Chemical Properties	411
1.3.	Production	412
1.3.1.	Chlorination, of Propene	412
1.3.2.	Other Production Processes	416
1.4.	Handling, Environmental Protection, Storage, and Transportation	417
1.5.	Quality and Analysis	418
1.6.	Uses	419
1.7.	Economic Aspects	419
2.	Allyl Alcohol	419
2.1.	Physical Properties	419
2.2.	Chemical Properties	420
2.3.	Production	421
2.3.1.	Hydrolysis of Allyl Chloride	422
2.3.2.	Isomerization of Propene Oxide	422
2.3.3.	Hydrolysis of Allyl Acetate	423
2.3.4.	Hydrogenation of Acrolein	424
2.4.	Quality and Analysis	424
2.5.	Uses	424
2.6.	Methallyl Alcohol	424
3.	Allyl Esters	425
3.1.	Properties	425
3.2.	Production	426
3.2.1.	Oxidation of Olefins	426
3.2.2.	Esterification	427
3.2.3.	Transesterification	427
3.2.4.	Other Production Methods	428
3.3.	Uses	428
3.3.1.	Polymer Production	428
3.3.2.	Other Uses	431
4.	Allyl Ethers	431
4.1.	Properties and Uses	431
4.2.	Production	432
5.	Allylamines	432
5.1.	Properties and Uses	432
5.2.	Production	433
6.	Toxicology and Occupational Health	433
7.	References	436

1. Allyl Chloride

Allyl chloride [107-05-1], the only chloropropene of industrial importance, was first produced in 1857 by A. CAHOURS and A. W. HOFMANN by reacting phosphorus chloride with allyl alcohol. The name allyl is derived from the latin *allium*, meaning garlic. Inhalation of even small amounts of allyl chloride produces, after a short time, the characteristic odor of garlic on the breath.

At the end of the 1930s, IG Farbenindustrie and the Shell Development Co. developed the high-temperature chlorination of propene, permitting large-scale production of allyl chloride with good yields. A significant part of the development was done by the Shell Chemical Co. when erecting a commercial plant in 1945. Dow, Solvay, and Asahi-Kashima developed their own processes.

1.1. Physical Properties

Allyl chloride, 3-chloropropene, $CH_2=CH-CH_2Cl$, is a colorless, mobile liquid with a penetrating, pungent odor, M_r 76.53, *mp* –134.5 °C, *bp* (101.3 kPa) 44.4 °C, n_D^{20} 1.416, n_D^{25} 1.413 [1]–[5]. Temperature-dependent physical data are given in Table 1.

Expansion coefficient (0–30 °C)	1.41 K^{-1}
Heat of combustion	
($CH_2=CH-CH_2Cl$ (g), HCl (g), CO_2 (g), H_2O (g))	1846 kJ/mol
Latent heat of vaporization at *bp*	29.1 kJ/mol
Critical temperature, t_{crit}	240 °C
Critical pressure, p_{crit}	4.79 MPa
Ratio of specific heat capacities c_p/c_v (at 14 °C)	1.137

Vapor pressure equation

$$\log p = 19.1403 - 2098.0/T - 4.2114 \cdot \log T;$$

(*p* in kPa, *T* in K)

[3].

Allyl chloride is miscible with most solvents in general use (e.g., octane, toluene, acetone); the solubility (mass fraction in %) of allyl chloride in water at 20 °C is 0.36, that of water in allyl chloride, 0.08. Azeotropic data for allyl chloride are presented in Table 2.

Flash point (closed cup)	-27 °C
Autoignition point	392 °C
Flammability limits in air	3.28 and 11.15 vol%

Table 1. Temperature dependence of the physical properties of allyl chloride

	Temperature, t, °C						
	10	15	20	25	30	40	50
Density, ϱ_4, g/cm^3	0.953		0.940		0.927	0.914	0.900
Specific heat capacity, c_p, kJ kg^{-1} K^{-1} at 101.3 kPa	1.633		1.666		1.700	1.733	1.771
Viscosity, η, µPa · s	368	347	336		307	282	
Surface tension, N/m		28.9	23.1		21.8		
Solubility in 15% hydrochloric acid, wt%				0.120			0.176

Table 2. Azeotropes of allyl chloride

Component	bp at 101.3 kPa, °C	Allyl chloride, mass fraction, %
Water	43	97.8
Methanol	40	90
Ethanol	44	95
2-Propanol	45	98
Formic acid	44	92.5

1.2. Chemical Properties

Allyl chloride is a very reactive compound undergoing the usual addition and polymerization reactions at the double bond. Also, because the chlorine atom can be exchanged readily with other groups, allyl chloride is a suitable starting material for the synthesis of a wide range of allyl derivatives.

Reactions of the Double Bond. The reaction of allyl chloride with oxygen in the liquid phase at ca. 120 °C and in the presence of metal acetates or hydrogen peroxide yields glycerol monochlorohydrin. Halogens add readily to the double bond yielding the corresponding trihalogeno compounds. The reaction with hypochlorous acid, yielding 2,3- and 1,3-glycerol dichlorohydrins (which are then dehydrochlorinated with alkali to give epichlorohydrin), is of great industrial importance. Allyl chloride reacts with hydrogen halides to form 1,2-dihalogeno compounds. In the presence of peroxides, the reaction with hydrobromic acid yields 1-bromo-3chloropropane (Kharasch effect), but in highly concentrated hydrogen peroxide solution, 1,2dibromo-3-chloropropane is formed [6]. Addition reactions of silanes [7], boranes [8], carboranes [9], and phosphorus trichloride [10], as well as cycloadditions of allyl cations with alkenes [11], are known. Allyl chloride polymerizes with sulfur dioxide to form polysulfones [12].

Reactions of the Chlorine Atom. Formerly, the most important reaction industrially was hydrolysis to allyl alcohol, in which small amounts of diallyl ether formed as byproduct. However, today allyl alcohol is produced increasingly by the isomerization of propene oxide (see Section 2.3.2). The chlorine atom is replaced easily by iodide, cyanide, isothiocyanate, sulfide, polysulfides (giving rubber-like condensation products [13]), and alkyl thiols [14]. The salts of carboxylic acids yield allyl esters (e.g., diallyl phthalate), which are easily polymerized to allyl resins or are copolymerized with other monomers. Sodium allyl sulfonate is obtained on reaction with sodium sulfite [15]. The reaction of allyl chloride with ammonia yields a mixture of mono-, di-, and triallylamines [16], [17], and the reaction with primary and secondary amines makes the corresponding alkylallylamines [18]. Organic polycarbonates are formed from a mixture of tertiary amines, sodium carbonate, alkylene glycol, carbon dioxide, and allyl chloride [19]. Phase-transfer-catalyzed carbonylation (e.g., with $Ni(CO)_4$ or $(Me_4N)_2[Ni_6(CO)_{12}]$) in the presence of sodium hydroxide leads to the sodium salt of vinylacetic acid [20], [21]. In the presence of alkalis, allyl chloride reacts with polyols to form the corresponding allyl ethers. Further reactions are described in [22].

1.3. Production

1.3.1. Chlorination, of Propene

Today, allyl chloride is produced on a large scale by the high-temperature (300 – 600 °C) chlorination of propene:

$CH_2=CH-CH_3 + Cl_2 \rightarrow CH_2=CH-CH_2Cl + HCl$
$\Delta H^0_{298} = -113$ kJ/mol

At these temperatures, chlorination proceeds by a free-radical chain mechanism, whereby the hydrogen atom in the allyl position is substituted preferentially by the chlorine, giving allyl chloride [23].

Byproducts. Below 200 °C propene reacts with chlorine mainly by addition to the double bond to give 1,2-dichloropropane; above 300 °C, this reaction is suppressed and the formation of allyl chloride predominates so that 1,2-dichloropropane is only a byproduct. The compounds *cis*- and *trans*-1,3-dichloropropene arise from a secondary reaction of allyl chloride, in which a further hydrogen atom is substituted by chlorine. Small amounts of other chlorination products are formed also:

$CH_3-CH=CH_2 + Cl_2 \rightarrow CH_3-CHCl-CH_2Cl$
$\Delta H^0_{298} = -184$ kJ/mol

$ClCH_2-CH=CH_2 + Cl_2 \rightarrow CH_2Cl-CH=CHCl + HCl$
$\Delta H^0_{298} = -101$ kJ/mol

$CH_3-CH=CH_2 + Cl_2 \rightarrow CH_3-C(Cl)=CH_2 + HCl$
$\Delta H^0_{298} = -121$ kJ/mol

$CH_3-CH=CH_2 + Cl_2 \rightarrow CH_3-CH=CHCl + HCl$

The most important variables in the industrial chlorination process of propene to allyl chloride are the temperature and the ratio of propene to chlorine, whereas pressure and residence time have only a slight effect on the allyl chloride yield [24]. Because the dominant reaction below 200 °C is addition to form 1,2-dichloropropane, the mixing temperature of propene and chlorine must be kept above 250 – 300 °C. The best yields of allyl chloride in industrial reactors are obtained at a maximum reaction temperature of 500 – 510 °C. If the reactor temperature is increased further, spontaneous pyrolysis occurs, with the formation of soot and high-boiling tars. Under laboratory conditions and temperatures of ca. 600 °C, benzene is formed and the yield of allyl chloride decreased [24], [25].

The *maximum reaction temperature* can be influenced by the temperatures of propene and chlorine entering the reactor and by the ratio of propene to chlorine. In general, only the propene is preheated. If the chlorine is also preheated, expensive construction materials must be used to avoid the danger of a "chlorine fire." For a given propene: chlorine ratio, the preheating temperature of the propene is fixed; for example, for $C_3H_6:Cl_2 = 3$ the temperature is set at ca. 300 °C and for $C_3H_6:Cl_2 = 5$, at ca. 400 °C [24].

Propene:Chlorine Ratio. The formation of byproducts decreases with increasing propene excess [26]. On the other hand, the cost of processing the propene rises simultaneously. The optimum reaction conditions are therefore strongly influenced by economic considerations, including the demand for the dichloride byproducts as nematicides [27]. The byproducts can be employed also as starting material for the production of C_1- or C_2-type solvents.

The *pressure* in the reactor has little influence on the yield or the distribution of products and is determined only by the pressure drop in the propene circulation system of the plant.

The *residence time* has only a small effect on the yield of allyl chloride. At high temperatures (300 – 600 °C), the chlorine has completely reacted after 1 – 3 s [24]. Too long a residence time, however, leads to thermal decomposition of the allyl chloride [25].

Reactor Construction. Industrial-scale reactors mainly operate adiabatically, even though higher yields would be expected when operating isothermally. Because the reaction is rapid and exothermic, the amount of heat lost through the reactor wall is not significant.

The technically simplest and oldest reactor type is the tube reactor, which sometimes is equipped with facilities for gas distribution and soot removal [25]. Many other reactor designs are known [26], [28] – [35], but all reactors are designed to achieve the mixing of the two reactants as rapidly and as thoroughly as possible in order to reduce the secondary reaction to form 1,3-dichloropropene. Under the good mixing conditions obtained in a highly turbulent flow, propene and chlorine are fed into the reaction zone at velocities up to 300 m/s. The importance of optimum mixing conditions is demonstrated by a cyclone reactor, in which, at a molar ratio of 3:1, similar yields are obtained as at a ratio of 5:1 with other designs [36]. Many reactors are equipped with a cooling jacket.

If the chlorine is distributed among several reactors arranged in a cascade, a chlorine conversion of up to 86% can be achieved [37], [38]. With this arrangement, higher preheating temperatures are possible because of a higher propene:chlorine ratio in the first reactor.

In another design of a cascade of reactors only the first reactor is charged with preheated propene and chlorine. Liquid propene and gaseous chlorine (molar ratio of 1:1) are fed into the other reactors. The advantages of this procedure are the cooling of the reaction gas and a quasiconstant ratio of the reactants over all stages. At an overall molar ratio of 3.2:1, the same yield is achieved as at a ratio of 7:1 in a single reactor [39].

Feed Preparation. The *purity* of the propene and the chlorine is important. Organic impurities in the propene cause the formation of byproducts and loss of chlorine. Especially propane leads to the formation of chlorine derivatives (1-chloropropane, 2-chloropropane) which are difficult to separate. Polymer-grade propene with a purity of 99.5% often is employed as the feedstock. The use of revaporized chlorine insures adequate purity. Small amounts of inert gases are thereby introduced into the process, the venting of which means propene losses. Both raw materials must contain as little water as possible. This is a particularly important condition for the choice of materials for the various stages of the plant (see below).

A process similar to that employed by Shell, for example [24], [40], is shown in Figure 1. Liquid propene is vaporized, then preheated to 350–400 °C (b), and fed, together with gaseous chlorine, into the reactor (c) via a mixing jet. The chlorine reacts completely, thereby increasing the temperature to 500–510 °C (under optimum conditions). Even under the best conditions, small amounts of carbon are formed; these catalyze the chlorination. A protective film of vitreous carbon deposits on the reactor walls. This material, which also contains highly chlorinated materials and tar, must be cleaned from the reactor walls at intervals of 4–8 weeks. Two parallel reactor chains often are in use so that partial production can be maintained during the cleaning. Another possibility is alternate operation of the two chains.

The gas stream leaving the chlorination reactor is precooled (d) and led to a prefractionator (e), the overhead temperature of which is maintained at ca. −40 °C by feeding liquid propene. This effectively separates all chlorinated hydrocarbons; the bottom product is free of propene and hydrogen chloride.

The gaseous mixture drawn off overhead is separated by absorption with water (i) into aqueous hydrogen chloride of commercial quality and propene. Then the propene is washed with caustic soda in a scrubber (k) to remove traces of hydrogen chloride. After compression to 1.2 MPa (12 bar), it is liquefied in a condenser (l, q). Water is separated (m) and liquid propene is dried by adsorption (n) and returned to the storage tank (a).

The bottom product of the prefractionator contains 80% allyl chloride, 3% 2-chloro-1-propene [557-98-2] and other low boilers, 16% dichlorides (mainly 1,2-dichloropropane [78-87-5] and *cis*- and *trans*-1,3-dichloro-1-propene [542-75-6]), and 1% 1,2,3-trichloropropane [96-18-4] and other heavy boilers [40]. These four fractions are separated by distillation (f, g, h).

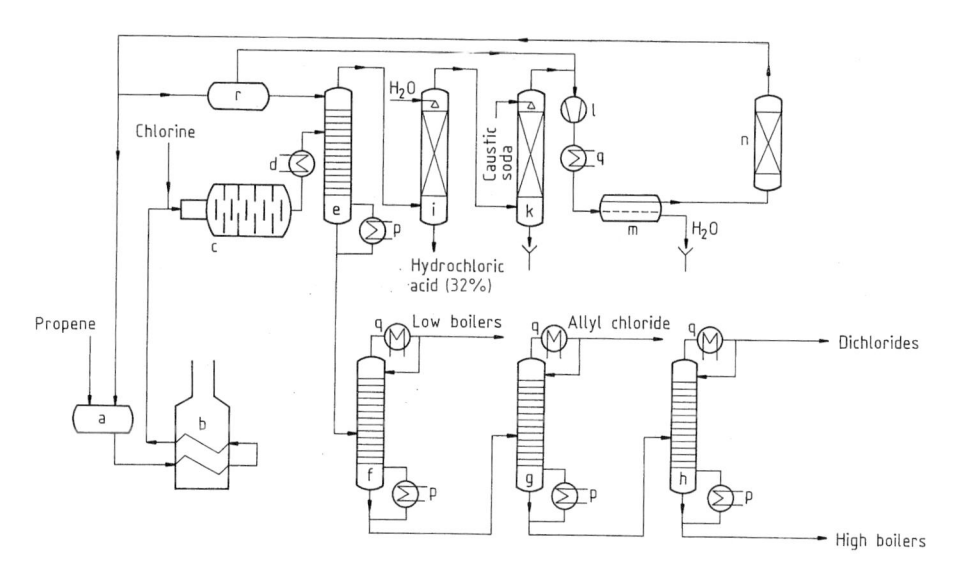

Figure 1. Allyl chloride production by high-temperature chlorination of propene
a) Storage vessel for liquid propene; b) Evaporator and superheater for propene; c) Reactor; d) Cooler; e) Prefractionator; f) Light-ends column; g) Allyl chloride purification column; h) Dichloropropene column; i) Hydrogen chloride absorber; k) Gas washer; l) Compressor; m) Decanter for removing water; n) Propene dryer; p) Evaporator; q) Condenser; r) Cold propene storage vessel

The basic principle of most production processes is as shown in Figure 1, although it has been modified for the synthesis and the fractionation processes [41]–[44]. For example, cooler (d) in Figure 1can be replaced by quench cooling [44], whereby the hot reaction gases are cooled by the evaporation of propene or of condensed reaction products.

Construction Materials. The choice of materials in the allyl chloride synthesis (propene circulation) depends on the individual process operations, temperature, and pressure. In principle, normal carbon steel is resistant where the water content of the product streams is low enough. If the chlorine stream is not preheated, carbon steel also can be utilized in the reactor area. However, materials resistant to chlorine at high temperatures (chromium–nickel steels, nickel, cupronickel alloys) often are employed in the reactor region. Graphite and PTFE coatings are favored for the absorption of the hydrogen chloride, whereas rubberized steel can be employed for the caustic soda scrubber. Carbon steel is suitable for the entire chlorinated hydrocarbon fractionation plant. However, in places when large amounts of water are present, expensive materials, such as nickel and cupronickel, are necessary [45]. Dry allyl chloride can be stored in steel vessels without any danger of corrosion.

1.3.2. Other Production Processes

Several other processes have been suggested for the production of allyl chloride, but none of these has been operated commercially.

Catalytic Chlorination of Propene. This process uses tellurium-containing catalysts [46]; allyl chloride yields of up to 82% are obtained. The main organic byproduct is 17% 2-chloropropane (isopropyl chloride). Unreacted propene and the hydrogen chloride produced in the reaction can be converted to allyl chloride in a second reaction step by oxychlorination.

The dehydrochlorination of 1,2-dichloropropane [47]–[49] produces only a 55% yield of allyl chloride plus a large amount of monochloropropenes. The process is of no commercial interest even though the starting material is dichloropropane, produced in large quantities during the production of propene oxide.

Oxichlorination. Oxichlorination was developed for the production of allyl chloride in order to utilize cheaply available hydrogen chloride instead of chlorine as the feedstock [50]–[62]. Palladium, vanadium, tellurium, copper, lithium, and manganese, as well as their chlorides and oxides and mixtures thereof, have been suggested as catalyst systems.

Most of these oxychlorination processes start from propene [50]–[60], although two of them start from propane [61], [62]. The following process was tested in a pilot plant by Hoechst [58].

Propene, hydrogen chloride, oxygen, and 2-chloropropane (either produced in a subsidiary reactor or introduced from outside) react in a main fluid-bed reactor at 200–260 °C and 0.1 MPa (1 bar) gauge pressure ($\Delta H = -218$ kJ/mol). The carrier catalyst contains tellurium, vanadium pentoxide, phosphoric acid, and a nitrogen compound as promotor. A part of the catalyst flow is treated in a side stream with air and nitric acid to remove coke and maintain constant reactivity. In a subsidiary reactor, 2-chloropropane is produced from propene, hydrogen chloride, and ferric chloride solution. The unreacted propene, 2-chloropropane, and hydrogen chloride are separated and recycled. The yield of allyl chloride is 88–94% based upon propene. The purity of the propene is not particularly critical.

Some of the suggested oxychlorination processes have disadvantages. For example, the activities of the catalysts deteriorate quickly as a consequence of the volatility of the metal salts employed and large volumes must be passed through the reactor because the conversion per pass is low. Also, it is difficult to remove the highly dilute allyl chloride from the reaction mixture without excessive loss of propene by reaction with oxygen. A recent suggestion [60] attempts to avoid these disadvantages by using manganese dioxide as both catalyst and carrier for oxygen. The following reactions take place in the main reactor:

$$MnO_2 + 4\ HCl \longrightarrow MnCl_2 + Cl_2 + 2\ H_2O$$

$$C_3H_6 + Cl_2 \longrightarrow C_3H_5Cl + HCl$$

The catalyst is then reoxidized and activated with oxygen:

$$MnCl_2 + O_2 \rightarrow MnO_2 + Cl_2$$

The process achieves an allyl chloride yield of between 71 and 81 % based on the consumption of propene.

1.4. Handling, Environmental Protection, Storage, and Transportation

Handling and Environmental Protection. Allyl chloride is a highly reactive, highly toxic, easily ignitable substance. Therefore, very stringent standards exist in many countries for atmospheric emission. Allyl chloride should be handled in closed systems in order to fulfill these requirements. Gases containing allyl chloride or byproducts of the production process must be purified by condensation in cold traps, by absorption, by adsorption, or in special cases by combustion before they are vented [63]. The compensation technique has proved of value during transfer from one vessel to another.

Any contamination of the soil, of underground water supplies, or of waste water must be avoided. Any danger of such contamination must be reported immediately to the responsible authorities.

Allyl chloride-containing wastes can be disposed of without difficulty in special incinerators, such as those used for the disposal of solvents, where the combustion gases are treated in absorbers or scrubbers to remove the hydrogen chloride formed.

Because of the high volatility and low flash point of allyl chloride, plants for its production and processing must comply with the relevant standards for fire and explosion protection.

Combustion of chlorinated hydrocarbons produces hydrogen chloride, so that fire fighters should be equipped with suitable protective suits and portable breathing apparatus. Combustion in the absence of adequate air supplies can lead to the formation of carbon monoxide and phosgene [5].

Allyl chloride is highly reactive. Therefore contact with other substances can lead to vigorous, exothermic, and even explosive reactions. This applies particularly to alkali and alkaline-earth metals, but also to aluminum and zinc and to strong oxidizing agents, such as concentrated sulfuric acid. The anhydrous halides (e.g., chlorides) of the metals mentioned above also react vigorously with allyl chloride [5]. In plants producing or processing allyl chloride, careful consideration must be given to the safety aspects of these exothermic reactions [64].

Storage. Pure, dry allyl chloride (water content ≤ 200 mg/kg) does not corrode iron and can be stored in ferrous containers for long periods at ambient temperatures without significant changes in quality. Lined vessels should be used if traces of iron chlorides are objectionable. Unwanted coloration can be prevented by the addition of a suitable stabilizer (e.g., propene oxide).

Transportation of the toxic and highly flammable allyl chloride is governed by many regulations [65]. International marine transportation is governed by the IMDG Code, D 3108, class 3.1, UN no. 1100; transportation in Europe is governed by RID, ADR, and ADNR class 6.1, no. 4 a, Rn 601, 2601, and 6601, respectively. European Economic Community: Yellow Book 78/79, EG-no. 602–029–00-X; Great Britain: Blue Book: Flammable Liquid, IMDG Code E 3023; United States: CFR 49: 172–189, Flammable Liquid (DOT Regulations) [5].

The most important regulations state that:

1) Vessels may be filled only to a maximum of 93 % of their volumes or the RID (Suppl. X and XI) must be complied with.
2) Road transport of loads of more than 1000 kg requires permission.
3) Containers with a capacity of more than 0.45 m^3 and tanks for railroad cars must be designed to carry an excess pressure of 1 MPa (10 bar). No pipe ducts or pipe connections are allowed below the surface of the liquid. Filling and emptying connections must be tightly closed and additionally secured by means of sealed caps. The tanks must be tested regularly to insure that they withstand total pressures of 0.5 MPa (5 bar).
4) Welded drums constructed of steel or stainless steel must be used for transport in drums. The authorities of some countries permit the use of one-way containers. Exceptions to these regulations are permitted in international traffic by bilateral agreement. The current national and international regulations for labeling packages, road vehicles, and rail tank cars must be followed [65].

1.5. Quality and Analysis

Commercial allyl chloride is at least 97.5 wt % pure and contains 1-chloropropene, 1-chloropropane, and 1,5-hexadiene as impurities. In addition, it contains a stabilizer, such as propene oxide, to scavenge any hydrogen chloride produced hydrolytically on long-term storage. Allyl chloride containing traces of iron has a pinkish tinge. The quantitative analysis of allyl chloride is performed exclusively by gas chromatography.

1.6. Uses

Allyl chloride [22], [66] is an important intermediate in the petrochemical industry, used chiefly for the production of epichlorohydrin, which in turn is used in the production of epoxy resins and as an intermediate in the synthesis of glycerol. Allyl chloride is a starting material in the synthesis of many esters (of which those of phthalic, phosphoric, and carboxylic acids are the most important (Section 3.2.4)) and some allyl ethers (Section 4.2) and allylamines (Section 5.2).

Further compounds made from allyl chloride are: *allyl isothiocyanate* (synthetic mustard oil), used in plant protection agents and pharmaceutical preparations; *allyl sulfonate*, used as an electroplating-bath additive [67] and in the production of carbon fibers [68]; *allylsilane*, used for the production of additives for the rubber industry; and *cyclopropane*, an anesthetic.

1.7. Economic Aspects

More than 90% (ca. 500 000 t worldwide in 1981 and 1982) of the allyl chloride produced is used for the production of epichlorohydrin. Less than 55 000 t worldwide went into other applications, of which ca. 25 000 t was in the United States and ca. 18 000 t in the EEC. In 1982, only about 1000 t was traded among the EEC countries [69], [70].

2. Allyl Alcohol

Allyl alcohol was first prepared in 1856 by A. CAHOURS and A. W. HOFMANN by saponification of allyl iodide. For general literature on allyl alcohol, see [71]–[76].

2.1. Physical Properties

Allyl alcohol [*107-18-6*], 2-propen-1-ol, $CH_2=CH-CH_2OH$, C_3H_6O, M_r 58.08, mp -129 °C, bp 96.9 °C (101.3 kPa), is a colorless, mobile liquid with an irritating odor.

Relative density,	d_4^{20} 0.8520; d_4^{25} 0.8476; d_{20}^{20} 0.8535
Refractive index,	n_D^{20} 1.4133; n_D^{25} 1.4111; n_D^{30} 1.4090
Critical temperature, t_{crit}	271.9 °C
Heat of vaporization at 101.3 kPa	39.98 kJ/mol
Specific heat capacity of vapor,	
c_p (g), (20 °C and 101.3 kPa)	2.428 kJ kg^{-1}K^{-1}
Specific heat capacity of liquid	
c_p (l), (20.5 – 95.5 °C)	2.784 kJ kg^{-1}K^{-1}
Heat of combustion at constant p	1853.8 kJ/mol

Vapor pressure vs. Temperature

t, °C	20	40	60	80
p, kPa	2.3	7.7	21.6	52.6

Viscosity	
at 15 °C	1.486 mPa · s
at 30 °C	1.072 mPa · s
Surface tension	
at 20 °C	25.68 mN/m
at 60 °C	22.11 mN/m
Dipole moment	1.63 D
Dielectric constant	
at 16.2 °C, $\lambda = 60$ cm	20.3
Ignition limits in air	
at 100 °C, 101.3 kPa	2.5 – 18.0 vol%
Flash point (closed cup)	22.2 °C

Allyl alcohol is miscible with water and organic solvents in all proportions at 20 °C. Table 3 contains the most important azeotropic data for binary and ternary allyl alcohol azeotropes.

Table 3. Azeotropes of allyl alcohol

Component	bp, °C	Allyl alcohol content, wt%	
Binary azeotropes			
Water	88.89	72.3	
Benzene	76.75	17.3	
Diallyl ether	89.8	30.0	
Allyl chloride	79.9	18	
Trichloroethylene	80.95	16	
Component I	Component II	bp, °C	Content, wt%
			Allyl alcohol / Component II
Ternary azeotropes			
Water	benzene	68.2	9.1 / 83.6
Water	diallyl ether	77.8	8.7 / 78.9

2.2. Chemical Properties

The allylic hydroxyl group and the olefinic double bond can undergo numerous reactions, such as oxidation, reduction, hydrogenation, condensation (formation of ethers and esters), and addition. At room temperature, allyl alcohol is a stable liquid. When heated to temperatures above about 100 °C, allyl alcohol forms water-soluble polymers (polyallyl alcohols) that react with alkenoic acids to give drying oils [77].

Allyl alcohol also can be grafted to polyimides [78] or copolymerized with styrene in the presence of oxygen [79]. Condensation of allyl alcohol with methyl glucoside polyethers, followed by bromination and addition of iso-cyanates, yields flame-resistant polyurethane foams [80].

Hydroxylation. The reaction of allyl alcohol with hydrogen peroxide in the presence of catalytic amounts of tungstic acid yields glycerol with excellent selectivity ($\rightarrow$ Glycerol). This commercially important process was developed by Shell [81]. The hydroxylation of allyl alcohol can be carried out also by organic hydroperoxides, such as ethylbenzene hydroperoxide, in the presence of a vanadium catalyst [82]. Organic hydroperoxides oxidize allyl alcohol selectively to glycidol (oxiranylmethanol) [556-52-5] [83]. Glycidol is also obtained by the reaction of allyl alcohol with peracids [84].

Allyl alcohol can be *dehydrogenated* with air in the gas phase over palladium, silver, or copper catalysts to give acrolein [85], [86]. The *oxidation* of allyl alcohol in the liquid phase (150–200 °C, Pd catalyst) yields acrylic acid and acrolein with 83% combined selectivity [87].

Catalytic *hydrogenation* of allyl alcohol gives 1-propanol [71-23-8] [88]. *Chlorination* in aqueous solution yields mono- and dichlorohydrins of glycerol, which can be hydrolyzed to glycerol [89]. *Bromination* proceeds very selectively in aqueous solution, buffered by calcium chloride [90]; 2,3-dibromo-1-propanol [96-13-9], used in flame-resistant materials [91], is formed.

Allyl alcohol reacts with allyl chloride to give diallyl ether [557-40-4] (*bp* 94 °C) and with methallyl chloride to give allyl methallyl ether [14289-96-4]. Carbon tetrachloride adds to allyl alcohol by radical initiation, leading to 2,4,4,4-tetrachloro-1-butanol [3290-70-8], which is a useful intermediate in flame-retardant technology [92].

Hydroformylation of allyl alcohol in the presence of cobalt carbonyl yields 4-hydroxybutyraldehyde [25714-71-0] [93] in 30% yield. With rhodium-complex catalysts, the 4-hydroxybutyraldehyde yield is improved to 80% [94]. The aldehyde then can be hydrogenated to 1,4-butanediol ($\rightarrow$ Butanediols, Butenediols, and Butynediols), a useful monomer for the production of polyurethanes [95]. When the hydroformylation of allyl alcohol is carried out in the vapor phase over rhodium triphenylphosphine complexes on porous carriers, tetrahydro-2-furanol [5371-52-8] is obtained in 94% yield [96]. Carbonylation of allyl alcohol in acetic acid solution with a palladium chloride catalyst gives 3-butenoic acid [97].

2.3. Production

The raw material in all commercial processes is propene. However, although allyl alcohol can be made via a variety of intermediates, only the processes proceeding via allyl chloride and propylene oxide are of major commercial importance.

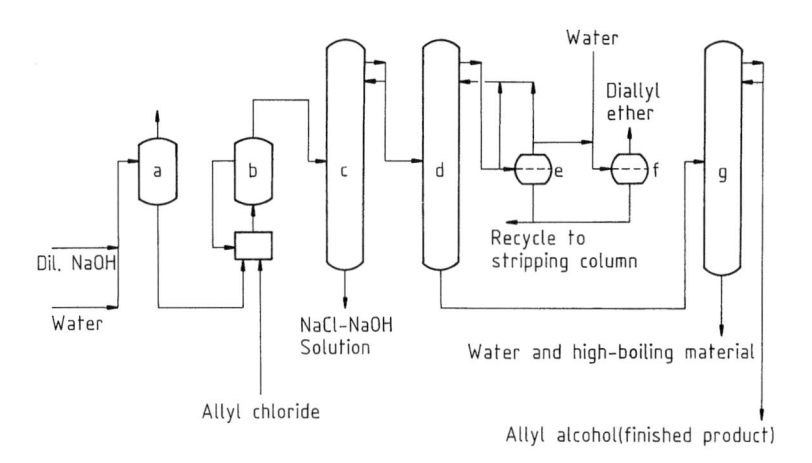

Figure 2. Allyl alcohol production
a) Preheater; b) Hydrolysis reactor with recycle system; c) Stripping column; d) Dehydration column; e) Separator; f) Diallyl ether washing tank; g) Allyl alcohol purification column

2.3.1. Hydrolysis of Allyl Chloride

Allyl alcohol is produced commercially by Shell and Dow from allyl chloride [24], [26]. Allyl chloride is hydrolyzed by a 5–10% sodium hydroxide solution at 150 °C and 1.3–1.4 MPa, yielding 85–95% allyl alcohol. Byproducts are diallyl ether (5–10%), chloropropenes, propionaldehyde, and high-boiling material. In order to obtain commercially satisfactory yields of allyl alcohol, it is necessary to mix the allyl chloride thoroughly with the aqueous alkaline solution, maintaining a constant pH value. The allyl chloride has to be converted nearly quantitatively, because its high corrosiveness does not allow economical recovery. The reaction therefore is carried out in a recycle reactor; the most suitable reactor material is nickel.

A commercial plant for the production of allyl alcohol by continuous allyl chloride hydrolysis [24] is shown in Figure 2. Water is removed from the raw allyl alcohol by azeotropic distillation with diallyl ether as entrainer.

For some reactions, the raw allyl alcohol – water azeotrope (72% allyl alcohol) may be used without further purification.

2.3.2. Isomerization of Propene Oxide

Currently, the catalytic rearrangement of propene oxide is of increasing commercial interest and has partially replaced the hydrolysis of allyl chloride. The most commonly used catalyst is lithium phosphate.

Vapor-Phase Process. Propene oxide vapor passes at 250–350 °C over a lithium phosphate catalyst containing up to 30% inert material. The catalyst is arranged in a fixed bed. Conversions are about 70–75%; allyl alcohol selectivity is 97%. Space-time yields of 0.5 kg allyl alcohol per liter of catalyst per hour are obtained. Olin Mathieson uses a lithium phosphate catalyst which contains 1% alkali hydroxide and which has to be regenerated after 40 h by washing with acetone [98]. The catalyst developed by Chemische Werke Hüls [99], containing 73.6% lithium phosphate and 17.5% silica, needs no regeneration. Even after 1200 h, an allyl alcohol selectivity of 97.3% is maintained. Lithium arsenate also has been proposed as an isomerization catalyst but has no essential advantage over lithium phosphate [100]. An older, chromium oxide-catalyzed process, developed by Wyandotte Chemical Corp. [101], gives much lower conversions and selectivities.

Liquid-Phase Process. The Progil process [102] is carried out with a finely ground lithium phosphate catalyst suspended in high-boiling solvents, such as terphenyl or alkyl benzenes. Propene oxide is bubbled through the suspension at 280 °C. Conversion is 60%, allyl alcohol selectivity 92%, and the space-time yield 1 kg allyl alcohol per liter per hour. Catalyst lifetimes of 500–1000 h are achieved. Jefferson Chemical Co. patented a similar process using a mixture of biphenyl and diphenyl ether as solvent [103].

2.3.3. Hydrolysis of Allyl Acetate

Propene, acetic acid, and oxygen react in the gas phase over palladium catalysts to form allyl acetate, which in turn is hydrolyzed to allyl alcohol:

$$C_3H_6 + CH_3COOH + 1/2\, O_2 \xrightarrow{Pd} CH_2=CHCH_2OCOCH_3 + H_2O$$

$$CH_2=CHCH_2OCOCH_3 + H_2O \rightarrow CH_2=CHCH_2OH + CH_3COOH$$

$$\overline{C_3H_6 + 1/2\, O_2 \rightarrow CH_2=CHCH_2OH}$$

The acetic acid is recovered and reused in the oxidation of propene. Therefore, only propene and oxygen are required as raw materials for the preparation of allyl alcohol.

This process needs no chlorine and therefore may be of commercial interest in the future. Allyl acetate can be hydrolyzed by heating at about 230 °C and 3 MPa, according to a process developed by Hoechst [104], [105]. In the Bayer process hydrolysis is carried out catalytically over an acid cation exchanger (sulfonated polystyrene) at 100 °C [106], [107]. The overall yield of allyl alcohol, relative to the propene feed, is 90%.

2.3.4. Hydrogenation of Acrolein

Allyl alcohol also can be obtained by catalytic hydrogenation of acrolein in the vapor phase. Use of cadmium–zinc catalysts gives yields up to 70% [108]. The hydrogenation also can be carried out over silver–cadmium alloys on inert carriers, such as alumina or silica, also giving yields up to 70% [109].

Ethanol or isopropyl alcohol can reduce acrolein to allyl alcohol in the presence of a mixture of magnesium oxide and zinc oxide [110], [111]. The reaction takes place in the vapor phase at 400 °C; allyl alcohol yields of 80% have been reported [110].

2.4. Quality and Analysis

Commercial allyl alcohol has the following specifications: content of allyl alcohol 98.0 wt% (minimum), content of water 0.3 wt% (maximum), boiling range 95–98 °C. The purity of allyl alcohol is determined by GC. In the absence of saturated alcohols, allyl alcohol can be determined also by acetylation with *N*-acetylpyridinium chloride and back titration of the unconverted *N*-acetylpyridinium chloride. Water is determined by Karl Fischer titration.

2.5. Uses

Allyl alcohol is an intermediate in the production of polymerizable allyl ethers and esters, especially diallyl phthalate. The bulk of the allyl alcohol produced commercially is consumed in the production of glycerol. Polymeric allyl alcohol reacts with unsaturated fatty acids to give drying oils [77]. Sulfur dioxide and allyl alcohol yield polymeric allylsulfonic acids, which have been proposed as intermediates for plasticizers and textile auxiliaries [112]. Allyl alcohol can be copolymerized with other monomers [78], [79] and then used as an intermediate in the production of flame-resistant materials [80], [91], [92] or as a nematicide, fungicide, or preservative.

2.6. Methallyl Alcohol

Methallyl alcohol [*513-42-8*], 2-methyl-2propen-1-ol, $CH_2=C(CH_3)-CH_2OH$, C_4H_8O, M_r 72.11, bp 114.5 °C (101.3 kPa), d_4^{20} 0.8515, n_D^{20} 1.4255, is only partially miscible with water (19.4 wt% methallyl alcohol in water, 33.8 wt% water in methallyl alcohol, at 20 °C). The methallyl alcohol-water azeotrope (*bp* 92 °C) contains 59.8 wt% methallyl alcohol.

Methallyl alcohol is prepared by hydrolysis of methallyl chloride in a 10 wt% aqueous sodium hydroxide solution [113], [114]. Pure methallyl alcohol can be obtained by methanolysis of methallyl acetate in the presence of catalytic amounts of bases [115]. Methallyl alcohol is used in the preparation of polymerizable esters.

3. Allyl Esters

3.1. Properties

Physical Properties. The lower allyl esters, such as allyl acetate, methallyl acetate, and allyl acrylate, are colorless, mobile liquids of low viscosity, with pungent odors, which cause violent irritation of the mucous membranes. The higher allyl esters, such as diallyl phthalate, have high viscosities and boiling points and are nearly odorless. The physical properties of the commercially most important allyl esters are summarized in Table 4.

Chemical Properties. Industrially, the most important property of allyl esters is their ability to polymerize in the presence of oxygen or peroxides. The allyl esters readily undergo hydrolysis and transesterification. Allyl acetate can be oxidized by oxygen and acetic acid over a palladium catalyst to 2-propene-1,1-diol diacetate [869-29-4] [116].

Addition of chlorine in the presence of catalytic amounts of transition-metal chlorides yields 2,3-dichloro-1-propanol acetate [589-96-8] [117]. Acetic acid can add to allyl acetate in the vapor phase (phosphoric acid catalyst, 200 °C) [118] or in the liquid phase with cation exchangers as catalysts [119], yielding 1,2-propanediol diacetate [623-84-7].

Substitution. Hydrogen chloride converts allyl acetate to allyl chloride; the reaction takes place at 100 °C in the liquid phase. A combination of copper(I) chloride and iron(III) chloride which is soluble in allyl acetate, is used as a catalytic system [105], [120], giving conversions of allyl acetate up to 94%, and an allyl chloride selectivity of 96%. Over copper catalysts, hydrocyanic acid converts allyl acetate to allyl cyanide (3-butenenitrile) [109-75-1] [121] – [123].

Hydroformylation of allyl acetate creates a mixture of acetoxybutyraldehydes [93], [124], [125], the main product being 4-acetoxybutyraldehyde [6564-95-0] in yields of about 70%. This aldehyde is an intermediate in a proposed production method for 1,4-butanediol and butyrolactone.

Recent investigations have proved that allyl acetate can undergo a metathesis reaction leading to 2-butene-1,4-diol diacetate (*cis*- and *trans*-isomers) [18621-75-5] and ethylene. With a catalyst system consisting of rhenium heptoxide on alumina promoted by tetramethyltin [594-27-4], allyl acetate conversions of 17% are obtained. The selectivity for 2-butene-1,4-diol diacetate is 96% [126].

Table 4. Physical data of important allyl esters

	CAS Registry No.	Mol. Formula, M_r	bp, °C/p, kPa	d_4^{20}	n_D^{20}
Allyl acetate (acetic acid 2-propenyl ester)	[591-87-7]	$C_5H_8O_2$ 100.12	103.5–104.5/101.3	0.9277	1.4050
Methallyl acetate (acetic acid 2-methyl-2-propenyl ester)	[820-71-3]	$C_6H_{10}O_2$ 114.14	124/101.3	0.9239 [a]	1.4050
Allyl acrylate (2-propenoic acid 2-propenyl ester)	[999-55-3]	$C_6H_8O_2$ 112.13	122/101.3	0.9441	1.4320
Allyl methacrylate (2-methyl-2-propenoic acid 2-propenyl ester)	[96-05-9]	$C_7H_{10}O_2$ 126.16	67/6.7	0.9335	1.4358
Diallyl maleate (cis-butenedioic acid di-2-propenyl ester)	[999-21-3]	$C_{10}H_{12}O_4$ 196.20	114–116/0.67 109–110/0.4	1.0773	1.4699
Diallyl fumarate (trans-Butenedioic acid di-2-propenyl ester)	[2807-54-7]	$C_{10}H_{12}O_4$ 196.20	142/3.7 93/0.4	1.0768	1.4675
Diallyl succinate (butanedioic acid di-2-propenyl ester)	[925-16-6]	$C_{10}H_{14}O_4$ 198.22	105/0.4	1.0510	1.4517
Diallyl adipate (hexanedioic acid di-2-propenyl ester)	[2998-04-1]	$C_{12}H_{18}O_4$ 226.27	141–142/1.1	1.0235	1.4540
Diallyl sebacate (decanedioic acid di-2-propenyl ester)	[3137-00-6]	$C_{16}H_{26}O_4$ 282.38	163/0.4	0.976	1.4551
Diallyl phthalate (1,2-benzene dicarboxylic acid di-2-propenyl ester)	[131-17-9]	$C_{14}H_{14}O_4$ 246.26	161/0.55 150/0.15	1.120	1.5190
Diallyl isophthalate (1,3-benzene dicarboxylic acid di-2-propenyl ester)	[1087-21-4]	$C_{14}H_{14}O_4$ 246.26	181/0.55	1.124	1.521 [c]
Diethylene glycol bis (allyl carbonate) (2,5,8,10-tetraoxa-9-oxotridec-12-enoic acid 2-propenyl ester)	[142-22-3]	$C_{12}H_{18}O_7$ 274.27	162/0.25	1.143	1.4503
Triallyl cyanurate (2,4,6-tris (2-propenyloxy)-1,3,5-triazine)	[101-37-1]	$C_{12}H_{15}O_3N_3$ 249.27	162/0.25	1.1133 [b]	1.540 [c]
Triallyl phosphate (phosphoric acid tri-2-propenyl ester)	[1623-19-4]	$C_9H_{15}O_4P$ 218.19	108–110/1.0 93–94/0.14	1.0815	1.4500

[a] d_{20}^{20}
[b] at 30 °C
[c] at 25 °C

3.2. Production

3.2.1. Oxidation of Olefins

Vapor-phase oxidation of olefins in the presence of lower alkanoic acids and over palladium catalysts yields the corresponding alkenyl esters.

The olefins are attacked in the allylic position. For example, propene is oxidized in the presence of acetic, propionic, or butyric acid to give allyl acetate, allyl propionate, or allyl butyrate, respectively. In a similar way, the methallyl esters are obtained by oxidation of isobutene.

Allyl Acetate. Propene is oxidized with oxygen in the presence of acetic acid in a tubular reactor [127]–[129]. The solid catalyst is located in a plurality of tubes and contains metallic palladium, alkali acetate, and promoters, such as compounds of iron or bismuth [127], [128]. The reaction takes place in the vapor phase at 50–250 °C and elevated pressure. Because of the high exothermicity of the reaction ($\Delta H = -1890$ kJ/mol) and to avoid explosion hazards, propene and acetic acid are converted only partially. The Bayer process [127], [130] yields allyl acetate with selectivities higher than 90 mol%, carbon dioxide being the sole byproduct.

Methallyl acetate is produced from isobutene, acetic acid, and oxygen by a similar process [131].

Liquid-Phase Oxidation. Olefins can be oxidized also in the liquid phase with a catalyst system consisting of palladium chloride and copper(II) chloride (redox system) [132]. This reaction is not selective; major byproducts are 1-propene-1-ol acetate, 1-propene-2-ol acetate, propionaldehyde, and acetone.

3.2.2. Esterification

A general method for the preparation of allyl esters is the reaction of allyl alcohol with the free acids, acid anhydrides, or acid chlorides. Catalysts frequently used are aromatic sulfonic acids (*p*-toluenesulfonic acid and naphthalene-2-sulfonic acid) [130]. Strong mineral acids are of less value because they can decompose the allyl alcohol. Polymerizable acids are esterified in the presence of phenolic inhibitors. The following allyl esters are prepared by this method: *diallyl phthalate, diallyl isophthalate, diallyl maleate, diallyl fumarate, triallyl cyanurate* (from cyanuric chloride and allyl alcohol), and *diethylene glycol bis* (*allyl carbonate*). The last of these compounds is obtained from allyl alcohol and the corresponding bischloroformate (prepared from diethylene glycol and phosgene).

3.2.3. Transesterification

Higher allyl esters can be prepared by transesterification of the corresponding methyl esters with allyl alcohol; sodium methanolate is the preferred catalyst [133], [134]. *Diallyl phthalate* and *diallyl sebacate* can be prepared in this way. The transesterification of methyl or ethyl esters with allyl acetate has been proposed [135], [136]. This method would be of industrial interest if allyl acetate were commercially available as a primary product, for example, by oxidation of propene [135]. Suitable transesterification catalysts are alkoxides or alkoxide complexes of elements of the first to third main groups of the periodic table or alkoxides of titanium and zirconium [136]. The ester is mixed with twice as much allyl acetate and catalytic amounts of the alkoxides and heated to reflux temperature. Methyl acetate (*bp* 57 °C) and ethyl acetate (*bp* 77 °C) have the lowest boiling points and are removed from the equilibrium by distillation. The desired allyl ester can be isolated in excellent yields and high purity.

3.2.4. Other Production Methods

The reaction of allyl chloride with the alkali salts of mono- or dicarboxylic acids to give the corresponding allyl esters is well known. These salts also may be formed during the reaction. In this way, *diallyl phthalate* is obtained from the reaction of allyl chloride and phthalic anhydride in the presence of sodium hydroxide, sodium carbonate, tertiary amines, or quaternary ammonium salts [137].

Allyl acrylate is prepared by the pyrolysis of allyl lactate, *allyl methacrylate* in a similar manner from allyl 2-hydroxyisobutyrate [138]. *Substituted tetrahydrophthalic acid allyl esters* are obtained by Diels-Alder reaction of the corresponding substituted butadienes and diallyl maleate [139]. The best method for the preparation of *triallyl phosphate* is the condensation of phosphorus trichloride with allyl alcohol to give *triallyl phosphite* [102-84-1] followed by oxidation with air [140]. *Diethylene glycol bis(allyl carbonate)* can be prepared in 87% yield by heating allyl chloride, diethylene glycol, and carbon dioxide in the presence of sodium carbonate and triethylamine in an autoclave [141].

3.3. Uses

3.3.1. Polymer Production

Polymer production is the most important application of allyl esters. They are especially suitable as components in copolymers. Currently, some allyl esters are being used for crosslinking or curing of polyolefins (graft copolymerization). Optimum crosslinking is obtained by adding specific peroxides or by high-energy irradiation. The crosslinked polymers and copolymers of allyl esters give thermoset articles of excellent heat resistance. Other applications of these polymers are in the production of casting sheets, molding material, electric and optical devices of high resistance, and flame-retardant materials.

Diethylene glycol bis(allyl carbonate) usually is referred to as *CR 39 monomer*. It is bulk polymerized by the addition of relatively large (compared to similar polymerizations) amounts of peroxides. The commercially preferred peroxide initiator is isopropyl percarbonate [105-64-6]. The homopolymer (*CR 39 polymer*) [25656-90-0] is suitable for the production of cast sheets, lenses, and other articles that have excellent resistance to scratching, impact, and heat as well as outstanding optical properties [142] – [144]. This polymer is resistant to common solvents with the exception of oxidizing acids. It is produced in the United States by PPG Ind. [142]. Even better qualities can be obtained by reacting CR 39 monomer with other monomers, such as methyl methacrylate [80-62-6] or triallyl isocyanurate [1025-15-6]. These copolymers are used for the production of light-weight lenses with very hard surfaces and high refractive indices [145].

Diallyl phthalate is an important monomer for the production of thermosetting molding compounds, which must have good dimensional stability and electrical properties, and be resistant to heat and solvents. Diallyl phthalate can be polymerized or copolymerized. The preferred technique is first to prepare a prepolymer or precopolymer in solution. This usually is done by dissolving the diallyl phthalate monomer in 2-propanol, adding 50% hydrogen peroxide at about 105 °C, and precipitating the prepolymer from the cooled, viscous solution with excess 2-propanol. Precopolymers can be prepared by adding comonomers, such as triallyl cyanurate, acrylates, vinyl compounds, acrylonitrile, styrene, or diallyl isophthalate. The prepolymers and precopolymers are bulk polymerized by mixing the prepolymer or precopolymer with large amounts of free-radical initiators (benzoyl peroxide, *tert*-butyl perbenzoate) and molding at temperatures up to 200 °C and 40–45 bar for about 15 min. Because of their outstanding mechanical and electrical properties, copolymers containing diallyl phthalate are suitable for specialty coating and for embedding, especially in the production of electronic devices. For example, the moisture-sensitive epoxy compounds now used in light-emitting diode (LED) displays can be replaced by stable diallyl phthalate epoxy encapsulating resins [146], [147].

By adding inorganic materials to diallyl phthalate prepolymer compositions, reinforced thermosetting molding compounds can be obtained. Glass cloth or paper can be impregnated with a solution of prepolymer, monomer, and peroxide initiator. After removal of the solvent, the glass cloth or paper is cured to give the desired film-protected material, which is used for decoration, stain-resistant overlays for household articles, and furniture.

Diallyl isophthalate (DAIP) has the advantage over the ortho isomer that it polymerizes faster and gives polymers of better heat resistance. The DAIP prepolymer is less stable than the diallyl phthalate prepolymer. Cured moldings from DAIP monomer-prepolymer compositions can be better processed than cured diallyl phthalate moldings because of their greater fluidity. Compositions containing DAIP prepolymers can be used for the production of hard, translucent, abrasive-resistant, and laser-trimmable coatings [148].

Diallyl esters of aliphatic dicarboxylic acids, e.g., diallyl succinate, adipate, and sebacate, are preferred for the preparation of soluble, thermoplastic precopolymers that sometimes contain vinyl monomers. These precopolymers can also be used for graft copolymerization with preformed polymers; heat- and solvent-resistant adhesives, plastics, and coatings are produced by this technique.

Diallyl maleate and diallyl fumarate have additional activated double bonds and copolymerize readily with styrenes, vinyl ethers, and acrylates. The less reactive allyl groups remain intact and can subsequently be crosslinked by heating with peroxides or treating with high-energy radiation. As in the case of the saturated diallyl esters, soluble precopolymers are obtained by heating the appropriate mixture of the monomers in the presence of peroxides [149]. Preferential uses are castings and moldings, plastisols, and adhesives.

The allyl acrylic monomer of most importance is allyl methacrylate; allyl acrylate is of secondary importance. Both are used in small amounts for the preparation of soluble, fusible copolymers that can be crosslinked in a subsequent reaction to form insoluble, heat-resistant plastics and coatings. Homopolymers also can be produced by radical-initiated polymerization but are too expensive and are of poorer quality than copolymers. About 10% of allyl methacrylate and allyl acrylate is copolymerized with other acrylic esters. Preformed vinyl polymers can be cross-linked and processed to form graft polymers. A copolymer of 10% allyl methacrylate and 90% vinyl chloride is a flame-retardant additive for polypropylene in amounts of up to 30% relative to the polyolefin [150]. Allyl acrylate is used in the production of dental plastics, optical lenses, reinforced plastic adhesives, coatings, rubbers, and components in printing processes. It can also be used to modify the properties of cotton [151], [152] or nylon 6 [153].

Triallyl cyanurate is used preferentially as a crosslinking agent in copolymers. On heating it may polymerize violently and then isomerize to the more stable triallyl isocyanurate [*1025-15-6*] with the allyl groups attached to the nitrogen atoms [154], [155].

At 30 °C viscous solutions of prepolymers form slowly. Triallyl cyanurate is used for the production of heat- and solvent-resistant coatings and moldings, reinforced plastics, and adhesives. Addition of 5–10% triallyl cyanurate to polyester–styrene or methyl methacrylate yields cast sheets of improved mechanical and thermal stability. Short-term heating of triallyl cyanurate with polymers in the presence of peroxides at 180 °C gives crosslinking; valuable copolymers are thus obtained from poly(vinyl chloride) elastomers [156] and fluoroelastomers [157]. Ethylene polymers and copolymers also may be crosslinked under similar reaction conditions [158]. High-impact plastics have been obtained by grafting butyl acrylate-triallyl cyanurate copolymer with a styrene–acrylonitrile mixture [159]. Other examples of peroxide-initiated curing with triallyl cyanurate (2–5%) at 150–160 °C are polyurethanes [160], nylons [161], cellulose [162], polyoxyethylene [163], vinyl-substituted polysiloxanes [164], and acrylate copolymers [165]. Crosslinking of polycarbonate with triallyl cyanurate by UV irradiation in the presence of polythiols gives scratch-resistant coatings [166].

Triallyl phosphate can explode on heating to about 130 °C and polymerizes rapidly on exposure to air. A possible commercial application of this reactive monomer is the crosslinking with polyolefins in the presence of peroxides and, if necessary, in the presence of foaming agents [167].

3.3.2. Other Uses

Some allyl esters find application as plasticizers, textile auxiliaries, and insecticides. Allyl cinnamate [1866-31-5] is a component of perfumes. 2,3-Dibromopropyldiallyl phosphate [33528-41-5] has been proposed as a flame retardant in poly(ethylene terephthalate) fibers [168]. Allyl esters also are suitable as diluents for the less reactive peroxides and as dispersion media for pigments and fillers.

4. Allyl Ethers

4.1. Properties and Uses

Simple allyl ethers, such as diallyl ether or alkyl allyl ethers, have only minor commercial importance. Valuable allyl ethers are polyol allyl ethers and allyl ethers that contain an epoxy group.

Allyl glycidyl ether [106-92-3], (allyloxymethyl)oxirane, M_r 114.14, bp 87.5–88 °C (10.9 kPa), d_4^{20} 0.9678, n_D^{20} 1.4345, is a toxic liquid used as an additive for epoxy resins and as a comonomer in polyglycols and polyolefins. Crosslinking is achieved by irradiation or by thermally induced peroxide decomposition. Copolymerization of allyl glycidyl ether with acrylamide or N,N'-methylenebis(acrylamide) yields water-soluble polymers containing epoxy groups. These can be condensed with pharmacologically active (controlled-release) compounds to form pharmacologically active polymers [169].

Ethylene glycol diallyl ether [7529-27-3], 1,2-Bis(allyloxy)ethane, $C_8H_{14}O_2$, M_r 142.20, bp 35–37 °C (0.13 kPa), d_4^{20} 0.8940, n_D^{20} 1.4340, is polymerized oxidatively with cobalt acetate and air in methanol solution soaked on the surface of filter paper. The resulting polymeric product confers a high degree of hydrophilicity to the filter paper [170].

Partial allyl ethers of trimethylolpropane and pentaerythritol are used to increase the drying rate of alkyd drying-oil coatings. By condensing a mixture of various pentaerythritol allyl ethers with polyesters and ethylene glycol monobutyl ether [111-76-2] and then curing with cobalt octanoate, a film of high tensile hardness and flexibility can be obtained [171].

Tetraallyl pentaerythritol ether [1471-18-7], $C_{17}H_{28}O_4$, 296.41, bp 124–125 °C (0.13 kPa), d_4^{20} 0.9497, n_D^{20} 1.4595, can be added to lubricating oil to improve its viscosity index (i.e., to reduce temperature effects) [172].

Allyl Ethers of Carbohydrates. Sucrose polyethers copolymerize with acrylic acid to yield branched, water-soluble polymers. Allyl ethers of starches have been proposed as air-drying protective coatings and as varnishes but have found only limited commercial interest because of their instability and water sensitivity.

4.2. Production

Allyl ethers of low molecular mass, such as diallyl ether and alkyl allyl ethers, are prepared by heating allyl alcohol or mixtures of allyl alcohol with monoalkanols in the presence of mineral acids. The more valuable polymerizable allyl ethers of polyols are prepared by reaction of allyl chloride with the polyols in the presence of sodium hydroxide [173], [174] or by treating an aqueous solution of the polyol with allyl chloride in a two-phase system to which a phase-transfer catalyst has been added. Using tetrabutylammonium bromide [1643-19-2] as the phase-transfer catalyst, up to 99.7 % tetraallyl pentaerythritol ether is obtained from the reactants [172].

5. Allylamines

5.1. Properties and Uses

Allylamine [107-11-9], $CH_2=CHCH_2NH_2$, C_3H_7N, M_r 57.10, bp 54.5 °C, d_4^{20} 0.7621, is a liquid with an ammoniacal odor. It can be converted to N-allylamides or -imides by reaction with esters, acid anhydrides, or acid chlorides. Allylamine is oxidized at 400 °C on a bismuth molybdate catalyst; acrylonitrile and propionitrile are the main products (85 % combined yield) [175]. Plasma polymerization of allylamine gives polymers that can be used as reverseosmosis membranes. A possible future application of these membranes is in closed environmental systems, such as manned spacecraft [176], [177]. Allylamine is useful as a corrosion inhibitor when pickling steel in acid [178]. The rhodium-catalyzed reaction of allylamine with carbon monoxide at 120 °C and 10 MPa yields 67 % γ-butyrolactam [616-45-5] [179].

Diallylamine [124-02-7], $C_6H_{11}N$, M_r 97.16, bp 111 – 112 °C, can be cyclized to give substituted pyridines [180]. It is an intermediate in the production of pharmaceuticals and resins.

Triallylamine [102-70-5], $C_9H_{15}N$, M_r 137.23, bp 150 – 151 °C, has been proposed as a catalyst for the production of polyesters [181] and as an initiator for the polymerization of butadiene [182].

Methallylamine [*2878-14-0*], $CH_2=C(CH_3)-CH_2NH_2$, C_4H_9N, M_r 71.12, bp 78.8 °C, d_4^{20} 0.782, is dehydrogenated to methacrylonitrile [*126-98-7*] over a silver catalyst. Copolymerization of methallylamine with acrylonitrile increases the affinity of polyacrylonitrile fibers for dyes [183].

5.2. Production

Allyl chloride reacts with aqueous ammonia at about 100 °C to give a mixture of the hydrochlorides of allyl-, diallyl-, and triallylamine, from which the pure amines are obtained by distillation [184]. Methallylamine is produced in a similar manner [185].

Pure allylamine can be prepared by hydrolysis of allyl isothiocyanate [186], thermal cleavage of allyl dithiocarbamate [187], hydrolysis of diallyl cyanamide [188], or ammonolysis of triallylamine [189]. Allylamine may be produced also by electroreduction of acrylonitrile on lead cathodes [190]. Polyallylamines are readily obtainable by catalytic hydrogenation of poly(acrylonitrile) over a Raney nickel catalyst [191].

6. Toxicology and Occupational Health

Allyl compounds are stronger irritants to the skin and mucous membranes and are more toxic than the corresponding alkyl compounds (Table 5). Many allyl compounds are absorbed easily through the skin and mucous membranes. After absorption, aliphatic allyl compounds can cause severe liver damage, in contrast to comparable alkyl compounds (Table 5). Other target organs are mainly the central nervous system, the kidneys, and the hematopoietic system.

Allyl Chloride [*107-05-1*]. *Acute and Subacute Toxicity.* LD_{50} = 460 mg/kg (rat, oral) [193]; LD_{50} = 3.7 mg/kg (rabbit, percutaneous) [5]; LC_{50} = 11 mg/L (rat, inhalation, 2 h) [193]. The inhalation of 3 ppm allyl chloride during 7 h/d on 5 days a week was tolerated by a group of rats, guinea pigs, and rabbits for 180 days without irreversible damage occurring. An analogous test using 8 ppm over a period of 35 days led to damage of the liver and kidneys [194]. Further experiments demonstrate a neurotoxic effect of allyl chloride, in particular to the peripheral nerves of cats and rabbits [193], [195], [196].

Carcinogenicity, Mutagenicity, Embryotoxicity. So far, little information on the *carcinogenic activity* of allyl chloride is available [197]–[201]. A carcinogenic effect is suspected, although a definite statement cannot be made at present [197]. The *mutagenicity* of allyl chloride has been confirmed in various tests [202]–[205]. A toxic effect on the development of rat embryos, as well as increased *embryo mortality*, was established

Table 5. Hepatotoxic effect* of some allyl and propyl compounds in rats

	Dosage, mg/kg	Degree of liver necrosis
Allyl alcohol	25	1.5
Allyl formate	40	2
Allyl acetate	45	2.5
Allyl butyrate	85	1.5
Allyl caproate	75	2.5
Allyl heptanoate	165	3
1-Propanol	2160	0
Propyl formate	1330	0
Propyl acetate	3120	0
Propyl butyrate	5000	0

* Degree of macroscopically visible liver necrosis on day 5 after oral application of 1/3 of the LD_{50} for four consecutive days. Degree 1: just clearly visible; degree 3: severe necrosis.

in rats that had inhaled air containing allyl chloride at a concentration of 3.1 mg/m^3. At 0.29 mg/m^3 no negative effects occurred [206].

General Characteristic Effects. Allyl chloride causes strong irritation of the skin and the mucous membranes. It is absorbed by inhalation, by ingestion, and through the skin. Direct contact and exposure to air–allyl chloride mixtures lead to strong irritation of the eyes, mucous membranes, and nasopharyngeal cavity. Acute symptoms are numbness, unconciousness, and — particularly upon repeated and lengthy exposure to higher concentrations — pulmonary edema. The heart and circulatory system, liver, and kidneys are endangered. At longer exposure times and higher concentrations, fatal poisoning is possible. After prolonged exposure, chronic damage of the liver and kidneys occurs [5], [207], [208].

Occupational Health. Because of its toxicity, allyl chloride is classified as a poisonous material. The threshold limit value (TLV) and the MAK are 1 ppm (8 h time weighted average) [209]–[211]. The short-term exposure limit is twice as high: 2 ppm [207], [210]. Because of its suspected carcinogenic potential, allyl chloride is classified in group III B by the MAK commission [209]. In the USSR, the maximum allowed workroom concentration is set at 0.1 ppm [212].

The particularly pungent odor of allyl chloride cannot serve as an adequate warning signal. The odor perception thresholds, at which 50 and 100 % of exposed persons perceive allyl chloride vapors in the air, are 3 to 6 and 25 ppm, respectively [5]. These values are considerably above the permissible working concentrations.

So far, few investigations in occupational medicine on the effects of allyl chloride exist [195], [207], [213].

Allyl Alcohol [*107-18-6*]. With respect to its irritating and toxic effects, allyl alcohol is the best investigated aliphatic allyl compound. The odor perception threshold is approximately 0.8 ppm [214], and 6–12 ppm causes irritation of the nose. At a concentration of 25 ppm, severe irritation of the eye occurs, with lacrimation, photophobia, blurred vision, and retrobulbar pain [214]. Corneal necrosis results in temporary blindness in persons exposed to higher concentrations [215]. Sensitive persons may react with slight irritation of the eyes at concentrations of 2–5 ppm [216], [217]. In animal experiments daily inhalation of concentrations between 2 and 7 ppm (7 h/d) has been tolerated without irritation in dogs, rabbits, guinea pigs, and rats during periods of up to 6 months [194].

The threshold limit value (TLV) and the MAK value are 2 ppm [210], [211]. The shortterm exposure limit is twice as high, namely 4 ppm.

A single oral application of allyl alcohol (0.02 mL/kg) caused liver necrosis in rats and the death of 7 out of 12 animals [218]. The oral and cutaneous LD_{50} for rabbits is 50–80 mg/kg, and for rats 64 mg/kg [219], [220]. A single dermal application of allyl alcohol to dogs (approximately 0.2 mL/kg) caused lethal gastrointestinal hemorrhages within a few hours [219]. In rats hepatotoxic effects with liver cell necrosis were observed after a single application of 62.5 µL/kg by gavage [221] and after a single intraperitoneal injection of 0.5 mL/kg [222]. On the other hand, the daily application of 50 ppm of allyl alcohol (4.8–6.2 mg/kg) with the drinking water was tolerated by rats for 15 weeks without effect [223]. In cases of skin contamination, the danger of systemic intoxication brought on by the high absorption rate of allyl alcohol through the skin (one drop of allyl alcohol equals approximately 50 mg) exists in addition to local irritation [220].

Allyl Esters. The toxicological effects of allyl esters are quite similar to those of allyl alcohol. Local irritation and damage to the central nervous system and the liver caused by absorption are the main effects. For *allyl acetate* [591-87-7], the dermal LD_{50} in rabbits is 1000 mg/kg and the LC_{50} in rats is 1000 ppm after 1 h of inhalation. Comparable values are reported for *diallyl maleate* [999-21-3] [214]. Table 5 shows the hepatotoxicity of simple allyl esters, compared with some propyl esters. The hepatotoxicity of *diallyl phthalate* [131-17-9] is probably in the same range as that of others in the table, but it causes only slight irritation of the skin or mucous membranes [224]. *Allyl methacrylate* [96-05-9] has an oral LD_{50} in rats of 430 mg/kg and the dermal LD_{50} in rabbits is 500 mg/kg. Allyl methacrylate is easily absorbed through the skin. The inhalational LCLo (lowest lethal concentration) in rats has been found to be 500 ppm [225].

Allyl Ethers. *Allyl glycidyl ether* [106-92-3] has an LD_{50} of 390 mg/kg (mouse, oral) and an LD_{50} of 1600 mg/kg (rat, oral). In rabbits, the dermal LD_{50} is 2550 mg/kg. Inhalational toxicity has been estimated: LC_{50} = 270 ppm (mice, 4 h) and 670 ppm (rats, 8 h) [226], [227].

Allyl glycidyl ether causes only slight irritation of the skin but is strongly irritating and corrosive to the eye. Following inhalation, the irritation of the respiratory tract can lead to pulmonary edema or to secondary bronchopneumonia [226], [227]. After absorption, systemic intoxication with disorders of the central nervous system and morphologic damage to the liver, kidneys, and spleen can occur [226], [227]. In humans, dermal sensitization has been observed besides the irritating effects [226],

[227]. The threshold limit (TLV) and the MAK values are 10 ppm or 45 mg/m^3 [210], [211].

Diallyl ether [*557-40-4*] smells like horseradish and severely irritates the mucous membranes; the LD$_{50}$ is approximately 270 mg/kg (rat, oral) [228]. Diallyl ether is absorbed easily through the skin; the LD$_{50}$ is 540 mg/kg (rabbits, dermal, estimate) [228].

Ethylene glycol diallyl ether [*7529-27-3*] has an oral LD$_{50}$ in rats of 1020 mg/kg. Exposure to it for 24 h causes severe irritation of the skin (500 mg) and eyes (250 μg) of rabbits [229].

Allyl ethyl ether [*557-31-3*] and *allyl vinyl ether* [*3917-15-5*] are irritants to the skin, mucous membranes, and eyes [230]. For allyl vinyl ether, the oral LD$_{50}$ in rats is 450 mg/kg [230].

Allylamines. Allylamines also are irritating to the skin and mucous membranes and are corrosive to the eye. In general, allylamines are absorbed easily through the skin. The dermal LD$_{50}$ of *allylamine* [*107-11-9*] in the rabbit is 35 mg/kg [231]. The oral LD$_{50}$ has been determined to be 106 mg/kg (rat) and 57 mg/kg (mouse) [231], [232]. For rats, the inhalational LC$_{50}$ is 286 ppm (4 h). In humans, the odor perception threshold of allylamine is approximately 2.5 ppm, whereas the TCLo (lowest toxic concentration) has been found to be 5 ppm (5 min). Above 2.5 ppm, irritating effects on the respiratory tract can occur; 14 ppm is intolerable [233].

In contrast to other allyl compounds, allylamine, diallylamine [*124-02-7*], and triallylamine [*102-70-5*] can cause myocardial degeneration as well as damage to the kidneys and liver. This has been observed in rats and rabbits [232]. According to recent investigations, allylamines cause a severe primary fibrosis of the myocardium [234], [235]. The oral application of allylamine in the drinking water over 81 – 104 days caused a dose-dependent myocardial degeneration in rats [235].

7. References

[1] *Ullmann*, 4th ed., **9**, 466.
[2] Solvay Note CMP 8049 V-2, 20 (internal document), 1976.
[3] *Kirk-Othmer*, 3rd ed., **5**,764.
[4] *Beilstein* **3** (1), 699.
[5] Shell Chemical Co.: Allyl Chloride Toxicity and Safety Bulletin 10/76, SC 195 – 76, 1976.
[6] Chemische Fabrik Kalk, DE 2913277, 1980 (H. Jenker, R. Strang).
[7] M. Capka, CS 187167, 1979 (M. Capka).
[8] M. J. Hawthorne, J. A. Dupont, *J. Am. Chem. Soc.* **80** (1958) 5830.
[9] J. Plesek, Z. Plzak, J. Stuchlik, T. Hermanek, *Collect. Czech. Chem. Commun.* **46** (1981) no. 8, 1748.
[10] L. Z. Soborowskii, Ju. M. Zinovev, L. I. Muler, *Zh. Obshch. Khim.* **29** (1959) 3907. *Beilstein* **1** (4), 739.

[11] H. Klein, H. Mayr, *Angew. Chem.* **93** (1981) 1069; *Angew. Chem. Int. Ed. Engl.* **20** (1981) 1027.
[12] E. M. Fettes, F. O. Davis, N. E. Gaylord (ed.): *Polyethers* part III, Interscience Publishers, New York 1962, pp. 225–270.
[13] H. Jacobi, W. Flemming, US 2259470, 1941.
[14] M. K. Gadzhiev, SU 810684, 1979.
[15] Kaustik Sterlitamak Industrial Enterprises USSR, SU 859357, 1981 (E. N. Denisov, Yu. D. Morozov).
[16] A. M. Mezheritskii, SU 578301, 1976 (A. M. Mezheritskii, M. M. Krivenko, N. N. Vdovenko).
[17] A. M. Mezheritskii, SU 654609, 1979 (A. M. Mezheritskii, M. M. Krivenko, N. N. Vdovenko).
[18] Akad. Wissenschaft DDR, DD 136497, 1978 (D. Ballschuh, R. Ohme, J. Rusche).
[19] Tokuyama Soda Corp., JP 8105442, 1981; JP 8105443, 1981 (S. Koyanagi, N. Hasegawa, T. Shimizu, S. Katsushima, I. Kaneho).
[20] M. Foa, L. Cassar, *Gazz. Chim. Ital.* **109** (1979) no. 12, 619.
[21] *Ullmann,* 4th ed., **9**, 163.
[22] C. E. Schildknecht: *Allyl Compounds and their Polymers, High Polymers,* vol. **XXVIII,** Wiley-Interscience, New York-London-Sydney-Toronto 1973.
[23] G. W. Hearne, T. W. Evans, H. L. Jale, M. C. Hoff, *J. Am. Chem. Soc.* **75** (1953) 1392. Shell Oil Co., US 2939879, 1957 (A. De Benedictis).
[24] A. W. Fairbairn, H. A. Cheney, A. J. Cherniavsky, *Chem. Eng. Prog.* **43** (1947) 280–290.
[25] L. M. Porter, *J. Am. Chem. Soc.* **78** (1956) 5571.
[26] E. C. Williams, *Trans. Am. Inst. Chem. Eng.* **37** (1941) 157–207.
[27] L. W. Chubb, *Ind. Chem.* **30** (1954) 491–496.
[28] Shell Development Co., US 2643272, 1947; US 2763699, 1956 (C. P. Van Dijk, F. J. F. van der Plas); GB 765764, 1955; GB 790166, 1956.
[29] Union Carbide Corp., US 3054831, 1958.
[30] Columbia-Southern Chemical Corp., GB 901680, 1962.
[31] Nissan Chemical Ind., BE 657 267, 1964.
[32] Asahi Denka Kogyo Co., JP 42/7366, 1966; JP 73/32087, 1968; JP 73/26732, 1969.
[33] Asahi Glass Co., JP 73/30250, 1968.
[34] Asahi Electrochem. Co., JP 72/13006, 1967; JP 72/ 13007, 1967; JP 74/39961, 1969; JP 75/16332:4, 1969.
[35] BASF, DE 1114184, 1959.
[36] P. Klucovsky, J. Dykyj, *Acta Chim. Acad. Sci. Hung.* **36** (1963) nos. 1–4, 145–155.
[37] Shell Development Co., GB 761831, 1954.
[38] G. A. Oshin, SU 706392, 1979.
[39] Solvay Cie., DE 1960063, 1969.
[40] D. L. Jabroff, J. Anderson, *World Pet. Congr. Proc.* **III,** *Sect. V* (1951) 22–30.
[41] Halcon International, DE 1200286, 1965; DE 1210806, 1966.
[42] L. Prochazka, CS 138599, 1968.
[43] Dow Chemical Corp., DE 2540336, 1975.
[44] Halcon International, DE 1215691, 1966.
[45] *DECHEMA-Werkstofftabellen A-Z DWT 53,* Dechema-Verlag, Frankfurt 1953.
[46] Hoechst, DE 1960158, 1969.
[47] Shell Development Co., US 2207193, 1940 (H. P. A. Groll).
[48] BASF, DE 926246, 1952.
[49] Bayer, DE 1210801, 1966.

[50] Hoechst, DE 1224302, 1964; DE 1230780, 1964; DE 1237554, 1965; DE 1237555, 1965; DE 1243670, 1965; DE 1274112, 1966; DE 1282637, 1967; DE 1283828, 1966; DE 1793132, 1968.
[51] Hoechst, DE 1222913, 1964; DE 1224301, 1964; DE 1227006, 1964; DE 1230781, 1965.
[52] Monsanto Chem. Co., US 2966525, 1959; GB 935088, 1960 (D. E. Steen).
[53] ICI, DE 1234705, 1963.
[54] Toyo Soda, GB 1016094, 1963.
[55] Deutsche Texaco, DE 1300930, 1967.
[56] El Paso Products Co., DE 1280847, 1966.
[57] Showa Denko Kabushiki Kaisha, DE 1817281, 1968.
[58] Hoechst, *Oil Gas J.* **68** (1970) 58.
[59] Shell Chemical Co., DE 2114302, 1971 (W. Rootsaert, R. Van Helden, F. Wattimina).
[60] Institut Neftechimitscheskich Processov Imeni Akademika Ju. G. Mamedalieva Akademii Nauk Azerbaidschanskoj SSR, Baku, SU 2585604, 1978.
[61] El Paso Products Co., DE 1293744, 1966.
[62] Lummus Co., DE 2426640, 1974 (H. D. Schindler, M. Sze, H. Riegel).
[63] C. A. Peterson, J. A. Key, F. D. Hobbs, J. W. Blackburn, H. S. Basdekis et al.: *Organic Chemical Manufacturing,* vol. **10,** Selected Processes, Report 3, Glycerin and its Intermediates, EPA Report 450/3-80-028 e (1980).
[64] N. Piccinini, U. Anatra, G. Malandrino, D. Barone, S. Donato: "Safety analysis for an allyl chloride," *Plant/Oper. Progr.* **1** (1982) 1,69.
[65] G. Hommel: *Handbuch der gefährlichen Güter* 1, Springer Verlag, Berlin-Heidelberg-New York 1980.
[66] F. Asinger: *Die petrolchemische Industrie,* vol. **II,** Akademie-Verlag, Berlin 1971, pp. 946–950 and pp. 953–960.
[67] Xantia National Corp., FR 2472161, 1981.
[68] Toray Industries Corp., JP 8137152, 1981; JP 8137153, 1981.
[69] Deutsche Solvay, Note CP 83017 (internal document), 1983.
[70] Comext Eurostat, Sept. 1983.
[71] *Beilstein,* **1** 436; **1** (1), 224; **1** (2), 474; **1** (3), 1873.
[72] F. Andreas, K. Gröbe: *Propylenchemie,* Akademie-Verlag, Berlin 1969.
[73] K. Weissermel, H.-J. Arpe: *Industrielle Organische Chemie,* Verlag Chemie, Weinheim-New York 1976.
[74] H. Warson: "Allyl alcohol and derivatives," *Polym. Paint Colour J.*1976, 698–699.
[75] R. K. Grasselli, J. D. Burrington, *Adv. Catal.* **30** (1981) 133–163.
[76] C. E. Schildknecht, *Encycl. Chem. Process. Des. 1977,* **2,** 460.
[77] American Cyanamid Co., GB 573757, 1945.
[78] Plastics Engineering Co., US 4168360, 1978 (G. F. D'Alelio).
[79] Osaka Soda Co., JP 7987788, 1977 (A. Suzui).
[80] F. H. Otey, C. A. Wilham, C. R. Russell, *Ind. Eng. Chem. Prod. Res. Dev.* **17** (1978) no. 2, 162–164.
[81] T. P. Forbeth, B. J. Gaffney, *Pet. Refiner* **34** (1955) no. 12, 160.
[82] Sumitomo Chemical Co., JP 8277636, 1980.
[83] Martinez de la Cuesta, E. Costa Novella, P. J. E. Rus Martinez, J. Torregrosa Anton, J. Augado Alonso. *An. Quim.* **76** (1980) no. 3, 380–386.
[84] K. Yamagishi, O. Kageyama, *Hydrocarbon Process.* **55**(1976) no. 11, 139–144.
[85] Shell Development Co., US 2042220, 1934 (H. P. A. Groll, H. W. de Jong).
[86] Bataafsche Petroleum Mij., FR 788921, 1935.

[87]   National Distillers and Chemical Corp., US 4051151, 1975 (J. H. Murib).
[88]   D. V. Sokol'skii, O. A. Tyurenkova, V. A. Dashevskii, E. I. Seliverstova, *Kinet. Katal.* **7** (1966) no. 6, 1032.
[89]   Olin Mathieson Chemical Corp., US 3037059, 1959 (D. W. Kaiser).
[90]   Chemische Fabrik Kalk, DE-OS 2526653, 1975 (H. Jenkner, O. Koenigstein).
[91]   Du Pont, US 3283013, 1962 (R. W. Rimmer).
[92]   Phillips Petroleum Co., US 3471579, 1967 (D. H. Kubicek).
[93]   H. Adkins, G. Krsek, *J. Am. Chem. Soc.* **71** (1949) 3051.
[94]   C. U. Pittman Jr., W. D. Honnick, *J. Org. Chem.* **45** (1980) no. 11, 2132.
[95]   M. Tamura, S. Kumano, *CEER Chem. Econ. Eng. Rev.* **12** (1980) no. 9, 32–35.
[96]   Stamicarbon B.V., EP-A 38609, 1980 (N. A. de Munck, J. J. F. Scholten).
[97]   Chevron Research Co., US 4189608, 1976 (V. P. Kurkov).
[98]   Olin Mathieson Chemical Corp., US 2986585, 1959 (W. I. Denton).
[99]   Chemische Werke Hüls, DE-AS 1271082, 1964 (W. Knepper, G. Hoeckele).
[100] Shell Oil Co., US 3209037, 1961 (R. W. Fourie, R. L. Maycock, G. H. Riesser).
[101] Wyandotte Chemical Corp., US 2479632, 1945 (L. H. Lundsted, E. C. Jacobs).
[102] Progil, DE-AS 1197077, 1961 (E. Charles, M. E. Degeorges, A. Thizy).
[103] Jefferson Chemical Co., US 3238264, 1961 (R. L. Rowten).
[104] Hoechst, DE-OS 1949537, 1969 (G. Roscher, H. Schmitz).
[105] H. J. Schmidt, G. Roscher, *Compend. Dtsch. Ges. Mineralölwiss. Kohlechem.* **75–76** (1975) 318–322.
[106] Bayer, DE-OS 1933538, 1969 (B. Engelhard, J. Grolig, M. Martin, K.-H. Reissinger, G. Scharfe, et al.).
[107] Bayer, DE-OS 2009742, 1970 (G. Scharfe, W. Schwerdtel, W. Swodenk, B. Engelhard, J. Grolig et al.).
[108] Degussa, DE 888691, 1952 (H. Brendlein).
[109] Celanese Corp., US 4127508, 1976 (T. H. Vanderspurt).
[110] Bataafsche Petroleum Mij., GB 619014, 1949 (S. A. Ballard, H. de V. Finch, E. A. Peterson).
[111] K. Yamagishi, *CEER Chem. Econ. Eng. Rev.* **6** (1974) no. 7, 40.
[112] IG-Farbenindustrie, FR 894673, 1943.
[113] Shell Development Co., US 2072015, 1932 (A. W.Tamele, H. P. A. Groll).
[114] Shell Development Co., US 2072016, 1932 (A. W. Tamele, H. P. A. Groll).
[115] Bayer, DE-OS 1939142, 1969 (W. Swodenk, G. Scharfe, J. Grolig).
[116] Bayer, DE-AS 1904236, 1969 (W. Swodenk, G. Scharfe, J. Grolig).
[117] Bayer, DE-OS 2121251, 1971 (J. Grolig, G. Scharfe, W. Swodenk).
[118] Hoechst, DE-OS 2219915, 1972 (H. Fernholz, D. Freudenberger).
[119] Bayer, DE-OS 2019428, 1970 (J. Grolig, G. Scharfe, W. Swodenk).
[120] Hoechst, DE-OS 1768242, 1968 (H. Fernholz, H. Wendt).
[121] Hoechst, DE-OS 2102263, 1971 (H. Krekeler, H. Fernholz, D. Freudenberger, H.-J. Schmidt, F. Wunder).
[122] Hoechst, DE-OS 2124755, 1971 (H. Fernholz, D. Freudenberger).
[123] Bayer, DE-OS 1960380, 1969 (P. Kurtz).
[124] General Electric Co., DE-OS 2425653, 1974 (W. E. Smith).
[125] Bayer, DE-AS 2430082, 1974 (C. Rasp, G. Scharfe, J. Grolig).
[126] J. C. Mol, E. F. G. Woerlee, *J. Chem. Soc. Chem. Commun.* 1979, 330.
[127] Bayer, FR 1346219, 1962 (H. Holzrichter, W. Krönig, B. Frenz).
[128] Bayer, DE-OS 1901289, 1969 (W. Krönig, G. Scharfe).

[129] Hoechst, DE-OS 1903954, 1969 (H. Fernholz, F. Wunder, H.-J. Schmidt).
[130] D. Swern, E. F. Jordan, *J. Am. Chem. Soc.* **70** (1948) 7.
[131] Bayer, DE-OS 1933537, 1969 (G. Scharfe, J. Grolig, W. Swodenk, M. Martin).
[132] Consortium, FR 1370867, 1963.
[133] Chem. Werke Witten, DE-AS 1211625, 1964.
[134] Du Pont, US 2218439, 1937 (H. S. Rothrock).
[135] Bayer, DE-OS 1933536, 1969 (W. Swodenk, G. Scharfe, J. Grolig).
[136] ICI, DE-OS 1568810, 1966 (D. K. V. Steel).
[137] FMC Corp., US 3250801, 1962 (H. Stange, W. B. Tuemmler).
[138] C. H. Fischer, C. Rehberg, L. Smith, *J. Am. Chem. Soc.* **65** (1943) 763 and 1003.
[139] Shell Development Co., US 2445627, 1944 (R. C. Morris, R. M. Horowitz).
[140] Hooker Chemical Corp., US 3136804, 1964 (J. J. Hodan, C. F. Baranaukas).
[141] Tokuyama Soda Co., JP 8102937, 1979.
[142] PPG, US 2379218, 1945 (W. R. Dial, C. Gould).
[143] PPG, US 4139578, 1975 (G. L. Baughman, H. C.Stevens).
[144] Deutsche Spezialglas, DE-OS 2938098, 1979 (H. Fricke, H. Schillert).
[145] Hoya Lens K. K., JP 8223901, 1980.
[146] A. M. Usmani, *J. Elastomers Plast.* **13** (1981) no. 3, 170–176.
[147] A. M. Usmani, I. O. Salyer, *J. Mater. Sci.* **16** (1981) no. 4, 915–926.
[148] North American Philips Corp., US 4350784, 1981 (K. E. Baum).
[149] Montecatini, IT 630731, 1961.
[150] Stauffer Chemical Co., US 4022849, 1971 (Jung Il Jin, P. Kraft).
[151] E. El-Alfy, M. I. Khalil, A. Hebeisch, *J. Polym. Sci. Polym. Chem. Ed.* **19** (1981) no. 12, 3137–3143.
[152] M. I. Khalil, F. I. Abdel-Hay, A. Hebeish, *Angew. Makromol. Chem.* 1982, 103 and 143.
[153] F. I. Abdel-Hay, M. I. Khalil, A. Hebeish, *J. Appl. Polym. Sci.* **27** (1982) no. 4, 1249–1258.
[154] J. K. Gillham, C. C. Mentzer, *Polym. Prepr. Am. Chem. Soc. Div. Polym. Chem.* **13** (1972) no. 1, 247.
[155] H. Boos, K. R. Hauschildt, *Angew. Makromol. Chem.* **25** (1972) 75.
[156] Ethyl Corp., US 3539488, 1970 (O. E. Klopfer, D. Hornbaker).
[157] Raychem., GB 1255493, 1971 (R. J. Penneck).
[158] G. Kerrutt, *Kautsch. Gummi Kunstst.* **24** (1971) 384.
[159] Hitachi Chem., JP 75124986, 1975 (F. Shoji et al.).
[160] E. N. Sotnikova et al., *Kozh. Oburvna Promst.* **13** (1971) no. 7, 55; *Chem. Abstr.* **75,** 141737.
[161] Yunichik, JP 7038735, 1970 (Y. Mori et al.).
[162] K. Nitzl, A. Dietl, DE-OS 1953121, 1971.
[163] General Electric Co., DE-OS 1815306, 1969 (F. F. Holub, M. M. Safford).
[164] Wacker-Chemie, FR 1546377, 1968.
[165] American Cyanamid Co., DE-OS 2119149, 1971 (N. P. Ermidis).
[166] General Electric Co., US 4199648, 1978 (H. L. Curry, W. L. Hall).
[167] K. Rauer, K. R. Adlassnig, J. Groepper, H. Hofmann, *Res. Discl.* 1981, no. 212, 436.
[168] Hoechst, DE-OS 1903954, 1969 (H. Fernholz, F. Wunder, H.-J. Schmidt).
[169] J. Pitha, S. Zawadzki, B. A. Hughes, *Macromol. Chem.* **183** (1982) no. 4, 781.
[170] Monsanto Co., US 4289864, 1980 (D. N. van Eenam).
[171] Esso Research and Eng. Co., FR 2071240, 1969.
[172] Elf France, BE 885670, 1979 (R. Leger, R. Nouguier, J. C. Fayard, P. Maldonado).
[173] Ciba-Geigy, EP-A 46731, 1980 (F. Lohse, C. E. Monnier).

[174] Bayer, DE-OS 2437789, 1974 (H. Haupt).
[175] J. D. Burrington, C. T. Kartisek, R. K. Grasselli, *J. Catal.* **75** (1982) no. 2, 225–232.
[176] D. Peric, A. T. Bell, M. Shen, *J. Appl. Polym. Sci.* **21** (1977) no. 10, 2661–2673.
[177] P. V. Hinman, A. T. Bell, M. Shen, *J. Appl. Polym. Sci.* **23** (1979) no. 12, 3651–3656.
[178] M. I. V. S. Florencio, *Rev. Port. Quim.* **22** (1980) no. 1–2, 48; *Chem. Abstr.* **96**, 55956.
[179] Texaco Dev. Co., DE-OS 2 750 250, 1976 (J. F. Knifton).
[180] C. Dauphin, L. David, B. Jamilloux, A. Kergomard, H. Veschambre, *Tetrahedron* **28** (1972) no. 4, 1055.
[181] M. F. Sorokin, E. L. Gershanova, SU 328124, 1969.
[182] Phillips Petroleum Co., US 3652456, 1969 (F. E. Naylor).
[183] A. Ageev, A. I. Ezvielev, E. S. Roskin, L. D. Mazo, SU 323409, 1966.
[184] Shell Development Co., US 2216548, 1938.
[185] M. Tamele, C. J. Ott, K. E. Marple, G. Hearne, *Ind. Eng. Chem.* **33** (1941), 115–120.
[186] *Org. Synth. Coll.* **2** (1943) 24.
[187] BASF, DE 845516, 1940 (W. Stade, W. Flemming).
[188] *Org. Synth. Coll.* **1** (1941) 201.
[189] Monsanto Co., US 4091019, 1976 (R. A. Keppel, J. S. McConaghy).
[190] Y. D. Smirnov, L. I. Saltikova, A. P. Tomilov, *Zh. Prikl. Khim. (Leningrad)* **43** (1970) no. 7, 1620.
[191] Shell Development Co., US 2456428, 1944 (J. H. Parker).
[192] J. M. Taylor et al., *Toxicol. Appl. Pharmacol.* **6** (1964) 378.
[193] L. Boquin, D. Shuwei, Y. Airu, X. Yinlin, G. Taibao, C. Tao, *Ecotoxicol. Environ. Saf.* **6** (1982) 19–27.
[194] T. Torkelson, M. A. Wolf, F. Oyen, V. K. Rowe, *Am. Ind. Hyg. Assoc. J.* **20** (1959) 217.
[195] H. Fengsheng, S. Dingguo, G. Yupu, L. Boquin, *Chin. Med. J. (Peking, Engl. Ed.)* **93** (1980) no. 3, 177.
[196] F. He, J. M. Jacobs, F. Scaravilli, *Acta Neuropathol.* **55** (1981) no. 2, 125.
[197] B. L. van Duuren: "Ethylene Dichloride, a Potential Health Risk," *Bambury Report*, vol. **5**, Cold Spring Harbor Laboratory, Cold Spring Harbor, New York 1980.
[198] B. L. van Duuren, B. M. Goldschmidt, G. Loewengart, A. C. Smith, S. Melchionne, I. Seldman, D. Roth, *JNCI J. Natl. Cancer Inst.* **63** (1979) no. 6, 1433.
[199] J. C. Theis, M. B. Shimkin, L. K. Poirier, *Cancer Res.* **39** (1979) 391.
[200] E. K. Weisburger, *EHP Environ. Health Perspect.* **21** (1977) 7.
[201] Carcinog. Tech. Rep. S – U.S., NCI-CG-TR-73, National Cancer Institute, USA 1978.
[202] L. Fishbein, *Sci. Total Environ.* **11** (1979) no. 2, 111.
[203] E. Eder, T. Neudecker, D. Lutz, D. Henschler, *Chem. Biol. Interact.* **38** (1982) 303.
[204] E. C. McCoy, L. Burrows, H. S. Rosenkranz, *Mutat. Res.* **57** (1978) 11.
[205] M. Bignami, G. Conti, L. Conti, R. Crebelli, F. Misuraca, *Chem. Biol. Interact.* **30** (1980) 9.
[206] S. M. Alizade, F. G. Guseinov, L. I. Denisko, *Azerb. Med. Zh.* **59** (1982) no. 2, 66.
[207] U.S. Department of Health, Education, and Welfare, HEW-Publication No. (NIOSH) 76-204.
[208] R. Kühn, K. Birett, *Merkblätter gefährliche Arbeitsstoffe*, 15th suppl., 12/81 A 47, Ecomed Verlagsgesellschaft, Landsberg 1981.
[209] Berufsgenossenschaft der chemischen Industrie: *Technische Regeln für gefährliche Arbeitsstoffe, MAK-Werte 1983*, UVV der BG-Chemie, Anlage 4, Jedermann Verlag, Heidelberg 1983.
[210] American Conference of Governmental Industrial Hygienists (ACGIH, ed.): *Threshold Limit Values 1982*, Cincinnati, Ohio 1982.

[211] Deutsche Forschungsgemeinschaft (ed.): *Maximale Arbeitsplatzkonzentrationen* (MAK) 1982, Verlag Chemie, Weinheim 1982.
[212] Deutsche Solvay, Note CMP-8049, V-1, 47 (internal document), 1976.
[213] M. Häusler, R. Lenich, *Arch. Toxicol.* **23** (1968) 209.
[214] M. K. Dunlap et al., *AMA Arch. Ind. Health* **18** (1958) 303.
[215] H. F. Smyth, *Am. Ind. Hyg. Assoc. J.* **17** (1956)129.
[216] Shell Chemical Corp., *Ind. Hyg. Bull. SC* 1957, 57–78.
[217] C. P. McCord, *JAMA J. Am. Med. Assoc.* **98** (1932) 2289.
[218] O. Strubelt et al., *Arch. Toxicol.* **22** (1967) 236.
[219] H. Oettel, *Hippokrates* **40** (1969) 285.
[220] American Conference of Governmental Industrial Hygienists, *Am. Ind. Hyg. Assoc. J.* **23** (1962) 419.
[221] P. G. Pagella, D. Faini, C. Turba, *Arzneim. Forsch.* **31** (1981) 1448.
[222] H. M. Maling, B. Highman, M. Williams, W. Saul, W. Butler, Jr., B. B. Brodie, *Toxicol. Appl. Pharmacol.* **27** (1974) 380.
[223] F. M. B. Carpanini, I. F. Gaunt, J. Hardy, S. D.Gangoli, K. R. Butterworth, *Toxicology* **9** (1978)29.
[224] F. A. Patty: *Industrial Hygiene and Toxicology,* vol. **2,** Wiley-Interscience, New York 1963, p. 1907.
[225] *Am. Ind. Hyg. Assoc. J.* **30** (1969) 470.
[226] C. H. Hine, J. K. Kodama, J. S. Wellington, M. K.Dunlap, H. H. Anderson, *AMA Arch. Ind. Health* **14** (1956) 250.
[227] C. H. Hine, V. K. Rowe: *Patty's Industrial Hygiene and Toxicology,* 2nd ed., vol. **2,** Wiley-Interscience, New York 1963, p. 1598.
[228] H. F. Smyth et al., *J. Ind. Hyg. Toxicol.* **31** (1949) 60.
[229] J. V. Marhold, in *Sborník výsledk toxikologickeho vyetení latex. pripravku,* Institut pro výchovu vedoucích pracovník Chemickeho prmyslu, Praha, Prague 1972, p. 38.
[230] C. J. Kirwin, E. E. Sandmeyer in G. D. Clayton, F. E. Clayton (eds.): *Patty's Industrial Hygiene and Toxicology,* 3rd ed., vol. **2 A,** Wiley-Interscience, New York 1981, p. 2515.
[231] C. H. Hine et al., *Arch. Environ. Health* **1** (1960) 343.
[232] R. J. Guzman, *Arch. Environ. Health.* **2** (1961)62.
[233] R. R. Beard, J. T. Noe in G. D. Clayton, F. E. Clayton (eds.): *Patty's Industrial Hygiene and Toxicology,* 3rd ed., vol. **2 B,** Wiley-Interscience, New York 1981, p. 3157.
[234] P. J. Boor, E. S. Reynolds, *Am. J. Pathol.* **86** (1977) 49 a.
[235] P. J. Boor, M. T. Moslem, E. S. Reynolds, *Toxicol. Appl. Pharmacol.* **50** (1979) 581.

Amines, Aliphatic

Individual keywords: → *Ethanolamines;* → *Propanolamines;* → *Methylamines*

GERD HEILEN, BASF Aktiengesellschaft, Ludwigshafen, Federal Republic of Germany (Chaps. 2–4)

HANS JOCHEN MERCKER, BASF Aktiengesellschaft, Ludwigshafen, Federal Republic of Germany (Chaps. 5, 6, and 8)

DIETER FRANK, Akzo Chemie America, Chicago, Ill, United States (Chaps. 7 and 9.2.2)

RICHARD A. RECK, Akzo Chemie America, Chicago, Ill, United States (Chaps. 7 and 9.2.2)

RUDOLF JÄCKH, BASF Aktiengesellschaft, Ludwigshafen, Federal Republic of Germany (Sections 9.1 and 9.2.1)

1.	Introduction	444	4.	Lower Alkylamines	453
2.	General Chemical Properties	444	4.1.	Physical Properties	453
2.1.	Salt Formation	445	4.2.	Storage and Transportation	453
2.2.	Conversion to Carboxamides	445	4.3.	Quality Specifications and Analysis	456
2.3.	Conversion to Sulfonamides	445	4.4.	Production	457
2.4.	Reaction with Carbonyl Compounds	445	4.5.	Uses	457
2.5.	Reaction with Carbon Dioxide and Carbon Disulfide	446	4.6.	Economic Aspects	459
2.6.	Reaction with Epoxides	446	5.	Cycloalkylamines	459
2.7.	Alkylation	447	6.	Cyclic Amines	464
2.8.	Reaction with Phosgene	447	7.	Fatty Amines	472
2.9.	Reaction with Acrylonitrile	447	7.1.	Properties	472
2.10.	Oxidation	447	7.2.	Production	475
2.11.	Dealkylation	448	7.3.	Analysis and Quality Control	477
3.	General Production Methods	448	7.4.	Uses	478
3.1.	Production from Alcohols	448	7.5.	Economic Aspects	480
3.2.	Production from Carbonyl Compounds	450	8.	Diamines and Polyamines	482
			8.1.	Diamines	482
			8.1.1.	1,2-Diaminoethane (Ethylenediamine)	482
3.3.	Production from Nitriles	450	8.1.2.	Diaminopropanes	484
3.4.	Production from Alkyl Halides	451	8.1.3.	Higher Diamines	486
3.5.	Production from Nitroalkanes	452	8.2.	Oligoamines and Polyamines	487
3.6.	Production from Olefins	452	8.2.1.	Physical Properties	487
3.7.	Other Processes	453	8.2.2.	Chemical Properties	488

8.2.3.	Production, Analysis, and Uses. .	489	9.2.	**Toxicology of Specific Amines**.	492
			9.2.1.	Alkylamines, Cyclic Amines, and	
9.	**Toxicology and Occupational**			Polyamines	492
	Health.	490	9.2.2.	Fatty Amines.	494
9.1.	**General Aspects**.	490	**10.**	**References**.	494

1. Introduction

Primary, secondary, and tertiary amines are distinguished on the basis of the number of hydrogen atoms in ammonia that have been replaced by organic groups. Substitution at the nitrogen atom by a fourth substituent gives quaternary ammonium compounds. The lower aliphatic amines are those with up to six carbon atoms per alkyl chain. Long-chain amines, with more than eight atoms per carbon chain, generally are known as fatty amines and are discussed in Chapter 7.

In 1849, Wurtz prepared methylamines and ethylamines by hydrolysis of the corresponding alkyl isocyanates, trialkyl cyanurates, and alkylureas [1]. However, Hofmann was the first to use the terms "primary," "secondary," and "tertiary" and to carry out fundamental work on the synthesis, properties, and structure of amines [2].

In general, naturally occurring amines are relatively complex compounds (e.g., alkaloids, vitamins, and amino acids). However, some lower alkylamines and diamines, such as tetramethylenediamine (putrescine) and pentamethylenediamine (cadaverine), are known as degradation or decomposition products of proteins.

With a current worldwide production of several 100 000 t/a, the aliphatic amines are among the most important organic intermediates in the chemical industry. The range of uses of these compounds is correspondingly wide. Major uses include those in the production of agrochemicals (in particular, herbicides), dyes, drugs, surfactants, and plastics; as auxiliaries for the rubber, textile, and paper industries; and as anticorrosion agents and process chemicals for gas scrubbing.

2. General Chemical Properties

The chemistry of the aliphatic amines is determined by the free electron pair on the nitrogen atom and by the tendency of the hydrogen atoms bonded to the nitrogen to be replaced by other substituents.

2.1. Salt Formation

Because they carry alkyl substituents, the aliphatic amines are stronger bases than ammonia (see Table 1); with acids, they form salts that are very soluble in water but insoluble in organic solvents. This property, combined with the difference in solubility between amine and salt, makes amines good acid acceptors and solvents for gas scrubbing and for certain extraction processes (e.g., in the synthesis of semisynthetic penicillins).

2.2. Conversion to Carboxamides

Amines react with carboxylic acids and their esters, chlorides, and anhydrides to give the corresponding substituted carboxamides:

$$R-\underset{\underset{O}{\|}}{C}-X + H_2N-R' \longrightarrow R-\underset{\underset{O}{\|}}{C}-NH-R' + HX$$

$$X = OH, OR, Cl$$

Yields of 90% or more are obtained, particularly with carboxylic acid chlorides or anhydrides. This type of reaction is utilized industrially, for example, in the synthesis of various herbicides with an acid amide structure.

2.3. Conversion to Sulfonamides

The reaction with benzenesulfonyl chloride is utilized for distinguishing among primary, secondary, and tertiary amines, as well as for their preparative separation (Hinsberg test). Whereas primary amines react forming alkali-soluble benzenesulfonamides, secondary amines give alkali-insoluble benzenesulfonamides, and tertiary amines do not react at all under these conditions.

$$RNH_2 + C_6H_5-SO_2Cl \xrightarrow{-HCl} C_6H_5-SO_2-NH-R \xrightarrow{NaOH}$$
$$C_6H_5-SO_2-NR^- + Na^+$$
$$R_2NH + C_6H_5-SO_2Cl \xrightarrow{-HCl} C_6H_5-SO_2-NR_2 \xrightarrow{NaOH} \!\!\!\!\!\!\not\rightarrow$$

2.4. Reaction with Carbonyl Compounds

Depending on the reaction conditions and the compound employed, carbonyl compounds react with amines to form Schiff bases (**1**) or enamines (**2**); these products can be hydrogenated to give more highly alkylated amines. This reaction is an important method for synthesizing higher amines.

$$RNH_2 + R'CH_2-\overset{O}{\underset{\|}{C}}-R'' \longrightarrow RN=C\overset{CH_2R'}{\underset{R''}{\diagdown}} + H_2O$$

$$\qquad\qquad\qquad\qquad\qquad\qquad\quad\mathbf{1}$$

$$\downarrow$$

$$\qquad\qquad\qquad\qquad\qquad RNH-C\overset{CHR'}{\underset{R''}{\diagdown}} + H_2O$$

R = alkyl
R', R'' = H, alkyl $\qquad\mathbf{2}$

2.5. Reaction with Carbon Dioxide and Carbon Disulfide

The carbamic acid or dithiocarbamic acid formed as a product in this reaction is unstable but can be obtained in a stable form as a salt or an ester.

$$RNH_2 + CX_2 \longrightarrow RNH-C\overset{X}{\underset{XH}{\diagdown}} \qquad X = O, S$$

Dithiocarbamates obtained from various amines play a key role as vulcanization accelerators in the rubber industry.

2.6. Reaction with Epoxides

Primary amines react with epoxides to give a mixture of mono- and dioxyalkylated derivatives, whereas secondary amines give monooxyalkylated compounds only, and tertiary amines give quaternary ammonium compounds.

$$R'R''NH + CH_2\underset{O}{-}CH_2 \longrightarrow R'R''N-CH_2-CH_2-OH$$

This oxyalkylation reaction is one of the most important industrial reactions of aliphatic amines and is utilized for the preparation of flocculents, surface coating resins, drug intermediates, and products for gas scrubbing (→ Ethanolamines and Propanolamines).

2.7. Alkylation

The reaction of amines with alkyl halides and dialkyl sulfates to give, ultimately, quaternary ammonium compounds is utilized in preparative drug chemistry and for the preparation of anticorrosion agents and biocides.

$$R^1-\underset{R^3}{N}-R^2 + R^4Cl \longrightarrow \left[R^1-\underset{R^4}{\overset{R^2}{N}}-R^3\right]^+ Cl^-$$

2.8. Reaction with Phosgene

The reaction of phosgene with primary amines leads first to a carbonyl chloride; subsequent cleavage of hydrogen chloride gives the alkyl isocyanate:

$$RNH_2 + COCl_2 \longrightarrow RNH-\underset{O}{\overset{\|}{C}}-Cl + HCl$$

$$RNH-\underset{O}{\overset{\|}{C}}-Cl \longrightarrow RNCO + HCl$$

This reaction is important in the preparation of various herbicides with urea, carbamate, or thiocarbamate structures and, particularly in the case of polyfunctional amines, in polyurethane chemistry.

The reaction of secondary amines with phosgene proceeds via an analogous indermediate to form N,N'-tetraalkylureas.

2.9. Reaction with Acrylonitrile

The addition of a primary or secondary amine to acrylonitrile to form an aminopropionitrile is utilized industrially on a large scale for the preparation of higher diamines and polyamines, because the nitrile can be readily hydrogenated to the amine.

$$RNH_2 + CH_2=CH-CN \longrightarrow RNH-CH_2-CH_2-CN$$

2.10. Oxidation

In contrast to their salts, the free amines are sensitive to oxidation, giving various products, depending on the oxidizing agent and the type of amine employed. Tertiary amines are oxidized by *hydrogen peroxide* to amine oxides, whereas the corresponding compounds formed from primary and secondary amines undergo further reaction to give the corresponding hydroxylamines or aldoximes:

$$R_3N \xrightarrow{H_2O_2} R_3N \rightarrow O$$

$$RCH_2NH_2 \xrightarrow{H_2O_2} RCH_2NH_2O \longrightarrow RCH_2NHOH$$
$$\longrightarrow RCH=NOH$$

Oxidation with *nitrous acid* can be used to distinguish among primary, secondary, and tertiary amines. Primary amines give alcohols with elimination of water und evolution of nitrogen. Secondary amines react to give *N*-nitrosoamines. Normally, tertiary amines do not react.

$$RCH_2NH_2 + HNO_2 \longrightarrow RCH_2OH + H_2O + N_2$$
$$(RCH_2)_2NH + HNO_2 \longrightarrow (RCH_2)_2N-N=O + H_2O$$

2.11. Dealkylation

Tertiary amines can be dealkylated by heating the quaternary ammonium salts; methyl groups are eliminated as methanol, all higher alkyl groups as alkenes, e.g.,

This reaction is utilized for the synthesis of amines and is an important degradation reaction in elucidating the structure of unknown amines.

3. General Production Methods

Depending on the availability of the starting materials, the most diverse processes have become accepted for the large-scale industrial production of aliphatic amines.

3.1. Production from Alcohols

The reaction of the appropriate alcohol with ammonia over a suitable catalyst is now the most common process for the preparation of lower alkylamines. In this reaction, the product is always a mixture of primary, secondary, and tertiary amines because the primary amine formed initially can react further with one or two molecules of alcohol. The product distribution can be controlled to a certain extent via the reaction conditions (temperature, excess of ammonia). Because the mixture of products usually obtained does not correspond to the requirements of the market, amines that are not marketable can be recycled. Thus, the yield of the preferred amine can be increased to above 90%.

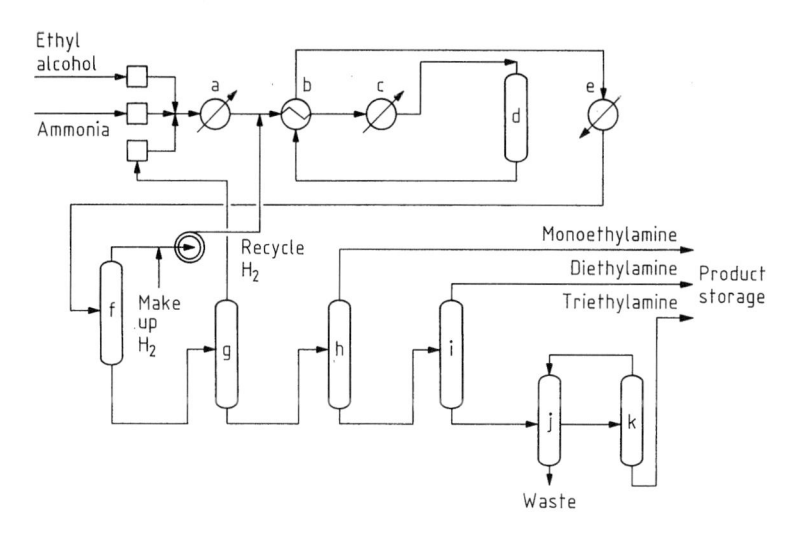

Figure 1. Continuous process for the production of ethylamines [5]
a) Vaporizer; b) Heat exchanger; c) Superheater; d) Catalytic converter; e) Product cooler; f) Gas separator; g) Ammonia column; h) Monoethylamine column; i) Diethylamine column; j) Decanter; k) Triethylamine column

Previously, this reaction was carried out using pure dehydrogenation catalysts (e.g., aluminum oxide, thorium oxide, tungsten oxide, chromium oxides, or various mixed oxides [3]); however, use of these catalysts is now restricted to the production of methylamines. In contrast, for the conversion of an alcohol containing two or more carbon atoms, catalysts possessing hydrogenating and dehydrogenating properties have become important. The catalysts used are mainly those based on nickel, cobalt, iron, or copper, and to a lesser extent those based on platinum or palladium. In this process, the alcohol, ammonia, and hydrogen are passed continuously over the catalyst in a fixed-bed reactor. The reaction takes place at about 0.5–20 MPa (5–200 bar) and about 100–250 °C, depending on the catalyst and on whether the liquid-phase or the gas-phase process is used. A twofold to eightfold excess of ammonia is used; hydrogen is not required as a direct reactant but is used to maintain the activity of the catalyst. The various versions of the process have been reviewed in detail [4]. A simplified flow diagram of the process for the synthesis of ethylamines and for working up the products is shown in Figure 1.

The same method is also useful for alkylating a primary or secondary amine instead of ammonia. When an amine and an alcohol having different aliphatic substituents react, the product is a mixed aliphatic amine. For example, *N,N*-dimethylethylamine is obtainable from dimethylamine and ethanol. To avoid transalkylation at the nitrogen atom in reactions of this type, copper catalysts are particularly recommended [6].

The ammonolysis of alcohols is a relatively complex process because various equilibrium reactions occur concurrently. Whereas the earlier literature postulates nitriles as intermediates in the reaction [7], more recent investigations indicate that the reaction proceeds via a carbonyl compound. The intermediate reacts with ammonia to form an azomethine, which then is hydrogenated to give the amine [8].

3.2. Production from Carbonyl Compounds

Aldehydes and ketones are used in preference to alcohols for amine synthesis if this is more economical.

Generally, this is the case only for the lower aldehydes (obtained from the oxo synthesis) and acetone obtained as a byproduct in the production of phenol.

The reaction between a carbonyl compound and ammonia or an amine occurs in two stages: the azomethine or Schiff base formed initially is hydrogenated in a second stage to give the amine. Other methods of carrying out this hydrogenation (for example, using formic acid according to the Leuckart-Wallach procedure) are of little industrial importance, so only the catalytic hydrogenation is discussed here. The usual procedure is similar to that for the conversion of alcohols; the reaction mixture, comprising the carbonyl compound, ammonia, and hydrogen, is passed over a fixed-bed catalyst.

In some cases, it can be advantageous to carry out the reaction in two stages. The carbonyl compound and ammonia or the amine react first, the water of reaction is removed, and only then is the hydrogenation carried out [9].

The essential difference between the amination of an aldehyde or ketone and that of an alcohol is that in the former hydrogen is the reactant and is consumed in a stoichiometric amount. The substantially higher heat of reaction of this process (e.g., 60.4 kJ/mol for acetone [10] versus 7.1 kJ/mol for isopropyl alcohol) necessitates a fundamentally different reactor design. In general, the process is carried out in the vapor phase at 100–160 °C and atmospheric or slightly above atmospheric pressure.

Generally, the same catalysts as for the amination of alcohols can be used; control of product distribution by adjusting the amount of excess ammonia and the workup by distillation also are much the same. Comprehensive reviews of industrial and preparative possibilities for synthesis of amines from carbonyl compounds are available [4], [11].

3.3. Production from Nitriles

Frequently, nitriles are obtainable more economically than the corresponding alcohols or carbonyl compounds and are synthesized by ammoxidation, the addition of an amine or an alcohol to acrylonitrile, or the conversion of the corresponding carboxylic acid. Nitriles are converted into the corresponding amines, preferably by catalytic hydrogenation:

$$\text{RCN} + \text{H}_2 \rightleftharpoons \text{RCH=NH} \xrightarrow{\text{H}_2} \text{RCH}_2\text{NH}_2$$

$$\Big\updownarrow \text{RCH}_2\text{NH}_2$$

$$\text{RCH=N-CH}_2\text{R} \xrightarrow{\text{H}_2} \text{RCH}_2\text{-NH-CH}_2\text{R}$$

$$\Big\updownarrow \text{RCH}_2\text{NH}_2$$

$$\underset{\underset{\text{CH}_2\text{R}}{|}}{\text{RCH}_2\text{-N-CH}_2\text{R}}$$

Noble metal (palladium, platinum, rhodium), nickel, or cobalt catalysts generally are employed, but recently, iron catalysts have been used to an increasing extent [12], largely for economic reasons. Whereas the noble metals permit the use of relatively mild reaction conditions (20–100 °C, 0.1–0.5 MPa (1–5 bar)), pressures as high as 25 MPa (250 bar) and temperatures up to 150 °C are required if nickel or cobalt catalysts are used. There are various versions of the process, but those most often used are the batch procedure using a suspended catalyst and the continuous procedure using a fixed-bed catalyst.

Side reactions resulting in the formation of secondary amines always occur, but these can be suppressed by the addition of ammonia, sodium hydroxide solution, or an acid [4]. Alternatively process conditions can be adjusted to make the formation of secondary or tertiary amines the principal reaction. This alternative is of interest, for example, in the synthesis of di- and triethylamines from acetonitrile [13].

The hydrogenation of acrylonitrile is a special case. Whereas saturated amines are normally the only products of this reaction, the use of copper chromite as a catalyst instead of noble metals, nickel, or cobalt leads to allylamine and diallylamine [14].

3.4. Production from Alkyl Halides

The reaction of an alkyl halide with ammonia or an amine gives an alkylammonium halide, from which the amine can be liberated using caustic alkali solution. Although a standard method of preparative chemistry, this route is industrially important only for the preparation of ethylenediamine, the homologous polyamines, and a few special amines such as allylamine and some small-volume pharmaceuticals. The reasons for this restriction in use are, first, the lack of cheap starting materials and, second, problems relating to corrosion and product quality that arise during the processing of halides.

3.5. Production from Nitroalkanes

Whereas the reduction of nitro compounds can be regarded as one of the most important methods for the synthesis of aromatic amines, this route has not become important for the production of aliphatic amines. Because the availability of appropriate nitroalkanes is restricted, this method is used only in a few special cases, e.g., the production of 2-amino-1-butanol, the precursor for the antituberculotic ethambutol.

The reaction proceeds via the nitroso or hydroxylamine intermediate, in some cases even at room temperature, and gives the amine in yields of 90% or more. Platinum, palladium, nickel, or copper is used as the hydrogenation catalyst [4]. The reaction is highly exothermic (~500 kJ/mol per nitro group), and therefore heat must be removed rapidly to keep the process under control.

3.6. Production from Olefins

Amines having a tertiary alkyl group adjacent to the nitrogen atom (e.g., *tert*-butylamine) are difficult to obtain by conventional synthetic methods. These compounds can be prepared readily by addition of hydrogen cyanide to an olefin, e.g., 2-methylpropene, in an acidic medium:

$$(CH_3)_2C=CH_2 + HCN + H_2O \xrightarrow{H^+} (CH_3)_3C-NH-CHO$$

$$\xrightarrow[OH^-]{H_2O} (CH_3)_3C-NH_2 + HCOOH$$

This process, known as the *Ritter reaction*, is carried out at 30 – 60 °C, and the resulting formamide intermediate is hydrolyzed when the reaction mixture is heated to ca. 100 °C. Because hydrocyanic acid is difficult to handle and causes corrosion problems, this process is restricted to those amines that cannot be obtained by one of the methods described above (e.g., *tert*-butylamine and a few amines of pharmaceutical interest).

Although advantageous thermodynamically, the direct addition of ammonia or an amine to an olefinic double bond is restricted to a few cases in which either an activated amine or a compound with an activated double bond is employed, for example, the addition of an amine to acrylonitrile. The patent literature, however, discloses a process in which, for example, ethylene reacts with ammonia over a zeolite to give ethylamines with good selectivity although with only low conversion rates [15].

3.7. Other Processes

The hydrogenation of aromatic amines to cycloalkylamines is restricted to the production of cyclohexylamines (see Chap. 5). A process for synthesizing amines from carbon monoxide, hydrogen, and ammonia or amines over an activated iron catalyst by a Fischer-Tropsch-type method was described [16], but this is not used on an industrial scale. All other synthetic routes described in the literature are essentially laboratory methods that are suitable for the preparation of small amounts and are carried out on an industrial scale only in isolated cases. For details see [3].

4. Lower Alkylamines

4.1. Physical Properties

The three methylamines and ethylamine are gaseous at room temperature; diethylamine, triethylamine, and higher amines up to about twelve carbon atoms per alkyl chain are liquid, and long-chain amines containing still higher alkyl groups are solid.

The short-chain amines are readily soluble in water, alcohol, ether, and conventional organic solvents. Amines containing more than five carbon atoms are only partially or very sparingly soluble in water. The solubility in water generally decreases with increasing temperature, and in some cases the miscibility gap vanishes at lower temperatures. For example, triethylamine is completely miscible with water below about 18 °C but is only partially miscible with water above this temperature.

All alkylamines, but particularly the lower members, which have high vapor pressures, have a characteristic ammonia odor. The odor decreases with increasing substitution.

Physical properties of importance for characterizing and handling the alkylamines are summarized in Table 1. For further data, reference may be made to the relevant tables and the producers data sheets.

4.2. Storage and Transportation

The aliphatic amines are stored usually in carbon steel or stainless steel containers; relatively small amounts can be kept in glass or ceramic vessels. Copper, aluminum, zinc, and their alloys are not resistant to amines. All three methylamines and ethylamine must be stored under pressure because they are vapors at room temperature. Isopropylamine (*bp* 32.4 °C) usually is stored under refrigeration or in pressurized containers. To facilitate storage and transport, these amines are produced and distributed also in the form of aqueous solutions (e.g., 50% or 70%). The aliphatic amines

Table 1. Physical properties of commercial alkylamines ($C \geq 2$)

Compound	CAS registry number	Formula	M_r	mp, °C	bp, °C	d_4^{20}	n_D^{20}	Flash point, °C	pk_b (25 °C)
Ethylamine	[75-04-7]	$C_2H_5NH_2$	45.09	−80.6	16.6	0.6829	1.3663	−52	3.25
Diethylamine	[109-89-7]	$(C_2H_5)_2NH$	73.14	−50	56.3	0.7056	1.3864	−23	2.88
Triethylamine	[121-44-8]	$(C_2H_5)_3N$	101.19	−115	89.3	0.7275	1.4010	−11	3.24
Propylamine	[107-10-8]	$C_3H_7NH_2$	59.11	−83	47.8	0.7173	1.3870	−30	3.41
Dipropylamine	[142-84-7]	$(C_3H_7)_2NH$	101.19	−63	109.2	0.7400	1.4050	7	3.09
Tripropylamine	[102-69-2]	$(C_3H_7)_3N$	143.26	−93.5	156	0.7558	1.4171	36	3.35
Isopropylamine	[75-31-0]	$(CH_3)_2CHNH_2$	59.11	−95.2	32.4	0.6886	1.3742	−37	3.37
Diisopropylamine	[108-18-9]	$[(CH_3)_2CH]_2NH$	101.19	−61	84	0.7169	1.3924	−17	3.43 (20 °C)
Butylamine	[109-73-9]	$C_4H_9NH_2$	73.14	−49.1	77.8	0.7414	1.4031	−8	3.39
Dibutylamine	[111-92-2]	$(C_4H_9)_2NH$	129.25	−60	159	0.7670	1.4177	39	3.04
Tributylamine	[102-82-9]	$(C_4H_9)_3N$	185.36	−70	213	0.7771	1.4291	70	3.11
Isobutylamine	[78-81-9]	$(CH_3)_2CHCH_2NH_2$	73.14	−85.5	68	0.7360	1.3988	−16	3.59
Diisobutylamine	[110-96-3]	$[(CH_3)_2CHCH_2]_2NH$	129.25	−73.5	139	0.7450	1.4090	25.5	3.18
Trisobutylamine	[1116-40-1]	$(CH_3)_2CHCH_2]_3N$	185.36	−21.8	191.5	0.7684	1.4252		3.68
1-Methylpropylamine (D,L)	[13952-84-6]	$C_2H_5CH(CH_3)NH_2$	73.14	−10.4	63.5	0.7246	1.3932	−20	3.43
Bis(1-methyl)propylamine (D,L)		$[C_2H_5CH(CH_3)]_2NH$	129.25	−70	135	0.7534	1.4111	21	
1,1-Dimethylethylamine	[75-64-9]	$(CH_3)_3CNH_2$	73.14	−67.5	44.4	0.6958	1.3784		3.39
Pentylamine	[110-58-7]	$C_5H_{11}NH_2$	87.17	−55	104.4	0.7547	1.4118	7	3.36
Dipentylamine	[2050-92-2]	$(C_5H_{11})_2NH$	157.30	−32	202	0.7771	1.4272	71	2.82
Tripentylamine	[621-77-2]	$(C_5H_{11})_3N$	227.44		240−5	0.7907	1.4366	88	
1-Methylbutylamine	[625-30-9]	$C_3H_7CH(CH_3)NH_2$	87.17		91.5	0.7384	1.4027		
3-Methylbutylamine	[107-85-7]	$(CH_3)_2CH(CH_2)_2NH_2$	87.17	<−60	95	0.7505	1.4083	4	3.98
Bis(3-methylbutyl)amine	[544-00-3]	$[(CH_3)_2CH(CH_2)_2]_2NH$	157.30	−44	188	0.7669	1.4235		
Tris(3-methylbutyl)amine	[645-41-0]	$[(CH_3)_2CH(CH_2)_2]_3N$	227.44		235	0.7848	1.4331		
Hexylamine	[111-26-2]	$C_6H_{13}NH_2$	101.19	−19	130	0.7660	1.4180	34	3.44

Table 1. (continued)

Compound	CAS registry number	Formula	M_r	mp, °C	bp, °C	d_4^{20}	n_D^{20}	Flash point, °C	pK_b (25 °C)
Octylamine	[111-86-4]	$C_8H_{17}NH_2$	129.25	0	179.6	0.7826	1.4294	58	3.35
2-Ethylhexylamine	[104-75-6]	$C_4H_9CH(C_2H_5)CH_2NH_2$	129.25	<−70	169.2	0.7894	1.4313	53	
Decylamine	[2016-57-1]	$C_{10}H_{21}NH_2$	157.30	17	220.5	0.7936	1.4369		3.36
N-Methylbutylamine	[110-68-9]	$C_4H_9NHCH_3$	87.17	−75	91	0.7341	1.4010	0	
N-Ethylbutylamine	[13360-63-9]	$C_4H_9NHC_2H_5$	101.19	−70	108−9	0.7398	1.4040	9	
N,N-Dimethylethylamine	[598-56-1]	$C_2H_5N(CH_3)_2$	73.14	<−70	36	0.6754		−45	
Allylamine	[107-11-9]	$CH_2=CH-CH_2-NH_2$	57.09	−88	53.2	0.7621	1.4205	−20	4.28
Diallylamine	[124-02-7]	$(CH_2=CH-CH_2)_2NH$	97.16	−88	111	0.7874	1.4387	21	4.87
N,N-Dimethyl-2-chloroethylamine	[107-99-3]	$ClCH_2CH_2N(CH_3)_2$	107.58		110		1.4286		
N,N-Diethyl-2-chloroethylamine	[100-35-6]	$ClCH_2CH_2N(C_2H_5)_2$	135.64		146−7	0.9166	1.4358		
3-Methoxypropylamine	[5332-73-0]	$CH_3O(CH_2)_3NH_2$	89.14		116−9	0.8727	1.4191		

have a virtually unlimited shelf life, but they should be stored under nitrogen to avoid contact with carbon dioxide (resulting in formation of carbonates) and atmospheric moisture.

All of the amines discussed here are inflammable. Some of them have very low flash points (see Table 1), and they form explosive mixtures with air. Hence, when these compounds are stored and transported, the relevant statutory regulations for inflammable liquids or pressurized gases must be observed. Because these vary greatly from country to country, they cannot be discussed in detail here.

To prevent odor nuisance (odor thresholds are below 0.1 ppm in some cases) and impermissible emission, special precautions must be taken during storage and, in particular, during transfer from one container to another. Where the process cannot be carried out in a closed system and relatively large amounts are being handled, absorption plants for washing waste gas must be provided. Scrubbing with water is frequently adequate, but better treatment of waste gas is achieved by scrubbing with an acid. For smaller plants, adsorption on active carbon filters is advisable under certain circumstances. This method has the additional advantage that the adsorbed amines can be recovered.

The toxicity (Chap. 9) of these amines also makes it necessary to take certain precautions with regard to occupational safety. These include, in particular, avoiding inhalation of amine vapors by using respiratory equipment and reliable prevention of skin and eye contact with fluid amines by wearing protective clothing.

4.3. Quality Specifications and Analysis

In the case of individual amines, the quality specifications are very high. Particularly for amines with a wide range of uses (e.g., ethylamine), specifications require a purity of above 99.7%, although usually 98.0–99.5% purity is considered adequate. Some end uses impose additional maximum values on the content of water, ammonia, and unsaturated amines or, in the case of tertiary amines, on the content of primary and secondary amines.

The amine content is determined by acid titration of the basic nitrogen, and the purity is determined preferably by gas chromatography. Also of importance is the determination of the water content by the Karl Fischer method, titration of the imine content by oximation, and the determination of physical properties.

4.4. Production

With few exceptions, aliphatic amines of industrial importance are produced by any one of the methods described in Section 2.1 or 2.2, i.e., by the reaction of the corresponding alcohol or carbonyl compound with another amine or with ammonia. These two alternatives are about equal with regard to technical feasibility, production costs, and product quality. The choice of process depends mainly on the following criteria:

1) Price of the starting materials
2) Availability of the starting materials
3) Conversion of any existing plants to allow for different pressures in the two processes
4) Availability of hydrogen (hydrogen consumption and pressure differ in the two processes)
5) Fiscal reasons (taxes or subsidies on particular materials, e.g., ethanol)

Because these factors vary greatly from country to country — in some cases even from region to region — and are changing continually, the reasons for the choice of a particular process from the two mentioned depend strongly on individual circumstances. Today, the situation is such that production in the United States is based almost exclusively on alcohols, whereas in Western Europe ethylamines are produced from ethanol and about equal amounts of the remaining aliphatic amines are produced by each of the two processes. In contrast, the emphasis in Japan is on carbonyl compounds as starting materials.

The aliphatic amines of large-scale industrial importance (mainly amines with two to four carbon atoms) can be produced in one production plant under very similar reaction conditions, and only in the workup are there small differences. For this reason, many plants are equipped to be multipurpose plants that can be used for the production of various amines.

Other production processes have become important only for a few amines. Examples are the production of *tert*-butylamine from isobutene and hydrogen cyanide (Section 3.6), the production of allylamines from allyl chloride (Section 3.4), the reaction of a dialkylethanolamine with thionyl chloride to give a dialkylchloroethylamine [17], and the hydrogenation of 3methoxypropionitrile to give 3-methoxypropylamine (Section 3.3).

4.5. Uses

Ethylamines. Of the amines discussed here, the ethylamines are by far the most important, accounting for about 35–40% of the annual world demand for alkylamines (excluding methylamines).

The principal use of *monoethylamine* is in the production of herbicides of the triazine type by reaction with cyanuric chloride. The most important commercial products are atrazine, ametryne, cyanazine, and simazine.

Diethylamine is used principally for the production of vulcanization accelerators. Reaction with carbon disulfide gives a dithiocarbamate that can be oxidized to a thiuram disulfide. The most important commercial products are zinc diethyldithiocarbamate and tetraethylthiuram disulfide.

Of the three ethylamines, *triethylamine* has the widest range of uses and is also the most expensive product (owing to the unfavorable distribution of monoethylamine, diethylamine, and triethylamine in the production). Much of the triethylamine is used as an organic acid acceptor in the most diverse syntheses, or as a salt former in precipitation or purification operations. Important examples of these uses are in the synthesis of semisynthetic penicillins and cephalosporins, and as solubilizers for 2,4-dichlorophenoxyacetic acid and 2,4,5-trichlorophenoxyacetic acid. Other important fields of use include polyurethane catalysts (for example, the hardening of core sands, i.e., cold-box casting), anticorrosion agents, textile and photographic auxiliaries, and anodic electrocoating.

Propylamine. The commercial demand for propylamines and isopropylamines is together about the same as that for the ethylamines. *Monoisopropylamine* is the most important product and is used particularly in the production of agrochemicals. *DipropylamineDipropylamine* is required principally as a starting material for the synthesis of herbicides. The remaining propylamines are of minor importance compared to monoisopropylamine and dipropylamine.

Butylamines. The most important of these is *diisobutylamine*, which is the principal starting material for the herbicide butylate. Although the demand for diisobutylamine is almost entirely restricted to this use, this product is among the most important of all the aliphatic amines, at least on the United States market. The second most important amine with four carbon atoms is *tert-butylamine*, which is used principally for the synthesis of vulcanization accelerators of the sulfenamide type. This application also constitutes the important outlet for *dibutylamine* (as dithiocarbamates or thiuram); the demand for this amine for other purposes (e.g., corrosion protection, flotation agents, cutting oils, and intermediates for insecticides and pharmaceuticals) is less important. *Monobutylamine* is an intermediate for the production of plasticizers, agrochemicals, and drugs (e.g., the antidiabetic tolbutamide). *Tributylamine* is also used commercially (a few 100 t/a) as a polymerization catalyst and as an acid acceptor.

Other Amines. Of the amines with more than four carbon atoms, only *octylamine* and *2-ethylhexylamine* are of importance. Octylamine is required, for example, for the production of the vasodilator suloctidil, whereas 2-ethylhexylamine is employed in the United States in anticorrosion formulations for engine oils and as a synergistic agent for pyrethrins. *Diallylamine* is the only unsaturated amine with a significant market; its principal uses are in crop protection and as an auxiliary in paper making. Although *chloroalkylamines* (mainly chloroethyl- or chloropropyl-*N*,*N*-dialkylamines) are not important in terms of the amounts employed, they are important starting materials for the production of a number of drugs (principally neuroleptics).

4.6. Economic Aspects

Worldwide production capacity for aliphatic amines (excluding methylamines) is estimated to be almost 400 000 t/a, of which about 40 % is attributable to the United States and about 40 % to Western Europe, the rest being distributed between Japan, the Eastern bloc, and other countries.

Because of the very large number of producers, exact figures for current annual worldwide production are not available. However, it may be assumed that the production capacity substantially exceeds demand, i.e., some production plants are not operating at full capacity.

Apart from those who are producing mainly for their own requirements, the most important producers are BASF, PCUK, Ruhrchemie, and ICI in Western Europe; Air Products, Union Carbide, Virginia, and Pennwalt in the United States; and Daicel in Japan. Important producers of special amines are Shell (allylamines) and Monsanto (*tert*-butylamine).

5. Cycloalkylamines

Cyclohexylamine [*108-91-8*], aminocyclohexane, $C_6H_{13}N$, M_r 99.18, is a colorless liquid, *mp* –17.8 °C, *bp* 134.5 °C (at 101.3 kPa), d_4^{25} 0.8647, n_D^{20} 1.4592.

$$\begin{array}{c} CH_2-CH_2 \\ CH_2 \qquad CH-NH_2 \\ CH_2-CH_2 \end{array}$$

Viscosity				
t, °C	0	20	50	75
η, Pa · s	3.73	2.10	1.14	0.77
Vapor pressure				
t, °C	20	60	100	120
p, kPa	1.43	8.66	35.99	65.31
Specific heat capacity				
t, °C		20	70	145
c, J g^{-1}K^{-1}		2.366	2.583	2.910

Heat of vaporization	399.86 J/g
Flash point (closed cups)	26.5 °C
Ignition temperature (class T3)	265 °C
Ignition range in air	1.6 – 9.4 vol%

Cyclohexylamine is infinitely miscible with water and the conventional organic solvents. With water it forms an azeotrope that contains 44.2% cyclohexylamine and boils at 96.4 °C.

Chemical Properties. Chemically cyclohexylamine has much in common with the acyclic aliphatic amines [3], [18]. Cyclohexylamine reacts with chlorine to form *N,N*-dichlorocyclohexylamine [19]. For conversion to azomethines and their chemistry, see [20]. *N*-Cyclohexylidenecyclohexylamine reacts with chloramine to give 1-cyclohexyl-3,3-pentamethylenediaziridine, which can be hydrolyzed to give cyclohexylhydrazine [21]. Cyclohexylamine and formaldehyde together react with peracetic acid to give 2-cyclohexyloxaziridine [22].

In addition to using alkyl halides, alkyl sulfates, or alkyl phosphates, cyclohexylamine can be alkylated with an alcohol in the presence of a catalyst, such as aluminum oxide, copper, nickel, cobalt, or platinum (reviewed in [23], [24]), or by the Leuckart-Wallach method [25].

Production. Cyclohexylamine has long been produced by catalytic hydrogenation of aniline under pressure. High yields are obtained using a nickel or cobalt catalyst treated with basic oxides [26]. Raney cobalt treated with calcium oxide and sodium carbonate gives a cyclohexylamine yield of more than 96% at 6 MPa (60 bar) and 230 °C [27]. Noble metals also are employed as catalysts [28].

The synthesis of cyclohexylamine from cyclohexanol and ammonia over supported cobalt catalysts [29] also is a very important method; calcium silicoaluminates also are suitable catalysts [30]. In this process, cyclohexanol reacts at 20 MPa (200 bar) and 220 °C with at least 3 mol of ammonia in the presence of circulating hydrogen over a fixed-bed catalyst. Using the same continuous procedure, cyclohexanone can be hydrogenated with ammonia under amination conditions [31]; nickel or cobalt catalysts are used for this process, which is carried out at 0.1 – 20 MPa (1 – 200 bar). Cyclohexylamine can also be obtained by reacting phenol with hydrogen and ammonia over a rhodium catalyst [32].

Quality. The purity of the commercial product is above 99.5% and is determined by gas chromatography; contaminants are ammonia and water.

Storage and Transportation. Cyclohexylamine can be stored and shipped in iron tanks. Nonferrous metals, particularly copper-containing materials, are attacked and are therefore unsuitable. The amine discolors on contact with air and therefore must be kept under nitrogen.

Uses. Cyclohexylamine is used primarily as a salt-forming and amide-forming component in many applications, for example, in vulcanization accelerators (benzothiazole-2-sulfonic acid cyclohexylamide [33]), plasticizers (salts with dodecanethiol and mercaptobenzothiazole [34]), and emulsifiers (emulsification of active ingredients in water with cyclohexylammonium alkylbenzenesulfonate [35]). Salts of fatty acids containing 10 to 14 carbon atoms prevent foam formation in mineral oils [36].

Cyclohexylamine is used also as a hardener for epoxy resins and as a catalyst for polyurethanes. Alone or mixed with other compounds, it has an anticorrosive action [37], for example, when used as an additive to heating oil or in the operation of steam boilers.

Sodium cyclohexylsulfamate and calcium cyclohexylsulfamate are important sweeteners.

N-Methylcyclohexylamine [100-60-7], $C_7H_{15}N$, M_r 113.2, is a colorless liquid, mp – 9 °C, bp 150 °C (at 101.3 kPa), d_4^{25} 0.8533, n_D^{25} 1.4530, η 5 mPa · s (at 25 °C).

Vapor pressure

t, °C	20	40	60	80	100	120	140
p, kPa	0.48	1.72	4.68	11.48	24.13	46.26	81.31

Flash point	36.1 °C
Ignition temperature (class T3)	255 °C
Explosion limits in air	2.2 and 10.5 vol% (corresponding to 103 and 494 g/m^3)
Specific heat capacity	2.14 J g^{-1} K^{-1} (at 20 °C)
	2.47 J g^{-1} K^{-1} (at 140 °C)

N-Methylcyclohexylamine is infinitely miscible with water, methanol, acetone, toluene, and cyclohexane.

Production. N-Methylcyclohexylamine can be prepared by a procedure similar to that used for cyclohexylamine, i.e., by hydrogenation of methylaniline over a supported nickel catalyst [38] or from cyclohexanone and methylamine under hydrogenation conditions [39]. Cyclohexylamine reacts with methanol over copper, zinc, or copper–calcium catalysts [40].

Uses. N-Methylcyclohexylamine is used as a component of vulcanization accelerators [41].

N,N-Dimethylcyclohexylamine [98-94-2], $C_8H_{17}N$, M_r 127.2, is a colorless liquid, mp <−50 °C, bp 159 °C (at 101.3 kPa), d_4^{25} 0.8467, n_D^{25} 1.4522, η 3 mPa · s (at 25 °C).

Vapor pressure

t, °C	20	40	60	80	100	120	140
p, kPa	0.29	1.05	3.21	8.29	18.13	35.99	63.72

Flash point	38.1 °C
Ignition temperature (class T4)	200 °C
Explosion limits in air	0.79 and 7.0 vol% (corresponding to 41.8 and 370 g/m^3)
Heat of vaporization	285.6 J/g
Specific heat capacity	1.88 J g^{-1}K^{-1} (at 20 °C)
	2.51 J g^{-1}K^{-1} (at 140 °C)

The compound is sparingly soluble (1%) in water, but the solubility of water in dimethylcyclohexylamine is ca. 20%. N,N-Dimethylcyclohexylamine is miscible with conventional organic solvents.

N,N-Dimethylcyclohexylamine is produced by hydrogenation of dimethylaniline at 180 °C and 6 MPa (60 bar) [42] or by amination of cyclohexanone under hydrogenation conditions in the presence of dimethylamine [43].

N,N-Dimethylcyclohexylamine is an excellent catalyst for the production of polyurethane, particularly for making foams, and for this purpose it is used as the free base [44] or as the salt of an organic acid [45]. The curing temperature of baking finishes comprising substances capable of polyurethane formation can be reduced by 50–80 °C by adding weakly acidic derivatives of the amine [46]. Like pyridine, dimethylcyclohexylamine catalyzes certain reactions and in the preparation of acid chlorides with thionyl chloride is slightly more efficient than pyridine [47].

N-Ethylcyclohexylamine [5459-93-8], C$_8$H$_{17}$N, M_r 127.2, is a colorless liquid, mp –43 °C, bp 165 °C (at 101.3 kPa), d_4^{20} 0.846, n_D^{20} 1.4525, η 1.39 mPa · s (at 20 °C).

Vapor pressure

t, °C	20	40	80	120
p, kPa	0.24	0.84	6.13	27.59

Flash point	46 °C
Ignition temperature (class T3)	245 °C
Explosion limit in air	1.2 and 7.6 vol%

N-Ethylcyclohexylamine is only slightly soluble (2.3%) in water, but the solubility of water in ethylcyclohexylamine is 25%. The amine is miscible with conventional organic solvents.

N-Ethylcyclohexylamine is produced by hydrogenation of ethylaniline at 180 °C and 6 MPa (60 bar) [42] or by amination of cyclohexanone under hydrogenation conditions in the presence of ethylamine [43]. The amine is a starting material for the beet herbicide Ro Neet (Stauffer), which is obtained by converting N-ethylcyclohexylamine into N-ethyl-N-cyclohexylthiocarbamate.

Dicyclohexylamine [101-83-7], C$_{12}$H$_{23}$N, M_r 181.3, is a colorless liquid, mp –0.1 °C, bp 256 °C (at 101.3 kPa), bp 133 °C (at 2.6 kPa), d_4^{25} 0.9104, n_D^{20} 1.4852.

Flash point	105 °C
Ignition temperature (class T3)	240 °C
Explosion limits in air	0.83 and 4.6 vol%
	(corresponding to 62 and 350 g/m^3)
Specific heat capacity	1.88 J g^{-1} K^{-1} (at 20 °C)
	2.47 J g^{-1} K^{-1} (at 140 °C)

Dicyclohexylamine is sparingly soluble in water (about 0.16% at 28 °C) but is readily soluble in the conventional solvents.

Dicyclohexylamine is formed as a byproduct in the production of cyclohexylamine from aniline or cyclohexanol/cyclohexanone [48]. When desired as the end product, dicyclohexylamine is obtained by amination of cyclohexanone under hydrogenation conditions [49] or, in higher yield, from cyclohexanone and cyclohexylamine over palladium/carbon at a hydrogen pressure of 0.4 MPa (4 bar) [50]. Dicyclohexylamine can be obtained also by amination of phenol under hydrogenation conditions, or the reaction of phenol with aniline over palladium/carbon at 0.5 MPa (5 bar) under hydrogenation conditions [51].

Dicyclohexylamine and its salts, especially the nitrite, have good anticorrosion properties in the vapor phase [52]. Paper treated with the nitrite is used as anticorrosion paper. The nitrite also imparts these properties when incorporated into alkyd resins [53]. Like the other cyclohexylamines, dicyclohexylamine is useful as a component of vulcanization accelerators and of pesticides.

N-Methyldicyclohexylamine [7560-83-0], C$_{13}$H$_{25}$N, M_r 195.3, is a liquid, bp 275 °C (at 101.3 kPa), d_4^{25} 0.9207, n_D^{25} 1.4881, η 10.2 mPa · s (at 25 °C), flash point 66 °C.

Vapor pressure

t, °C	63	90	150	190	248
p, kPa	0.133	0.599	2.66	13.33	66.65

N-Methyldicyclohexylamine is sparingly soluble in water and tetrahydrofuran but readily soluble in acetone, methanol, toluene, cyclohexane, and chloroform. The compound is prepared by methylating dicyclohexylamine with dimethyl sulfate or by the Leuckart-Wallach reaction [54] and has uses similar to those of the other cyclohexylamines.

2,2-Bis(4-aminocyclohexyl)-propane [3377-24-0], C$_{15}$H$_{30}$N$_2$, M_r 238.42, mp 51–53 °C, bp 155–160 °C (at 0.65 kPa), d_4^{20} 0.99.

$$H_2N-\bigcirc-\underset{\underset{CH_3}{|}}{\overset{\overset{CH_3}{|}}{C}}-\bigcirc-NH_2$$

The amine is obtained by amination of the corresponding bishydroxy compound in the liquid phase under superatmospheric pressure. Product yields ca. 90% of theoretical are obtained over hydrogenation catalysts containing ruthenium [55] or cobalt, manganese, and phosphoric acid [56]. The industrial product is a mixture of three stereoisomers. Concentration of the trans, trans isomer by treating the isomer mixture with phosphoric acid under reduced pressure at 230 °C or by recrystallization from branched octanes [57] is very important because polymerization of the pure isomer gives polyamides of better quality.

Cyclooctylamine [*5452-37-9*], $C_8H_{17}N$, M_r 127.23, is a liquid; *bp* 80 °C (at 1.3 kPa), *mp* of hydrochloride 244 – 45 °C, *mp* of picrate 193 – 94 °C. Its chemical behavior is analogous to that of other primary amines.

The amine can be produced by a Ritter reaction of cyclooctanol with hydrocyanic acid [58], by reduction of cyclooctanone oxime with sodium and an alcohol [59], by a Ritter reaction of cyclooctene with hydrocyanic acid [60], or by catalytic hydrogenation of cyclooctanone under amination conditions [61]. Cyclooctylamine is used as a component of urea herbicides.

Cyclododecylamine [*1502-03-0*], $C_{12}H_{25}N$, M_r 183.34, *mp* 31 – 32 °C, *bp* 95 – 96 °C (at 0.25 kPa), $n_D^{32.5}$ 1.4849; *mp* of hydrochloride 268 – 270 °C, *mp* of picrate 232 – 234 °C. The compound behaves chemically as a primary amine.

Cyclododecylamine is produced by reduction of cyclododecanone oxime with sodium and an alcohol [62], by catalytic reduction of nitrocyclododecane with hydrogen [63], by Ritter reaction of cyclododecene with hydrocyanic acid [64], or by hydrogenation of cyclododecanone under amination conditions [65]. The reaction of cyclododecylamine with propylene oxide leads to a product that can undergo cyclization to give the fungicide Badilin (BASF).

6. Cyclic Amines

Pyrrolidine [*123-75-1*], C_4H_9N, M_r 71.08, is a colorless liquid, *mp* –60 °C, *bp* 87 – 88 °C (at 101.3 kPa), d_4^{20} 0.8576, n_D^{20} 1.4428, η 1.1 mPa · s (at 20 °C).

$$\begin{array}{c} H \\ | \\ H_2C^{\nwarrow N \nearrow} CH_2 \\ | \quad\quad | \\ H_2C \!-\!\!\!-\! CH_2 \end{array}$$

Flash point	3 °C
Ignition temperature (class T2)	345 °C
Explosion limits in air	1.6 and 10.6 vol% (corresponding to 47 and 314 g/m³)
Specific heat capacity	2.20 J g⁻¹ K⁻¹ (at 21 °C)
	2.25 J g⁻¹ K⁻¹ (at 65 °C)
	2.27 J g⁻¹ K⁻¹ (at 77 °C)
Heat of vaporization	529 J/g (at 0.1 MPa)

Pyrrolidine is infinitely miscible with water and the conventional solvents, such as methanol, acetone, ether, and chloroform.

Chemically, pyrrolidine is like a secondary amine in every respect. For example, it undergoes Leuckart-Wallach or Mannich reactions and is very readily converted into an enamine. In the presence of a catalyst, such as platinum at 360 °C or rhodium at 650 °C [66], pyrrole is formed. In the presence of a copper catalyst, N-methylpyrrolidone is converted into N-methylpyrrolidine.

Production. Pyrrolidine can be produced from butanediol and ammonia over an aluminum–thorium oxide catalyst at 300 °C [67], or from tetrahydrofuran and ammonia over aluminum oxide at 275–375 °C [68].

Uses. Pyrrolidine is used virtually exclusively for the synthesis of drugs and antibiotics. In addition, it is used in vulcanization accelerators.

Piperidine [*110-89-4*], $C_5H_{11}N$, M_r 85.15, derives its name from the alkaloid piperine, the pepper flavoring, in which piperidine is present as piperic acid piperidide.

Piperidine, a colorless liquid, *mp* –10.5 °C, *bp* 106.4 °C (at 101.3 kPa), with a sharp ammonia odor, is hygroscopic and fumes in air; d_4^{20} 0.8613 (the temperature dependence of the density is given in [69]), n_D^{20} 1.4532.

Vapor pressure

t, °C	13.0	25.0	40.0	51.0	67.8	78.8	87.2
p, kPa	2.0	4.0	8.0	13.3	26.7	40.0	53.3

The vapor pressure curve is given in [70]; flash point 3 °C, dissociation constant 1.6×10^{-3} (at 25 °C).

Piperidine is infinitely soluble in water, lower alcohols and ketones, ethers, aliphatic and aromatic hydrocarbons, ethyl acetate, and dimethylformamide. It forms an azeotrope with water:

p, kPa	100.8	53.3	26.7
bp, °C	93.7	76.9	61.4
Piperidine, mass fraction in %	63.7	68.1	76.0

The preparation of enamines from piperidine takes place particularly smoothly [71]–[73]. Unlike open-chain amines, cyclic amines react with aldehydes to give aminals as intermediates [74]. Only pyrrolidine is comparably versatile.

Production. Pyridine obtained from tar distillation is contaminated by sulfur-containing compounds and therefore can be hydrogenated economically only in a two-stage process. The crude pyridine is first hydroraffined over a sulfidic metal catalyst at

280–310 °C and then hydrogenated quantitatively to piperidine at 120–160 °C over an Ni–Al$_2$O$_3$ catalyst [75]. Sulfur-free pyridine can be hydrogenated over a ruthenium [76] or cobalt catalyst.

The preparation of piperidine by aminolysis of 1,5-pentanediol under hydrogenation conditions at 225–260 °C and 20 MPa (200 bar) over a cobalt catalyst has been described [77]; a similar preparation from tetrahydrofurfuryl alcohol also has been described [78]. When piperidine is prepared by hydrogenating glutarates [79], glutaric acid [80], or glutaraldehyde [81], the reaction is carried out in the presence of a large excess of ammonia.

Uses. The secondary amine piperidine is highly reactive and is therefore frequently employed as an intermediate for drugs [82]. It is used also as an accelerator in rubber manufacture, and as an oil or fuel additive [83].

Piperidine and, in many cases, piperidine acetate are useful catalysts for condensation reactions, e.g., the Knoevenagel reaction [84], aldol condensation [85], or the condensation of a nitroparaffin with an aldehyde [86]. However, for the last of these reactions, diethylamine is the preferred catalyst. The use of piperidine is particularly advisable where the reactants or products are unstable in the presence of stronger bases.

Hexamethyleneimine [*111-49-9*], azacycloheptane, hexahydroazepine, homopiperidine, C$_6$H$_{13}$N, M_r 99.18, is a liquid, bp 139 °C (at 101.3 kPa), d_4^{20} 0.88, flash point 30 °C, ignition temperature 255 °C.

Hexamethyleneimine and water form a one-to-one azeotrope that boils at 95–95.5 °C. The hexamethyleneimine–water solubility curve is given in [87]. The chemical behavior of hexamethyleneimine is determined entirely by the secondary amine functional group.

Production. Hexamethyleneimine is produced in 84 % yield by heating hexamethylenediamine at 350 °C in a stream of hydrogen. The catalyst, ammonium vanadate on activated alumina, is prereduced with hydrogen at 500 °C [88]. Residues from the industrial distillation of hexamethylenediamine can be converted into hexamethyleneimine over aluminum silicate or aluminum oxide in a stream of nitrogen [89].

For the preparation from caprolactam, see [90]. Hexamethyleneimine can be prepared by dimerizing acrolein, reducing the product to 2-hydroxymethyltetrahydropyran, expanding the ring to give oxepane, and reacting it with ammonia over aluminum oxide at 350 °C [91].

Uses. The most important use is the conversion of hexamethyleneimine into *S*-ethylhexahydro-1*H*-azepine-1-carbothioate, the rice herbicide Ordram, C$_2$H$_5$-S-CO-NC$_6$H$_{12}$ (Stauffer). Minor amounts of hexamethyleneimine are used for the production of drugs and textile auxiliaries; a large number of patents describe the anticorrosive action of hexamethyleneimine and its derivatives.

Morpholine [*110-91-8*], C_4H_9NO, M_r 87.12, is a colorless hygroscopic liquid; mp −3.1 °C, bp 128.2 °C (at 101.3 kPa), d_4^{20} 1.007, n_D^{20} 1.4542, viscosity 2.3 mPa · s (at 20 °C).

Vapor pressure

t, °C	10	20	40	60	80	100	120
p, kPa	0.57	1.1	3.2	8.3	20.5	40.9	81.8

Critical pressure	5.302 MPa
Critical temperature	344 °C
Flash point	43 °C
Ignition temperature	275 °C
Explosion limits in air	1.8 vol % and 15.2 vol %
Specific heat capacity	1.2 J g^{-1} K^{-1} (at 25 °C)
	2.15 J g^{-1} K^{-1} (at 100 °C)
Heat of vaporization	505 J/g (at 25 °C)
	425 J/g (at 128 °C)

Morpholine is infinitely miscible with water and the conventional solvents, such as methanol, acetone, benzene, and glycol. The mixture comprised of 88 % morpholine and 12 % water has a freezing point below −50 °C and a viscosity of 7.45 mPa · s at 20 °C.

Chemically, morpholine behaves as a secondary amine in every respect. It forms complexes with many cations. In addition to the Leuckart-Wallach and Mannich reactions, the conversion to enamines has become important [71], [72]. Morpholine, pyrrolidine, and piperidine are preferred to the open-chain amines for these reactions because they react faster with the carbonyl component.

Production. Morpholine can be obtained from diethanolamine by cyclization with sulfuric acid or via diethanolamine hydrochloride [92] or from di(2-chloroethyl) ether and ammonia [93]. In an industrially more important method, diglycol is converted under aminating and hydrogenating conditions, under pressure and over a cobalt or nickel catalyst [94]. Other catalysts used are described in [95]. A byproduct frequently obtained is 2-(2-aminoethoxy)ethanol, which is the result of simple amination without subsequent cyclization; after isolation, this compound can be recycled.

Quality and Storage. When produced from diglycol, commercial morpholine is more than 99 % pure, being contaminated particularly by *N*-ethylmorpholine and ethylenediamine.

Morpholine can be stored for an unlimited time in iron containers if it is protected from atmospheric moisture and carbon dioxide. Copper, zinc, and their alloys are unstable to morpholine.

Uses. Morpholine is an intermediate for a large number of drugs, crop protection agents, dyes, and optical brighteners. An important use is the conversion to vulcanization accelerators, which are based on 2-mercaptobenzothiazole, dithiocarbamic acid, thiuram polysulfides, or sulfenamides.

Salts of morpholine with long-chain fatty acids, such as oleic or stearic acid, have waxlike properties and are used as emulsifiers.

Because water and morpholine have similar volatilities and the latter is an anticorrosion agent, it is useful in steam cycles [96] and for aqueous hydraulic liquids [97] and similar systems. Up to 550 °C, no substantial loss occurs as a result of decomposition.

An aqueous solution of morpholine (2 M) can be used for removing CO_2, H_2S, or HCN from gases [98]. In the extraction of aromatics from hydrocarbon mixtures, morpholine is highly selective [99].

Morpholine Derivatives. *N-Methyl-morpholine* and *N-ethylmorpholine* are used as catalysts for the production of polyurethane foams. *N-Formylmorpholine* is used as a selective solvent for the extraction of very pure aromatic compounds in the Formex process (SNAM Progetti).

N-Methylmorpholine [109-02-4], $C_5H_{11}NO$, M_r 101.15, *mp* −65 °C, *bp* 114 °C (at 101.3 kPa), d_4^{20} 0.919, η 0.90 mPa · s (at 20 °C), is obtained by reacting morpholine with formaldehyde to give *N*-hydroxymethylmorpholine and reducing the product with excess formaldehyde [100], with formic acid [101], or with hydrogen over cobalt or manganese catalysts and under pressure [102]. Another important reaction is the alkylation of morpholine with methanol at a total pressure above 3.5 MPa (35 bar) and a hydrogen partial pressure above 1 MPa (10 bar) [103].

Alkylmorpholines are obtained also by cyclization of a bis(2-chloroethyl) ether with an amine [104] or of diethylene glycol with an amine in the presence of hydrogen and a hydrogenation catalyst [105].

N-Ethylmorpholine [100-74-3], $C_6H_{13}NO$, M_r 115.18, *mp* −63 °C, *bp* 134 – 137.5 °C (at 101.3 kPa), d_4^{20} 0.913, η 1.05 mPa · s (at 20 °C).

N-Formylmorpholine [4394-85-8], $C_5H_9NO_2$, M_r 115.13, *mp* 20 – 21 °C (anhydrous), 9 °C (5 % water), −3 °C (10 % water), *bp* 244 °C (anhydrous), 172 °C (2 % water), 151 °C (5 % water), 132 °C (8 % water) (all at 101.3 kPa), d_4^{20} 1.1528, d_4^{50} 1.1266, η 3.61 c St (at 50 °C), flash point 124 °C.

Piperazine [110-85-0], $C_4H_{10}N_2$, M_r 86.14, forms strongly hygroscopic, colorless lamellar crystals with a typical amine odor, *mp* 109.6 °C, *bp* 148 °C (at 101.3 kPa), *mp* of the hexahydrate 42.0 °C.

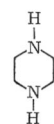

Vapor pressure

t, °C	110.8	121.7	125	140.6
p, kPa	30.3	44.6	49.6	79.7

Flash point	98 °C
Ignition temperature	340 °C
Explosion limits in air	3.93 kPa and 12.0 kPa (corresponding to 139 g/m³ and 430 g/m³)
	These are the saturated vapor pressures above piperazine at 64.0 °C and 90.5 °C.
Specific heat capacity	2.94 J g⁻¹ K⁻¹ (at 130 °C)
	2.99 J g⁻¹ K⁻¹ (at 140 °C)
Heat of vaporization	47.3 J/g

Piperazine is readily soluble in water, methanol, and ethanol but only slightly soluble in diethyl ether, benzene, and heptane.

Piperazine has the chemical properties of a secondary amine. Vapor-phase dehydrogenation over copper chromite or palladium gives pyrazine.

Production. Piperazine is obtained as a byproduct in the production of ethylenediamine. If ethanolamine is used as a starting material [106] and reacts with ammonia at 150–220 °C and 10–25 MPa (100–250 bar), piperazine can be distilled from the reaction mixture, which also contains ethanolamine, ethylenediamine, diethylenetriamine, aminoethylethanolamine, and polyamines.

Piperazine is obtained with a purity of 99 %, and can be converted to flakes, even after conversion to the hexahydrate. Frequently, it is transported as an approximately 65 % strength aqueous solution, which has a low melting point of about 45 °C and hence corresponds roughly to a eutectic mixture. The purity is determined in most cases by gas chromatography. If it can be assumed that organic impurities are absent, the purity can be determined titrimetrically (HCl or HClO₄) in aqueous solution.

Storage. The flakes are stored in barrels lined with a polyethylene sack. To avoid yellowing, the barrels should be air tight and not exposed to direct sunlight. The aqueous solution is stored at 50–60 °C in insulated iron tanks that can be heated.

Uses. Piperazine, in the form of the hexahydrate or of salts, is used as an anthelmintic agent in human and veterinary medicine. Furthermore, the piperazine ring is a constituent of a large number of other drugs.

The production of polyamides from piperazine and aliphatic dicarboxylic acids has long been known [107], but has remained unimportant. The polymers so formed have a heat resistance superior to that of conventional polyamides but stabilizers must be used [108].

The use of 1,4-bis(1-aziridinylcarbonyl)piperazine or 1,4-bis(chloroacetyl)piperazine for curing gelatine for photographic purposes was proposed [109]. 1-Aminoethylpiperazine is used as a hardener for epoxy resins.

Triethylenediamine [280-57-9], 1,4-diazabicyclo[2.2.2]octane, "Dabco", $C_6H_{12}N_2$, M_r 112.18, is a highly symmetrical molecule with a cage structure. The colorless hygroscopic crystals melt at 159.8 °C; bp 174 °C, pK_{A1} 2.95, pK_{A2} 8.60.

Triethylenediamine reacts virtually quantitatively with bromine to give a one-to-one adduct. With alkyl halides it forms quaternary salts, even in nonpolar solvents. Apart from its highly nucleophilic nature, triethylenediamine exhibits catalytic activity in base-catalyzed reactions.

Production. Triethylenediamine can be produced from ethylenediamine or ethanolamine [110], diethanolamine [111], or diethylenetriamine [112]. All processes give only moderate yields. For example, in the presence of a catalyst consisting of 86 % silicon dioxide and 12 % aluminum oxide, at 360 °C and atmospheric pressure, diethylenetriamine gives about equal amounts of piperazine, alkylpiperazines, pyrazines, and ethylenediamine and homologues in addition to the bicyclic compound. Fractional distillation, subsequent recrystallization from acetone, and washing with petroleum ether give 95 % pure triethylenediamine; sublimation gives a 98 % pure product.

Processes starting from piperazine derivatives (*N*-hydroxyethylpiperazine or *N*-aminoethylpiperazine) also were described [113].

Uses. The only industrial use of triethylenediamine is as a hardener for polyurethane foams [114]. Recently, however, more effective aza- and diazabicyclooctane derivatives intended for the same purpose have been the subject of patent applications [115].

Hexamethylenetetramine [100-97-0], hexamine, $C_6H_{12}N_4$, M_r 140.19, is a highly symmetrical molecule with an adamantane structure. Trade names: Urotropin, Formin, Aminoform. It crystallizes from ethanol as colorless rhombododecahedra.

At reduced pressure (2 kPa), hexamethylenetetramine sublimes at 230–270 °C, virtually without decomposition, mp above 270 °C (decomp.), ϱ 1.331 g/cm^3 at −5 °C, enthalpy of formation 124.1 ± 0.75 kJ/mol. For further thermodynamic data, see [116].

For the IR spectrum, see [117]. The vapor pressure (in kPa) between 20 and 280 °C can be calculated from the formula $\log p = -524.8/T + 1.334$.

At 12 °C, 81.3 g of hexamethylenetetramine dissolves in 100 g of water; the solubility decreases slightly with temperature. The dissociation constant is 1.4×10^{-9} (temperature not given). The solubility of hexamethylenetetramine in 100 g of various solvents at 20 °C is as follows: 13.4 g in chloroform, 7.25 g in methanol, 2.89 g in absolute ethanol, 0.65 g in acetone, 0.23 g in benzene, 0.14 g in xylene, 0.06 g in ether, and near zero in petroleum ether.

Hexamethylenetetramine gives a monobasic reaction, but with strong acids it is dibasic. The equilibrium $C_6H_{12}N_4 + 6 H_2O \rightleftharpoons 6 CH_2O + 4 NH_3$ in aqueous solution permits the use of hexamethylenetetramine as a formaldehyde or ammonia donor. Thermolysis leads to hydrocyanic acid in a yield of 73 % at 800 °C, and 92 % at 1200 °C [118]. Nitration gives 1,2,3-trinitro1,2,3-triazine, which is an important explosive (hexogen, cyclonite, RDX). The intact tricyclic structure can be alkylated and arylated at the nitrogen atom and halogenated at the carbon atom. It undergoes a large number of addition reactions: adduct formation with phosgene, disulfur dichloride, and many inorganic salts.

Production. Formaldehyde and ammonia can be converted into hexamethylenetetramine in the gas phase, in aqueous solution, or in a suspension in an inert solvent, although the most important of these procedures is that carried out in water. The usual process gives a pure product in yields near 90 % and includes an economical drying step.

If aqueous solutions of ammonia (27 %) and formaldehyde (30 %) react continuously at up to 95 °C in a V4A stainless steel tube reactor, it is necessary to dewater a 14 % solution of hexamethylenetetramine [119]. If formaldehyde and ammonia are reacted in the vapor phase, the product is precipitated with cold water [120] or is discharged with an aqueous solution of hexamethylenetetramine [121].

The heat of reaction in aqueous solution is 230 kJ/mol and 745 kJ/mol in the vapor phase. Therefore, the presence of a minimum amount of water is desirable. In this method [122], ammonia gas and formaldehyde gas are passed continuously into the boiling reaction solution. The boiling point is brought to 50 – 70 °C. The heat of hydration and heat of reaction vaporize the water of condensation. The hexamethylenetetramine is removed in crystalline form or as a concentrated solution.

In another method, ammonia gas and formaldehyde solution react at pH 7.5 – 8 and 50 – 90 °C, and the product is granulated by spraydrying in a hot inert gas stream (220 – 250 °C), or by spraying it onto hexamethylenetetramine particles in a fluidized bed in an air stream at 80 – 115 °C [123].

Quality. As a rule, the industrial product is 99 % pure. The compound is purified by stepwise crystallization (see [124]). The addition of paraffin, benzoic acid, acid amides, or diatomaceous earth gives free-flowing products [125].

Uses. Hexamethylenetetramine is used mainly as an ammonia or formaldehyde donor, for example, in the production of phenol resins, urea – formaldehyde resins and in fuel tablets. Further uses are those in the production of the high explosives

Table 2. Mixture composition of primary fatty amines (mass fraction in %)

	Number of carbon atoms in alkyl chain													
	Saturated alkyl									Unsaturated alkyl *				
	6	8	10	12	14	15	16	17	18	14'	16'	18'	18"	Other
Amine														
Coco	0.5	8	7	50	18		8		1.5			6		1
Hydrogenated-tallow				1	4	0.5	30	1.5	60	0.5	0.5	2		
Tallow				1	3	0.5	29	1	23	1	3	37	1.5	
Oleyl				0.5	3.5	0.5	4	1	5	1.5	5	76	3	
Soya					0.5	1	16		15		1	49.5	13	4

*A single prime indicates one double bond; a double prime, two double bonds.

hexogen and octogen, as vulcanization accelerator, and as anticorrosion agent. Proposed uses relate to the activation of chlorites as bleaches [126] and the preservation of foodstuffs [127], paints, finishes, latices, etc. [128]. Because of its bactericidal action hexamethylenetetramine is employed in medicine, but this use is of very minor importance.

7. Fatty Amines

Straight-chain primary, secondary, and tertiary amines with chain lengths between 8 and 24 carbon atoms are commonly known as fatty amines. This group of compounds also includes derivatives, such as the *N*-alkyl-1,3-propanediamines. Of commercial importance are fatty amine mixtures, such as coco amine, tallow amine, hydrogenated-tallow amine, oleylamine, and soya amine, which are derived from naturally occurring fatty acids. More recently, alkylamines of similar structure were produced from synthetic feed stocks, e.g., olefins or paraffins, and these also are called fatty amines (β-amines, branched-chain amines, aminoalkanols).

7.1. Properties

Physical Properties. Table 2 shows typical mixture compositions of fatty amines. The properties of these formulations can be varied to a certain degree by blending chain lengths having the desired physical characteristics.

Solubilities of selected fatty amines in polar and nonpolar organic solvents are given in Table 3. Only the short-chain fatty amines with eight to ten carbon atoms per alkyl chain are of limited solubility in water.

Vapor pressures at ambient temperature are generally very low and decrease from 26.5 Pa (0.26 mbar) for dodecylamine to 2.7×10^{-2} Pa (0.27×10^{-3} mbar) for eicosyl-

Table 3. Solubility (g per 100 mL at 30°C) of fatty amines in organic solvents

	2-Propanol	Hexane
Primary amines		
Dodecylamine	∞	∞
Tetradecylamine	458	216
Hexadecylamine	169	64.8
Octadecylamine	86	27.9
Secondary amines		
Dioctylamine	∞	∞
Didodecylamine	55	27.5
Dioctadecylamine	1.2 (50 °C)	2.1 (40 °C)
Tertiary amines		
Trioctylamine	∞	∞
Tridodecylamine	23.9	∞
Trioctadecylamine		36.4

Table 4. Physical properties of fatty amines

	Formula	M_r	mp, °C	bp *, °C	Hydrochloride decomp. temp., °C	Acetate mp, °C
Pure primary amines						
Octylamine	$C_8H_{17}NH_2$	129.25	-0.1	180.0	198	
Decylamine	$C_{10}H_{21}NH_2$	157.30	15.9	221.8	194	
Dodecylamine	$C_{12}H_{25}NH_2$	185.35	28.2	259.1	179 – 181	56.0 – 56.3
Tetradecylamine	$C_{14}H_{29}NH_2$	213.41	38.0	292.3	150 – 167	66.0 – 66.8
Hexadecylamine	$C_{16}H_{33}NH_2$	241.46	46.0	321.9	148 – 166	74.0 – 74.2
Octadecylamine	$C_{18}H_{37}NH_2$	269.52	53.1	348.5	158 – 161	79.8 – 80.0
Eicosylamine	$C_{20}H_{41}NH_2$	297.57	59.5	372.5	151 – 154	85.0 – 85.2
Docosylamine	$C_{22}H_{45}NH_2$	325.63	65.3	394.6	150 – 153	88.8 – 89.0
Commercial mixtures of primary amines						
Coco amine			12 – 17	130 – 227		ca. 50
Oleylamine				200 – 210		
Tallow amine			32 – 40	200 – 230		ca. 55
Hydrogenated-tallow amine			48 – 56			ca. 60
Soya amine			27 – 30			

* At 101.3 kPa (1.01 bar) for pure compounds, at 3.6 kPa (36 mbar) for mixtures

amine. The more important physical properties for pure compounds and technical mixtures are given in Tables 4 and 5.

The oscillation of *melting points* between the series of odd-chain and even-chain homologues is shown in Figure 2. This effect disappears in derivatives of primary fatty amines. Diamines and 2-(alkylimino)diethanols (N,N-bis(2-hydroxyethyl) alkylamines) are low-melting pasty solids.

Fatty amines and their derivatives are cationic surfactants and show strong substantivity to the negatively charged surface of most solids, e.g., cotton, minerals, and metals, thus changing physical properties of the substrates [135], [136].

Chemical Properties. Fatty amines, like short-chain amines, are stronger bases than ammonia. Their ionization constants [137] do not change significantly with chain

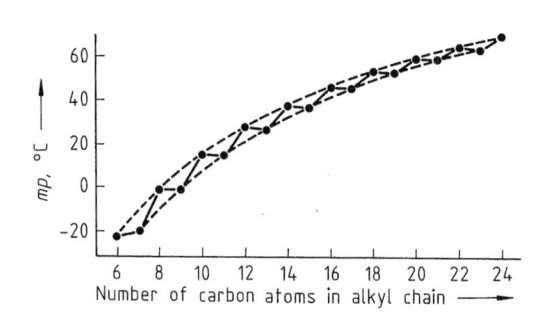

Figure 2. Melting points of primary amines

Table 5. Physical properties of secondary and tertiary fatty amines

	Formula	M_r	mp, °C (α)	mp, °C (β)	bp, °C (p,Pa)
Secondary amines					
Dioctylamine	$(C_3H_{17})_2NH$	241.46	14.8	26.5	115 (133)
Didecylamine	$(C_{10}H_{21})_2NH$	297.57	33.6	41.4	152 (133)
Didodecylamine	$(C_{12}H_{25})_2NH$	353.68	47.2	51.0	183 (133)
Ditetradecylamine	$(C_{14}H_{29})_2NH$	409.79	58.1	58.8	209 (133)
Dihexadecylamine	$(C_{16}H_{33})_2NH$	465.90	67.0		231 (133)
Dioctadecylamine	$(C_{18}H_{37})_2NH$	522.01	74.7		250 (133)
Dicoco amine			40–47		
Dihydrogenated-tallow amine			60–65		
Tertiary amines					
N-Methyldioctadecylamine	$(C_{18}H_{37})_2NCH_3$	536.03	40		252–259 (6.7)
N,N-Dimethyldodecylamine	$(C_{12}H_{25})N(CH_3)_2$	213.41	−20.5		135 (1333)
N,N-Dimethyloctadecylamine	$(C_{18}H_{37})N(CH_3)_2$	287.49	22.5		202 (1333)

Table 6. pK_b Values of amines in water at 25°C *

R	RNH_2	R_2NH	R_3N
Methyl	3.38	3.29	4.24
Ethyl	3.37	3.02	
Octyl	3.35	3.00	
Octadecyl	3.40	3.00	

* The pK_b of NH_3 in water at 25 °C is 4.79.

length (see Table 6), and their reactions are similar to those known from the lower homologues.

The following are industrially important reactions:

1) *Salt formation* with inorganic and organic acids. The hydrochlorides are slightly more soluble, the acetates considerably more soluble in water than the free amines. Acetates and salts of higher fatty acids are readily soluble in organic solvents [130]. As would be expected, melting points of the salts are higher than those of the free amines.

2) *Conversion* into N-alkyl carboxamides with carboxylic acids or their derivatives.

3) Reaction with *carbon dioxide* to give carbamates, which are formed as contaminants in amines exposed to air.

4) *Alkylation* with excess of strong nucleophiles, such as methyl chloride or dimethyl sulfate, leading eventually to quaternary ammonium compounds. This reaction is normally carried out in aqueous alcohol as the solvent and in the presence of alkali as the proton acceptor [138], [139].
5) *Oxyalkylation* with ethylene oxide, a more selective reaction than alkylation. The uncatalyzed reaction with primary amines at 150 to 200 °C yields almost exclusively 2-(alkylimino)diethanols (N,N-bis(2-hydroxyethyl) fatty amines). Base catalysis even at temperatures as low as 100 °C leads to bis(polyoxyethyl) amines [140]. Under mild conditions (80 °C, 350 Pa) and in the presence of water or inorganic or organic acids, secondary and tertiary amines react readily with ethylene oxide to form the corresponding bis- or mono(2-hydroxyethyl) quaternary ammonium salts [141] – [143]. Long-chain epoxides undergo a similar reaction [144].
6) Reaction of fatty amines with *halogenated carboxylic acids* or with *lactones*, yielding amphoteric compounds (betaines).
7) *Michael addition* to activated double bonds. For example, the important 1,3-propanediamines are synthesized by the reaction of a primary fatty amine with acrylonitrile followed by hydrogenation of the intermediate 2-(alkylimino)ethane nitrile [145].
8) *Mannich reactions* with formaldehyde and nucleophiles, such as aldehydes, nitroparaffins, or phenols, to form polyfunctional derivatives [146] – [148].
9) *Reaction with phosgene* to form isocyanates and ureas.
10) *Oxidation* by hydrogen peroxide, peracids, or ozone to give amine oxides.

7.2. Production

Fatty amines can be produced from natural fats and oils or from synthetic raw materials.

Production from Fatty Nitriles. The most important source of fatty amines is still fatty nitriles, which are formed from carboxylic acids and ammonia over dehydrating catalysts (Al_2O_3, ZnO, or salts of Mn or Co) in liquid-phase reactors or liquid- and vapor-phase reactors at 280 – 360 °C.

Depending on the process used, nitriles are purified by distillation or used as such. Hydrogenation of nitriles can lead to all three types of amines, depending on reaction conditions.

Saturated primary amines are formed by hydrogenation of fatty nitriles, typically at 80 – 140 °C and 1 – 4 MPa (10 – 40 bar), over nickel catalysts. Raney cobalt or copper chromite catalysts have to be used to obtain *unsaturated amines* [149] – [151]. In order to avoid the formation of a secondary amine, ammonia is commonly used as a suppressant.

Secondary Amines. If ammonia is continuously vented from the reactor and temperatures of 160–210 °C and pressures of 5–20 MPa (50–200 bar) are maintained, saturated and unsaturated secondary amines can be obtained in yields of greater than 90 % [152], [153].

Tertiary Amines. Symmetrical trialkylamines are produced from nitriles via the imine, RCH=NH, and Schiff base, RCH=N–CH$_2$, using supported nickel catalysts and hydrogen sparge at 230 °C and 0.7 MPa (7 bar) [154].

More important are the *N*-alkyl-*N*-methylalkylamines and the *N,N*-dimethylalkylamines, which can be synthesized by the Leuckart reaction of primary or secondary amines [155], [156]:

$$R_2NH + CH_2O + HCOOH \longrightarrow R_2NCH_3 + CO_2 + H_2O$$

N,N-dimethylalkylamines can be produced also from fatty nitriles and dimethylamine [157]. Both types of methyl-substituted tertiary amines are accessible also from primary alcohols [158], [159].

Production from Alcohols or Carbonyl Compounds. Fatty amines are produced by reductive alkylation of ammonia or substituted amines with primary alcohols at 90–190 °C under low pressure (atmospheric up to 0.7 MPa) in the presence of a Raney nickel catalyst. If water is continuously removed from the reactor, only secondary and tertiary amines are obtained [160]–[162]. Similarly, aldehydes can be converted into primary amines by reductive amination at 110–120 °C and 1.5 MPa (15 bar) using rare earth-promoted cobalt catalysts [163]. Aldehydes produced in the oxo process and methylamine react to form imines, which are then hydrogenated to give *N*-methylalkylamines at 115 °C and 3 MPa (30 bar) using Raney nickel catalyst [164]. Similarly, both alcohols and aldehydes can be converted into *N,N*-dimethyl tertiary amines using dimethylamine. Alcohols are converted into amines at 230 °C at atmospheric pressure using copper chromite catalysts, whereas fatty aldehydes require noble metal, copper chelate, or copper carboxylate catalysts [165]–[168].

Production from Olefins. Straight-chain primary amines can be produced from ammonia and 1-bromoalkanes, which are made readily by radical addition of hydrogen bromide to α-olefins [169]. Under hydrocarbonylation conditions, α-olefins and dimethylamine form *N,N*-dimethylalkylamines if highly selective noble metal catalysts are used [170]. Use of the less specific tributylphosphinecobalt catalysts leads to mixtures of amines and alcohols [171].

In a *Ritter reaction*, α-olefins react with hydrogen cyanide or acetonitrile in concentrated sulfuric acid or aqueous hydrogen fluoride, via the intermediate 2-acylaminoalkane, to give branched alkyl primary amines (*β*-amines). The reaction is not specific and positional isomers also form [172], [173].

Aluminum alkoxides, as formed in the Ziegler process, can be used to convert secondary amines into tertiary amines [174].

Production from Alkane Derivatives. *Nitroalkanes* with a statistical distribution of nitro groups can be hydrogenated over palladium–charcoal at 190 °C and 3.9 MPa (39 bar) to form secondary alkyl primary amines in yields up to 93 % [175]. Selectively *chlorinated alkanes* react with dimethylamine to yield up to 65 % terminal tertiary amine [176].

Production from Other Starting Materials. *Glycerides* and also *methyl esters* of fatty acids can be converted into fatty nitriles at 220–300 °C provided ammonia is vented rapidly. Yields of nitrile and glycerine in excess of 80 % were reported using zinc carboxylate or aryl sulfate catalysts [177]. Under hydrogenation conditions and using Zn-Al catalysts at high pressures (10–30 MPa) and temperatures (200–350 °C), coconut glycerides can be converted into primary coco amine. Glycerine, however, decomposes completely under these conditions [178].

Primary fatty amides can be converted in high yields into odd-chain primary amines by *Hofmann degradation* of the chloroamide intermediate [179]:

$$RCH_2-CONH_2 \xrightarrow[-HCl]{Cl_2} RCH_2CONHCl \xrightarrow[\substack{-CO_2 \\ -NaCl}]{NaOH} R-CH_2-NH_2$$

7.3. Analysis and Quality Control

Because of their relatively high basicity, fatty amines can be titrated easily with acids. Most of the amines are insoluble in water; therefore, organic solvents, such as isopropyl alcohol, are commonly used. End points are determined colorimetrically (bromophenol blue) or potentiometrically. In the second case, perchloric acid and glacial acetic acid are preferred titrant and solvent, respectively.

Industrial mixtures of amines normally contain up to 90 % of the main compound. These mixtures are analyzed by stepwise titrations with acid. The total basicity is determined first. Primary amines then are removed with salicylaldehyde, and primary and secondary amines can be removed with phenylisothiocyanate or acetic acid, so that the remaining free amine can be titrated with acid. Unsaturated amines are determined from the iodine number. Because amine groups also consume iodine, the original Wijs method must be slightly modified by the use of acetic acid as the solvent. The amine salt formed absorbs the halide at a much slower rate than the double bonds.

Quaternary ammonium compounds usually are determined by solvent partition titration with anionic surfactants or sodium tetraphenylboron [180]. Iodine numbers of quaternary ammonium compounds are obtained in sodium lauryl sulfate containing chloroform to prevent the free iodine from being retained by the nonaqueous phase. A summary of approved wet chemical test methods (both AOCS and ASTM) is given in [181].

Chain-length distribution and hydrocarbon content in fatty amines are determined by gas chromatography [182]–[186]; nitrile and amide impurities are determined by IR analysis [187], [188].

More recently, NMR spectroscopy has been used for determination of quaternary ammonium compounds in the presence of free amines [189] and also for establishing the relative amounts of primary, secondary, and tertiary amines in mixtures [190].

For environmental information, analysis for volatile reagents used in the production of various types of amines (formaldehyde, ethylene oxide, methyl chloride, acrylonitrile) and found in the head spaces of reactors (head space analysis) is most often used in industry.

7.4. Uses

Fabric Softeners. By far the largest use of fatty amines is in fabric softening. Prior to the early 1950s, fabric softeners were not needed because most detergents were based on tallow-derived soap formulations and some of the active detergent remained on the textile, which resulted in a softer dried fabric. However, the advent of synthetic detergents created the market for fabric softeners in home care and institutional laundries.

Most of these softeners are used in *rinse cycles* where a 5–7% formulated dispersion is added leaving 0.1–0.2% active quaternary ammonium compound on the surface of the clothes. Recently, formulations have been designed in which the fabric softener is applied to a substrate, such as paper, to woven or nonwoven cloth, or to a foam and then added to the *drying cycle*. The formulation contains a quaternary ammonium compound and a transfer agent that allows the active ingredient to be transferred to the cloth during the drying cycle.

The fatty quaternary ammonium compounds used are dimethyldihydrogenated-tallow ammonium chloride or the corresponding methyl sulfate; tallow amidotallowimidazolinium methyl sulfate; and ditallowdiamidohydroxyethyl methyl sulfate. Compounds of the first type have the largest market share and are the most effective [191].

Flotation. One of the first commercial uses of primary fatty amines was in the beneficiation of potash [192]. In this process, fatty amine salts adhered to the surface of potassium chloride, allowing it to be removed by froth flotation and leaving the sodium chloride in the tailings. The technique spread to other mineral beneficiation processes [193]. In general, any negatively charged surface can be removed from a system with an amine salt. Lower grades of phosphate rock are concentrated by the use of amines and their salts [194]. Frequently, fatty amine salts are added to mineral products to prevent caking of the powdered mineral.

Corrosion Inhibition. Large quantities of amines are used also in corrosion inhibition. Again, the cationic amino function adheres to the surface of metals and thereby prevents attack on the metal by corrosive liquids and gases. Primary amines and N-alkyl-1,3-propanediamines are used most frequently for this purpose. The amines may be neutralized with oleic, naphthenic, and acetic acids to adjust their solubilities. A typical oil-soluble inhibitor formulation contains N-tallow-1,3-propanediamine dioleate.

Asphalt Emulsification. The cationic nature of fatty amines makes them very useful as emulsifiers for asphalt. Primary amines, diamines, alkoxylated amines, and fatty quaternary ammonium compounds can be used in this application. The formulation used depends on the properties of the emulsion desired. The cationic emulsion can be rapid, medium, or slow set. In all cases a cationic emulsifier is used at a low rate, 0.5 – 1.0 wt % of the emulsion. When the emulsion is mixed with the negatively charged aggregate, the emulsion breaks, thereby setting the road-covering formulation. Fatty amines, because of their positive charge, also increase the adherence of the asphalt to the aggregate.

Other Applications. Dimethyldihydrogenated-tallow ammonium chloride is also used in *sugar-refining processes* [195]. The quaternary ammonium salt removes anionic impurities from the sugar solution by forming complexes that precipitate. The precipitate then can be removed by filtration or centrifugation.

There are many applications for fatty amines in *agriculture* as emulsifiers, adjuvants, and intermediates in pesticide production. Much of their usefulness relies on the controllable lipid solubility, which aids penetration or absorption of many active pesticides. The emulsifiers used are ethoxylated amines and quaternary ammonium compounds. Some of the surfactants also can act as foaming agents or foam stabilizers [196].

Neutralization of various herbicidal acids with fatty amines enhances the activity of the herbicides because the corresponding salts are more soluble in oil and less volatile than the free amines [197] – [200]. In some cases, the N,N-dimethylalkylamines were found to control the growth of suckers or unwanted leaves, thereby producing a better grade of product, for example, in tobacco [201]. The reaction of dodecylamine and cyanamide leads to dodecylguanidine, which is an excellent fungicide for fruit trees.

Quaternary ammonium compounds are also used extensively as *bactericides, sanitizers, and disinfectants*. The three types of compounds used for these applications are alkyl(benzyl)dimethylammonium chlorides, alkyltrimethylammonium chlorides, and dialkyldimethylammonium chlorides. For an extensive review of the biological activity of fatty nitrogen compounds, see [202]. Generally, alkyl(benzyl)dimethyl and alkyltrimethyl compounds in which the fatty alkyl group contains 12 – 14 carbon atoms are most effective against a broad range of organisms. The dialkyldimethyl compounds are most effective when the fatty alkyl group contains 8 – 10 carbon atoms.

Many fatty alkylamine derivatives are used as additives in plastic formulations. For example, 2-(alkylimino)diethanols (bis(2-hydroxyethyl)alkylamines) are added to polypropylene or polyethylene as internal *antistatic agents*, either as 100 % amines or as a 75 % solid masterbatch to prevent static charge buildup at low humidities.

Fatty amines, diamines, and ethoxylated amines also can be used as *pigment-grinding aids* and as *dispersants* for pigments in paints, coatings, and magnetic tape. The most

Table 7. Markets for fatty amines, 10^3 t/a,

	Western Europe 1982	USA & Canada 1982	Japan 1980	Total 1983
Fabric softeners	74	42	10	126
Petroleum	6	20	1	27
Intermediates	14	15	–	29
Asphalt	8	11	1	20
Mining	10	9	–	19
Others	18	16	16	50
Total	130	113	28	271

widely used compound is N-tallow-1,3-propanediammonium dioleate. The amine compound coats the pigment particle making it more readily and evenly dispersible in the formulation.

The *organo-clay* market also is a very large user of fatty amines. Most of this market is satisfied with methyl or benzyl quaternary ammonium chlorides produced from methyldihydrogenated-tallow amine. In either case, the quaternary ammonium chloride reacts with an anionic clay, such as bentonite, to produce clay platelets separated with an oleophilic compound. These organo-clay products are used as thickening agents in oleophilic mixtures, mainly in drilling muds and oil-based coatings.

7.5. Economic Aspects

The economics of fatty amines production generally depend on the worldwide fat and oil market process. Most of the fatty amine products are made from tallow and coconut oils; others are made from oils such as soya bean, palm, and cottonseed. Minor amounts of tall-oil acids are used. When pure chain lengths are required, these can be obtained by fractionating the acid mixture found in the initial triglyceride. Pure fractions are much more expensive than mixtures.

Estimated sales of fatty amines by application and territory are given in Table 7. Installed capacity worldwide in 1983 may be 30 % higher. Growth has been remarkable: from 120 000 t in 1971 to 271 000 t in 1983.

The most important producers of fatty amines are Henkel, Hoechst-Gendorf, and Akzo-Chemie in the Federal Republic of Germany; Akzo-Chemie in the United Kingdom; Ceca in France; Oleofina, Akzo-Chemie, and Kemangard in Belgium; Sinor-Kao in Spain; Kemangard in Sweden; Sherex, Akzo Chemie, Humko, and Jetco in the United States; Armak in Canada; and Kao Soap and Lion-Akzo in Japan. Because most fatty amines offered commercially contain mixed alkyl chain lengths, trade names are common. A number of these are listed in Table 8.

Table 8. Trade names of fatty amines

Amine	CAS registry number	Trade name (company)
Primary amines		
Octylamine	[111-86-4]	Armeen 8D (Armak)
Decylamine	[2016-57-1]	Armeen 10D (Armak)
Dodecylamine	[124-22-1]	Adogen 163D (Sherex), Alamine 4D (Henkel), Armeen 12D (Armak), Jetamine 12 (Jetco)
Tetradecylamine	[2016-42-4]	Alamine 5D (Henkel), Armeen 14D (Armak),
Hexadecylamine	[143-27-1]	Alamine 6D (Henkel), Armeen 16D (Armak), Kemamine P-880 (Humko)
Octadecylamine	[124-30-1]	Adogen 142D (Sherex), Alamine 7D (Henkel) Armeen 18D (Armak), Kemamine P-990D (Humko), Jetamine 18D (Jetco)
Coco amine	[61788-46-3]	Adogen 160D (Sherex), Alamine 21D (Henkel), Armeen CD (Armak)
Tallow amine	[61790-33-8]	Adogen 170 (Sherex), Alamine 26D (Henkel), Armeen T (Armak)
Hydrogenated-tallow amine	[61788-45-2]	Adogen 140D (Sherex), Alamine H26D (Henkel), Armeen HT (Armak)
Oleylamine	[112-90-3]	Armeen O (Armak)
Soya amine	[61790-18-9]	Armeen S (Armak)
Secondary amines		
Dioctylamine	[1120-48-5]	Armeen 2–8 (Armak)
Didecylamine	[1120-49-6]	Armeen 2–10 (Armak)
Didodecylamine	[3007-31-6]	Alamine 204 (Henkel)
Dihexadecylamine	[16724-63-3]	Alamine 206 (Henkel)
Dioctadecylamine	[112-99-2]	Alamine 207 (Henkel)
Dicoco amine	[61789-76-2]	Adogen 260 (Sherex), Alamine 221 (Henkel), Armeen 2C (Armak), Kemamine S-650 (Humko)
Dihydrogenated-tallow amine	[61789-79-5]	Adogen 240 (Sherex), Alamine H226 (Henkel), Armeen 2HT (Armak)
Disoya amine		Armeen 2S (Armak)
Tertiary amines		
Trioctylamine	[1116-76-3]	Alamine 336 (Henkel), Adogen 363 (Sherex)
Tridodecylamine	[102-87-4]	Alamine 304 (Henkel)
N-Methyldihydrogenated-tallow amine	[61788-63-4]	Armeen M2HT (Armak)
N,N-Dimethyldodecylamine	[112-18-5]	Armeen DM12D (Armak), Kemamine T6702D (Humko), Onamine 12 (Onyx)
N,N-Dimethyltetradecylamine	[112-75-4]	Armeen DM14D (Armak), Barlene 14S (Lonza), Jetamine T-14D (Jetco)
N,N-Dimethylhexadecylamine	[112-69-6]	Armeen DM16D (Armak), Kemamine T8702D (Humko), Jetamine T-16D (Jetco), Onamine 16 (Onyx)
N,N-Dimethyloctadecylamine	[124-28-7]	Armeen DM18D (Armak), Adogen 342D (Sherex), Barlene 18 (Lonza), Kemamine T9902D (Humko)
N,N-Dimethylcoco amine	[61788–93-0]	Armeen DMCD (Armak), Kemamine T6502D (Humko), Jetamine TCD (Jetco)
N,N-Dimethylsoya amine	[61788-91-8]	Armeen DMSD (Armak), Kemamine T9972D (Humko), Jetamine T-SD (Jetco)
N,N-Dimethylhydrogenated-tallow amine	[61788-95-2]	Armeen DMHTD (Armak), Adogen 345D (Sherex), Kemamine T9702D (Humko)
N-Octadecenyl-1,3-propanediamine	[7173-62-8]	Duomeen O (Armak)
N-Coco-1,3-propanediamine	[61791-63-7]	Diam 21 (Henkel), Duomeen C (Armak), Duomeen CD (Armak)
N-Tallow-1,3-propanediamine	[61791-55-7]	Diam 26 (Henkel), Duomeen T (Armak)
N-Soya-1,3-propanediamine	[61791-67-1]	Duomen S (Armak), Kemamine D-997 (Humko)

8. Diamines and Polyamines

8.1. Diamines

8.1.1. 1,2-Diaminoethane (Ethylenediamine)

(For triethylenediamine, see Chap. 6)

Physical Properties. 1,2-Diaminoethane [*107-15-3*], $H_2NCH_2CH_2NH_2$, $C_2H_8N_2$, M_r 60.10, is a colorless liquid, *mp* 10.9 °C, *bp* 117 °C (at 101.3 kPa), d_4^{20} 0.8966, n_D^{20} 1.4571.

Viscosity

t, °C	25	60	100
η, mPa · s	1.35	0.7	0.4

Vapor pressure

t, °C	20	40	60	80	100
p, kPa	1.21	3.87	11.13	26.53	55.99

Flash point	39.4 °C
Ignition temperature (class T2)	390 °C
Explosion limits in air	2.7 and 16.6 vol% (corresponding to 67.4 and 415 g/m^3)
Specific heat capacity	3.41 J g^{-1} K^{-1} (at 100 °C)
	3.45 J g^{-1} K^{-1} (at 110 °C)
	3.50 J g^{-1} K^{-1} (at 120 °C)
Heat of vaporization	636.4 J/g (at 0.1 MPa)
	660 J/g (at 40 kPa)

Vapor pressures, thermodynamic data, densities, and viscosities have been shown in graphical form [205]. Diaminoethane is infinitely miscible with water, methanol, diethyl ether, acetone, and benzene but is only slightly soluble in lower hydrocarbons.

Chemical Properties. As a primary diamine, diaminoethane forms stable mono- and diacid salts. The free base can be liberated from these salts by inorganic alkali solutions. The mononitrate and dinitrate are highly explosive in the dry state. The reaction of diaminoethane with chloroacetic acid or with formaldehyde and hydrogen cyanide or an alkali-metal cyanide to give ethylenediaminetetraacetic acid or its salts is particularly important (→ Ethylenediaminetetraacetic Acid and Related Chelating Agents). Diaminoethane forms complex salts with salts of copper, manganese, cobalt, and other subgroup elements.

If diaminoethane is heated with a fatty acid under conditions favoring dehydration, the prod- duct is a mono- or diamide, which can be converted into an imidazoline by cyclization. At high temperatures and pressures and in the presence of a catalyst, ammonia is eliminated from diaminoethane, and the resulting product undergoes cyclization to give piperazine. Diaminoethane reacts with aldehydes and ketones to give Schiff bases, which in turn can be hydrogenated to give secondary amines. Condensation with 1,2-diketones results in the elimination of water to give the corresponding 2,3-dihydropyrazines, which can be readily dehydrogenated to pyrazines [206]. Phosgene reacts with diaminoethane to give the hydrochloride of N,N'-ethyleneurea (2-imidazolidone) [207]. One mole of diaminoethane reacts with two moles of carbon disulfide in the presence of a basic catalyst to give ethylene bisdithiocarbamate [208]. With two moles of acrylonitrile, diaminoethane undergoes a smooth cyanoethylation reaction [209].

Production. 1,2-Diaminoethane is produced mainly by reacting ethylene chloride with aqueous or liquid ammonia at about 100 °C [210]. This process has been modified frequently [211].

In the past few decades, the reaction of ethanolamine with ammonia has developed into an important process [212]. In a continuous procedure, ethanolamine, ammonia, and hydrogen are passed over a cobalt catalyst at 20 MPa (200 bar) and 150–230 °C. The yield of 1,2-diaminoethane is 74%, based on an ethanolamine conversion of 93%.

A third important production process is the reaction of ethylene oxide with ammonia [213], as carried out by Berol in Sweden. Among the byproducts of this process is piperazine (ca. 10%). Union Carbide (USA) has recently brought a similar process into operation. In another commercial process, formaldehyde reacts catalytically with hydrocyanic acid, ammonia, and hydrogen [214]. Amination of diglycol also has been proposed. The production of diaminoethane by the reaction of chloroacetyl chloride with ammonia [215] is of less importance, whereas production by hydrogenation of aminoacetonitrile [216] seldom is carried out today.

Quality, Analysis, and Storage. The purity of diaminoethane, determined by titration and gas chromatography, is usually above 99%.

For storage, containers made of stainless steel and aluminum are preferred to avoid deterioration of the color number when the product is stored for long periods. Because it is corrosive, diaminoethane should not be stored in containers made of copper, copper alloys, or uncoated iron. Frequently containers made of iron coated with plastics, such as Lupolen (BASF), are used for shipping and storing diaminoethane. Because amines are hygroscopic and react with carbon dioxide in the air, tanks must be flushed with nitrogen before use. This also avoids discoloration during storage.

Uses. Diaminoethane has a number of uses and possible uses in a very large variety of areas. For example, it is converted into zinc and manganese salts of ethylene-bisdithiocarbamic acid, which are used as fungicides. Ethylenediaminetetraacetic acid is a further secondary product.

Diaminoethane-polyester condensates that are hydroxymethylated with formaldehyde can be used as plasticizers for phenolformaldehyde resins [217]. The use of condensates of diaminoethane, epoxides, and urea as nitrogen-containing polyol components for the production of polyurethane foams was proposed [218].

The addition of diaminoethane to viscose fiber spinning baths is said to improve the tensile strength of the fiber [219]. In the wet spinning of polyurethane fibers, diaminoethane is a rapid-action curing agent [220]. The incorporation of diaminoethane into diisocyanate-polyester prepolymers results in useful polymers for the production of elastic polyurethane fibers [221]. Ethylene–urea derivatives prepared from diaminoethane are used for textile finishing.

Tetraacetylethylenediamine, which is obtained from diaminoethane, is used in the detergent industry. Diaminoethane is employed as a stabilizer for rubber latex.

Thermoplastic adhesives are produced by condensing oligomeric fatty acids with diaminoethane [222]. In the mineral oil industry, diaminoethane acts as a stabilizer for halogen-containing high-pressure lubricating oils [223] and, in the form of Schiff bases with ketones, is also used as a metal deactivator [224].

Economic Aspects. After an annual increase of about 5–10% in the demand for ethylenediamine and its homologues up to about 1975, the market appears to have leveled off at present and is unlikely to increase in the future. In 1981, the free-market demand for diaminoethane worldwide, i.e., not including the producer's own requirements, was about 60 000 t/a. A substantial part is converted into fungicides and other products for agricultural purposes.

8.1.2. Diaminopropanes

In the past few years, the demand for diaminopropanes and in particular for *N*-substituted 1,3-diaminopropanes has increased sharply in a large variety of fields.

1,2-Diaminopropane [*78-90-0*], propylene-1,2-diamine, $H_2NCH_2CH(NH_2)CH_3$, $C_3H_{10}N_2$, M_r 74.13, *mp* −37.2 °C, *bp* 120.5 °C (at 101.3 kPa), d_4^{25} 0.8584, is a colorless liquid that is very hygroscopic and miscible with water and most organic solvents. It forms azeotropes with butanol, 2-methylpropanol, and toluene.

Because 1,2-dichloropropane does not react smoothly with ammonia, 1,2-diaminopropane is usually prepared by aminating amino-2-propanol, which is obtained from propylene oxide and ammonia [225]. The process differs from corresponding processes for the production of diaminoethane in that the temperature in the catalytic amination is higher (190–230 °C). Producers of 1,2-diaminopropane include Bayer and BASF.

1,2-Diaminopropane is used as an intermediate for crop-protection agents, e.g., Antracol (Bayer) or Basfungin (BASF). Especially in the United States, derivatives of the amine are required for use as fuel and lubricant additives.

1,3-Diaminopropane [*109-76-2*], propylene-1,3-diamine, trimethylenediamine, $C_3H_{10}N_2$, $H_2NCH_2CH_2CH_2NH_2$, M_r 74.13, *mp* −11.8 °C, *bp* 139.7 °C (at 101.3 kPa) d_4^{25} 0.884, n_D^{20} 1.4600. 1,3-Diaminopropane is a colorless liquid that is miscible with water and most organic solvents.

1,3-Diaminopropane is produced in a two-stage continuous process, under a pressure of 10–20 MPa (100–200 bar). In a first reactor, acrylonitrile reacts with excess ammonia at 70–100 °C to give 2-aminopropionitrile and bis(cyanoethyl)amine. The mixture of the two aminonitriles is then hydrogenated in a downstream reactor over a fixed-bed catalyst (cobalt or nickel) at 60–120 °C to give 1,3-diaminopropane and bis(aminopropyl)amine [226]. The yield of 1,3-diaminopropane can be increased by adding a polar solvent and water or by using a large excess of ammonia. Most of the 1,3-diaminopropane produced is converted into the textile finishing agent 1,3-dihydroxymethylhexahydropyrimid-2-one. 1,3-Diaminopropane is also used for ion exchangers, anticorrosion agents, etc.

N-Substituted 1,3-diaminopropanes can be prepared from acrylonitrile and ammonia or an amine, also with such a two-stage procedure. It is most efficient to produce the substituted aminopropionitrile batchwise and then hydrogenate it continuously in the presence of ammonia [227]. In some cases, hydrogenation also can be carried out batchwise over Raney nickel or Raney cobalt. A primary amine or ammonia reacts with acrylonitrile in a molar ratio of about 1:2 to give alkyl-bis(aminopropyl)amine, $RN[(CH_2)_3NH_2]_2$, as the principal product, whereas when the molar ratio exceeds 1:1, the principal product is a 1-amino-3-alkylaminopropane, $NH_2(CH_2)_3NHR$. Dialkylamines give 1-amino-3-dialkylaminopropanes.

Major producers of diamines, oligoamines, and polyamines of acrylonitrile include the Jefferson Chemical Co. and BASF.

The following liquid, water-miscible *N*-substituted diaminopropanes are industrially important. Their chief use is as components of epoxy resins:

1-Amino-3-methylaminopropane [*6291-84-5*], $H_2N(CH_2)_3NHCH_3$, $C_4H_{12}N_2$, M_r 88.1, *bp* 140–141 °C (at 101.3 kPa), d_4^{20} 0.852.

1-Amino-3-dimethylaminopropane [*109-55-7*], $H_2N(CH_2)_3N(CH_3)_2$, $C_5H_{14}N_2$, M_r 102.2, *mp* <−60 °C, *bp* 135 °C (at 101.3 kPa), n_D^{20} 1.4355, d_4^{20} 0.817, η (at 20 °C) 1.6 mPa · s, flash point 32 °C, ignition temperature 305 °C (class T2).

1-Amino-3-diethylaminopropane [*104-78-9*], $H_2N(CH_2)_3N(C_2H_5)_2$, $C_6H_{18}N_2$, M_r 130.2, *mp* <−50 °C, *bp* 169 °C (at 101.3 kPa), n_D^{20} 1.4424, d_4^{20} 0.829, η (at 20 °C) 1.4 mPa · s.

1-Amino-3-cyclohexylaminopropane [*3312-60-5*], $H_2N(CH_2)_3NHC_6H_{11}$, $C_9H_{20}N_2$, M_r 156.3, *mp* −16 °C, *bp* 113 °C (at 1.3 kPa), n_D^{20} 1.4810, d_4^{20} 0.9154, η (at 20 °C) 9.4 mPa · s, flash point 103 °C, ignition temperature 205 °C (class T3).

8.1.3. Higher Diamines

1,3-Diaminobutane [*590-88-5*], M_r 88.15, $H_2N(CH_2)_2CH(NH_2)CH_3$, $C_4H_{12}N_2$, can be produced from crotonaldehyde by amination under conditions of hydrogenation [228]. Diamines with four to five carbon atoms are important intermediates for drugs. However, the production of these diamines is restricted by the facts that the starting materials are difficult to obtain and that there is a strong tendency to cyclization during preparation.

1,4-Diaminobutane [*110-60-1*], tetramethylenediamine, putrescine, $H_2N(CH_2)_4NH_2$, $C_4H_{12}N_2$, M_r 88.15, mp 27 – 28 °C, bp 158 – 159 °C (at 101.3 kPa), d_4^{25} 0.877, n_D^{20} 1.4569, occurs as colorless crystals soluble in water and the conventional organic solvents.

1,4-Diaminobutane can be produced from 1,4-dichloro-2-butene or 1,4-dihalobutane (obtainable from the reaction of butadiene or tetrahydrofuran with a halogen, e.g., chlorine), preferably via the phthalimido compounds. Alternatively, 1,4-diaminobutane can be prepared by hydrogenating succinodinitrile [229].

1,5-Diaminopentane [*462-94-2*], pentamethylenediamine, cadaverine, $H_2N(CH_2)_5NH_2$, $C_5H_{14}N_2$, M_r 102.18, mp 9 °C, bp 178 – 180 °C (at 101.3 kPa), d_4^{25} 0.867, n_D^{25} 1.4561, can be prepared by methods similar to those described for 1,4-diaminobutane. The starting materials are 1,5-dichloropentane (obtainable from tetrahydropyran), glutarodinitrile, or glutaraldehyde (e.g., Relugan, BASF).

1-Diethylamino-4-aminopentane [*140-80-7*], $(C_2H_5)_2N(CH_2)_3CH(NH_2)CH_3$, $C_9H_{22}N_2$, M_r 158.29, bp 191 °C (at 101.3 kPa), d_4^{20} 0.821, is a colorless liquid that is soluble in water and the conventional organic solvents. The diamine is produced by a conventional method, for example, by aminating 1-diethylaminopentan-4-one under conditions of hydrogenation, by hydrogenating the oxime, or by aminating 1-diethylaminopentan-4-ol. The ketone is obtained from (2-chloroethyl) (diethyl)amine by synthesis of the acetoacetate, and 1-diethylaminopentan-4-ol is obtained by Mannich condensation of diethylamine, formaldehyde, and 1-butyn-3-ol followed by hydrogenation of the product, or is prepared from 1-methyltetrahydrofuran [230], [231]. Producers of 1-diethylamino-4-aminopentane are BASF, Bayer, and Rhône-Poulenc.

Derivatives of 1-diethylamino-4-aminopentane are particularly important as antimalarial agents [231] (e.g., Atebrin and Resochin, Bayer).

1,8-Diaminooctane [*373-44-4*], $H_2N(CH_2)_8NH_2$, $C_8H_{20}N_2$, M_r 144.26, mp 52 – 53 °C, bp 122 – 123 °C (at 2.4 kPa).

1,10-Diaminodecane [*646-25-3*], $H_2N(CH_2)_{10}NH_2$, $C_{10}H_{24}N_2$, M_r 172.32, mp 61 – 62 °C, bp 140 °C (at 1.6 kPa).

1,12-Diaminododecane [2783-17-7], $H_2N(CH_2)_{12}NH_2$, $C_{12}H_{28}N_2$, M_r 200.37, mp 66–67 °C, bp 162–164 °C (at 19 kPa).

Diamines containing more than six carbon atoms can be obtained readily by hydrogenating the corresponding dinitriles over a Raney nickel catalyst in a batch procedure under pressure at up to 125 °C. A five- to tenfold excess of liquid ammonia is used, with or without methanol solvent [232].

Higher diamines can be prepared continuously, as described for the production of hexamethylenediamine. For methods of preparing specific higher diamines, see [3].

8.2. Oligoamines and Polyamines

8.2.1. Physical Properties

Diethylenetriamine [111-40-0], bis(2-aminoethyl)amine, $(H_2NCH_2CH_2)_2NH$, $C_4H_{13}N_3$, M_r 103.17, is a colorless liquid, mp −39 °C, bp 207 °C (at 101.3 kPa), d_4^{20} 0.9542, n_D^{20} 1.4859, η (at 20 °C) 7 mPa · s.

Vapor pressure

t, °C	20	89	174
p, kPa	0.049	1.333	40.0

Flash point	102 °C
Ignition temperature	330 °C (class T2)
Specific heat capacity	3.44 J g^{-1} K^{-1} (at 190 °C)
	3.48 J g^{-1} K^{-1} (at 200 °C)
	3.53 J g^{-1} K^{-1} (at 210 °C)
Heat of vaporization	456.4 J/g (at 0.1 MPa)
	485.7 J/g (at 40 kPa)
Dissociation constant	7.07 × 10^{-5}
Surface tension	43.8 mN/m (at 25 °C)
	40.8 mN/m (at 50 °C)
Coefficient of expansion	9.1 × 10^{-4} (at 55 °C).

The amine is miscible with water, methanol, acetone, ether, and benzene, but insoluble in heptane.

Triethylenetetramine [112-24-3], bis(2-amino-ethyl)ethylenediamine, $H_2NCH_2CH_2NH-CH_2CH_2NHCH_2CH_2NH_2$, $C_6H_{18}N_4$, M_r 146.24, is a slightly yellowish viscous liquid, mp −35 °C, bp 277 °C (at 101.3 kPa), d_4^{20} 0.9819, n_D^{20} 1.4986, η 26.7 mPa · s (at 20 °C).

Vapor pressure

t, °C	20	144	240
p, kPa	0.001	1.33	40.0

Flash point 144 °C
Heat of vaporization 374.7 J/g (at 0.1 MPa)
 397.8 J/g (at 40 kPa)
Dissocation constant 6.7×10^{-5}

Tetraethylenepentamine [112-57-2], $H_2N(CH_2CH_2NH)_3CH_2CH_2NH_2$, $C_8H_{23}N_5$, M_r 189.31, is a slightly yellowish viscous liquid, bp 340 °C (at 101.3 kPa), d_4^{20} 0.9980, n_D^{25} 1.5076, η 96.2 mPa · s (at 20 °C).

Vapor pressure

t °C	20	195	300
p, kPa	0.0013	13.5	40

Dissociation constant 6.5×10^{-5}
Flash point 163 °C

Tetraethylenepentamine is infinitely miscible with water, methanol, acetone, benzene, and diethyl ether but is insoluble in heptane.

Dipropylenetriamine [56-18-8], 1-amino-3(3-aminopropyl)-aminopropane, $C_6H_{17}N_3$, M_r 131.2, $H_2N(CH_2)_3NH(CH_2)_3NH_2$, mp −15.1 °C; bp 238 °C (at 101.3 kPa), d_{20}^{20} 0.9290, vapor pressure < 0.001 kPa at 20 °C, dissociation constant 3.96×10^{-4}, η (at 20 °C) 9.6 mPa · s, n_D^{20} 1.4821, is miscible with water, methanol, and the conventional organic solvents.

N-Methyldipropylenetriamine [105-83-9], 1-amino-3-N-methyl-N-(3-aminopropyl)-aminopropane, $H_2N(CH_2)_3N(CH_3)(CH_2)_3NH_2$, $C_7H_{19}N_3$, M_r 145.3, mp −32 °C, bp 107 °C (at 1.7 kPa), n_D^{20} 1.4732, d_4^{20} 0.9040, η 7.4 mPa · s (at 20 °C), flash point 107 °C, ignition temperature 250 °C (class T3), is miscible with water and the conventional organic solvents.

3-(2-Aminoethyl)-aminopropylamine [13531-52-7], N-(2-aminoethyl)–1,3-diaminopropane, $H_2NCH_2CH_2NH(CH_2)_3NH_2$, $C_5H_{15}N_3$, M_r 117.2, mp −10 °C, bp 80 °C (at 0.3 kPa).

N,N'-Bis(3-aminopropyl)–1,2-diaminoethane [79554-59-9], N,N'-bis(3-aminopropyl)ethylenediamine, $C_8H_{22}N_4$, M_r 174.3, $H_2N(CH_2)_3NHCH_2CH_2NH(CH_2)_3NH_2$, mp −1.5 °C, bp 170 °C (at 0.3 kPa).

8.2.2. Chemical Properties

Because they possess terminal primary amino groups, condensates of 1,2-diaminoethane react very much like 1,2-diaminoethane itself. The higher condensates of 1,3-diaminopropane have similar properties. Dibasic acids react with oligoamines and polyamines to give polyamides [233]. A number of urea compounds can be obtained from oligoamines, e.g., diethylenetriamine, or from polyamines [234]. Theses amines react with aliphatic dihalides to give water-soluble cationic products [235] and with epichlorohydrin to give anionic polymers [236].

In a manner similar to diaminoethane, diethylenetriamine reacts with formaldehyde and sodium cyanide to give diethylenetriaminepentaacetic acid, which also has complex-forming properties. Diethylenetriamine undergoes also a smooth cyanoethylation reaction [209]. The oligoamines react directly with formaldehyde to give permethylated products. At elevated temperatures, diethylenetriamine can be cyclized to piperazine; if the temperature is about 400 °C, the cyclization reaction gives pyrazine directly [237].

8.2.3. Production, Analysis, and Uses

Diethylenetriamine is obtained as a byproduct in the synthesis of 1,2-diaminoethane from ethanolamine and ammonia. Diethylenetriamine also is formed, in addition to higher amines, when dichloroethane reacts with ammonia. A number of polyamines are obtained by cyanoethylation of diaminoethane or a diaminopropane with acrylonitrile, followed by hydrogenation of the product [108]. The regulations governing storage and transportation of the oligoamines and polyamines are the same as those for 1,2-diaminoethane.

Purity is determined preferably by titration, and analysis is made by fractional distillation. The oligoamines, e.g., diethylenetriamine, can be analyzed also by gas chromatography, but this method is unsuccessful for the higher polyamines, owing to their low vapor pressures.

Polyamines are used for textile finishing and are said to increase the water-repellent properties of many products, e.g., paper. The polyamine amides possessing terminal primary amino groups, which are prepared from fatty acids and polyamines, are used as curing agents for epoxy resins.

Distillation cuts of higher polyamines are being employed to an increased extent as asphalt additives in road building [238]. The use of polyamines as anticorrosion agents is also known [239]. The addition of diethylenetriamine to printing inks increases their resistance to moisture [240].

Table 9. MAK and TLV values of aliphatic amines

	MAK, ppm	TLV (TWA), ppm
Methylamine	10	10
Ethylamine	10	10
Dimethylamine	10	10
Diethylamine	25	10
Ethyldimethylamine	25	–
Trimethylamine	25	10
Butylamines	5	5
2-Propylamine	5	5
Diisopropylamine	–	5
2-Aminoethanol	3	3
Ethylenediamine	10	10
Cyclohexylamine	10	10
Morpholine	20	20
N-Ethylmorpholine	–	5
Piperazine dihydrochloride	–	5
Diethylenetriamine	–	1

9. Toxicology and Occupational Health

9.1. General Aspects

The aliphatic amines are highly irritating and corrosive. In their nonprotonated form they are lipid soluble and therefore penetrate rapidly into the lower layers of the tissue. Alkylamines — including the long-chain species — as well as alkylenediamines cause extremely severe corrosion with widely spread and deep necroses. Even in low concentration the vapors cause swelling and damage of the mucous membranes of the eye and the respiratory tract; "halos" are formed on the eye, and subsequently the cornea becomes turbid.

Acute Toxicity. Generally, the LD_{50} (rat, oral or dermal) is on the order of 100 mg/kg [241]. The symptoms are caused chiefly by high penetrability, corrosion, alkalosis, and, in some cases, by functional interference with the nerve-impulse conduction. The LD_{50} values of the hydrochlorides are usually higher because they do not cause corrosion and alkalosis: 1600 – 3200 mg/kg (rat, oral) [241]. The MAK and TLV values of aliphatic amines (1983) are compiled in Table 9.

β-Chloroalkylamines have an alkylating potential in addition to their corrosiveness. Therefore, they are suspected carcinogens. When inhaled, they cause severe burns of the respiratory tract. Secondary amines (including those with cyclic structures) can be transformed to the extremely carcinogenic N-nitrosodialkylamines in the presence of nitrites, or nitrates under reducing conditions.

Human Exposure. Irritation and corrosion of the skin and mucous membranes were observed frequently [241], [244], also on exposure to the vapor [245]. Accidental inhalation of large amounts can cause headache, nausea, excitation, and short convulsions [241]. Some aliphatic amines — especially di- and polyamines — can cause sensitization of the skin and the respiratory tract (allergies) [241], [246]. In this respect, ethylenediamine, diethylenetriamine, and triethylenetetramine are well-known sensitizers; cyclohexylamine and piperazine are suspected sensitizers. Less than 0.9 ppm of 3-(dimethylamino)- propylamine in the workplace air may adversely affect the respiratory function [247].

Systemic Toxicity. In neutral media the amines are almost completely protonated to the ammonium compounds. Therefore, interaction with the cation channels of the nerve-cell membranes is possible. This may result in a change of permeability for physiologically important cations and thus in inhibition of impulse conduction along the nerve fibers.

The essential structural element of most local anesthetics is a tertiary amino group. If significant amounts of these substances are present in the blood, effects adverse to the impulse conduction of the heart and especially to the inhibitory neurons of the central nervous system result. The clinical symptoms range from excitation to clonic convulsions. Consequently, excitation of the central nervous system with increased blood pressure and short spasms were observed along with irritation and injury to the respiratory tract on accidental inhalation of large amounts of aliphatic amines.

Long-chain mono- and dialkylamines are potential histamine liberators and may produce reddening and edema of the skin and mucous membranes, itching, decreased blood pressure, tachycardia, and bronchioconstriction [241], [248]; the maximum effect occurs with C_{10} monoamines and C_{14} diamines.

Certain aliphatic di- and polyamines, such as putrescine [*110-60-1*] ($H_2N(CH_2)_4NH_2$), spermidine [*124-20-9*] ($H_2N(CH_2)_4NH(CH_2)_3NH_2$), and spermine [*71-44-3*] ($H_2N(CH_2)_3NH(CH_2)_4NH(CH_2)_3NH_2$), are found characteristically in proliferating cells and are formed under the influence of hormones, natural growth stimuli, specific tumor promoters, and other factors. Their physiological action is obviously involved in cell division and tissue proliferation [249]. Whether these and related compounds exhibit tumorigenic potential at higher dosage and exposure cannot be evaluated at present.

Metabolism. Little is known about the metabolism of alkylamines. Generally, monoalkylamines are alkylated to dialkylamines, whereas trialkylamines are dealkylated to dialkylamines [241], [250]. In addition, primary, secondary, and tertiary amines are oxidatively dealkylated stepwise by monoamine oxidase to the aldehydes (or carboxylic acids), ammonia, and hydrogen peroxide. The oxidation rate increases with chain length; methylamine is not transformed at all by monoamine oxidase [250], [251]. Short-chain mono- and diamines are oxidized also by diamine oxidase (histaminase): those most easily oxidized are 1,4-butanediamine and the C_5-C_6 monoamines. Cyclohexylamine is eliminated unchanged with the urine [252].

9.2. Toxicology of Specific Amines

9.2.1. Alkylamines, Cyclic Amines, and Polyamines

Cyclohexylamine [*108-91-8*], LD_{50} 710 mg/kg (rat, oral), LD_{50} 320 mg/kg (rabbit, dermal) [253], is strongly irritating to the skin and mucous membranes [241]. This compound has been studied intensively because it is a metabolite of the sweetener saccharin [*81-07-2*]. Intake of 200 mg/kg daily in the feed of rats for two years did not lead to higher incidence of tumors [254]. Reduced feed intake, retarded development of body weight, and reduced spermatogenesis were observed with rats after 90 days [255] at daily doses of 2000 and 6000 ppm, but no such symptoms were evident at 600 ppm (30 mg/kg) daily intake. According to the same dosage scheme, cyclohexylamine was not found to be carcinogenic for rats in a two-year study [256]. A dose of 3000 ppm administered over 80 weeks to mice led neither to tumors nor to atrophy of the testicles [257].

In a multigeneration study (two years, 150 mg/kg daily of cyclohexylammonium chloride), the fertility of rats was not affected adversely in spite of slight atrophy of the testicles [258]. Based on these and other results [259], [260], a review concludes that no mutagenicity, carcinogenicity, or teratogenicity is suspected for cyclohexylamine [261]. However, a skin-sensitizing potential cannot be excluded at present [241].

Piperazine [*110-85-0*], 1,4-diethylenediamine, LD_{50} ca. 2200 mg/kg (rat, oral) [262], is strongly irritating and corrosive; the TLV is given in Table 9.

Human exposure. Headache, nausea, coordination disorders, and exanthems have been observed occasionally during repeated intake of piperazine salts (as anthelmintics) in daily doses of 30–75 mg/kg [263]. Apparently sensitizations of the skin and also the respiratory tract can occur; these are also caused by the dihydrochloride [264]–[266].

1-(2-Aminoethyl)piperazine [*140-31-8*], LD_{50} ca. 1500 mg/kg (rat, oral) [262], is corrosive to the skin and mucous membranes. Sensitization was observed in animal experiments [267].

Pyrrolidine [*123-75-1*], LD_{50} 300–670 mg/kg (rat, oral) [268], LC_{50} 1.3 mg/L (447 ppm; mouse, inhalation, 2 h); however, 6-h inhalation of 120, 444, and 2600 ppm was survived by rats [269]. For *N*-methylpyrrolidine an LC_{50} of 4.5–10.1 mg/L (rat, aerosol inhalation, 1 h) was found [262]. Depending on its concentration, the compound can irritate or corrode the skin and mucous membranes. As with other amines, sympathomimetic action is observed, resulting in increased blood pressure and a tendency to convulsions [268].

Ethylenediamine [*107-15-3*], LD$_{50}$ ca. 0.7 – 1.4 g/kg (rat, oral), is strongly corrosive to skin and eyes [241], [262]. Dermatitis related to primary irritation and also to sensitization occurs frequently [241]. Sensitization of the mucous membranes of the respiratory tract with asthmatic symptoms has also been described [270], [271].

Diethylenetriamine [*111-40-0*], LD$_{50}$ 1080 mg/kg (rat, oral) [241], [272], is strongly irritating; sensitization of the skin and respiratory tract occurs frequently [241]. Therefore, exposure should be as small as possible and below the odor threshold [273]. Derivatives (e.g., *N*-hydroxyethyl) of diethylenetriamine are less toxic and less irritating [274].

Triethylenetetramine [*112-24-3*], LD$_{50}$ 4340 mg/kg (rat, oral), LD$_{50}$ 820 mg/kg (rabbit, dermal) [253], is strongly irritating to corrosive to the skin and mucous membranes. The compound is also an effective sensitizer [241]. In addition, it is absorbed in significant amounts by the skin of guinea pigs and may lead to toxic effects in the kidneys, liver, brain, and placenta and possibly cause abortion [275], [276]. (Data on effective doses are not given.)

Hexamethylene imine, hexahydro-1*H*-azepine [*111-49-9*], LD$_{50}$ ca. 350 mg/kg (rat, oral), corrodes skin and mucous membranes [262].

Hexamethylenetetramine [*100-97-0*], HMTA, Urotropin, has an LD$_{50}$ of 9200 mg/kg (rat, i.v.). Daily oral intake of 5 g/kg by mice for ten days did not lead to toxic effects [268].
Chronic Toxicity. Several long-term tests of carcinogenicity did not reveal any evidence of a carcinogenic potential [268], [277] – [279]. During one experiment involving repeated subcutaneous injection, sarcomas developed at the site of injection [280]; however, this kind of response can also be caused by noncarcinogenic chemicals or insoluble materials.
Teratogenicity, Mutagenicity. In a study with female beagles (600 and 1250 ppm of HMTA in the feed; 4th to 56th day after mating), slight embryotoxic effects were found at higher dosage, but no malformations were observed [281]. In sublethal doses a weak mutagenic effect is found in male mice [282]. The compound is described as strongly mutagenic for larvae of *Drosophila* [283].
Sensitization. Repeated skin contact with HMTA can lead to allergies. Sensitized persons can suffer from asthmatic attacks on inhaling HMTA [283]; the sensitization is attributed to formaldehyde liberated from HMTA.

Morpholine [*110-91-8*], LD$_{50}$ 1600 – 1900 mg/kg (rat, oral) [262], [268], LD$_{50}$ ca. 500 mL/kg [268], irritates skin and mucous membranes strongly [262].
In a subacute feeding test (four weeks, rats) a daily dose of 323 mg/kg led to increased weight of the suprarenal gland accompanied by retarded development of body weight; daily doses of 27 and 93 mg/kg were without effect [268].

In vitro experiments of purified morpholine did not give any evidence of a mutagenic or carcinogenic potential [262], [284]. However, there is a possibility of nitrosation in the presence of traces of nitrous acid or nitrogen oxides [268], [285]: *N*-nitrosomorpholine is a potent carcinogen [284].

9.2.2. Fatty Amines

Skin and Eye Irritation. Fatty amines are local irritants of eyes, skin, and mucous membranes. Both the primary amines and the propanediamines are generally regarded as severely irritating and possibly corrosive. Even dilute solutions (5%) of fatty amines may cause irreversible damage to the eye; cornea inflammation caused by 1% solutions heals without permanent effects. Concentrated solutions of salts of fatty amines and quaternary ammonium compounds are almost neutral but they have an irritating potential similar to the free bases. Ethoxylation of fatty amines reduces the irritating effects.

A small number of workers became sensitized to selected fatty amines. Symptoms included skin rash, dermatitis, eye swelling, and a sensation of the skin, described as "crawling". Appropriate skin and eye protection must be worn when handling fatty amines.

Inhalation. At ambient temperatures only the short-chain amines, especially if branched, have sufficient vapor pressure to present any hazard by inhalation. The LC_{100} of 2-ethylhexylamine is 1.3 mg/L (rat, 4 h) [286], whereas exposure to air saturated with *N,N*-di-(tridecyl)tridecanamine for eight hours resulted in no deaths [287].

Acute and Chronic Oral Toxicity. None of the fatty amines are highly toxic by ingestion. The more corrosive ones, such as coco-1,3-diaminopropane, have oral LD_{50} values as low as 147 mg/kg, whereas compounds of higher molecular weight, such as di(hydrogenated-tallow) dimethylammonium chloride, are relatively nontoxic, with oral LD_{50} values of 7000 mg/kg and higher [288].

Chronic feeding studies have shown that rats fed with 500 ppm of 1-octadecylamine in the diet for two years showed no observable adverse health effects, whereas dogs fed the same level for twelve months showed mainly nonpathologic irritant effects on the mucosa of the gastrointestinal tract [289].

10. References

[1] A. Wurtz, *C. R. Hebd. Seances Acad. Sci.* **28** (1849) 223.
[2] E. Fischer, *Ber. Dtsch. Chem. Ges. Sonderheft* 1902, 197.
[3] *Houben-Weyl*, vol. **9/1**.
[4] J. Falbe, U. Hasserodt (eds.): *Katalysatoren, Tenside und Mineralöladditive*, Thieme Verlag, Stuttgart 1978.

[5] *Hydrocarbon Process.* **60** (Nov. 1981) 146.
[6] BASF, DE-OS 2 838 184, 1980 (W. Schroeder, H. Toussaint, W. Franzischka).
[7] Ruhrchemie, FR 1 080 359, 1954.
[8] J. Paek, P. Kondelik, P. Richter, *Ind. Eng. Chem. Prod. Res. Dev.* **11** (1972) 333.
[9] Ruhrchemie, DE-OS 2 048 750, 1970 (H. Goethel, B. Cornils, H. Feichtinger, H. Tummes, J. Falbe).
[10] J. Paek, P. Richter, *Chem. Prum.* **19** (1969) 63.
[11] W. S. Emerson in R. Adams: *Organic Reactions,* vol. **4**, J. Wiley & Sons, New York 1948, p. 174.
[12] ICI, DE 2 429 293, 1975 (T. Dewdney, D. Dowden, B. Hawkins, W. Morris).
[13] Erdölchemie, DE 2 839 134, 1978 (K. Mainusch, R. Schorsch, B. Schleppinghoff).
[14] C. Barnett, *Ind. Eng. Chem. Prod. Res. Dev.* **8** (1969) 145.
[15] Air Products, US 4 307 250, 1981 (J. Peterson, H. Fales).
[16] H. Kölbel, I. Abdulahad, M. Raleh, *Erdöl Kohle Erdgas Petrochem.* **28** (1975) no. 8, 385.
[17] H. Gilman, D. Shirley, *J. Am Chem. Soc.* **66** (1944) 888.
[18] S. Patai: *The Chemistry of the Amino Group,* Interscience, London-New York-Sydney 1968.
[19] L. K. Jackson, G. N. R. Smart, G. F. Wright, *J. Am. Chem. Soc.* **69** (1947) 1539.
[20] S. Patai: *The Chemistry of the Carbon Nitrogen Double Bond,* Interscience, London 1970.
[21] E. Schmitz, *Angew. Chem.* **73** (1961) 23; *Houben-Weyl,* vol. **10/2** (1967) 71.
[22] H. Krimm, *Chem. Ber.* **91** (1957) 1057; *Houben-Weyl,* vol. **10/4** (1968) 449.
[23] L. Spialter, J. A. Pappalardo: *The Acyclic Aliphatic Tertiary Amines,* MacMillan Publ. Co., New York 1965, p. 29.
[24] V. A. Nekrasova, N. I. Shuikin, *Russ. Chem. Rev. (Engl. Transl.)* **34** (1965) 843.
[25] Monsanto, GB 716 649, 1952.
[26] *Houben-Weyl,* vol. **11/1** (1957) 687; ICI, GB 969 542, 1964 (J. D. Seddon); K. Smekal, G. Esser, DD 27 348, 1964. BASF, DE 1 921 467, 1969 (H. Corr, W. Friedrichsen).
[27] Mitsui Chem. Ind., JP 3 180, 1968 (T. Sasa, H. Hirai).
[28] BASF, DE 3 045 719, 1980 (N. Goetz, P. Jacobs); Monsanto, US Appl. 271 869, 1981 (H. L. Merten, G. R. Wilder); US 3 196 179, 1962 (R. M. Robinson).
[29] BASF, DE 1 543 377, 1966 (K. Adam, E. Haarer). Inst. Cercetari, RO 63 243, 1975 (E. Dutkay).
[30] Halcon Internat., NL 6 401 010, 1964.
[31] Inst. Français du Pétrole, GB 1 050 589, 1965 (B. Choffe, R. Stern). Quaker Oats, FR 1 446 554, 1965 (A. P. Dunlop, D. G. Manly). Mitsubishi Chem. Ind., JP 4 332, 1968 (R. Matsura, T. Otaki). Toa Gosei Chem. Ind., JP 19 897, 1970 (T. Kawaguchi, T. Matsubara); JP 19 898, 1970 (T. Kawaguchi, T. Matsubara).
[32] BASF, DE 3 045 719, 1980 (N. Goetz, P. Jacobs).
[33] Monsanto, US 2 191 657, 1937 (M. W. Harman).
[34] US Rubber, GB 898 319, 1962.
[35] Monsanto, US 2 731 338, 1951 (E. A. Fike, H. L. Morrill).
[36] Esso, GB 793 737, 1958 (J. Scott, R. J. Westland).
[37] Dearborn Chem., US 2 956 889, 1956 (W. L. Denman); *Chem. Eng. (N.Y.)* **62** (1955) no. 4, 140.
[38] BASF, DE 1 921 467, 1970 (W. Corr, W. Friedrichsen).
[39] W. S. Emerson, *Org. React. (N.Y.)* **4** (1948) 218.
[40] Shell Development, US 2 580 284, 1949 (T. J. Deahl, F. H. Stross).
[41] Goodyear, US 3 086 018, 1957 (A. F. Hardman). Monsanto, US 2 930 777, 1955 (H. M. Leeper).
[42] Du Pont, US 3 376 341, 1964 (C. R. Bauer). BASF, DE 851 189, 1944 (O. Stichnoth).
[43] W. S. Emerson, *Org. React. (N.Y.)* **4** (1948) 174. BASF, DE 1 543 354, 1966 (H. Corr, E. Haarer).

[44] Bayer, DE 826 641, 1949 (E. Windemuth).
[45] Bayer, DE 962 113, 1953 (F. Brochhagen, A. Höchtlen).
[46] Bayer, DE 946 173, 1953 (W. Bunge, K. H. Mielke).
[47] J. F. Norris, A. E. Bearse, *J. Am. Chem. Soc.* **62** (1940) 953.
[48] BASF, DE 805 518, 1949 (O. Stichnoth). Ugine Kuhlmann, FR 1 530 477, 1967.
[49] Witco Chem., US 3 551 486, 1968 (J. M. Solomon).
[50] Abbott Lab., FR 1 333 693, 1963 (R. M. Robinson).
[51] Dow, US 2 571 016, 1949 (J. Dankert). Abbott Lab., FR 1 343 391, 1963 (R. M. Robinson); US 3 351 661, 1967 (F. H. Van Munster).
[52] I. Garz, K. Schwabe, *Werkst. Korros.* **14** (1965) 842. Nagynyomasu Kiserleti Intezet, HU 16 386, 1976 (J. Gemes).
[53] R. Johannsen, *Farbe Lack* **70** (1964) 189.
[54] S. Hünig, M. Kiessel, *J. Prakt. Chem.* **277** (1958) 224.
[55] Du Pont, DE 1 468 779, 1964 (L. D. Brake).
[56] BASF, DE 1 272 919, 1966 (P. Raff, H. G. Peine).
[57] BASF, DE 2 036 502, 1970; DE 2 038 023, 1970 (K. Adam, H. Hoffmann).
[58] R. Jacquier, H. Christol, *Bull. Soc. Chim. Fr.* 1954, 556.
[59] L. Ruzicka, M. W. Goldberg, M. Hürbin, *Helv. Chim. Acta* **16** (1933) 1339.
[60] L. Schuster, *Tetrahedron Lett.* **1963**, 2001. BASF, DE 1 178 846, 1964 (L. Schuster).
[61] BASF, FR 1 468 354, 1966 (K. Adam, E. Haarer).
[62] V. Prelog, M. Fausy El-Neweihy, O. Häfliger, *Helv. Chim. Acta* **33** (1950) 365.
[63] H. Meister, *Justus Liebigs Ann. Chem.* **679** (1964) 83.
[64] L. Schuster, *Tetrahedron Lett.* **1963**, 2001. BASF, NL 6 408 389, 1965.
[65] BASF, FR 1 528 400, 1967.
[66] J. M. Patterson, P. Drenchko, *J. Org. Chem.* **24** (1959) 878.
[67] I. G. Farbenind., DE 701 825, 1938 (W. Reppe).
[68] Du Pont, US 2 525 584, 1948 (C. A. Bordner).
[69] J. Timmermans, *J. Chim. Phys. Phys. Chim. Biol.* **34** (1937) 693.
[70] S. A. G. Osborn, D. R. Douslin, *J. Chem. Eng. Data* **13** (1968) 543.
[71] A. G. Cook: *Enamines*, Marcel Dekker, New York 1969.
[72] I. Szmuszkovicz, *Adv. Org. Chem.* **4** (1963) 1.
[73] K. Blàha, O. Cesvinka, *Adv. Heterocycl. Chem.* **6** (1966) 147.
[74] G. Laban, R. Bayer, *Z. Chem.* **8** (1968) 165.
[75] K. Smeykal, K. K. Moll, *Chem. Tech. (Leipzig)* **19** (1967) 92.
[76] A. A. Ponomarev, A. S. Ĉegolja, *Khim. Geterotsikl. Soedin.* 1966, 239.
[77] Celanese, DE 1 802 122, 1968 (D. R. Larkin).
[78] Quaker Oats Co., US 3 163 652, 1962 (D. G. Manly). Inst. of Petrochem. Synth., SU 2 478 887, 1977 (A. N. Bashkirov).
[79] ICI, GB 971 187, 1962 (C. Gardner, G. A. Silverstone).
[80] Rhoâne-Poulenc, FR 1 475 961, 1966 (G. Chichery, P. Perras).
[81] ICI, NL 6 605 495, 1966.
[82] Ciba, US 2 826 583, 1954 (K. Hoffmann, J. Heer).
[83] Phillips Petroleum, US 2 771 737, 1951 (C. R. Scott, A. L. Ayers).
[84] G. Jones, *Org. React. (N.Y.)* **15** (1967) 204.
[85] A. T. Nielsen, W. J. Honlihan, *Org. React. (N.Y.)* **16** (1968) 1.
[86] G. B. Bachmann, M. T. Atwood, *J. Am. Chem. Soc.* **78** (1956) 484.
[87] E. N. Zilbermann, E. D. Skorikowa, *Zh. Obshch. Khim.* **23** (1953) 1659.

[88]   Bayer, DE 738 448, 1941 (H. Raab).
[89]   BASF, DE 1 138 780, 1960 (A. Palm, H. Stanger).
[90]   Artemev, GB 1 570 647, 1980 (D. M. Popov, E. V. Genkina). BASF, DE 2 837 290, 1978 (W. Schröder, W. Franzischka). Mitsubishi, JP 82 183 771, 1982.
[91]   Monsanto, US 3 635 952, 1969 (D. A. Tyssee).
[92]   Soc. Prod. Chim. Marles Kuhlmann, FR 1 583 461, 1969.
[93]   *Houben-Weyl*, vol. **11/1** (1957) 45, 129.
[94]   Wyandotte, DE 1 049 864, 1957 (W. K. Langdon, E. Jaul).
[95]   Goodyear, DE 3 002 342, 1979 (K. J. Frech). BASF, DE 3 125 662, 1981 (W. Schröder, W. Lengsfeld, G. Heilen). Jefferson, US 3 155 657, 1961 (W. C. Bedoit).
[96]   M. F. Obrecht, *Effluent Water Treat. J.* **4** (1964) 279.
[97]   Chem. Werke Hüls, BE 642 387, 1964.
[98]   H. Koppers GmbH, US 3 555 782, 1968 (H. Deringer).
[99]   E. Cinelli, P. L. Girotti, R. Tesei, *Hydrocarbon Process. Pet. Refiner* **42** (1963) no. 8, 141.
[100] Wyandotte, US 3 167 551, 1961 (E. A. Weipert). Agency Ind. Sci. Techn., JP 80 43 008, 1980 (Y. Sugi).
[101] Jefferson, US 3 210 349, 1961 (N. B. Godfrey).
[102] BASF, DE 1 793 380, 1968 (E. Fürst, H. Toussaint).
[103] Jefferson, US 3 087 928, 1961 (N. B. Godfrey).
[104] Pennsalt Chem., US 3 155 656, 1960 (R. H. Goshorn, R. Ferren).
[105] Jefferson, US 3 151 113, 1962 (P. S. Advani, G. P. Speranza). ICI, GB 1 106 084, 1964 (L. L. Dufty).
[106] BASF, FR 1 575 059, 1968.Union Carbide, US 3 112 318, 1957 (R. C. Lemmon, R. C. Myerly).
[107] Du Pont, US 2 130 948, 1937 (W. H. Carothers).
[108] M. M. Epstein, C. W. Hamilton, *Mod. Plast.* 37 (1960) 142.
[109] Eastman Kodak, US 2 950 197, 1960 (C. F. H. Allen, E. R. Webster). AGFA, DE 1 130 283, 1962 (W. Himmelmann, H. Ulrich).
[110] Air Products and Chemicals, US 4 362 886, 1981 (M. E. Ford).
[111] Air Products and Chemicals, US 3 166 558, 1961 (R. L. Mascioli).
[112] Houdry Process Corp., US 2 937 176, 1956 (E. C. Herrick).
[113] Air Products and Chemicals, EP 69 322, 1981 (J. E. Wells). Chem. Werke Hüls, DE 1 445 578, 1963 (W. Thomas). Bayer, DE 2 846 813, 1978 (L. Imre, W. Horstmann).
[114] Houdry Process Corp., DE 1 120 691, 1958 (M. Orchin).
[115] National Starch and Chem. Corp., DE 2 040 607, 1970 (H. H. Stockmann).
[116] S. S. Chang, E. F. Westrum, *J. Phys. Chem.* **64** (1960) 1547.
[117] A. Chentin, I. P. Mathieu, *J. Chem. Phys.* **53** (1956) 106.
[118] VEB Gärungschemie Dessau, DE 1 076 104, 1959 (H. Plischke, W. Zettl). H. Plischke, *Z. Chem.* **1** (1961) 217.
[119] Stachowski, PL 106 934, 1980. Celanese, US 2 640 826, 1949 (A. F. Maclean, A. L. Stautzenberger).
[120] Bayer, DE 977 337, 1953 (R. Ludwig, F. Halle).
[121] Montecatini, BE 601 096, 1960. Soc. Belge de l'Azote et des Prod. Chim., GB 962 925, 1962 (R. Englebert, A. Genkenne, A. Lefebre).
[122] F. Meissner, E. Schwiedessen, D. F. Othmer, *Ind. Eng. Chem.* **46** (1954) 724.
[123] Tenneco, US 3 538 199, 1968 (S. Weiss, D. X. Klein); BE 743 974, 1969 (S. Weiss).
[124] Phillips Petroleum, US 3 547 597, 1967 (G. E. Hayes).
[125] Degussa, DE 1 104 967, 1961 (H. J. Mann). BASF, US 2 912 435, 1957 (H. Scholz).

[126] Degussa, DE 1 085 491, 1959 (A. Lehn).
[127] Bayer, FR 895 645, 1962. Food Machinery and Chem. Corp., US 2 833 656, 1953 (A. F. Kalmar, H. F. Fitzpatrick).
[128] Dow, GB 960 732, 1964; US 3 268 399, 1964 (R. R. Langner); US 3 524 854, 1968 (S. J. Kuhn).
[129] "Armeen Aliphatic Amines", *Armak Product Data Bulletin* 72-8 (1972).
[130] *Kirk-Othmer,* 2nd ed., vol. **2**, pp. 130, 132.
[131] Melting points were calculated using references [132], [133] , and internal Armak data converted into a polynomial. Vapor pressures were correlated in a similar way using the Antoine equation given in [134].
[132] R. R. Dreisbach: "Physical Properties of Chemical Compounds, III", *Adv. Chem.* 1961, no. 29. 1.
[133] *Selected Values of Properties of Chemical Compounds,* Thermodynamics Research Center, Department of Chemistry, Texas A & M University, June 30, 1971.
[134] J. A. Dean: *Lange's Handbook of Chemistry,* 11th ed., McGraw-Hill, New York 1973, p. 10–31.
[135] G. Tauber, A. May, *Tenside Deterg.* **19** (1982) no. 3, 151–156.
[136] A. Weiss, *Tenside Deterg.* **19** (1982) no 3, 157.
[137] C. W. Hoerr, M. R. McCorkle, A. W. Ralston, *J. Am. Chem. Soc.* **65** (1943) 328, 329.
[138] Kao Soap Co., JP Kokai 7 616 603, 1976 (U. Nishimoto, K. Okabe, M. Takaku).
[139] Akzona Inc., US 4 237 064, 1980 (R. A. Reck).
[140] H. L. Sanders, J. B. Braunwarth, R. B. McConnell, R. A. Swenson, *J. Am. Oil Chem. Soc.* **46** (1969) no. 3, 167–170.
[141] Henkel DE-OS 2 117 427, 1972 (G. Demmering).
[142] Armour, GB 1 223 730, 1967 (R. A. Reck).
[143] Akzona Inc., US 4 368 127, 1983 (J. M. Richmond).
[144] Union Carbide, US 3 636 114, 1972 (E. Tobler, D. J. Foster).
[145] Armour, US 3 222 402, 1965 (M. C. Cooperman).
[146] R. S. Egly, Nitroparaffin Symposium, Commercial Solvents Corp., Chicago, Jan. 25, 1956.
[147] Akzona Inc., US 4 334 887, 1982 (D. Frank, L. D. Metcalfe).
[148] Akzona Inc., US 4 357 181, 1982 (D. Frank, L. D. Metcalfe).
[149] P. Jowett, GB 1475 689, 1973.
[150] W. R. Grace & Co., GB 1 388 053, 1975 (R. A. Diffenbach).
[151] Chemcell, US 3 574 754, 1971 (G. Specken).
[152] Hoechst, DE-OS 1 941 290, 1971 (H. Mueller, W. Kuehn, G. Mueller-Schiedmayer, J. Strauss).
[153] Hoechst, DE-AS 1 280 243, 1968 (H. Oberrauch, W. Froehlich, H. Lewinsky); *Chem. Abstr.* **70** (1969) 11 108.
[154] Archer-Daniels-Midland Co., US 3 264 254, 1966 (E. J. Sawyer).
[155] Armour, GB 860 922, 1961 (S. H. Shapiro, F. Pilch).
[156] Lion Fat and Oil, US 4 248 801, 1981 (S. Tomidokoro, M. Sato, D. Saika).
[157] Hoechst, US 3 444 205, 1969 (W. Froehlich, H. Oberrauch, K.Fischer).
[158] Kao Soap Co., JP Kokai 81 152 441, 1981.
[159] BASF, DE-OS 2 639 648, 1978 (H. Hoffmann, H. Mueller, H. Toussaint, A. Wittwer).
[160] Archer-Daniels-Midland Co., US 3 223 734, 1965 (H. T. Fallstad, A. E. Rheineck).
[161] Gulf Research & Development Co., US 2 953 601, 1960 (A. C. Whitaker).
[162] BASF, DE-OS 2 645 712, 1978 (H. Hoffmann, H. Mueller, H. Toussaint, A. Wittwer); US 4 207 263, 1980.
[163] Japan Catalytic Chem. Ind., JP Kokai 7 424 905, 1974 (H. Koike, N. Kurata, Y. Okuda).

[164] Imperial Chem. Ind., GB 2 048 866, 1979 (C. B. Hanson, G. E. B. Smith).
[165] Gulf Research and Development Co., US 4 251 465, 1981 (H. E. Swift, R. A. Innes, P. Adams).
[166] UOP, DE-OS 3 014 455, 1981.
[167] Kao Soap Co., US 4 254 060, 1980 (H. Kimura, K. Matsutani, S. Tsutsumi).
[168] Kao Soap Co., US 4 210 605, 1980 (F. Hoshino, H. Kimura, K. Matsutani).
[169] Gulf Research and Development Co., US 3 742 060, 1973 (R. W. Lagally, G. D. Johann).
[170] UOP, US 4 250 115, 1981 (T. Imai).
[171] Shell Oil Co., US 3 234 283, 1966 (H. v. Finch, R. E. Meeker).
[172] M. R. McCorkle, P. L. Dubrow, B. E. Marsh, *J. Am. Oil Chem. Soc.* **45** (1968) 10A.
[173] Armour, US 3 338 967, 1966; US 3 530 153, 1966 (R. H. Potts, E. J. Miller, A. Mais).
[174] Ethyl Corp., GB 1 115 238, 1968.
[175] Texaco Development Corp., US 3 739 027, 1971.
[176] Continental Oil Co., US 3 371 118, 1968 (A. J. Lundeen, K. L. Motz).
[177] Hoechst, US 4 234 509, 1980; DE-OS 2 813 294, 1978; DE-OS 2 737 607, 1979 (S. Billenstein, B. Kukla, H. Stuehler).
[178] Henkel, DE 1 288 595, 1969 (H. Rutzen); US 3 579 585, 1971 (H. Rutzen).
[179] SNIA Viscosa, US 4 198 348, 1980 (C. A. Pauri).
[180] C. Jungermann: *Cationic Surfactants*, Marcel Dekker, New York 1970, Chap. 13.
[181] L. D. Metcalfe, *J. Am. Oil Chem. Soc.* **56** (1979) 786A.
[182] K. G. van Senden, *Recl. Trav. Chim. Pays Bas* **84** (1965) 1459.
[183] DGF-Einheitsmethoden M-V-6 (57).
[184] G. Grossi, *J. Gas Chromatogr.* **3** (1965) no. 5, 170.
[185] L. D. Metcalfe, A. A. Schmitz, *J. Gas Chromatogr.* **2** (1964) no. 1, 15.
[186] L. D. Metcalfe, R. J. Martin, *Anal. Chem.* **44** (1972) 403.
[187] R. B. Stage, J. B. Stanley, P. B. Moseley, *J. Am. Oil Chem. Soc.* **49** (1972) no. 2, 87.
[188] G. F. Okunev, L. G. Morozova, E. M. Shmeleva, *Zavod. Org. Sint.* **2** (1972) 173; *Chem. Abstr.* **83** (1975) 178 196 v.
[189] A. E. Merbach, *Mitt. Geb. Lebensmittelunters. Hyg.* **66** (1975) 176.
[190] F. Mozayeni, *Appl. Spectrosc.* **33** (1979) 520.
[191] W. P. Evens, *Chem. Ind.(London)* 1969, 893–903.
[192] Du Pont, US 2 088 325, 1937 (J. E. Kirby).
[193] Du Pont, US 2 132 902, 1938 (S. Lenhar).
[194] J. J. Barr, Jr, *Rock Prod.* **49** (1946) 88.
[195] Tate & Lyle, US 3 698 651, 1972 (M. C. Bennett).
[196] L. L. Johnsen, *Weeds* **13** (1965) 123–130.
[197] Armour, US 3 246 015, 1966 (H. L. Lindaberry, W. W. Abramitis).
[198] Akzona, US 3 506 433, 1970 (W. W. Abramitis, R. A. Reck).
[199] Sherwin-Williams, US 3 166 469, 1965 (H. A. Pass, M. S. H. Nurse, B. J. Watt).
[200] Armour, US 3 135 656, 1964 (W. W. Abramitis, R. A. Reck).
[201] Armour, US 3 223 517, 1965 (W. W. Abramitis, R. A. Reck).
[202] H. D. Hueck, D. M. M. Adema, J. R. Wiegman, *Appl. Microbiol.* **14** (1966) 308.
[203] *Fatty Acids*, North America, Hull and Co., New York 1975.
[204] Akzo-Chemie, internal market studies.
[205] R. W. Gallant, *Hydrocarbon Process.* **48** (1969) 143.
[206] I. Flament, M. Stoll, *Helv. Chim. Acta* **50** (1967) 1754.
[207] N. Puschin, R. Mitič, *Justus Liebigs Ann. Chem.* **532** (1937) 300.
[208] Rohm & Haas, US 2326643, 1941 (W. F. Hester).

[209] H. Bruson, *Org. React. (N.Y.)* **5** (1949) 79.
[210] IG Farbenind., *BIOS Final Rep.* no. 1154, no. 5; Toyo Soda, JP 7926288, 1979 (S. Otsuki).
[211] Simon Carves, GB 1147984, 1965 (J. G. Blears, P. Simpson). Nippon Soda, JP 6325, 1965.
[212] Dow, US 2861995, 1956 (G. F. Mac Kenzie). BASF, DE 1172268, 1962 (S. Winderl, E. Haarer).Union Carbide, US 3112318, 1957 (R. C. Lemon, R. C. Myerly).
[213] *Eur. Chem. News* **20** (1971) no. 495, 16.
[214] Rohm & Haas, JP 1909, 1964.
[215] Prod. Chim. Pechiney-St. Gobain, US 3281471, 1963 (P. Chassaing, J. Coillard).
[216] Du Pont, US 2429876, 1944 (W. F. Gresham).
[217] Monsanto, FR 1426289, 1964 (J. R. Le Blanc).
[218] Bayer, FR 1411269, 1964 (H. J. Hennig, G. Braun).
[219] Phrix, DE 1185331, 1959 (F. Kaiser, P. Weber).
[220] Thiocol Chem., GB 1038355, 1964.
[221] Polythane, BE 633253, 1963 (A. D. Schneider, T. V. Peters).
[222] Schering, NL 6506239, 1965.
[223] Texas Corp., US 2696473, 1951 (S. J. Sokol).
[224] Shell Dev., US 2285878, 1942 (E. R. White).
[225] BASF, FR 1347648, 1963 (S. Winderl, E. Haarer).
[226] BASF, DE 1034184, 1957 (H. Scholz, H. Günthert).
[227] *Houben-Weyl,* vol. **11/1** (1957) 272, 565.
[228] *Houben-Weyl,* vol. **11/1** (1957) 606.
[229] *Houben-Weyl,* vol. **11/1** (1957) 92, 95, 558. Nitto Chemical, JP 7940524, 1979 (S. Tomita).
[230] A. Wingler, *Angew. Chem.* **61** (1949) 52.
[231] *Houben-Weyl,* vol. **11/1** (1957) 61, 72, 131, 498.
[232] *Houben-Weyl,* vol. **11/1** (1957) 558, 567.
[233] Cyanamid, US 2371104, 1943 (R. H. Kienle).
[234] Cyanamid, US 2554475, 1947 (T. J. Suen).
[235] Talbott Dev. Assoc., US 2694633, 1950 (D. K. Pattiloch).
[236] Cyanamid, US 2469683, 1945 (J. R. Dudley).
[237] Cyanamid, US 2414552, 1942 (H. F. Pfann).
[238] *Chem. Week* **80** (1957) May 4, 54.
[239] Nat. Aluminate, US 2580923/4, 1947 (A. L. Jacoby).
[240] A. F. Schmutzler, D. F. Othmer, US 2606123, 1950.
[241] R. R. Beard, J. T. Noe: "Aliphatic and Alicyclic Amines," in C. D. Clayton, F. Clayton (eds.): *Patty's Industrial Hygiene & Toxicology,* vol. **II B**, Wiley Interscience, New York 1982.
[242] Deutsche Forschungsgemeinschaft (ed.): *Maximale Arbeitsplatzkonzentrationen (MAK) 1983,* Verlag Chemie, Weinheim 1983.
[243] Am. Conf. of Governmental Industrial Hygienists (ed.): *Threshold Limit Values (TLV) 1983,* Cincinnati, Ohio, 1983.
[244] L. B. Bourne, F. J. M. Miller, L. B. Alberman, *Br. J. Ind. Med.* **16** (1959) 81.
[245] C. P. Carpenter, H. F. Smyth, jr., *Am. J. Ophthalmol.* **29** (1946) 1363–1372.
[246] R. E. Eckardt, *EHP Environ. Health Perspect.* **17** (1976) 103–106.
[247] R. E. Brubaker, H. J. Muranko, D. B. Smith, G. J. Beck, G. Scovel, *J. Occup. Med.* **21** (1979) 688–690.
[248] T. Sollman: *A Manual of Pharmacology,* 8th ed.,Saunders, Philadelphia 1957.
[249] D. R. Morris, L. J. Marton: *Polyamines in Biology and Medicine,* Marcel Dekker, New York-Basel 1981.

[250] H. Blaschko, *Pharmacol. Rev.* **4** (1952) 415.
[251] M. L. Simenhoff: "Metabolism of Aliphatic Amines," *Kindney Int. Suppl.* **3** (1975) 314–317.
[252] D. Henschler (ed.): *Gesundheitsschädliche Arbeitsstoffe,* Verlag Chemie, Weinheim 1979.
[253] H. Smyth, C. Carpenter, C. Weil, U. Pozzani, J. Striegel, J. Nycum, *Am. Ind. Hyg. Assoc. J.* **30** (1969) 470.
[254] D. Schmähl, *Arzneim. Forsch.* **23** (1973) 1466–1470.
[255] J. F. Gaunt, M. Sharrat, P. Grasso, A. B. Landsdown, S. D. Gangolli, *Food Cosmet. Toxicol.* **12** (1974) 609–624.
[256] J. F. Gaunt, J. Hardy, P. Grasso, S. D. Gangolli, K. R. Butterworth, *Fed. Cosmet. Toxicol.* **14** (1976) 255–267.
[257] J. Hardy, J. F. Gaunt, J. Hooson, R. J. Hendy, K. R. Butterworth, *Fed. Cosmet. Toxicol.* **14** (1976) 269–276.
[258] D. L. Oser, S. Carson, G. E. Cox, E. E. Vogin, S. S. Sternberg, *Toxicology* **6** (1976) 47–65.
[259] M. S. Legator, K. A. Palmer, S. Green, K. W. Peterson, *Science* **165** (1969) 1139–1140.
[260] J. H. Turner, D. H. H. Hutchinson, *Mutat. Res.* **26** (1974) 407.
[261] P. Cooper, *Food Cosmet. Toxicol.* **15** (1977) 69–70.
[262] BASF Aktiengesellschaft, unpublished results.
[263] J. R. Berger, M. Globus, E. Melamed, *Arch. Neurol.* **36** (1979) 180–181.
[264] S. Fregert, *Manual of Contact Dermatitis,* Scandinavian University Books, Munksgaard 1974.
[265] J. Pepys, C. A. C. Pickering, H. W. G. London, *Clinical Allergy* **2** (1972) 189–196.
[266] L. Hagmar, T. Bellander, B. Bergbö, B. G. Simonsson, *JOM J. Occup. Med.* **24** (1982) 193–197.
[267] A. Thorgeirson, *Acta Derm. Venereol.* **58** (1978) 332–336.
[268] C. F. Reinhardt, M. S. Britelly in G. D. Clayton, F. Clayton (eds.): *Patty's Industrial Hygiene and Toxicology,* vol. **II A,** Wiley-Interscience, New York 1981, p. 2671–2822.
[269] G. N. Zaeva, N. N. Ordynskaya, L. J. Duninina, N. J. Osipenko, V. N. Ivanov, *Gig. Tr. Prof. Zabol.* **2** (1974) 29–32.
[270] C. U. Dernehl, *Ind. Med. Surg.* **20** (1951) 541.
[271] S. Lam, M. Chang-Yeung, *Am. Rev. Resp. Dis.* **121** (1980) 151–155.
[272] F. S. Malette, E. v. Haan, *Arch. Ind. Hyg. Occup. Med.* **5** (1952) 311.
[273] Amer. Ind. Hyg. Assoc.: "Hygienic Guide Series," *Am. Ind. Hyg. Assoc. J.* **21** (1960) 266, 268.
[274] C. H. Hine, J. K. Kodoma, H. H. Anderson, D. W. Simonson, J. S. Wellington, *AMA Arch. Ind. Health* **17** (1958) 129–144.
[275] W. Dobryszycka, J. Kulpa, A. Woyton, J. Woyton, J. Szacki, A. Dziboa, *Arch. Immunol. Ther. Exp.* **33** (1975) 867–870.
[276] H. Szacki, J. Woyton, A. Dzioba, J. Rabczynski, A. Woyton, *Arch. Immunol. Ther. Exp.* **22** (1974) 123–128.
[277] R. Brendel, *Arzneim. Forsch.* **14** (1964) 51.
[278] G. Della Porta, *Food Cosmet. Toxicol.* **4** (1966) 362–363.
[279] H. Netvig, J. Anderson, W. Wulff-Rasmussen, *Food Cosmet. Toxicol.* **9** (1971) 491–500.
[280] F. Watanabe, S. Sugimuto, *Gann* **46** (1955) 365–366.
[281] H. Hurni, H. Ohder, *Food Cosmet. Toxicol.* **11** (1973) 459–462.
[282] G. Röhrborn, F. Vogel, *Dtsch. Med. Wochenschr.* **92** (1967) 2315–2321.
[283] I. A. Rapoport, *C. R. Acad. Sci. USSR* **54** (1946) 65–67.
[284] P. N. Magee, J. M. Barnes, *Adv. Canc. Res.* **10** (1967) 163–246.
[285] K. Brunnemann, S. Hecht, D. Hoffmann, *J. Toxicol. Clin. Toxicol.* **19** (1982/83) 661–688.
[286] H. F. Smyth, C. P. Carpenter, C. S. Weil, *J. Ind. Hyg. Toxicol.* **31** (1949) 60–63.

[287] H. F. Smyth, C. P. Carpenter, C. S. Weil, *Toxicol. Appl. Pharmacol.* **28** (1974) 313–319.
[288] *Armak Product Data Bulletin* 81-6 (1981), 80-8 (1980), and 73-6 (1973), Armak Co., Chicago.
[289] W. B. Deichmann, J. L. Radomski, W. E. MacDonald, R. L. Kascht, R. L. Erdmann, *AMA Arch. Ind. Health* **18** (1958) 483–487.

Amines, Aromatic

Individual keywords: → *Amines, Aliphatic;* → *Aminophenols;* → *Aniline;* → *Benzidine and Benzidine Derivatives;* → *Naphthalene Derivatives;* → *Phenylenediamines and Toluenediamines;* → *Toluidines and Xylidines*

PETER F. VOGT, Mobay Chemical Corp., Pittsburgh, Pennsylvania 15205, United States (Chap. 1–4)

JOHN J. GERULIS, Mobay Chemical Corp., Pittsburgh, Pennsylvania 15205, United States (Chap. 5)

1.	Introduction	503
2.	Physical and Chemical Properties	506
2.1.	Physical Properties	506
2.2.	Chemical Properties	506
2.2.1.	Acetylation and Similar Reactions	506
2.2.2.	N-Alkylation and Condensations	507
2.2.3.	C-Alkylation	508
2.2.4.	Cyclization	508
2.2.5.	Diazotization	509
2.2.6.	Halogenation	510
2.2.7.	Nitration	510
2.2.8.	Oxidation	511
2.2.9.	Sulfonation	511
2.2.10.	Reduction	512
2.2.11.	Other Reactions	512
3.	Production	513
3.1.	Reduction	513
3.1.1.	Chemical Reduction	513
3.1.2.	Catalytic Hydrogenation	516
3.2.	Nucleophilic Substitution	521
3.2.1.	Exchange of Halide	521
3.2.1.1.	Activated Halides	521
3.2.1.2.	Unactivated Halides	522
3.2.2.	Exchange of Hydroxyl and Ether Groups	523
3.2.3.	Exchange of Sulfo Groups	524
3.3.	Other Processes	525
4.	Economic Aspects	525
5.	Toxicology, Occupational Health, and Environmental Protection	525
5.1.	Toxicity	525
5.2.	Exposure Limits	528
5.3.	Protective Measures	528
5.4.	Environmental Monitoring	528
6.	References	530

1. Introduction

Definitions. *Aromatic amines* are organic nitrogen compounds that may be considered derivatives of ammonia (NH_3) with at least one of the hydrogen atoms replaced by an aryl group. The nitrogen must be bound directly to the aromatic ring and so be able to interact with the aromatic π-electron system. The amine can be primary, secondary, or tertiary depending on whether one, two, or three of the protons are replaced by alkyl or aryl groups. The simplest aromatic amine is derived from benzene and is called aniline or benzenamine ($C_6H_5NH_2$). The amines from toluene are referred to as toluidines and from naphthalene as naphthylamines. Pyridine, pyrrole, and others in which nitrogen forms part of the ring are treated not as aromatic amines but as heterocyclic compounds.

Nomenclature. Aromatic amines most often are named by adding the suffix "amine" to the name of the radical (or radicals) replacing one or more of the hydrogen atoms in ammonia, except when some trivial name exists. According to IUPAC nomenclature, the amino (NH_2) or modified amino group (NHR, NRR') is considered a substituting group of the aromatic hydrocarbon. Therefore, these compounds are named as derivatives of benzene, toluene, naphthalene, and others, e.g., *N,N*-dimethylaminobenzene, $C_6H_5N(CH_3)_2$.

Aniline is generally used as the parent structure for its derivatives unless a carbon atom is attached to the benzene ring or unless a function of higher seniority than amino is present in the molecule. Important derivatives of aniline are named as such, e.g., *N*-methylaniline, *p*-nitroaniline, chloroaniline, whereas the sulfonic acids are named as derivatives of benzene (aminobenzenesulfonic acid) or by their trivial names (metanilic acid for *m*-aminobenzenesulfonic acid). If two of the hydrogen atoms of ammonia are replaced by aryl groups, the compounds are generally described as diarylamines, with diphenylamine (DPA) as the parent compound. Aromatic amines with two amino groups on the same benzene ring are named phenylenedi-amines. *Chemical Abstracts* called these compounds benzenediamines in 1972.

Historical Development. In 1826 O. UNVERDORBEN first obtained aniline by the dry distillation of indigo and called it "krystallin." In 1834 F. RUNGE found it in coal tar; in 1841 C. F. FRITZSCHE prepared the same oily liquid by heating indigo with potash and gave it the name "aniline," from "anil" (the Portuguese word for indigo), which was identical to the "benzidam" of N. N. ZININ (1841). The structure was proved by A. W. VON HOFMANN in 1843, who showed that it was obtained from the reduction of nitrobenzene. This is still the most common method of production. Diphenylamine, the simplest diarylamine, was first prepared by VON HOFMANN in 1863 by destructive distillation of triphenylmethane dyes.

Major Uses and Importance. Aromatic amines are widely used as dye intermediates, especially for azo dyes, pigments, and optical brighteners; as intermediates for photographic chemicals, pharmaceuticals, and agricultural chemicals; in polymers via isocyanates for polyurethanes; and as antioxidants. Recently they have been listed as corrosion inhibitors of mild steel in the pickling process [1].

The oldest use of aromatic amines is as intermediates for *dyes*. Substituted anilines and especially naphthylamines have been precursors of azo dyes since the middle of the 1800s. Even with the replacement of the once predominant anthraquinone dyes with more and more azo dyes derived from aromatic amines, the previous strong demand of aniline in the dye industry has decreased markedly in the United States to less than 4%

of its former magnitude. Once over 700 dyes listed in the Colour Index (C.I.) [2] were prepared from aniline and its derivatives; now only a few are produced in commercial quantities.

The production and sales volume for the aniline derivatives as a group is generally growing only at the average rate of expansion of the chemical market as a whole. United States production of many of them has ceased, and the supply is shifting to Japan, Korea, China, and other East Asian countries along with textile and dye production.

p-Nitroaniline [100-01-6], with a United States production of 6200 t in 1981 [3] and an import volume of 630 t in 1983 [4], is probably the largest in this group. *p*-Nitroaniline is used mainly in the production of dyes, antioxidants, and pharmaceuticals.

Anisidines, especially *o*-anisidine [90-04-0], with a limited United States production and imports of 570 t in 1983 [4], is an important intermediate for pigments and azo dyes. Japan reported a production of only 320 t in 1981 [5], down from a high of 997 t in 1974. *Chloroanilines* are used mostly for agricultural products. *4,4'-Diamino-2,2'-stilbenedisulfonic acid* [81-11-8] has maintained its strength as the largest volume raw material for optical brighteners (United States use 5000 t in 1984).

By far the largest share of the *aniline* consumption (67% of total 1984 consumption) is used for the manufacture of isocyanates, primarily for 4,4'-methylenebis(phenylisocyanate), (MDI) to make polyurethanes. About 20% of the aniline consumption in the United States and in Western Europe versus 27% in Japan is used in the rubber industry for the manufacture of antioxidants; vulcanization accelerators, such as 2-mercaptobenzothiazoles (MBT) [149-30-4]; diphenylguanidines; for condensation products of aniline with aldehydes and ketones, e.g., 2,2,4-trimethyl-1,2-dihydroquinoline (TMDHQ) [147-47-7] and its polymer [26780-96-1]. Many herbicides, fungicides, insecticides, and animal repellents are made from aniline or its derivatives. Between 1960 and 1980, United States aniline consumption grew at an average rate of 8.8% annually and then slowed down to an average rate of 3.1% in the period 1980–1984. Consumption is expected to stay at that rate at least until 1990.

Analgesics and sulfonamides are examples of important pharmaceuticals derived from aromatic amines. Many sulfa drugs start from acetanilide (antifebrin) [103-84-4], which itself has been used as an antipyretic and analgesic agent.

Diarylamines readily donate their amine hydrogen atom to terminate various free-radical reactions. For this reason they are widely used as antioxidants for various polymers and elastomers, and as stabilizers for nitrocellulose [6]. Nearly all commercial synthetic rubbers are protected with alkylated diphenylamines. Because they darken on oxidation, diphenylamines cannot be used in light colored rubber and polymer compositions. In this case N,N'-di-β-naphthyl-p-phenylenediamines are used instead. Some di-arylamines retard the effect of intense gamma or neutron radiation [7]. Diarylamines have been used to stabilize sulfur trioxide, are used as corrosion inhibitors in roofing asphalt, and are essential in promoting crosslinking in poly(vinyl chloride) [8] and in polyethylene [9]. Diaryl-amines containing other functional groups, such as amino or hydroxy, are used as dye intermediates [10], antiozonants, in color photography, and for hair dyeing and have been patented for use in the image formation process in photographic films [11], [12].

2. Physical and Chemical Properties

2.1. Physical Properties

All aromatic amines are either solids or liquids at room temperature. The ones without other functional groups are colorless, high-boiling materials having a characteristic odor. Most oxidize on exposure to air and darken. They are only slightly soluble in water.

In contrast to the aliphatic amines, which are stronger bases than ammonia, the aromatic amines are only weakly basic with basicity constants $K_B \leq 10^{-10}$ in water. Their onium salts, therefore, are ionized in water and act as Brønsted acids.

The effect of the phenyl groups becomes even more dominant in diphenylamine ($K_B = 10^{-14}$), which can form salts only with strong acids that are completely ionized in water, whereas triphen-ylamine rarely forms salts and has no basic properties.

The physical properties of the most important aromatic amines are given under → Aniline.

2.2. Chemical Properties

The chemical reactions of aromatic amines are similar to those of aliphatic amines. These reactions are influenced greatly by the basicity of the amine nitrogen. Therefore, such electron-withdrawing substituents as nitro and cyano reduce the base strength of the amine nitrogen when substituted in the ortho or para positions of the benzene ring(s). Aromatic amines undergo a variety of chemical reactions. The ones of greatest industrial significance are described in this section.

2.2.1. Acetylation and Similar Reactions

Primary and secondary aromatic amines react with carboxylic acids, acid chlorides, anhydrides, and esters to form a wide variety of commercially important amides. For example, *p*-acetanisidine, *N*-(4-methoxyphenyl)acetamide [51-66-1], is obtained in nearly theoretical yield, especially if the water formed is removed (e.g., by distillation).

$$2\ CH_3O\text{-}C_6H_4\text{-}NH_2 + (CH_3CO)_2O \longrightarrow 2\ CH_3O\text{-}C_6H_4\text{-}NHCOCH_3 + H_2O$$

Acid chlorides, in general, give the best yields of pure product. Polymeric aromatic amides have been made from *m*- and *p*-phenylenediamines by reaction with iso- or terephthaloyl chloride. These so-called aramid fibers are marketed by Du Pont (United States) under the trade names Nomex, poly(1,3-phenyleneisophthalamide) [*24938-60-1*], and as Kevlar, poly(1,4-phenyleneterephthalamide) [*24938-64-5*] [13], [14].

The reaction with phosgene gives isocyanates or ureas depending on the reaction conditions.

$$Cl\text{-}C_6H_4\text{-}NH_2 + COCl_2 \longrightarrow Cl\text{-}C_6H_4\text{-}NCO + 2\ HCl$$

Commercially important isocyanates are obtained from toluenediamine, methylenedianiline, and other aromatic primary diamines. The monoisocyanates are used in such agricultural products as herbicides, insecticides, bacteriostats, fungicides, and pesticides. The major use of the diisocyanates is as intermediates for polyurethanes.

Reaction with dimethylformamide in the presence of sodium methoxide yields *N*-arylformamides in 45–88% yield [15], [16]. With aromatic aldehydes, Schiff bases (azomethines, Ar–CH=N–Ar') are formed, and with chloral and hydroxylamine, oximinoacetanilides result in 80–91% yield [17]. Dithiocarbamic acid ammonium salts are formed with CS_2 and NH_3 in toluene. These salts give aryl isothiocyanates (70–90% yield) when treated with phosgene [18].

$$ArNH_2 + CS_2 + NH_3 \xrightarrow{toluene} ArNH\text{-}\underset{\underset{S}{\|}}{C}\text{-}S^-\ NH_4^+ \xrightarrow{COCl_2} Ar\text{-}N\text{=}C\text{=}S$$

Instead of toluene, water also can be used as the solvent, followed by treatment with heavy metal salts, e.g., $Pb(NO_3)_2$, $CuSO_4$, $FeCl_3$ [19]. Using CS_2 alone in ethanol or acetone, aromatic amines react to yield *N,N'*-diarylthioureas.

With cyanogen chloride, *N,N*-diarylguanidines are formed in 90% yield.

$$Cl\text{-}C\text{≡}N + 2\ ArNH_2 \xrightarrow{H_2O,\ 95\text{-}100°C} ArNH\text{-}\underset{\underset{NH}{\|}}{C}\text{-}NHAr \cdot HCl$$

2.2.2. N-Alkylation and Condensations

Alkylation of aromatic amines to the corresponding *N*-alkyl- and *N,N*-dialkylamines is generally of importance only for aniline (→ Aniline), [20]–[23].

Substituted mono- or diethanolamines can be produced in high yields from ethylene or propylene oxide and an aromatic amine.

[Reaction scheme: 2-methoxy-4-acetamidoaniline + 2 ethylene oxide (CH₂–CH₂–O), 20–60°C, 90%, giving the N,N-bis(2-hydroxyethyl) derivative with OCH₃ and NHCOCH₃ substituents]

The resulting products, as such, or after further acetylation of the hydroxy groups, are important coupling compounds for azo dyes.

Aldehydes, ketones, acetals, and orthoformates give a variety of condensation products depending on the reaction conditions. For example the reaction of aromatic amines with ketones gives Schiff bases that can be hydrogenated to the alkyl-substituted amines. In the case of phenylenediamine derivatives, important anti-oxidants and antiozonants for rubber are produced, such as N-(1,3-dimethylbutyl)-N′-p-phenylenediamine [793-24-8], which is sold under various *trade names:* Vulkanox 4020 (Mobay, Bayer), Flexone 7L (Uniroyal), Santoflex 13 (Monsanto), UOP 588 (UOP), Wingstay 300 (Goodyear).

[Reaction scheme:
Ph–NH–C₆H₄–NH₂ + CH₃–CO–CH₂CH(CH₃)₂
→ (−H₂O) Ph–NH–C₆H₄–N=C(CH₃)–CH₂CH(CH₃)₂
→ (H₂/cat.) Ph–NH–C₆H₄–NH–CH(CH₃)–CH₂CH(CH₃)₂]

2.2.3. C-Alkylation

Under Friedel-Crafts conditions, the aromatic ring can be alkylated with alkenes to form the corresponding alkylanilines at temperatures of 200–280 °C and 20 MPa (200 bar). The alkylation takes place in the positions ortho to the amino group and if one is occupied only monoalkylation results [24]–[28].

These compounds are used as chain extenders for polyurethanes (e.g., 3,5-diethyl-2,4-diaminotoluene), and for herbicides (e.g., 2,6-diethylaniline and 2-methyl-6-ethylaniline).

2.2.4. Cyclization

Aromatic amines undergo numerous cyclization reactions to give a variety of N-heterocyclic compounds. Quinoline derivatives are obtained with aldehydes and ketones. Aniline condenses with acetone in the presence of hydrochloric acid, p-toluenesulfonic acid, acidic clays, or acidic ion-exchange resins to give 2,2,4-trimethyl-1,2-dihydroquinoline (TMDHQ) [147-47-7], which on oligomerization is used as an antioxidant in the rubber industry.

[Reaction scheme: aniline + 2 CH₃COCH₃ → 2,2,4-trimethyl-1,2-dihydroquinoline + 2 H₂O]

With aldehydes and styrenes, 1,2,3,4-tetrahydroquinolines result [29], and with glycerol or acrolein, quinoline [91-22-5] is formed in the Skraup synthesis [30]. Many of the substituted quinolines are intermediates for antimalarial agents. In the Combes quinoline synthesis [31], aniline and 1,3-diketones are condensed to make 2,4-disubstituted quinolines. With hydroxyketones or with haloketones, aromatic amines yield substituted indoles [32]. Aromatic amines and 2-aminophenols condense with CS_2 to form 2-mercapto-1,3-benzothiazole. With sulfur monochloride (S_2Cl_2) 1,2,3-benzodithiazole is formed and can be hydrolyzed to the corresponding 2-aminothiophenol in yields of up to 93% [33].

[Reaction scheme: o-toluidine·HCl + 3 S₂Cl₂ → benzodithiazolium chloride (30–60 °C, –5 HCl, –4 S); then 1. H₂O, 2. NaOH, 3. H⁺ → 2-amino-4-chloro-3-methylthiophenol]

Benzofuroxane [480-96-6] results from the treatment of *o*-nitroaniline with sodium hypochlorite [34].

[Reaction scheme: o-nitroaniline → benzofuroxane (NaOCl–NaOH–H₂O, 98%)]

Diphenylamines form carbazoles when heated with iodine as the catalyst, whereas heating diphenylamine with sulfur in the presence of various catalysts results in phenothiazine [92-84-2].

2.2.5. Diazotization

Primary aromatic amines react with nitrous acid to form diazonium salts.

[Reaction: PhNH₂ + NaNO₂ + 2 HCl → Ph–N⁺≡N Cl⁻ + NaCl + 2 H₂O]

This reaction has considerable general utility and forms the basis for azo dye production by coupling of the diazonium salt with aromatic amines and phenols. In the Sandmeyer reaction the diazo group is replaced by chlorine from a CuCl solution. The reaction also is used for the introduction of Br, CN, I, and sometimes also aryl and alkyl groups. Secondary aromatic amines react with nitrous acid to form *N*-nitroso compounds in the same way as do aliphatic amines.

Tertiary aromatic amines do not give *N*-nitroso compounds but undergo *C*-nitrozation.

$$\text{C}_6\text{H}_5-\text{N}(\text{CH}_3)_2 \xrightarrow{\text{HNO}_2} \text{ON}-\text{C}_6\text{H}_4-\text{N}(\text{CH}_3)_2$$

Diaryldiazenes are obtained by the oxidation of aromatic amines with $KMnO_4$ in aqueous KOH [35] or by the reaction of aromatic amines with aromatic nitrosamines [36].

$$\text{Ar}-\text{NO} + \text{H}_2\text{N}-\text{Ar}' \xrightarrow[70-95\%]{\text{CH}_3\text{COOH}} \text{Ar}-\text{N}=\text{N}-\text{Ar}' + \text{H}_2\text{O}$$

o-Phenylenediamines react with nitrite at neutral to slightly acidic pH to undergo ring closure to the corresponding benzotriazoles [37], [38]; these are important corrosion inhibitors for Cu-containing alloys.

2.2.6. Halogenation

The chlorination of aniline salts in aqueous solutions is so fast that only the trichloroaniline can be isolated [39]. Chlorine in the presence of hydrogen chloride in an anhydrous solvent yields 2,4,6-trichloroaniline [*634-93-5*] [40]. In aqueous sulfuric acid at 110 °C aniline gives chloranil, 2,3,5,6-tetrachloro-2,5-cyclohexadiene-1,4-dione [*118-75-2*], in 91% yield [41].

2.2.7. Nitration

Generally, nitration of aromatic amines with nitric acid results in concomitant oxidation of the amino function. If it is protected, predominantly the para- and ortho-nitro isomers are formed. Nitration of *p*-acetanisidine [*51-66-1*] in sulfuric acid yields 2-nitro-4-acetaminoanisole [*50651-39-3*] as the major product. This compound can be hydrolyzed subsequently to the free amine.

The nitration of *p*-acetaminotoluene [*103-89-9*], however, leads predominantly to substitution ortho to the acetamino group.

2.2.8. Oxidation

Aromatic amines are very susceptible to oxidation and most turn brown or black if exposed to air for extended periods. The course of oxidation depends on the nature of the oxidizing agent. With hydrogen peroxide and percarboxylic acids (e.g., peracetic acid), oxidation to the corresponding *N*-phenylhydroxylamine or to nitroso or nitrobenzene may occur, depending on temperature and the amount of oxidizing agent present [42]. 2,6-Dichloroaniline [*608-31-1*] is oxidized to 2,6-dichloronitrobenzene by trifluoroperacetic acid [43].

Oxidation with MnO_2 or $Na_2Cr_2O_7$ in sulfuric acid leads to quinones; e.g., 1-naphthylamine is oxidized to 1,4-naphthoquinone. Oxidation of aniline with dichromate or chlorate in the presence of copper or vanadium yields aniline black.

2.2.9. Sulfonation

Reaction of aromatic amines with sulfuric acid, oleum, or sulfur trioxide leads to the introduction of the sulfonic acid group mainly in the positions ortho or para to the amino group. 2-Methoxy-5-methylaniline [*120-71-8*] is sulfonated with fuming sulfuric acid to 4-amino-5-methoxy-2-methylbenzenesulfonic acid [*6471-78-9*] (the intermediate to make the dye FD & C Red 40 [*25956-17-6*] [2]).

1-Naphthylamine [*134-32-7*] is sulfonated with oleum at room temperature to give 1-amino-4- [*84-86-6*] and 5-naphthalenesulfonic acid [*84-89-9*].

Many aromatic amines also sulfonate via the *baking process* in which the amine sulfate is heated to 220 °C under vacuum to yield either the *para-* or *ortho-*sulfonic acid. The process is very selective and gives high yields of the *p*-sulfonic acid, e.g., aniline leads to sulfanilic acid (4-aminobenzenesulfonic acid [*121-57-3*]). If the para position is occupied, sulfonation takes place ortho to the amino group; e.g., *p*-toluidine [*106-49-0*] gives 2-amino-5-methylbenzenesulfonic acid [*88-44-8*] [44]. 1-Naphthylamine yields only 1-amino-4-naphthalenesulfonic acid [*84-86-6*], also known as naphthionic acid [10]. This compound is used as a dye intermediate for Congo Red, Fast Red A, Azo Rubin, and others.

2.2.10. Reduction

Hydrogenation of aromatic amines results in the corresponding cycloaliphatic amines. This process is of greatest importance with amines that do not contain other easily reducible functional groups. Such catalysts as Raney nickel, cobalt-alumina, and more exotic ones are used [45]. Hydrogenation of aniline using an alumina–silicate catalyst [46] leads to 80% cyclohexyl-amine [*108-91-8*] and 15% dicylohexylamine [*101-83-7*]; bis(4-amino-3-methylphenyl)methane is reduced in 89% yield on an alumina catalyst [47]. Hydrogenation of aniline over a Raney nickel catalyst gives 95% dicyclohexylamine [48]. Recently good results have been obtained with rhodium catalysts. 4,4′-Methylenedianiline (MDA) [*101-77-9*] has been hydrogenated on a 5% rhodium–aluminum catalyst to give bis(4-aminocyclohexyl)methane in good yield [47].

2.2.11. Other Reactions

In a reversal of the Bucherer reaction, naphthylamines can be hydrolyzed to the corresponding naphthol in yields approaching 100%. Sulfonic acid groups on the ring accelerate the reaction [49], [50].

3. Production

Generally, three types of reactions are used for the production of aromatic amines:

1) *Reductions:* using such metals as Fe, Zn, Sn, and Al or their salts; sulfur compounds; electrochemical methods; and catalytic hydrogenation
2) *Nucleophilic substitutions:* replacement of such substituents as halogen, hydroxyl, alkoxy, and sulfo groups
3) *Rearrangements and degradations:* including the benzidine and Beckmann rearrangements, and the Schmidt and Hofmann degradations.

Only the first two reaction types are of general importance. Chemical rearrangements and degradations seldom lead to clean reaction products in high yield.

3.1. Reduction

All aromatic carbon–nitrogen compounds with nitrogen oxidation numbers ≥ -2 can be reduced to aromatic amines (for a summary see [51]). Only the reduction of nitro compounds has gained widespread industrial acceptance, be-cause the starting materials can be prepared with a high degree of specificity for a wide variety of compounds. Other aromatic nitrogen compounds cannot be produced economically enough to be used as starting materials for aromatic amines.

3.1.1. Chemical Reduction

In the chemical reduction of aromatic nitro compounds to aromatic amines, the hydrogen that becomes attached to the nitrogen normally comes from the solvent, in many cases water, or from the added acid. The most important reductants are metals, such as iron, tin, and zinc; phosphorus, sulfides, sulfites, and sulfur dioxide are used also. Because oxidation of the reducing agent may result in a useless waste product, which has to be removed in an environmentally safe way, chemical reduction has lost its importance in favor of catalytic reduction for the large-scale production of products.

Reduction with Iron. The reduction with iron in dilute acid was described by BÉCHAMP in 1854 and the process still carries his name. Iron is still the most important metal for the reduction of nitro compounds to aromatic amines, and generally, all nitro compounds can be reduced to the corresponding aromatic amines with iron in water and acid. Side products are expected only if the molecule contains other substituents that can be reduced easily (e.g., nitroso, azo, hydrazine, sulfoxide, and other nitro groups) or can be saponified [51]. Carbon–carbon multiple bonds are not attacked. Bayer (Federal Republic of Germany) and Mobay (United States) both use this iron reduction for the production of iron oxides with aniline as the resulting side product. For a further discussion, → Aniline, Section 2.3, and [51]–[54].

Another example of industrial importance is the reduction of 4,4'-dinitro-2,2'-stilbenedisulfonic acid to 4,4'-diamino-2,2'-stilbenedisulfonic acid [*81-11-8*], which is still done exclusively using iron in neutral medium. Also, many small-volume dye intermediates are reduced with iron.

Reduction with Other Metals. Aromatic nitro groups also can be reduced with zinc, tin, and aluminum, or their salts in acidic, neutral, or alkaline medium [51].

These reducing agents are generally very mild and do not interfere with, e.g., –OH, –OR, –COOH, –CO–Ar, halogen, or –CN groups. However, except for laboratory applications they are not of any industrial significance. Instead, catalytic hydrogenation is used because of wastewater regulations.

Reductions with Sulfide, Hydrogensulfite, and Sulfur Dioxide. Compounds with several nitro groups can be reduced to nitro amines if a stoichiometric amount of *sulfide* is used [55], [56]. In this case, sodium or ammonium sulfide is the reagent of choice. *m*-Nitroaniline [*99-09-2*] can be obtained in 87% yield from 1,3-dinitrobenzene [55]:

$$4 \text{ C}_6\text{H}_4(\text{NO}_2)_2 + 6 \text{ Na}_2\text{S} + 7 \text{ H}_2\text{O} \xrightarrow{60-70°C} 4 \text{ C}_6\text{H}_4(\text{NH}_2)(\text{NO}_2) + 6 \text{ NaOH} + 3 \text{ Na}_2\text{S}_2\text{O}_3$$

(87%)

The addition of sulfur reduces the required amount of sodium sulfide. 4-Nitrotoluene is reduced nearly quantitatively to 4-aminobenzaldehyde [*556-18-3*] with sodium sulfide and sulfur [57], [58].

Azo groups stay unchanged if the reduction with sodium sulfide is carried out at 40–70 °C, if the reaction times are kept short, and if sulfide is not in excess. However, halogen substituents are replaced easily by –SH [55], [56]. When excess reducing agent is not a problem, ammonium hydrogensulfide in the presence of ammonia can be used [56], [59].

The reduction of aromatic nitro, nitroso, or azo compounds with *sulfite* or *hydrogensulfite* to the corresponding aromatic amines is known as the Piria reaction (R. PIRIA, 1851). In aqueous or alcoholic medium, mainly *N*- and *C*-sulfonic acids result [60]. On treatment of these with mineral acid, a mixture of aromatic amines and aminobenzenesulfonic acids is formed.

The quinone oxime intermediate facilitates substitution of hydrogensulfite onto the aromatic ring in the reduction of *ortho*- and *para*-nitrosophenols or naphthols:

$$\underset{\text{NO, OH}}{\text{naphthalene}} + \text{NaHSO}_3 + \text{SO}_2 + \text{H}_2\text{O} \xrightarrow[\text{up to 60\%}]{0-20°\text{C}} \underset{\text{NH}_2, \text{OH}, \text{SO}_3\text{H}}{\text{naphthalene}} + \text{NaHSO}_4$$

If the hydroxyl group is in the para position, the sulfo group will enter in the 2 position. Substituents in the 3 position (e.g., Cl or COOH) are replaced by H [61].

Only after methods were found to reduce the amount of C-sulfonation did reduction with SO_2 gain industrial importance. This reduction is best carried out in a closed system in strongly acidic media (e.g., 15–40% aqueous sulfuric acid) at temperatures of 80–180 °C with iodine, hydrogen iodide, or iodine salts as catalysts. Glass-lined reactors are the best [62].

Sodium dithionite reduces the aromatic nitro and nitroso group nearly quantitatively [63]:

$$\text{R-C}_6\text{H}_4\text{-NO}_2 + 3\ \text{Na}_2\text{S}_2\text{O}_4 + 4\ \text{H}_2\text{O} \xrightarrow{\text{up to 100\%}} \text{R-C}_6\text{H}_4\text{-NH}_2 + 6\ \text{NaHSO}_3$$

When possible, reductions with sulfide, sulfite, and other sulfur compounds have been replaced by catalytic hydrogenations, resulting in cleaner reaction mixtures, less corrosion (resulting when SO_2 is used), and fewer environmental problems.

Electrochemical Reduction of Nitro Compounds. Electrochemical reduction is generally a special case of the chemical reduction. An inorganic compound is used as the reducing agent; the oxidized form is then reduced at the cathode and can react again.

$$\text{R-C}_6\text{H}_4\text{-NO}_2 + 6\ \text{H}^+ + 6\ e^- \longrightarrow \text{R-C}_6\text{H}_4\text{-NH}_2 + 2\ \text{H}_2\text{O}$$

The cathode is made of Pb, Sn, Ni, or Cu, and 15–20% hydrochloric acid is used on the cathode side of a semi-permeable membrane. On the anode side 30% sulfuric acid is employed [64].

Although electrochemical reductions have been known for more than 100 years, they have remained mostly curiosities and have not been used commercially to any great extent in the past. However, interest is growing as better cells and membranes become available. Some products have been produced electrochemically, especially in India; e.g., p-aminobenzoic acid [150-13-0], p-aminophenol [123-30-8], and many others have reached pilot-plant scale or beyond [65], [66].

3.1.2. Catalytic Hydrogenation

Reduction with Hydrazine. The reduction of aromatic nitro compounds with hydrazine represents, in most cases, a special variation of catalytic hydrogenation, where hydrazine is the source of the hydrogen. The mechanism of decomposition of hydrazine on precious-metal catalysts differs depending on the pH value, with the amount of hydrogen generated per mole of hydrazine increasing with higher pH [67]. At weakly alkaline or neutral conditions, 1 mol H_2 is generated:

$$3\ N_2H_4 \xrightarrow{Pt,\ Pd,\ or\ Ni} 2\ NH_3 + 2\ N_2 + 3\ H_2$$

If barium hydroxide or calcium carbonate is added, 2 mol of H_2 are formed:

$$N_2H_4 \xrightarrow{Pd\ or\ Pt} N_2 + 2\ H_2$$

The reduction with hydrazine sometimes occurs without a catalyst but in most cases is done in the presence of a hydrogenation catalyst, such as Raney nickel, Pd on carbon, or Pd on $CaCO_3$ [68], [69]. Ideally 1 mol of hydrazine gives four reduction equivalents depending on the pH and catalyst, but in practice fewer are formed, and an excess of hydrazine must be used. Carbon–carbon double bonds and carbonyl groups are not attacked [70]. Depending on the reaction conditions, the intermediates of the reduction can be isolated.

$$ArNO_2 \longrightarrow Ar-\overset{\overset{O}{\uparrow}}{N}=N-Ar \longrightarrow Ar-N=N-Ar \longrightarrow Ar-NH-NH-Ar \longrightarrow ArNH_2$$

In the reduction using hydrazine without a catalyst, carbonyl-containing compounds are reduced to the corresponding hydroxy compounds via the hydrazone intermediate [71]. The reduction of aromatic nitro compounds with hydrazine has no advantage over catalytic reduction with much cheaper hydrogen, except for laboratory work and very small product volumes, where it allows use of normal pressure. Examples of catalytic reduction with hydrazine hydrate on Pd - C and on Raney nickel can be found in [69], [72] – [74].

$$R-C_6H_4-NO_2 + \tfrac{3}{2} N_2H_4 \xrightarrow{Pd/C\ or\ Ni} R-C_6H_4-NH_2 + \tfrac{3}{2} N_2 + 2\ H_2O$$

50–100%

Cyclohexene also has been used on a laboratory scale as a hydrogen source for the catalytic hydrogenation of polynitro compounds [75].

Reduction with Hydrogen. Primary aromatic mono- or polyamines generally are produced by the catalytic hydrogenation of the corresponding nitro compound, either in the vapor or in the liquid phase, with or without a solvent [45], [76]–[78]. All large-scale products, such as aniline (→ Aniline), or *o*- and *p*-toluidine (→ Toluidines), 2,4- and 2,6-diaminotoluene (→ Phenylenediamines and Toluenediamines), and 1-naphthylamine (→ Naphthalene Derivatives), are produced in this way. For the reaction meachanism see [79], [80].

Catalytic reduction of nitro compounds is very exothermic. In the hydrogenation of nitrobenzene in the liquid phase, 553.5 kJ/mol are released, whereas hydrogenation in the vapor phase at 200 °C releases 493.2 kJ/mol. Unless this heat is dissipated properly, explosions can result, especially if thermal decomposition of the nitro compound occurs or condensation reactions are initiated as with chloro–nitro compounds [51]. To reduce these hazards, the concentration of nitro compound, the amount and partial pressure of hydrogen, the temperature, and the activity of the catalyst are controlled.

Vapor-Phase Hydrogenation. Industrial use of vapor-phase hydrogenation is limited by the boiling point and the thermal stability of the nitro compounds. Most aniline produced in the United States is made by vapor-phase hydrogenation of nitrobenzene over copper–silica catalyst in $\geq$ 99% yield, whereas Pd–alumina is used by Bayer in Europe and Brazil.

Liquid-Phase Hydrogenation. It is preferable to hydrogenate most aromatic nitro compounds in the liquid phase. In this case the pressure and temperature can be changed independently. The temperature is limited by the hydrogenation reaction of the aromatic ring which occurs above 170–200 °C. Normally the reduction is carried out at 100–170 °C. Some compounds (e.g., 1-bromo-4-nitrobenzene) must be hydrogenated at lower temperatures (20–70 °C) to avoid cleavage of sensitive groups. Pressures of 1–15 MPa (10–150 bar) are used industrially. For the hydrogenation of sensitive nitro aromatic compounds lower pressures are advisable (0.1–5 MPa, 1–50 bar) [81]–[83]. Generally the pressure only influences the speed of the reaction by increasing the phase transfer and saturation of the catalyst with hydrogen. The reaction time depends on many parameters, such as hydrogen pressure, concentration, temperature, activity and concentration of the catalyst, and mixing. Often an induction period is observed, which is independent of the activity of the catalyst [84], [85]. Generally the reaction time is from a few minutes to several hours. Considerably longer times are necessary if these parameters are not optimized.

Because of the exothermic nature of the reaction, many safety precautions must be observed, especially for the industrial hydrogenation of aromatic polynitro compounds in the liquid phase without solvents [86]–[88]. The exothermic reaction is controlled by the continuous addition of small quantities of the nitro compound, thus keeping its concentration below 2% [89]. Deionized water can be added to remove the heat of reaction by continuous evaporation. The water also affects the activity of the catalyst;

e.g., the more water present in the batch hydrogenation of dinitrotoluene without solvent, the lower the activity of the Pd on carbon catalyst [90], [91]. If the amine shows good water solubility, water can be used as the solvent. Water also can be used in cases where the nitro compound forms water-soluble salts with alkali, such as with nitrocarbonic or sulfonic acids. Sometimes 30–40% solutions of the amine in water can be produced directly in the hydrogenation reactor without an additional concentration step [51], [92].

Solvents. Methanol and 2-propanol are preferred; also dioxane, tetrahydrofuran, and N-methylpyrrolidone have been used; e.g., o-nitrophenol is hydrogenated in good yields in methanol with Pt on carbon. The concentration of the nitro compound can vary substantially. According to [93] a 33 wt% solution of dinitrotoluene in methanol is fed into the hydrogenation reactor, whereas Olin recommends a 15–35 wt% solution [94]. In the hydrogenation with a water-immiscible solvent, such as toluene, the water must be removed, as in solvent-free hydrogenation, in order to maintain the activity of the catalyst. Very polar nitro compounds, e.g., bis(4-amino-3-nitrophenyl)sulfone, are hydrogenated in liquid ammonia [95].

Hydrogenation Catalysts. For vapor-phase hydrogenation only metals or metal derivatives on supports are used in fixed beds or fluidized beds, whereas metals with large surfaces (Raney nickel) or metals and metal derivatives on supports with large surfaces (carbon, alumina) are used for the liquid-phase hydrogenation. In practice only Raney nickel, Raney nickel–iron, Raney cobalt, and Raney copper are used as pure-metal catalysts because of their relatively low cost. Precious-metal catalysts, such as Pt and Pd, generally are used at concentrations of 0.5–5 wt% on support material with large surfaces, such as charcoal, silica, aluminum oxide, or alkaline-earth carbonates. The powdered form of the catalyst is used in slurries and the pellet form in fixed beds [96]. Because the support material may activate or deactivate the catalyst [80], it is important to find the optimum conditions that not only allow the catalyst to be used repeatedly and to be regenerated but also are economical for large-scale (> 500 t/a) products [87], [97]. Decomposition and oxidation products generally deactivate catalysts by blocking the active surfaces [80]. Even in small quantities, compounds containing sulfur, arsenic, or antimony are catalyst poisons. However, with increased oxidation state (e.g., $As^{3-} \rightarrow As^{5+}$ or $S^{2-} \rightarrow S^{4+}$) these compounds have less effect as catalyst poisons. With noble metal catalysts, such ions as halide, sodium, and magnesium, cobalt metal, and CO_2 reduce the activity. Nitro aromatic compounds containing sulfur are better reduced with molybdenum or tungsten sulfide [98], or with nickel catalysts that contain zinc or calcium carbonate [101]. In the hydrogenation with Pd on carbon, copper salts are catalyst poisons, whereas in the hydrogenation of dinitrotoluene, both nitrophenol and nitrocresol are strong catalyst poisons and decomposition activators [45], [100].

Hydrogenation of Halide-Substituted Nitro Compounds. Whereas most aromatic nitro compounds are hydrogenated at 100–170 °C, lower temperatures are used with halogen-substituted nitro compounds to avoid halide–hydrogen exchange. The catalysts have reduced activity, e.g., sulfide-poisoned metal catalysts. 1-Chloro-2nitrobenzene can be reduced at 80 °C and 1.0–5.0 MPa (10–50 bar) with a 5 wt% Pt on carbon catalyst, which is treated at 22 °C with dilute sulfuric acid, then with H$_2$ followed by H$_2$S [101]. The catalyst can be recycled up to 30 times without any further purification. The reaction time is 30 min per batch and only 0.1% chlorine is cleaved. Therefore, a product of high purity and stability can be obtained in nearly quantitative yield. Also 3-chloroaniline [108-42-9], 4-chloroaniline [106-47-8], 2,5-dichloroaniline [95-82-9], 3,4-dichloroaniline [95-76-1], 2,6-dichloroaniline [608-31-1], 6-chloro-2-aminotoluene [87-60-5], and 4-chloro-2-aminotoluene [95-79-4] can be prepared in high yield and purity by this method.

A similar sulfide-poisoned catalyst has been described [102], in which 5% platinum on carbon catalyst was treated with 1% aqueous sulfuric acid, hydrogen, and sodium sulfide solution. 2,5-Dichloronitrobenzene was hydrogenated with this catalyst at 90–100 °C, 1.0–3.0 MPa (10–30 bar), 40 min per batch. The catalyst was recycled at least 30 times without loss of activity or selectivity while chlorine cleavage was kept $\leq$ 0.1%; yield of 2,5-dichloroaniline [95-82-9] was 97%. Also a 1% Pt on carbon catalyst was treated with dimethylsulfoxide in water before it was used in the hydrogenation of chloronitro aromatic compounds in 98.4% yield [103]. Additional treatment with hydrazine hydrate increased the activity of the catalyst with yields of 99.6% for 2-chloroaniline [95-51-2]. Modification of Pt on carbon with lead, bismuth, or silver salts for the hydrogenation of chloronitrobenzene at 100 °C gave chloroaniline in 99% yield, whereas untreated catalyst at 77 °C resulted in 100% hydrolysis [104]. 3-Bromonitrobenzene has been hydrogenated using Pt on barium or strontium carbonate at 20 °C, 0.1 MPa (1 bar) [105]. The catalytic activity of Raney nickel can be modified also. For the hydrogenation of 3-chloronitrobenzene the addition of 0.1–20% of thiocyanate to the catalyst has proved beneficial [83].

Equipment and Construction Materials. Hydrogenation in the vapor phase is carried out only continuously, but in the liquid phase it can be done in batch or continuous operation de-pending on volume. Continuous liquid-phase hydrogenation is used exclusively for a relatively few large-scale products, such as dinitrotoluene, nitrotoluene, and 1-nitronaphthalene [91]. The vast majority of aromatic amines have small annual volumes (< 500 t) and are produced by batch hydrogenation with catalyst slurries. This allows fast changeover from one product to another, which is not possible when fixed-bed catalysts are used. The intensive mixing of the three phases (hydrogen gas, nitro compound solution, and solid catalyst) is of great importance. The transport of the reactants toward the active centers of the catalyst, and of the products away from them, must be rapid for good reaction.

Although batch catalytic hydrogenations traditionally are done in stirred, steel or stainless steel autoclaves (Fig. 1), improvements in performance are being realized increasingly through loop reactor technology. Production units employing the loop reactor principle (Fig. 2) have been available for the last two decades and have been applied successfully to the catalytic hydrogenation of aromatic nitro compounds by Buss (Basel, Switzerland) [106]–[109]. The major advantages reported over stirred autoclaves are increased heat and mass transfers and improved reaction selectivity. In

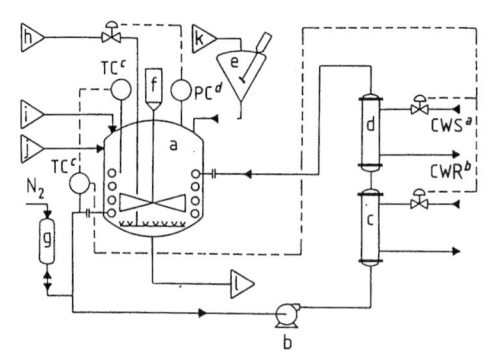

Figure 1. Conventional stirred autoclave system [106]–[108]
a) Reactor with internal coil; b) Recirculation pump; c) Heater; d) Cooler; e) Catalyst preparation vessel; f) Agitator; g) Expansion vessel; h) Hydrogen; i) Reactant; j) Solvent; k) Catalyst; l) Product;
[a] Cold water supply; [b] Cold water return; [c] Temperature control; [d] Pressure control

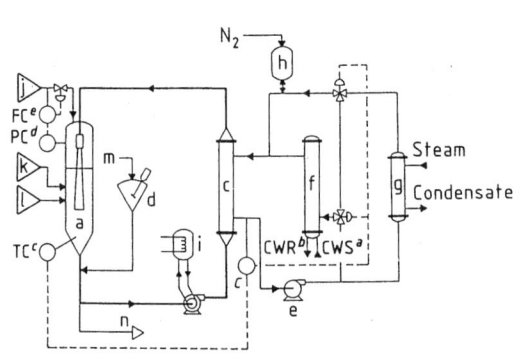

Figure 2. Loop reactor system [106]–[108]
a) Reactor with mixing nozzle; b) Primary loop recirculation pump; c) Primary loop heat exchanger; d) Catalyst preparation vessel; e) Secondary loop recirculation pump; f) Secondary loop cooler; g) Secondary loop heater; h) Secondary loop expansion vessel; i) Seal fluid cooler; j) Hydrogen; k) Reactant; l) Solvent; m) Catalyst; n) Product;
[a] Cold water supply; [b] Cold water return; [c] Temperature control; [d] Pressure control; [e] Flow control

continuous hydrogenation, generally shorter batch cycle times and higher product yields result. In addition catalyst usage is often lower.

The basic operating principle of a loop reactor is shown in Figure 3 [106]–[109]. The batch charge and slurry-type catalyst are pumped continuously through an external heat exchanger to a mixing nozzle in the top head of the autoclave. The nozzle acts as an eductor, and the action of the slurry passing through it creates intimate gas–liquid–solid mixing. The chemical reaction primarily takes place in a well-defined zone inside the nozzle prior to gas–liquid disengagement at the nozzle exit. The autoclave acts as a holding vessel. A loop reactor allows for simple and reliable transfer from pilot plant to production, because the scaleup is based only on the nozzle configuration and pump-around rate. In the case of a stirred autoclave, scaleup is more difficult because of the ratio between a small pilot vessel and a large production vessel (e.g., 1:1500 to 1:12 500). Temperature and concentration gradients as well as longer reaction times in the production vessel often result in undesired side reactions and a decreased product yield. Loop reactors minimize these problems by limiting the reaction to a highly agitated, well-defined zone, with a considerably decreased cycle time.

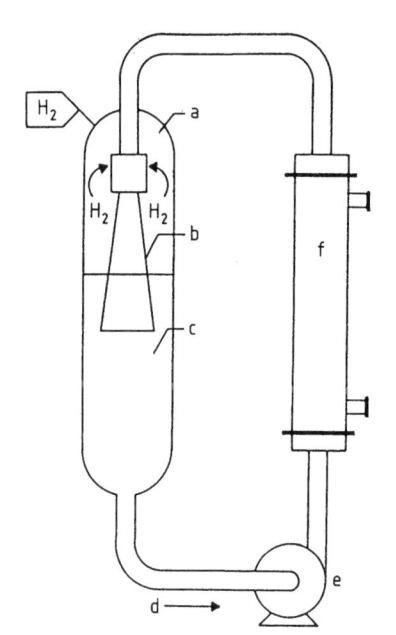

Figure 3. Operating principle of a loop reactor
a) Autoclave; b) Mixing and reaction zone; c) Reactant, solvent, catalyst suspension; d) Flow; e) Primary loop re-circulation pump; f) Primary loop heat exchanger

3.2. Nucleophilic Substitution

Practically all non-carbon substituents on the aromatic ring can be exchanged for amino groups. However, only a few of these substitution reactions are of industrial importance, including:

1) Exchange of halide
2) Exchange of hydroxyl and ether groups
3) Exchange of sulfo groups

3.2.1. Exchange of Halide

Exchange of halide is universally useful for the production of amines with mono-, poly-, or heterocyclic substituents as well as for diaryl-amines [110]–[112], and arylpolyamines [113], [114]. Many of the products are used as dye or pigment intermediates. The substitution of a halide on the aromatic ring with ammonia or with amines occurs via two different mechanisms, depending on whether the halide is activated or unactivated.

3.2.1.1. Activated Halides

Aromatic halides containing ortho and/or para electron-withdrawing substituents, e.g., $-NO_2$ or $-CN$, undergo nucleophilic aromatic substitution [115], [116]. Both *o*- and *p*-chloronitrobenzene react with ammonia at 170 °C, whereas 2,4-dinitrochlorobenzene requires only 70 °C and no catalyst [117], [118].

The more basic the amine (unless it is sterically hindered), the faster the reaction with the haloaromatic compound. In the reaction of methylamine and ammonia with a solution or suspension of 2,4-dinitrochlorobenzene at higher temperature, good yields of *N*-methyl-2,4-dinitroaniline are obtained, rather than 2,4-dinitroaniline [119].

Reaction Conditions. *4-Nitroaniline* [*100-01-6*] is produced in an agitated titanium reactor in 99.3 % yield by reacting 4-chloronitrobenzene [*100-00-5*] with a tenfold excess of aqueous ammonia at 175 °C and 4.2 MPa (42 bar) for 10 h [120].

4-(4-Methylanilino)–3-nitrobenzenesulfonamide is prepared in 90.1 % yield by heating 4-chloro-3-nitrobenzenesulfonamide and *p*-toluidine [*106-49-0*] for 5 h at 130 °C [110]. *4-Nitrodiphenylamine* [*836-30-6*] (precursor for antiozonants for rubber) is made by the condensation of 4-nitrochlorobenzene with aniline in chlorobenzene. *N*-Alkylated *p*-phenylenediamine derivatives, however, cannot be produced in aqueous medium.

3.2.1.2. Unactivated Halides

Unactivated halides undergo elimination – addition reaction through a benzyne intermediate. This requires higher temperatures than nucleophilic aromatic substitution and results in isomeric mixtures if other substituents are present. The industrial usefulness of this reaction is therefore limited.

Chlorobenzene and ammonia react at a reasonable rate only above 200 °C [121], generally with the addition of a catalyst, such as copper(I) and copper(II) salts or their amine complexes [114], [122], [123]. Also metal amides (e.g., $NaNH_2$, $LiNEt_2$) are used. The energy needed to exchange the halide is the reverse of the order for activated halides:

$F \gg Cl \gg Br > I$

Reaction Conditions. The reaction is generally carried out in aqueous suspension or in solution at elevated temperature and pressure. Because the reaction of arylhalides with amines is exothermic, it is especially important in the batch process to watch the temperature while heating to reaction conditions. This normally poses no problem in the continuous reaction. The hydrogen halide released during the reaction is neutralized either by excess ammonia or by the addition of bases, such as sodium carbonate or calcium oxide [112]. With excess amine the reaction generally proceeds with nearly quantitative yield. With an insufficient excess of strongly basic amines in aqueous solution, side reactions such as hydrolysis or the formation of diarylamines, can occur. Cyano groups are hydrolyzed to amides in aqueous solution, but this reaction can be avoided by using an inert solvent. Carboxylic acids are converted easily to the corresponding amide by reaction in nonaqueous media.

Material of Construction. The presence of halides and amines at elevated temperature and pressure can cause severe corrosion; therefore, the selection of materials and equipment becomes difficult. Autoclaves of normal steel can be used for laboratory work, but considerable reduction in the wall thickness must be expected (several millimeters per year). In addition, stress corrosion can lead to much faster deterioration of the reactor. Low-alloy steels are not useful for production units, especially with continuous working reactors. Stainless steels (chromium- and nickel-containing austenitic steels of the 316 types) were used with success for reactions in aqueous media. Because even these materials become pitted, other more corrosion-resistant alloys, such as Hastelloy and Inconel, have come into use. In addition, zirconium, titanium, and tantaclad (explosion clad tantalum on steel) have been used as reactor materials [120].

3.2.2. Exchange of Hydroxyl and Ether Groups

The displacement of an aromatic hydroxyl group by an amine is not only possible with phenols and cresols but also with quinoline and isoquinoline derivatives. In addition ammonia can replace alkoxy groups that are activated by nitro, halo, or cyano in the ortho or para positions. For example, 2,4-dinitroanisole [119-27-7] reacts with ammonia at 50–200 °C to form 2,4-dinitroaniline [97-02-9] in 50–90% yield [124]. Interest in the reaction of phenols with amines is often a function of the cost of the phenol raw material. The principal exchange processes are:

1) *Reaction in aqueous medium without a catalyst.* 3-Nitrosalicylic acid reacts with aqueous ammonia at increased temperature and pressure to yield 2-amino-3-nitrobenzoic acid [125].
2) *Reaction in the presence of an acid catalyst.* m-Aminophenols are produced in the reaction of resorcinol [108-46-3] and aqueous ammonium hydroxide or alkylamines with boric acid as the catalyst [126]. Under anhydrous conditions with the same catalyst, 3-(β- and γ-hydroxyalkylamino)phenols are obtained from resorcinol and

ethanolamine or 3-amino-1propanol [127]. A similar process uses an acid activated alumina–silica catalyst [128].

3) *In a multiphase system*, nitrophenols and ammonia react to yield nitroanilines, preferably using a solvent in which the reaction product has good solubility [129]. Many of the new phase-transfer catalysts are being applied to these reactions.

4) *p-Nitrosoaniline* and *p*-nitroso-*N*-phenylaniline [156-10-5] have been produced in liquid ammonia in the presence of either an ammonium salt or a tertiary amine, respectively, and an organic solvent [130].

5) *p-Nitrosophenols* are reacted with alkyl-amines via the *p*-nitrosophenylethers to give the corresponding *N*-substituted *p*-nitrosoanilines. The water that is eliminated in this process must be removed to drive the reaction to completion [131].

6) *The Halcon process* is the industrially important vapor-phase amination of phenol [132], [133]. Phenols or naphthols react in the gas phase with ammonia at 250 °C on metallic or nonmetallic oxides, e.g., MgO, B_2O_3, Al_2O_3, SiO_2, TiO_2, or mixtures of these. The Halcon process has been used by U.S. Steel Chemicals since 1982 in its plant at Haverhil, Ohio for the production of aniline from phenol (aniline capacity 100 000 t/a) [134].

7) *The Bucherer reaction* is the reaction of 1- or 2-naphthols with aqueous amine or ammonium solutions using hydrogen sulfite as the catalyst. It is nearly exclusively used for the naphthalene series [115]. The use of hydrogensulfite reduces the reaction temperature by 50–150 °C and results in high yield (85–95%) and purity. The reaction mechanism has been elucidated [135].

3.2.3. Exchange of Sulfo Groups

Ammonia can replace the sulfo groups in benzene, naphthalene, and anthraquinone derivatives. However, only in the anthraquinone series is this exchange of industrial importance. Until recently 1aminoanthraquinones were made by reaction of 1-sulfoanthraquinones with amines. To destroy the released hydrogensulfite, stoichiometric amounts of an oxidizing agent, e.g., sodium 3-nitrobenzenesulfonate, were added. Increasingly 1-aminoanthraquinone [82-45-1] is made by reduction of the corresponding nitroanthraquinone. This method avoids the mercury-catalyzed sulfonation to the 1-sulfoanthraquinone. 1-Cyclohexylaminoanthraquinone-5-sulfonic acid is obtained from anthraquinone-1,5-disulfonic acid and cyclohexyl-amine in aqueous medium under pressure at 100–200 °C [137]. The hydrogensulfite is oxidized with sodium perchlorate or 3-nitrobenzenesulfonic acid.

3.3. Other Processes

Only a few of all other possible methods for the production of aromatic amines described in the literature should be mentioned here. 4,4′-Methylenedianiline is produced on a large scale by condensation of aniline with formaldehyde in aqueous or aqueous–methanolic solution. Diphenylamine is made by vapor-phase condensation of aniline over alumina or titanium catalyst at 450–500 °C [138] or in the liquid phase at 175–450 °C [139]. The rearrangement of arylhydroxylamines in dilute sulfuric acid leads to o- and p-aminophenols [140]. 3,3′-Dichlorobenzidine [91-94-1], an important pigment intermediate, is produced by Bofors (plant in Muskegan, MI) using the benzidine rearrangement starting from o-chloronitrobenzene.

4. Economic Aspects

The economically most important aromatic amines are discussed in separate keywords. For the others, data are generally unavailable.

5. Toxicology, Occupational Health, and Environmental Protection

5.1. Toxicity

The toxicity of aromatic amines has long been of concern to the chemical industry because of the serious nature of both acute and chronic effects of exposure to aromatic amines. Despite this, the toxicology of many of the individual compounds has not been studied thoroughly. Therefore, a discussion of aromatic amine toxicity in general is more appropriate than a discussion of individual compounds.

The primary hazards associated with aromatic amine exposure are methemoglobinemia and carcinogenesis. Both of these effects are attributed to activation products of metabolic N-oxidation [141]. Acute exposure, from either inhalation or skin contact, results in anoxia because of an impairment of the oxygen transport system of the blood.

In aromatic amine exposure the heme iron of hemoglobin is oxidized from Fe(II) to Fe(III). The resulting product, methemoglobin, cannot combine with oxygen or carbon monoxide. The development of *methemoglobinemia* after exposure is often insidious; i.e., the onset of symptoms may be delayed for hours. Usually the first symptom is headache; anoxia, including cyanosis of lips, nose, and earlobes, develops when the methemoglobin blood level reaches 15%. At levels of 40%, weakness and dizziness are experienced, and at > 70%, coma and death may occur.

The extent to which various aromatic amines can cause methemoglobin formation varies widely. However, the arylhydroxylamines are recognized as the strongest methemoglobin formers in this group. The reason for their potency is that they are involved in a cyclic reaction. Phenylhydroxylamine reacts with hemoglobin to form methemoglobin and nitrosobenzene [142]. Nitrosobenzene is then reduced back to phenylhydroxylamine, which then is available to oxidize more hemoglobin. A small amount of nitrosobenzene appears to be reduced all the way to its amine, which eventually stops the reaction. As much as 50 mmol of methemoglobin may be produced by 1 mmol of hydroxylamine [143].

The *carcinogenic potential* of aromatic amines is of greatest concern. A number of them are known or are suspected of being carcinogens. Indeed it has been reported that the epidemiology of aromatic amine carcinogenesis is essentially the epidemiology of human cancer of industrial origin [144]. The first suspicion of aromatic amine carcinogenesis was raised in 1895 when cases of bladder cancer in Swiss dye workers were reported [145]. Since then a number of epidemiologic investigations and animal studies have identified which aromatic amines are carcinogenic [146]–[148]. Because most aromatic amines have not been examined thoroughly for carcinogenic potential, researchers are attempting to develop theoretical models for determining which ones are carcinogenic [149]–[151]. Structural requirements, such as the type of aromatic ring system, the position of the amine group on the ring, and the presence and position of other substituents, are believed to be important to carcinogenic potential. Chemical carcinogens exert their effect by causing DNA damage, which leads to tumor formation. The metabolic reaction products, rather than the aromatic amines themselves, are believed to damage the DNA. These reaction products are electrophilic and are covalently bonded to the nucleic acid bases of DNA. The following reaction sequence has been proposed, where $ArNH_2$ is the aromatic amine [152].

$$ArNH_2 \xrightarrow{oxid.} ArNHOH \xrightarrow[H]{pH<7} Ar\overset{+}{N}OH_2 \rightarrow Ar\overset{+}{N}H \rightarrow \text{covalent bonding to nucleophilic sites on critical macromolecules}$$

This reaction sequence implies that structural factors in the aromatic amine favoring the cation formation (e.g., electron-donating substituents ortho and para to the amino group) enhance carcinogenicity. Figure 4 is a sampling of some carcinogenic amines [153].

In addition to the effects cited in Table 1 [143], aromatic amines can cause a wide variety of other symptoms, such as eye and skin irritation. By drawing correlations from

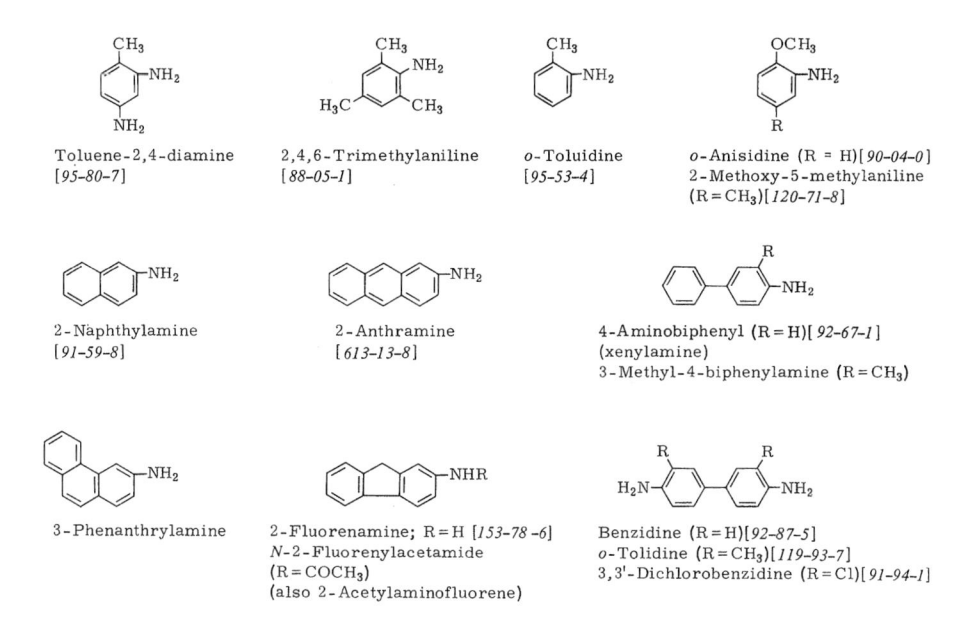

Figure 4. Examples of carcinogenic amines

Table 1. Effects of the chronic administration of primary amines on the blood, liver, and bladder of dogs

	Methemoglobin formation	Toxicity and/or carcinogenesis (liver)	Bladder carcinogenesis
4-Aminobiphenyl [92-67-1]	+++	−	+++
2-Naphthylamine [91-59-8]	+	−	++
1-Naphthylamine [134-32-7]	−	−	−
Benzidine [92-87-5]	−	−	+
Aniline [62-53-3]	++	−	−
Dichlorobenzidine [91-94-1]	−	+	++
Methylenedianiline [101-77-9]	−	+++	+−b*
4,4′-Methylene-bis(2-chloroaniline) [101-14-4]	−	+	++
4,4′-Methylene-bis(2-methylaniline) [838-88-0]	−	+++	−
Bis aniline A [2479-47-2]	−	+−	+
4-Aminobiphenylamine [101-54-2]	−	++	+−b*

* Indicates only preneoplastic changes, no tumors.

animal studies, exposure to some of these compounds may lead to liver and kidney damage.

5.2. Exposure Limits

Various countries have established their own exposure limits for workers [155]–[159]. As an illustration, Table 2 lists those aromatic amines with exposure limits established in the United States (ACGIH) [160] and the Federal Republic of Germany (MAK) [161]. These recommendations are in addition to the U.S. Occupational Safety and Health Administration's Subpart Z exposure limits [162].

5.3. Protective Measures

Exposure to aromatic amines occurs primarily through two routes: skin contact and inhalation. Because of the potential health hazards, strict control measures are necessary to prevent exposure. From an industrial hygiene standpoint, exposure is best prevented through the use of engineering controls. Whenever possible, processes containing aromatic amines should be enclosed. In a closed process exposure can occur only when equipment fails or when process streams are sampled. If a process is not enclosed, ventilation (both general diluted and local exhaust) should be used to prevent or reduce exposure. Ventilation specifications for a number of industrial processes are available [163]. When ventilation controls are not sufficient to lower exposure, respirators should be used until effective controls are installed. When respirators are used, a complete respiratory protection program should be instituted that includes regular training, maintenance, inspection, cleaning, and evaluation. Where skin contact may occur, workers should wear impervious clothing, including gloves and boots. Contaminated clothing should be stored in closed containers until discarded or cleaned. Safety showers should be available in the immediate area where exposure can occur so that the body can be washed quickly.

5.4. Environmental Monitoring

Routine monitoring of the airborne aromatic amine concentration should be performed to evaluate worker exposure. Measurements should be taken on a single 8-h sample or two 4-h samples to determine the 8-h TWA exposure. In addition, short-term samples should be taken to determine peak exposure when workers are doing specific tasks with a potential for exposure. Air samples should be taken in the worker's breathing zone. Several methods are available for airborne aromatic amine monitoring [164]–[166].

Table 2. ACGIH aromatic amines workplace exposure limit (mg/m^3)

Substance	TWA^a	STEL^b	MAK^c
4-Aminobiphenyl (skin) [92-67-1]	A/b^d	A/b^d	III A 1^f
Aniline & homologues (skin) [62-53-3]	10	20	A 2^e
Anisidine (o-, p-isomers) (skin) [29191-52-4]	0.5	-	0.5
ANTU (1-Naphthylthiourea) [86-88-4]	0.3	0.9	0.3
Benzidine (skin) [92-87-5]	A/b^d	A/b^d	III A 1^f
3,3′-Dichlorobenzidine (skin) [91-94-1]	A 2^e	A 2^e	III A 2^g
N,N-Dimethylaniline (skin) [121-69-7]	25	50	25
Diphenylamine [122-39-4]	10	20	-
N-Isopropylaniline (skin) [643-28-7]	10	20	-
N-Methylaniline (skin) [100-61-8]	2	5	9
4,4′-Methylene-bis(2-chloroaniline) (skin) [101-14-4]	0.22, A 2^e	-	III A 2^g
4,4′-Methylenedianiline (skin) [101-77-9]	0.8	4	-
2-Naphthylamine [91-59-8]	A/b^d	A/b^d	III A 1^f
p-Nitroaniline (skin) [100-01-6]	3	-	6
N-Phenyl-2-naphthylamine [135-88-6]	A 2	-	A 2^e
p-Phenylenediamine [106-50-3]	0.1 (skin)	-	0.1
o-Tolidine [119-93-7]	A 2^e	-	A 2^e
o-Toluidine (skin) [95-53-4]	9, A 2^e	-	-
Triphenylamine [603-34-9]	5	-	-
m-Xylene-α,α′-diamine [1477-55-0]	C^{h}0.1	-	-
Xylidene (all isomers except 2,4-xylidene) [1300-73-8]	10 (skin)	-	25

^{a} Time-weighted average (8-h workday and a 40-h week);
^{b} Short-term exposure limit (15-min TWA exposure);
^{c} Maximum concentration values in the workplace;
^{d} A recognized carcinogen. No assigned TLV. No exposure permitted by any route;
^{e} A suspect carcinogen for man. Exposures should be controlled carefully;
^{f} Compounds capable of inducing malignant tumors as shown by experience with humans;
^{g} Unmistakably carcinogenic in animal experimentation only;
^{h} Ceiling limit; TLV should not be exceeded even instantaneously, avoid skin contact, exposures are likely by skin absorption.

6. References

[1] M. N. Desai, *Trans. Saest* **16** (1981) no. 2, 77–87. *Chem. Abstr.* **95** (1981) 105208 p.
[2] *Colour Index,* 3rd ed., Society of Dyers and Colourists, Bradford, England, and American Association of Textile Chemists and Colorists, North Carolina, United States, 1971.
[3] United States International Trade Commission, *Publication 1588,* U.S. Government Printing Office, Washington, D.C., 1984.
[4] United States International Trade Commission, *Publication 1548,* "Import of Benzenoid Chemicals and Products 1983," Graphics Branch, Washington, D.C., July 1984, p. 10, 23.
[5] Japan Dyestuff Industry Assn., *Japan Chem.* 1982 (April 15) 8.
[6] W. L. Semon in Davis, Bake (ed.): *The Chemistry and Technology of Rubber,* Reinhold Publ. Co., New York 1937, p. 414.
[7] R. G. Bauman, J. W. Born, *J. Appl. Polym. Sci.* **1**, (1959) 351.
[8] G. S. Petrov, V. F. Prosvirkina, *Zh. Prikl. Khim. (Moscow)* **30**, (1956) 1660. *Chem. Abstr.* **52** (1958) 5028 f.
[9] G. Oster, US 3014799, 1961.
[10] H. E. Fierz-David, L. Blangey: *Fundamental Processes of Dye Chemistry,* 5th ed., Interscience, New York 1949, pp. 99, 140, 179, 306.
[11] Horizon Inc., US 3082086, 1963 (R. H. Sprague).
[12] Horizon Inc., US 3056673, 1962 (E. Wainer).
[13] Du Pont, US 3673143, 1971 (T. I. Bair, P. W. Morgan).
[14] R. E. Wilfong, J. Zimmerman, *Rubber World* **166** (1972) 40, 47.
[15] A. Galat, G. Elion, *J. Am. Chem. Soc.* **65** (1943) 1566.
[16] G. R. Petit, E. G. Thomas, *J. Org. Chem.* **24** (1959) 895.
[17] C. S. Marvel, G. S. Hiers, *Org. Synth. Coll.* **1** (1941) 327.
[18] K. H. Slotta, H. Dressler, *Ber. Dtsch. chem. Ges.* **63** (1930) 889, 894. *Houben-Weyl* **9**, 871.
[19] F. B. Dains, R. Q. Brewster, C. P. Olander, *Org. Synth. Coll.* **1** (1941) 447.
[20] Shell Development Co., US 2580284, 1951 (T. J. Deahl, F. H. Stross, M. D. Taylor).
[21] Du Pont, US 3649693, 1975 (D. R. Coulson).
[22] Biller, Michealis & Co., US 2991311, 1961 (M. Thoma).
[23] A. K. Bhattacharyya, D. K. Naudi, *Ind. Eng. Chem. Prod. Res. Dev.* **14** (1975) no. 3, 162.
[24] R. Stroh, J. Ebersberger, H. Haberland, W. Hahn, *Angew. Chem.* **69** (1957) 124.
[25] Ethyl Corp., US 4219502, 1980 (K. G. Ihrmann, M. Brandt).
[26] Ethyl Corp., US 4128582, 1978 (L. J. Governale, J. C. Wollensak).
[27] Ethyl Corp., US 3923892, 1974 (O. H. Klopfer).
[28] Ethyl Corp., US 3843698, 1974 (J. H. Dunn).
[29] K. D. Hesse, *Justus Liebigs Ann. Chem.* **741** (1970) 117.
[30] C. W. Smith: *Acrolein,* J. Wiley & Sons, New York 1962, p. 105.
[31] R. C. Elberfield: *Heterocyclic Compounds,* vol. **4,** J. Wiley & Sons, New York 1952, pp. 6–80.
[32] H. J. Roth, P. Lepke, *Arch. Pharm. (Weinheim, Germ.)* **305** (1972) 159.
[33] *Houben-Weyl* **9**, 40.
[34] F. B. Mallory, *Org. Synth. Coll.* **4** (1963) 74.
[35] *Houben-Weyl* **4 (1 b),** 662.
[36] H. D. Anspon, *Org. Synth. Coll.* **3** (1955) 711. *Houben Weyl* **10 (3),** 332.
[37] Sherwin-Williams Co., US 4363914, 1982 (J. W. Long, L. Vacek).
[38] Bayer, US 4424360, 1984 (F. Hagedorn, W. Evertz).

[39] *Houben-Weyl* **5 (3)**, 705.
[40] Ethyl Corp., US 2675406, 1954 (H. D. Orloff, J. F. Napolitano).
[41] U. S. Rubber Co., GB 588642, 1947.
[42] *Houben-Weyl* **10** (1), 1054.
[43] A. S. Pagano, W. D. Emmond, *Org. Synth. Coll.* **5** (1973) 367.
[44] C. F. H. Allen, J. A. van Allen, *Org. Synth. Coll.* **3** (1955) 824.
[45] R. J. Peterson: *Hydrogenation Catalysts,* Noyes Data Corp., Park Ridge, N.J., 1977, pp. 139–144, 254–267.
[46] BASF, US 3799983, 1974 (H. Corr et al.).
[47] Upjohn Co., US 3856862, 1974 (T. H. Chung et al.).
[48] BASF, DE 805518, 1951 (O. Stichnoth).
[49] N. L. Drake, *Org. React.* **1,** (1942) 105.
[50] W. W. Hartmann, J. R. Byers, J. B. Dickey, *Org. Synth. Coll.* **2** (1948) 451.
[51] *Houben-Weyl* **11 (1)**, 341, 394.
[52] G. W. Gray et al., *J. Chem. Soc. B.* 1952, 1959.
[53] D. C. Owsley, J. J. Bloomfield, *Synthesis* 1977 118.
[54] S. A. Cogswell: *Chemical Economics Handbook,* SRI International, Menlo Park, Calif., 1984, p. 611.5032 K.
[55] *Houben-Weyl* **11 (1)** 409, 474.
[56] H. K. Porter, *Org. React.* **20** (1973) 455.
[57] *Houben-Weyl* **7 (1),** 156.
[58] Monsanto Chemical Co., US 2745614, 1957 (O. DeGarmo, E. J. McMullen).
[59] G. G. Henderson, M. M. J. Sutherland, *J. Chem. Soc. Trans.* **97** (1910) 1616.
[60] J. F. Bunnett, R. E. Zahler, *Chem. Rev.* **49** (1951) 398.
[61] *Houben-Weyl* **9,** 519.
[62] Bayer, US 3417090, 1965 (H. Pelster, C. König, R. Pütter); FR 1531889, 1967 (H. Pelster, C. König, R. Pütter).
[63] *Houben-Weyl* **11 (1),** 437.
[64] *Houben-Weyl* **11 (1),** 472.
[65] R. Jansson, *Chem. Eng. News* **62** (1984) no. 47, 43–57.
[66] M. M. Baizer, H. Lund: *Organic Electrochemistry,* 2nd ed., Marcel Dekker, New York 1983, pp. 295–313.
[67] T. Kaufmann et al. in W. Foerst (ed.): *Neure Methoden der präparativen organischen Chemie,* vol. **4,** Verlag Chemie, Weinheim 1966, p. 62.
[68] D. Balcom, A. Fürst, *J. Am. Chem. Soc.* **75** (1953) 4334.
[69] A. Fürst et al., *Chem. Rev.* **65** (1965) 51–69.
[70] T. L. Fletcher et al., *J. Org. Chem.* **25** (1960) 996.
[71] E. J. Colter, S. S. Wang, *J. Org. Chem.* **27** (1962) 1517.
[72] Filiala Cluj a Academiei R. S. Romania, DE-OS 2007656, 1971 (M. Ionesca, I. Mester).
[73] V. V. Korshak et al., SU 218898, 1965. *Ullmann,* 4th ed., 7 : 398.
[74] P. M. Bavin, *Org. Synth. Coll.* **5** (1973) 30.
[75] R. A. Raphael, E. C. Taylor, H. Wynberger, *Advan. Org. Chem.* **2** (1960) 350.
[76] American Cyanamid Co., US 2891094, 1951 (O. C. Kaskalitis et al.).
[77] Monsanto, US 3944615, 1976 (A. F. M. Iqbal).
[78] Du Pont, DE-AS 1187243, 1965 (J. R. Kosak).
[79] S. Tsutsumi, H. Terade, *Kogyo Kagaku Zasshi* **54** (1951) 527.
[80] L. F. Albright et al., *Chem. Eng. (New York)* **74** (1967) 197–202, 251–256.

[81] Engelhard-Minerals & Chem. Corp., DE 2042368, 1971 (S. G. Hindin, D. L. Blair, D. R. Steele).
[82] Du Pont, US 3499034, 1970 (R. A. Gonzales).
[83] Bayer, DE-OS 1643379, 1971 (W. Böhm, A. Wissner).
[84] P. N. Rylander: *Catalytic Hydrogenation over Platinum Metals*, Academic Press, New York 1967, p. 169.
[85] B. R. James: *Homogenous Hydrogenation*, J. Wiley & Sons, New York-London 1973, p. 180.
[86] Du Pont, US 3328465, 1967 (L. Spiegler).
[87] Du Pont, US 3127356, 1964 (J. M. Hamilton, L. Spiegler).
[88] Olin Corp., US 3935264, 1976 (S. K. Bhutani).
[89] ICI, US 3154584, 1964 (C. Gardner, R. G. New).
[90] Allied Chem. Corp., US 2976320, 1961 (L. O. Winstrom, R. D. Samdahl, P. S. Perch).
[91] Olin Mathieson Chem. Corp., US 3356729, 1967 (W. I. Denton, P. D. Hammond).
[92] Du Pont, US 2619503, 1952 (R. G. Benner, A. C. Stevenson).
[93] Mobay Chem. Corp., US 3517063, 1970; BE 645406, 1964 (B. R. Nason).
[94] Olin Mathieson Chem. Corp., BE 661047, 1965 (J. J. Cimerol, W. M. Clark, W. I. Denton).
[95] Monomers & Des. Inst., SU 233660, 1968 (A. V. Ivanov et al.).
[96] *Houben-Weyl* **4 (2),** 137.
[97] Allied Chem. & Dye Corp., US 2671763, 1954 (L. O. Winstrom, W. B. Harris).
[98] BASF, DE-AS 1233408, 1967 (K. Ehrmann).
[99] Office Official Industrial L'azote, BE 646607, 1964.
[100] M. Sittig: *Amines, Nitrites and Isocyanates, Processes and Products*, Noyes Dev. Corp., Park Ridge, N.J., 1969.
[101] Farbwerke Hoechst AG, US 3761425, 1973 (K. Baessler, K. Mayer).
[102] Farbwerke Hoechst AG, US 3929891, 1975; US 3803054, 1974 (K. Habig, K. Baessler).
[103] Degussa, US 3941717, 1976 (G. Volheim et al.).
[104] Engelhard Minerals & Chem. Corp., DE 2042368, 1970; US 3666813, 1972 (S. G. Hindlin, D. L. Bair).
[105] Engelhard Industries Inc., FR 1440419, 1965.
[106] G. M. Leuteritz, *Chemie-Anlagen & Verfahren* **3** (March 1971) 49–54.
[107] G. M. Leuteritz, *Process Eng.* Dec. 1973 **62,** 63.
[108] G. M. Leuteritz et al., *Hydrocarbon Process.* June 1976 **99,** 100.
[109] R. J. Malone, *Chem. Eng. Prog.* **53** (1980) 53–59.
[110] M. Day, A. T. Peters Jr., *J. Soc. Dyers Colour.* **83** (1967) no. 4, 137–143.
[111] S. K. Bhattacharyya et al., *Indian J. Technol.* **5** (1967) no. 7, 232–233.
[112] Universal Oil, FR 1471866, 1967 (M. Earl, A. Nielsen).
[113] Phillips Petroleum Co., US 3484487, 1969 (J. S. Dix).
[114] Du Pont, DE-OS 1695590, 1971 (J. Pikl).
[115] *Houben-Weyl* **11 (1),** 26.
[116] J. F. Bunnet, R. F. Zahler, *Chem. Rev.* **49** (1951) 273.
[117] F. B. Wells, C. F. H. Allen, *Org. Synth. Coll.* **2** (1948) 221.
[118] F. Pietra, *Q. Rev. Chem. Soc.* **23** (1969) 504–521.
[119] Etat Francais, FR 1332501, 1963 (J. Issoire).
[120] Monsanto, DE-OS 1768518, 1968 (T. A. Garabedian).
[121] Mobil Oil Co., US 3231616, 1966 (D. J. Jones).
[122] Kanto Elektrochem. Ind. Co., JP-Applic. 671864, 1967; 6726289, 1967.
[123] J. F. Bunnet, *Chem. Rev.* **49** (1951) 395.

[124]   *Houben-Weyl* **11 (1)**, 189.
[125]   General Aniline, GB 1115535, 1966 (C. H. Chang). GAF Corp., US 3468941, 1969 (C. H. Chang).
[126]   Sterling Drug Inc., US 3102913, 1963 (R. E. Werner).
[127]   BASF, BE 703695, 1967.
[128]   Monsanto, US 3170956, 1965 (J. F. Olin).
[129]   Sherwin-Williams Co., NL 7017339, 1970.
[130]   Hercules, US 3338966, 1967 (J. C. Snyder); US 3340302, 1967 (H. L. Young).
[131]   Hercules Powder Co., GB 926897, 1961.
[132]   Halcon International Inc., FR 1 405 816, 1965 (M. R. S. Barker); BE 740758, 1969 (J. L. Russell); US 3578714, 1971 (J. L. Russell); BE 744936, 1970 (M. Becker, R. Fernandez).
[133]   Mitsui Petrochemical Ind. Ltd., JP 7123052, 1968.
[134]   I. McKechnie, F. Bayer, J. Drennan, *Chem. Eng. (N.Y.)* **87** (Dec. 29, 1980) 26–27.
[135]   A. Rieche, H. Seeboth, *Justus Liebigs Ann. Chem.* **638** (1960) 43–110.
[136]   N. L. Drake, *Org. React.* **1** (1942) 110.
[137]   Bayer, DE 1222051, 1965 (H. J. Schulz, H. S. Bien).
[138]   Amer. Cyanamid Co., US 2943111, 1960 (G. L. Wiesner); US 3118944, 1964 (G. I. Addis).
[139]   B. F. Goodrich Co., US 3071619, 1959 (H. J. Kehe, R. T. Johnson et al.).
[140]   E. D. Hughes, C. K. Ingold, *Quart. Rev.* **6** (1952) 45.
[141]   H. G. Neumann, E. W. Wieland, *Ind. Med. Surg.* **42** (Nov-Dec 1973) no. 11, 15–19.
[142]   M. Kiese, *Pharmacol. Rev.* **18** (1966) 1091–1161.
[143]   J. L. Radomski, *Annu. Rev. Pharmacol. Toxicol.* **19** (1979) 129–157.
[144]   H. G. Parkes in C. E. Scarle (ed.): *Chemical Carcinogens,* American Chemical Society Monograph No. 173, American Chemical Society, Washington, D.C., 1976, pp. 462–480.
[145]   L. Rehn, *Arch. Klin. Chir.* **50** (1895) 588–600.
[146]   *Second Annual Report on Carcinogens,* U.S. Dept. of Health and Human Services, U.S. Govt. Printing Office, December 1981.
[147]   International Agency for Research on Cancer (IARC): *Monographs on the Evaluation of Carcinogenic Risk of Chemicals to Man (4),* Lyon 1974.
[148]   International Agency for Research on Cancer (IARC): *Monographs on the Evaluation of Carcinogenic Risk of Chemicals to Man* (16), Lyon 1978.
[149]   S. S. Thorgeirsson, H. A. J. Schut, P. J. Wirth, E. Dybing (ed.): *Mutagenicity and Carcinogenicity of Aromatic Amines, Molecular Basis of Environmental Toxicity,* Ann Arbor Science Publishers, Inc., Ann Arbor, Mich., 1980, pp. 275–292.
[150]   R. A. Thuraisingham, S. H. M. Nilar, *J. Theor. Biol.* **86** (1981) 557–580.
[151]   K. Yuta, P. C. Jurs, *J. Med. Chem.* **24** (1981) 241–251.
[152]   E. C. Miller, *Can. Res.* **38** (1978) 1479.
[153]   J. H. Weisburger, G. M. Williams (ed.): *Casarett and Doull's Toxicology, The Basic Science of Poisons,* 2nd ed., Macmillan Publ. Co., New York 1980, p. 93.
[154]   G. W. Thorn et al. (ed.): *Harrison's Principles of International Medicine,* 8th ed., McGraw-Hill, New York 1977, p. 1710–1713.
[155]   Deutsche Forschungsgemeinschaft (ed.): *Report No. 11 of the Commission for Investigation of Health Hazards of Chemical Compounds in the Work Area,* Bonn, Federal Republic of Germany, 1979.
[156]   Ontario Ministry of Labor, *Exposure Criteria for Potentially Harmful Agents and Substances in Workplaces,* Toronto, Ontario, 1981.
[157]   Z. B. Smelyanskiy, I. P. Ulanova, *Gig. Tr. Prof Zabol* **3** (1959) no. 5, 7–15.

[158] *Recommendation for Permissible Concentration, etc. (Japan)*, Sangyo Igaku **13** (1971) 475–484.
[159] *Permissible Levels of Occupational Exposure to Airborne Toxic Substances*, WHO Technical Report Series No. 415, WHO, France 1968.
[160] American Conference of Governmental Industrial Hygienists (ACGIH) (ed.): Threshold Limit Values for Chemical Carcinogens in Work Air (TLV), Cincinnati, Ohio, 1982.
[161] Deutsche Forschungsgemeinschaft (DFG) (ed.): *Maximum Concentrations at the Workplace (MAK)*, Verlag Chemie, Weinheim 1982.
[162] Occupational Safety and Health Administration, *OSHA Safety and Health Standards*, 29 CFR 1910, U.S. Government Printing Office, Washington, D.C., Nov. 7, 1978.
[163] American Conference of Governmental Industrial Hygienists (ACGIH) (ed.): *Industrial Ventilation-A Manual of Recommended Practice*, 17th ed., Cincinnati, Ohio, 1982.
[164] M. A. Pinches, R. F. Walker, *Ann. Occup. Hyg.* **23** (1980) 335–352.
[165] E. E. Campbell, G. O. Wood, R. G. Anderson, *IARC Sci. Publ.* **40** (1981) 109–118.
[166] D. N. Meddle, *Analyst (London)* **106** (1981) Issue 1261, 1088–1095.

Amino Acids

Axel Kleemann, Degussa AG, Hanau-Wolfgang, Federal Republic of Germany
Wolfgang Leuchtenberger, Degussa AG, Hanau-Wolfgang, Federal Republic of Germany
Bernd Hoppe, Degussa AG, Hanau-Wolfgang, Federal Republic of Germany
Herbert Tanner, Degussa AG, Hanau-Wolfgang, Federal Republic of Germany

1.	Introduction and History	536
2.	Properties	540
2.1.	Physical Properties and Structure	540
2.2.	Chemical Properties	541
3.	Production	548
3.1.	General Methods	548
3.1.1.	Methods for D,L-Amino Acids	549
3.1.2.	Methods for L-Amino Acids	550
3.1.3.	Methods for D-Amino Acids	553
3.2.	Production of Specific Amino Acids	553
3.2.1.	Alanine	553
3.2.2.	Arginine	553
3.2.3.	Aspartic Acid and Asparagine	554
3.2.4.	Cystine and Cysteine	554
3.2.5.	Glutamic Acid and Glutamine	555
3.2.6.	Glycine	555
3.2.7.	Histidine	555
3.2.8.	Isoleucine	556
3.2.9.	Leucine	556
3.2.10.	Lysine	556
3.2.11.	Methionine	557
3.2.12.	Phenylalanine	558
3.2.13.	Proline	559
3.2.14.	Serine	559
3.2.15.	Threonine	560
3.2.16.	Tryptophan	560
3.2.17.	Tyrosine	562
3.2.18.	Valine	562
4.	Biochemical and Physiological Significance	562
5.	Uses	565
5.1.	Human Nutrition	565
5.1.1.	Supplementation	567
5.1.2.	Flavorings, Taste Enhancers, and Sweeteners	570
5.1.3.	Other Uses in Foodstuff Technology	572
5.2.	Animal Nutrition	573
5.3.	Pharmaceuticals	575
5.3.1.	Nutritive Agents	575
5.3.2.	Therapeutic Agents	577
5.4.	Cosmetics	583
5.5.	Agrochemicals	584
5.6.	Industrial Uses	584
6.	Chemical Analysis	585
7.	Economic Significance	587
8.	Toxicology	587
9.	References	590

1. Introduction and History

The proteins, although they occur in an almost infinite variety, are composed of a relatively small number of basic building blocks, all α-amino acids. In addition, the amino acids fulfill certain regulatory functions in the metabolism and are required for the biosynthesis of other functional structures. This review is limited, for the most part, to the protein-forming α-amino acids, because they are by far the most widely distributed in nature and are of considerable economic interest.

The ca. 20 different α-amino acids found in proteins are rather simple organic compounds, in which an amino group and a side chain (R) are attached alpha to the carboxyl function. The R group may be aliphatic, aromatic, or heterocyclic and may possess further functionality.

$$R-\underset{NH_2}{\underset{|}{\overset{H}{\overset{|}{C}}}}-COOH \quad \text{Amino acid}$$

$$\downarrow -H_2O$$

$$H_2N-\underset{R}{\underset{|}{\overset{H}{\overset{|}{C}}}}-CO-NH-\underset{R}{\underset{|}{\overset{H}{\overset{|}{C}}}}-COOH \quad \text{Dipeptide}$$

$$\downarrow -H_2O$$

$$H_2N-\underset{R}{\underset{|}{\overset{H}{\overset{|}{C}}}}-CO(NH-\underset{R}{\underset{|}{\overset{H}{\overset{|}{C}}}}-CO)_{n-2}-NH-\underset{R}{\underset{|}{\overset{H}{\overset{|}{C}}}}-COOH$$

$$\text{Protein } (n > 50)$$

At present over 200 naturally occurring α-amino acids are known [1]–[3], [11]. Table 1 shows the structures of the α-amino acids found in proteins, where they occur exclusively as the L-enantiomers. D-Amino acids have been found only in the cell walls of some bacteria, in peptide antibiotics, and in the cell pools of some plants [6], [12], [13]. Table 2 lists some amino acids and derivatives that do not occur in proteins.

History. The history of amino acid chemistry began in 1806, when two French investigators, VAUQUELIN and ROBIQUET, isolated asparagine from asparagus juice. It was not until 1925 that SCHRYVER and BURTON isolated threonine from oat protein, the last discovered of the ca. 20 protein-forming amino acids. STRECKER synthesized alanine in 1850 from acetaldehyde and hydrogen cyanide. ESCHER established the hypothesis of essential amino acids. EMIL FISCHER discovered that the amino acids were building blocks of the proteins. ABDERHALDEN synthesized threonine from acrylic acid derivatives and methanol. ROSE et al. recognized threonine as the last of the eight essential amino acids. D,L-Methionine was produced industrially in Germany in 1948, and in 1956 L-glutamic acid was produced by fermentation in Japan.

Origin of Amino Acids. The first amino acids were probably produced on the earth more than 3×10^9 years ago via "prebiotic synthesis" in the primordial atmosphere. The concept of prebiotic synthesis is based on laboratory experiments in which glycine, alanine, aspartic acid, glutamic acid,

Table 1. Common amino acids of proteins (CAS Registry Numbers and physical properties are given in table 3)

Trivial name	IUPAC abbreviation	Systematic name and synonyms	Structural formula	Empirical formula	M_r	Protein	Content, g/100 g [a]	[b]
L-Alanine	Ala	2-aminopropionic acid	$CH_3CH-COOH$ $\quad\quad\ \ \|$ $\quad\quad\ \ NH_2$	$C_3H_7NO_2$	89.09	silk fibroin	25	29.7
L-Arginine	Arg	2-amino-5-guanidinovaleric acid 2-amino-5[(aminoiminomethyl)amino]pentanoic acid	$H_2N-C-NH-CH_2CH_2CH_2CH-COOH$ $\quad\ \ \|\|\quad\quad\quad\quad\quad\quad\quad\quad\quad\ \|$ $\quad\ \ NH\quad\quad\quad\quad\quad\quad\quad\quad\ \ NH_2$	$C_6H_{14}N_4O_2$	174.20	salmin edostin wool gelatin rat liver histone	87 17 10	 8.3 15.9
L-Asparagine	Asn	2-aminosuccinamic acid 2,4-diamino-4-oxobutanoic acid	$H_2N-C-CH_2-CH-COOH$ $\quad\quad\ \|\|\quad\quad\quad\quad\quad\ \|$ $\quad\quad\ O\quad\quad\quad\quad\quad NH_2$	$C_4H_8N_2O_3$	132.13			
L-Aspartic acid	Asp	aminosuccinic acid 2-amino-1,3-butanedioic acid	$HOOC-CH_2CH-COOH$ $\quad\quad\quad\quad\quad\ \|$ $\quad\quad\quad\quad\quad NH_2$	$C_4H_7NO_4$	133.10	edestin hemoglobin barley globulin	12 9–10	 10.3
L-Cysteine	Cys	2-amino-3-mercaptopropionic acid 3-mercaptoalanine	$HS-CH_2CH-COOH$ $\quad\quad\quad\quad\ \|$ $\quad\quad\quad\quad NH_2$	$C_3H_7NO_2S$	121.16	wool keratin human hair keratin feather keratin	11.9 14.4 8.2	
L-Cystine	Cys-Cys, (Cys)$_2$, Cyss	2,2'-diamino-3,3'-dithiobis(propionic acid) 3,3'-dithiobis(2-aminopropanoic acid)	$HOOC-CH-CH_2S-S-CH_2CH-COOH$ $\quad\quad\quad\ \|\quad\quad\quad\quad\quad\quad\quad\ \|$ $\quad\quad\quad\ NH_2\quad\quad\quad\quad\quad\ NH_2$	$C_6H_{12}N_2O_4S_2$	240.30			
L-Glutamic acid	Glu	2-aminoglutaric acid 2-amino-1,4-pentanedioic acid	$HOOC-CH_2CH_2CH-COOH$ $\quad\quad\quad\quad\quad\quad\quad\ \|$ $\quad\quad\quad\quad\quad\quad\quad NH_2$	$C_5H_9NO_4$	147.13	gliadin zein wheat gliadin maize zein	47 31	 39.2 22.9
L-Glutamine	Gln	2-aminoglutaramic acid 2,5-diamino-5-oxopentanoic acid	$H_2N-C-CH_2CH_2CH-COOH$ $\quad\quad\ \|\|\quad\quad\quad\quad\quad\quad\ \|$ $\quad\quad\ O\quad\quad\quad\quad\quad\quad NH_2$	$C_5H_{10}N_2O_3$	146.15			
Glycine	Gly	aminoacetic acid	H_2N-CH_2COOH	$C_2H_5NO_2$	75.07	gelatin silk fibroin	26 44	
L-Histidine	His	α-amino-1H-imidazole-4-propionic acid 1H-imidazole-4-alanine	imidazole-$CH_2CH-COOH$ $\quad\quad\quad\quad\quad\quad\ \ \|$ $\quad\quad\quad\quad\quad\ \ NH_2$	$C_6H_9N_3O_2$	155.16	hemoglobin	7	
L-Hydroxyproline	Hyp	trans-4-hydroxy-2-pyrrolidinecarboxylic acid	pyrrolidine-COOH (HO-substituted)	$C_5H_9NO_3$	131.13	gelatin	15	
L-Isoleucine	Ile	2-amino-3-methylvaleric acid 2-amino-3-methylpentanoic acid	$CH_3CH_2CH-CH-COOH$ $\quad\quad\quad\quad\ \|\quad\ \|$ $\quad\quad\quad\quad CH_3\ NH_2$	$C_6H_{13}NO_2$	131.18	edestin hemoglobin serum proteins oat globulin beef serum albumin	21 29 20	 4.3 2.6

Table 1. continued

Trivial name	IUPAC abbreviation	Systematic name and synonyms	Structural formula	Empirical formula	M_r	Protein	Content, g/100 g [a]	[b]

L-Leucine	Leu	2-amino-4-methylvaleric acid
		2-amino-4-methylpentanoic acid
		2-amino-isocaproic acid
			$CH_3-CH-CH_2-CH-COOH$
			       $\|$ $\|$
			      CH_3 NH_2
				$C_6H_{13}NO_2$	131.18	edestin
						hemoglobin
						serum proteins
						maize zein
							21
							29
							20
								
								19

L-Lysine	Lys	2,6-diaminohexanoic acid
		2,6-diaminocaproic acid
			$H_2N-CH_2CH_2CH_2CH_2CH-COOH$
			                                      $\|$
			                                   NH_2
				$C_6H_{14}N_2O_2$	146.19	serum albumin
						serum globulin
						horse myoglobin
							13
							6
							
								15.5

L-Methionine	Met	2-amino-4-(methylthio)butyric acid
		2-amino-4-(methylthio)butanoic acid
			$CH_3-S-CH_2CH_2CH-COOH$
			                              $\|$
			                           NH_2
				$C_5H_{11}NO_2S$	149.21	egg albumin
						casein
						β-lactoglobulin
							5
							3
							
								4.1
								3.2

L-Phenylalanine	Phe	2-amino-3-phenylpropionic acid
		α-aminobenzenepropanoic acid
			⌬—$CH_2CH-COOH$
			           $\|$
			        NH_2
				$C_9H_{11}NO_2$	165.19	zein
						egg albumin
						serum albumin
							8
							5
							
								7.7
								7.8

L-Proline	Pro	2-pyrrolidinecarboxylic acid 2-carboxypyrrolidine	⟨N⟩—COOH H	$C_5H_9NO_2$	115.13	gelatin gliadin salmin casein	17 13 	 6.9 10.6

L-Serine	Ser	2-amino-3-hydroxypropionic acid
		2-amino-3-hydroxypropanoic acid
			$HO-CH_2CH-COOH$
			                   $\|$
			                NH_2
				$C_3H_7NO_3$	105.09	silk fibroin
						trypsinogen
						pepsin
							13	16.2
								16.7
								12.2

L-Threonine	Thr	2-amino-3-hydroxybutyric acid
		2-amino-3-hydroxybutanoic acid
			$CH_3CH-CH-COOH$
			         $\|$ $\|$
			       OH NH_2
				$C_4H_9NO_3$	119.12	casein
						human hair keratin
						avidin
							4	
								8.5
								10.5

L-Tryptophan	Trp	2-amino-3-(3'-indolyl)propionic acid
		α-amino-1H-indole-3-propanoic acid
			⌬—$CH_2CH-COOH$
			              $\|$
			           NH_2
			 (indole)
				$C_{11}H_{12}N_2O_2$	204.23	fibrin
						egg lysozyme
							3	
								10.6

L-Tyrosine	Tyr	2-amino-3-(4-hydroxyphenyl)-propionic acid
		[3-(4-hydroxyphenyl)]alanine
		2-amino-3-(p-hydroxyphenyl)-propionic acid
		α-amino-4-hydroxybenzenepropanoic acid
			HO—⌬—$CH_2CH-COOH$
			                         $\|$
			                      NH_2
				$C_9H_{11}NO_3$	181.19	fibrin
						silk fibroin
						papain
							6
							13
							
								14.7

L-Valine	Val	2-amino-3-methylbutyric acid
		2-aminoisovaleric acid
			$CH_3-CH-CH-COOH$
			         $\|$ $\|$
			       CH_3 NH_2
				$C_5H_{11}NO_2$	117.15	casein
						beef sinew
						beef aorta
							8	
								17.4
								17.6

Table 2. Important natural nonproteinogenic amino acids

Amino acid	Formula	Occurrence
Neutral		
β-Alanine [107-95-9]	$H_2N-CH_2CH_2COOH$	apple, constituent of pantothenic acid, carnosine, anserine
γ-Aminobutyric acid (GABA) [56-12-2]	$H_2N-CH_2CH_2CH_2COOH$	citrus fruits, sugar beet, brain
Betaine [107-43-7]	$CH_3-\overset{CH_3}{\underset{CH_3}{N^{\pm}}}-CH_2COO^-$	sugar beet
Carnitine [461-06-3]	$CH_3-\overset{CH_3}{\underset{CH_3}{N^{\pm}}}-CH_2\underset{OH}{CH}-CH_2COO^-$	Lys metabolite, muscle
Citrulline [372-75-8]	$H_2N-\underset{O}{\overset{\|}{C}}-NH-CH_2CH_2CH_2\underset{NH_2}{CH}-COOH$	urea cycle, watermelon
Basic		
Creatine [57-00-1]	$H_2N-\underset{NH}{\overset{CH_3}{\overset{\|}{C}}-N-CH_2COOH}$	muscle of vertebrates
Ornithine [70-26-8]	$H_2N-CH_2CH_2CH_2\underset{NH_2}{CH}-COOH$	urea cycle, shark liver
Saccharopine [997-68-2]	$HOOC-\underset{NH_2}{CH}-(CH_2)_4-NH-\underset{COOH}{CH}-CH_2CH_2COOH$	baker's and brewer's yeast
Aromatic		
3,4-Dihydroxyphenylalanine (DOPA) [59-92-7]	HO—C$_6$H$_3$(OH)—$CH_2\underset{NH_2}{CH}-COOH$	Tyr metabolite, bean
5-Hydroxytryptophan [56-69-9]	HO-indole-$CH_2\underset{NH_2}{CH}-COOH$	serotonin precursor
Thyroxine [51-48-9]	HO—C$_6$H$_2$I$_2$—O—C$_6$H$_2$I$_2$—$CH_2\underset{NH_2}{CH}-COOH$	thyroid gland
Sulfur containing		
Homocysteine [6027-13-0]	$HS-CH_2CH_2\underset{NH_2}{CH}-COOH$	Met metabolite, mushrooms
S-Methylmethionine (vitamin U) [10332-17-9]	$CH_3-\overset{+}{\underset{CH_3}{S}}-CH_2CH_2\underset{NH_2}{CH}-COO^-$	cabbage, asparagus
Penicillamine [52-67-5]	$CH_3\underset{HS}{\overset{CH_3}{\overset{\|}{C}}}-\underset{NH_2}{CH}-COOH$	hydrolysis product of penicillin
Imino acid		
Pipecolic acid [535-75-1]	piperidine-2-COOH	legumes, metabolite of Lys

and other compounds were produced by the action of an electrical discharge on a simulated primordial atmosphere consisting of methane, hydrogen, water, and ammonia [14]. Since then, traces of amino acids have been detected in moon rocks, meteorites, and interstellar space.

2. Properties

2.1. Physical Properties and Structure

α-Amino acids are nonvolatile, white, crystalline compounds with no defined melting points. They are relatively stable on heating, generally decomposing at 250–300 °C. Both the low volatility and the thermal stability result from the low-energy dipolar structure (zwitterion, inner salt, betaine), which the amino acids assume in the solid state.

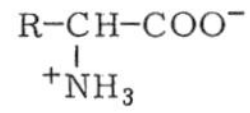

$$R-CH-COO^- \\ |\\ ^+NH_3$$

Evidence for this structure is provided by infrared and Raman spectra in which the bands typical of $-NH_2$ and $-COOH$ moieties are absent. Equilibrium in solution also lies almost exclusively on the side of the dipolar form; therefore, amino acids are insoluble in nonpolar solvents and usually not very soluble in polar ones. The only amino acids that exhibit any appreciable solubility in alcohol are proline and hydroxyproline. Solubility in water depends on the pH: the minimum is at the isoelectric point.

This solubility minimum at the isoelectric point is quite useful for purifying and recrystallizing amino acids. The analytical technique for separating amino acid mixtures by electrophoresis is based on the fact that a specific amino acid does not migrate in an electric field at its isoelectric point, pI, a physical constant for each amino acid.

The physical properties of the most important α-amino acids are listed in Table 3.

Stereochemistry. With the exception of glycine, the simplest amino acid (R = H), all natural α-amino acids are chiral compounds occurring in two enantiomeric (mirror-image) forms.

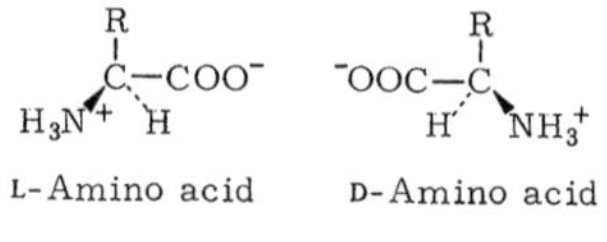

L-Amino acid   D-Amino acid

The prefixes L and D express the absolute configuration about the α-carbon atom by means of the formal stereochemical relationship to L- or D-glyceraldehyde, the reference substance introduced by EMIL FISCHER in 1891. In addition to the spacial representations shown above, the so-called Fischer projections are also universally recognized and used:

```
        COOH              COOH
   H₂N──┼──H          H──┼──NH₂
        R                 R
     L-Amino acid    D-Amino acid
```

Polarimetric determination of the specific rotation $[\alpha]_D^t$ can be used to differentiate between the two enantiomers and to check their optical purity. The molecular rotation $[M]_D^t$ is less common:

$$[M]_D^t = \frac{M_r}{100} \cdot [A]_D^t$$

M_r molecular mass; t temperature; D 589.3 nm (wavelength of the sodium D line)

Further methods for investigating the structure of amino acid enantiomers include the Cotton effect (change in molecular rotation as a function of the wavelength of plane-polarized light), optical rotational dispersion (reversal of the direction of the molecular rotation at the wavelength of the absorption maximum), and circular dichroism (differing absorption for left- and right-handed circularly polarized light). L-Amino acids exhibit a positive carbonyl Cotton effect, D-amino acids a negative one.

Isoleucine, threonine, and hydroxyproline contain two chiral carbon atoms each; therefore, they appear in four stereoisomeric forms. Cystine, which likewise contains two chiral carbons, has only three stereoisomers: L-, D-, and *meso*-cystine, the meso form having a plane of symmetry. According to the Cahn-Ingold-Prelog rule (R, S rule) [17], all proteinogenic L-α-amino acids, with the exception of L-cysteine and L-cystine, are S; the D-α-amino acids are R. According to this system, L-threonine, for example, is termed (2S, 3R)-threonine. However, the R,S system has not attained wide acceptance for the simple L-α-amino acids.

Absorption Spectra. Aliphatic amino acids exhibit no absorption in the UV region above 220 nm, with the exception of cystine (240 nm). The aromatic amino acids, phenylalanine, tyrosine, and tryptophan, absorb between 250 and 300 nm [1, vol. 2]. The exact position of the maximum and the molar extinction coefficient ε are affected by the pH of the aqueous solution.

Two infrared bands are especially characteristic of amino acids: 1560 – 1600 cm^{-1} (–COO$^-$) and ca. 3070 cm^{-1} (–NH$_3^+$). These bands are also evidence of the bipolar nature of the amino acids.

2.2. Chemical Properties

Acidity and Basicity. The chemical properties of the α-amino acids are primarily the properties of the amino and carboxyl groups. Amino acids react with strong acids as proton acceptors (bases) and with strong bases as proton donors (acids). In acidic medium they are present predominantly as cations; in basic medium they are present predominantly as anions. Figure 1 shows the titration curves of glycine, glutamic acid, and lysine. The pK$_1$ and pK$_2$ values correspond to the inflection points of the titration

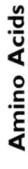

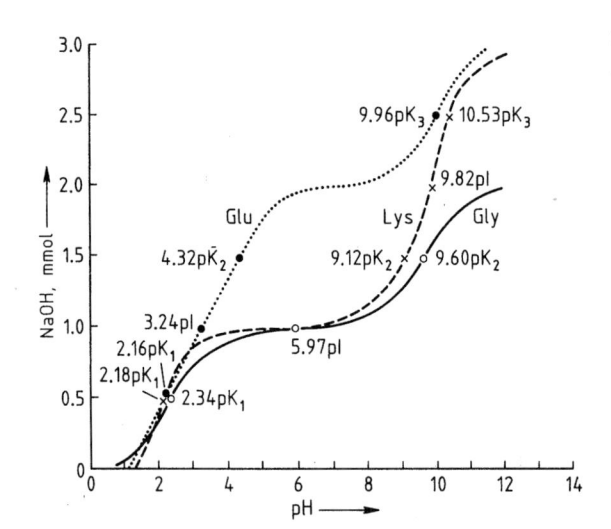

Figure 1. Titration curves of glycine, glutamic acid, and lysine [5]

curve, the pH value where the concentrations of the zwitterion form and the cationic or anionic form are equal. The pK_1, pK_2, and pI values can be calculated from the titration curves with the Henderson-Hasselbach equation:

$$pK_1 = pH - \log \frac{\begin{bmatrix} R-CH-COO^- \\ {}^+NH_3 \end{bmatrix}}{\begin{bmatrix} R-CH-COOH \\ {}^+NH_3 \end{bmatrix}}$$

$$pK_2 = pH - \log \frac{\begin{bmatrix} R-CH-COO^- \\ NH_2 \end{bmatrix}}{\begin{bmatrix} R-CH-COO^- \\ {}^+NH_3 \end{bmatrix}}$$

$$pI = \frac{pK_1 + pK_2}{2}$$

Amino acids with additional basic or acidic groups (arginine, lysine, glutamic acid, cysteine) exhibit additional pK values. Amino acids act as buffers in the region of their pK values.

The pK_1 values show the amino acids to be considerably stronger acids than acetic acid. However, because of intramolecular protonation of the amine moiety by the carboxyl group, aqueous solutions of amino acids are only weakly acidic. The pH values of aqueous monoaminomonocarboxylic acids lie between 5.5 and 6.0. Solutions of the acidic amino acids aspartic acid and glutamic acid have pH values of ca. 2. The weakly basic amino acid histidine has a pH of 7.5 in aqueous solution; the more strongly basic amino acids lysine and arginine have pH values of ca. 11–12. These

Table 3. Properties of important α-amino acids [1], [4], [15], [16]

Amino acid	Decomposition point, °C	Specific rotation $[\alpha]_D^{25}$	pK_1	pK_2	pK_3	pI	Solubility, g/100 g H_2O
L-Alanine [56-41-7]	314	+ 14.47 ° (c = 10.03 in 6 M HCl)	2.34^b	9.69^b		6.01^b	16.51 (25 °C)
D,L-Alanine [302-72-7]	295		2.35	9.87		6.11	16.72 (25 °C)
L-Arginine [74-79-3]	244	+ 27.58 ° (c = 2 in 6 M HCl)	2.01	9.04 (α-NH_2)	12.48 (guanidyl)	10.76	14.87 (20 °C)
L-Arginine hydrochloride [1119-34-2]	220						75.1 (20 °C)
L-Asparagine [70-47-3]	236	+ 32.6 ° (c = 1 in 0.1 M HCl, 20 °C)	2.02^b	8.80^b		5.41^b	3.11 (28 °C)
L-Aspartic acid [56-84-8]	270^a	+ 25.4 ° (c = 2 in 5 M HCl)	2.10	3.86 (β-COOH)	9.82	2.98	0.5 (25 °C)
D,L-Aspartic acid [617-45-8]	275^a						0.775 (25 °C)
L-Citrulline [372-75-8]	234–7, 222	+ 24.2 ° (c = 2 in 5 M HCl)	2.43^b	9.41^b		5.92^b	10.3 (20 °C)
L-Cysteine [52-90-4]	240	+ 9.7 ° (c = 8 in 1 M HCl)	1.71^b	8.27^b(–SH)	10.78^b	5.02^b	28^d (25 °C)
L-Cysteine hydrochloride monohydrate [7048-04-6]	178 (anhydr.)	+ 6.53 ° (calculated as cysteine) (c = 2 in 5 M HCl)					16 (20 °C) > 100 (20 °C)
L-Cystine [56-89-3]	260	-212 ° (c = 1 in 1 M HCl)	1.04	2.05 (–COOH)	8.0 (–NH_2)^c	5.02	0.011 (25 °C)
L-Glutamic acid [56-86-0]	224–5, 249	+ 31.5 ° (c = 2 in 5 M HCl, 20 °C)	2.10	4.07^b	9.47	3.08	0.843 (25 °C)
L-Glutamine [56-85-9]	185–6	+ 31.8 ° (c = 2 in 1 M HCl)	2.17^b	9.13		5.65^b	3.6 (18 °C)
Glycine [56-40-6]	262, 292		2.35	9.78		6.06	24.99 (25 °C)

Table 3. (continued)

Amino acid	Decomposition point, °C	Specific rotation $[\alpha]_D^{25}$	pK_1	pK_2	pK_3	pI	Solubility, g/100 g H_2O
L-Histidine [71-00-1]	277, 287	+ 13.0 ° (c = 1 in 6 M HCl)	1.77	6.10 (imidazolyl)	9.18	7.64	4.29 (25 °C)
L-Histidine hydrochloride monohydrate [5934-29-2]	259	+ 13.0 ° (calculated as histidine) (c = 1 in 6 M HCl)					16.99 (20 °C)
L-Hydroxyproline [51-35-4]	274	-75.2 ° (c = 2 in H_2O, 22 °C)	1.92	9.73		5.82	36.11 (25 °C)
L-Isoleucine [73-32-5]	285–6	+ 40.6 ° (c = 2 in 6 M HCl, 20 °C)	2.36[b]	9.68[b]		6.02[b]	4.117 (25 °C)
D,L-Isoleucine [443-79-8]	292		2.32	9.76		6.04	2.011 (25 °C)
L-Leucine [61-90-5]	293–5, 314–5	+ 15.6 ° (c = 2 in 5 M HCl)	2.36[b]	9.60[b]		5.98[b]	2.19 (25 °C)
D,L-Leucine [328-39-2]	293–5, 332		2.33	9.74		6.04	1.00 (25 °C)
L-Lysine [56-87-1]	224–5	+ 25.9 ° (c = 2 in 5 M HCl)	2.18	8.95 (α-NH_2)	10.53	9.47	> 100 (25 °C)
L-Lysine hydrochloride [657-27-2]	253–6	+ 25.9 ° (c = 2 in 5 M HCl) (calculated as lysine)					72.5[d] (25 °C)
D,L-Lysine hydrochloride [70-53-1]	264						35.98 (20 °C)
L-Methionine [63-68-3]	283	+ 23.4 ° (c = 5 in 3 M HCl, 28 °C)	2.28[b]	9.21[b]		5.74[b]	5.37 (20 °C)
D,L-Methionine [59-51-8]	281		2.28	9.21		5.74	3.35 (25 °C)
L-Ornithine [70-26-8]	140	+ 16.5 ° (c = 4.6 in H_2O)	1.94[b]	8.65[b] (α-NH_2)	10.76[b]	9.70[b]	
L-Ornithine hydrochloride [3184-13-2]	215	+ 28.3 ° (c = 2 in 5 M HCl) (calculated as ornithine)					54.36 (20 °C)
L-Phenylalanine [63-91-2]	283–4	-35.1 ° (c = 2 in H_2O, 20 °C)	1.83[b]	9.13[b]		5.48[b]	2.965 (25 °C)
D,L-Phenylalanine [150-30-1]	284–8, 320		2.58	9.24		5.91	1.29 (25 °C)
L-Proline [147-85-3]	220–2	-85.0 ° (c = 1 in H_2O)	2.00	10.60		6.30	162.3 (25 °C)
D,L-Proline [609-36-9]	205						
L-Serine [56-45-1]	228	-6.8 ° (c = 10 in H_2O, 26 °C)	2.21[b]	9.15[b]		5.68[b]	35.97 (20 °C)

Table 3. (continued)

Amino acid	Decomposition point, °C	Specific rotation $[\alpha]_D^{25}$	pK_1	pK_2	pK_3	pI	Solubility, g/100 g H_2O
D,L-Serine [302-84-1]	246[a]		2.21	9.15		5.68	5.023 (25 °C)
L-Threonine [72-19-5]	253	−28.6 ° (c = 2 in H_2O, 26 °C)	2.71[b]	9.62[b]		6.16[b]	9.03 (20 °C)
D,L-Threonine [80-68-2]	234–5						20.5 (25 °C)
L-Tryptophan [73-22-3]	290–2, 281	−32.15 ° (c = 1 in H_2O)	2.38	9.39		5.88	1.14 (25 °C)
D,L-Tryptophan [54-12-6]	285						0.25 (30 °C)
L-Tyrosine [60-18-4]	342-4[a], 297–8	−7.27 ° (c = 4 in 6 M HCl)	2.20	9.11	10.07 (−OH)	5.63	0.045 (25 °C)
D,L-Tyrosine [556-03-6]	340, 318						0.351 (25 °C)
L-Valine [72-18-4]	315	+ 26.7 ° (c = 3.4 in 6 M HCl, 20 °C)	2.32[b]	9.62[b]		5.96[b]	8.85 (25 °C)
D,L-Valine [516-06-3]	298[a]		2.29	9.72		6.00	7.09 (25 °C)

[a] Sealed tube.
[b] Stereochemistry not specified.
[c] pK_4 (second NH_2 group) = 10.25.
[d] Grams per 100 mL of solution.

differences in acidities and basicities are utilized in the separation of amino acid mixtures by ion-exchange chromatography and electrophoresis.

Reactions. Because of their bifunctional and sometimes trifunctional character, the α-amino acids are capable of taking part in a variety of chemical reactions. Comprehensive treatments of these may be found in monographs and reviews [1], [5]–[8], [18].

The introduction of groups protecting the amino, carboxyl, and side-chain functions is especially important for peptide synthesis [9]. Additionally, α-amino acids play a prominent role as intermediates in the synthesis of heterocycles [19]. The chiral α-amino acids and their derivatives are inexpensive and, for the most part, readily available synthons for numerous natural products and pharmaceuticals. In many cases optically active amino acids can also be used to induce chirality during the course of a synthetic process [20].

α-Amino acids form *chelate-like complexes* with heavy metal ions. The best known are the dark blue, easily crystallized copper chelates:

Bis (glycinato) copper(II) hydrate

This complex formation can be used to protect both the α-amino and the carboxyl group during synthesis of ε-N-acetyl derivatives or carbamates of lysine or other side-chain derivatives. In these copper or cobalt complexes the α-carbon atom is activated sufficiently to react with aldehydes. A well-known example is the alkaline condensation of the glycine–copper complex with acetaldehyde, resulting in threonine. Especially important is the reaction of α-amino acids with ninhydrin to form a blue-violet dye, the basis of a sensitive optical method for detecting amino acids (see Chap. 6).

Free amino acids react with *nitrous acid* to yield α-hydroxycarboxylic acids, with retention of configuration. Volumetric measurement of the nitrogen gas set free is the basis of Van Slyke's method for the quantitative analysis of amino acids. Reaction of amino acid esters with nitrous acid gives the acid-labile diazocarboxylic acid esters. Treatment of N-alkyl- or N-arylamino acids with nitrous acid yields the N-nitroso derivatives, which can be dehydrated to sydnones in the presence of acetic anhydride:

$$R^1-\underset{H}{N}-\underset{COOH}{CH}-R^2 \xrightarrow{HNO_2} R^1-\underset{ON}{N}-\underset{COOH}{CH}-R^2$$

$$\xrightarrow{(CH_3CO)_2O}$$

Oxidizing agents attack the amino group, converting the amino acid into an iminocarboxylic acid. These are unstable and either hydrolyze to α-oxocarboxylic acids or decompose, after decarboxylation, into ammonia and the aldehyde containing one carbon atom less. Diketones, triketones, N-bromosuccinimide, or silver oxide may serve as the oxidizing agent. The best known example is the reaction with ninhydrin. Oxidative deamination also can be carried out enzymatically with D- or L-amino acid oxidases. This also proceeds via the α-iminocarboxylic acids, which subsequently are hydrolyzed to α-oxo acids:

$$\text{R-CH(NH}_2\text{)-COOH} \xrightarrow[-2\text{H}]{\text{amino acid oxidase}} \text{R-C(=NH)-COOH} \xrightarrow[-\text{NH}_3]{\text{H}_2\text{O}} \text{R-C(=O)-COOH}$$

This enantioselective enzymatic oxidative deamination is the basis of analytical methods for the determination of amino acid enantiomers.

The *N-acylation* of α-amino acids with acyl chlorides or anhydrides under Schotten-Baumann conditions produces *N*-acyl-α-amino acids, which have numerous uses. For example, the *N*-acetyl derivatives of D,L-amino acids (e.g., alanine, valine, methionine, phenylalanine, tryptophan) are intermediates in the production of L-amino acids by enzymatic resolution using aminoacylases (see p. 551). Amino acids acylated with naturally occurring fatty acid residues are used industrially as easily degradable surfactants.

If free amino acids are heated above 200 °C, especially in the presence of soda lime or metal ions, they readily *decarboxylate* to form amines. The enzymatic decarboxylation of amino acids gives biogenic amines. Some of these amines are physiologically active neurotransmitters (histamine, tyramine, dopamine, serotonin).

The *esters* usually are prepared by direct esterification, e.g., reaction of the α-amino acid with anhydrous alcohol in the presence of anhydrous hydrogen chloride. The initial product is the hydrochloride of the ester, which is set free by addition of base. On standing or warming, the free esters of α-amino acids eliminate alcohol to form 2,5-diketopiperazines:

$$2\ \text{R-CH(NH}_2\text{)-COOR} \xrightarrow{-2\text{ROH}} \text{2,5-diketopiperazine}$$

Esters of amino acids are useful for peptide synthesis because the carboxyl function is protected. The esters may be converted to amino alcohols by treatment with a strong reducing agent, such as lithium aluminum hydride.

Several cyclic derivatives of the α-amino acids are of industrial importance. The *hydantoins* or imidazolidine-2,4-diones, which have been discussed as raw materials for the synthesis of α-amino acids, are most conveniently prepared by treatment of aldehyde cyanohydrins with ammonium carbonate or urea. They also may be obtained by reacting amino acids with cyanates or isocyanates.

The initial product of this reaction is the ureidocarboxylic acid (hydantoic acid derivative):

$$R^1\text{-CH(NH}_2)\text{-COOH} \xrightarrow{R^2\text{-N=C=O}} R^1\text{-CH(NH-CO-NH-R}^2)\text{-COOH} \xrightarrow{H^+} \text{Hydantoin derivative}$$

The thiohydantoins are obtained by the reaction with isothiocyanates.

Dehydration of *N*-acylamino acids with acetic anhydride or carbodiimide yields *1,3-oxazolin-5-ones* (azlactones):

$$R^1\text{-CH(NH-CO-R}^2)\text{-COOH} \xrightarrow{(CH_3CO)_2O} \text{azlactone}$$

Racemization occurs readily during the reaction. The azlactones are intermediates in the synthesis of amino acids.

The *N-carboxylic acid anhydrides* (Leuchs anhydrides, 1,3-oxazolidine-2,5-diones) are often used in peptide synthesis, especially to prepare poly-α-amino acids. These reactive amino acid derivatives are obtained by treating the amino acid with phosgene in the presence of tertiary amines

$$R\text{-CH(NH}_2)\text{-COOH} \xrightarrow{COCl_2, NR_3} \text{Leuchs anhydride}$$

or by elimination of benzyl chloride from *N*-benzyloxycarbonylamino acid chlorides.

3. Production

3.1. General Methods

There are four processes suitable for the production of amino acids: extraction, chemical synthesis, fermentation, and enzymatic catalysis. In the extraction process the natural amino acids are isolated from protein hydrolysates; the raw materials are protein-containing animal and vegetable products. Synthetic amino acids generally are racemic mixtures; in some cases, the production of L-amino acids is possible from prochiral precursors, so-called asymmetric syntheses. Fermentation and enzymatic catalysis utilize whole cells (bacteria, yeast) and active cell components (enzymes), respectively. Generally, starting materials are natural products (sugar, molasses) for

Table 4. Production methods for amino acids

Amino acid	Chemical synthesis	Extraction	Fermentation	Enzymatic catalysis
L-Alanine		+		+
D,L-Alanine	+			
L-Arginine		+	+	
L-Aspartic acid		+		+
L-Asparagine		+		
L-Cystine		+		
L-Cysteine	+*			
L-Glutamic acid (Na)		(+)**	+	
L-Glutamine		(+)	+	
Glycine	+			
L-Histidine (· HCl)		+	+	
L-Isoleucine		+	+	(+)
L-Leucine		+		
L-Lysine (· HCl)			+	+
L-Methionine				+
D,L-Methionine	+			
L-Phenylalanine		(+)	+	+
L-Proline		+	(+)	
L-Serine		+	+	
L-Threonine		+	+	
L-Tryptophan			+	+
L-Tyrosine		+		
L-Valine		+	(+)	+

* Reduction of L-cystine.
** Parenthesis indicates minor importance.

fermentation and suitable amino acid precursors for enzymatic catalysis. For certain amino acids (L-Glu, D,L-Met), one of these processes is clearly the most advantageous. Usually, however, the different processes are quite competitive. Table 4 shows which processes are used to produce individual amino acids.

Numerous publications and monographs [10], [21], [22] provide detailed descriptions of the methods of synthesis and related technological developments.

3.1.1. Methods for D,L-Amino Acids

In principle, all amino acids may be prepared by synthesis. Only a few processes, however, have attained large-scale industrial production, because only a few racemic amino acids are of economic interest. These amino acids are methionine, alanine, and the *N*-acetyl derivatives of D,L-valine, D,L-phenylalanine, and D,L-tryptophan. These *N*-acetyl derivatives are intermediates in the production of the L-amino acids. The Strecker synthesis (or the Bucherer variation) often is of use commercially. Aminonitriles, obtainable from aldehydes, ammonia, and hydrogen cyanide, are either directly hydrolyzed or first converted to hydantoins by treatment with carbon dioxide and then saponified:

$$\text{RCHO} + \text{NH}_3 + \text{HCN} \longrightarrow \text{R-CH(NH}_2\text{)-CN}$$

$$\text{R-CH(NH-CO-NH)-C=O} \xrightarrow{\text{OH}^-} \text{R-CH(NH}_2\text{)-COOH}$$

The condensation of aldehydes with hydantoin and subsequent hydrogenation and hydrolysis or the reductive amination of α-oxocarboxylic acids can be advantageous for the preparation of aromatic amino acids such as phenylalanine.

The racemization of amino acids is of interest in the commercial production of L-amino acids via the resolution of racemic mixtures. The undesirable D-enantiomer is racemized in aqueous solution, either thermally or in the presence of strong bases or acids, and then recycled. A particularly mild method is catalytic racemization in acetic acid in the presence of salicylaldehyde [23]. Acetyl derivatives may be effectively racemized thermally in the presence of acetic anhydride.

3.1.2. Methods for L-Amino Acids

With a few exceptions, only the L-amino acids are nutritionally effective. Therefore, most processes concentrate on the production of L-amino acids.

Extraction. For several amino acids (cystine, tyrosine, proline), isolation from natural raw materials (extraction) is still the most economical production method. The typical amino acid compositions of various protein starting materials are shown in Table 5. The standard procedure for isolating an amino acid from an aqueous protein hydrolysate consists of passing the hydrolysate over the H^+ form of a strongly acidic ion-exchange resin, binding the amino acids to the resin. The resin is then washed with water. Elution with aqueous ammonia frees the amino acids, which are collected in fractions. Amino acids of pharmaceutical quality are obtainable by this procedure. Washing, regeneration with a strong acid, and renewed washing ready the ion exchanger for the next cycle.

Synthetic Processes. The asymmetric synthesis of L-amino acids from prochiral starting materials, e.g., with optically active rhodium complexes, is a promising method of preparing aromatic amino acids [24]. Industrially, this process is currently limited to the production of L-dopa, 3,4-dihydroxyphenylalanine, used to treat Parkinson's disease. The more common methods are optical resolution by direct crystallization of enantiomeric mixtures [25] or by fractional crystallization of diastereomeric salt pairs, e.g., D- or L-mandelates [26], and significant progress has been made in chromatographic resolution with chiral phases or chiral eluents. Finally, there is asymmetric

Table 5. Typical raw materials for the isolation of amino acids (content in percentage of dry substance, TS=total solid)

Amino acid	Swine bristles (90% TS)	Blood meal (95% TS)	Collagen (60% TS)	Horn chips (90% TS)
L-Ala	4	7.1	10	4.2
L-Arg	8.1	3.6	2.9	8.9
L-Asp	6.3	9.3	4.2	6.6
L-(Cys)$_2$	8.2	—	—	—
L-Glu	13.6	8.5	9	12.4
Gly	3.7	3.8	19.1	5.1
L-His	0.9	5.8	0.5	4.2
L-Hyp	—	—	8.4	—
L-Ile	2.3	0.9	0.9	2.8
L-Leu	6.4	10.2	2.7	7.5
L-Lys	2.9	7.6	2.9	3.4
L-Met	0.4	1.1	0.7	0.5
L-Orn	—	—	2.9	—
L-Phe	2.1	5.8	1.8	2.8
L-Pro	7.6	3.9	12.9	4.4
L-Ser	7.7	4.4	0.8	7.2
L-Thr	5.4	3.5	0.4	4.4
L-Tyr	2.6	2.2	–	4.8
L-Val	3.8	6.8	2.3	4.2
Total	86	84.5	82.4	88.6

synthesis, in which a new center of chirality is induced by a chiral group already present in the molecule.

Microbiological Processes. Although it is now possible to prepare all natural amino acids by microbiological methods, fermentative processes using special, high-performance mutants find large-scale industrial application only for the production of L-glutamic acid and L-lysine (see Sections 3.2.5 and 3.2.10). The raw materials for these processes are molasses or glucose and sometimes acetic acid. For the fermentative preparation of certain amino acids, so-called precursors are also used as starting materials. These are substances added to the nutrient medium of a bacteria culture to be converted by the microbes into the desired L-amino acid. Table 6 lists several bacteria strains that produce L-amino acids and reports the final product concentration. Modern developments in gene technology promise a tremendous increase in the productive capacity of microorganisms, and a considerable reduction in the cost of L-amino acids.

Enzymatic Processes. The microbiological preparation of amino acids is a biosynthesis using the entire enzyme complement of the microorganism. Enzymatic catalysis, in contrast, is the use of specific enzymes or enzyme systems. Enzymatic catalysis is advantageous for cases where L-amino acids can be produced from inexpensive precursors with high specificity in a continuously operating process. The acylase-catalyzed stereospecific hydrolysis of N-acetyl-D,L-amino acids is used to produce alanine, valine,

Table 6. Amino acid production by fermentation

L-Amino acid	Final concentration, g/L	Bacteria strain*
Alanine	40	c
Arginine	29	a
Glutamic acid	39	a
Histidine	10	a
Isoleucine	15	a
Leucine	28	d
Lysine	32	a
	44	b
Phenylalanine	2.2	a
	6	e
Ornithine	26	b
Proline	29	a
Threonine	18	a
Tryptophan	1.9	a
	6.1	e
Valine	23	d

* a) *Brevibacterium flavum*, b) *Corynebacterium glutamicum*, c) *Corynebacterium gelatinosum*, d) *Brevibacterium lactofermentum*, e) *Bacillus subtilis*. Newly developed strains produce even higher concentrations, see Section 3.2.2

methionine, phenylalanine, and tryptophan. The enzyme is recycled either by binding it to a carrier and carrying out the reaction in a fixed-bed column [28] or by performing the enzymatic resolution in a loop reactor and employing an ultrafilter membrane to retain the dissolved enzyme in the system (enzyme membrane reactor) [29]. Enzymatic resolution can be competitive only when coupled with efficient racemization, either catalytically (e.g., with acetic anhydride) or enzymatically.

$$\text{Acetyl-D,L-amino acid} \xrightarrow{\text{L-acylase}} \text{L-Amino acid} + \text{Acetyl-D-amino acid} \xrightarrow{\text{racemization}}$$

Enzyme-catalyzed addition can directly convert prochiral starting materials into chiral products. A practical example is the aspartase-catalyzed addition of ammonia to fumaric acid in the production of L-aspartic acid. L-Aspartate-β-decarboxylase-catalyzed CO_2 elimination readily affords L-alanine.

$$\text{HOOC-CH=CH-COOH} + NH_3 \xrightarrow{\text{L-aspartase}} \begin{array}{c} COOH \\ | \\ H\cdots C\blacktriangleleft NH_2 \\ | \\ CH_2 \\ | \\ COOH \end{array}$$

$$\xrightarrow{\text{L-aspartate-}\beta\text{-decarboxylase}} \begin{array}{c} COOH \\ | \\ H\cdots C\blacktriangleleft NH_2 \\ | \\ CH_3 \end{array}$$

The enzymatic conversion of α-oxo acids to L-amino acids, e.g., using L-amino acid-specific dehydrogenases with cofactor regeneration [30] or using transaminases, is quite promising, provided that these enzyme systems prove to be stable in a continuous process.

3.1.3. Methods for D-Amino Acids

With the exception of D-phenylglycine and D-*p*-hydroxyphenylglycine, which are starting materials for the semisynthetic antibiotics ampicillin and amoxicillin, the D-amino acids make up only a modest part of the amino acid market. There are, however, quite simple and elegant procedures for their preparation, including the hydrolysis of the acetyl-D-amino acids remaining after L-acylase resolution and the enzymatic ring cleavage of 5-substituted hydantoins by active microbes containing D-hydantoinase [31], [32]. The enzymatic resolution with D-acylase leads directly to D-amino acids [33].

3.2. Production of Specific Amino Acids

3.2.1. Alanine

Production of D,L-alanine from acetaldehyde via aminopropionitrile or 5-methylhydantoin (Strecker-Bucherer synthesis) has been optimized [34], [35], making acetyl-D,L-alanine an inexpensive starting material for the production of L-alanine. L-Alanine is produced via asymmetric hydrolysis with microbial acylase, and it is still isolated from protein hydrolysates on an industrial scale.

L-Alanine may be prepared more elegantly by enzymatic decarboxylation of L-aspartic acid with an immobilized microorganism such as *Pseudomonas dacunhae* (IAM 1152). The enzyme is L-aspartic acid-β-decarboxylase [36], [37]. Although direct fermentation is of no importance for the preparation of L-alanine, it may be used to produce D-alanine with, e.g., *Corynebacterium fascians* (ATCC 21950). The D-alanine is produced in a concentration of 7 g/L [38].

3.2.2. Arginine

L-Arginine is produced commercially by isolation from acidic hydrolysates of protein (gelatins). It can be produced by conversion of fermentatively produced L-ornithine on treatment with guanidizing agents, such as cyanamide, *S*-methylthiourea, or *O*-methylisourea. Direct fermentative procedures are gaining importance. Good L-arginine-producing microbes are the mutant strains of *Bacillus subtilis* (ATCC 31183–9,

ATCC 31002) and *Brevibacterium flavum* (FERM-P 5641) [39]. Yields of more than 40 g/L have been attained.

3.2.3. Aspartic Acid and Asparagine

After commercialization of the asymmetric, aspartase-catalyzed synthesis of L-aspartic acid, production of D,L-aspartic acid by amination of fumaric acid under pressure declined. The production of L-aspartic acid by hydrolysis of asparagine or by isolation of protein hydrolysates plays only a minor role.

Suitable microbes for the industrial bioconversion of fumaric acid to L-aspartic acid include mutant strains of *Pseudomonas* (ATCC 21973), *Brevibacterium flavum* (ATCC 14067), *Brevibacterium lactofermentum* (ATCC 13869), *Corynebacterium glutamicum* (ATCC 13032), and *Corynebacterium acetoacidophilum* (ATCC 15870) [40]. By immobilizing the microorganism, the production of L-aspartic acid may be carried out continuously in a fixed-bed reactor, making it particularly economical [41].

L-Asparagine can be isolated as a byproduct from the production of potato starch. Because L-aspartic acid is readily available from microbial processes and can be easily esterified to the β-methyl ester, the synthesis of L-asparagine by treatment of the ester with ammonia is a simple, economical method of production [42].

3.2.4. Cystine and Cysteine

L-Cystine can be isolated easily from hydrolysates of hair or other keratins because it is less soluble than all other amino acids (except tyrosine). The reduction of cystine, especially electrolytic reduction, continues to be the most important method of producing L-cysteine.

The search for alternative enzymatic and synthetic processes for cysteine, just as L-cystine is becoming scarce, has led to a series of new, interesting, and potentially promising synthetic routes. L-Cysteine can be prepared from β-chloro-D,L-alanine and sodium sulfide with cysteine desulfhydrase, an enzyme obtained from, e.g., *Citrobacterium freundii* (IFO 12681) [43]. 2-Aminothiazoline-4-carboxylic acid (ATC), which is readily available from methyl α-chloroacrylate and thiourea, can be converted enzymatically in high yield to Lcystine and L-cysteine [44].

For the synthesis of D,L-cysteine, processes that proceed via β-chloroalanine [45] or via α-chloro-β-alanine and aziridinecarboxylic acid [46] compete with the thiazoline process [47]. The thiazoline process begins with the inexpensive starting materials chloroacetaldehyde, acetone, ammonia, sodium hydrosulfide, and hydrogen cyanide and can be practiced on an industrial scale:

$$\begin{array}{c} \text{CH}_2\text{O} \quad \text{NH}_3 \\ | \\ \text{CH}_2\text{Cl} \quad \text{O}{\diagdown}\text{CH}_3 \\ \text{NaHS} \quad \text{CH}_3 \end{array} \longrightarrow \underset{\text{S}}{\overset{\text{N}}{\diagup}}\underset{\text{CH}_3}{\overset{\text{CH}_3}{\diagdown}}$$

$$\xrightarrow{\text{HCN}} \text{N}\equiv\text{C} \diagdown \underset{\text{S}}{\overset{\text{NH}}{\diagup}}\underset{\text{CH}_3}{\overset{\text{CH}_3}{\diagdown}} \xrightarrow{\text{HCl, H}_2\text{O}} \begin{array}{c} \text{COOH} \\ | \\ \text{CH–NH}_2 \cdot \text{HCl} \cdot \text{H}_2\text{O} \\ | \\ \text{CH}_2\text{SH} \end{array}$$

3.2.5. Glutamic Acid and Glutamine

Glutamic acid is the most important amino acid in terms of quantity and economic value. Fermentation is firmly established as the only production process of the monosodium glutamate industry, although a series of attractive glutamic acid syntheses — based on acrylonitrile [48] or acrolein [49] and the resolution of D,L-glutamic acid by selective crystallization [50] — has been developed.

For fermentation, organisms of the genera *Micrococcus*, *Brevibacterium*, and *Corynebacterium* can be used. They are cultured in aqueous media that contain beet molasses, cane molasses, raw sugar, or starch hydrolysate. Numerous refinements in the culturing of the microorganisms have been discovered [51]. In some cases methanol or ammonium acetate can be used as a partial source of carbon [52]. Accumulation of L-glutamic acid in the broth can exceed 100 g/L.

L-Glutamine can be obtained by extraction of natural raw materials, such as sugar beets. However, synthesis starting from L-glutamic acid via the γ-methyl ester or fermentative processes are better suited for large-scale production.

The sodium salt of L-glutamic acid is widely used as a flavor-enhancing factor. For more details on the manufacture of glutamic acid and its sodium salt.

3.2.6. Glycine

For glycine, the simplest natural amino acid and the only achiral one, chemical synthesis is the most convenient method of preparation. Amination of chloroacetic acid [53] and the hydrolysis of aminoacetonitrile, which is available from formaldehyde, hydrogen cyanide, and ammonia, are the favored production methods [54].

3.2.7. Histidine

L-Histidine is produced primarily by isolation of blood-meal hydrolysates [55]. Chemical synthesis is relatively difficult on account of the imidazole ring system. It is hoped that efficient fermentation processes will be developed for histidine production. Mutants belonging to the genera *Brevibacterium* and *Corynebacterium* are capable of producing L-histidine [56], [57], and fermentation media can contain more than 10 g/L L-histidine. Recombinant DNA methods are being used to develop more efficient L-histidine-producing microorganisms [58].

3.2.8. Isoleucine

Because isoleucine contains two chiral carbon atoms, synthesis by the Strecker method, starting from 2-methylbutyraldehyde, results in a mixture of four stereoisomers: D- and L-(threo)-isoleucine and D- and L-alloisoleucine. D,L-Isoleucine can be isolated by fractional crystallization. Fractional crystallization can be carried out also on the aminonitrile intermediate [59]. Isolation of L-isoleucine, the only nutritional isomer, from the racemic mixture is quite difficult. Isolation from protein hydrolysates is also problematic because of the difficulty in separating the L-isoleucine from the other branched-chained amino acids, L-leucine and L-valine.

Fermentation is therefore the method of choice for producing L-isoleucine [60]. Genetically modified L-isoleucine-producing microorganisms [61] may greatly exceed the productive capacities of existing strains.

3.2.9. Leucine

Isolation from protein hydrolysates is currently the method used for the production of L-leucine. However, because the demand for leucine seems likely to increase in the future, either chemical synthesis combined with enzymatic resolution or fermentation must be considered to supplement extraction. For example, N-acetyl-D,L-leucine can be hydrolyzed by L-acylase to provide L-leucine and acetyl-D-leucine [62]. Alternatively, N-acetyl-D,L-leucine ester may be treated with proteolytic enzymes to yield a mixture of N-acetyl-L-leucine and N-acetyl-D-leucine ester, separation and hydrolysis providing L-leucine and D-leucine [63]. Suitable bacteria for a fermentative production of L-leucine include *Brevibacterium* and *Corynebacterium* [64].

3.2.10. Lysine

L-Lysine, a feed additive, is one of the most important amino acids economically. Because synthetic D,L-lysine is only 50% nutritionally active, an achiral synthesis has no chance of industrial realization unless coupled with effective resolution. So far no chemical process has been able to challenge fermentative production.

The synthetic process suggested by DSM starts with condensation of acetaldehyde and acrylonitrile to yield γ-cyanobutyraldehyde and proceeds via the hydantoin with subsequent hydrogenation and hydrolysis to D,L-lysine [65]. The racemate is resolved through crystallization by seeding with L-lysine sulfanilate [66]:

$$N\equiv C-CH=CH_2 + CH_3CHO \rightarrow N\equiv C-CH_2CH_2CH_2CHO$$

$$\rightarrow N\equiv C-CH_2CH_2CH_2-\underset{HN\underset{\|}{}NH}{\overset{O}{\diagup}} \rightarrow \text{D,L-Lysine}$$

A process combining chemical and microbiological techniques for the production of L-lysine was introduced by Toray Industries [67]. α-Amino-ε-caprolactam (ACL) is formed from cyclohexanol:

[reaction scheme: cyclohexanol → cyclohexene → (NOCl) → 1-chloro-2-nitrosocyclohexane → 2-aminocyclohexanone oxime → (Beckmann rearrangement) → ACL]

ACL

This intermediate is then hydrolyzed with L-ACL-hydrolase to yield L-lysine and D-α-amino-ε-caprolactam. Added with the L-ACL-hydrolase is a racemase that converts D-α-amino-ε-caprolactam to the D,L mixture for subsequent reaction with the L-ACL-hydrolase. The enzymes are employed in immobilized form. This process is competitive with fermentation.

The usual raw material for the fermentative production of L-lysine is molasses. High-performance microbes belong to the species *Corynebacterium glutamicum* or *Brevibacterium lactofermentum* [68], [69]. They are capable of producing up to 70–80 g/L lysine (calculated as hydrochloride) in a 60-h batch fermentation. New, more productive strains may be developed by gene technology, specifically by cell fusion [70].

3.2.11. Methionine

D,L-Methionine is produced on a large scale by chemical synthesis; it is second only to L-glutamic acid in total production. Because D- and L-methionine are of equal nutritive value (see Chap. 4), the racemate can be used directly as a feed additive. However, L-methionine is required, in much smaller quantities, for pharmaceutical and special applications.

All industrial producers of D,L-methionine start with the same raw materials: acrolein, methanethiol (methyl mercaptan), hydrogen cyanide, and ammonia or ammonium carbonate. The multistep process can be carried out in batches or continuously. The first step is addition of methyl mercaptan to acrolein to form β-methylthiopropionaldehyde, which reacts with hydrogen cyanide in the second step to give α-hydroxy-γ-methylthiobutyronitrile. Treatment with ammonia leads to α-amino-γ-methylthiobutyronitrile, which is hydrolyzed to methionine:

$$CH_3SH + CH_2{=}CH{-}CHO \rightarrow CH_3S{-}CH_2CH_2CHO$$

$$\xrightarrow{HCN} CH_3S{-}CH_2CH_2{-}\underset{OH}{CH}{-}CN \xrightarrow{NH_3}$$

$$CH_3S{-}CH_2CH_2{-}\underset{NH_2}{CH}{-}CN$$

Better is the treatment of α-hydroxy-γ-methylthiobutyronitrile with ammonia and carbon dioxide or ammonium carbonate to form 5-(βmethylthioethyl)hydantoin in a Bucherer-type reaction:

$$CH_3S{-}CH_2CH_2{-}\underset{OH}{CH}{-}CN \xrightarrow{NH_3, CO_2} CH_3S{-}CH_2CH_2{-}\underset{HN\underset{O}{}NH}{\overset{O}{\diagup}}$$

$$\xrightarrow{hydrolysis} CH_3S{-}CH_2CH_2{-}\underset{NH_2}{CH}{-}COOH$$

The hydantoin is subjected to alkaline hydrolysis at elevated temperature and pressure. The amino acid can be isolated by acidification of the saponified solution to the isoelectric point of methionine (pH = 5.7). Environmentally, a process practiced by Degussa [71] is especially elegant. Potassium carbonate is the base for saponification of the hydantoin. The potassium carbonate, excess ammonia, and carbon dioxide are continuously recycled. Methionine yields of up to 95 % (based on acrolein) are attained.

L-Methionine is produced by the acylase-catalyzed hydrolysis of N-acetyl-D,L-methionine. Other enzymatic procedures, such as the cleavage of N-acetyl-D,L-methionine esters by proteolytic enzymes [72] or the asymmetric hydrolysis of N-carbamoyl-D,L-methionine by enzymes [73], have not proved to be successful production processes.

3.2.12. Phenylalanine

All of the four principal methods for preparing amino acids have been described and developed as production processes for L-phenylalanine: protein hydrolysis, asymmetric synthesis, enzymatic catalysis, and fermentation. The expectations that have arisen from the marketing of the dipeptide sweetener Aspartame (the methyl ester of L-aspartyl-L-phenylalanine) [74] have resulted in careful evaluation of processes for large-scale industrial production. Because the dipeptide ester need not necessarily be made from L-phenylalanine but can be prepared enzymatically starting from D,L-phenylalanine [75], syntheses of D,L-phenylalanine are also significant.

D,L-Phenylalanine can be prepared by the Bucherer process in 90% yield from phenylacetaldehyde [76]. A newer process is based on the reductive amination of phenylpyruvic acid [77], which can be obtained by carbonylation of benzyl chloride. The carbonylation of benzyl chloride in the presence of acetamide catalyzed by octacarbonyldicobalt leads directly to N-acetyl-D,L-phenylalanine in a one-pot process [78].

The condensation product of benzaldehyde with hydantoin, benzylidene hydantoin, can be converted to D,L-phenylalanine in two steps by hydrogenation and hydrolysis. An optically active hydrogenation catalyst produces L-phenylalanine directly [79]. The Erlenmeyer method, starting from benzaldehyde and acetylglycine via the azlactone (see 546), has been investigated as a method of phenylalanine production. An optically active hydrogenation catalyst gives N-acetyl-L-phenylalanine [80].

The enzymatic hydrolysis of N-acetyl-D,L-phenylalanine by microbial acylase has proved useful for the production of L-phenylalanine, especially the pharmaceutical amino acid, although racemization of the acetyl-D-phenylalanine and recycling the racemate are necessary. Newer processes, e.g., the enzymatic addition of ammonia to *trans*-cinnamic acid [81] or the transamination of phenylpyruvate salts using L-aspartate [82], can compete with purely fermentative processes [83]. The introduction of new genetic information, via plasmids, into microorganisms producing amino acids suggests the future direction of the fermentative production of L-phenylalanine [84].

3.2.13. Proline

L-Proline is still produced predominantly by isolation from protein hydrolysates. Chemical and fermentative processes are likely to gain significance as demand increases. L-Proline can be obtained from L-glutamic acid via reduction of L-pyroglutamic acid. Syntheses of D,L-proline, e.g., starting from glutamic acid γ-semiacetal, $(C_2H_5O)_2CH\text{-}CH_2CH_2CH(NH_2)COOH$, require subsequent resolution [85], either enzymatically via amidase-catalyzed cleavage of D,L-prolinamide [86] or via diastereomeric salts [87]. For the fermentative production of L-proline, microorganisms of the genera *Brevibacterium*, *Corynebacterium*, or *Microbacterium* [88] and *Serratia* [89] have been suggested. The preparation of plasmids capable of carrying the gene for the biosynthesis of L-proline is likely to revolutionize proline synthesis [90].

3.2.14. Serine

Extraction of protein hydrolysates and fermentation of glycine substrate are the preferred methods for producing L-serine. Both L-serine and D,L-serine have attracted the attention of numerous research groups as a starting material for the microbial synthesis of L-tryptophan from indole (Section 2.2.16). Convenient starting materials for the synthesis of D,L-serine include glycolaldehyde precursors, such as vinyl acetate or vinyl ether, glycolaldehyde derivatives [91], and aziridine-2-carboxylic acid [92]. Microorganisms suggested for the fermentative production of L-serine from glycine include mutants of the genera *Pseudomonas* [93], *Corynebacterium glycinophilum* [94], and *Nocardia* [95].

3.2.15. Threonine

The essential amino acid threonine contains two chiral centers. Achiral synthesis, e.g., from the copper complex of glycine plus acetaldehyde [96] or from α-isocyanoacetamide plus acetaldehyde [97], results in a mixture of four stereoisomers: the D- and L-threo forms and the D- and L-allo forms. L-Threonine can be isolated from the mixture by enzymatic or chemical resolution. All of these procedures, however, including the recently suggested treatment with D-threonine aldolase and L-allothreonine aldolase [98], are most complicated.

The only biologically active form, the L-threo isomer, is obtained directly by extraction of protein hydrolysates or by fermentation. Fermentative processes are likely to play a larger role in the future. With a mutant strain of *Serratia marcescens* and a reaction time of 120 h at 30 °C, concentrations of up to 45.4 g/L were accumulated in the broth [99]. The success with cloning organisms to increase the rate of L-threonine production [100] raises hopes that larger quantities of less expensive L-threonine will be available soon.

3.2.16. Tryptophan

L-Tryptophan is an essential amino acid. Because it is a limiting amino acid in feedstuffs, it is a potential feed additive. However, its price must be reduced dramatically. Much effort has been spent in developing chemical, enzymatic, and fermentative processes for economical large-scale industrial production. Basically, all of these processes may be divided into two categories: those using indole as starting material and those in which the indole ring is constructed during the synthesis. The D,L-tryptophan syntheses from acrolein belong to the latter category. In the classical Warner-Moe synthesis [101], [102] acrolein is reacted with acetamidomalonic ester. The resulting aldehyde is converted to the phenylhydrazone, which, after Fischer rearrangement, is hydrolyzed to tryptophan.

$$\text{OCH-CH=CH}_2 + \underset{\underset{\text{NH-COCH}_3}{|}}{\overset{\overset{\text{COOR}}{|}}{\text{CH-COOR}}} \longrightarrow$$

$$\underset{\underset{\text{NH-COCH}_3}{|}}{\overset{\overset{\text{COOR}}{|}}{\text{OCH-CH}_2\text{CH}_2\text{C-COOR}}} \xrightarrow{\text{Phenyl-hydrazine}}$$

$$\text{C}_6\text{H}_5\text{-NH-N=CH-CH}_2\text{CH}_2\underset{\underset{\text{NH-COCH}_3}{|}}{\overset{\overset{\text{COOR}}{|}}{\text{C-COOR}}} \xrightarrow{\text{H}^+}$$

$$\underset{\text{indole}}{\text{[indolyl]}}\text{-CH}_2\underset{\underset{\text{NH-COCH}_3}{|}}{\overset{\overset{\text{COOR}}{|}}{\text{C-COOR}}} \xrightarrow[\text{decarboxylation}]{\text{saponification}}$$

$$\underset{\text{indole}}{\text{[indolyl]}}\text{-CH}_2\underset{\underset{\text{NH}_2}{|}}{\text{CH-COOH}}$$

D,L - Tryptophan

In two other procedures, acrolein acetate [103] (→ Acrolein and Methacrolein) or acrolein acetal [104] are hydroformylated with carbon monoxide and converted to the acetal of β-(hydanto-5-yl)propionaldehyde under Bucherer conditions. Condensation with phenylhydrazine followed by Fischer indole rearrangement leads to tryptophan hydantoin, which is hydrolyzed to D,L-tryptophan in the presence of bases. A variation begins with β-cyanopropionaldehyde, which is converted to 5-(β-cyanoethyl)hydantoin by the Bucherer procedure and then catalytically reduced to β-(hydanto-5-yl)propionaldehyde. This variation was reported as a synthetic route to tryptophan hydantoin [105].

Another proposed process is based on 4formyl-2-aminobutyric acid (glutamic acid γ-semialdehyde) or its derivatives, which, after conversion to phenylhydrazones and Fischer rearrangement, lead to tryptophan [106], [107]. Pure L-tryptophan can be obtained from L-glutamic acid γ-semialdehyde, which is synthesized from L-glutamic acid, or by fermentation [108], [109].

L-Tryptophan currently is prepared by the enzymatic resolution of N-acetyl-D,L-tryptophan with mold acylase. Another enzymatic procedure consists of the reaction of indole with serine or pyruvic acid in the presence of tryptophan synthetase or tryptophanase. Numerous variations have been reported. As a rule, the enzymes are left in the microorganisms, which are either immobilized or cultured in a batch process [110]–[113]. L-Tryptophan also may be prepared by the microbial reaction of indole with β-chloroalanine [114] or methanol [115].

There are also promising fermentations not requiring indole as starting material. They use L-tryptophan-producing mutants of *Bacillus subtilis* that grow in a medium containing anthranilic acid [116]; overproduction of L-tryptophan is obtained by methanol-utilizing yeast [117] and by mutants of the genera *Serratia* [118] or

Bacillus [119]. Recently developed microorganisms carrying recombinant plasmids controlling L-tryptophan production [120] offer hope for less expensive L-tryptophan.

3.2.17. Tyrosine

L-Tyrosine is produced exclusively from protein hydrolysates. Its low solubility in water makes isolation of the amino acid quite simple. Tyrosine can be synthesized by procedures analogous to phenylalanine syntheses; however, these procedures have no industrial significance.

3.2.18. Valine

L-Valine is produced industrially in pharmaceutical quality by enzymatic resolution of *N*-acetyl-D,L-valine. L-Valine is also obtained by extraction of protein hydrolysates; however, this process generally is used to produce a technical-grade product.

The best fermentative processes for L-valine use mutants of *Brevibacterium* and *Corynebacterium* [121], [122]. The Strecker-Bucherer synthesis from isobutyraldehyde and hydrogen cyanide is the most economical route to D,L-valine or *N*-acetyl-D,L-valine, and recent variations promise to improve this process [123].

D-Valine is the starting material for the valine derivative fluvalinate, a new insecticide [124]. The D-valine is available from the hydrolysis of *N*-acetyl-D-valine, a byproduct in the enzymatic resolution of acetyl-D,L-valine. Alternatively, 5-isopropylhydantoin can be cleaved stereospecifically with veal liver dihydropyrimidase [125] or with microorganism hydantoinase [126].

4. Biochemical and Physiological Significance

The biosynthesis of amino acids begins with atmospheric nitrogen, which is reduced to ammonia by bacteria and plants. Ammonia is used by plants, by bacteria, and, to a limited extent, by ruminants as a raw material for amino acids. Amino acids, in turn, serve as starting materials for the synthesis of proteins and a variety of other nitrogen-containing compounds, such as the purine and pyrimidine bases in nucleic acids. Bacterial degradation leads, once again, to ammonia and nitrogen [127].

α-Ketoglutaric acid plays the central role in the assimilation of ammonia. Its transamination product, glutamic acid, in turn, can provide its amino group for the synthesis of other amino acids, e.g., alanine (Fig. 2).

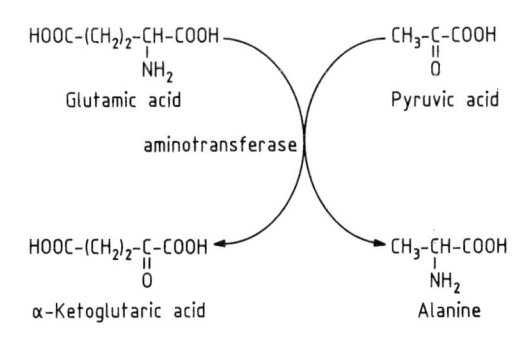

Figure 2. The transamination of amino acids

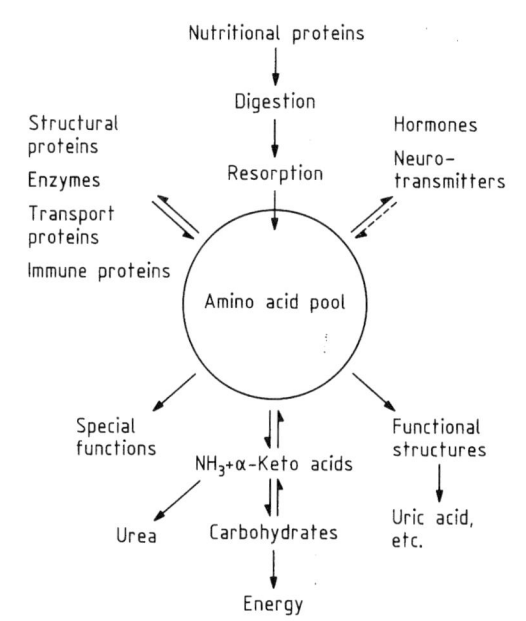

HOOC–(CH$_2$)$_2$–CH–COOH → CH$_3$–C–COOH
         | ||
         NH$_2$ O
   Glutamic acid   Pyruvic acid

aminotransferase

HOOC–(CH$_2$)$_2$–C–COOH ← CH$_3$–CH–COOH
         || |
         O   NH$_2$
  α-Ketoglutaric acid   Alanine

Figure 3. The amino acid pool and functions of amino acids in the intermediary metabolism

Nutritional proteins → Digestion → Resorption → Amino acid pool

Structural proteins
Enzymes
Transport proteins
Immune proteins

Hormones
Neuro-transmitters

Special functions → Urea
NH$_3$+α-Keto acids ⇅ Carbohydrates → Energy
Functional structures → Uric acid, etc.

Humans, animals, and some bacteria are incapable of synthesizing all the necessary amino acids in their own intermediary metabolism; i.e., they are heterotrophs and are therefore dependent on the biosynthetic capability of plants. Proteins that are consumed as foodstuffs by humans and animals are hydrolyzed to amino acids by the digestive enzymes. The amino acids are resorbed in the upper part of the intestine and enter the liver by way of the portal vein. The liver is the central organ for metabolism and homoeostasis of the plasma amino-acid level. The body's various requirements are met from the pool of free amino acids (Fig. 3), ca. 50 g in adult humans.

The lion's share of the amino acids ($\approx$ 300 g/day for adults) is required for synthesis of proteins [128]: structural proteins, enzymes, transport proteins, and immune proteins. Additionally, amino acids are required for the synthesis of oligopeptides and polypeptides that fulfill regulatory functions in the body, i.e., hormones. Some amino acids or their metabolites are directly active as hormones or facilitate the transmission of nerve impulses (neurotransmitters), e.g., serotonin. Furthermore, there are amino acids that serve special functions, such as methionine, which is a methyl group donor. Finally, a series of amino acids serves as precursors for the biosynthesis of other structures. For

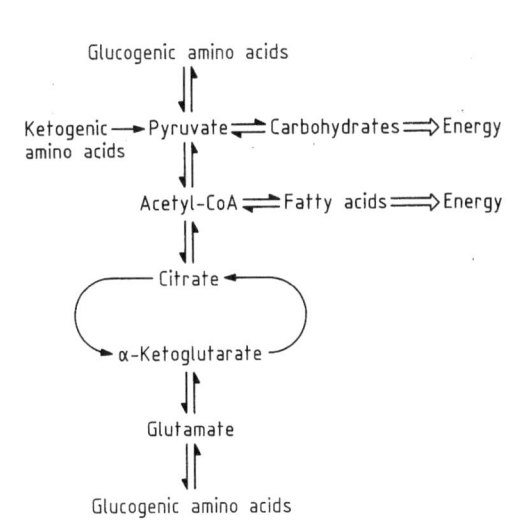

Figure 4. Intermediary metabolism of amino acids (simplified)

example, glycine is used in the construction of the porphyrin skeleton. Amino acids are metabolized to produce energy in the case of a protein-deficient or a protein-excess diet.

The free amino acids in the amino acid pool undergo numerous transformations (Fig. 4), involving transamination and oxidative deamination, by which other amino acids can be synthesized. The α-keto acids are intermediates and in addition allow amino acids entrance into the carbohydrate (through pyruvate) and fatty acid (through acetylcoenzyme A) metabolisms. A distinction is therefore drawn between glucogenic and ketogenic amino acids.

D-Amino acids occur in the cell pool of plants and gram-positive bacteria and as building blocks in peptide antibiotics and bacterial cell walls [6], [12]. They do not occur in human or animal metabolism; proteins are made exclusively of L-amino acids. The traces of D-amino acids detected in metabolically inert protein (teeth, eye lenses) are believed to originate from racemization. Orally ingested D-amino acids are resorbed from the intestinal lumen more slowly than the L-form. The D-enantiomers cannot be utilized or can be utilized only to a slight extent as essential amino acids [129]. The one important exception is D-methionine. Animals and adult humans convert D-methionine into L-methionine by transamination. The α-keto acid of methionine is an intermediate. Otherwise, D-amino acids are degraded with the help of D-amino acid oxidases [130] to be used as an energy source [131].

The major end products of amino acid metabolism are urea, uric acid, ammonium salts, creatinine, and allantoin. The loss of nitrogen via these metabolites stabilizes at about 22 g protein per day after a few days on a protein-free diet.

Inborn disorders in amino acid metabolism [132] can lead to marked alterations in the excretion profile. These disorders usually take the form of an enzyme or transport deficiency [127], [133]. The most common example is phenylketonuria, a disruption of the normal metabolic pathway from phenylalanine to tyrosine caused by severe limitation in the activity of the phenylalanine hydroxylase [134].

Humans and animals are not capable of producing all the required L-amino acids in their intermediary metabolism. Therefore, they are dependent on an external source of these *essential* amino acids (Table 7). In situations of increased requirements (rapid growth, stress, trauma), histidine and arginine also become essential for humans.

Table 7. Essential (+) and semiessential (±) amino acids

	Baby	Adult	Rat	Chicken	Hen	Cat	Salmon
L-Arginine	±	±	±	+	±	+	+
L-Cysteine	+ (?)						
Glycine				+			
L-Histidine	+	±	+	+	±	±	+
L-Isoleucine	+	+	+	+	+	+	+
L-Leucine	+	+	+	+	+	+	+
L-Lysine	+	+	+	+	+	+	+
L-Methionine	+	+	+	+	+	+	+
L-Phenylalanine	+	+	+	+	+	+	+
L-Threonine	+	+	+	+	+	+	+
L-Tryptophan	+	+	+	+	+	+	+
L-Tyrosine	+ (?)						
L-Valine	+	+	+	+	+	+	+

Cysteine and tyrosine may be essential for infants during their first few weeks, because their intermediary metabolism does not yet function well enough to produce these from methionine and phenylalanine in sufficient quantities.

5. Uses

The uses of amino acids have been treated in recent review articles [19], [135] – [137].

5.1. Human Nutrition

In addition to their nutritive value, amino acids are important flavor precursors and taste enhancers. In foods for humans, the flavor uses of amino acids represent the dominant factor in total market value. In animal nutrition, amino acids are used almost exclusively for their nutritive value.

Addition of small amounts of amino acids to improve the nutritive value of proteins is known as supplementation. Both supplementation and the combination of proteins with complementary amino acids are used to increase the biologic value of proteins. Usually the supply of at least one of the essential amino acids lies below the requirement. This, the limiting amino acid, determines what percentage of the protein (or, more precisely, its amino acids) can be used to meet the body's amino acid requirements. In most cases, methionine is the first limiting amino acid. Sometimes it is lysine; now and then it is both together.

The contents of essential amino acids found in several animal and vegetable foodstuffs are compiled in Table 8. Considerable variations may be present in the amino acid contents of a given foodstuff.

Amino Acids

Table 8. Average amino acid content of some foodstuffs (mg/100 g)

Food	Ile	Leu	Lys	Cyss	Met	Phe	Tyr	Thr	Trp	Val	Arg	His	Protein, %	Moisture content, %
Maize, grain	350	1190	254	147	182	464	363	342	67 M	461	398	258	9.5	12.0
Rice, husked	300	648	299	84	183	406	275	307	98 M	433	650	197	7.5	13.0
Wheat, whole grain	426	871	374	332	196	589	391	382	142 M	577	602	299	12.2	12.0
Wheat, flour, 70–80% extr. rate	435	840	248	304	174	581	277	321	128 M	493	422	248	10.9	12.0
Potato (*Solanum tuberosum*)	76	121	96	12	26	80	55	75	33 M	93	100	30	2.0	78.0
Bean (*Phaseolus vulgaris*)	927	1685	1593	188	234	1154	559	878	223 M	1016	1257	627	22.1	11.0
Soybean, milk	171	278	195	57	50	175	133	128	48 M	165	253	84	3.2	92.0
Soy protein, isolate [139]	4147	7119	5777	1008	1092	4644	3458	3211	1080	4210	6767	2378	75.7	4.7
Lettuce, leaves (*Lactuca sativa*)	50	83	50	24	–	67	35	54	10 M	71	59	21	1.3	94.8
Tomato (*Solanum lycopersicum*)	20	30	32	7	7	20	14	25	9 M	24	24	17	1.1	93.8
Apple (*Malus silvestris*)	13	23	22	5	3	10	6	14	3	15	10	7	0.4	84.0
Orange (*Citrus sinensis*)	23	22	43	10	12	30	17	12	6	31	52	12	0.8	87.4
Beef, veal, edible flesh	852	1435	1573	226	478	778	637	812	198 M	886	1118	>603	17.7	61.0
Fish, fresh, all types	900	1445	1713	220	539	737	689	861	211 M	1150	1066	665	18.8	74.1
Milk, cows, untreated	162	328	268	28	86	185	163	153	48 M	199	113	92	3.5	87.3
Milk, human	48	104	81	16	19	41	39	53	20 M	54	46	30	1.2	87.6
Cheese, all types	956	1864	1559	76	530	950	973	725	217 M	1393	651	556	18.0	51.0
Egg, whole	778	1091	863	301	416	709	515	634	184 M	847	754	301	12.4	74.0

[a] Chemical determination. M Microbiological determination.

Table 9. Essential amino acid requirements of humans

Amino acid	Suggested patterns of requirements [140], g/100 g protein			Adult requirement, mg kg^{-1} d^{-1}			
	Infant child	School-age	Adult 1973	FAO/WHO 1980 (144)	NAS/NCR	Rose^a	Hegsted^b
His	1.4	–	–	–	–	–	–
Ile	3.5	3.7	1.8	10	12	10	10
Leu	8.0	5.6	2.5	14	16	11	13
Lys	5.2	7.5	2.2	12	12	9	10
Met + Cys^c	2.9	3.4	2.4	13	10	14	13
Phe + Tyr^d	6.3	3.4	2.5	14	16	14	13
Thr	4.4	4.4	1.3	7	8	6	7
Trp	0.85	0.46	0.65	3.5	3	3	3
Val	4.7	4.1	1.8	10	14	14	11
Total	37.3	32.6	15.2	83.5	91	81	80

^{a} For men.
^{b} For women.
^{c} Cys can partly cover the total S-amino acid requirement.
^{d} Tyr can partly cover the total aromatic amino acid requirement.

The published requirements for the individual essential amino acids differ. The values (Table 9) usually contain safety factors and therefore are higher than the minimum requirement. Requirement values were first determined by Rose [141]; those published by Hegsted [142] are considered the most reliable at present. The amino acid requirement pattern suggested by the FAO/WHO [140] is considered optimal for the greatest part of the population.

The "average safe level of daily protein intake for men and women," based on these amino acid requirement figures, is given as 0.55 g/kg body weight. The acute daily protein requirement, however, varies between 0.5 and 2.5 g/kg with age and constitution [143]. The Deutsche Gesellschaft für Ernährung (DGE) recommends a daily consumption of 0.9 g/kg body weight [144]. The Committee on Dietary Allowances, Food and Nutrition Board of the National Academy of Sciences (NAS), USA, cites 0.8 g/kg as a desirable level of daily protein comsumption [145].

5.1.1. Supplementation

In general, animal protein contains the essential amino acids in larger quantities and in a more favorable ratio than vegetable protein, which is often deficient in essential amino acids. Lysine is the first limiting amino acid in wheat, rye, barley, oats, maize, and millet, whereas methionine is the first limiting amino acid in meat, milk, soybeans, and other beans. The second limiting amino acids are usually threonine (wheat, rice) and tryptophan (maize, rice, casein). The limiting amino acids for several foodstuffs are listed in Table 10.

Improving the biologic value of vegetable protein in human nutrition is practiced for economic and dietary reasons. Combining different protein types is not always prac-

Table 10. Limiting amino acids in foodstuffs

Proteins	First limiting amino acid	Second limiting amino acid(s)
Peanut	Thr	Lys and Met
Fish	Met	Lys
Casein	Met	Trp
Torula yeast	Met	
Sesame	Lys	
Skim milk	Met	
Beans	Met	
Sunflower seed	Lys	Thr
Soy protein	Met	Lys
Wheat	Lys	Thr
Rice	Lys	Thr and Trp
Rye	Lys	Thr and Trp
Gelatin	Trp	
Maize	Lys	Trp and Thr

tical. Often a complementary protein is unavailable, too expensive, or not of acceptable taste. In these cases supplementation with amino acids is the simplest method of increasing the biologic value of proteins. There is a monumental volume of literature on the subject of amino acid supplementation [146]–[153].

The biologic value is an important criterion for the evaluation of proteins or amino acid mixtures. It can be determined experimentally [154], [155]. In principle, all methods measure the ability of the nutritional protein to replace body protein. Table 11 lists the biologic values of nutritional proteins as determined by the minimum requirement [143]. Whole egg protein is the reference in this scale. Another scale of evaluation is the protein efficiency ratio (PER). This is the daily weight gain of young animals, usually rats, under standard feeding conditions (see Table 11). In an improved method, the animals are fed diets of various protein levels, and the protein efficiency is then determined using regressional analysis.

The net protein retention (NPR) method and the net protein utilization (NPU) method are more accurate than the PER method because they consider the protein (NPR) or total nitrogen (NPU) requirement for maintenance. The "chemical score," in which the availability of the amino acids is not considered, is suitable for a gross estimation of the biologic protein quality. In this method the amino acid content is determined analytically and compared with the amino acid pattern of a reference protein, e.g., the FAO/WHO provisional scoring pattern (Table 12). This method provides an immediate picture of the size of amino acid gaps and the sequence of limiting amino acids.

$$\text{Chemical Score} = \frac{\text{Content of amino acid in test protein}}{\text{Content of amino acid in reference protein}} \times 100$$

As can be seen in Table 13, addition of ca. 0.1–0.5% of the limiting amino acid to such basic foodstuffs as wheat, rice, maize, and soybeans raises the protein efficiency in rat growth tests impressively.

Table 11. Protein quality of food and minimum requirements (human)

Food	Biologic value, g kg^{-1} d^{-1} [143]	Minimum requirement*, ratio (rat) [143]	Protein efficiency [138]
Whole egg	100	35	3.92
Beef	92	39	2.30
Cow's milk	88	40	3.09
Potato	86 [156]	41 [156]**	3.0 [157]
Fish			3.55
Casein			2.50
Soybean	84	42	2.32
Rice	81	44	2.18
Rye flour	76	46	
Maize	72	49	1.18
Beans	72	49	1.48
Wheat flour	56	63	0.6

* Of the protein part.
** Calculated.

Table 12. FAO/WHO provisional scoring pattern 1973

Amino acid	g/100 g protein
L-Isoleucine	4.0
L-Leucine	7.0
L-Lysine	5.5
L-Methionine + L-cystine *	3.5
L-Phenylalanine + L-tyrosine **	6.0
L-Threonine	4.0
L-Tryptophan	1.0
L-Valine	5.0
Total	36.0

* Cys can partly cover the total S-amino acid requirement.
** Tyr can partly cover the total aromatic amino acid requirement.

Clinical studies [158]–[160] and field trials have solidified the evidence of the benefits to human nutrition of amino acid supplementation [161]–[164]. However, the general supplementation of basic foodstuffs, such as bread and rice, is not yet practiced extensively. In dietary nutrition, however, supplementation already plays an important role (Table 14).

Some infants exhibit lactose or cow's milk protein incompatibility. The formulas marketed for this condition often are based on isolated soybean protein and are supplemented with L-methionine to increase the biologic value. The advantageous effects of L-methionine supplementation on the physical development of infants has been demonstrated in a series of clinical studies [165], [166]. Additionally, the food for pregnant and nursing women, seniors, overweight persons, and athletes can also be supplemented. Extruded soy protein, which is already used in large quantities as a meat extender and vegetarian meat substitute, can be supplemented with N-acetyl-L-methionine [167].

Table 13. Increase of the protein efficiency ratio (PER) by supplementation with amino acids

Food (protein content)	L-Lys·HCl, %	L-Thr, %	D,L-Trp, %	D,L-Met, %	PER*
Wheat flour (10%)					0.65
	0.2				1.56
	0.4				1.63
	0.4	0.15			2.67
Rice (7.8%)					1.50
	0.2		0.1		2.61
Maize (8.75%)					1.41
	0.4		0.07		2.33
Soybean milk (10%)					2.12
				0.3	3.01
Extruded soy protein (10%)					1.99
				0.23	2.62

*Reference protein: casein, PER=2.50.

Table 14. Amino acids in dietetic products

Protein/protein hydrolysate	Supplemented amino acid	Use	Indication
Cow's milk, casein, whey protein	L-Cyss and/or L-Lys · HCL	infant nutrition	adapted nutrition
Soy protein	L-Met	infant nutrition	lactose incompatibility and milk protein allergy
Casein/yeast	L-Lys · HCl	meal supplement	protein malnutrition, in place of conventional nutrition

5.1.2. Flavorings, Taste Enhancers, and Sweeteners

Free amino acids occur in almost all protein-based foods. In some foods their concentration is several percent. Foodstuffs having a relatively high concentration of free amino acids include fruit juices [168], cheese [169], [170], beer [171], and seafood [172]. Approximately 85% of the free amino acids in orange juice is proline, arginine, asparagine, γ-aminobutyric acid, aspartic acid, serine, and alanine [173].

Amino acids are relatively tasteless. Nonetheless, they contribute to the flavor of foods. They have characteristic synergistic flavor-enhancing and flavor-modifying properties, and they are precursors of natural aromas [174], [176]. Amino acids and protein hydrolysates are therefore useful additives in the food industry. The sodium salt of L-glutamic acid (MSG) exhibits a particularly pronounced flavor-enhancing effect [177] and has been recognized as a flavoring factor for seaweed, sake, miso, and soy sauce since 1908. The substance is used in concentrations of 0.1–0.4% as an additive for spices, soups, sauces, meat, and fish, usually in combination with nucleotides [176].

Table 15. Tastes of - and -amino acids*

	L-Amino acid	D-Amino acid
Alanine	sweet (12–18)	sweet (12–18)
Arginine	bitter	neutral
Asparagine	neutral	sweet (3–6)
Aspartic acid	acidic/neutral	acidic/neutral
Cysteine	sulfurous	sulfurous
Glutamine	neutral	sweet (8–12)
Glutamic acid	acidic/"glutamate-like"	acidic/neutral
Glycine	sweet (25–35)	sweet (25–35)
Histidine	bitter	sweet (2–4)
Isoleucine	bitter	sweet (8–12)
Leucine	bitter	sweet (2–5)
Lysine	sweet/bitter	sweet
Methionine	sulfurous	sweet/sulfurous (4–7)
Phenylalanine	bitter	sweet (1–3)
Proline	sweet/bitter (25–40)	neutral
Serine	sweet (25–35)	sweet (30–40)
Threonine	sweet (35–45)	sweet (40–50)
Tryptophan	bitter	sweet (0.2–0.4)
Tyrosine	bitter	sweet (1–3)
Valine	bitter	sweet (10–14)

* Threshold values for sweet taste in parenthesis, μmol/ml; the threshold value for sucrose is 10–12 μmol/ml.

L-Cysteine especially enhances the aroma of onion [178] and is therefore used to rearomatize dried onions. Glycine, which has a refreshing, sweetish flavor, occurs abundantly in mussels and prawns. It is considered to be an important flavor component of these products. When used as an additive for vinegar, pickles, and mayonnaise, it attenuates the sour taste and lends a note of sweetness to their aroma. D,L-Alanine is used for the same purpose in the Far East [179]. Glycine is used to mask the aftertaste of the sweetener saccharin [180], [181].

The L- and D-amino acids usually exhibit pronounced flavor differences. Many L-enantiomers taste weakly bitter, whereas their optical antipodes, the D-amino acids, taste sweet [182]–[184] (Table 15). For the most part, dipeptides and oligopeptides have bitter flavors. One of the few exceptions is the methyl ester of the dipeptide L-aspartyl-L-phenylalanine (Aspartame) [185], [187], which is 150–200 times sweeter than sucrose.

The free amino acids are used widely in foodstuff technology as precursors for aromas and brown food colors [188]. The flavors are formed during foodstuff production, e.g., during the ripening of cheese [189], [190], the fermentation of alcoholic beverages [191], [192], or the leavening of dough [193], [194], or foodstuff cooking, e.g., frying, roasting, boiling, by the Maillard reaction between amino acids and reducing sugars (nonenzymatic browning) [174], [195], [196]. The Strecker degradation of amino acids plays a central role in this process. A broad spectrum of aroma-intensive, volatile compounds forms [197], [198]. The most important classes are aliphatic carbonyl compounds and heterocycles, such as furans, pyrones, pyrroles, pyrrolidines, pyridines,

Table 16. Amino acids for Maillard flavors *

Meat, poultry [199]	Cys , Cyss, Gly, Glu, Ala, Met, His, Ser, Asp, Pro
Bread, cracker, biscuit [193], [200]	Pro , Lys, Arg, Val, His, Leu, Glu, Phe, Asp, Gly, Gln
Chocolate, cocoa [201]	Leu, Phe, Val, Glu, Ala
Honey [202]–[204]	Phe
Cream, butter [205]	Pro , Lys, Ala, Gly, His
Nut, peanut [206]	Leu, Val, Ile, Pro, Glu, Gln, His, Phe, Asp, Asn
Potato [207]	Met
Tobacco [203], [208]	Asn , Arg, GABA, Gln, Ala, Gly, Orn, Glu, Asp, Leu, Val, Thr, Pro, Tyr, Phe

* Key amino acids underlined.

imidazoles, pyrazines, chinoxalines, thiophenes, thiolanes, trithianes, thiazoles, and oxazoles.

It is often possible to assign certain aromas to specific amino acids [198]. For example, the sulfur-containing amino acid cysteine is primarily responsible for the formation of meat flavor. Proline seems to be important for the aroma of bread crust. Phenylalanine, as well as the branched-chain amino acids leucine and valine, is important for the characteristic flavor of chocolate. Valine and leucine are also involved in the aroma of roasted nuts. Methionine plays a key role in the aroma of french fries. The flavors of such products as precooked foods, snack articles, and spices may be improved by addition of the proper Maillard aromas. One variation is adding the precursors of the Maillard aromas, i.e., amino acid plus sugar, to the foodstuff and allowing the fragrance to form in situ. Some aroma profiles that can be prepared from amino acids are compiled in Table 16.

5.1.3. Other Uses in Foodstuff Technology

Amino acids are used in the foodstuff industry for purposes other than supplementation and flavoring. L-Cysteine, for example, is used by the baked goods and pasta industry as a flour additive [209], [210]. As a reducing agent, it relaxes wheat gluten proteins (by cleavage of the disulfide linkages), homogenizes the dough, accelerates dough development, and improves the structure of the baked product, while allowing shorter kneading times.

Because they are capable of forming complexes with metals, amino acids act as antioxidants for fats and fat-containing foodstuffs [211]. This effect is strengthened by primary antioxidants, such as α-tocopherol. Melanoidines, which are formed during the Maillard reaction, are stronger antioxidants than the amino acids themselves [212]. Maillard products are also reported to be preservatives [213]. Glycine apparently exhibits a special preservative effect [214].

Table 17. Amino acid composition of feedstuffs, wt%

Feedstuff	Dry matter	Crude protein	Met	Met+ Cys	Lys	Thr	Trp	Arg	Leu	Ile	Val	His
Alfalfa	93	16.0	0.22	0.40	0.75	0.67	0.26	0.73	1.15	0.66	0.85	0.32
Barley	88	10.5	0.16	0.40	0.36	0.36	0.10	0.49	0.72	0.33	0.52	0.23
Blood meal	91	87.6	1.15	2.19	8.25	4.15	1.16	3.67	11.16	0.94	7.59	5.42
Corn gluten meal	90	61.7	1.64	2.76	0.96	2.09	0.29	2.01	10.06	3.23	2.75	1.32
Fish meal, menhaden	91	60.0	1.66	2.18	4.58	2.51	0.46	3.51	4.31	2.24	2.81	1.28
Tapioca, 60–65% starch	90	2.4	0.03	0.06	0.08	0.07	0.02	0.11	0.12	0.07	0.09	0.03
Meat and bone meal	93	50.0	0.58	0.95	2.24	1.42	0.22	3.19	2.61	1.20	1.81	0.80
Oats	89	11.6	0.18	0.56	0.40	0.37	0.12	0.74	0.80	0.42	0.57	0.23
Rapeseed meal, extracted	90	38.3	0.82	1.81	2.19	1.68	0.45	2.36	2.70	1.54	2.03	1.09
Sorghum, grain	89	9.0	0.18	0.36	0.22	0.33	0.07	0.34	1.10	0.34	0.46	0.22
Soybean meal, extracted, 44% crude protein	89	44.0	0.64	1.33	2.82	1.78	0.56	3.22	3.50	1.97	2.11	1.31
Soybean meal, extracted, 48% crude protein	89	48.0	0.68	1.41	3.06	1.91	0.61	3.55	3.80	2.13	2.28	1.42
Sunflower seed meal, with hulls	93	29.8	0.70	1.24	1.07	1.05	0.38	2.29	1.88	1.19	1.48	0.72
Wheat	87	12.5	0.22	0.51	0.34	0.38	0.12	0.59	0.84	0.41	0.54	0.29
Wheat, winter	87	10.5	0.19	0.44	0.31	0.32	0.11	0.50	0.70	0.35	0.50	0.25
Wheat middlings	88	16.6	0.26	0.59	0.66	0.55	0.25	1.07	1.02	0.50	0.75	0.45

5.2. Animal Nutrition

The use of amino acids for the nutrition of monogastric animals is based on the same foundation as the supplementation of human foodstuffs and the clinical experience with humans. In practice, the enrichment of animal feeds and formulated feeds with amino acids, especially methionine and lysine, represents far greater quantities than does human nutrition. By supplementing feeds or formulated feeds with the first limiting amino acid an obvious cost reduction can be achieved, while maintaining the quality of the ration.

Of the ca. 20 amino acids found in feed protein, about one half are essential for monogastric animals (see Table 7). Most natural feeds are relatively poor in methionine and lysine (Table 17). The requirement of our livestock, however, is comparatively high [216].

When formulating a feed mix for a given animal type, the manufacturer has two choices for meeting the requirement of a particular amino acid. He may use either an excess of feed protein that contains large amounts of this amino acid or a minimum of natural protein and supplement it with synthetic amino acid. Because methionine and lysine are commercially available and inexpensive, they are often used in formulated feed [217]. L-Threonine and L-tryptophan, which in many cases are the next limiting amino acids after methionine and lysine, are still too expensive to be used for supplementation.

Amino Acids Content of Feedstuffs. Effective supplementation requires an exact knowledge of the natural amino acid content of both the individual feedstuffs and the formulated feed mix: the desired rates of supplementation must always be capable for being measured analytically. Ion-exchange and high-pressure liquid chromatography are reliable and proved methods for this.

The amino acid contents of individual feedstuffs are published internationally in a large series of tabulations (see, e.g., Table 17). However, such data must be current and reliable.

Amino Acid Requirements of Livestock. Determining the amino acid requirements of animals requires difficult, time-consuming experiments, and the values are not constants valid for all times [169]. Any alteration in breed, age, environmental conditions, required performance, energy content of feed, vitamin dosage, etc., also changes the amino acid requirement.

There are essentially three methods for determining these requirements: synthetic rations, semi-synthetic rations, and carcass analysis. In the first method, the animal is fed a synthetic mixture of all amino acids along with an otherwise balanced ration. The content of a single amino acid is reduced stepwise until a measurable reduction in growth or feed utilization occurs. In the second method, only a single amino acid is present in an amount less than the requirement for the animal. (This requires exact analysis of the amino acid contents of the various feed components.) The limiting amino acid is then supplemented stepwise until growth and feed utilization do not improve further.

In the carcass method, the amino acid content of the carcass is taken as a first-order approximation to the amino acid requirement of the animal.

The amino acid requirements of monogastric domestic animals are listed in Table 18.

Economics of Amino Acid Supplementation. The purpose of modern, formulated feed mixes is to meet all nutritional requirements of the animal at a minimum cost, and amino acids and proteins are usually among the most expensive components of a feed mix. The performance of a feed mix is measured primarily by feed utilization and weight gain. Both feed utilization and weight gain, as well as other factors such as laying performance or feather or hair growth, are directly dependent on a sufficient supply of ami no acids. Methionine and lysine play the major roles. Methionine in the form of D,L-methionine and lysine in the form of L-lysine · HCl are used in place of or as a supplement to natural methionine and lysine sources [217].

Linear programming is the method of choice for simultaneously optimizing a ration and minimizing cost [222]. This method allows simultaneous consideration of all demands that are made on the ration. A prerequisite is exact data on the nutrient content of all available feedstuffs and additives as well as their prices and availabilities. Additionally, the restrictions, i.e., the dietary requirements that

Table 18. Amino acid recommendations, wt%, for poultry [218]–[220] and pigs [218], [221]

Species	Metabolizable energy, MJ/kg	Crude protein	Met	Met+Cys	Lys	Thr	Trp
Broiler							
starter	13.2	21	0.56	0.96	1.18	0.77	0.21
grower	13.4	20	0.52	0.92	1.06	0.70	0.19
finisher	13.6	18	0.43	0.82	0.93	0.65	0.17
Laying hen	11.3	14	0.35	0.65	0.70	0.48	0.13
Turkey (weeks of age)							
0– 4	11.7	28	0.65	1,12	1.60	1.05	0.25
5– 8	12.2	25	0.58	1.00	1.43	0.94	0.23
9–12	12.6	22	0.49	0.85	1.25	0.82	0.20
13–16	13.0	19	0.44	0.75	1.01	0.71	0.17
> 16	13.4	16	0.41	0.70	0.90	0.63	0.15
Pig (kg live weight)							
10– 19	12.8 *	18	0.38	0.68	1.05	0.67	0.21
20– 35	12.8 *	17	0.35	0.65	1.00	0.64	0.20
36– 55	12.8 *	16	0.30	0.55	0.85	0.54	0.17
56–100	12.8 *	14	0.26	0.49	0.75	0.48	0.15
Sow							
lactating	12.8	15	0.25	0.46	0.70	0.45	0.15
pregnant	12.0	12	0.20	0.36	0.55	0.35	0.12

* Elevation of metabolizable energy by 1 MJ/kg requires elevation of amino acid content by 8%.

the ration has to fulfill, must be known. Table 19 shows two examples of feed mixes formulated by the linear programming method commonly used.

5.3. Pharmaceuticals

The pharmaceutical industry requires amino acids at a rate between 2000 and 3000 t/a worldwide. More than half of this is used for infusion solutions. During the last few years the potential of amino acids and their derivatives as active ingredients for pharmaceuticals has been recognizedclearly, and considerable growth can be predicted.

5.3.1. Nutritive Agents

Infusion Solutions. Parenteral nutrition with L-amino acid infusion solutions is a well-established component of clinical nutrition therapy. A standard infusion solution contains the eight classical essential amino acids, the semiessential amino acids L-arginine and L-histidine, and several nonessential amino acids, generally glycine, L-alanine, L-proline, L-serine, and L-glutamic acid.

Also available are special infusion solutions tailored to the requirements of particular groups, such as newborn infants, seniors, or patients with an extreme negative nitrogen balance. Solutions rich in

Table 19. Two examples of formulated feeds

Broiler feed composition, wt%		Pig fattening feed composition, wt%	
Yellow corn	28.00	Feed grain (barley, wheat, corn)	35.00
Wheat	20.00	Soybean meal (44% crude protein)	19.00
Soybean meal (48% crude protein)	30.00	Tapioca meal	20.00
Tapioca meal	10.00	Corn gluten feed	15.00
Fat	7.00	Meat and bone meal (45% crude protein)	3.00
Meat and bone meal (45% crude protein)	3.00	Fat	3.00
Mineral premix	1.25	Beet molasses	2.00
Vitamin-trace element premix	0.50	Mineral premix	2.43
D,L-Methionine	0.25	Vitamin-trace element premix	0.50
	100.00	D,L-Methionine	0.07
			100.00

the branched-chained amino acids leucine, isoleucine, and valine and poor in methionine and aromatic amino acids are available for liver-disease patients. Solutions containing only essential amino acids are available for kidney patients. Enzymatic protein hydrolysates, which were used as infusion solutions until a few years ago, have disappeared almost completely from the market. They were not available in the optimal composition, and there were often compatibility problems. Only pure, crystalline L-amino acids are used in modern infusion solutions. The solutions (up to 10%), which also contain electrolytes in addition to amino acids, are sterile and pyrogen-free. The simultaneous administration of carbohydrates is necessary for optimal utilization of the amino acids. Glucose is normally a separate infusion. Some commercially available amino acid infusion solutions contain an energy source in the form of sugar alcohols (sorbitol, xylitol), which do not enter into a Maillard reaction with the amino acids.

Normally, parenteral nutrition is only practiced over a limited time. In principle, however, total parenteral nutrition over many years is possible. In such a case, all essential nutrients (unsaturated fatty acids, vitamins, and trace elements) must be provided.

Elemental Diets. Enteral nutrition is also a means of providing the essential nutrients [223]. Elemental diets, which were developed originally for the astronauts [224], contain chemically defined nutritive components. In addition to free amino acids the mixtures generally contain carbohydrates, fats, minerals, and vitamins in a combination adapted to the requirements. In many cases, elemental diets are used as an alternative and supplement to parenteral nutrition. They have high nutritional value and are totally resorbable. They are largely independent of the digestive function of the pancreas and reduce the intestinal bacteria flora. Amino acid elemental diets generally are used in cases of anatomic, functional, or enzymatic defects [225].

Formula diets based on peptides currently are gaining ground as an alternative to elemental diets based on L-amino acids. According to recent studies [226], short-chained peptides are resorbed rapidly via a peptide transport system in the gut, therefore in a process that is independent of amino acid transport. Recently, compositions of nitrogen-free amino acid analogues (keto acids and hydroxy acids) have come into use for the special case of kidney insufficiency (chronic renal failure).

Elemental diets or formula diets are administered orally or via a nasogastric tube directly into the gastrointestinal tract.

5.3.2. Therapeutic Agents

Many therapeutic agents are derivatives of natural or nonnatural amino acids. Examples are benserazide, captopril, and dextrothyroxine. They are described under keywords such as Spasmolytics, Blood Pressure Affecting Agents, or Thyrotherapeutic Agents. Only therapeutically useful amino acids and simple derivatives are treated here.

Amino Acids and Salts. The amino acids and their simple salts that are currently important therapeutic agents are compiled in Table 20. The proprietary names listed represent only a selection.

N-Acetylcysteine [616-91-1], $C_5H_9NO_3S$, M_r 163.2, mp 109–110 °C, $[\alpha]_D^{20}$ + 5 ° (c = 3, H_2O), is a mucolytic and secretolytic agent.

$$HS-CH_2-\underset{NHCOCH_3}{\overset{H}{C}}-COOH$$

It is prepared by reaction of cysteine hydrochloride monohydrate with acetic anhydride in the presence of sodium acetate [227], [228].

Trade names: Fluimucetin (Inpharzam, FRG), Fluimucil (Inpharzam, FRG; Zambon, Italy), Mucolyticum "Lappe" (Lappe, FRG), Mucomyst (Allard, France; Mead Johnson, USA), Airbron, Parvolex (Duncan Flockhart, UK).

Carbocisteine (carbocysteine) [638-23-3], S-carboxymethyl-L-cysteine, $C_5H_9NO_4S$, M_r 179.2, mp 204–207 °C (decomp.), $[\alpha]_D^{20}$ −34.0 to −36.0 ° (c = 10, H_2O), is used to treat disorders of the respiratory tract associated with excessive mucus.

$$HOOC-CH_2S-CH_2-\underset{NH_2}{\overset{H}{C}}-COOH$$

Synthesis involves S-alkylation of L-cysteine with chloroacetic acid in the presence of sodium hydroxide [229], [230].

Trade names: Transbronchin (Degussa Pharma, FRG), Mucodyne (Berk, UK), Mucolex (Warner, UK), Mucopront (Mack, FRG), Rhinathiol (Joullié, France).

Levodopa [59-92-7], (S)-3-(3,4-dihydroxyphenyl)alanine, $C_9H_{11}NO_4$, M_r 197, mp 285.5 °C (decomp.), $[\alpha]_D^{20}$ −12.15 ° (c = 4 in 1 M HCl).

Amino Acids

Table 20. Amino acids and their salts as drugs

Compound		Formula	M_r	Medical use	Trade names
L-Alanine	[56-41-7]	$C_3H_7NO_2$	89.09	parenteral nutrition, stimulant of glucagon secretion	
L-Arginine	[74-79-3]	$C_6H_{14}N_4O_2$	174.20	parenteral nutrition, stimulation of pituitary release of growth hormone and prolactin, stimulation of pancreatic release of glucagon and insulin	
L-aspartate	[7675-83-4]	$C_{10}H_{21}N_5O_6$	307.31	treatment of hyperammonemia	Argihepar, Laevil, Potentiator, Sargenor, Sorbenor
L-glutamate hydrochloride	[4320-30-3]	$C_{11}H_{23}N_5O_6$	321.34	treatment of hyperammonemia	Modumate Argivene, R-Gene,
	[1119-34-2]	$C_6H_{15}ClN_4O_2$	210.66	treatment of hyperammonemia, electrolyte concentrate for i.v. solutions	Polilevo
L-pyroglutamate		$C_{11}H_{21}N_5O_5$	303.27		
2-oxoglutarate	[16856-18-1]	$C_{11}H_{20}N_4O_7$	320.3	treatment of hepatic disorders	Leberam, Anetil, Argiceto, Eucol
L-Asparagine monohydrate	[70-47-3]	$C_4H_8N_2O_3$	132.13	parenteral nutrition	
	[5794-13-8]	$C_4H_{10}N_2O_4$	150.13		
D,L-Aspartic acid	[617-45-8]	$C_4H_7NO_4$	133.10		
magnesium, tetrahydrate	[52101-01-6]	$C_8H_{20}N_2O_{12}Mg$	360.54	cardiac agent, management of fatigue, mineral supplement	Magnesium Verla,
potassium, semihydrate	[923-09-1] (anhydrous)	$C_4H_6NO_4K \cdot 1/2\ H_2O$	180.20	cardiac agent, management of fatigue, mineral supplement	Trommcardin, Trophicard-Köhler
sodium, monohydrate		$C_4H_8NO_5Na$	173.10		
L-Aspartic Acid	[56-84-8]	$C_4H_7NO_4$	133.10	parenteral nutrition	
ferrous, tetrahydrate		$C_8H_{12}ON_2O_{12}Fe$	392.1	treatment of iron deficiency	Sideryl, Spartocine
magnesium, dihydrate	[2068-80-6]	$C_8H_{16}N_2O_{10}Mg$	324.52	management of fatigue and heart conditions	
potassium, semihydrate	[1115-63-5] (anhydrous)	$C_4H_6NO_4K \cdot 1/2\ H_2O$	180.20	management of fatigue and heart conditions	Magnesiocard, Corroverlan
sodium, monohydrate	[3792-50-5] (anhydrous)	$C_4H_8NO_5Na$	173.10	parenteral nutrition	

Table 20. (continued)

Compound		M_r	Formula	Medical use	Trade names
Betaine			$C_5H_{11}NO_2$		
citrate	[107-43-7]	117.15	$C_{11}H_{19}NO_9$	lipotropic	Flacar, Panstabil
hydrochloride	[17671-50-0]	309.27	$C_5H_{12}ClNO_2$	lipotropic, gastric acidifier	Aciventral, Acidol, Pesimuriat
	[590-46-5]	153.61			
monohydrate	[17146-86-0]	135.15	$C_5H_{13}NO_3$	lipotropic	Hepaderichol
L-Citrulline	[372-75-8]	175.19	$C_6H_{13}N_3O_3$	treatment of hepatic disorders	Polilevo
L-Cysteine	[52-90-4]	121.16	$C_3H_7NO_2S$	treatment of damaged skin, topically in ophthalmology, detoxicant	Reducdyn, Hepa-Loges, Irradian, Cicatrex, Felacomp
hydrochloride	[52-89-1]	157.62	$C_3H_8ClNO_2S$		Cheihepar, Choldestal
hydrochloride monohydrate	[7048-04-6]	175.64	$C_3H_{10}ClNO_3S$		
L-Cystine	[56-89-3]	240.30	$C_6H_{12}N_2O_4S_2$	parenteral nutrition, lipotropic agent, treatment of hair and nail damage	Cystin "Brunner", Gerontamin, Pantovipar, Priorin
L-Glutamic acid	[56-86-0]	147.13	$C_5H_9NO_4$	parenteral nutrition, dietary supplement, treatment of hyperammonemia	Aciglut, Glutamin-Verla
calcium, dihydrate	[5996-22-5]	368.2	$C_{10}H_{20}N_2O_{10}Ca$		Vivacalcium
hydrochloride	[138-15-8]	183.54	$C_5H_{10}ClNO_4$	gastric acidifier	Bioprotein-Holzinger, Pansan, Pepsalara
monosodium, monohydrate	[142-47-2]	187.13	$C_5H_{10}NO_5Na$	flavoring and seasoning of food	
magnesium, tetrahydrate	[19238-50-7] (anhydrous)	388.62	$C_{10}H_{24}N_2O_{12}Mg$	tonics, mineral supplements	Magnesium Verla, Glutergen
potassium, monohydrate	[19473-49-5] (anhydrous)	203.24	$C_5H_{10}NO_5K$	tonics, mineral supplements	
magnesium, hydrobromide		397.5	$C_{10}H_{17}BrN_2O_8Mg$	tranquilizer	Psicosoma, Psychoverlan

Table 20. (continued)

Compound		Formula	M_r	Medical use	Trade names
L-Glutamine	[56-85-9]	$C_5H_{10}N_2O_3$	146.15	parenteral nutrition, treatment of mental disorders and alcoholism	
Glycine	[56-40-6]	$C_2H_5NO_2$	75.07	parenteral nutrition, antacid in conjunction with calcium carbonate	Cicatrex, Felacomp, Gastripan-K, Mirogastrin
aluminum, hydrate	[13682-92-3]	$C_2H_6NO_4Al\ (+ x\ H_2O)$	135.1	antacid	Acidrine, Al-Glycin
L-Histidine	[71-00-1]	$C_6H_9N_3O_2$	155.16	parenteral nutrition, essential amino acid for infants	
acetate		$C_8H_{13}N_3O_4$	215.21		
hydrochloride monohydrate	[5934-29-2]	$C_6H_{12}ClN_3O_3$	209.63		Anti-rheuma, Rollkur-Ankermann, Laristine, Plexamine
L-Isoleucine	[73-32-5]	$C_6H_{13}NO_2$	131.18	parenteral nutrition, treatment of hepatic coma, dietary supplement	
L-Leucine	[61-90-5]	$C_6H_{13}NO_2$	131.18	parenteral nutrition, treatment of hepatic coma, dietary supplement	numerous combinations
D,L-Lysine	[70-54-2]	$C_6H_{14}N_2O_2$	146.19	formation of salts with acidic drugs (to enhance solubility)	
monohydrochloride	[70-53-1]	$C_6H_{15}ClN_2O_2$	182.65	geriatric	Jestrosemin
acetylsalicylate	[34220-70-7]	$C_{15}H_{22}N_2O_6$	326.34	soluble form of acetylsalicylic acid (aspirin) for injection	Aspisol, Delgesic
L-Lysine	[56-87-1]	$C_6H_{14}N_2O_2$	146.19	parenteral nutrition, dietary supplement, prophylaxis of herpes simplex infection (?)	numerous combinations
acetate	[57282-49-2]	$C_8H_{18}N_2O_4$	206.24	dietary supplement	
L-aspartate	[27348-32-9]	$C_{10}H_{11}N_3O_6$	279.30	dietary supplement	
L-glutamate	[5408-52-6]	$C_{11}H_{23}N_3O_6$	293.32	dietary supplement	
L-malate	[71555-10-7]	$C_{10}H_{20}N_2O_7$	280.28	dietary supplement	
monohydrate	[39665-12-8]	$C_6H_{16}N_2O_3$	164.21		
monohydrochloride	[657-27-2]	$C_6H_{15}ClN_2O_2$	182.65	treatment of hypochloremia, alkaloses	Aktivanad, Athensa, Omnival, Vivioptal

Table 20. (continued)

Compound		Formula	M_r	Medical use	Trade names
D,L-Methionine	[59-51-8]	$C_5H_{11}NO_2S$	149.21	lipotropic and choleretic agent	numerous combinations
L-Methionine	[63-68-3]	$C_5H_{11}NO_2S$	149.21	parenteral nutrition, dietary supplement, lipotropic agent, treatment of paracetamol poisoning	
L-Ornithine acetate	[60259-81-6]	$C_7H_{16}N_2O_4$	192.22	parenteral nutrition	
L-aspartate	[3230-94-2]	$C_9H_{19}N_3O_6$	265.27	treatment of hepatic disorders	Hepa-Merz
monohydrochloride	[3184-13-2]	$C_5H_{13}ClN_2O_2$	168.62	parenteral nutrition	Ornitaine, Polilevo
2-oxoglutarate	[5191-97-9]	$C_{10}H_{18}N_2O_7$	278.14	treatment of hepatic disorders (hyperammonemia)	Ornicetil
D-Phenylalanine	[673-06-3]	$C_9H_{11}NO_2$	165.19	antidepressant	
L-Phenylalanine	[63-91-2]	$C_9H_{11}NO_2$	165.19	parenteral nutrition	
L-Proline	[147-85-3]	$C_5H_9NO_2$	115.13	parenteral nutrition, dietary supplement, starting material for captopril and enalapril	
L-Pyroglutamic acid	[98-79-3]	$C_5H_7NO_3$	129.07	formation of salts with basic drugs	
D,L-Serine	[302-84-1]	$C_3H_7NO_3$	105.09	starting material for benserazide	Aktiferrin
L-Serine	[56-45-1]	$C_3H_7NO_3$	105.09	parenteral nutrition, dietary supplement	Sulfolitruw
L-Threonine	[72-19-5]	$C_4H_9NO_3$	119.12	parenteral nutrition, dietary supplement	
L-Tryptophan	[73-22-3]	$C_{11}H_{12}N_2O_2$	204.23	parenteral nutrition, antidepressant, sleep inducer, dietary supplement	Optimax, Kalam, Pacitron
L-Tyrosine	[60-18-4]	$C_9H_{11}NO_3$	181.19	parenteral nutrition, dietary supplement	
L-Valine	[72-18-4]	$C_5H_{11}NO_2$	117.15	parenteral nutrition, dietary supplement, treatment of hepatic coma	

Uses Levodopa is widely used for treatment of Parkinson's disease, most often in combination with peripheral decarboxylase inhibitors such as benserazide and carbidopa.

Mecysteine hydrochloride [*18598-63-5*], methyl cysteine hydrochloride, methyl L-2-amino-3-mercaptopropionate hydrochloride, $C_4H_{10}ClNO_2S$, M_r 171.66, mp 140–141 °C, is used in the treatment of disorders of the respiratory tract associated with excessive mucus. It is prepared by esterification of L-cysteine hydrochloride monohydrate with methanol in the presence of hydrogen chloride [231].

$$H_2N-\underset{\underset{CH_2SH}{|}}{\overset{\overset{COOCH_3}{|}}{C}}-H \cdot HCl$$

Trade names: Acthiol (Joullié, France), Actiol (Lirca, Italy), Visclair (Sinclair, UK).

Methiosulfonium chloride (iodide) [*1115-84-0*], L-methylmethionine sulfonium chloride, vitamin U, $C_6H_{14}ClNO_2S$, M_r 199.7; mp 134 °C (decomp.). Iodide [*3493-11-6*], $C_6H_{14}INO_2S$, M_r 291.1.

$$H_2N-\underset{\underset{CH_2CH_2\overset{+}{S}\underset{CH_3}{\overset{CH_3}{<}}}{|}}{\overset{\overset{COOH}{|}}{C}}-H \quad Cl^- \text{ (I}^-)$$

Methiosulfonium chloride is used for its protective effect on the liver and gastrointestinal mucosa, whereas the iodide finds use for rheumatic disorders. The compounds are made by heating L-methionine with methyl chloride or methyl iodide [232].

Trade names: Chloride: Ardesyl (Beytout, France); Cabagin (Kowa, Japan). Iodide: Lobarthrose (Opodex, France).

Oxitriptan [*4350-09-8*], (S)-5-hydroxytryptophan, $C_{11}H_{12}N_2O_3$, M_r 220, mp 273 °C (decomp.), $[\alpha]_D^{22}$ −32.5 ° (c = 1 in H_2O), $[\alpha]_D^{22}$ + 16.0 ° (c = 1 in 4 M HCl).

This intermediate in mammalian biosynthesis of serotonin is used as an antidepressant. It is produced either by total synthesis (analogous to L-tryptophan via the 5-benzyloxy derivative) [233]–[235] or by fermentation with *Chromobacterium violaceum* [236].

Trade names: Levothym (Karlspharma, FRG), Prétonine (Arkodex, France), Quietim (Nativelle, France).

D-Penicillamine [52-67-5], D-3-mercaptovaline, D-β,β-dimethylcysteine, $C_5H_{11}NO_2S$, M_r 149.21, mp 198.5 °C, $[\alpha]_D^{25}$ −63 ° (c = 0.1, pyridine). Hydrochloride [2219-30-9], $C_5H_{12}ClNO_2S$, M_r 185.7, mp 177.5 °C (decomp.) $[\alpha]_D^{25}$ −63 ° (c = 1 in 1 M NaOH).

$$\begin{array}{c} \text{COOH} \\ | \\ \text{H---C}\blacktriangleleft\text{NH}_2 \\ | \\ \text{H}_3\text{C--C--SH} \\ | \\ \text{CH}_3 \end{array}$$

D-Penicillamine is a chelating agent that aids the elimination of toxic metal ions, e.g., copper in Wilson's disease. It is used, as an alternative to gold preparations, in the treatment of severe rheumatoid arthritis. It is useful in treating cystinuria because it reacts with cystine to form cysteine-penicillamine disulfide, which is much more soluble than cystine.

D-Penicillamine is produced by hydrolysis of benzylpenicillin via its Hg(II) complex [237] or by total synthesis. In the synthesis, isobutyraldehyde, sulfur, and ammonia are condensed to 5,5-dimethyl-2-isopropyl-Δ^3-thiazoline, which, on reaction with hydrogen cyanide, gives 5,5-dimethyl-2-isopropyl-thiazolidine-4-carbonitrile.Hydrolysis with boiling hydrochloric acid yields D,L-penicillamine hydrochloride. Cyclization with acetone and formylation leads to D,L-3-formyl-2,2,5,5-tetramethylthiazolidine-4-carboxylic acid, which can be resolved with (-)-phenylpropanolamine via the diastereomeric salts. Hydrolysis with hydrochloric acid leads to D-penicillamine hydrochloride [238].

Trade names: Cuprimine (Merck Sharp & Dohme, USA), Depen (Carter-Wallace, USA), Distamine (Dista, UK), Metalcaptase (Heyl/Knoll, FRG), Trolovol (Bayer/Homburg, FRG).

5.4. Cosmetics

Amino acids are a component of the "natural moisturizing factor" (NMF) that protects the skin surface from dryness, brittleness, and a deleterious environment [239]. The epidermis of the skin contains about 15% water, which, in the presence of amino acids, forms a stable water-in-oil emulsion with the skin lipids in the form of a thin layer. The amino acids simultaneously stabilize the pH of the skin (the acidic layer). Because of these properties, amino acids [240], [241], protein hydrolysates [242], and proteins [240] are widely utilized in skin and hair cosmetics, e.g., in mild skin creams, skin cleansing lotions, and hair shampoos.

The sodium salts of the reaction products of fatty acids with amino acids, such as glutamic acid [243]–[246], or short-chain peptides (protein hydrolysates) [247], [248] are surfactants. They are effective, skin-compatible cleaners and emulsifiers, which are used in shampoos, shower gels, baby baths, medicinal skin cleansers, cold-wave preparations, etc. [244], [249], [250]. The sulfur-containing amino acids exhibit a special normalizing effect on skin metabolism, e.g., in cases of excess skin lipid production (serborrhea), dandruff, or acne. Substances utilized for this purpose include derivatives of cysteine (e.g., *S*-carboxymethylcysteine), homocysteine (2amino-4-mercaptobutyric acid), and methionine [251]. Amino acids are also used in hair lotions, where they are reported to have a nutritive effect [252]. Cysteine, which acts as a reducing agent, is gaining importance, especially in Japan, as a substitute for thioglycolic acid in permanent wave preparations that are less damaging to hair [253]. Use of aluminum, tin, and zirconium complexes of amino acids, especially glycine, as deodorants [254] and antiperspirants [254], [255] has been reported.

5.5. Agrochemicals

Amino acids are building blocks in certain growth regulators, herbicides, pesticides, and fungicides. In many cases, synthesis begins not with the amino acids but with other raw materials. The total herbicide glyphosphate, for example, is not prepared from glycine but from iminodiacetic acid or from iminodiacetonitrile. The herbicide Suffix and the fungicide metalaxyl, which contain the alanine moiety, are prepared from the substituted anilines.

Agrochemical applications for free amino acids, although relatively rare, also are reported in the literature. Examples include D,L-amino acids as nematocides [256], L-phenylalanine as a fungicide [257], L-tryptophan in combination with gibberellin as a growth promoter [258], and L-proline as a promoter for blossom formation [259].

5.6. Industrial Uses

Amino acids and derivatives, as polyfunctional compounds, have potential for special industrial uses. The primary reason is their physicochemical properties, such as high thermal stability, low volatility, amphoterism, buffering capacity, and complex-forming capability. Increasingly important, however, are such aspects as environmental acceptability and low toxicity, properties that are characteristic for most amino acids and derivatives. Potential uses include acylamino acid monomers for epoxy resins [260]; amino acid dispersing agents for pigments in coloring polyester fibers [261]; *N*-acylamino acid dispersants for polyurethanes in water [262]; amino acid setting retarders for cement [263]; zinc salts of *o*-*N*-acyl-derivatives of basic amino acids and *N*-acylamino acids for the thermal stabilization of PVC [264]; polyglutamic acid esters and polyaspartic acid esters coatings for natural and synthetic leather [265]; amino acid hardening agents for methacrylate resins [266]; *N*-acylamino acid, amino acid ester, and amino acid amide gel-forming agents for oils [267]; basic amino acid vulcanization accelerators for natural rubber [268]; amino acid [269] and *N*-acylamino acid [270] corrosion inhibitors for metals; amino acids to stabilize the latent image of photographic emulsions [271]; and amino acid brighteners in galvanic baths [272], [273].

6. Chemical Analysis

Amino acids do not have defined melting points but decompose over a broad range between 250 and 300 °C. Therefore, they must be transformed into derivatives before melting points are useful for identification. Phenylisothiocyanate is used to yield the phenylthiohydantoin amino acid (PTH amino acid) [4], or 2,4-dinitrofluorobenzene (Sanger's reagent) is used to yield the dinitrophenyl amino acid [4].

Spectroscopic methods for the identification of amino acids include infrared [274], Raman, ^{1}H NMR, and ^{13}C NMR spectroscopy [275] of free amino acids or PTH derivatives and mass spectrometry of PTH derivatives [276]. Ultraviolet spectroscopy is important only for aromatic amino acids.

Various *staining methods* may be used for the qualitative identification of α-amino acids [277]. Some of these dye-forming reactions are suitable for quantitative analysis within the validity range of the Lambert-Beer law. By far the most important is the reaction with ninhydrin, which yields a red-violet to blue-violet dye (λ_{max} = 570 nm).

Ninhydrin   blue-violet dye

The imino acids proline and hydroxyproline form a structurally different dye, with an absorption maximum at 440 nm.

Fluorescent reagents also have been used successfully for quantitative analysis. The amino acid is converted into a strongly fluorescent derivative, which increases the sensitivity by orders of magnitude. Typical fluorescent reagents are *o*-phthalaldehyde–2-mercaptoethanol [278], 1-dimethylamino-naphthalene-5-sulfonyl chloride (dansyl chloride) [279], and 4-phenylspiro [furan-2(3*H*)–1′-phthalan]-3,3′-dione (fluorescamine) [280].

The *separation* of amino acid mixtures is possible by electrophoresis or chromatography. The latter is especially useful, techniques including paper, thin layer, ion-exchange, high-pressure liquid, and gas chromatography. Paper chromatography is generally carried out in two dimensions. A number of eluents are available for this purpose. Quite often the amino acid is converted to its dinitrophenyl derivative.

More common than paper chromatography is *thin layer chromatography* (TLC) of either the free amino acids [281] or their PTH derivatives [282]. Silica gel, aluminum oxide, cellulose powder, or polyacrylamide may be used as carrier, but silica is preferred. An indicator reagent, most often ninhydrin, is used for detection [283]. For detection of very small quantities the amino acids separated on an TLC plate can be converted to the fluorescamine derivatives [284]. The detection limit is 10 pmol. The time required for the analysis of dinitrophenyl amino acids can be reduced by using high-pressure thin layer chromatography [285].

Ion-exchange chromatography [286], [287] on organic resins (Dowex, Amberlite, etc.) has proved to be the most exact and reliable method for the separation and quantitative analysis of amino acids. Before the automatic analysis technique was introduced [288], complete analysis of an amino acid mixture required 24 h. Today 2 h is the rule. Sodium citrate or lithium citrate buffer solutions are the eluents. The eluate is reacted with ninhydrin [286], [289] or *o*-phthalaldehyde [290]. With ninhydrin,

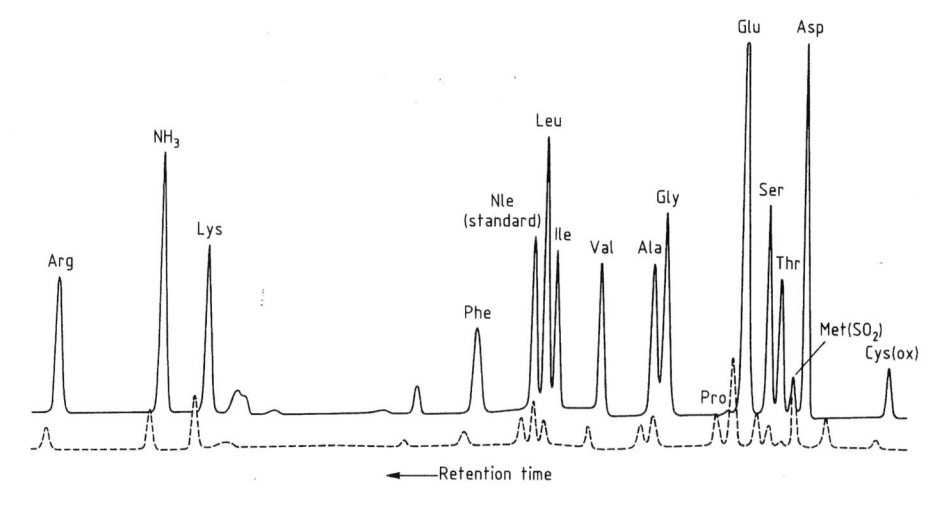

Figure 5. Amino acid chromatogram of a broiler feed
Internal standard: norleucine. Solid curve: UV detection at $\lambda = 570$ nm. Dotted curve: UV detection at $\lambda = 440$ nm for Pro (and Hyp).

1–20 nmol amino acid can be measured with an accuracy of ± 1–5 %. The detection limit with o-phthalaldehyde is in the picomole range.

Ion-exchange chromatography is currently the method of choice for analyzing amino acids in feeds, foodstuffs, and biologic fluids. In general, analysis is preceeded by a hydrolysis, which degrades the proteins and peptides to their component amino acids. Figure 5 shows a sample aminogram of broiler feed.

Utilization of *high-pressure liquid chromatography* (HPLC) allows a further reduction in analysis time. However, not all amino acids can be separated cleanly. Furthermore, the amino acids generally must be converted to derivatives, e.g., to dansyl amino acids [291], PTH amino acids [292], or dinitrophenyl derivatives [293], before analysis. Reversed phases are the preferred stationary phases. Ninhydrin, o-phthalaldehyde–mercaptoethanol [294], or fluorescamine [295] is the usual reagent for detection. An HPLC method with direct UV detection has also been described [296] for analysis of infusion solutions.

Gas chromatographic methods [297] are useful for the analysis of complex amino acid mixtures. However, the amino acids must be converted into volatile derivatives, e.g., PTH amino acids [298], methyl esters of N-trifluoroacetylamino acids [299], n-butyl esters of N-trifluoroacetylamino acids [300], N-trimethylsilylamino acids [301], or N,O-bis-trimethylsilylamino acids [302].

Electrophoresis [303], which employs the differing rates of migration in an electric field, is relatively unimportant. Capillary isotachophoresis is a new high-resolution electrophoresis technique for amino acids.

The chromatographic separation of amino acid enantiomers is the subject of intensive investigation. Separation is currently possible by gas chromatography [304] and high-pressure liquid chromatography [305], using optically active phases or chiral solvents [306].

The *microbiological analysis* of amino acids is based on the fact that several L-amino acids are essential for certain bacteria strains. The growth of the bacteria cultures under standard conditions can be quantitatively evaluated (acidimetry or turbidimetry) and related to the amino acid concentration. Lactic acid bacteria [307] (Lactobacteriaceae) can be used to analyze 19 L-amino acids. Typical test microbes include *Leuconostoc mesenteroides* (ATCC 8042), *Lactobacillus arabinosus 17–5* (ATCC 8014), and *Streptococcus faecalis* R (ATCC 9790, 8043). The conventional microbiological methods are quite complicated but can be simplified by automation [308].

A series of L- or D-amino acids can be analyzed by *enzyme methods*. L-Amino acids, or the enantiomeric purity of D-amino acids, can be determined with bacteria decarboxylases by measurement of CO_2 formed. D-Amino acids, or the enantiomeric purity of L-amino acids, can be determined with kidney D-amino acid oxidases by measurement of the O_2 consumption. Enzymes that react only with a single amino acid allow determination of that amino acid, e.g., arginine using liver arginase. An improved enzymatic method consists of the use of enzyme electrodes that contain the enzymes [309], or microorganisms having a special enzyme in a fixed form. Enzyme electrodes, however, are relatively unstable [310].

The *quantitative determination of crystallized amino acids* is carried out by acidimetric titration in nonaqueous medium [311], [312]. Glacial acetic acid is a suitable solvent. Formic acid may be added to improve solubility. The titrant is perchloric acid. Formol titration by the method of Sörensen can be used for aqueous solutions but is less accurate.

Standards of purity for individual L- and D,L-amino acids that are used in drugs or as food additives are published in pharmacopeias [311], [313] and food codices [314].

7. Economic Significance

The current world market for amino acids is estimated at ca. 450 000 t/a (Table 21). The "big three" amino acids, sodium L-glutamate, D,L-methionine, and L-lysine · HCl, account for more than 95% of the volume (Table 22). The other amino acids play only a small role. The dominant amino acid, sodium glutamate, is used almost exclusively as a taste enhancer. D,L-Methionine and L-lysine · HCl are used almost exclusively to improve the nutritive value of animal feeds. The other amino acids have diversified applications. With the exception of glycine, they are more expensive than the big three amino acids. Table 23 shows the market volume and value broken down by use.

The main manufacturers of amino acids are located in the Far East (e.g., Ajinomoto, Kyowa Hakko, Tanabe) and in Europe (Degussa, Alimentation Equilibrée de Commentry).

8. Toxicology

Excess amino acids are rapidly disposed of by increased metabolic degradation. Should the amino acid dose be suddenly increased, e.g., by extremely high protein consumption, within about two days the liver adaptively increases the levels of amino acid-catabolizing enzymes — transaminases, enzymes of the urea cycle, cystathionase, tryptophan pyrrolase, etc. The excess amino acids are to a large extent used to provide energy. The nitrogen is eliminated as urea. A smaller portion is used in protein synthesis, mainly liver protein and plasma albumin.

Table 21. World market for amino acids, 1982

Amino acid	Quantity, t/a [315] ([27])
L-Alanine	130
D,L-Alanine	2000 (700)
L-Arginine[a]	500
L-Aspartic acid[b]	450 (250)[d]
L-Asparagine	50
L-Cysteine[a]	}700
L-Cystine	
L-Citrulline	50
Glycine	6000
L-Glutamic acid[c]	340 000 (270 000)
L-Glutamine	500
L-Histidine[a]	200
L-Hydroxyproline	50
L-Isoleucine	150
L-Leucine	150
L-Lysine[a]	34 000 (32 000)
L-Methionine	150
D,L-Methionine	110 000
L-Ornithine	50
L-Phenylalanine	150[d]
L-Proline	100
L-Serine	50
L-Threonine	160
L-Tryptophan	200[d]
L-Tyrosine	50 (100)
L-Valine	150

[a] Hydrochloride and free amino acid.
[b] Free acid and salts.
[c] Sodium salt.
[d] Quantities in 1984: L-Asp 5000 t, L-Phe 3000 t, L-Trp 300 t.

Table 22. Percentage of individual amino acids as a part of the total market, 1979

Amino acid	Quantity, %	Value, %
L-Glutamic acid (Na)	64	56
D,L-Methionine	26	23
L-Lysine (HCl)	7	8
Glycine	1.4	–
Other amino acids	1.6	13
	100 = ca. 422 000 t/a	100 = $ 1 645 000 000 per year

When too little protein or no protein is consumed or when the component amino acids are imbalanced, alteration of the ribosome profile occurs in the liver, and ribonucleic acids are catabolized. The manifestation of chronic protein deficiency is known as marasmus (slight deficiency) and kwashiorkor (extreme deficiency). Protein deficiency is usually coupled with calorie deficiency (protein-calorie malnutrition). This manifests itself on the biochemical level as a negative nitrogen balance, indicating a reduction in protein inventory. Initially, the labile enzyme and plasma proteins are

Table 23. Market volume and value by field of use, 1982

Use	Volume, %	Value, %
Human nutrition	66	52
Animal nutrition	33	30
Specialty *	1	18
	100 = 455 000 t/a	100 = $ 1 150 000 000 per year

* Pharmaceuticals, cosmetics, agrochemicals, and industrial uses.

consumed, the greatest losses occurring first in the liver, then in the musculature. Brain, heart, and kidneys suffer only minimal protein loss. Symptoms of protein synthesis disorders include disturbances in wound healing and bone growth, lowered resistance to infection and stress, loss of fertility and appetite, and anorexia. The urinary excretion of 3-methylhistidine is a common indicator of the catabolism of muscle protein.

The total absence of an essential amino acid in the diet is more serious than protein deficiency. In this case the proteins or amino acids in the diet are totally worthless because protein synthesis can occur only by degradation of body protein. A general interruption of the protein synthesis results. This manifests itself rapidly as a drop in enzyme activity and an impoverishment of the plasma proteins. Noticeable symptoms are loss of appetite and weight, alteration of cornea and lens, anatomic organ alterations, and an increased rate of mortality. In addition there appear specific deficiency symptoms characteristic of the missing amino acid or acids.

The metabolic disturbances brought about by gross divergences from the optimal amino acid pattern have three different causes [317], [318]: imbalance, antagonism, and toxicity. *Amino acid imbalance* manifests as an appearance of deficiency symptoms for the first limiting amino acid when the other amino acids are consumed in great excess. The symptoms of imbalance are eliminated by administration of the first limiting amino acid in sufficient quantities. The main symptom is a severe reduction of food or feed assimilation and depression of growth. The depression of growth in rats has been investigated intensively by adding individual L- and D,L-amino acids to basal diets of various protein levels [319], [320].

Amino acid antagonism is caused by competition for common transport systems. An example is antagonism of the branched-chain amino acids isoleucine, leucine, and valine. The symptoms are reversible. In a study with young rats, addition of 5% L-leucine to a low-protein diet (9% casein) reduced the plasma levels of isoleucine and valine, depressed growth, and reduced feed consumption [321]. These effects were eliminated after a latent period of three days by small doses of L-isoleucine (0.16%) and L-valine (0.15%).

Amino acid toxicity occurs when very large quantities of one or more amino acids are consumed and is characterized by total failure of the adaptive mechanisms. The toxic level has been studied by adding increasing quantities of individual amino acids to a protein basal diet [318]. The toxicity of individual amino acids depends on the total protein consumption. Imbalance, antagonism, and toxicity are less pronounced when overall protein consumption is sufficient but become more severe at lower levels of protein consumption. The consumption of toxic amounts of amino acids increases their concentration in the plasma and brain. Because of the blood-brain barrier, however, the increase in the brain is not as great [322]. Failure of the adaption mechanisms during consumption of large excess of an amino acid can lead to accumulation of the amino acids or certain metabolites in the organism, leading directly or indirectly (e.g., by influencing hormone secretion) to anatomic or functional damage.

The toxicity of amino acids has been reviewed [318], [321]. The acute toxicities of most L-amino acids and some derivatives have been determined [323]–[325]. The toxicology of D-amino acids is discussed in review articles [318] and other publications [324], [325]. There is no evidence to date that the D-enantiomers of the α-amino acids found in proteins exhibit specific toxic effects. Their LD_{50} values are generally higher than those of the L-amino acids.

9. References

General References

[1] J. P. Greenstein, M. Winitz: *Chemistry of the Amino Acids,* vol. **1, 2,** and **3,** J. Wiley & Sons, New York – London 1961.

[2] A. Meister:*Biochemistry of the Amino Acids,* 2nd ed., vol. **1,** Academic Press, New York – London 1965.

[3] I. Wagner, H. Musso: "Neue natürliche Aminosäuren" *Angew. Chem.* **95** (1983), 827–920; *Angew. Chem. Int. Ed. Eng.* **22** (1983) 816 (review of the literature since 1956).

[4] Th. Wieland et al.: "Methoden zur Herstellung und Umwandlung von Aminosäuren und Derivaten," *Houben-Weyl,* 11/2.

[5] H. D. Jakubke, H. Jeschkeit: *Aminosäuren, Peptide, Proteine,* Verlag Chemie, Weinheim 1982.

[6] B. Weinstein (ed.): *Chemistry and Biochemistry of Amino Acids, Peptides and Proteins,* vol. **1–5,** Marcel Dekker, New York 1971–1978.

[7] D. Barton, W. D. Ollis: *Comprehensive Organic Chemistry,* vol. **2,** Chap. 9.6, p. 815; vol. **5** Pergamon Press, New York – Toronto – Sydney – Paris – Frankfurt 1979.

[8] G. Krügers:"Aminocarbonsäuren" in: *Methodicum Chimicum,* vol. **6,** p. 611.

[9] E. Wünsch: "Synthese von Peptiden," *Houben-Weyl,* **15/1** and **2.**

[10] I. C. Johnson: *Amino Acids Technology,* Noyes Data Corp., Parke Ridge, N.J., 1978.

Specific References

[11] IUPAC Commission on Nomenclature of Organic Chemistry (CNOC) and IUPAC-IUB Commission on Biochemical Nomenclature (CBN), *Biochemistry* **14** (1975) 449.

[12] S. K. Bhattacharyya, A. B. Banerjee, *Folia Microbiol. (Prague)* **19** (1974) 43.

[13] T. Robinson, *Life Sci.* **19** (1976) 1097.

[14] R. E. Dickerson, *Spektrum der Wissenschaften* 1979, no. 9, 98.

[15] *Handbook of Chemistry and Physics,* 62nd ed., CRC Press, Boca Raton, Florida, 1981–1982.

[16] D. M. Greenberg: *Amino Acids and Proteins,* Charles C. Thomas Publ., Springfield, Illinois, 1951.

[17] R. S. Cahn, C. Ingold, V. Prelog, *Angew. Chem.* **78** (1966) 413;*Angew. Chem. Int. Ed. Engl.* **5** (1966) 385, 511.

[18] K. Lübke,E. Schröder,G. Kloss: *Chemie und Biochemie der Aminosäuren, Peptide und Proteine,* vol. **I** and **II,** Thieme Verlag, Stuttgart 1975.

[19] A. Kleemann, *Chem. Ztg.* 106 (1982) 151–167.

[20] K. H. Drauz, A. Kleemann, J. Martens, *Angew. Chem.* **94** (1982) 590; *Angew. Chem. Int. Ed. Engl.* **21** (1982) 584.

[21] M. S. Sadovnikova, V. M. Belikov, *Russ. Chem. Rev. (Engl. Transl.)* **47** (1978) 199.
[22] Y. Izumi, J. Chibata, T. Itoh, *Angew. Chem.* **90** (1978) 187; *Angew. Chem. Int. Ed. Engl.* **17** (1978) 176.
[23] S. Yamada, C. Hongo, R. Yoshioka, J. Chibata, *J. Org. Chem.* **48** (1983) 843.
[24] Degussa, US 4356324, 1982 (W. Bergstein, A. Kleemann, J. Martens).
[25] A. Collet, M.-J. Birenne, J. Jacques, *Chem. Rev.* **80** (1980) 215.
[26] Nippon Kayaku, US 4224239, 1980 (Y. Tashiro, T. Nagashima, S. Aoki, R. Nishizawa).
[27] *Finechemical (Jpn.)* 1982, no. 3, 4–26.
[28] J. Chibata: *Immobilized Enzymes,* Kodansha – Halsted Press, Tokyo – New York 1978.
[29] C. Wandrey:*Advances in Biochemical Engineering,* vol. **12,** Springer Verlag, New York 1979.
[30] Degussa, US 4304858, 1981 (C. Wandrey et al.).
[31] Snamprogetti, DE 2621076, 1976 (F. Cecere et al.).Ajinomoto, US 4211840, 1980 (S. Nakamori et al.). AEC (Alimentation Equilibrée de Commentry), BE 883322, 1980. BASF, EP 46186, 1981 (R. Lungershausen et al.).
[32] M. Guivarach et al., *Bull. Soc. Chim. Fr.* 1980, no. 1/2, II-91–II-95.
[33] M. Sugie, H. Suzuki, *Agric. Biol. Chem.* **44** (1980) 1089.
[34] Dow Chemical, US 2700054, 1955 (H. C. White).
[35] Ajinomoto, GB 908735, 1962 (J. Kato et al.).
[36] Tanabe Seiyaku, US 3898128, 1975 (I. Chibata et al.).
[37] K. Yamamoto et al., *Biotechnol. Bioeng.* **22** (1980) 2045.
[38] Tanabe Seiyaku, US 3871959, 1975 (I. Chibata et al.).
[39] Kyowa Hakko Kogyo, US 4086137, 1978 (K. Nakayama, K. Araki, H. Yoshida). Tanabe Seiyaku, US 3902967, 1975 (I. Chibata, M. Kisumi, J. Kato).Ajinomoto, GB 2084566, 1982 (K. Akashi et al.).
[40] Prod. Organ. de Santerre Orsan, US 3933586, 1976 (Nguyen-Cong Puc). Ajinomoto, US 4000040, 1976 (T. Tsuchida, K. Kubota, Y. Hirose).
[41] Tanabe Seiyaku, DE 2252815, 1972 (I. Chibata, T. Tosa, T. Sato)
[42] Maggi, DE 2449711, 1979 (P. Hirsbrunne, R. Bertholet).
[43] Mitsubishi Chemical Ind., US 3974031, 1976 (H. Yamada, K. Kumagai, H. Ohkishi).
[44] Ajinomoto, US 4006057, 1977 (K. Sano et al.).
[45] Showa Denko, DE 3021566, 1979 (K. Nakayasu, O. Furuya, C. Inouè).Showa Denko, JP-Kokai 80164669, 1980.
[46] Mitsui Toatsu Chemicals, EP 79966, 1983.
[47] J. Martens, H. Offermanns, P. Scherberich, *Angew. Chem.* **93** (1981) 680; *Angew. Chem. Int. Ed. Engl.* **20** (1981) 668.
[48] Ajinomoto, DE 1196630, 1970 (J. Kato et al.). Ajinomoto, DE 1122538, 1962 (H. Kageyama, M. Sato, T. Inouè).
[49] Shell Oil, US 3131211, 1964 (G. A. Kurhajec, D. S. La France).
[50] T. Akashi, *Nippon Kagaku Zasshi* **81** (1960) 421.
[51] Ajinomoto, US 3971701, 1976 (K. Takinami et al.). Ajinomoto, US 4347317, 1982 (M. Yoshimura et al.).
[52] Kyowa Hakko Kogyo, US 3939042, 1976 (K. Nakayama et al.). Commercial Solvents Corp., US 3929575, 1975 (G. M. Miescher).
[53] Chattan Drug & Chemical, US 3510515, 1970 (C. S. Colburn).
[54] Degussa, DE 1134683, 1963 (H. Wagner et al.). Showa Denko, DE 2343599, 1982 (N. Mihara, O. Furuya, K. Wada).
[55] M. B. Vickery, *J. Biol. Chem.* **143** (1942) 77.

[56]   Tanabe Seiyaku, US 3902966, 1975 (I. Chibata et al.).
[57]   Ajinomoto, US 3875001, 1975 (K. Kubota et al.).
[58]   Ajinomoto, EP 82637, 1982 (T. Tsuchida et al.).
[59]   Degussa, DE 2048790, 1976 (H. Wagner, F. Schäfer). Tanabe Seiyaku, JP 82156448, 1982.
[60]   Tanabe Seiyaku, JP-Kokai 7922513, 1979.W. R. Grace, US 4329427, 1982 (M. H. Updike, G. J. Calton).
[61]   Ajinomoto, EP 71023, 1982 (T. Tsuchida, K. Miwa, S. Nakamori).
[62]   T. Tosa, T. Mori, N. Fuse, I. Chibata, *Enzymologia* **32** (1967) 153.
[63]   Ethyl Corp., US 4259441, 1981 (D. P. Bauer).
[64]   Ajinomoto, US 3865690, 1975 (S. Okumura et al.).
[65]   Stamicarbon, DE 2005515, 1970 (I. A. Thoma, J. F. Klein, L. M. Geurts). Stamicarbon, DE 2010696, 1970 (I. A. Thoma, W. Reichrath).
[66]   Stamicarbon, DE 1949585, 1969 (W. K. van der Linden, G. M. Surerkropp).
[67]   Toray Ind., US 3770585, 1973, US 3796632, 1974 (T. Fukumura).
[68]   O. Tosaka, K. Takinami, Y. Hirose, *Agric. Biol. Chem.* **42** (1978) 745.
[69]   Kyowa Hakko Kogyo, DE 2531999, 1979 (K. Inuzuka, S. Hamada, Y. Hofu).
[70]   Ajinomoto, FR 2482622, 1981 (K. Miwa et al.) Ajinomoto, US 4346170, 1982 (K. Sano, T. Tsuchida).
[71]   Degussa, DE 1906405, 1969 (H. Wagner et al.).
[72]   Procter & Gamble, US 3963573, 1976 (C. E. Stauffer).
[73]   Kanegafuchi Kagaku Kogyo, US 4148688, 1979 (H. Yamada et al.).
[74]   R. H. Mazur, J. M. Schlatter, A. H. Goldkamp, *J. Am. Chem. Soc.* **91** (1969) 2684.
[75]   Y. Isowa et al., *Tetrahedron Lett.* 1979, 2611.
[76]   Ajinomoto, US 3867436, 1975 (M. Nakamura et al.).
[77]   Dynamit Nobel, DE 2931224, 1979 (H. aus der Fünten, K. Schrage).
[78]   Ajinomoto, DE 2364039, 1973 (T. Yukawa et al.).
[79]   Degussa, GB 2078218, 1981 (T. Lüssling, P. Scherberich).
[80]   Anic, EP 77099, 1982 (M. Fiorini, M. Riocci, M. Giongo).
[81]   S. Yamada et al., *Appl. Environ. Microbiol.* **42** (1981) 773.
[82]   Tanabe Seiyaku, JP 7020556, 1966.
[83]   Ajinomoto, US 3909353, 1975.
[84]   Ajinomoto, GB 2053906, 1980 (T. Tsuchida, K. Sano). Genex Corp., EP 77196, 1982 (R. A. Synenki).
[85]   Degussa, EP 52201, 1981 (K. Drauz, A. Kleemann, M. Samson).
[86]   V. E. Price et al., *Arch. Biochem.* **26** (1950) 92.
[87]   S. Yamada, C. Hongo, I. Chibata, *Agric. Biol. Chem.* **41** (1977) 2413.
[88]   Ajinomoto, US 4224409, 1980 (S. Nakamori, H. Morioka, F. Yoshinaga).
[89]   Tanabe Seiyaku, EP 76516, 1982 (I. Chibata et al.).
[90]   Schering, DE 3127361, 1981 (G. Siewert).
[91]   L. Bassignani et al., *Chem. Ber.* **112** (1979) 148.
[92]   Mitsui Toatsu Chemical, EP 30474, 1980 (R. Mita et al.).
[93]   Ges. für Biotechnol. Forschung, US 4060455, 1977 (F. Wagner, H. Sahm, W. H. Keune).
[94]   Ajinomoto, US 3880741, 1975 (K. Kageyama et al.).
[95]   Kyowa Hakko Kogyo, US 4183786, 1980 (K. Nakayama, K. Araki, Y. Tanaka).
[96]   M. Sato, K. Okawa, S. Akabori, *Bull. Chem. Soc. Jpn.* **30** (1957) 937.
[97]   Y. Ozaki et al., *Synthesis* 1979, 216.
[98]   Denki Kagaku Kogyo, DE 3247703, 1982 (M. Kato, T. Miyoshi, I. Kibayashi).

[99] Tanabe Seiyaku, GB 2072185, 1981 (I. Chibata et al.).
[100] V. G. Debabov et al., US 4321325, 1980. Ajinomoto, US 4347318, 1982 (K. Miwa et al.). Ajinomoto, EP 66129, 1982 (T. Tsuchida, K. Miwa, S. Nakamori).
[101] O. T. Warner, O. A. Moe, *J. Am. Chem. Soc.* **70** (1948) 2765.
[102] Degussa, DE 2647255, 1978 (H. Offermanns, H. Weigel).
[103] I. Maeda, R. Yoshida, *Bull. Chem. Soc. Syn.* **41** (1968) 2975.
[104] Degussa, EP 52199, 1981 (A. Kleemann, M. Samson).
[105] Ajinomoto, GB 982727, 1965.
[106] Ajinomoto, JP 7903021, 1979; JP 7903059, 1979; JP 7903010, 1979.
[107] Degussa, EP 52200, 1981 (A. Kleemann, M. Samson).
[108] Ajinomoto, DE 2819148, 1978 (S. Nakamori, K. Yokozeki, K. Mitsugi).
[109] F. Masumi et al., *Chem. Pharm. Bull.* **30** (1982) 3831.
[110] Tanabe Pharmaceutical, JP 78001836, 1978.
[111] DE 2841642, 1978 (F. Wagner, J. Klein).
[112] Mitsui Toatsu Chemicals, US 4335209, 1982 (Y. Asai, M. Shimada, K. Soda).
[113] Asahi Kasei Kagyo, US 4349627, 1982; US 4360594, 1982 (A. Mimura et al.).
[114] Ajinomoto, US 3929573, 1975 (S. Konosuke, K. Mitsugi).
[115] AB Bofors, US 3963572, 1976 (S. V. Gatenbeck, P. O. Hedman).
[116] Showa Denko, US 4363875, 1982 (T. Akashiba, A. Nakayama, A. Murata).
[117] E. O. Denenu, A. L. Demain, *Appl. Environ. Microbiol.* **42** (1981) 497.
[118] Mitsubishi Petrochemical, US 4271267, 1981 (H. Yukawa et al.).
[119] Ajinomoto, EP 81107, 1982 (O. Kurahashi, M. Kamada, H. Enei).
[120] Ajinomoto, US 4371614, 1983 (D. M. Anderson, K. M. Hermann, R. L. Somerville). Ajinomoto, EP 80378, 1982 (T. Tsuchida et al.).
[121] Ajinomoto, DE 2411209, 1982 (T. Tsuchida, F. Yoshinaga).
[122] Prod. Organ. de Santerre Orsan, GB 1578057, 1980.
[123] W. D. Jefferson, *J. Am. Chem. Soc.* **43** (1978) 3980.
[124] Zoecon Corp., US 4260633, 1981 (R. J. Anderson, K. G. Adams, C. A. Henrick).
[125] Snamprogetti, CH 620943, 1980 (D. Dinelli, F. Morisi, D. Zaccardelli).
[126] Kanegafuchi Kagaku Kogyo, DE 2757980, 1977 (H. Yamada, S. Takahashi, K. Yoneda). Kanegafuchi Kagaku Kogyo, JP-Kokai 8158493, 1981.
[127] D. A. Bender: *Amino Acid Metabolism*, J. Wiley & Sons, London – New York – Sydney – Toronto 1975, p. 2.
[128] H. N. Munro, *Drug Intell. Clin. Pharm.* **6** (1972) 216.
[129] M. L. Sunde, *Poult. Sci.* **51** (1972) 44.
[130] R. F. Barker, D. A. Hopkinson, *Ann. Hum. Genet.* **41** (1977) 27.
[131] L. D. Stegink: *Clinical Nutrition Update: Amino Acids*, Am. Med. Assoc. Publ., Chicago 1977, pp. 198–205.
[132] K. Schreier, *Ärztl. Fortbildung* **19** (1971) 107.
[133] U. Porath, K. Schreier, *Med. Ernähr.* **11** (1970) 229.
[134] H. Bickel, S. Kaiser-Grubel, *Dtsch. Med. Wochenschr.* **96** (1971) 1415.
[135] Y. Izumi, I. Chibata, T. Itoh, *Angew. Chem.* **90** (1978) 187; *Angew. Chem. Int. Ed. Eng.* **17** (1978) 176.
[136] M. S. Sadovnikova, V. M. Belikov, *Russ. Chem. Rev. (Engl. Transl.)* **47** (1978) 199.
[137] K. Drauz, A. Kleemann, J. Martens, *Angew. Chem.* **94** (1982) 590; *Angew. Chem. Int. Ed. Engl.* **21** (1982) 584.

[138] FAO: *Nutritional Studies No. 24, Amino Acid Content of Foods and Biological Data on Proteins,* Interprint (Malta), Rome 1970.
[139] Degussa, unpublished
[140] Report of a Joint FAO/WHO Ad Hoc Expert Committee: *Energy and Protein Requirements,* Rome 1973.
[141] W. C. Rose, *Nutr. Abstr. Rev.* **27** (1957) 631.
[142] D. M. Hegsted, *Fed. Proc. Fed. Am. Soc. Exp. Biol.* **22** (1963) 1424.
[143] E. Kofrányi, *Ernähr. Umsch.* **23** (1976) 205.
[144] Dtsch. Ges. für Ernährung: *Empfehlungen für die Nährstoffzufuhr,* 3rd ed., Umschau-Verlag, Frankfurt 1975.
[145] *Recommended Dietary Allowances,* 9th ed., Nat. Acad. of Sci., Washington, D.C., 1980.
[146] G. K. Parman, *J. Agric. Food Chem.* **16** (1968) 169.
[147] H. H. Ottenheym, P. J. Jenneskens, *J. Agric. Food Chem.* **18** (1970) 1010.
[148] N. S. Scrimshaw, A. M. Altschul: *Amino Acid Fortification of Protein Foods,* MIT Press, Cambridge, Mass., and London 1971.
[149] E. E. Howe, G. R. Jansen, E. W. Gilfilian, *Am. J. Clin. Nutr.* **16** (1965) 315.
[150] A. M. Altschul, *Nature (London)* **248** (1974) 643.
[151] J. Kato, N. Muramatsu, *J. Am. Oil Chem. Soc.* **48** (1971) 415.
[152] J. Mauron, *Z. Ernährungswiss. Suppl.* no. **23** (1979) 10.
[153] D. M. Hegsted, *Am. J. Clin. Nutr.* **21** (1968) 688.
[154] L. D. Satterlee, H. F. Marshall, J. M. Tennyson, *J. Am. Oil Chem. Soc.* **56** (1979) 103.
[155] H. W. Staub, *Food Technol. (Chicago)* **32** (1978) 57.
[156] F. Jekat, *Fette Seifen Anstrichm.* **79** (1977) 273.
[157] J. C. Somogyi in: *Die Bedeutung der Eiweiße in unserer Ernährung,* Issue 48 a, Schriftenreihe der Schweizer Vereinigung für Ernährung, 1982, p. 3.
[158] G. G. Graham et al., *Am. J. Clin. Nutr.* **22** (1969) 1459.
[159] G. R. Jansen, C. F. Hutchison, M. E. Zanetti, *Food Technol. (Chicago)* **20** (1966) 323.
[160] R. Bressani et al., *J. Nutr.* **79** (1963) 333.
[161] J. L. Iwan, *Cereal Sci. Today* **13** (1968) 202.
[162] D. Rosenfield, F. J. Stare, *Mod. Gov.* **11** (1970) 47.
[163] S. N. Gershoff et al., *Am. J. Clin. Nutr.* **30** (1977) 1185.
[164] H. N. Parthasarathy et al., *Can. J. Biochem.* **42** (1964) 385.
[165] S. J. Fomon et al., *Am. J. Clin. Nutr.* **32** (1979) 2460.
[166] A. L. Jung, S. L. Carr, *Clin. Pediatr. (Philadelphia)* **16** (1977) 982.
[167] Procter & Gamble, US 3878305, 1975 (R. A. Damico, R. W. Boggs).
[168] S. Wallrauch, *Flüss. Obst* **44** (1977) 386.
[169] H. D. Pruss, I. P. G. Wirotama, K. H. Ney, *Fette Seifen Anstrichm.* **77** (1975) 153.
[170] C. Ambrosino et al., *Minerva Pediatr.* **18** (1966) 759.
[171] C. A. Masschelein, J. Van de Meerssche, *Tech. Q. Master Brew. Assoc. Am.* **13** (1976) 240.
[172] T. Take, H. Otsuka, *Chem. Abstr.* **70** (1969) 46270 m.
[173] J. Koch, *Flüss. Obst* **46** (1979) 212.
[174] G. Baumann, K. Gierschner, *Dtsch. Lebensm. Rundsch.* **70**(1974) 273.
[175] A. Askar, H. J. Bielig, *Alimenta* **15** (1976) 3.
[176] W. Hashida, *Food Trade Rev.* **44** (1974) 21.
[177] *Food Technol. (Chicago)* **34** (1980) 49.
[178] S. Schwimmer, D. G. Guadagni, *J. Food Sci.* **32**(1967) 405.
[179] Riken Kagaku, JP 7249707, 1972 (H. Watanabe et al.); *Chem. Abstr.* **79** (1973) 114219 q.

[180] Pillsbury Comp., US 3510310, 1970.
[181] C. Colburn, *Am. Soft Drink J.* **126** (1971) 16.
[182] R. S. Shallenberger, T. E. Acree, C. Y. Lee, *Nature (London)* **221** (1969) 556
[183] J. Solms, L. Vuataz, R. H. Egli, *Experientia* **21** (1965) 693.
[184] H. Wieser, H. Jugel, H.-D. Belitz, *Z. Lebensm. Unters. Forsch.* **164** (1977) 277.
[185] R. H. Mazur, J. M. Schlatter, A. H. Goldkamp, *J. Am. Chem. Soc.* **91** (1969) 2684
[186] G. A. Crosby, *CRC Crit. Rev. Food Sci. Nutr.* **7** (1976) 297.
[187] L. A. Pavlova et al., *Russ. Chem. Rev. (Engl. Transl.)* **50** (1981) 316.
[188] W. J. Herz, R. S. Shallenberger, *Food Res.* **25** (1960) 491.
[189] Lever Brothers, US 3922365, 1975 (K. H. Ney et al.).
[190] H. Tanaka, Y. Obata, *Agric. Biol. Chem.* **33** (1969) 147.
[191] M. Giaccio, L. Surricchio, *Quad. Merceol.* **16** (1977) 151.
[192] H. Valaize, G. Dupont, *Ind. Agric. Aliment.* **68** (1951) 245.
[193] E. L. Wick, M. DeFigueiredo, D. H. Wallace, *Cereal Chem.* **41** (1964) 300.
[194] A. A. M. El-Dash, Dissertation, Kansas State University 1969 (Univ. Microfilms Inc., Ann Arbor, Michigan, No. 69-21123).
[195] W. Baltes, *Ernähr. Umsch.* **20** (1973) 35.
[196] M. Angrick, D. Rewicki, *Chem. Unserer Zeit* **14** (1980) 149.
[197] H. E. Nurstein, *Food Chem.* **6** (1980) 263.
[198] T. A. Rohan, *Food Technol. (Chicago)* **24** (1970) 29.
[199] Maggi, DE 2246032, 1973 (R. J. Gasser). Z. Mielniczuk et al., *Acta Aliment. Pol.* **2** (1976) 213.Chas. Pfizer & Co., US 3365306, 1968 (M. A. Perret).Y.-P. C. Hsieh et al., *J. Sci. Food Agric.* **31** (1980) 943.R. A. Wilson, *J. Agric. Food Chem.* **23** (1975) 1032. P. van de Rovaart, J.-J. Wuhrmann, US 3930044, 1975. R. Schroetter, G. Woelm, *Nahrung* **24** (1980) 175.
[200] H. Kisaki, *Chem. Abstr.* **69** (1968) 9824 d. Ajinomoto, FR 2005896 (1969). G. Rubenthaler, Y. Pomeranz, K. F. Kinney, *Cereal Chem.* **40** (1963) 658. I. R. Hunter, M. K. Mayo, US 3425840, 1969. Y. H. Liau, C. C. Lee, *Cereal Chem.* **47** (1970) 404. Y.-Y. Linko, J. A. Johnson, B. S. Miller, *Cereal Chem.* **39** (1962) 468. G. L. Bertram, *Cereal Chem.* **30** (1953) 126. Research Corp., US 3268555, 1966 (L. Wiseblatt). Hoffmann-La Roche, US 3547659, 1970 (W. Cort, L. Neck). M. Rothe, *Nahrung* **24** (1980) 185. A. A. El-Dash, A. A. Johnson, *Cereal Chem.* **47** (1970) 247. L. Wiseblatt, H. F. Zoumut, *Cereal Chem.* **40** (1963) 162. M. Rothe, *Ernährungsforschung* **5** (1960) 131.
[201] G. Ziegleder, D. Sandmeier, *Dtsch. Lebensm. Rundsch.* **78** (1982) 315. W. Mohr, E. Landschreiber, Th. Severin, *Fette Seifen Anstrichm.* **78** (1976) 88. S. Turos, US 4346121, 1982.
[202] E. Cremer, M. Riedmann, *Monatsh. Chem.* **96** (1965) 364.
[203] Yuki Gosei Kogyo, US 3478015, 1969 (I. Onishi, A. Nishi, T. Kakizawa).
[204] Fuji Oil Co., GB 1357511, 1974; GB 1488282, 1977.
[205] Naarden Int., NL 7712745, 1979.
[206] J. A. Newell, M. E. Mason, R. S. Matlock, *J. Agric. Food Chem.* **15** (1967) 767.
[207] S.-C. Lee, B. R. Reddy, S. S. Chang, *J. Food Sci.* **38** (1974) 788. Research Corp., US 3814818, 1974 (S. S. Chang, B. R. Reddy). T. Y. Fan, M. H. Yueh, *J. Food Sci.* **45** (1980) 748. P. T. Arroyo, D. A. Lillard, *J. Food Sci.* **35** (1970) 769.
[208] Yuki Gosei Kogyo, DE 1593733, 1972 (I. Onishi et al.).Japan Monopoly, Tanabe Seiyaku, US 3722516, 1973 (K. Suwa et al.). Philip Morris, US 4306577, 1981 (D. L. S. Wu, J. W. Swain).Reynolds Tobacco, US 3996941, 1976 (C. W. Miller, J. P. Dickerson, C. E. Rix).
[209] R. G. Henika, N. E. Rodgers, *Cereal Chem.* **42** (1965) 397.

[210] J. M Bruemmer, W. Seibel, H. Stephan, *Getreide Mehl Brot* **34** (1980) 173. Patent Technology, US 3803326, 1974.J. Geittner, *Gordian* **79** (1979) 202.
[211] R. Marcuse, *Fette Seifen Anstrichm.* **63** (1961) 940.
[212] H. Iwainsky, C. Franzke, *Dtsch. Lebensm. Rundsch.* **52** (1956) 129. H. Lingnert, C. E. Eriksson, *J. Food Process. Preserv.* **4** (1980) 161.
[213] N. Watanabe et al., JP 7314042, 1973; *Chem. Abstr.* **79** (1973) 145052 j.
[214] A. G. Castellani, *Appl. Microbiol.* **1** (1953) 195. Nisshin Flour Milling, JP 7319945, 1973 (G. Ogawa, K. Taguchi); *Chem. Abstr.* **81** (1974) 76689 z. Nippon Kayaku, JP-Kokai 81109580, 1981; *Chem. Abstr.* **95** (1981) 202313 b.
[215] H.-L. Bertram, H. Schmidtborn, *Feed International* 1984 (May) 37.
[216] M. Kirchgeßner: *Tierernährung*, 5th ed., DLG-Verlag, Frankfurt 1982, p. 88.
[217] M. L. Scott, M. C. Nesheim, R. J. Young: *Nutrition of the Chicken*, 3rd ed., M. L. Scott, Ithaka, New York, 1982.
[218] *D,L-Methionine, The Amino Acid for Animal Nutrition*, Degussa, Frankfurt 1980.
[219] *Nutrient Requirements of Poultry*, Nat. Res. Council, no. 1, 7th ed., Washington, D.C., 1977.
[220] AEC (Alimentation Equilibrée de Commentry): *Alimentation Animale*, Doc. 4, Commentry, France, 1978.
[221] *Nutrient Requirements of Swine*, Nat. Res. Council, 1979. Agricultural Research Council, Commonwealth Agricultural, Burlaux, London 1981. P. L. M. Berende, H.-L. Bertram, *Z. Tierphysiol. Tierernähr. Futtermittelk.* **49** (1983) 30. H.-L. Bertram, P. L. M. Berende, *Kraftfutter* (1983) 46.
[222] W. Prinz, A. Becker, *Arch. Geflügelk.* **29** (1965) no. 2, 135.
[223] R. Chernoff, *J. Am. Diet. Assoc.* **79** (1981) 426. M. R. Polk, *Am. Pharm.* **NS 22** (1982) 25.
[224] M. Winitz et al., *Nature (London)* **205** (1965) 741.
[225] R. J. Russell, *Gut* **16** (1975) 68. W. F. Caspary, *Dtsch. Ärztebl.* 1978 (2. Febr.) 243. H. Kasper, *Aktuel. Ernährung* **1** (1978) 22.
[226] D. M. Matthews, S. A. Adibi, *Gastroenterology* **7 1** (1976) 151. M. T. Lis, R. F. Crampton, D. M. Matthews, *Br. J. Nutr.* **27** (1972) 159.
[227] Mead Johnson, US 3091569, 1963; US 3184505, 1965.
[228] H. A. Smith, G. Gorin, *J. Org. Chem.* **26** (1961) 820.
[229] Rech. et Propagande Scientif., FR 1288907, 1962.
[230] Degussa, US 4129593, 1978.
[231] M. Bergmann, G. Michalis, *Ber. Dtsch. Chem. Ges.* **63** (1930) 987.
[232] Degussa, DE 1239697, 1963.
[233] May & Baker, GB 845 034, 1957.
[234] B. Witkop et al., *J. Am. Chem. Soc.* **76** (1954) 5579.
[235] A. J. Morris et al., *J. Org. Chem.* **22** (1957) 306.
[236] E. A. Bell et al., *Nature (London)* **210** (1966) 529.
[237] Distillers, GB 854339, 1957. Squibb, US 3281461, 1966. Heyl & Co., DE 2114329, 1971; DE 2413185, 1974.
[238] Degussa, DE 1795299, 1968; DE 1795297, 1968; DE 2032952, 1970; DE 2123232, 1971; DE 2156601, 1971; DE 2335990, 1973; DE 2138122, 1971; DE 2258411, 1972; DE 2304055, 1973.
[239] K. Schrader, *Am. Cosmet. Perfum.* **87** (1972) 49. S. Tatsumi, *Am. Cosmet. Perfum.* **87** (1972) 61.O. K. Jacobi, *Am. Cosmet. Perfum.* **87** (1972) 35. G. Hopf, J. König, G. Padberg, *Kosmetologie* **4** (1971) 132. A. Szakall, *Arch. Klin. Exp. Dermatol.* **201** (1955) 331. K. Laden, R. Spitzer, *J. Soc. Cosmet. Chem.* **18** (1967) 351. H. W. Spier, G. Pascher, *Klin. Wochenschr.* **31** (1953) 997.

[240] Y. Kumano et al., *J. Soc. Cosmet. Chem.* **28** (1977) 285.
[241] L'Oreál, DE 2807607, 1978 (J.-C. Ser et al.). Unilever, DE 2337342, 1973 (G. F. Johnston et al.). Orlane Paris, DE 2524297, 1975 (A. Meybeck, H. Noel). Shiseido, US 4035513, 1977 (Y. Kumano).
[242] E. S. Cooperman, *Am. Cosmet. Perfum.* **87** (1972) 65. Colgate Palmolive, GB 1573529, 1980.
[243] M. Takehara et al., *J. Am. Oil Chem. Soc.* **50** (1973) 227; **51** (1974) 419.
[244] Ajinomoto, Kawaken Fine Chemicals, US 4273684, 1981 (T. Nagashima et al.)
[245] Ajinomoto, GB 1483500, 1977.
[246] Kawaken Fine Chemicals. JP-Kokai 75117806, 1975 (K. Nakazawa); *Chem. Abstr.* **84** (1976) 8858 r.
[247] G. Schuster, H. Modde, E. Scheld, *Seifen Öle Fette Wachse* **38** (1965) 477. G. Schuster, H. Modde, *Parfüm + Kosmet.* **45** (1964) 337.
[248] Estee Lauder, US 4005210, 1977 (J. Gubernick).
[249] American Cyanamid, US 3988438, 1976. Ajinomoto, DE 2010303, 1970 (R. Yoshida et al.).
[250] R. Yoshida, M. Takehara, *Chem. Abstr.* **84** (1976) 6843 h.
[251] Dominion Pharmacal., US 4176197, 1979 (B. N. Olson). L'Oréal, US 4002634, 1977 (G. Kalopissis, C. Bouillon). L'Oréal, GB 1397623, 1975 (G. Kalopissis, G. Manoussos). L'Oréal, DE 1492071, 1965 (G. Kalopissis).
[252] Mare Corp., US 4201235, 1980 (V. G. Ciavatta). Unilever, EP 8171, 1981 (G. P. Mathur et al.).
[253] Kyowa Hakko Kogyo, US 4139610, 1979 (Y. Miyazaki et al.) Hans Schwarzkopf, DE 958501, 1957 (J. Saphir, E. Kramer). K. Yoneda et al., DE 2951923, 1979.
[254] Schuylkill Chem., GB 1516890, 1978.
[255] Procter & Gamble, EP 47650, 1982. Unilever, GB 1597498, 1981 (K. Gosling, M. R. Hyde).
[256] K. S. K. Prasad, K. G. H. Setty, H. C. Govindu, *Indian J. Plant Prot.* **5** (1977) 153.
[257] Roussel-Uclaf, FR 2405650, 1979 (A. Boudet).
[258] T. Kato, A. Iio, JP-Kokai 7629229, 1976; *Chem. Abstr.* **85** (1976) 88519 q.
[259] A. Bouniols, J. Margara, *C. R. Hebd. Séances Acad. Sci. Ser. D* **273** (1971) 1193.
[260] Ciba Geigy, EP 18948, 1980.
[261] Emori Shoji, JP 78097022, 1978.
[262] Dainichi Seika Kogy, JP 78079990, 1978.
[263] L. Mueller, *Zem. Kalk Gips* **27** (1974) 69.
[264] Ajinomoto, DE 2533136, 1975 (R. Yoshida et al.).
[265] Kyowa Fermentation, GB 1400741, 1975. Honny Chemicals, GB 1402758, 1975 (Y. Nakagoshi).Kyowa Hakko Kogyo, DE 2229488, 1972 (Y. Fujimoto et al.).Toyo Cloth, US 3676206, 1972 (K. Nishitani et al.).
[266] Sanyo Trading Co., DE 2716758, 1977 (M. Onizawa, S. Ohmiya).
[267] Ajinomoto, JP 81035179, 1981. Ajinomoto, JP-Kokai 7522801, 1975 (T. Saito et al.); *Chem. Abstr.* **84** (1976) 33494 b.
[268] Sanyo Trading Co., DE 2602988, 1976; US 4069213, 1978; DE 2604053, 1976 (M. Onizawa); DE 2658693, 1976 (M. Onizawa, S. Ohmiya).
[269] R. M. Saleh, A. M. Shams El Din, *Corros. Sci.* **12** (1972) 688.
[270] Ajinomoto, JP-Kokai 7426145, 1974 (Y. Kita et al.); *Chem. Abstr.* **81** (1974) 53195 w.
[271] Ilford, GB 1378354, 1974 (A. D. Ezekiel). Ilford, DE 2316632, 1973 (R. Jefferson).
[272] Fr. Blasberg & Co., DE 2050870, 1973 (W. Immel, W. Adams).Yokozawa Chemical Ind., JP 7410572, 1974 (K. Aoya et al.); *Chem. Abstr.* **81** (1974)57631 h. Sony Corp., DE 2325109, 1973 (S. Fueki et al.).
[273] A. Steponavicius, R. Visomirskis, *Electrodepos. Surface Treat. Lausanne* **1** (1972) 37.

[274] F. S. Parker, D. M. Kirschenbaum, *Spectrochim. Acta* **16** (1960) 910.M. Tsuboi, T. Takenishi, A. Nakamura, *Spectrochim. Acta* **19** (1963) 271. J. F. Pearson, M. A. Slifkin, *Spectrochim. Acta Part A* **28** (1972) 2403.

[275] C. S. Tsai et al., *Can. J. Biochem.* **53** (1975) no. 9, 1005.

[276] H. Hagenmeyer et al., *Z. Naturforsch. B: Anorg. Chem. Org. Chem.* **25 B** (1970), 681.

[277] E. Scoffone, A. Fontana, *Mol. Biol. Biochem. Biophys.* **8** (1970) 185.

[278] M. Roth, *Anal. Chem.* **43** (1971) 880. J. R. Benson, P. E. Hare, *Proc. Natl. Acad. Sci. USA* **72** (1975) 619.

[279] E. Bayer et al., *Anal. Chem.* **48** (1976) 1106.

[280] S. Udenfried et al., *Science (Washington, D.C.)* **178** (1972) 871.

[281] C. Haworth, R. W. A. Oliver, *J. Chromatogr.* **64** (1972) 305. E. Stahl: *Dünnschichtchromatographie*, Springer Verlag, Berlin – Heidelberg – New York 1967, p. 701. A. R. Fahmy et al., *Helv. Chim. Acta* **44** (1959) 245.

[282] M. Kubota et al., *Anal. Biochem.* **64** (1975) no. 2, 494. K. D. Kulbe, *Anal. Biochem.* **44** (1970) 548. P. A. Laursen, *Biochem. Biophys. Res. Commun.* **37** (1969) 663.

[283] A. Wolf, *Prax. Naturwiss. Chem.* **23** (1974) no. 3, 74.

[284] H. Nakamura, J. J. Pisano, *J. Chromatogr.* **121** (1976) 33.

[285] K. Macek, Z. Deyl, M. Smr, *J. Chromatogr.* **193** (1980) 421.

[286] P. B. Hamilton, *Anal. Chem.* **35** (1963) 2055.

[287] S. Blackburn: *Amino Acid Determination,* Marcel Dekker, New York – Basel 1978.

[288] D. H. Spackman, W. H. Stein, S. Moore, *Anal. Chem.* **30** (1958) 1190.

[289] S. Moore, W. H. Stein, *J. Biol. Chem.* **211** (1954) 907.

[290] H.-M. Lee et al., *Anal. Biochem.* **96** (1979) 298.

[291] E. Bayer et al., *Anal. Chem.* **48** (1976) 1106. J. M. Wilkinson, *J. Chromatogr. Sci.* **16** (1978) 547. K.-T. Hsu, B. L. Currie, *J. Chromatogr.* **166** (1978) 555. T. Jamabe, N. Takei, H. Nakamura, *J. Chromatogr.* **104** (1975) 359. A. Khayat, P. K. Redenz, L. A. Gorman, *Food Technol. (Chicago)* **36** (1982) 46.

[292] A. P. Graffeo, A. Haag, B. L. Karger, *Anal. Lett.* **6** (1973) no. 6, 505. J. K. De Vries, R. Frank, C. Birr, *FEBS Lett.* **55** (1975) no. 1, 65. P. Frankhauser et al., *Helv. Chim. Acta* **57** (1974) 271. C. C. Zimmermann, E. Appella, J. J. Pisano, *Anal. Biochem.* **77** (1977) 569. G. Frank, W. Strubert, *Chromatographia* **6** (1973) no. 12, 522.

[293] H. Beyer, U. Schenk, *J. Chromatogr.* **89** (1969) 483.

[294] T. A. Kan, W. F. Shipe, *J. Food Sci.* **47** (1981) 338. H. Umagat, P. Kucera, L. F. Wen, *J. Chromatogr.* **239** (1982) 463.

[295] W. Voelter, K. Zech, *J. Chromatogr.* **112** (1975) 643.

[296] R. Schuster, *Anal. Chem.* **52** (1980) 617.

[297] A. Darbre, *Biochem. Soc. Trans.* **2** (1974) 70. B. M. Nair, *J. Agric. Food Chem.* **25** (1977) 614.

[298] J. J. Pisano, T. J. Bronzert, H. B. Brewer, *Anal. Biochem.* **45** (1972) 43.

[299] A. Darbre, A. Islam, *Biochem. J.* **106** (1968) 923.

[300] C. W. Gehrke, R. W. Zumwalt, K. Kuo, *J. Agric. Food Chem.* **19** (1971) 605.

[301] K. Ruhlmann, W. Giesecke, *Angew. Chem.* **73** (1961) 113.

[302] D. L. Stalling, C. W. Gehrke, R. W. Zumwalt, *Biochem. Biophys. Res. Commun.* **31** (1968) 4.

[303] W. Grassmann, K. Hannig, *Houben-Weyl,* **1/1,** 708.

[304] I. Abe, S. Musha, *J. Chromatogr.* **200** (1980) 195. G. J. Nicholson, H. Frank, E. Bayer, *HRC CC J. High Resolut. Chromatogr. Chromatogr. Commun.* **2** (1979) 411. W. A. König, G. J. Nicholson, *Anal. Chem.* **47** (1975) 951.

[305] W. Lindner, *Chimia* **35** (1981) 294. V. A. Davankov et al., *Chromatographia* **13** (1980) no. 11, 677.
[306] P. E. Hare, E. Gil-Av, *Science (Washington, D.C.)* **204** (1979) 1226.
[307] L. M. Henderson, E. E. Snell, *J. Biol. Chem.* **172** (1947) 15. I. Grote, *Mühle Mischfuttertech.* **116** (1979) 465.
[308] H. Itoh, T. Morimoto, I. Chibata, *Anal. Biochem.* **60** (1974) 573.
[309] Ch. Calvot et al., *FEBS Lett.* **59** (1975) no. 2, 258. S. J. Updike, G. P. Hicks, *Nature (London)* **214** (1967) 986.
[310] Ajinomoto, FR 2421380, 1979.
[311] *Europäische Pharmacopoe*, vol. **1–3**, Deutscher Apotheker-Verlag, Stuttgart, Govi-Verlag, Frankfurt 1974, vol. **1**, p. 104.
[312] W. Seaman, E. Allen, *Anal. Chem.* **23** (1951) no. 4, 592. H. P. Deppeler, G. Witthans, *Fresenius Z. Anal. Chem.* **305** (1981) 273.
[313] *United States Pharmacopeia* USP XX, Convention, Rockville, Md., 1979.
[314] *Food Chemicals Codex*, 3rd ed., National Academy Press, Washington, D.C., 1981.
[315] T. Akashi: "Amino Acid Production and Use to Improve Nutrition of Foods and Feeds," *Chemrawn II Conference* Manila 1982.
[316] *Chem. Eng. News* **61** (1983) Jan. 3, 18.
[317] H. N. Munro, *Adv. Exp. Med. Biol.* **105** (1978) 119.
[318] A. E. Harper, N. J. Benevenga, R. M. Wohlhueter, *Physiol. Rev.* **50** (1970) 428.
[319] R. G. Daniel, H. A. Waisman, *Growth* **32** (1968) 255.
[320] H. E. Sauberlich, *J. Nutr.* **75** (1961) 61.
[321] K. Lang: *Biochemie der Ernährung*, 4th ed.,Steinkopff, Darmstadt 1979.
[322] Y. Peng et al., *J. Nutr.* **103** (1973) 608.
[323] Degussa, unpublished, 1973 and 1983. R. J. Breglia, C. O. Ward, C. I. Jarowski, *J. Pharm. Sci.* **62** (1973) 49. P. Gullino et al., *Arch. Biochem. Biophys.* **58** (1955) 253. O. Strubelt, C.-P. Siegers, A. Schütt, *Arch. Toxicol.* **33** (1974) 55. W. Braun, *Strahlentherapie* **108** (1959) 262. H. Gutbrod et al., *Acta Hepatol.* **5** (1957) no. 1/2, 1. I. Petersone et al., *Eksp. Klin. Farmakoter.* **3** (1972) 5. D. G. Gallo, A. L. Sheffner, DE 2018599, 1971. Transbronchin, Homburg Pharma, Frankfurt 1971. L. Bonanomi, A. Gazzaniga, *Therapiewoche* **30** (1980) 1926. W. F. Riker, H. Gold, *J. Am. Pharm. Assoc.* **31** (1942) 306.
[324] G. Maffii, G. Schott, M. G. Serralunga, *Res. Prog. Org. Biol. Med. Chem.* **2** (1970) 262.
[325] Y. Kawaguchi et al., *Iyakuhin Kenkyu* **11** (1980) 635. Kaken Chemical, JP 52083940, 1976 (S. Suzuki). P. Gullino et al., *Arch. Biochem. Biophys.* **64** (1956) 319. Degussa, unpublished, 1981 and 1982. E.-J. Kirnberger et al., *Arzneim. Forsch.* **8** (1958) 72.

Aminophenols

STEPHEN C. MITCHELL, University of Birmingham, Birmingham, United Kingdom
ROSEMARY H. WARING, University of Birmingham, Birmingham, United Kingdom

1.	Aminophenols	601	1.6.	Environmental Considerations	609
1.1.	Physical Properties	602	1.7.	Uses	610
1.2.	Chemical Properties	603	1.8.	Toxicology	610
1.3.	Production	606	2.	Aminophenol Derivatives	611
1.3.1.	Reduction	606	2.1.	Derivatives of 2-Aminophenol	611
1.3.2.	Substitution	608			
1.3.3.	Purification	608	2.2.	Derivatives of 3-Aminophenol	617
1.4.	Storage and Handling	609	2.3.	Derivatives of 4-Aminophenol	618
1.5.	Analysis	609	3.	References	622

1. Aminophenols

Aminophenols and their derivatives are of steadily increasing commercial importance, both in their own right and as intermediates in the chemical and dye industries. They are amphoteric and can behave either as weak acids or weak bases, but the basic character usually pre-dominates. 2-Aminophenol (**1**) and 4-aminophenol (**3**) are oxidized easily. This is the basis for their main applications, for example, as photographic developers. In contrast, 3-aminophenol (**2**) is fairly stable in air and is not oxidized rapidly.

Table 1. Solubility* of aminophenols in common solvents

	2-Aminophenol	3-Aminophenol	4-Aminophenol
Acetone	3	3	2
Acetonitrile	3	3	2
Benzene	1	1	0
Chloroform	1	1	0
Diethyl ether	2	2	1
Dimethylsulfoxide	3	3	3
Ethanol	3	3	1
Ethyl acetate	3	3	2
Toluene	1	1	1
Water			
Cold	1	2	1
Hot	2	3	2

* 0, insoluble; 1, slightly soluble; 2, soluble; 3, very soluble

Table 2. Spectral characteristics of the aminophenol isomers

	2-Aminophenol	3-Aminophenol	4-Aminophenol
IR[a], cm^{-1}	3380, 3300, 1600, 1510, 1470,	3370, 3310, 1600, 1470, 1390,	3050–2580, 1500, 1470, 1240,
	1270, 900, 740	1260, 1180, 910	970, 830, 750
UV[b], nm	233, 285 (methanol)	287 (methanol)	234, 301 (methanol)
	229, 281 (water)	270 (0.1 M HCl)	229, 294 (water)
	235, 288 (cyclohex.)	234, 284 (cyclohex.)	235, 304 (cyclohex.)
NMR[c], ppm	6.9–7.5, 8.6 (TFA)	4.7, 6.0, 6.1, 6.8, 8.8 (DMSO)	7.1, 7.4, 8.7 (TFA)

[a] Only IR absorption bands reported as very strong are included (accuracy: ± 10 cm^{-1}) [4], [5].
[b] UV spectra: cyclohex. = cyclohexane containing 2 vol% diethyl ether [6]–[8].
[c] Proton chemical shift spectra over the range of 0–15 ppm (± 0.1 ppm); TFA, trifluoroacetic acid; DMSO, dimethylsulfoxide. When complex spectra caused by second-order effects or overlapping resonances were encountered the range was recorded [10], [11].

1.1. Physical Properties

The simple aminophenols exist in three isomeric forms depending on the relative positions of the hydroxyl and amino groups around the benzene ring. At room temperature they are solid crystalline compounds. The commercial-grade compounds are usually impure because of contamination with oxidation products and may take on a yellow-brown or pink-purple hue, the 2- and 4-aminophenols being more susceptible to this phenomenon than the 3-isomer. The solubilities of these compounds in common solvents and their spectral characteristics are given in Tables 1 and 2, respectively [3]–[12].

2-Aminophenol (**1**) [95-55-6], 2-hydroxyaniline, 2-amino-1-hydroxybenzene, C.I. Oxidation Base 17, C.I. 76520, C_6H_7NO, M_r 109.13, forms white orthorhombic bipyramidal needles when crystallized from water or benzene, mp 174 °C. The crystals have eight molecules to the ele-mentary cell and a density of 1.328 g/cm³ (1.29 also quoted) [13]–[15]. At reduced pressure (1.47 kPa), 2-aminophenol sublimes rapidly at 153 °C without decomposition [16]. Acid–base dissociation constants are pK_1 4.72 (water at 21 °C), 4.66 (1 vol% ethyl alcohol in water at 25 °C); pK_2 9.66 (water at 15 °C), 9.71 (water at 22 °C) [17], [18]. Salts: hydrochloride [51-19-4], needles, mp 207 °C; formate, mp 120 °C; oxalate, mp 167.5°C (decomp.); acetate, mp 150 °C.

3-Aminophenol (**2**) [591-27-5], 3-hydroxyaniline, 3-amino-1-hydroxybenzene, C_6H_7NO, M_r 109.13, forms white prisms when crystallized from water or toluene, mp 122–123 °C. The orthorhombic crystals have a tetramolecular unit and a density of 1.195 g/cm^3 (1.206 and 1.269 also quoted) [14], [15]. 3-Aminophenol boils at 164 °C at 1.47 kPa with slight decomposition [16]. Acid–base dissociation constants are: pK_1 4.17 (water at 21 °C), 4.31 (1 vol% aqueous ethyl alcohol at 25 °C); pK_2 9.87 (water at 22 °C) [17]. Salts: hydrochloride [51-81-0], prisms, mp 229 °C; hydrobromide, prisms, mp 224 °C; hydroiodide, prisms, mp 209 °C; sulfate [66671-80-5], plates or needles, mp 152 °C; oxalate, mp 275 °C.

4-Aminophenol (**3**) [123-30-8], 4-hydroxyaniline, 4-amino-1-hydroxybenzene, C_6H_7NO, M_r 109.13, forms white plates when crystallized from water, mp 189–190 °C (decomp.). The crystals exist in two forms. The α form (from alcohol, water, or ethyl acetate) is the more stable and has an orthorhombic pyramidal structure containing four molecules per unit cell. It has a density of 1.290 g/cm^3 (1.305 also quoted). The less stable β form (from acetone) exists as acicular crystals that turn into the α form on standing; they are orthorhombic bipyramidal or pyramidal and have a hexamolecular unit [14], [15], [19]. At reduced pressure (40 Pa) 4-aminophenol sublimes at 110 °C with slight decomposition. Boiling points quoted at various pressures are as follows: 284 °C (101.3 kPa), 174 °C (1.47 kPa), 167 °C (1.07 kPa), 150 °C (0.4 kPa), 130.2 °C (0.04 kPa), decomposition usually occurring. Acid–base dissociation constants are: pK_1 5.50 (water at 21 °C), 4.86 (water at 30 °C), 5.48 (1 vol% aqueous ethyl alcohol at 25 °C); pK_2 10.30 (water at 22 °C), 10.60 (water at 30 °C) [17], [20]. Salts: hydrochloride [51-78-5], prisms, mp 306 °C (decomp.); hydrosulfate [15658-52-3], needles, mp 272 °C; oxalate, mp 183 °C; acetate [13871-68-6], mp 183 °C; chloroacetate, needles, mp 148 °C; trichloroacetate, needles, mp 166 °C.

1.2. Chemical Properties

The chemical properties and reactions of the aminophenols are to be found in detail in many standard chemical texts and only a summary is given here [21].

The acidity of the phenols is depressed by the presence of an amino group on the benzene ring; this phenomenon is most pronounced with 4-aminophenol. They also behave as weak bases, giving salts with both mineral and organic acids. The aminophenols are true ampholytes, with no zwitterion structure; hence they exist either as neutral molecules (**4**) or as ammonium cations (**5**) or phenolate anions (**6**), depending on the pH value of the solution.

$$\underset{5}{\text{HO-C}_6\text{H}_4\text{-NH}_3^+} \underset{H^+}{\rightleftharpoons} \underset{4}{\text{HO-C}_6\text{H}_4\text{-NH}_2} \underset{H^+}{\rightleftharpoons} \underset{6}{^-\text{O-C}_6\text{H}_4\text{-NH}_2}$$

However, deviations from theoretical curves observed during acid–base titrations have led to postulation of the existence of half-salt complex cations B_2^+, formed by the association of an ammonium cation, B^+, with a neutral molecule, B. This association phenomenon is most apparent with 4-aminophenol but is also displayed by the other isomers [20].

The aminophenols are chemically reactive, undergoing reactions involving both the aromatic amino group and the phenolic hydroxyl moiety, as well as substitution on the benzene ring. Oxidation leads to the formation of highly colored polymeric quinoid structures. 2-Aminophenol undergoes a variety of cyclization reactions.

Alkylation. All of the possible mono-, di-, and trimethylated aminophenols are known. *N*-Monoalkylation occurs when the aminophenol is heated with the appropriate alkyl halide or with an alcohol and Raney nickel; equal or even better results can be achieved using aldehydes or ketones in place of the alcohol. Specific alkylation of the hydroxyl group to form methoxyanilines (anisidines) or ethoxyanilines (phenetidines) is difficult because of the reactivity of the amino group; mixed alkylated products usually are obtained. 3-Methoxyaniline may be prepared by methylation of 3-aminophenol under alkaline conditions, but it is more usual to protect the amino group and to methylate 3-acetylaminophenol, followed by hydrolysis. The other anisidines and phenetidines are prepared indirectly by reduction of the nitro analog:

$$\text{Ar-NO}_2\ (o,p\text{-OR}) \xrightarrow{\text{red.}} \text{Ar-NH}_2\ (o,p\text{-OR})$$

where R is alkyl

Acylation. Acylation of the aminophenol (using acetic anhydride in alkali or pyridine, acetyl chloride and pyridine in toluene, or ketene in ethanol) usually leads to *N*-acylated products. If an excess of reagent is used, however, especially with 2-aminophenol, *O,N*-diacylated products will be formed. Aminophenyl carboxylates (*O*-acylated aminophenols) normally are prepared by the reduction of the corresponding nitrophenyl carboxylates, which is of particular importance with the 4-aminophenol derivatives. A migration of the acyl group from the *O* to the *N* position is known to occur for both 2- and 4-aminophenol acylated products. 2-Aminophenyl ethyl carbonate (**7**) has been shown to rearrange slowly in dilute acid to ethyl 2-hydroxyphenyl carbamate (**8**); the corresponding 4 derivative does not undergo this particular reaction [22].

[Structures 7 and 8: 7 is a benzene ring with OCO₂C₂H₅ and NH₂ substituents; 8 is a benzene ring with OH and NHCO₂C₂H₅ substituents]

Diazonium Salt Formation. The aromatic amino group of aminophenols can be converted to the diazonium salt using sodium nitrite in aqueous acid, although difficulties may be encountered when the aminophenol is oxidized easily or of low solubility. Crystalline diazonium salts have been isolated using the hydrochloride or sulfate of the appropriate aminophenol under anhydrous conditions. Such diazo derivatives find extensive use in the dye industry [23], [24].

Cyclization Reactions. Because of the close proximity of the amino and hydroxyl groups in 2-aminophenol, this isomer is particularly susceptible to cyclization and condensation reactions. Oxidation by iron(III) chloride, enzymes, or light or autoxidation on silica thin-layer plates gives 2-aminophenoxazin-3-one (**9**). Further oxidation with iron(III) cyanide or heating with potassium hydroxide in ethanol gives a five-ring structure, triphenoxdioxazine (benzoxazinophenoxazine) (**10**). 2-Aminophenol and its derivatives are useful starting materials for the synthesis of phenoxazones, phenoxazines, benzoxazoles, and thiobenzoxazoles. Most of these condensation reactions involve heating at 200–300 °C with a suitable catalyst [21].

[Structures 9 (2-aminophenoxazin-3-one) and 10 (triphenoxdioxazine)]

Condensation Reactions. Substituted diphenylamines or diphenyl ethers are obtained from aminophenols and suitable reactants by elimination of ammonia or hydrogen chloride:

$$\text{HO-C}_6\text{H}_4\text{-NH}_2 + \text{H}_2\text{N-C}_6\text{H}_5 \rightarrow \text{HO-C}_6\text{H}_4\text{-NH-C}_6\text{H}_5 + \text{NH}_3$$

$$\downarrow \text{NH}_3$$

$$\text{H}_2\text{N-C}_6\text{H}_4\text{-NH-C}_6\text{H}_5$$

$$\text{O}_2\text{N-C}_6\text{H}_4\text{-Cl} + \text{HO-C}_6\text{H}_4\text{-NH}_2 \rightarrow \text{O}_2\text{N-C}_6\text{H}_4\text{-O-C}_6\text{H}_4\text{-NH}_2$$

$$\downarrow \text{red.}$$

$$\text{H}_2\text{N-C}_6\text{H}_4\text{-O-C}_6\text{H}_4\text{-NH}_2$$

Reactions of the Benzene Ring. Both the amino and hydroxyl groups are electron-donating moieties and many substituted derivatives are known. The controlled interaction of aminophenols with chlorine or bromine in glacial acetic acid can give a variety of mono-, di-, tri-, or tetra-halogenated products. The use of concentrated sulfuric acid or oleum, with or without heat, gives aromatic sulfonic acids. The sulfonic acid group enters the ortho or para position relative to the hydroxyl group. Further treatment with oleum leads to the formation of disulfonated compounds. The carboxylation of *m*-aminophenol leads to the formation of *p*-aminosalicylic acid.

1.3. Production

Aminophenols are made either by reduction of nitrophenols or by substitution. Reduction is accomplished with iron or with hydrogen in the presence of a catalyst. The last mentioned is the method of choice today for the production of 2- and 4-aminophenol.

1.3.1. Reduction

Iron Reduction [25]. The reduction of nitrophenols with iron turnings takes place in weakly acidic solution or suspension. Before the iron – iron oxide sludge is separated from the solution [26] the product aminophenol must be made water soluble by adding sodium hydroxide. The resulting sodium aminophenolate is very susceptible to oxidation in aqueous solution; various methods are recommended for its purification (see Section 1.3.3). Subsequent to this, the aminophenols are precipitated from acidic solution by neutralization with base and in the absence of air; reducing agents usually must be added during this procedure.

When 2-nitrophenol is reduced with iron, red insoluble color lakes are formed as byproducts that decrease the yield. Therefore, the iron reduction of 2-nitrophenol is of minor industrial importance today.

Catalytic Reduction [27] – [29]. Catalytic reduction usually takes place in solution, emulsion, or suspension in autoclaves or pressurized vessels; after the catalyst is added, the vessel is pressurized with hydrogen. Water and methanol are the preferred solvents; in water the addition of alkali hydroxide [30] – [33], alkali carbonate [30], [33], or acid [33], [34] has been recommended. Nickel [35] – preferably Raney nickel [31], [32], [36] – or supported precious metals [30], [37], such as platinum or palladium [33], [38] on activated carbon, or the oxides of these metals [34], [39] are used as catalysts. The catalyst life can be extended, the catalyst consumption decreased, and the product quality enhanced by adding organic solvents that are not miscible with water [31], [33]. The preferred hydrogen pressure is 2 MPa; the hydrogenation can can also be performed at atmospheric pressure or at higher pressure up to 6 MPa. The reaction temperature does not exceed 100 – 110 °C.

Reduction of Nitrobenzene. When reducing nitrobenzene in acidic medium [40], the intermediate phenylhydroxylamine [41] rearranges to 2- and 4-aminophenol before it is further reduced to aniline. The main product of this reaction is 4-aminophenol; byproducts are 2-aminophenol, aniline, and 4,4′-diaminodiphenyl ether:

In the past, metals in dilute sulfuric acid were used as reducing agents [42]. Today, the reducing agent is hydrogen in the presence of preciousmetal catalysts [41], [43], e.g., palladium or platinum. Other catalysts have been suggested: molybdenum and platinum sulfide [44], [45], and a platinum–ruthenium mixed catalyst [46]. Either the catalysts are used as their oxides, or they are supported on activated carbon.

Dilute aqueous mineral acid is used as reaction medium, for example, dilute sulfuric acid; acidic salts also can be added to the reduction medium [44].

In a two-step process, nitrobenzene first is selectively reduced to phenylhydroxylamine with hydrogen in the presence of Raney copper and in an organic solvent, such as 2-propanol [41]. The product rearranges to 4-aminophenol after addition of dilute sulfuric acid [47].

The addition of *wetting agents* increases the aminophenol yield [48]; these agents must be water soluble and stable in the presence of sulfuric acid. Quaternary ammonium salts that contain at least one alkyl group with at least ten carbon atoms are suitable, e.g., dodecyltrimethylammonium chloride. The reaction usually is performed below 100 °C, either at atmospheric or at higher pressure. Hydrogen is added during the reaction as consumed. The addition of inert organic solvents further increases the yield of 4-aminophenol and the product quality [49].

In another variant, only 88 % of the nitrobenzene is reduced [50]; after that, the reaction mixture consists of two phases with the precious-metal catalyst (palladium on activated carbon) remaining in the unreacted nitrobenzene phase. Therefore, phase separation is sufficient as workup, and the nitrobenzene phase can be recycled directly to the next batch. The aqueous sulfuric acid phase contains 4-aminophenol and byproduct aniline. After neutralization, the aniline is stripped, and 4-aminophenol is obtained by crystallization after the aqueous phase is purified with activated carbon.

Electrolytic reduction also is possible [42], [51]–[53]; this method causes less concern over pollution than metal–acid reduction systems, but it has not yet found industrial application. Electrolysis of nitrobenzene or phenylhydroxylamine in the presence of sulfuric acid or of azoxybenzene in acid solution yields specifically 4-aminophenol [54].

1.3.2. Substitution

Substitution of various groups by amino or hydroxyl groups is industrially unimportant for the production of 2- and 4-aminophenol [55], but this type of reaction is used for the synthesis of 2- or 4-aminophenol derivatives.

However, 3-aminophenol cannot be obtained easily by reduction. It is made mainly by the reaction of 3-aminobenzene sulfonic acid with sodium hydroxide or by the reaction of resorcinol with ammonia. Substitution of the sulfonic acid group in 3-aminobenzene sulfonic acid is accomplished by caustic soda fusion (5–6 h; 240–245 °C). The product is purified by vacuum distillation [56].

Alternatively, resorcinol reacts with ammonia [57], for example, in the presence of diammonium phosphate and arsenic pentoxide [58] or ammonium sulfite to form 3-aminophenol. The compound also may be made by hydrolysis of 3-aminoaniline [59], [60].

1.3.3. Purification

Generally, aminophenols can be purified by sublimation at reduced pressure and higher temperature. 3-Aminophenol may be purified by vacuum distillation; in order to obtain a colorless product sulfur dioxide is added during distillation [61] or the distillate is collected under a blanket of an unreactive liquid of lower density, such as water [62].

Another method for purifying aminophenols is the treatment of their aqueous solutions with activated carbon [63], [64]. During this treatment, sodium sulfite, sodium dithionite, or disodium ethylenediaminotetraacetate [64] is added to increase the quality and stability of the products and to chelate heavy-metal ions that would catalyze oxidation. Addition of sodium dithionite, hydrazine [64], or sodium hydrosulfite [65] also is recommended during precipitation or crystallization of aminophenols.

Contaminants, which are usually present in the 4-aminophenol made by catalytic reduction, can be reduced or even removed completely by a variety of procedures: treatment with 2-propanol [66]; with aliphatic, cycloaliphatic or aromatic ketones [67]; with aromatic amines [68]; with toluene or low molecular mass alkyl acetates [69]; with phosphoric acid, hydroxyacetic acid, hydroxypropionic acid, or citric acid [70]; or by extraction with methylene chloride, chloroform [71], or nitrobenzene [72].

1.4. Storage and Handling

Under atmospheric conditions, 3-aminophenol is the most stable of the three isomers. Both 2- and 4-aminophenol are unstable; they darken on exposure to air and light and should be stored in brown glass containers, preferably in an atmosphere of nitrogen. The use of activated iron oxide in a separate cellophane bag inside the storage container inhibits the discoloration of 4-aminophenol [73]. The salts, especially the hydrochlorides, are more resistant to oxidation and should be used where possible.

1.5. Analysis

Aminophenols have been detected in waste water by investigating UV absorptions at 220, 254, and 275 nm [74] and also in parts per thousand by using potentiometric titrations in a two-phase chloroform–water medium [75]. More specifically, 2-aminophenol can be identified using an iron(II) sulfate–hydrogen peroxide reagent [76]. 4-Aminophenol can be analyzed by voltammetry [77]. A colorimetric method using 3-cyano-N-methoxypyridinium perchlorate as reagent detects 4-aminophenol in the presence of N-acetyl-4-aminophenol [78]. 3-Aminophenol has been analyzed colorimetrically by oxidation in base and subsequent extraction of a violet quinoneimide dye [79].

Chromatographic methods for the separation and identification of aminophenols also have been described. Thin layer chromatography provides a rapid and convenient method of separating the isomers from many derivatives, and subsequent spraying with a variety of chromogenic reagents gives additional information [80], [81]. Several gas–liquid [82] and high-pressure liquid [83], [84] chromatographic separation techniques have also been reported.

1.6. Environmental Considerations

Because the aminophenols are oxidized easily they tend to remove oxygen from solutions. Hence, if they are released from industrial waste waters into streams and rivers they will deplete the capacity of these environments to sustain aquatic life. Biologic degradation of 3-aminophenol has been carried out in aeration tanks using adapted microflora from active sludge; the process took 4 months [85]. An enzymatic method for removing aminophenols from waste waters has been described using horseradish peroxidase to crosslink and precipitate the compounds as insoluble polymers [86].

1.7. Uses

Both 2- and 4-aminophenols are strong reducing agents and are employed as photographic developers under the trade names of Atomal and Ortol (2-aminophenol); Activol, Azol, Certinal, Citol, Paranol, Rodinal, Unal, and Ursol P (4-aminophenol); they may be used alone or in combination with hydroquinone. The oxalate salt of 4-aminophenol is marketed under the name of Kodelon.

The aminophenols are versatile intermediates and are employed in the synthesis of virtually every class of stain and dye. In addition, 2-aminophenol (Oxidation Base 17; C.I. 76520) is specifically used for shading leather, fur, and hair from grays and browns to yellowish brown. 3-Aminophenol has found application as a hair colorant and as a coupler molecule in hair dyes [87], [88]. 4-Aminophenol is used as an intermediate in the synthesis of pharmaceuticals, as a wood stain imparting a roselike color to timber [89], and as a dyeing agent for fur and feathers.

As a result of the close proximity of the amino and hydroxyl groups on the benzene ring and their ease of condensation with suitable reagents, 2-aminophenol is a principal intermediate in the synthesis of such heterocyclic systems as oxyquinolines, phenoxamines, and benzoxazoles. The last-named compounds have been used as inflammation inhibitors [90]. 3-Aminophenol has found use as a stabilizer of chlorine-containing thermoplastics [91], although its major use is as an intermediate in the production of 4-amino-2-hydroxybenzoic acid, a tuberculostat. Similarly, nitrogen-substituted 4-aminophenols have long been known as antipyretics and analgesics, and the production of these derivatives is a major use of 4-aminophenol.

1.8. Toxicology

In general, aminophenols are irritants. Their toxic hazard rating is slight to moderate, but repeated contamination may cause general itching, skin sensitization, dermatitis, and allergic reactions [92]. Immunogenic conjugates are spon-taneously produced on exposure to 2- and 4-aminophenol [93]. Methemoglobin formation with subsequent cyanosis is another possible complication. Inhalation of 4-aminophenol may precipitate this event and also cause bronchial asthma. 3-Aminophenol is less hazardous than the others. The sulfonated derivatives are less irritating than the unsulfonated compounds.

4-Aminophenol is nephrotoxic and stongly inhibits proximal tubular function [94]. Respiration, oxidative phosphorylation, and ATPase activity are inhibited in rat kidney mitochondria [95]. According to recent work 2- and 4-aminophenol, but not 3-aminophenol, are teratogenic in the hamster [96]. 4-Aminophenol is known to inhibit DNA synthesis and alter DNA structure in human lymphoblasts. The aminophenols also have been shown to be genotoxic, as evidenced by the induction of sister chromatid exchanges [97], [98].

Obviously, care should be taken in handling these compounds; prolonged exposure should be avoided. Contaminated clothing should be removed immediately and the affected area washed thoroughly with running water for at least 10 min.

2. Aminophenol Derivatives

The derivatives of the aminophenols have important uses in both the photographic and the pharmaceutical industries. They are also exten-sively applied as precursors and intermediates in the synthesis of more complicated molecules, especially those used in the staining and dye industry. All of the major classes of dyes have representatives that incorporate substituted aminophenols. The varying degrees of ease with which the aminophenols can be diazotized and coupled has led to their use in the manufacture of azo compounds. The sulfonated aminophenols also are employed in this context; these compounds, however, are treated elsewhere (→ Benzenesulfonic Acids and Their Derivatives). Details concerning which aminophenol derivatives and sulfonated compounds are produced commercially as dye intermediates have been reviewed and are given in the publications listed in the reference section [99]. The more commonly encountered simple derivatives of the aminophenols can be found in the review articles cited in the references and also in standard organic chemistry texts [100]. A few examples, which have specific uses or are manufactured in large quantities, have been selected for detailed discussion.

2.1. Derivatives of 2-Aminophenol

2-(N-Methylamino)phenol (11) [611-24-5], C_7H_9NO, M_r 123.15, mp 99 °C, forms colorless plates; the aqueous hydrochloric acid solution turns reddish brown when $FeCl_3$ solution is added.

Compound **11** is synthesized by reaction of 2-aminophenol with phosgene, alkylation of the intermediate benzoxazolone, and cleavage of N-methylbenzoxazolone with hydrochloric acid [101]. Alternatively, 2-N-acetyl-N-methylaminophenol is heated at 140 °C in the presence of hydrochloric acid [102]:

2-Methoxyaniline (**12**) [*90-04-0*], *o*-anisidine, C_7H_9NO, M_r 123.15, *mp* 6.22 °C, *bp* 224 °C (102.1 kPa), *bp* 105–106 °C (1.85 kPa), d_4^{20} 1.0923, $d_4^{98.7}$ 1.0263, n_D^{20} 1.5738, dynamic viscosity η 2.211 mPa · s (55 °C), η 1.051 mPa · s (98.7 °C), dipole moment 1.62 D (vapor), is slightly soluble in water (1.38 wt% at 25 °C).

The compound is produced from 2-methoxynitrobenzene by reduction with iron [103] or by hydrogenation in the presence of precious-metal catalysts [104] (platinum or palladium on activated carbon; solvent: toluene) or of Raney nickel [105] (solvent: methanol or 2-propanol).

2-Ethoxyaniline (**13**) [*94-70-2*], *o*-phenetidine, $C_8H_{11}NO$, M_r 137.18, d_{20}^{20} 1.0513, *bp* 232.5 °C (101.3 kPa), is produced from 2-ethoxynitrobenzene by reduction with iron or by catalytic hydrogenation in the presence of precious-metal catalysts or Raney nickel [106].

2-Acetamidophenol) (**14**) [*614-80-2*], 2-hydroxyacetanilide, $C_8H_9NO_2$, M_r 151.16, forms colorless crystals, *mp* 209 °C, that are quite soluble in ethanol and hot water. Production is by acetylation of 2-aminophenol with acetic anhydride [107].

2-Amino-5-nitrophenol (**15**) [*121-88-0*], $C_6H_6N_2O_3$, M_r 154.13, forms orange needles, *mp* 207–208 °C, that are sparingly soluble in water. Production is by nitration of benzoxazolone (**16**) and separation from the 2-amino-4-nitrophenol isomer after treatment with hydrochloric acid [108].

2-Amino-4-nitrophenol (**17**) [*99-57-0*], $C_6H_6N_2O_3$, M_r 154.13, 2-hydroxy-5-nitroaniline, forms orange prisms from water that are hydrated with one molecule of water of crystallization, *mp* 80–90 °C. They can be dehydrated over sulfuric acid to the anhydrous form, *mp* 143–145 °C, soluble in alcohol, ether, acetic acid, and warm benzene and slightly soluble in water.

Compound **17** is produced commercially by the partial reduction of 2,4-dinitrophenol; this may be achieved electrolytically using vanadium [109] or chemically with polysulfide, sodium hydrosulfide, or hydrazine and copper [110]–[112]. Alternatively, 2-acetamidophenol or 2-methylbenzoxazole may be nitrated in sulfuric acid to yield a mixture of 4- and 5-nitro derivatives that are then separated and hydrolyzed with sodium hydroxide [113].

The major use of this compound is in the production of mordant and acid dyes. 2-Amino-4-nitrophenol also has found limited use as an antioxidant and light stabilizer in butyl rubbers and as a catalyst in the manufacture of hexa-diene.

2-Amino-4-nitrophenol has been shown to be a skin irritant and continuous exposure should be avoided. Toxicologic studies have shown it to be nonaccumulative [114].

2-Amino-4,6-dinitrophenol (18) [96-91-3], $C_6H_5N_3O_5$, M_r 199.13, 4,6-dinitro-2-aminophenol, picramic acid, forms dark red needles from alcohol and prisms from chloroform, mp 169–170 °C. The compound flashes at 210 °C in contact with an open flame, ignites rapidly, and burns relatively fast. 2-Amino-4,6-dinitrophenol is soluble in glacial acetic acid, water, alcohol, benzene, and aniline and sparingly soluble in ether and chloroform.

The compound can be prepared from 2,4,6-trinitrophenol (picric acid) by reduction with sodium hydrosulfide [115], with ammonia – hydrogen sulfide followed by acetic acid neutralization of the ammonium salt [116], with ethanolic hydrazine and copper [112], or electrolytically with vanadium sulfate in alcoholic sulfuric acid [109]. Heating 4,6-dinitro-2-benzamidophenol in concentrated HCl at 140 °C also yields picramic acid [117]. It is a major intermediate in the manufacture of colorants, especially mordant dyes. It has also been used as an indicator dye in titrations (yellow with acid, red with alkali) and as a reagent for albumin determination.

2,4-Diaminophenol (19) [95-86-3], 4-hydroxy-m-phenylenediamine, $C_6H_8NO_2$, M_r 124.14, forms leaflets that darken on exposure to air and melt at 78–80 °C with decomposition. The compound is soluble in acid, alkali, alcohol, and acetone and sparingly soluble in chloroform, ether, and ligroin. 2,4-Diaminophenol usually is sold as the sulfate [72556-58-2] (Diamol) or dihydrochloride salt [137-09-7] (Acrol, Amidol).

2,4-Diaminophenol can be prepared from 2,4-dinitrophenol by catalytic hydrogenation or, less conveniently, by metal reduction in acid solution (Béchamp method) [118] – [123]. Alternatively, electrolytic reduction and subsequent hydroxylation of 1,3-dinitrobenzene or 3-nitroaniline in sulfuric acid can be undertaken [124] – [126].

The dihydrochloride salt is used as a photographic developer. It also is applied as an intermediate in the manufacture of fur dyes, in hair dyeing, as a reagent in testing for ammonia and formaldehyde, and as an oxygen scavenger in water to prevent boiler corrosion [127].

2-Amino-4-chlorophenol [95-85-2], C_6H_6ClNO, M_r 143.58, forms colorless plates, *mp* 139–140 °C, that are extremely sensitive to oxidation when wet. For its production 2,5-dichloronitrobenzene is converted to 4-chloro-2-nitrophenol by reaction with sodium hydroxide; this product is then reduced either with iron [128], with hydrazine, or with hydrogen [129] in the presence of Raney nickel or platinum catalysts.

2-Amino-4-methylphenol [95-84-1], C_7H_9NO, M_r 123.15, forms crystals, *mp* 135 °C (decomp.), that are almost insoluble in cold water. The compound is quite soluble in ethanol, diethyl ether, and chloroform but is almost insoluble in benzene. 2-Amino-4-methylphenol is obtained by reduction of 2-nitro-4-methylphenol either with hydrazine or catalytically [130].

2-Methoxy-5-methylaniline (20) [120-71-8], 2-amino-4-methylanisole, 3-amino-4-methoxytoluene, $C_8H_{11}NO$, M_r 137.18, forms needles or plates, *mp* 52 °C, *bp* 235 °C (101.3 kPa), that can be steam distilled. The compound is quite soluble in ethanol, diethyl ether, and benzene, but not very soluble in cold water. Production is by alkylation of 4-methyl-2-nitrophenol and subsequent reduction with iron [131].

2-Methoxy-acetanilide (21) [93-26-5], $C_9H_{11}NO_2$, M_r 165.19, crystallizes as needles, *mp* 86.5 °C, *bp* 303–305 °C (101.3 kPa). The compound is quite soluble in acetic acid and hot water. Production is by acetylation of 2-methoxyaniline with acetic anhydride.

2-Methoxy-5-chloroaniline (22) [2401-24-3], C_7H_8ClNO, M_r 157.61, *mp* 84 °C, is quite soluble in ethanol and can be steam distilled. Production is by reduction of the corresponding nitro compound with iron [132].

2-Amino-4,6-dichlorophenol (23) [527-62-8], $C_6H_5Cl_2NO$, M_r 178.03, is sparingly soluble in water and ethanol and decomposes when heated above 109 °C. Industrial production is by reduction of the corresponding nitro compound with iron [133] or with hydrazine.

2-Methoxy-4-nitroaniline (24) [97-52-9], $C_7H_8N_2O_3$, M_r 168.15, *mp* 140–142 °C, is prepared by acetylation and subsequent nitration of 2-methoxyaniline to form intermediate **33**, which is then deacetylated to **24**.

2-Methoxy-5-nitroaniline (25) [*99-59-2*], $C_7H_8N_2O_3$, M_r 168.15, mp 118.5 – 119 °C, is prepared either by partial reduction of 2,4-dinitroanisole (**26**) with ammonium or sodium sulfide [134] or by nitration of 2-methoxyaniline (**27**) [135].

2-Amino-4-chloro-5-nitrophenol (28) [*6358-07-2*], $C_6H_5ClN_2O_3$, M_r 188.58, forms yellow needles that darken when heated above 200 °C and decompose above 225 °C. The compound is scarcely soluble in hot water but quite soluble in ethanol. Industrial production is by reaction of 2-amino-4-chlorophenol with phosgene, nitration of the product 5-chlorobenzoxazolone (**29**) with nitric acid to form 5-chloro-4-nitrobenzoxazolone (**30**), and subsequent hydrolysis with sodium hydroxide [136].

2-Amino-6-chloro-4-nitrophenol (31) [*6358-09-4*], $C_6H_5ClN_2O_3$, M_r 188.58, crystallizes from water in yellow needles that contain 1 mol of water. Dehydration occurs when the needles are heated above 100 °C; the crystals turn scarlet and finally melt at 160 °C. The compound is sparingly soluble in water and quite soluble in ethanol and diethyl ether. Industrial production is by partial reduction of 6-chloro-2,4-dinitrophenol with sodium hydrosulfide or ammonium sulfide [137].

3-Amino-4-methoxybenzoic acid (32) [*2840-26-8*], $C_8H_9NO_3$, M_r 167.16, mp 204 – 205 °C, forms needles or prisms that are sparingly soluble in cold water and also in boiling water (0.12 wt%). The crystals are sparingly soluble in diethyl ether and quite soluble in hot ethanol. Production is by reduction of the corresponding nitro compound.

4-Amino-2-methoxy-acetanilide (34) [*5329-15-7*], 4-acetamido-3-methoxyaniline, $C_9H_{12}N_2O_2$, M_r 180.2, mp 120 – 121 °C, is quite soluble in ethanol, ethyl acetate, acetone, and chloroform. The solubility in diethyl ether is low. Production is by acetylation of 2-methoxyaniline and subsequent nitration with nitric acid [138]. Nitro compound **33** is then reduced with iron in dilute acetic acid [139].

2-Methylbenzoxazole (35) [95-21-6], C_8H_7NO, M_r 133.16, mp 8.5 – 10 °C, bp 178 °C, n_D^{20} 1.5497, is made by cyclization of 2-acetamidophenol. Its chief use is for the production of 2-amino-5-nitrophenol (**15**).

Orthocaine [536-25-4], 3-amino-4-hydroxybenzoic acid methyl ester, methyl 3-amino-4-hydroxybenzoate, Orthoform New, $C_8H_9NO_3$, M_r 167.16, mp 143 °C, is almost insoluble in cold, but soluble in hot water, and very soluble in alcohol, ether, and dilute alkali. For preparation, see [140], [141].

The compound is used as a topical anesthetic.

Acetarsone [97-44-9], 3-acetamido-4-hydroxyphenylarsonic acid, acetarsol, stovarsol, $C_8H_{10}AsNO_5$, M_r 275.08, mp 240 – 250 °C (decomp.), is practically insoluble in alcohols or dilute acids, slightly soluble in water, and freely soluble in dilute aqueous alkali. For its preparation, see [142], [143].

Salts of acetarsone are used in the treatment of intestinal amoebiasis, trade name Acetarsol, of vaginal trichomoniasis, trade name S.V.C. (May and Baker), and as a constituent of a mouth wash, trade name Pyrex (Bengué). Because of acetarsone's toxicity, safer drugs have been developed; lethal dose LD_{50} 150 mg/kg (rabbit, oral).

2.2. Derivatives of 3-Aminophenol

3-(N,N-Dimethylamino)phenol (36) [*99-07-0*], 3-hydroxydimethylaniline, $C_8H_{11}NO$, M_r 137.18, forms white needles, *mp* 87 °C. Boiling points quoted at various pressures are 265–268 °C (101.3 kPa), 206 °C (13.3 kPa), 194 °C (6.7 kPa), 153 °C (0.7 kPa). The compound is soluble in alkali, mineral acid, ethanol, ether, acetone, and benzene and practically insoluble in water.

OH
N(CH₃)₂
36

3-(N,N-Dimethylamino)phenol can be prepared by heating resorcinol with an aqueous solution of dimethylamine and its hydrochloride at 200 °C under pressure for 12 h [144], [145]. Alternatively, the treatment of dimethylaniline with oleum at 55–60 °C, followed by fusion with sodium hydroxide at 270–300 °C, gives 3-(N,N-dimethylamino)phenol [146]. In addition, 3-aminophenol may be methylated with dimethyl sulfate under neutral conditions [147] or its hydrochloride salt heated with methanol at 170 °C under pressure for 8 h to give the desired product [148]. The compound is used primarily as an intermediate in the production of basic (Red 3 and Red 11) and mordant (Red 77) dyes.

3-(N-Methylamino)phenol (37) [*14703-69-6*], C_7H_9NO, M_r 123.15, *bp* 170 °C (1.6 kPa), is easily soluble in ethyl acetate, ethanol, diethyl ether, and benzene; it is sparingly soluble in cold water, better in hot water. Industrial synthesis is by heating 3-(N-methylamino)benzenesulfonic acid with sodium hydroxide at 200–220 °C [149] or by the reaction of resorcinol with methylamine in the presence of aqueous phosphoric acid at 200 °C [150].

3-(N,N-Diethylamino)phenol (38) [*91-68-9*], $C_{10}H_{15}NO$, M_r 165.23, forms rhombic bipy-ramidal crystals, *mp* 78 °C, *bp* 276–280 °C (101.3 kPa), *bp* 209–211 °C (1.6 kPa). Industrial synthesis is analogous to the previously described synthesis of 3-(N,N-dimethylamino)phenol: from resorcinol and diethylamine, by reaction of 3-(N,N-diethylamino)benzenesulfonic acid with sodium hydroxide, or by alkylation of 3-aminophenol hydrochloride with ethanol.

OH   OH   OH
NHCH₃   N(C₂H₅)₂   NH(C₆H₅)
37   38   39

3-(N-Phenylamino)phenol (39) [*101-18-8*], $C_{12}H_{11}NO$, M_r 185.22, *mp* 81.5–82 °C, *bp* 340 °C (101.3 kPa), is slightly soluble in ethanol, diethyl ether, acetone, benzene, and water. The compound is made by heating resorcinol and aniline at 200 °C in the presence of aqueous phosphoric acid [58] or calcium chloride [151]. In another process, 3-aminophenol is heated with aniline hydrochloride at 210–215 °C [152].

4-Amino-2-hydroxybenzoic acid (41) [65-49-6], 4-aminosalicylic acid (PAS), $C_7H_7NO_3$, M_r 153.13, white crystals from alcohol, melts at 150–151 °C with effervescence and darkens on exposure to light and air. A reddish-brown crystalline powder is obtained on recrystallization from ethanol–ether. 4-Amino-2-hydroxybenzoic acid is soluble in dilute solutions of nitric acid and sodium hydroxide, alcohol, and acetone; slightly soluble in water and ether; and virtually insoluble in benzene, chloroform, or carbon tetrachloride. It is unstable in aqueous solution and decarboxylates to form 3-aminophenol. Because of the instability of the free acid it is usually prepared as the hydrochloride salt, mp 224 °C (decomp.), dissociation constant pK_a 3.25.

4-Amino-2-hydroxybenzoic acid is manufactured by carboxylation of 3-aminophenol under pressure with ammonium carbonate at 110 °C [153] or with potassium bicarbonate and carbon dioxide at 85–90 °C [154] and subsequent acidification.

The major use of this compound is as a bacteriostatic agent against tubercle bacilli. The compound also is used as an adjunct to streptomycin and isoniazid. The free acid and its sodium, potassium, and calcium salts are marketed under many trade names. Up to 10–15 g of the sodium salt may be administered each day, although prolonged use may give rise to toxic symptoms. Oral LD_{50} of the free acid in mice is 4 g/kg.

2.3. Derivatives of 4-Aminophenol

4-(N-Methylamino)phenol (41) [150-75-4], 4-hydroxy-N-methylaniline, C_7H_9NO, M_r 123.15, forms needles from benzene, mp 87 °C, bp 168–169 °C (2 kPa); it is slightly soluble in alcohol and insoluble in ether.

Industrial synthesis involves the decarboxylation of N-(4-hydroxyphenyl)glycine (47) at elevated temperature in such solvents as chlorobenzene–cyclohexanone [155]. It also can be prepared by the methylation of 4-aminophenol [156] or from methylamine by heating with 4-chlorophenol and copper sulfate at 135 °C in aqueous solution [157] or with hydroquinone at 200–250 °C in alcoholic solution [158], [159].

Its chief use is as a component in photographic developers. Because the free compound is unstable in air and light, it usually is marketed as the sulfate salt [55-55-0], Metol, mp 260 °C (decomp.). It also finds application as an intermediate for fur and hair dyes and, under certain circumstances, as a corrosion inhibitor for steel. Prolonged exposure to 4-(N-methylamino)phenol has been associated with the development of dermatitis and allergies.

4-(N,N-Dimethylamino)phenol (42) [*619-60-3*], 4-hydroxydimethylaniline, $C_8H_{11}NO$, M_r 137.18, forms large rhombic crystals from ether–hexane or ether–ligroin that melt at 75–76 °C, *bp* 101–103 °C (66.7 Pa). The compound forms a salt with sulfuric acid, *mp* 208–210 °C [160], [161].

Methylation of 4-aminophenol with a methyl halide under pressure produces 4-(*N*,*N*dimethylamino)phenol. The major competing product, 4-hydroxyphenyltrimethyl ammonium halide (or the corresponding base), also yields 4-(*N*,*N*-dimethylamino)phenol on distillation. Alternatively, it can be synthesized by dealkylation of 4-methoxydimethylaniline with hydroiodic acid at reflux temperature for 10 h [162] or by the photodecomposition of 4-dimethylaminobenzenediazonium tetrafluoroborate [160].

The compound is an intermediate in several synthetic reactions and recently has found extensive use in experimental toxicity studies in animals. It has been shown to cause methemoglobinemia; its metabolism in humans has been discussed in [163]–[168].

OH	OH	OH
NHCH₃	N(CH₃)₂	NHCOCH₃
41	42	43

4-Hydroxyacetanilide (43) [*103-90-2*], 4-acetamidophenol, acetaminophen, paracetamol, forms large white monoclinic prisms from water that melt at 169–171 °C. The compound is odorless and has a bitter taste. 4-Hydroxyacetanilide is insoluble in petroleum ether, pentane, and benzene; slightly soluble in ether and cold water; and soluble in hot water, alcohols, dimethylformamide, 1,2-dichloroethane, acetone, and ethyl acetate. The dissociation constant, pK_a, is 9.5 (25 °C).

Production is by the acetylation of 4-aminophenol. This can be achieved with acetic acid and acetic anhydride at 80 °C [169], [170], with acetic anhydride in pyridine at 100 °C [171], with acetyl chloride and pyridine in toluene at 60 °C [172], or by the action of ketene in alcoholic suspension [173]. 4-Hydroxyacetanilide also may be synthesized directly from 4-nitrophenol. The available reduction–acetylation systems include tin with acetic acid [174], hydrogenation over Pd–C in acetic anhydride [175], and hydrogenation over platinum in acetic acid [176]. Other routes include rearrangement of 4-hydroxyacetophenone hydrazone with sodium nitrite in sulfuric acid [177] and the electrolytic hydroxylation of acetanilide [178].

4-Hydroxyacetanilide is used as an intermediate in the manufacture of azo dyes and photographic chemicals. The compound possesses antipyretic and analgesic properties and is used widely in this context. The oral LD_{50} in rats is 3.7 g/kg.

4-Methoxyaniline [*104-94-9*], C_7H_9NO, has M_r 123.55, *mp* 58.5 °C, *bp* 123.2–123.5 °C (1.9 kPa), dipole moment 1.8 D (benzene), d_4^{55} 1.092, n_D^{67} 1.5559, viscosity η 3.215 mPa · s (55 °C). For its industrial production 4-nitroanisole is reduced either with sodium sulfide [179] or with hydrogen in the presence of precious-metal catalysts or Raney nickel.

4-Ethoxyaniline [*156-43-4*], $C_8H_{11}NO$, M_r 137.18, *mp* 2.4 °C, *bp* 249.9 °C (101.3 kPa), $d_4^{14.9}$ 1.0652, $n_\alpha^{15.9}$ 1.5572, is a colorless compound at room temperature when pure; in the light or in the presence of air, the liquid turns brown. 4-Ethoxyaniline is slightly soluble in water.

Industrial production of 4-ethoxyaniline is by reduction of 4-nitrophenetole with iron [180] or with hydrogen under pressure in the presence of a catalyst. Alternatively, 4-chlorophenetole is heated with aqueous ammonia in the presence of copper(I) oxide at 225 °C; the yield of 4-ethoxyaniline is 85% [181].

4-Ethoxyacetanilide (44) [*62-44-2*], phenacetin, *p*-acetophenetidine, 4-acetamidophenetole, $C_{10}H_{13}NO_2$, M_r 179.21, is a white crystalline powder, *mp* 134–135 °C. The compound is odorless and has a slightly bitter taste; it is sparingly soluble in cold water and more soluble in hot water, alcohol, ether, and chloroform. At relative humidities between 15 and 90% the equilibrium moisture content is about 2% (25 °C).

$$\underset{44}{\underset{\text{NHCOCH}_3}{\overset{\text{OC}_2\text{H}_5}{\bigcirc}}}$$

The main production route to 4-ethoxyacetanilide is by catalytic reduction of 4-nitrophenetole with hydrogen and subsequent acetylation using acetic anhydride. The compound also can be synthesized by ethylating 4-nitrophenol with ethyl sulfate in alkali, reducing the nitro group to an amino group with iron in acid, and then acetylating by boiling with glacial acetic acid [182]. Alternatively, 4-aminophenol may be ethylated with ethyl iodide in alcoholic alkali; the resulting 4-phenetidine is then acetylated. The acetylation also may be carried out first, followed by the ethylation.

4-Ethoxyacetanilide possesses both antipyretic and analgesic properties but is of little value for the relief of severe pain. Its use for prolonged periods should be avoided because one of its minor metabolites (2-hydroxyphenetidine) is nephrotoxic and may be involved in the formation of methemoglobinemia. The oral LD_{50} in rats is 1.65 g/kg [183].

5-Aminosalicylic acid (45) [*89-57-6*], 5-amino-2-hydroxybenzoic acid, $C_7H_7NO_3$, M_r 153.13, forms colorless crystals that turn brown on heating above 250–260 °C and melt at 283 °C (decomp.). The crystals are insoluble in ethanol and cold water and sparingly soluble in hot water.

The compound is made by reduction of 3-nitrobenzoic acid at 115–145 °C and 3.5 MPa in dilute sulfuric acid and in the presence of platinum catalyst. The intermediate hydroxylamine derivative (**46**) rearranges to **45** under these conditions [184]. Alternatively, 5-nitrosalicyclic acid is reduced, for example, with zinc in hydrochloric acid [185].

N-(4-Hydroxyphenyl)glycine (47) [122-87-2], 4-hydroxyphenylaminoacetic acid, 4-oxyanilinoacetic acid, photoglycine, $C_8H_9NO_3$, M_r 167.16, forms aggregate spheres or shiny leaflets from water; it turns brown at 200 °C, begins to melt at 220 °C, and melts completely with decomposition at 245–247 °C. The compound is soluble in alkali and mineral acid and sparingly soluble in water, glacial acetic acid, ethyl acetate, alcohol, ether, acetone, chloroform, and benzene.

N-(4-Hydroxyphenyl)glycine can be prepared from 4-aminophenol and chloroacetic acid [186], [187] or by alkaline hydrolysis of the corresponding nitrile with subsequent elimination of ammonia [188].

N-(4-Hydroxyphenyl)glycine is used as a photographic developer under the trade names of Glycine, Iconyl, and Monazol. It also is applied as a photoresist in the dye industry and serves as an intermediate in the production of 4-(N-methylamino)phenol (Metol) by liberation of CO_2. N-(4-Hydroxyphenyl)glycine is used in analytical chemistry for the determination of iron, phosphorus, and silicon and as an acid indicator in bacteriology. Prolonged use of this compound may result in kidney damage [189].

4-Amino-2,6-dichlorophenol (48) [5930-28-9], $C_6H_5Cl_2NO$, M_r 178.03, can be sublimed and melts at 167 °C. The compound is soluble in ethanol, diethyl ether, acetone, and acetic acid; it is sparingly soluble in benzene and almost insoluble in water.

3. References

General References

[1] K.-F. Wedemeyer, in"Phenole,"*Houben-Weyl* Teil 1 and 2 (1976).

[2] *Beilstein*, **13**, 354–549; **13 (2)**, 164–308.

Specific References

[3] E. G. Smith: *The Wiswesser Line-Formula Chemical Notation*, McGraw-Hill, New York 1968.

[4] G. Varsanyi: *Assignment for Vibrational Spectra of Seven Hundred Benzene Derivatives*, Akademiai Kiado, Budapest – Adam Hilger, London 1974, pp. 136, 206, 253, 473, 476, 478.

[5] C. J Pouchert (ed.): *The Aldrich Library of Infrared Spectra*, 3rd ed., vol. **3,** Aldrich Chemical Co., Milwaukee, Wisc. 1981, pp. 718 B, 720 D, 725 D.

[6] W. F. Forbes, I. R. Leckie, *Can. J. Chem.* **36** (1958) 1371–1380.

[7] J. C. Dearden, W. F. Forbes, *Can. J. Chem.* **37** (1959) 1294–1304.

[8] *Sadtler Catalog of Ultraviolet Spectra*, The Sadtler Research Co., Philadelphia, Pa., SAD no. 236, 1894, 3509.

[9] *Aldermaston Eight-peak Index of Mass Spectra*, Imperial Chemical Industries, Organics Division, Manchester, & Mass Spectrometry Data Centre, Atomic Weapons Research Establishment (AWRE), Aldermaston, U.K., p. 46.

[10] C. J. Pouchert, J. R. Campbell (eds): *The Aldrich Library of NMR Spectra*, vol. **5.**, Aldrich Chemical Company, Milwaukee, Wisc., 1974–75, pp. 45 C, 51 B, 57 A.

[11] *Sadtler Catalog of NMR Spectra*, The Sadtler Research Co., Philadelphia, Pa., SAD no. 717, 1176, 10220.

[12] J. G. Grasselli (ed.): *CRC Atlas of Spectral Data and Physical Constants for Organic Compounds*, CRC Press, Cleveland, Ohio, 1973, p. B 754.

[13] S. Ashfaquzzaman, A. K. Pant: *Acta Crystallogr., Sect. B.* **B 35** (1979) no. 6, 1394–1399; *Chem. Abstr.* **91** (1979) 66707.

[14] W. A. Caspari, *Phil. Mag.* **4**, 1927, no. 7, 1276–1285.

[15] J. D. H. Donnay, H. M. Ondik in: *Crystal Data; Determinative Tables*, 3rd ed., U.S. Dept. of Commerce National Bureau of Standards & Joint Committee on Powder Diffraction Standards, Washington, D.C., 1972,pp. 0–23, 0–73, 0–97, 0–106.

[16] N. V. Sidgwick, R. K. Callow, *J. Chem. Soc. (Trans.)* **125** (1924) 522–527.

[17] R. Kuhn, A. Wassermann, *Helv. Chim. Acta* **11** (1928) 1–30.

[18] V. H. Veeley, *J. Chem. Soc.* **93** (1908) 2122–2144.

[19] R. W. G. Wyckoff, in: *Crystal Structures*, 2nd ed., vol. **6**, part 1, Interscience, New York 1969, pp. 186–190.

[20] G. Chuchani, J. A. Hernández, J. Zabicky, *Nature* **207** (1965) 1385–1386.

[21] S. Coffey (ed.): *Rodd's Chemistry of Carbon Compounds*, 2nd ed., vol. **3 A**, Elsevier Publishing Co., Amsterdam 1971, pp. 352–363.

[22] J. H. Ransom, *Am. Chem. J.* **23** (1900) 1–50; *J. Am. Chem. Soc.* **22** (1900) 89–91.

[23] K. H. Saunders: *The Aromatic Diazo Compounds and Their Technical Application*, Arnold, London 1949,pp. 17–20.

[24] H. Zollinger: *Azo and Diazo Chemistry*, Interscience, New York 1961, pp. 51–53 and 250–265.

[25] *BIOS Report* **986** (1946) 45.

[26] CS 159564, 1975, (D. Kulda, J. Fuka, J. Ott, Z. Misar).

[27] CPC International, US 3535382, 1970 (B. Brown, F. Schilling).
[28] Mallinckrodt Inc., EP 41837, 1980 (W. R. Clingan, E. L. Derrenbacker, T. J. Dunn).
[29] Mallinckrodt Inc., US 4264529, 1975 (T. J. Dunn).
[30] Universal Oil, FR 1354430, 1964.
[31] BASF, DE 1244196, 1967.
[32] Du Pont, US 2183019, 1939.
[33] Seiko Chem. Ind. Co. Ltd., JP 75135042, 1975.
[34] Abbott Lab., US 3079435, 1963.
[35] Sumitomo, JP 30294/68, 1968.
[36] Monsanto, US 2035292, 1933.
[37] Sumitomo, JP 2701/69, 1969.
[38] Abbott Lab., FR 1307415, 1962.
[39] Du Pont, US 2947781, 1960.
[40] Constructors John Brown Ltd., DE-OS 2026039, 1970. Continental Oil Co., US 3338806, 1967. Miles Laboratories, Inc., DE 1066589, 1959. Miles Lab. Inc., GB 856436, 1960. Ionics Incorp., US 3103473, 1963. Warner-Lambert Pharmaceutical Co., US 2998450, 1958. National Carbon Co., US 2427433, 1947.
[41] Du Pont, FR 1392098, 1965.
[42] Hoechst, DE 96853, 1897. Belvedere Chem. Ltd., US 2132454, 1935. Eastman Kodak Co., US 2446519, 1948, US 2525515, 1950.
[43] Soc. des Usines Chimiques Rhoâne-Poulenc, FR 1559841, 1969. Frontier Chem. Co., US 3265735, 1966. ICI, GB 856366, 1960. Prochim S. A., FR 1338899, 1963. Du Pont, GB 713622, 1954.
[44] Koppers Co., US 3953509, 1976.
[45] Du Pont, US 198249, 1938.
[46] International Nickel Ltd., NL 6814316, 1969.
[47] GB 802619, 1958. (R. Debus, P. Leprince).
[48] CPC Internat. Inc., US 3535382, 1970.
[49] Engelhard Minerals & Chem. Corp., DE-OS 2118334, 1971. Seiko Chem. Co. Ltd., JP Kokai 75142525, 1975.
[50] R. G. Benner, DE-OS 1643255, 1971.
[51] M. Noel, P. N. Anantharama, H. V. K. Udupa, *J. Appl. Electrochem.* **12** (1982) no. 3, 291–298.
[52] E. Theodoridou, D. Jannakondakis, *Z. Naturforsch., B* **36 B** (1981) no. 7, 840–845.
[53] K. S. Udupa, G. S. Subramanian, H. V. K. Udupa, *Trans. Soc. Adv. Electrochem. Sci. Technol.* **7** (1972) no. 2, 49–50.
[54] Continental Oil, US 3338806, 1967 (W. Harwood).
[55] Trimb. Ltd., AU 205525, 1957. Akt.-Ges. f. Anilinfabr., DE 205415, 1908.
[56] *BIOS Report 986*, British Intelligence Objectives Subcommittees, Eng. 1946, p. 45–46, 412.
[57] Hoechst, DE-OS 2140786, 1973; Leonhardt & Co., DE 49060, 1889.
[58] Eastman Kodak Co., US 2376112, 1945.
[59] Universal Oil Products, FR 1354430, 1964 (J. Levy).
[60] Mitsui Toatsu Chem. Inc., JP Kokai 8120553, 1981.
[61] Toaka Dyestuffs MFG. Co. Ltd., JP Kokai 7611, 722, 1976.
[62] BASF, DE 1104970, 1961.
[63] Bayer, DE-OS 1902418, 1970.
[64] ICI, GB 1038005, 1966. (Z. Grzymalski, F. Mirek).
[65] PL 46829, 1963.

[66] ICI, GB 1038078, 1966.
[67] Soc. des Usines Chimiques Rhoâne-Poulenc, FR 1564882, 1969.
[68] Howard Hall & Co., DE-OS 2050943, 1971.
[69] Howard Hall & Co., DE-OS 2050927, 1971.
[70] Howard Hall & Co., DE-OS 2054282, 1971.
[71] MacFarlan Smith Ltd., DE-OS 2103548, 1971.
[72] CPC Internat. Inc., US 3876703, 1975.
[73] Mitsui Toatsu Chem. Inc., JP Kokai 8081843, 1980.
[74] K. Urano, K. Kawamoto, K. Hayashi, *Suishitsu Odaku Kenkyu* **4** (1981) 43–50.
[75] R. A. Hux, Su Puon, F. F. Cantwell, *Anal. Chem.* **52** (1980) 2388–2392.
[76] C. S. P. Sastry, K. V. S. S. Murty, *Natl. Acad. Sci. Lett. (India)* **5** (1982) 15–17.
[77] D. J. Miner, J. R. Rice, R. M. Riygin, P. T. Kissinger, *Anal. Chem.* **53** (1981) 2258–2263.
[78] M. A. Korany, D. Heber, J. Schnekenburger, *Talanta* **29** (1982) 332–334.
[79] A. Mazzeo-Farina, M. A. Ionio, A. Laurenzi, *Ann. Chim. (Rome)* **71** (1981) 103–109.
[80] L. Reio, *J. Chromatogr.* **1** (1958) 338–373.
[81] S. C. Mitchell, R. H. Waring, *J. Chromatogr.* **151** (1978) 249–251.
[82] A. Ya. Yakobi, L. A. Oberemok, *Khim. Promst. Ser. Metody Anal. Kontrolya Kach. Prod. Khim. Promst.* **2** (1981) 46–47.
[83] N. T. Bernabei, V. Ferioli, G. Gamberini, R. Cameroni, *Farmaco Ed. Prat.* **36** (1981) 249–255.
[84] L. A. Stevenson, *IARC Sci. Publ.* **40** (1981) 219–228.
[85] L. N. Zayidullina, E. A. Korneeva, A. S. Lukyanova, *Vodosnabzh. Sanit. Tekh.* **8** (1981) 12–13.
[86] A. M. Klibanov, B. N. Alberti, E. D. Morris, L. M. Felshin, *J. Appl. Biochem.* **2** (1980) 414–421.
[87] Alberto Culver Co., US 4297098, 1980 (G. F. Dasher, T. J. Schamper).
[88] Wella AG, GB 2085483, 1981 (E. Konrad, H. Mager).
[89] Matsushita Electric Works Ltd., JP Kokai 8120162, 1981.
[90] M. Terashima, M. Ishi, Y. Kanaoka, *Synthesis* **6** (1982) 484–485.
[91] Ciba-Geigy AG, EP 48222, 1982 (H. O. Wirth, J. Buessing, H. H. Friedrich).
[92] *Patty*, 3rd ed., vol. **2A,** 2443–2445.
[93] M. Cirstea, G. Suhaciu, M. Cirje, G. Petec, A. Vacariu, *Rev. Roum. Morphol. Embryol. Physiol. Physiol.* **17** (1980) 91–96.
[94] J. D. Tange, B. D. Ross, J. G. G. Ledingham, *Clin. Sci. Mol. Med.* **53** (1977) 485–492.
[95] C. A. Crowe, I. C. Calder, N. P. Madsen et al., *Xenobiotica* **7** (1977) 345–356.
[96] J. V. Rutkowski, V. H. Ferm, *Toxicol. Appl. Pharmacol.* **63** (1982) 264–269.
[97] N. K. Hayward, M. F. Lavin, P. W. Craswell, *Biochem. Pharmacol.* **31** (1982) 1425–1429.
[98] G. Kirchner, U. Bayer, *Hum. Toxicol.* **1** (1982) 387–392.
[99] *Colour Index*, 3rd ed., The Society of Dyers and Colorists, Bradford, Yorkshire, England, vol. 4, 1971, p. 4001–4689, 4691–4863; vol. 6, 1975, p. 6391–6404, 6407–6410.
[100] *Beilstein*, **13**, 354–549; **13 (2)**, 164–308.
[101] J. Ransom, *Am. Chem. J.* **23** (1900) 34.
[102] F. Lees, F. Shedden, *J. Chem. Soc.* **83** (1903) 756.
[103] J. Schwyzer: *Die Fabrikation pharmazeutischer und chemisch-technischer Produkte*, Springer Verlag, Berlin 1931, p. 203.
[104] General Aniline & Film Corp., FR 1321689, 1963.
[105] E. Profft, *Dtsch. Chem. Z.* **2** (1950) 194.
[106] J. R. Reasenberg, E. Lieber, G. B. L. Smith, *J. Am. Chem. Soc.* **61** (1939) 384.
[107] R. Meldola, G. H. Woolcott, E. Wray, *J. Chem. Soc.* **69** (1896) 1321.
[108] R. Nodzu et al., *Yakugaku Zasshi* **79** (1959) 1378.

[109] H. Hofer, F. Jakob, *Ber. Dtsch. Chem. Ges.* **41** (1908) 3187–3199.
[110] *P. B. Report 12272* U.S. Department Commission Office Technology Service, Washington, D.C., 1945, p. 39–46.
[111] W. W. Hartmann, H. L. Silloway, *Org. Synth.* **25** (1945) 5–7.
[112] S. Kubota, K. Nara, S. Onishi, *Yakugaku Zasshi* **76** (1956) 801.
[113] *P.B. Report 74197* U.S. Department Commission Office Technology Service, Washington, D.C., 1947, p. 695–697.
[114] M. P. Slyusar, *Nauchn. Tr. Ukr. Nauchno Issled. Inst. Gig. Tr. Profzabol.* **27** (1958) 103; *Chem. Abstr.* **54** (1960) 13435h.
[115] *P. B. Report 85172* U.S. Department Commission Office Technology Service, Washington, D.C., 1948, p. 242.
[116] G. Egerer, *J. Biol. Chem.* **35** (1918) 565.
[117] H. Hubner, *Justus Liebigs Ann. Chem.* **210** (1881) 328–396.
[118] W. E. Bradt, *J. Phys. Chem.* **34** (1930) 2711–2718.
[119] E. A. Braude, R. P. Linstead, K. R. H. Wooldridge, *J. Chem. Soc.* (1954) 3586–3595.
[120] T. Nelson, H. C. S. Wood, A. G. Wylie, *J. Chem. Soc.* 1962, part I, 371–372.
[121] L. Gattermann, *Ber. Dtsch. Chem. Ges.* **26** (1893) 1844–1856.
[122] W. Hemilian, *Ber. Dtsch. Chem. Ges.* **8** (1875) 768.
[123] H. Pomeranz, DE 269542, 1914.
[124] Friedrich Bayer & Co., DE 75260, 1894.
[125] K. Udupa, G. Subramanian, H. Udupa, *Bull. Acad. Pol. Sci. Ser. Sci. Chim.* **9 (6)** (1961) 419.
[126] Eastman Kodak Co., US 2525515, 1950 (F. R. Bean).
[127] Betz Laboratories Inc., US 4279767, 1980 (J. A. Muccitelli).
[128] H. E. Fierz-David, L. Blangey: *Grundlegende Operationen der Farbenchemie,* Springer Verlag, Wien 1943, p. 105.
[129] A. M. Popow, *Anilinokras. Promst.* **3** (1933)391; *Chem. Zentralbl.* 1935, I, 226.
[130] W. J. Close, B. D. Tiffany, M. H. Spielmann, *J. Am. Chem. Soc.* **71** (1949) 1265.
[131] *BIOS Final Report* **986** M. O. de Vries, *Recl. Trav. Chim. Pays-Bas* **28** (1909) 288.
[132] Akt.-Ges. f. Anilinf., DE 137956, 1902.
[133] *P. B. Report 74051*, U.S. Department Commission Office Techn. Service, Washington, D.C., 1946, p. 281.
[134] A. Cahours, *Justus Liebigs Ann. Chem.* **74** (1850) 301.C. Niemann, J. F. Mead, A. A. Benson, *J. Am. Chem. Soc.* **63** (1941) 609.
[135] Du Pont, US 1998794, 1932.
[136] Bayer, DE 186655, 1905.
[137] *P.B. Report 74197* U.S. Department Commission Office Techn. Service, Washington, D.C., 1947 610. A. Faust, H. Müller, Justus Liebigs Ann. Chem. 173 (1874) 315.
[138] Fabr. de Prod. Chimiques de Thann, DE 98637, 1898. Amer. Cyanamid, US 2459002, 1944.
[139] IG Farbenind., DE 552267, 1932.
[140] A. Einhorn, B. Pfyl, *Justus Liebigs Ann. Chem.* **311** (1900) 34–66.
[141] K. Auwers, H. Rohrig, *Ber. Dtsch. Chem. Ges.* **30** (1897) 988–998.
[142] G. W. Raiziss, J. L. Gavron, *J. Am. Chem. Soc.* **43** (1921) 582–585.
[143] G. W. Raiziss, B. C. Fisher, *J. Am. Chem. Soc.* **48** (1926) 1323–1327.
[144] A. Leonhardt & Co. DE 49060, 1889.
[145] Badische Anilin- und Soda-Fabrik, DE 121683, 1901.
[146] Ges. F. Chem. Ind., DE 44792, 1888.

[147] M. L. Crossley, P. F. Dreisbach, C. M. Hoffmann, R. P. Parker, *J. Am. Chem. Soc.* **74** (1952) 573–578.
[148] Badische Anilin- und Soda-Fabrik, DE 44002, 1888.
[149] BASF, DE 48151, 1889.
[150] Eastman Kodak Co., US 2376112, 1942.
[151] A. Calm, *Ber. Dtsch. Chem. Ges.* **16** (1883) no. 2, 787.
[152] BASF, DE 46869, 1888.
[153] J. T. Sheehan, *J. Am. Chem. Soc.* **70** (1948) 1665–1666.
[154] H. Erlenmeyer, B. Prijs, E. Sorkin, E. Suter, *Helv. Chim. Acta* **31** (1948) 988–992.
[155] Industrial Dyestuff Co., US 1993253, 1934, US 2101749, US 2101750, 1937.
[156] K. Fricker, DE 449047, 1927.
[157] Akt.-Ges. f. Anilin-Fabr., DE 205414, 1908.
[158] E. Merck AG, DE 260234, 1913.
[159] R. N. Harger, *J. Am. Chem. Soc.* **41** (1919) 270–276.
[160] J. DeJonge, R. Dijkstra, *Recl. Trav. Chim. Pays-Bas* **68** (1949) 426–429.
[161] H. Wieland, *Ber. Dtsch. Chem. Ges.* **43** (1910) 712–728.
[162] F. G. Bordwell, P. J. Boutan, *J. Am. Chem. Soc.* **78** (1956) 87–91.
[163] J. E. Bright, T. C. Marrs, *Arch. Toxicol.* **50** (1982) 57–64.
[164] T. C. Marrs, J. E. Bright, D. W. Swanston, *Arch. Toxicol.* **50** (1982) 89–92.
[165] R. Elbers, S. Soboll, H. G. Kampffmeyer, *Biochem. Pharmacol.* **29** (1980) 1747–1753.
[166] P. Jansco, L. Szinicz, P. Eyer, *Arch. Toxicol.* **47** (1981) 39–45.
[167] P. Eyer, H. G. Kampffmeyer, *Biochem. Pharmacol.* **27** (1978) 2223–2228.
[168] P. Eyer, H. Gaber, *Biochem. Pharmacol.* **27** (1978) 2215–2221.
[169] H. Fierz-David, W. Kuster, *Helv. Chim. Acta* **22** (1939) 82–112.
[170] Warner-Lambert Pharmaceutical Co., US 2998450, 1961.
[171] A. L. LeRosen, E. D. Smith, *J. Am. Chem. Soc.* **70** (1948) 2705–2709.
[172] V. R. Olsen, H. B. Feldman, *J. Am. Chem. Soc.* **59** (1937) 2003–2005.
[173] F. Stern, DE 453577, 1927 (M. Bergmann).
[174] H. N. Morse *Ber. Dtsch. Chem. Ges.* **11** (1878) 232–233.
[175] Nepera Chemical Co., US 3341587, 1967 (B. Duesel, G. Wilbert).
[176] J. H. Burckhalter, F. H. Tendick, E. M. Jones, P. A. Jones, W. F. Holcomb, A. L. Rawlins, *J. Am. Chem. Soc.* **70** (1948) 1363–1373.
[177] D. E. Pearson, K. Carter, C. M. Greer, *J. Am. Chem. Soc.* **75** (1953) 5905–5908.
[178] C. F. Boehringer & Soehne GmbH, DD 1259344, 1968 (H. Staudinger, V. Ullrich).
[179] *BIOS Final Report* 986.
[180] J. Schwyzer: *Die Fabrikation pharmazeutischer und chemisch-technischer Produkte*, Springer Verlag, Berlin 1931, p. 204.
[181] Dow, US 1932653, 1928.
[182] Monsanto Chemicals Co., US 2887513, 1959 (C. M. Eaker, J. R. Campbell).
[183] P. K. Smith in: *Acetophenetidin*, a Monograph, Interscience, New York 1958.
[184] Du Pont, US 2198249, 1938.
[185] H. Weil, M. Traun, S. Marcel, *Ber. Dtsch. Chem. Ges.* **55** (1922) 2664.
[186] H. Vater, *J. Prakt. Chem.* **29** (NF) (1884) 286–299.
[187] R. Medola, H. S. Foster, R. Brightman, *J. Chem. Soc.* **111** (1917) 551–553.
[188] L. Galatis, *Helv. Chim. Acta* **4** (1921) 574–579.
[189] B.-N. Li, T. Kemeny, J. Sos, *Acta Med. Acad. Sci. Hung.* **19** (1963) suppl. 19, 111.

Aniline

F. R. LAWRENCE, E. I. du Pont de Nemours & Co., Deepwater, New Jersey 08023, United States

W. J. MARSHALL, E. I. du Pont de Nemours & Co., Deepwater, New Jersey 08023, United States

1.	Introduction	627
2.	Physical and Chemical Properties	628
3.	Production	629
3.1.	Catalytic Liquid-Phase Hydrogenation of Nitrobenzene	630
3.2.	Catalytic Vapor-Phase Hydrogenation of Nitrobenzene	631
3.3.	Reduction of Nitrobenzene with Iron and Iron Salts	632
3.4.	Amination of Phenol	632
4.	Quality Specifications	633
5.	Handling, Storage, and Transportation	634
6.	Aniline Derivatives and Uses	634
6.1.	Methylene Diphenylene Isocyanate (MDI)	634
6.2.	Other Uses	635
6.3.	Derivatives	638
7.	Economic Aspects	639
8.	Environmental Protection, Toxicology, and Occupational Health	640
9.	References	641

1. Introduction

Aniline, $C_6H_5NH_2$ [62-53-3], NIOSH Access No. BW 665000, is a member of the aromatic amines chemical family (→ Amines, Aromatic). It is also known by the synonyms aniline oil, aminobenzene, phenylamine, benzenamine (CAS name), aminophen, and by the trade name Blue Oil. It was first obtained from the destructive distillation of indigo in 1826 by O. UNVERDORBEN, who named it krystalline. F. RUNGE identified it as a constituent of coal tar in 1834 and named it kyanol. In 1841 C. J. FRITZSCHE heated indigo with caustic potash and called the product aniline. In the same year, N. N. ZININ obtained an amine by reduction of nitrobenzene and called it benzidam. A. W. VON HOFMANN proved these products to be identical in 1843 and fixed the name of the compound as aniline. The Béchamp process, discovered in 1854, in which nitrobenzene and iron borings were reacted in the presence of acetic acid, was the first economical commercial process. Since 1854 aniline has become an important commercial intermediate for isocyanates, rubber chemicals, agricultural chemicals, pharmaceuticals, and dyes. World capacity in 1983 was 1.324×10^6 t.

2. Physical and Chemical Properties

Aniline, C_6H_7N, M_r 93.13, bp 184.4 °C, (101.3 kPa), when freshly distilled, is a colorless, oily liquid that turns brown on exposure to air and light.

The most important physical data are listed below:

Density,	
liquid (20 °C)	1.022 g/cm³
vapor (at bp, air = 1)	3.30
Refractive index n_D^{20}	1.5863
Viscosity,	
at 20 °C	4.35 mPa · s
at 60 °C	1.62 mPa · s
Solubility (30 °C),	
aniline in water	3.7 wt%
water in aniline	5.4 wt%
Specific heat (25 °C)	2.06 J g⁻¹ K⁻¹
Heat of vaporization (at bp)	478.55 J/g
Critical temperature	425.6 °C
Critical pressure	5.30 MPa
Heat of combustion	36.45 kJ/g
Flash point	
closed cup	70 °C
open cup	75.5 °C
Ignition temperature	615 °C
Lower flammable limit	1.3 vol%
Odor threshold	0.5 ppm

Aniline is miscible with most organic sol-vents. It is a primary aromatic amine with chemical properties [1] characteristic of compounds having a benzene nucleus activated by an amino group. Aniline is a weak base ($K_b = 3.8 \times 10^{-10}$) and forms stable, water-soluble salts with strong acids, such as hydrochloric (aniline hydrochloride, $C_6H_5NH_3^+Cl^-$) and sulfuric (aniline acid sulfate, $C_6H_5NH_3^+OSO_3H^-$). Because free aniline oxidizes easily, many syntheses start with the aniline salt in order to protect the amino group. The nitration of aniline acid sulfate in sulfuric acid, followed by treatment with sodium hydroxide, produces m-nitroaniline, $O_2NC_6H_4NH_2$, as the major product. The o-, m-, and p-isomers are separable by fractional precipitation. Reaction of aniline with glacial acetic acid or acetic anhydride gives acetanilide, $C_6H_5NHCOCH_3$. Orthoarsenic acid, H_3AsO_4, reacts with aniline to produce arsanilic acid, $AsO(OH)_2 \cdot C_6H_4NH_2$. In the presence of hydrochloric acid, aniline and formaldehyde yield methylenedianiline (4,4′-diaminodiphenylmethane). Alkylidene or arylalkylidene amines (Schiff bases) are formed when aniline reacts with aliphatic and aromatic aldehydes. Alkyl halides combine with aniline to form the expected secondary or tertiary amines or quaternary ammonium salts. Aniline is highly activated toward substitution by electrophilic reagents; for example, aniline is easily halogenated directly to trichloro- or tribromoaniline (isomer distribution depends on process conditions). The reaction pathway involving oxidation depends on the nature of the oxidizing agent and reaction conditions. Manufacture of p-benzoquinone (→ Benzoquinone) involves the oxidation of aniline with manganese dioxide and sulfuric acid. Aniline can be hydrogenated catalytically to cyclohexyl-amine, $C_6H_{11}NH_2$, at elevated temperature and pressure.

3. Production

Numerous chemical routes to aniline are described in the literature but only those that start with nitrobenzene (Eq. 1) or phenol (Eq. 2) are currently economical.

$$C_6H_6 + HNO_3 \xrightarrow[-H_2O]{H_2SO_4} \underset{\text{Nitrobenzene}}{C_6H_5NO_2} \xrightarrow[-2H_2O]{3H_2,\text{ cat.}} C_6H_5NH_2 \quad (1)$$

$$C_6H_6 + CH_2\!=\!CHCH_3 \xrightarrow{\text{cat.}} \underset{\text{Cumene}}{C_6H_5CH(CH_3)_2} \xrightarrow[-CH_3COCH_3]{\begin{array}{c}1)\,O_2\\2)\,H_2SO_4\end{array}}$$

$$\underset{\text{Phenol}}{C_6H_5OH} \xrightarrow[-H_2O]{NH_3,\text{ cat.}} C_6H_5NH_2 \quad (2)$$

The nitrobenzene route is used by all world producers with the exception of Mitsui Petrochemical Co. (Japan) and United States Steel Chemicals, both of whom use the phenol route. One other route, used commercially but now abandoned, involved the amination of chlorobenzene.

Commercial aniline producers find it economical to manufacture the nitrobenzene or phe nol also. Therefore, companies using nitrobenzene as the starting material have facilities to carry out a mixed ($HNO_3 + H_2SO_4$) acid nitration of benzene to nitrobenzene at 70–80 °C followed by separation from spent acid, water wash, neutralization, water wash, and distillation (Fig. 1). The spent acid is reconcentrated and recycled. Similarly, producers using phenol can oxidize cumene.

The nitration of benzene necessitates dis-posal or reconcentration of sulfuric acid. Drum concentrators generally are used but nitration processes have been developed that reconcen-trate the sulfuric acid in situ. An example is the adiabatic nitration process developed jointly by Amer. Cyanamid and Canadian Industries [2], [3] in which a large volume of acid is used to absorb the exothermic heat of reaction. Water is removed by flash distillation at reduced pressure.

Du Pont has patented two energy-efficient nitration processes. The first is an azeotropic nitration process [4] in which the water produced by reaction of benzene

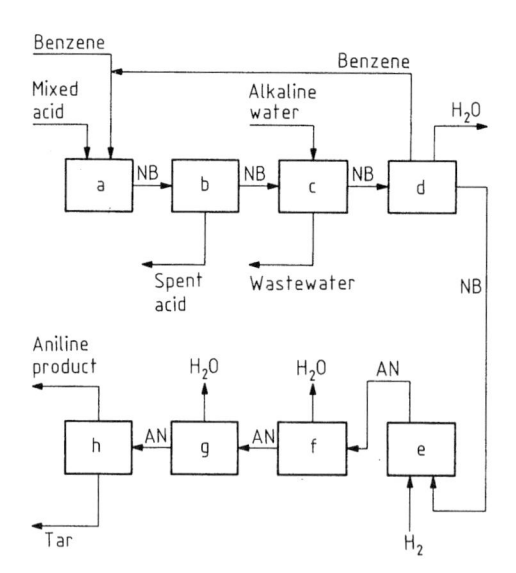

Figure 1. Benzene to nitrobenzene (NB) to aniline (AN) simplified flow sheet
a) Nitration; b) Separation; c) Washing neutralization; d) Topping; e) Hydrogenation (catalyst); f) Decanter; g) Dehydration; h) Rectification

with nitric acid, as well as water that enters the system with the nitric acid, is distilled from the reactor by a benzene-rich stream. The vapors are condensed; the organic phase is separated, revaporized, and returned to the reactor. The water phase, containing some nitric acid, is discarded. This avoids the need to remove and reconcentrate the sulfuric acid, which is a process that is wasteful in energy and prone to maintenance difficulties. The second, dehydrating nitration [5] also strips the water with a vapor stream but recycles inert gas along with the benzene vapor. This reduces the energy requirement for the process because the stripping gas need only be cooled in a partial condenser before recycle to the reactor.

3.1. Catalytic Liquid-Phase Hydrogenation of Nitrobenzene

Many industrial aniline processes involve the liquid-phase hydrogenation of nitrobenzene using suspended, highly active catalysts. The hydrogenation of nitrobenzene (Eq. 1) is highly exothermic (about 545 kJ/mol) and heat transfer becomes an important consideration in plant-scale reactor design. Liquid-phase hydrogenation processes are operated at 80–250 °C under pressure. For a given reactor size, a liquid-phase reactor usually has a greater production capacity than the corresponding vapor-phase system. Conversions of nitrobenzene normally are complete on a single reactor pass with yields of 98–99%. Aniline is purified using vacuum distillation.

In the 1960s, ICI developed a continuous, catalytic, liquid-phase hydrogenation process [6] that used aniline as the solvent in a proportion $\geq$ 95 wt% of the liquid phase and preferably as close to 100% as possible. By operating at or near the boiling point of the solvent (usually at pressures < 100 kPa), some or all of the heat of reaction is dissipated by allowing the reaction mixture to evaporate. Water is removed with the effluent vapors and sufficient aniline is returned to the reaction vessel to maintain steady-state conditions. One of the preferred catalysts is finely divided nickel on kieselguhr. Du Pont hydrogenates nitrobenzene in the liquid phase using a supported precious metal catalyst, which provides good catalyst life, high activity, and protection against hydrogenation of the aromatic ring.

3.2. Catalytic Vapor-Phase Hydrogenation of Nitrobenzene

Nitrobenzene is reduced to aniline, usually in about 99% yield, using fixed-bed or fluidized-bed vapor-phase processes. The catalyst is generally a nonnoble metal supported on a carrier suitable for the required high temperature reaction system. Copper is one of the preferred metals, giving good activity and selectivity. Nickel catalysts usually are formulated in combination with other metals or modifiers in order to control the high activity of the nickel and to protect against ring hydrogenation.

In the *Lonza process* [7], which is operated by First Chemical Co. [8], a homogenized feed of hydrogen and benzene is passed over a fixed-bed catalyst of copper on pumice at 150–300 °C. A molar ratio of nitrobenzene to hydrogen of 1:2.5 to 1:6 is employed at a gauge pressure of 200–1500 kPa. In a preferred process, the circulating gas stream is constricted and the nitrobenzene is atomized into the hot gas stream at the constricted position. The heat of reduction is absorbed by the recycling hydrogen in the system. The aniline is purified by distillation. Carbon steel can be used throughout the plant.

BASF operates a *vapor-phase, fluidized-bed process* [9]. Nitrobenzene is partially evaporated by atomizing it with the aid of a hot stream of gas comprised substantially of hydrogen. The stream of gas is circulated in the presence of a fluidized catalyst, the reaction products are condensed, and the aniline is separated from the isolated crude reaction products. One type of preferred catalyst is copper ($\approx$ 15 wt%) on a silica support promoted with chromium, zinc, and barium, with particle sizes of 0.2–0.3 mm. The two-phase mixture of hydrogen and nitrobenzene is injected through nozzles located at several heights in the fluid bed and the hydrogenation is carried out at 250–300 °C and 400–1000 kPa of pressure in the presence of excess hydrogen. The hot product gas is cooled by passing it through a heat exchanger, and aniline is isolated in a liquid–gas separator. The reaction heat is used for steam production. The yield of aniline is > 99%.

3.3. Reduction of Nitrobenzene with Iron and Iron Salts

A commercial variation of the nitrobenzene route is the *Béchamp process*, which uses iron and acid. This process coproduces iron oxide colored pigment. It is practiced by Mobay Chemical for part of their aniline production and by Bayer in the Federal Republic of Germany.

In the Béchamp process, nitrobenzene is reduced in an agitated reaction vessel with ferrous chloride solution and ground iron filings. The reduction is exothermic so the components must be mixed carefully to avoid excessive temperature and pressure buildup. After mixing is complete, the temperature is maintained at approximately 100 °C with steam. The reaction is completed in about 8 h.

$$4\, C_6H_5NO_2 + 4\, H_2O + 9\, Fe \xrightarrow{FeCl_2} 4\, C_6H_5NH_2 + 3\, Fe_3O_4$$

The reaction mixture is neutralized with lime; it is then transferred to a separator, and the organic phase containing aniline is withdrawn. Aniline is recovered from the organic phase by water stripping and distillation. Residual aniline is recovered from the material remaining in the separator before the iron oxide powder slurry is processed into a fine particle, colored pigment. The aniline yield based on nitrobenzene is > 98 %. The color of the iron oxide byproduct can be controlled by additives to the reaction medium (e.g., aluminum salts and phosphoric acid), by the use of different types of iron, and by subsequent calcination conditions [10], [11].

The advantage of the Béchamp process is the production of byproduct iron oxide. Disadvan-tages include a slow reaction rate, difficulties in separating product and impurities, and equipment corrosion. The reducer vessel usually is fabricated of cast iron with replaceable lining plates.

3.4. Amination of Phenol

In the commercial phenol route developed by Halcon [12]–[17], phenol is aminated in the vapor phase using ammonia in the presence of a silica–alumina catalyst. The reaction is reversible, so high conversion (95 %) is obtained by the use of excess ammonia (mole ratio of 20:1) and a low reaction temperature, which also reduces the dissociation of ammonia. Byproduct impurities include diphenylamine, triphenylamine, and carbazole. Their formation is inhibited by the use of excess ammonia. Yields based on phenol and ammonia are $\geq$ 96 % and 80 %, respectively. The reaction is mildly exothermic (8.374 kJ/mol).

In the process (Fig. 2), phenol and ammonia are vaporized separately (to prevent yield losses) and combined in the fixed-bed amination reactor (a) containing silica–alumina catalyst. After reaction at ca. 370 °C and 1.7 MPa, the gas is cooled and passed through a column (b) for ammonia recovery and

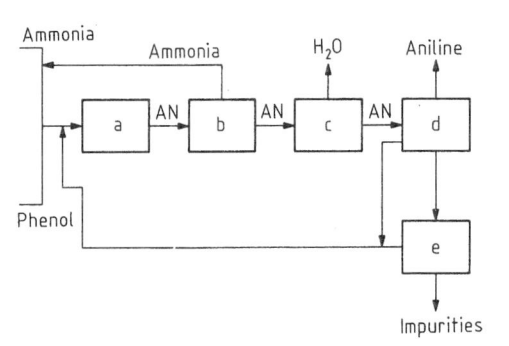

Figure 2. Phenol to aniline (AN) simplified flow sheet
a) Amination (catalyst); b) Ammonia recovery; c) Drying; d) Rectification; e) Residuals

recycle. The reaction mixture is separated into a water and an organic layer. The organic portion is passed through a drying column (c) to remove water and then through a finishing column (d) to separate aniline from residual phenol and impurities. The phenol, containing some aniline, is recycled.

Process steps preceding the amination of phenol include benzene → cumene → phenol (Eq. 2, Chap. 3). Cumene is prepared by alkylating benzene with propene using a carrier catalyst containing absorbed phosphoric acid. Temperature is ca. 230 °C and pressure 3.5 MPa. After purification, cumene is oxidized to cumene hydroperoxide at ca. 100 °C and 550 kPa using air and an aqueous sodium carbonate catalyst. The hydroperoxide is separated from cumene and converted into phenol and acetone by adding sulfuric acid. Small amounts of α-methyl styrene and other byproducts are formed in the reaction.

Comparison of the phenol route to the nitrobenzene route, both starting from benzene, shows four steps for the former (cumene to phenol involves two steps via the intermediate cumene hydroperoxide) versus two for the latter. The nitrobenzene route has an overall advantage in yield and lower total energy requirements. The phenol route has an advantage in catalyst life and is the preferred method for producers.

4. Quality Specifications

Technical-grade aniline usually meets the following specifications: aniline, min. 99.9%; nitrobenzene, max. 2×10^{-6} parts; water, max. 0.05%; color, APHA 100 max.; fp (dry basis) −6.2 °C min. The infrared spectrum, refractive index, and boiling range (184.4 °C±0.1 °C) are useful methods for identifying aniline. The organic purity usually is measured by gas chromatography using a flame ionization detector. Standard specifications and methods are described in ASTM D3264–76.

5. Handling, Storage, and Transportation

Aniline sampling and handling procedures in the United States are described by ASTM D3436–75. It is classified by the U.S. Interstate Commerce Commission (ICC) as a poisonous liquid, Class B (Regulation 173347) and as a Class 6 poison by the United Nations. As such, it must be packaged in ICC specification containers when shipped by rail, water, or highway, and all of the ICC regulations regarding loading, handling, and labeling must be followed. Aniline ordinarily is transported in tank cars, tank trucks, or metal drums. Carbon steel or cast iron are considered materials of choice for handling aniline except when discoloration must be kept to a minimum. Stainless steels (400 series) are recommended for color-critical applications. Aniline attacks copper, brass, and other copper alloys. It is rated as a Class 111A combustible liquid (NFPA Std. No. 30) and usually can be handled with little danger of fire.

6. Aniline Derivatives and Uses

Since the original discovery of aniline as an important raw material, the major uses have shifted from dyes to rubber chemicals in the 1960s and to isocyanates in the 1970s (Table 1).

6.1. Methylene Diphenylene Isocyanate (MDI)

More than half of the 1982 aniline production was consumed in the manufacture of MDI, 4,4'-methylene bis(phenyl isocyanate) [*101-68-8*], and poly-MDI [*9016-87-9*].

when $n = 0$, MDI
when $n > 0$, poly-MDI

These, in turn, are reacted with polyols to manufacture rigid polyurethane foams and elastomers. The polyurethane foams provide insulation in construction and transportation and have had a marked impact on demand for aniline. In the United States, aniline demand for isocyanates grew at an 8–10% per year rate during the 1970s.

United States 1983 MDI capacity was 263×10^3 t/a and the major producers included Mobay Chemical, Rubicon Chemicals, BASF, and Upjohn. Mobay and Rubicon

Table 1. Estimated United States consumption of aniline by use (1982)

Use	Percentage of production
MDI	55
Rubber chemicals	20
Agricultural chemicals and fibers	10
Pharmaceuticals	3
Dye intermediates and other uses	9
Hydroquinone	3

produce their own aniline; however, Mobay's requirements surpassed their production capacity so they were also net buyers. Upjohn and BASF purchased their total aniline requirements.

A typical MDI process used by commercial producers is as follows:

A mixture of polymeric polyphenylamines is first prepared by condensing aniline with formaldehyde in the presence of HCl. The reaction involves intermediate formation of amine hydrochlorides, which subsequently are neutralized with caustic soda. The condensation step is the critical one in the reaction sequence because it controls the number of rings in the molecule and the isomer distribution. The reaction usually is run with a large excess of aniline at subatmospheric pressure and at temperatures between 70 and 105 °C. The yield is about 96 %.

The polyamine product is converted to a crude mixture of isocyanates by reaction with phosgene in chlorobenzene solvent. Reaction temperature is about 120 °C, pressure is 345 kPa, and the yield ca. 97 %. Chlorobenzene is removed by distillation. The crude is separated by vacuum distillation into pure MDI and poly-MDI. Hydrogen chloride is produced as a byproduct of the reaction.

An alternative process for MDI was developed by Arco and avoids the use of aniline. It starts with nitrobenzene, which is carbonylated with CO in ethanol solvent in the presence of selenium catalyst to produce urethane (ethyl phenyl carbamate). The urethane is condensed with formaldehyde to form polyurethane, which is then decomposed by heat in the presence of catalyst to the isocyanate mixture. Ethanol also is formed and can be recycled to the carbonylation reaction. The Arco process has not been commercialized, but any large-scale future development is likely to have an impact on the aniline market.

6.2. Other Uses

Aniline's second largest use is as a raw material for *rubber-processing chemicals*, including antioxidants, antiozonants, accelerators, stabilizers, vulcanizers, and activators. With a trend toward smaller automotive tires and tires with longer life, future use of aniline in the rubber industry is expected to grow at the low rate of $1-2\%$ per year.

Other uses include agricultural chemicals (e.g., Monsanto's herbicide Lasso), pharmaceuticals (e.g., sulfonamides), dyes, and hydroquinone. Table 2 shows substituted anilines, their properties, and their uses.

Table 2. Substituted anilines

Class and name	Formula	CAS registry number	M_r	Appearance[a]	mp, °C	bp, °C	d_4^{20} [b]	n_D^{20} [b]	Solubility[a]	Use[c]
Salts										
Aniline hydrochloride	$C_6H_5NH_2 \cdot HCl$	[142-04-1]	129.60	w. cr.	198	245	1.2215^4	1.5881	s. w, alc	(1)
Aniline sulfate	$(C_6H_5NH_2)_2 \cdot H_2SO_4$	[542-16-5]	284.34	w. cr.	dec.		1.377^4		s. w	
Chloro										
o-Chloroaniline	$ClC_6H_4NH_2$	[95-51-2]	127.58	c. liq.	−14	209	1.2125	1.5881	s. alc, eth, be, ace	(1)
m-Chloroaniline		[108-42-9]	127.58	c. liq.	−10	230	1.2161	1.5941	s. alc, eth, be	(1)
p-Chloroaniline		[106-47-8]	127.58	c. cr.	72.5	232	1.429^{19}	1.5546^{87}	s. hot w, alc, eth, ace	(1) (8)
2,3-Dichloroaniline	$Cl_2C_6H_3NH_2$	[608-27-5]	162.02	cr.	24	252	1.383^{25}		s. alc, eth, ace, be	
2,4-Dichloroaniline		[554-00-7]	162.02	cr.	63	245	1.567		s.s. w; s. alc, eth	
2,5-Dichloroaniline		[95-82-9]	162.02	cr.	50	251			s. alc, eth, be	(1)
2,6-Dichloroaniline		[608-31-1]	162.02	cr.	39				s. alc, eth	
3,4-Dichloroaniline		[95-76-1]	162.02	w. cr.	72	272	1.33^{80}		s. alc, eth	
3,5-Dichloroaniline		[626-43-7]	162.02	cr.	51	260			s. alc, eth, be	(1) (8)
2,4,5-Trichloroaniline	$Cl_3C_6H_2NH_2$	[636-30-6]	196.46	cr.	97	270			s. alc, eth	
Alkyl										
2,4,5-Trichloroaniline	$H_3CC_6H_4NH_2$	[95-53-4]	107.17	y. liq.	−16.4	200	0.9994	1.5725	s.s. w; s. alc, eth	(1)
m-Toluidine		[108-44-1]	107.17	c. liq.	−30	203	0.9889	1.5681^{22}	s.s. w; s. alc, eth, ace, be	(1)
p-Toluidine		[106-49-0]	107.17	w. cr.	44–45	200.6	0.9619	1.5636	s.s. w; s. alc, eth, ace, be	(1)
2,3-Xylidine	$(H_3C)_2C_6H_3NH_2$	[87-59-2]	121.20	liq.	<−15	221–222	0.9931	1.5684	s. alc, eth	
2,4-Xylidine		[95-68-1]	121.20	liq.	−14	215	0.9723	1.5569	s. alc, eth, be	(1)
2,5-Xylidine		[95-78-3]	121.20	oily liq.	16	214	0.9790^{21}	1.5593	s. eth	(1)
2,6-Xylidine		[87-62-7]	121.20	liq.	11	216	0.9842	1.5610	s. alc, eth	(4)
3,4-Xylidine		[95-64-7]	121.20	cr.	51	226	1.076^{18}		s. eth	(1)
3,5-Xylidine		[108-69-0]	121.20	oily liq.	9.8	220–221	0.9706	1.5581	s. eth	
Alkoxy										
o-Anisidine	$H_3COC_6H_4NH_2$	[90-04-0]	123.17	y. liq.	6.2	225	1.0923	1.5793	s. alc, eth, ace, be; s.s. w	(1)
m-Anisidine		[536-90-3]	123.17	oily liq.	<−12	251	1.096	1.5794	s. alc, eth, ace, be; s.s. w	
p-Anisidine		[104-94-9]	123.17	w. cr.	57.2	243	1.0605^{67}	1.5559^{67}	s. alc, eth, ace, be; s.s. w	(1)
o-Phenetidine	$H_5C_2OC_6H_4NH_2$	[94-70-2]	137.20	oily liq.	<−21	232		1.5560	s. alc, eth,	(1) (2) (3)
p-Phenetidine		[156-43-4]	137.20	liq.	2–3	254	1.0652^{15}	1.5528	s. alc, eth; s.s. w	(1)
p-Cresidine	$H_3CO(CH_3)$–$C_6H_3NH_2$	[120-71-8]	137.20	w. cr.	52	235			s. alc, eth, ace, be; s.s. w.	(1)

Table 2. (continued)

Class and name	Formula	CAS registry number	M_r	Appearance[a]	mp, °C	bp, °C	d_4^{20} [b]	n_D^{20} [b]	Solubility[a]	Use[c]
N-Alkyl, N-Aryl										
N-Methylaniline	$C_6H_5NHCH_3$	[100-61-8]	107.17	y. liq.	−57	196	0.9891	1.5684	s. alc; s.s. w	(1)
N,N-Dimethylaniline	$C_6H_5N(CH_3)_2$	[121-69-7]	121.20	y. liq.	2	194	0.9557	1.5582	s. alc, eth, ace, be	(1) (5)
N-Ethylaniline	$C_6H_5NHC_2H_5$	[103-69-5]	121.20	liq.	−63.5	205	0.9625	1.5559	s. alc, eth, ace, be	(1)
N,N-Diethylaniline	$C_6H_5N(C_2H_5)_2$	[91-66-7]	149.26	y. liq.	−38.8	216	0.9351	1.5409	s.s. alc, eth, ace	(1)
N-Ethyl-N-benzylaniline	$C_6H_5N(C_2H_5)-CH_2C_6H_5$	[92-59-1]	211.20	y. liq.		314	1.034		s. alc, eth	(1)
N,N-Diphenylamine	$C_6H_5NHC_6H_5$	[122-39-4]	169.24	w. cr.	54–55	302	1.159		s. alc, eth, ace, be	(3)
N-Acyl										
Formanilide	C_6H_5NHCOH	[103-70-8]	121.15	w. cr.	50	271	1.1437^{17}		s. alc; s.s. w	(1) (6)
Acetanilide	$C_6H_5NHCOCH_3$	[103-84-4]	135.18	cr.	114	304	1.219^{15}		s. alc, eth, ace, be, hot w	(1) (3) (6)
Acetoacetanilide	$C_6H_5NHCOCH_2COCH_3$	[102-01-2]	177.22	w. cr.	86				s. alc, eth, ace, be	(1)
Nitro										
o-Nitroaniline	$O_2NC_6H_4NH_2$	[88-74-4]	138.14	y. cr.	71–72	284	1.442^{15}		s. alc, eth, ace, be, hot w	(1)
m-Nitroaniline		[99-09-2]	138.14	y. cr.	114	306			s. alc, eth, ace; s.s. w	(1)
p-Nitroaniline		[100-01-6]	138.14	y. cr.	148–149	332	1.424		s.s. alc	(1) (3)
2,4-Dinitroaniline	$(O_2N)_2C_6H_3NH_2$	[97-02-9]	183.14	y. cr.	187–188		1.615^{14}			(1)
2,4,6-Trinitroaniline	$(O_2N)_3C_6H_2NH_2$	[489-98-5]	228.12	y. cr.	192–195		1.762^{12}			(7)
Sulfonated										
Orthanilic acid (o)	$H_2NC_6H_4SO_3H$	[88-21-1]	173.9	cr.	>320					(1)
Methanilic acid (m)		[121-47-1]	173.9	w. cr.	dec.				s.s. alc, w	(1)
Sulfanilic acid (p)		[121-57-3]	173.9	w. cr.	288		1.485^{25}		s.s. w	(1) (6)

[a] c. = colorless; w. = white; y. = yellow; liq. = liquid; cr. = crystals; dec. = decomposes; s. = soluble; s.s. = slightly soluble; w = water; alc = alcohol; eth = ether; ace = acetone; be = benzene
[b] Density and refractive index superscripts refer to temperature (°C)
[c] (1) dyes, pigments; (2) synthetic sweetener; (3) rubber antioxidant, accelerator; (4) vitamin B_2(riboflavin); (5) explosive stabilizer; (6) medication (analgesic), pharmaceuticals; (7) explosive; (8) agricultural chemicals

6.3. Derivatives

o-Chloroaniline [95-51-2], 1-amino-2-chlorobenzene, $ClC_6H_4NH_2$, is a water-white to tan liquid, M_r 127.6, Tagliabue open cup (TOC) flash point 90 °C, bp 208 °C, fp –2.3 °C. o-Chloroaniline is *used* as an intermediate for rubber chemicals, pigments, pesticides, and dyes. *Production* is by low-pressure hydrogenation of 2-chloronitrobenzene with noble metal and/or noble metal sulfide catalysts at temperatures of ca. 50–100 °C in inert organic solvents. The yield is almost quantitative.

m-Chloroaniline [108-42-9], 1-amino-3-chlorobenzene, $ClC_6H_4NH_2$, is a water-white to light-amber liquid, M_r 127.6, flash point (TOC) 90 °C, bp 230 °C, fp –10.5 °C. This compound is *used* as an intermediate for pesticides, pharmaceuticals, and dyes. It is *produced* by low-pressure hydrogenation of 3-chloronitrobenzene in the liquid phase with noble metal and/or noble metal sulfide catalysts. The addition of metal oxides to the reaction helps avoid dehalogenation. The yield is approximately 98%.

p-Chloroaniline [106-47-8], 1-amino-4-chlorobenzene, $ClC_6H_4NH_2$, is a white to light amber, crystalline solid at room temperature, M_r 127.6, bp 230 °C, mp 70 °C. It is *used* as an intermediate for pesticides, pharmaceuticals, pigments, and dyes. *Production* is similar to that of m-chloroaniline [18].

3,4-Dichloroaniline [95-76-1], 1-amino-3,4-dichlorobenzene, $Cl_2C_6H_3NH_2$, is a gray to dark-brown crystalline solid at room temperature, M_r 162.0, fp 66–71 °C, bp 272 °C. It is *used* as an intermediate for pesticides and dyes and is *produced* by catalytic hydrogenation of 3,4-dichloronitrobenzene with noble metal catalysts under pressure. Various types of additives prevent dehalogenation during production and reactors must be fabricated with special steel alloys to inhibit corrosion [18], [19].

N,N-Dimethylaniline [121-69-7], N,N-di-methylphenylamine, $(H_3C)_2NC_6H_5$, is a light yellow (straw colored) to brown oily liquid, freezing to a crystalline solid at low temperatures, M_r 121.2, fp 2.1 °C, flash point, Tagliabue closed cup (TCC) 63 °C, bp 193 °C. It is *used* as a polymerization catalyst and as an intermediate for pharmaceuticals and dyes. Dimethylaniline is *produced* from aniline and methanol under pressure in the presence of acidic catalysts or by passing dimethyl ether and aniline vapor over highly activated aluminum oxide at 230–295 °C [20].

N,N-Diethylaniline [91-66-7], N-phenyl-diethylamine, $(H_5C_2)_2NC_6H_5$, is a yellow to brown, oily liquid, M_r 149.2, flash point (TOC) 85 °C, bp 216 °C. Its *production* is analogous to methods used for dimethylaniline production except that ethanol instead of methanol is reacted with aniline. *Uses* are also similar to those for the dimethyl analogue.

N-Phenylacetamide [*103-84-4*], acetanilide, acetylaniline, $C_6H_5NHCOCH_3$, is a colorless, glossy, crystalline material, M_r 135.2, *fp* 114 °C, *bp* 304 °C. It is *used* as a dye intermediate, plasticizer and stabilizer for cellulose esters, and as an intermediate for the synthesis of sulfonamides. It is *produced* by heating aniline with acetic acid, acetic anhydride, or acetyl chloride [21].

2,4-Xylidine [*95-68-1*], *m*-xylidine, 2,4-di-methylaniline, $(H_3C)_2C_6H_3NH_2$, is a colorless to tan, oily liquid, M_r 121.2, flash point (TCC) > 93 °C, *bp* 213 °C. It is *used* as an intermediate for photographic chemicals, pesticides, and dyes. It is *produced* by the catalytic hydrogenation of the corresponding nitro derivative.

2,6-Xylidine [*87-62-7*], 2,6-dimethylaniline, $H_2NC_6H_3(CH_3)_2$, is a colorless to reddish-yellow, clear liquid, M_r 121.2, flash point (TCC) > 94 °C, *bp* 217 °C. This compound is *used* as an intermediate for dyes, pesticides, and pharmaceuticals. Production is by the catalytic hydrogenation of the corresponding nitro derivative [22].

N-Phenylaminobenzene [*122-39-4*], diphenylamine, $C_6H_5NHC_6H_5$, is isolated in the form of colorless leaflets, M_r 212.2, flash point (TCC) 153 °C, *bp* 302 °C, *mp* 53 °C. Diphenylamine is *used* as a stabilizer for elastomers, nitrocellulose, and nitroglycerine and as an intermediate for dyes. It is *produced* by passing aniline over various types of acidic catalysts at elevated temperatures and pressures [23].

7. Economic Aspects

World aniline capacity was estimated at 1324×10^3 t for 1983. Capacity ($t \times 10^3$) by geographic area is:

North America	
United States	580
Mexico	5
Western Europe	
Federal Republic of Germany	200
United Kingdom	190
Belgium	100
Portugal	50
France	25
Eastern Europe	
German Democratic Republic	25
Yugoslavia	14
Poland	11
Far East	
Japan	105
India	18
Africa	
Egypt	1

Table 3. Major world aniline producers

Company	Location	Estimated 1983 capacity $t \times 10^3$
Bayer	Belgium and Federal Republic of Germany	250
ICI	United Kingdom	188
Du Pont	United States	190
Rubicon Chemicals	United States	127
USS Chemicals	United States	90
First Chemical	United States	114
Mobay Chemical	United States	59
BASF	Belgium and Federal Republic of Germany	52

The major world producers of aniline are listed in Table 3.

United States producers were operating at approximately 50% of capacity in 1982. Average capacity utilization for other world locations are estimated at the same level because of depressed economic conditions. World aniline capacity should be adequate to meet projected demand until the late 1980s. Approximately 60–70% of the aniline is used captively by the producer.

8. Environmental Protection, Toxicology, and Occupational Health

In the United States, the current OSHA [24] permissible exposure limit (PEL) averaged over an 8-h work shift is 5 parts of aniline per 10^6 parts of air (19 mg/m^3). The ACGIH has recommended a TLV of 2 ppm (8 mg/m^3) with a skin notation (possibility of cutaneous absorption). The MAK is 8 mg/m^3 and the compound is listed as being justifiably suspected of having carcinogenic potential. Aniline can affect the body if it is inhaled, comes in contact with the eyes or skin, or is swallowed. It is readily absorbed through the skin, either as a liquid or vapor. Aniline affects the ability of the blood to carry oxygen, and moderate overexposure may cause only a bluish discoloration of the skin. As oxygen deficiency increases, the blue discoloration may be associated with headache, weakness, irritability, drowsiness, shortness of breath, and unconsciousness. If treatment is not given promptly, death can occur. Absorption of aniline into the body leads to the formation of methemoglobin which, in sufficient concentration, causes cyanosis. Because reversion of methemoglobin occurs spontaneously after exposure ceases, moderate degrees of cyanosis need be treated only by supportive measures, such as bed rest and oxygen inhalation. Workers should be provided with and required to use impervious clothing, gloves, face shields, and other appropriate protective clothing to prevent skin contact with liquid aniline where contact may occur. Contaminated clothing should be removed and the skin should be washed with soap or mild detergent to remove any aniline.

The United States EPA classifies aniline and aniline waste as hazardous [25]. All aniline labels should carry the "poison" designation. No criterion has been set, but the EPA has suggested an ambient limit in water of 262 µg/L based on health effects. Included in the list of hazardous aniline wastes are: distillation bottoms from aniline production, process residues from aniline extraction, and combined wastewater streams generated from nitrobenzene – aniline production.

Based on tests with laboratory animals, aniline may cause cancer. Both the National Cancer Institute and the Chemical Industry Institute of Toxicology (CIIT) conducted lifetime rodent feeding studies. Both studies found tumors of the spleen at high dosage levels of aniline hydrochloride (100 – 300 mg kg^{-1} day^{-1}) whereas the CIIT found a no-tumor level of 10 – 30 mg kg^{-1}day^{-1}. The latter level is roughly equivalent to a human 8-h inhalation level of 17 – 50 parts of aniline vapor per 10^6.

The 1982 International Agency for Research on Cancer (WHO) report [26] indicates that the high risk of bladder cancer observed originally in workers in the aniline dye industry was probably a result of exposure to chemicals other than aniline. Studies of individuals exposed to aniline but to no other known bladder carcinogens have shown little evidence of increased risk (one death from bladder cancer among 1223 men producing or using aniline with 0.83 deaths expected from population rates).

Based on current information of risk assess-ment, observance of the acceptable exposure limit of 2 parts per 10^6, 8 – 12-h TWA, contin-ued avoidance of skin contact with aniline, and observance of good industrial hygiene practices will provide a sub-stantial margin of safety in protecting against any potentially adverse health effect from aniline.

The toxicology of aniline derivatives is discussed under → Amines, Aromatic.

9.   References

[1]   "Aniline – Properties, Uses, Storage, and Handling". *Du Pont Product Bulletin*, 1983.
[2]   Amer. Cyanamid, US 4021498, 1977 (V. Alexanderson, J. Trecek, C. Vanderwaart).
[3]   Amer. Cyanamid, US 4091042, 1978 (V. Alexanderson, J. Trecek, C. Vanderwaart).
[4]   Du Pont, US 3928475, 1975 (M. W. Dassel).
[5]   Du Pont, US 4331819, 1982 (R. McCall).
[6]   ICI, US 3270057, 1966 (E. V. Cooke, H. J. Thurlow).
[7]   Lonza Ltd., DE-OS 1809711 (1968), US 3636152, 1972 (L. Szigeth).
[8]   Lonza/First Chemical Co., *Hydrocarbon Process* **59** (Nov. 1979) no. 11, 136.
[9]   BASF, US 3136818, 1964 (H. Sperber, G. Poehler, H. J. Pistor, A. Wegerich).
[10]   P. H. Groggins: *Unit Processes in Organic Synthesis*, 5th ed., McGraw-Hill, New York 1958, p. 135.
[11]   Bayer AG, US 4234348, 1980 (H. Brunn, H. Bade, F. Hund).
[12]   Halcon International, Inc., US 3272865, 1966 (R. S. Barker).

[13] Halcon International, Inc., DE-OS 2003842, 1970 (M. Becker et al.).
[14] Halcon International, Inc., DE-OS 2026053, 1970 (M. Becker et al.).
[15] Halcon International, Inc., US 3682782, 1972 (C. Y. Choo).
[16] Halcon International, Inc., US 3578714, 1971 (J. L. Russell).
[17] Halcon International, Inc., US 3860650, 1975 (M. Becker, S. Khoobiar).
[18] Du Pont, US 3361819, 1968 (J. R. Kosak, L. Spiegler).
[19] Dow, US 3067253, 1962 (A. J. Dietzler, T. R. Kell).
[20] Biller, Michaelis, US 2991311, 1961 (M. Thoma).
[21] British Industrial Solvents, US 2462221, 1949 (E. S. Pemberton).
[22] Standard Oil Development, US 2421608, 1947; US 2620356, 1952; US 2481245, 1949.
[23] Amer. Cyanamid, US 2968676, 1961 (A. G. Potter, R. G. Weyker).
[24] *NIOSH/OSHA Occupational Health Guideline for Aniline*, U.S. Dept. of Health and Human Services, U.S. Dept. of Labor., U.S. Government Printing Office, Washington, D.C., September 1978.
[25] Marshall Sittig: *Handbook of Toxic and Hazardous Chemicals,* Noyes Publications, Park Ridge, N.J., 1981.
[26] International Agency for Research on Cancer (IARC): *Monograph on the Evaluation of the Carcinogenic Risk of Chemicals to Humans,* Lyon 1981 (published April 1982, vol. 27).